AF896324

Springer Proceedings in Physics

Volume 194

The series Springer Proceedings in Physics, founded in 1984, is devoted to timely reports of state-of-the-art developments in physics and related sciences. Typically based on material presented at conferences, workshops and similar scientific meetings, volumes published in this series will constitute a comprehensive up-to-date source of reference on a field or subfield of relevance in contemporary physics. Proposals must include the following:

- name, place and date of the scientific meeting
- a link to the committees (local organization, international advisors etc.)
- scientific description of the meeting
- list of invited/plenary speakers
- an estimate of the planned proceedings book parameters (number of pages/articles, requested number of bulk copies, submission deadline).

More information about this series at http://www.springer.com/series/361

Indrani Bhattacharya · Satyajit Chakrabarti
Haricharan Singh Reehal
Vasudevan Lakshminarayanan
Editors

Advances in Optical Science and Engineering

Proceedings of the Third International Conference, OPTRONIX 2016

Springer

Editors
Indrani Bhattacharya
Research and Development Cell
University of Engineering & Management
Kolkata, West Bengal
India

Satyajit Chakrabarti
University of Engineering & Management
Kolkata, West Bengal
India

Haricharan Singh Reehal
School of Engineering
London South Bank University
London
UK

Vasudevan Lakshminarayanan
School of Optometry and Vision Science
University of Waterloo
Waterloo, ON
Canada

ISSN 0930-8989 ISSN 1867-4941 (electronic)
Springer Proceedings in Physics
ISBN 978-981-10-3907-2 ISBN 978-981-10-3908-9 (eBook)
DOI 10.1007/978-981-10-3908-9

Library of Congress Control Number: 2017932402

© Springer Nature Singapore Pte Ltd. 2017
This work is subject to copyright. All rights are reserved by the Publisher, whether the whole or part of the material is concerned, specifically the rights of translation, reprinting, reuse of illustrations, recitation, broadcasting, reproduction on microfilms or in any other physical way, and transmission or information storage and retrieval, electronic adaptation, computer software, or by similar or dissimilar methodology now known or hereafter developed.
The use of general descriptive names, registered names, trademarks, service marks, etc. in this publication does not imply, even in the absence of a specific statement, that such names are exempt from the relevant protective laws and regulations and therefore free for general use.
The publisher, the authors and the editors are safe to assume that the advice and information in this book are believed to be true and accurate at the date of publication. Neither the publisher nor the authors or the editors give a warranty, express or implied, with respect to the material contained herein or for any errors or omissions that may have been made. The publisher remains neutral with regard to jurisdictional claims in published maps and institutional affiliations.

Printed on acid-free paper

This Springer imprint is published by Springer Nature
The registered company is Springer Nature Singapore Pte Ltd.
The registered company address is: 152 Beach Road, #21-01/04 Gateway East, Singapore 189721, Singapore

This volume is dedicated to the students and researchers who dream and work for a better tomorrow

Preface

Optics and Photonics are related to the science of light and it can be harnessed for utilization in an enormous variety of applications, as evidenced by its increasing pervasiveness in our everyday lives. We use photonic technologies to light up our homes, offices and cities, to harvest renewable energy from the sun, to make telephone calls or surf the Internet, to enable early medical diagnosis and treatments, to establish clean and efficient manufacture of a multitude of everyday products, or to provide reliable security systems to protect us as we travel. Already, over 90% of all the data transferred for telecommunications is transmitted by optical fibre. Green photonics is the study and development of applied optical systems for generating clean, renewable energy. This includes solar cells and photovoltaic devices, creating energy-efficient optical sources for lighting and display applications and developing environmental friendly materials for optoelectronic devices and components.

The 3rd International Conference on Opto-Electronics and Applied Optics, OPTRONIX 2016 (www.optronix.iem.edu.in) has been organized by Research and Development Council, University of Engineering and Management, UEM, New Town, Kolkata, India, in association with Institute of Engineering and Management, IEM, Salt Lake, Kolkata, India, on 18–20 August, 2016 at UEMK Campus. The Conference has got technical collaboration with Optical Society of India, OSI and American Journal of Advanced Computation Technology (AJACT), Canada and was supported by SPIE (Society of Photonics and Instrumentation Engineers, USA) and OSA (Optical Society of America, USA) approved IEM–UEMK Group Student Chapters.

The theme of the Conference is selected as the emerging research area, "Green Photonics: The Scope of Research and Development in harnessing Clean Energy". In addition to the theme, diversified areas, namely Silicon Photonics, Advanced Photovoltaics, Application of Solar Energy, Fibre and Integrated Optics, Organic Photovoltaics, Nano-Photonics, Bio-Photonics and Bio-Medical Optics, Opto-Electronic Materials, Optical and Digital Data and Image Processing, Photonics for Space Applications, Adaptive Optics, Optical Design, Fabrication and Testing, E.M. Radiation Theory and Antenna, Interferometry, Micro-Electronics and VLSI, Quantum Optics and Information Processing, Digital Holography, Terahertz Technology, etc., have been selected on which researchers have disseminated their contributions. The conference is organized to promote the International Solar Alliance of over 120 countries initiated by Hon'ble Prime Minister of India, Mr. Narendra Modi with the French President, François Hollande, at the Paris COP21 climate summit. "Solar technology is evolving, costs are coming down and grid connectivity is improving", as said by Hon'ble Prime Minister "The dream of universal access to clean energy is becoming more real. This will be the foundation of the new economy of the new century". Twenty seven eminent Scientists and Professors from India and abroad had been invited to deliver keynote, plenary as well as invited addresses in the related field. More than seventy research papers have been registered and presented in the Conference. Among distinguished International Invited Speakers were Prof. Toyohiko Yatagai, Director, Centre for Optical Research and Education (CORE), Utsunomiya University, JAPAN and Former President SPIE (2015), San Diego, USA; Prof. Vasudevan Lakshminarayanan, Professor of Optometry (Vision Science), Physics, Electrical and Computer Engineering and Systems Design Engineering, University of Waterloo, Canada; Prof. Motoharu Fujigaki, Professor, Human and Artificial Intelligent Systems, Graduate School of Engineering, University of Fukui, Japan; Prof. Takeo Sasaki, Professor, Department of Chemistry, Tokyo University of Science, Japan; and Prof. Tigran Galstian (over Skype), Professor, Department of Physics, Centre d'optique, photonique et laser (COPL), University of Laval, Quebec, Canada.

The distinguished National Invited Speakers include Prof. Ajoy Ghatak, Prof. Kehar Singh, both are eminent Physicists and former professors of IIT Delhi; Prof. Rajpal Sirohi, Professor of Eminence, Tezpur University, Assam and Former Director, IIT Delhi; Prof. Samit Kr. Ray, Dean of Postgraduate Studies and Research, IIT Kharagpur; Prof. Lakshminarayan Hazra, Emeritus Professor, Department of Applied Optics and Photonics, Calcutta University, India; Dr. Amitava Roy, Scientist 'F', Department of Science and Technology, Govt. of India; Prof. Rajesh Gupta, IIT Bombay; Prof. Navin Nishchal, IIT Patna; Prof. Rajan Jha, IIT Bhubaneswar; Prof. Partha Roy Chaudhuri, IIT Kharagpur; Prof. Samudra Roy, IIT Kharagpur; Prof. Asim Kar, Former Professor, Jadavpur University; Prof. Shyamal Bhadra, Emeritus Scientist, Raman Centre for Atomic, Molecular and Optical Sciences, Indian Association for the Cultivation of Science (IACS), Jadavpur, Kolkata; Prof. Kallol Bhattacharya, Department of Applied Optics and Photonics, Calcutta University; Prof. P.T. Ajithkumar, CEO and Leading Scientists, Light Logics Holography and Optics, Trivandrum, Kerala; Prof. Tarun K Gangopadhyay, Senior Principal Scientist, CSIR-Central Glass and Ceramics Research Institute, Jadavpur, Kolkata; Prof. Somnath Ghosh, INSPIRE Professor, Calcutta University; and Prof. A.K. Chakraborty, Prof. K.K. Ghosh, Prof. K.P. Ghatak and Prof. Rajiv Ganguly, all professors of UEMK and IEM Group.

I am extremely thankful to SERB, Symposia Section, Department of Science and Technology, Govt. of India, for providing partial grants-in-aid to support the

TA and DA of Senior Scientists and young researchers to participate in the conference. My sincere gratitude to the International Advisory Committee and Technical Program Committee for providing valuable guidance and extended support in reviewing the submitted contributions continuously. I would like to thank Prof. Dr. Mrimnoy Chakraborty, Assistant Professor, Dr. B.C. Roy Engineering College, Durgapur, for his support and cooperation in the review process.

My special thanks to Prof. Dr. Mina Ray, Assistant Professor, Department of Applied Optics and Photonics, University of Calcutta, Kolkata and Prof. Dr. Debotosh Guha, Professor, Institute of Radio Physics and Electronics, Calcutta University, eventually both are my seniors at Rajabazar Science College Campus of University of Calcutta, for their presence and chairing technical sessions in their area of expertise. I would like to express my sincere thanks and gratitude to the respected faculty members and staffs as well as the Student Organizing Committee of UEMK–IEM Group who has provided relentless support and cooperation to make the Conference successful.

Kolkata, India

Indrani Bhattacharya

Optronix 2016 Organizing Committee

Patron	**Dr. Satyajit Chakrabarti** Chancellor University of Engineering and Management, UEM, Jaipur and Kolkata, India Founder Director Institute of Engineering and Managemnt, IEM, Salt Lake, Kolkata, India
	Banani Chakrabarti Director Institute of Engineering and Management, IEM, Salt Lake, Kolkata, India
	Gopa Goswami Director, Placement Cell Institute of Engineering and Management, IEM, Salt Lake, Kolkata, India
General Chair	**Dr. Satyajit Chakrabarti (Alex)** **Pro-Vice Chancellor** University of Engineering and Management, UEM, Jaipur and Kolkata, India Director, Institute of Engineering and Management, IEM, Salt Lake, Kolkata, India
Convener & TPC Chair	**Indrani Bhattacharya** Head, Research and Development Cell University of Engineering & Management, Kolkata, India

Advisory Committee

Davinder Kumar Banwet
Vice Chancellor
University of Engineering & Management, Kolkata, India

Vasudevan Lakshminarayanan
Professor of Optometry (Vision Science), Physics, Electrical and Computer Engineering and Systems Design Engineering
University of Waterloo, Canada, and University of Michigan, USA

Haricharan Singh Reehal
Professor and Director of Research, School of Engineering
London South Bank University, 103, Borough Road, London, UK

Brian Culshaw
Emeritus Professor, Electronic and Electrical Engineering
University of Strathclyde, Glasgow, UK

Graham Reed
Professor, Optoelectronic Research Centre
University of Southampton, UK

Tigran Galstian
Professor, Department of Physics, Engineering Physics, Optical Centre
University of Laval, Quebec, Canada

Lorenzo Pavesi
Professor, Department of Physics
University of Trento, Italy

Michael Lengden
Department of Electronic and Electrical Engineering
University of Southampton, Glasgow, UK

Takeo Sasaki
Department of Chemistry, Faculty of Science
Tokyo University of Science, Tokyo, Japan

Yasuhiro MIZUTANI
Department of Mechanical Engineering
Osaka University, Japan

Yukitoshi Otani
Professor, Centre for Optical Research and Education, CORE
Japan

Ajoy Ghatak
Eminent Physicist and Formerly Professor, Department of Physics
Indian Institute of Technology, Delhi, India

Samit Kr. Ray
Professor, Department of Physics
Dean, Postgraduate Studies and Research
Indian Institute of Technology, Kharagpur, India

Kehar Singh
ITM University, Gurgaon and Former Professor, Department of Physics
Indian Institute of Technology, Delhi, India

B.P. Pal
Eminent Physicist and Former Professor, Department of Physics
Indian Institute of Technology, Delhi, India

Rajpal S Sirohi
Professor of Eminence, Department of Physics,
Tejpur University, Assam, India

A.K. Chakraborty
Professor, IEM, Kolkata, India
Former Professor, Department of Applied Optics and Photonics
University of Calcutta, Kolkata, India

Lakshminarayan Hazra
Emeritus Professor, Department of Applied Optics and Photonics University of Calcutta, Kolkata, India

Anurag Sharma
Professor, Department of Physics
Indian Institute of Technology, Delhi, India

Nikhil Ranjan Das
Dean, Faculty of Engineering and Technology
Director, Centre for Research in Nanoscience and Nanotechnology Professor and Head, Institute of Radio-Physics and Electronics, University of Calcutta, Kolkata, India

Shyamal Bhadra
Former Chief Scientist, CSIR-Central Glass and Ceramic Research Institute, Jadavpur, Kolkata

Rajiv Ganguly
Professor and Dean
University of Engineering & Management, New Town, Kolkata, India

Senthil Kumar M.
Head, Reflective Optics Division, SAC(ISRO), Ahmedabad

C.S. Narayan Murthy
Professor of Physics
Indian Institute of Space Science and Technology, Thiruvananthapuram, Kerala, India

Naveen Kumar Nishchal
Associate Professor
Indian Institute of Technology, Patna, India

Partha Roy Chaudhari
Department of Physics
Indian Institute of Technology, Kharagpur, India

Amitava Roy
Scientist 'F'
Department of Science and Technology (DST), Delhi, India

Kallol Bhattacharya
Associate Professor, Department of Applied Optics and Photonics, University of Calcutta, Kolkata, India

Tarun Kumar Gangopadhyay
Senior Principal Scientist, Fibre Optics & Photonics Division
Head, Computer Divison, CSIR-Central Glass and Ceramic Research Institute, Govt. of India, Kolkata

Technical Chairs

Ajoy Ghatak
Eminent Scientist, IIT, Delhi
Ajoy Chakraborty
Institute of Engineering & Management, Kolkata, India

Abir Chattopadhyay
University of Engineering & Management, Kolkata, India

Debotosh Guha
Department of RPE, University of Calcutta, Kolkata

Kehar Singh
ITM University, Gurgaon, India

M.R. Shenoy
Department of Physics, IIT, Delhi, India

Niranjan Mondal
University of Engineering & Management, Kolkata, India

R.S. Sirohi
Tejpur University, Assam, India

Motoharu Fujigaki
University of Fukui, Japan

Rajan Jha
Department of Physics, IIT, Bhubaneswar, Orissa, India

Rajesh Gutpa
Department of Energy Science, IIT, Powaii, Mumbai, India

Malay Ganguly
Department of ECE, Institute of Engineering & Management, Kolkata, India

Mina Ray
Department of AOP and Photonics, Calcutta University, Kolkata, India

Navin Nishchal
Department of Physics, IIT, Patna

Partha Roy Chaudhari
Dept. of Physics, IIT Kharagpur, India

Samudra Roy
Dept. of Physics, IIT Kharagpur, India

Takeo Sasaki
Tokyo University of Science, Japan

Yushiro Mizutani
University of Tokushima, Japan

Technical Program Committee

Biswajoy Chatterjee
University of Engineering & Management, Kolkata, India

K.K. Ghosh
Institute of Engineering & Management, Kolkata, India

K.P. Ghatak
University of Engineering & Management, Kolkata, India

Rajiv Ganguly
University of Engineering & Management, Kolkata, India

Sukalyan Goswami
University of Engineering & Management, Kolkata, India

Ratna Chakrabarty
Institute of Engineering & Management, Kolkata, India

Mili Sarkar
Institute of Engineering & Management, Kolkata, India

Soumen Banerjee
University of Engineering & Management, Kolkata, India

Madhusree Chatterjee
University of Engineering & Management, Kolkata, India

Souvik Chatterjee
University of Engineering & Management, Kolkata, India

Sohini Mukherjee
University of Engineering & Management, Kolkata, India

Bipasha Mukhopadhyay
University of Engineering & Management, Kolkata, India

Debkumar Chowdhury
University of Engineering & Management, Kolkata, India

Joydip Ghosh
University of Engineering & Management, Kolkata, India

Raj Nag
University of Engineering & Management, Kolkata, India

Arundhati Chakraborty
University of Engineering & Management, Kolkata, India

Sangita Dutta
University of Engineering & Management, Kolkata, India

Dibyendu Bikash Seal
University of Engineering & Management, Kolkata, India

Souvik Ghosh
University of Engineering & Management, Kolkata, India

Sharmistha Nag
University of Engineering & Management, Kolkata, India

Subhalaxmi Chakraborty
University of Engineering & Management, Kolkata, India

Student Organizing Committee

IEM B.Tech Outgoing

Kaustav Dutta
Archisman Saha (Web Master of OPTRONIX)
Sahana Lahiri
Subhajit Dutta Chowdhury

IEM B.Tech 4th Year

Kuntal Sengupta
Ritam Ghosh
Sombuddha Chatterjee
Sampoorna Das Mazumdar

UEM B. Tech 2nd Year

Sayak Dhar
Aparna Singh

Sudipta Sikder
Sumit Kumar Ghosh
Sourav Nandy
Mounadeep Poddar
Shubham Roy
Subhashis Paul
Jayita Chatterjee
Gourav Saha
Aiatur Rahman Mallick
Atreyee Hazra
Raya Chowdhury
Rahul Paul
Ankush Pal
Debrup Sarkar
Dwip Biswas
Gunjan Chakraborty
Moinak Nandi
Soumik Dey
Aniruddha Das
Bodhisatya Banerjee
Biplab Gorai
Bishal Chakraborty
Charls Ghosh
Gourab Dipta Ghosh
Souradeep Ganguli
Dwipayan Gupta

Contents

Editors and Contributors

About the Editors

Indrani Bhattacharya is Associate Professor and Head, Research and Development Cell, University of Engineering & Management, Kolkata, India, the newly formed University by well-known Institute of Engineering & Management Group (www.iem.edu.in). Ms. Bhattacharya has secured her M.Sc. (Tech) degree from the Department of Applied Physics, University of Calcutta in Applied Physics with specialization in Opto-electronics in 1988. Ms. Bhattacharya is pursuing her Ph.D. (Tech) from Department of Applied Optics and Photonics, University of Calcutta under the guidance of Emeritus Professor Prof. Lakshminarayan Hazra. Ms. Bhattacharya has a long professional experience ($\sim$20 years) working as a Research Scientist and Engineer in a leading Telecom Industry in India. She has experience in study, development and characterization of Fused Fibre Optics Biconical Taper Couplers, Optical Fibre Joint Boxes, Slotted core and loose tube fibre-optic cables. Ms. Bhattacharya is trained in UNIX, SYBASE, ORACLE with Developer 2000. Ms. Bhattacharya is the Faculty Adviser of SPIE and OSA approved IEM-UEMK Group Student Chapters and she has been enlisted as Women in Optics in the year planner 2016–17 published by SPIE. Prof. Bhattacharya is the Co-coordinator of the Department of Science and Technology (DST), Govt. of India sponsored INSPIRE (INnovation in Science Pursuit for Inspired Research) Science Internship Program, "The Enchanting World of Science" in IEM/UEMK. Ms. Bhattacharya has published ten research works in scientific journals including Conference publications starting from 2011. Her interest of research area includes PSF Engineering, Optical Macro- and Micro-Trapping, Bio-Medical Applications, Light-Matter Interactions, Quantum Computing, Optical Fibre Technology, Nano-Photonics, Solar Concentrator Optics, and so on. Ms. Bhattacharya is the member of Optical Society of India, Optical Society of America, Society of Photonics and Instrumentation Engineering (SPIE) and has been selected as Steering Committee member of International Society of Opto-mechatrinics (ISOM) in Tokyo, Japan. Ms. Bhattacharya is the Convener and TPC chair of International Conference on Opto-Electronics and Applied Optics, OPTRONIX 2014, 2015 and 2016.

Prof. Dr. Satyajit Chakrabarti Pro-Vice Chancellor, University of Engineering & Management, Kolkata and Jaipur Campus, India and Director of Institute of Engineering & Management, IEM, a premier organization in education sector in India, started many entrepreneurial ventures starting from Aug, 2002 to till date. As the Director of one of the most reputed organization in Engineering & Management in Eastern India of Budget—$10 Million with Reporting Team Size 250, he started PGDM Programme to run AICTE approved Management courses, Toppers Academy to train students for certificate courses, Software Development in the field of ERP solutions. Dr. Chakrabarti was Project Manager in TELUS, Vancouver, Canada from Feb, 2006 to Sept, 2009 where he was involved with extensive experience in planning, execution, monitoring, communicating with stakeholders, negotiating with vendors and cross-functional teams, motivating members. He managed Projects a combined budget of $3 million along with 50 people in his team.

Prof. Dr. Chakrabarti has got his B.Tech in Computer Science & Engineering in May 2002 from University of Kalyani, West Bengal, India and got his MS in Computer Science from University of British Columbia (UBC), Vancouver, Canada in Sept 2004. He won scholarships and graduated in the top 1% of the class and led a team of four Computer Engineers to successfully complete the Research Project: "Advanced OS (AVOS): A Real-Time 32 bit Operating System". He is the Convener and General Chair of 7th International Conference on IEMCON which is going to be held in Canada and New York, USA from 15–17 October, 2016 and 20–22 October, 2016.

Vasudevan Lakshminarayanan (aka Vengu) received his Ph. D. from the University of California at Berkeley and is currently a professor of Vision Science, Physics, Electrical and Computer Engineering and Systems Design Engineering at the University of Waterloo. He has held visiting and faculty appointments at the universities of California at Berkeley, at Irvine, University of Missouri, the University of Michigan, Glasgow Caledonian University, Ryerson University and the Indian Institute of Technology, Delhi. He has been selected as a GIAN Professor of Physics at the Indian Institute of Technology, Madras (2016), and has been named as the International Visiting Professor of Biophotonics at the University of Brescia, Italy in 2017. He has had numerous honours, including Fellow of AAAS, APS, IoP, OSA, SPIE, etc., as well as Senior Member status of IEEE and the recipient of various awards, including most recently, the Esther Hoffman Beller medal of OSA (2013), and the Optics Educator award of SPIE (2011). He has been a KITP Scholar at the Kavli Institute of Theoretical Physics at the University of California at Santa Barbara. He is currently the Chair of

Education of SPIE—The International Society for Optics and Photonics and is deeply involved with education and outreach in developing nations. He is a consultant to the ophthalmic and medical devices section of the FDA (from 2013) and is technical editor or section editor of a number of journals including Journal of Modern Optics and Optics Letters. He has authored or edited about 13 books, and over 300 publications in topics ranging from optical physics and engineering, neuroscience, bioengineering, applied math, cognitive science and ophthalmology/optometry and serves on grant review panels for a number of funding agencies both in the US and abroad.

Prof. Haricharan Singh Reehal is head of the Advanced Materials group in the School of Engineering at London South Bank University (LSBU). His research is focused on PV materials and devices, particularly thin-film silicon, low temperature processing and novel concepts for PV devices including nanowires and plasmonics. His interests include improving PV module performance and PV systems. He has wide expertise in thin film growth and in the characterization of materials and devices. Prior to joining LSBU, Prof. Reehal was a lecturer in Physics at Bradford University working on electroluminescent display devices and then Team Leader of the Solid State and Surface Science Groups at BP Research UK. His work at BP ranged from photovoltaics (PV) and novel devices to advanced materials, coatings and surface science. In PV, his team made significant contributions to the development of BP's laser buried grid solar cell technology as well as thin-film crystalline Si and GaAs technologies. Prof. Reehals' work has been supported by EPSRC, Industry and charities and he has been a partner in several research consortia and international collaborations. He has supervised more than 30 Ph.D. students and research fellows and acted as the external Ph.D. examiner in more than 20 national and international vivas. He has organized several successful UK conferences and meetings on PV and is a member of various national and international PV conference committees. He is a member of EPSRC's peer review college and acted as a reviewer for many bodies. He contributed to the UK Energy Research Centre (UKERC) task force on developing a roadmap for PV research. He is a chartered engineer (CEng) and a Fellow of the Institute of Physics. He is on the editorial board of the Journal of Solar Energy and has published more than 100 papers, 4 patents as well as contributed to more than 100 confidential BP technical reports. Prof. Reehal has carried out several senior academic management roles at LSBU including Director of Research in the Faculty of Engineering, Science and the Built Environment.

Contributors

Mona Aggarwal Department of EECE, The NorthCap University, Gurgaon, Haryana, India

Swaran Ahuja Department of EECE, The NorthCap University, Gurgaon, Haryana, India

P.T. Ajith Kumar Light Logics Holography and Optics, Thiruvallom, Thiruvananthapuram, Kerala, India

Sajan Ambadiyil Center for Development of Imaging Technology, Thiruvanathapuram, Kerala, India

Subhash C. Arya Lightwave Communication Systems Laboratory, Department of Electronics and Communication Engineering, North-Eastern Hill University, Shillong, Meghalaya, India

Nagarajan Asvin Centre for Nonlinear Science and Engineering (CeNSE), School of Electrical & Electronics Engineering, SASTRA University, Thanjavur, Tamil Nadu, India

M. Bala Subramanian Department of Electronics and Communication Engineering, SSN College of Engineering, Chennai, Tamil Nadu, India

Narayanan Balaji Centre for Nonlinear Science and Engineering (CeNSE), School of Electrical & Electronics Engineering, SASTRA University, Thanjavur, Tamil Nadu, India

Aritra Banerjee Department of Physics, University of Calcutta, Kolkata, West Bengal, India

Jayeta Banerjee Department of Applied Optics and Photonics, University of Calcutta, Kolkata, India

Jyoti Sekhar Banerjee Department of ECE, Bengal Institute of Technology, Kolkata, West Bengal, India

Meenakshi Banerjee Department of Electronics and Communication Engineering, RCC Institute of Information Technology, Kolkata, West Bengal, India

Rudraishwarya Banerjee Indian Institute of Engineering Science and Technology, Shibpur, Howrah, West Bengal, India

Soumen Banerjee Department of ECE, University of Engineering & Management, Kolkata, West Bengal, India

Sudipta Banerjee Department of ECE, NSHM Knowledge Campus, Durgapur, West Bengal, India

A.M. Basil Department of Photonics, Society of Applied Microwave Electronics Engineering and Research (SAMEER), Mumbai, Maharashtra, India

Rahul Basu Adarsha Institute of Technology, Bangalore, Karnataka, India

Mahua Bera Department of Applied Optics and Photonics, University of Calcutta, Kolkata, West Bengal, India

Sanjay Bhadra Department of EE, University of Engineering and Management, Saltlake, Kolkata, West Bengal, India

Shantanu Bhadra Department of Computer Application, RAMT, MAKAUT, Kolkata, West Bengal, India

Shyamal K. Bhadra Fiber Optics and Photonics Division, CSIR-Central Glass and Ceramic Research Institute, Kolkata, West Bengal, India; Raman Centre for Atomic, Molecular and Optical Sciences, Indian Association for the Cultivation of Science (IACS), Kolkata, West Bengal, India

Saurabh Bhardwaj Department of Photonics, Society for Applied Microwave Electronics Engineering and Research (SAMEER), Mumbai, Maharashtra, India

Vimal Bhatia Signals and Software Group, Discipline of Electrical Engineering, Indian Institute of Technology Indore, Simrol, Madhya Pradesh, India

Anuj Bhatnagar Department of Photonics, Society of Applied Microwave Electronics Engineering and Research (SAMEER), Mumbai, Maharashtra, India

Madhumita Bhattacharya Department of Physics, Gushkara Mahavidyalaya, Gushkara, West Bengal, India

Shanmugam Boomadevi Department of Physics, National Institute of Technology, Tiruchirappalli, Tamil Nadu, India

Litesh Bopche Photonics Laboratory, Institute of Engineering & Technology, Devi Ahilya University, Indore, Madhya Pradesh, India

Gokul Boro Lightwave Communication System Laboratory, Department of Electronics & Communication Engineering, North-Eastern Hill University, Shillong, Meghalaya, India

Maruthi M. Brundavanam Department of Physics, Indian Institute of Technology Kharagpur, Kharagpur, West Bengal, India

Stefania Campopiano Department of Engineering, Centro Direzionale di Napoli Isola C4, University of Naples "Parthenope", Naples, Italy

A.K. Chakraborty Institute of Engineering & Management, Kolkata, West Bengal, India

Ajoy Chakraborty Institute of Engineering and Management, Kolkata, West Bengal, India

Arpita Chakraborty Department of ECE, Bengal Institute of Technology, Kolkata, West Bengal, India

Ipsita Chakraborty Department of Applied Optics and Photonics, University of Calcutta, Kolkata, West Bengal, India

Ritabrata Chakraborty Department of Electronics and Communication Engineering, RCC Institute of Information Technology, Kolkata, West Bengal, India

S. Chakraborty Department of ECE, CSET, Barasat, Kolkata, West Bengal, India

Suprava Chakraborty National Institute of Solar Energy, Gurugram, Haryana, India

Tamalika Chakraborty Department of Electronics and Communication Engineering, RCC Institute of Information Technology, Kolkata, West Bengal, India

Madhuria Chanda Department of Electrical Engineering, University of Engineering & Management, Kolkata, West Bengal, India

V. Charan Teja Department of Electronics Engineering, Madras Institute of Technology Campus, Anna University, Chennai, Tamil Nadu, India

Amit Chatterjee Photonics Laboratory, Department of E & I, Institute of Engineering and Technology, DAVV, Indore, Madhya Pradesh, India

Sombuddha Chatterjee Department of ECE, Institute of Engineering & Management, Kolkata, West Bengal, India

Abir Chattopadhyay Department of ECE, University of Engineering & Management, Kolkata, West Bengal, India

Rik Chattopadhyay Fiber Optics and Photonics Division, CSIR-Central Glass and Ceramic Research Institute, Kolkata, West Bengal, India; Raman Centre for Atomic, Molecular and Optical Sciences, Indian Association for the Cultivation of Science (IACS), Kolkata, West Bengal, India

Sanatan Chattopadhyay Department of Electronic Science, University of Calcutta, Kolkata, West Bengal, India

Partha Roy Chaudhuri Indian Institute of Technology Kharagpur, Kharagpur, West Bengal, India

B.S. Chauhan Instrument Research & Development Establishment (IRDE), Dehradun, Uttar Pradesh, India

Siva Chellappa Centre for Nonlinear Science and Engineering (CeNSE), School of Electrical & Electronics Engineering, SASTRA University, Thanjavur, Tamil Nadu, India

Suyog Choudhari Department of Electronics and Telecommunication Engineering, Ramrao Adik Institute of Technology, Navi Mumbai, Maharashtra, India

Rahul K. Choudhary Department of Physics, Tezpur University, Tezpur, Assam, India

Sheli Sinha Choudhury Department of ETCE, Jadavpur University, Kolkata, West Bengal, India

Basudev Nag Chowdhury Centre for Research in Nanoscience and Nanotechnology (CRNN), University of Calcutta, Kolkata, West Bengal, India

S. Chowdhury Department of ECE, CSET, Barasat, Kolkata, West Bengal, India

Snigdha Chowdhury (Kolay) Department of ECE, CSET, MAKAUT, Kolkata, West Bengal, India

Vijay R. Dahake Department of Electronics and Telecommunication Engineering, Ramrao Adik Institute of Technology, Navi Mumbai, West Bengal, India

Antara Das Department of Electronics and Communication Engineering, RCC Institute of Information Technology, Kolkata, West Bengal, India

S. Das Department of CSE, Indian Institute of Engineering Science and Technology, Shibpur, Kolkata, West Bengal, India

Sneha Das Department of ECE, University of Engineering & Management, Kolkata, West Bengal, India

Soumen Das Department of CSE, UIT, Burdwan University, Durgapur, West Bengal, India

Tania Das Department of Electronics and Communication Engineering, Heritage Institute of Technology, Kolkata, West Bengal, India

U. Das Centre for Lasers and Photonics (CELP), IIT Kanpur, Kanpur, Uttar Pradesh, India

Somshuddha Datta Department of Electronics and Communication Engineering, RCC Institute of Information Technology, Kolkata, West Bengal, India

T. Datta Department of Basic Science and Humanities, Institute of Engineering and Management, Kolkata, West Bengal, India

Moutusi De Optical Fiber Laboratory, Department of Applied Physics, Indian Institute of Technology (Indian School of Mines), Dhanbad, Jharkhand, India

Romi Dey Department of Electronics and Communication Engineering, RCC Institute of Information Technology, Kolkata, West Bengal, India

Sandip Dey Department of Electronics and Communication Engineering, RCC Institute of Information Technology, Kolkata, West Bengal, India

Tumpa Dey Department of IT, Women's College, Agartala, Tripura, India

Arpan Deyasi Department of Electronics and Communication Engineering, RCC Institute of Information Technology, Kolkata, West Bengal, India

Jitendra Dhanotia Signals and Software Group, Discipline of Electrical Engineering, Indian Institute of Technology Indore, Simrol, Madhya Pradesh, India

Reena Disawal Photonics Laboratory, Department of E & I, Institute of Engineering and Technology, DAVV, Indore, Madhya Pradesh, India

Manoj Dongre Department of Electronics and Telecommunication Engineering, Ramrao Adik Institute of Technology, Navi Mumbai, Maharashtra, India

Manoj Kr. Dutta Department of Physics, Birla Institute of Technology, Ratanpur, Jasidih, Jharkhand, India

Ruma Dutta Department of Electronics and Communication Engineering, RCC Institute of Information Technology, Kolkata, West Bengal, India

Flavio Esposito Department of Engineering, Centro Direzionale di Napoli Isola C4, University of Naples "Parthenope", Naples, Italy

Areeba Fatima Department of Physics, Indian Institute of Technology Patna, Patna, Bihar, India

Motoharu Fujigaki Human and Artificial Intelligent Systems, Graduate School of Engineering, University of Fukui, Fukui, Japan

Bani Gandhi Amity Institute of Telecom Engineering and Management, Amity University, Noida, Uttar Pradesh, India

Arvind Ganesh Centre for Nonlinear Science and Engineering (CeNSE), School of Electrical & Electronics Engineering, SASTRA University, Thanjavur, Tamil Nadu, India

M. Ganesh Madhan Department of Electronics Engineering, Madras Institute of Technology Campus, Anna University, Chennai, Tamil Nadu, India

Malay Gangopadhyaya Department of ECE, Institute of Engineering & Management, Kolkata, West Bengal, India

Tarun Kumar Gangopadhyay Fiber Optics & Photonics Division, CSIR-CGCRI, Kolkata, West Bengal, India

Arijit Ganguly Department of Electrical Engineering, University of Engineering & Management, Kolkata, West Bengal, India

Rahul Kumar Gangwar Optical Fiber Laboratory, Department of Applied Physics, Indian Institute of Technology (Indian School of Mines), Dhanbad, Jharkhand, India

K.P. Ghatak Department of ECE, University of Engineering and Management, Kolkata, West Bengal, India

S. Ghatak Department of BCA and M.Sc., Institute of Engineering and Management, Kolkata, West Bengal, India

K.K. Ghosh Department of Electronics and Communication Engineering, Institute of Engineering and Management, Kolkata, West Bengal, India

Solanki Ghosh Department of Electronics and Communication Engineering, RCC Institute of Information Technology, Kolkata, West Bengal, India

Dibyendu Ghoshal Department of ECE and EIE, NIT Agartala, Jirania, Agartala, Tripura, India

J. Gope Department of ECE, CSET, Barasat, Kolkata, West Bengal, India

Jayanta Gope Department of ECE, CSET, MAKAUT, Kolkata, West Bengal, India

Utpal Goswami Department of Electrical Engineering, University of Engineering & Management, Kolkata, West Bengal, India

Rajesh Gupta Department of Energy Science and Engineering, Indian Institute of Technology Bombay, Mumbai, Maharashtra, India

Arindam Haldar Fiber Optics and Photonics Division, CSIR-Central Glass and Ceramic Research Institute, Kolkata, West Bengal, India; Optoelectronics Research Centre, University of Southampton, Southampton, UK

Takumi Hayashi Graduate School of Systems Engineering, Wakayama University, Wakayama, Japan

L.N. Hazra Department of Applied Optics & Photonics, University of Calcutta, Kolkata, West Bengal, India

Sunit M. Hazarika Department of Physics, Tezpur University, Tezpur, Assam, India

Archana Hegde Department of Photonics, Society of Applied Microwave Electronics Engineering and Research (SAMEER), Mumbai, Maharashtra, India

Agostino Iadicicco Department of Engineering, Centro Direzionale di Napoli Isola C4, University of Naples "Parthenope", Naples, Italy

Jijo V. Ittiarah Fiber Optics & Photonics Division, CSIR-CGCRI, Kolkata, West Bengal, India

Vijay Janyani Department of Electronics and Communication Engineering, Malaviya National Institute of Technology, Jaipur, Rajasthan, India

Avinash Kumar Jha Hindustan Institute of Technology & Science, Chennai, Tamil Nadu, India

Pranali Jouras Department of IT, Universal College of Engineering, Thane, Maharashtra, India

Dhiman Kakati Lightwave Communication Systems Laboratory, Department of Electronics and Communication Engineering, North-Eastern Hill University, Shillong, Meghalaya, India

Anupam Karmakar Department of Electronic Science, University of Calcutta, Kolkata, West Bengal, India

Parvinder Kaur Department of Physics, Indian Institute of Technology Delhi, New Delhi, Delhi, India

Triveni Keskar Department of Electronics and Telecommunication Engineering, Ramrao Adik Institute of Technology, Navi Mumbai, Maharashtra, India

Himanshu Khanna Department of EECE, The NorthCap University, Gurgaon, Haryana, India

Yash Khot Department of IT, Universal College of Engineering, Thane, Maharashtra, India

Dhirendra Kumar Department of Physics, Indian Institute of Technology Patna, Bihta, Patna, Bihar, India

Harsh Kumar Department of Electronics and Communication Engineering, Malaviya National Institute of Technology, Jaipur, Rajasthan, India

Laxman Kumar Department of Electronics and Communication Engineering, Malaviya National Institute of Technology, Jaipur, Rajasthan, India

Pradeep Kumar Department of Physics, Indian Institute of Technology, New Delhi, Delhi, India

Sagarika Kumar Department of Energy Science and Engineering, Indian Institute of Technology Bombay, Mumbai, Maharashtra, India

M. Kundu Department of EEE, CSET, Barasat, Kolkata, West Bengal, India

Vasudevan Lakshminarayanan School of Optometry and Vision Science, University of Waterloo, Waterloo, ON, Canada

Khoa Van Le Department of Chemistry, Faculty of Science, Tokyo University of Science, Tokyo, Japan

V.P. Mahadevan Pillai Department of Optoelectronics, University of Kerala, Thiruvanathapuram, Kerala, India

Mainuddin Department of Electronics & Communication Engineering, Jamia Millia Islamia, New Delhi, Delhi, India

Roshan Makkar Photonics Division, Society of Applied Microwave Electronics Engineering and Research (SAMEER), Mumbai, Maharashtra, India

Yogita Mane Department of IT, Universal College of Engineering, Thane, Maharashtra, India

Sampoorna Das Mazumdar Department of ECE, Institute of Engineering & Management, Kolkata, West Bengal, India

Toijam Sunder Meetei Centre for Nonlinear Science and Engineering (CeNSE), School of Electrical & Electronics Engineering, SASTRA University, Thanjavur, Tamil Nadu, India

Sanjana Meshram Department of Electronics and Telecommunication Engineering, Ramrao Adik Institute of Technology, Navi Mumbai, Maharashtra, India

M. Mitra Department of ECE, Indian Institute of Engineering Science and Technology, Shibpur, Kolkata, West Bengal, India

Kshitij Mittholiya Department of Photonics, Society for Applied Microwave Electronics Engineering and Research (SAMEER), Mumbai, Maharashtra, India

Twinkle Mohanty Department of ECE, University of Engineering & Management, Kolkata, West Bengal, India

S. Mondal Department of EEE, CSET, Barasat, Kolkata, West Bengal, India

Tanmoy Mondal Department of CSE, Kalyani Government Engineering College, Kalyani, West Bengal, India

Shouta Morino Department of Chemistry, Faculty of Science, Tokyo University of Science, Tokyo, Japan

S. Mukhopadhyay Department of Applied Optics & Photonics, University of Calcutta, Kolkata, West Bengal, India

Yorinobu Murata Faculty of Systems Engineering, Wakayama University, Wakayama, Japan

Arun V.S. Nair Fiber Optics & Photonics Division, CSIR-CGCRI, Kolkata, West Bengal, India

Prita Nair Department of Physics, SSN College of Engineering, Chennai, Tamil Nadu, India

R. Navamathavan Division of Physics, School of Advanced Sciences, VIT University Chennai, Chennai, Tamil Nadu, India

Shrabani Nayak Department of Electronics and Communication Engineering, RCC Institute of Information Technology, Kolkata, West Bengal, India

Naveen K. Nishchal Department of Physics, Indian Institute of Technology Patna, Bihta, Patna, Bihar, India

Buryy Oleh Institute of Telecommunication Radioelectronics and Electronic Engineering, Lviv Polytechnic National University, Lviv, Ukraine

Divya Pandey Department of Physics, National Institute of Technology, Jamshedpur, Jharkhand, India

G.N. Pandey Department of Physics, Amity Institute of Applied Science, Amity University, Noida, Uttar Pradesh, India

J.P. Pandey Department of Physics, M.L.K. P.G. College, Balrampur, Uttar Pradesh, India

Krishnamoorthy Pandiyan Centre for Nonlinear Science and Engineering (CeNSE), School of Electrical & Electronics Engineering, SASTRA University, Thanjavur, Tamil Nadu, India

P.K. Panigrahi Centre for Lasers and Photonics (CELP), IIT Kanpur, Kanpur, Uttar Pradesh, India

Mohana Kuppuswamy Parthasarathy School of Optometry and Vision Science, University of Waterloo, Waterloo, ON, Canada

Susanta Kumar Parui Indian Institute of Engineering Science and Technology, Shibpur, Howrah, West Bengal, India

Mukesh Patil Department of Electronics and Telecommunication Engineering, Ramrao Adik Institute of Technology, Navi Mumbai, Maharashtra, India

Arkadeep Paul Department of Electronics and Communication Engineering, RCC Institute of Information Technology, Kolkata, West Bengal, India

Mukul C. Paul Fiber Optics and Photonics Division, CSIR-Central Glass and Ceramic Research Institute, Kolkata, West Bengal, India

R. Paul Department of CSE, University of Engineering and Management, Kolkata, West Bengal, India

Somdatta Paul Centre for Research in Nanoscience and Nanotechnology (CRNN), University of Calcutta, Kolkata, West Bengal, India

Sushmita Paul Department of Applied Optics and Photonics, University of Calcutta, Kolkata, West Bengal, India

S. Piramasubramanian Department of Electronics Engineering, Madras Institute of Technology Campus, Anna University, Chennai, Tamil Nadu, India

Shiva Prabhakar Centre for Nonlinear Science and Engineering (CeNSE), School of Electrical & Electronics Engineering, SASTRA University, Thanjavur, Tamil Nadu, India

Somarpita Pradhan Indian Institute of Technology Kharagpur, Kharagpur, West Bengal, India

Shashi Prakash Photonics Laboratory, Department of E & I, Institute of Engineering and Technology, DAVV, Indore, Madhya Pradesh, India

Navin Infant Raj Centre for Nonlinear Science and Engineering (CeNSE), School of Electrical & Electronics Engineering, SASTRA University, Thanjavur, Tamil Nadu, India

Rishabh Raj SENSE, VIT University Chennai, Chennai, Tamil Nadu, India

Biswarup Rana Department of ETCE, Indian Institute of Engineering Science & Technology, Shibpur, Howrah, West Bengal, India

Rajeev Ranjan Department of Engineering, Centro Direzionale di Napoli Isola C4, University of Naples "Parthenope", Naples, Italy; Department of Physics, National Institute of Technology, Jamshedpur, Jharkhand, India

R. Ranjith Department of Electronics Engineering, Madras Institute of Technology Campus, Anna University, Chennai, Tamil Nadu, India

Mina Ray Department of Applied Optics and Photonics, University of Calcutta, Kolkata, West Bengal, India

K.S. Resmi Department of Physics, SSN College of Engineering, Chennai, Tamil Nadu, India

Abhijit Roy Department of Physics, Indian Institute of Technology Kharagpur, Kharagpur, West Bengal, India

Kyamelia Roy Department of ECE, University of Engineering & Management, Kolkata, West Bengal, India

Samudra Roy Department of Physics, IIT Kharagpur, Kharagpur, West Bengal, India

Srinjini Roy Department of Applied Optics and Photonics, University of Calcutta, Kolkata, West Bengal, India

Subinoy Roy Department of Energy Science and Engineering, Indian Institute of Technology Bombay, Mumbai, Maharashtra, India

Tanvi Sachdev Amity Institute of Telecom Engineering and Management, Amity University, Noida, Uttar Pradesh, India

P.K. Sadhu Indian School of Mines, Dhanbad, Jharkhand, India

Soham Saha Department of CSE, Kalyani Government Engineering College, Kalyani, West Bengal, India

Sanyogita Centre for Lasers and Photonics (CELP), IIT Kanpur, Kanpur, Uttar Pradesh, India

Debasmita Sarkar Department of Electronics and Communication Engineering, RCC Institute of Information Technology, Kolkata, West Bengal, India

Debasree Sarkar Department of CSE, UIT, Burdwan University, Burdwan, West Bengal, India

Koushik Sarkar Department of ECE, SVIST, Kolkata, West Bengal, India

Koustuv Sarkar Department of EE, University of Engineering and Management, New Town, Kolkata, West Bengal, India

Takeo Sasaki Department of Chemistry, Faculty of Science, Tokyo University of Science, Tokyo, Japan

Ubizskii Serhij Institute of Telecommunication Radioelectronics and Electronic Engineering, Lviv Polytechnic National University, Lviv, Ukraine

T.L. Shaji Sam Light Logics Holography and Optics, Crescent Hill, Thiruvallom, Thiruvananthapuram, Kerala, India; Optical Image Processing Division and Holography Studio, Centre for Development of Imaging Technology (C-DIT), Thiruvananthapuram, Kerala, India

Pradeep Shankhwar Instrument Research & Development Establishment (IRDE), Dehradun, India

Varsha Shaw Department of Electronics and Communication Engineering, RCC Institute of Information Technology, Kolkata, West Bengal, India

Himanshu Shekhar Hindustan Institute of Technology & Science, Chennai, Tamil Nadu, India

M.R. Shenoy Department of Physics, Indian Institute of Technology Delhi, New Delhi, Delhi, India

Anil Kumar Shukla Amity Institute of Telecom Engineering and Management, Amity University, Noida, Uttar Pradesh, India

Subhrajit Sikdar Department of Electronic Science, University of Calcutta, Kolkata, West Bengal, India

A.K. Singh Instrument Research & Development Establishment (IRDE), Dehradun, Uttarakhand, India

Ghanshyam Singh Department of Electronics and Communication Engineering, Malaviya National Institute of Technology, Jaipur, Rajasthan, India

Kehar Singh The NorthCap University, Gurgaon, Haryana, India

Rakesh K. Singh Department of Physics, Indian Institute of Space Science and Technology, Thiruvananthapuram, Kerala, India

Vinod Kumar Singh Optical Fiber Laboratory, Department of Applied Physics, Indian Institute of Technology (Indian School of Mines), Dhanbad, Jharkhand, India

Pintu Singha Department of Physics, University of Calcutta, Kolkata, West Bengal, India

Gaurav Singhal Laser Science and Technology Centre, New Delhi, Delhi, India

Aloka Sinha Department of Physics, Indian Institute of Technology, New Delhi, Delhi, India

Archana Sinha Department of Energy Science and Engineering, Indian Institute of Technology Bombay, Mumbai, Maharashtra, India

Santa Sircar Department of Applied Optics and Photonics, University of Calcutta, Kolkata, West Bengal, India

Rajpal S. Sirohi Department of Physics, Tezpur University, Tezpur, Assam, India

Jenifar Sultana Centre for Research in Nanoscience and Nanotechnology (CRNN), University of Calcutta, Kolkata, West Bengal, India

Yash Thaker Department of IT, Universal College of Engineering, Thane, Maharashtra, India

Khem. B. Thapa Department of Physics, B.B.A. University, (A Central University), Lucknow, Uttar Pradesh, India

Anukriti Tyagi SCSE, VIT University Chennai, Chennai, Tamil Nadu, India

R.K. Tyagi Laser Science and Technology Centre, New Delhi, Delhi, India

Indrajeet Vadgama Department of IT, Universal College of Engineering, Thane, Maharashtra, India

A.K. Varshney Laser Science and Technology Centre, New Delhi, Delhi, India

Sharda Vashisth North Cap University, Gurgaon, Haryana, India

Anshul Vats North Cap University, Gurgaon, Haryana, India

Avinash C. Verma Laser Science and Technology Centre, New Delhi, Delhi, India

Part I
Keynote Address

Investigation of Degradation in Photovoltaic Modules by Infrared and Electroluminescence Imaging

Archana Sinha, Subinoy Roy, Sagarika Kumar and Rajesh Gupta

Abstract Photovoltaic (PV) modules suffer from a variety of degradation that reduces their long-term performance and reliability. This paper presents a comprehensive investigation of various outdoor degradation in PV modules by spatially-resolved infrared (IR) thermography and electroluminescence (EL) characterization techniques. Active IR thermography has been implemented for the investigation of delamination, corroded interconnects and other electrical losses in a PV module, while EL characterization technique has been exploited for the quantification of discoloration extent in encapsulant material, and detection of finger and cell breakages. Due to fast and non-destructive nature, these imaging techniques can be employed for large-area inspection of PV modules in shorter time.

1 Introduction

Solar photovoltaic (PV) is a promising solution to harness the freely available solar energy and to reduce the dependency on the conventional fossil fuels. In the last few years, large-scale PV deployment has attracted huge attention worldwide after improvement in energy-driven policies and considerable reduction in module cost. The structure of wafer-based crystalline-silicon PV module consists of front transparent tempered glass, polymeric encapsulant (ethyl vinyl acetate, EVA), solar cell string, EVA and a backsheet. Degradation in any of these component layers deteriorates the overall module's performance and its service lifetime. Some of the field studies show that PV modules suffer from a variety of degradation due to the impact of environmental conditions [1, 2]. It mainly includes delamination, discoloration, corrosion, interconnect failures, cracks and broken fingers. However, the type and extent of degradation depend on location, exposure time, module

A. Sinha · S. Roy · S. Kumar · R. Gupta (✉)
Department of Energy Science and Engineering, Indian Institute of Technology Bombay, Mumbai, Maharashtra 400076, India
e-mail: rajeshgupta@iitb.ac.in

© Springer Nature Singapore Pte Ltd. 2017

I. Bhattacharya et al. (eds.), *Advances in Optical Science and Engineering*, Springer Proceedings in Physics 194, DOI 10.1007/978-981-10-3908-9_1

technology and packaging materials used in the module. One such important degradation is delamination, which is the loss of adhesion strength between encapsulant and other layers of module [3]. It decreases the heat transfer efficiency at delaminated regions, leading to hotspot formation. It also increases moisture ingression and accumulation capacity inside the module, which increases the possibility of corrosion of cell metallic parts. Encapsulant discoloration is another common degradation, which directly decreases light transmission to the solar cells, reducing short-circuit current and maximum output power [4]. Other than degradation of encapsulant layers, the electrical defects such as micro-cracks, finger breakage, shunts are also present in the module that decreases its performance. Thus, it is important to detect and characterize such degradation in PV modules to improve their performance and sustainability.

I-V characterization is used to measure the change in maximum power and other electrical parameters. However, it only provides the global module parameters from which it is difficult to find the exact reason of degradation. In order to identify the location of degradation and its induced effects in the module, spatial characterization techniques are required. Camera-based infrared (IR) and electroluminescence (EL) characterization techniques have received high importance due to their inherent advantages of large area inspection in a shorter time while providing good spatial information along with their non-destructive nature and capability to be employed in field conditions by means of some portable equipment. The IR imaging technique had been widely used for solar cell characterization, mainly for shunts investigation [5, 6]. In the present study, it has been exploited for investigation of delamination and other degradation effects in the module which has been less explored. EL imaging is another popular technique, which has been extensively used for mapping of electrical parameters (diffusion length, series resistance) and detection of localized defects in the solar cells and modules [7, 8]. In this study, it has been utilized for quantification of encapsulant discoloration and investigation of severity of some common electrical defects in the PV modules.

2 Experimental Methodology

Experiments were performed on some degraded mono/multi c-Si photovoltaic modules for investigating various kinds of outdoor degradation and corresponding induced effects on the module electrical parameters. For IR thermographic measurements, a cooled IR camera having InSb detector has been used. Two different approaches of IR thermography were applied: by irradiating the module with a heat flux using a halogen lamp, and by passing the electrical current using a programmable power supply under lock-in mode. The IR images of module were taken from the backside of module, since the front glass was opaque to IR radiation.

For EL measurements, a cooled Si-CCD camera has been used to measure EL signal. Under dark surrounding, the current equivalent to I_{sc} under forward bias was injected into the module and the luminescence emission (near-IR region) of solar cells has been captured by an EL camera through the encapsulant and glass.

3 Results and Discussion

The characterization results obtained from IR and EL imaging techniques for investigation of delamination, discoloration and electrical cell defects in the degraded PV modules are presented in this section.

3.1 Delamination

Figure 1a shows the schematic of PV module assembly showing a delamination at solar cell/ EVA interface. In order to detect and characterize this delamination in PV module, a novel method based on active IR thermography approach has been exploited. In this method, a discrete part of module was thermally excited by subjecting a constant heat flux and the corresponding temperature evolution at the back surface was measured by IR camera. The delamination defect perturbs the heat propagation inside the module, which develops a temperature difference on the module surface. The delaminated region has low thermal conductivity compared to non-delaminated region, which build up the heat over the delaminated region that leads to an increase in surface temperature. Figure 1b shows an IR image of PV module where the excess hot area was identified as a delaminated region [9]. In this particular case, a larger portion of cell area is affected by delamination that can deteriorate the cell performance and cause considerable mismatch losses in the module. Figure 1c displays the experimental plot of thermal contrast evolution, whose shape is like a charging curve of capacitor. This plot has a characteristic parameter similar to 'time constant' of RC circuit. This parameter would not be influenced by the heat flux intensity and mainly affected by the thickness of

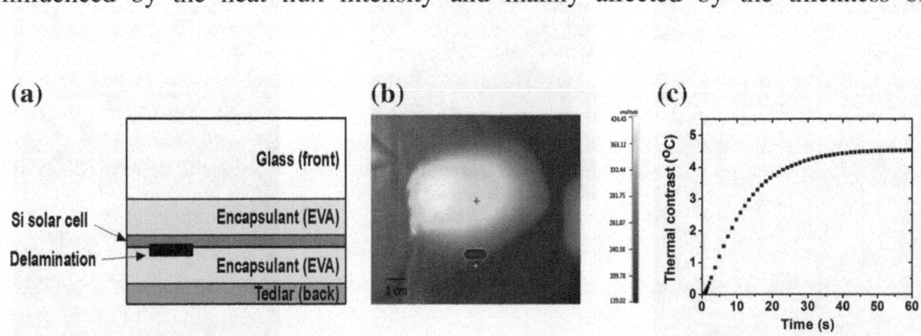

Fig. 1 **a** Schematic of a delaminated PV module, **b** IR image of PV module showing delamination, and **c** thermal contrast evolution plot

delamination. The evolution plot of thermal contrast can be utilized for estimation of delamination size and for determination of optimum detection parameters. It has been realized that active IR thermography provides a fast, non-destructive and non-contact approach to investigate delamination defect in PV modules that can be easily implemented for quality control in the production lines and predictive maintenance in the outdoor fields.

3.2 Encapsulant Discoloration

Encapsulant discoloration is characterized by the color appeared over the cells of module, extending from light yellow shade to dark brown shade. However, visual inspection is not an efficient way to determine discoloration since the interpretation of color may vary for different people. For quantification of discoloration effect, a fast and non-destructive EL characterization method has been presented. In this method, emitted EL signal from the solar cell has been measured by an EL camera. Figure 2a shows an EL image of a discolored module consisting of 36 mono c-Si cells where a peculiar dark circular pattern was observed at the central region of each solar cell. The appearance of dark pattern was ascribed to EVA discoloration, which minimizes the transmission intensity of EL signal. EL image also shows that the boundary of cells is not affected by this discoloration, which is due to photo-bleaching of discolored encapsulant [10]. The extent of reduction in EL intensity over the discolored region can be used to quantify the discoloration extent over the cell. Line scan of EL intensity along the middle of a solar cell from

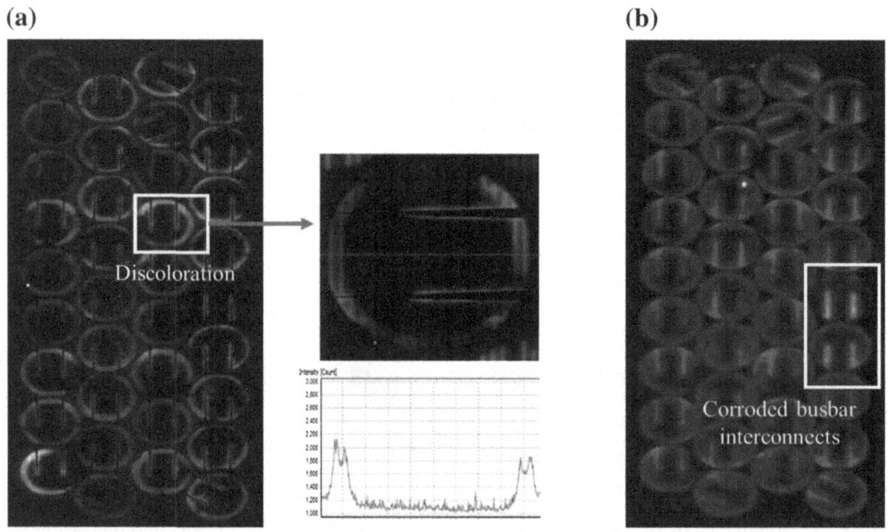

Fig. 2 **a** EL and **b** IR images of a discolored c-Si PV module

discolored module shows that the EL intensity was considerably reduced in the dark circular area as compared to the cell periphery, which can be correlated to the extent of discoloration. This information is useful for quantitative assessment of discoloration in PV modules, which cannot be obtained from visual inspection. Above results demonstrated that EL technique can be used for quantification of discoloration extent in a module as an alternative of direct-light conventional method, which usually requires a large experimental setup.

Generally, discoloration is non-uniform over the module, which produces uneven photocurrent in the cells. This non-uniform generation of current directly leads to mismatch losses in solar module. However, discoloration also leads to uneven electrical defects in module which contribute in mismatch losses. These mismatch losses generate temperature variation over the module, which has been imaged and investigated by IR imaging technique. Figure 2b shows an IR image of same discolored module, which revealed non-uniformity in temperature over the module indicating the variation in electrical parameter of each cell. High temperature was observed over the busbar-interconnects, which was mainly due to series resistance (R_s) increase of corroded metallization parts of cells. The generated heat would accelerate the other degradation effects, such as hotspots and interconnect failures. These results indicate that IR thermography can serve as effective diagnostic technique for non-destructive assessment of adverse effects of discoloration in photovoltaic modules by imaging the temperature variation.

3.3 Degradation-Induced Electrical Defects

Apart from manufacturing faults, the outdoor degradation also generate various kinds of defects in the solar cells such as cracks, finger breakage, finger cuts, corroded front metallization and electrical isolation, which cause significant reduction in the output power. EL imaging has been utilized to find the location and severity of these defects in the module. Figure 3a shows an EL image of degraded module consisting of 36 multi c-Si cells. It shows some dark regions (marked by region-A) that signify electrically inactive regions of the cells. The distributed dark line strips between the busbar of cells (marked by region-B) were ascribed to finger breakages, due to thermo-mechanical stress caused by difference in temperature coefficient of PV materials.

IR imaging has also been used to detect the electrical defects commonly found in the solar cells. Figure 3b shows an IR image of same module in which the region-A also appeared dark. This confirms that these regions were electrically inactive regions that do not contribute in the power generation. The defects associated to finger breakage also appeared dark in IR image as shown in region-B. The current does not flow through these regions, the power dissipation decreases causing reduced IR signal. This effect can also be ascertained by comparing IR image with EL image, which show similar patterns. In addition, some localized bright spots (marked by region-C) corresponding to shunts appeared in the vicinity of busbar

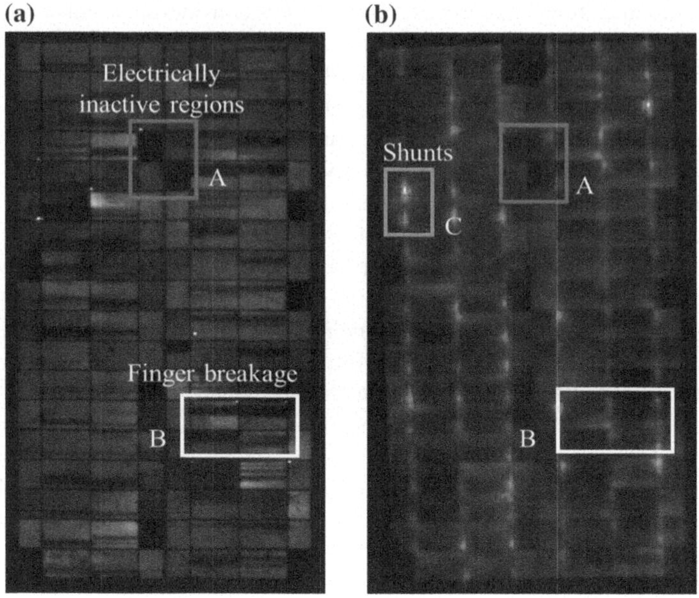

Fig. 3 **a** EL and **b** IR images of degraded c-Si PV module showing electrical defects

and cell edges, which were not easily identified in EL image. Shunt sinks the cell current from its neighboring region that causes localized heating and hence appeared as a hotspot. The above results exemplified that both EL and IR imaging techniques are capable of investigating the electrical problems in a solar cell that cause significant loss in its electrical parameters and overall module performance.

4 Conclusions

In this paper, a variety of outdoor degradation in PV modules has been investigated and their adverse effects on different electrical parameters have been discussed. An approach based on active IR thermography to detect the delamination in a PV module has been presented, which enabled the estimation of delamination size. For investigation of encapsulant discoloration, EL imaging appeared to be a good technique that allowed fast and non-destructive approach for quantification of discoloration extent over the cells in module, whereas IR thermography imaged the presence of electrical mismatch and localized defects induced in the module due to discoloration. Other than these degradation defects, EL and IR imaging techniques have been used for finding the location and severity of electrical defects in the degraded modules. Both these imaging techniques can be used as diagnostic tools for holistic investigation of wide range of degradation and defects in PV modules. These techniques have the capability to be implemented in the production line for quality control and for predictive maintenance of installed PV modules.

References

1. Ndiaye, A., Charki, A., Kobi, A., Kebe, C. M. F., Ndiaye, P. A. and Sambou, V., "Degradations of silicon photovoltaic modules: A literature review," Sol. Energy, 96, 140–151 (2013).
2. Ferrara, C. and Philipp, D., "Why do PV modules fail?," Energy Procedia, 15, 379–387 (2012).
3. Quintana, M. A., King, D. L., McMahon, T. J. and Osterwald, C. R., "Commonly observed degradation in field-aged photovoltaic modules," Proc. 29th IEEE PVSC, 1436–1439 (2002).
4. Holley, W., Agro, S., Galica, J., Thomas, L. and Yorgensen, R., "Investigation into the causes of browning in EVA encapsulated flat plate PV modules," Proc. 4th WCPEC, 893–896 (1994).
5. Kasemann, M., Kwapil, W., Schubert, M. C., Habenicht, H., Walter, B., The, M., Kontermann, S., Rein, S., Breitenstein, O., Bauer, J., Lotnyk, A., Michl, B., Nagel, H., Schutt, A., Carstensen, J., Foll, H., Trupke, T., Augarten, Y., Kampwerth, H., Bardos, R. A., Pingel, S., Berghold, J., Warta, W. and Glunz, S. W., "Spatially resolved silicon solar cell characterization using infrared imaging methods," Proc. 33rd IEEE PVSC, 1–7 (2008).
6. Gupta, R. and Breitenstein, O., "Unsteady-state lock-in thermography – Application to shunts in solar cells," Quant. Infr. Therm. J., 4, 85–105 (2007).
7. Fuyuki, T. and Kitiyanan, A., "Photographic diagnosis of crystalline silicon solar cells utilizing electroluminescence," Appl. Phys. A: Mater. Sci. Process., 96, 189–196 (2009).
8. Camino-Villacorta, M., Coello, J., Perez, L. and Dominguez, F., "Comparative Analysis of Electroluminescence, IR Thermography and Visual Inspection on c-Si Modules," Proc. 27th EUPVSEC, 3363–3365 (2012).
9. Sinha, A., Sastry, O. S. and Gupta, R., "Detection and characterization of delamination in PV Modules by active infrared thermography," Nondestruct. Test. Eval., 31, 1–16 (2016).
10. Czanderna, A. W. and Pern, F. J., "Encapsulation of PV modules using ethylene vinyl acetate copolymer as a pottant: A critical review," Sol. Energy Mat. Sol. Cells, 43, 101–181 (1996).

Solar Holography—A Potential High Efficiency Green Energy Solution

P.T. Ajith Kumar

Abstract Two impediments that acted globally against the progress and popularity of solar electricity were low efficiency and high price of traditional Silicon photovoltaic (PV) panels. Though mass manufacturing and large scale competition between manufacturers have resulted in per Watt price of about 0.3–0.5 USD, low efficiency of these panels still remains as a major hurdle. Concentrator Photovoltaics (CPV) with spectral splitting ability gained worldwide research interest during the recent decades. This paper presents an overview of activities in this highly emerging green energy avenue, some of our work and future design possibilities of relevance.

1 Introduction

Globally there is an increasing and pressing need for the production of green energy. Energy from Sun is not only green but is bestowed free of cost and perpetual also. This is especially relevant for tropical countries that are blessed with abundant sunshine during major part of a year. Due to drastic reduction in price per Watt and aggressive promotional efforts of governments of various countries, recent years witnessed an increase in global popularity of one Sun solar panels. Matured technology, mass manufacturing and increased global competition were other reasons that fuelled the above reduction in price. These PV panels are made of either monocrystalline or polycrystalline Silicon materials. Optimum conversion efficiency under standard test conditions claimed by leading Silicon cell manufacturers is about 18% and this figure reduces to about 12% under true field level conditions. Second generation multi-junction thin film panels are also becoming

P.T. Ajith Kumar (✉)
Light Logics Holography and Optics, Crescent Hill, Thinavila, Thiruvallom, Trivandrum, Kerala 695 027, India
e-mail: ceo@lightlogics.in

© Springer Nature Singapore Pte Ltd. 2017

I. Bhattacharya et al. (eds.), *Advances in Optical Science and Engineering*, Springer Proceedings in Physics 194, DOI 10.1007/978-981-10-3908-9_2

popular now. Leading commercial brands of multi-junction thin film panels have Copper, Indium and Selenium (CIS) as the main ingredients, and include Gallium and Sulphur in the semiconductor layer (CIGS). These panels claim about 14% conversion efficiency and better overall performance under real world conditions where fluctuations in both temperature and solar insolation are realities. Under peak sunshine hours, if we assume an insolation of about 1 kW/m^2, both of the above commercially available panels, in real field conditions, will practically yield only about 120 W/m^2 output. Thus, though the price is affordable, low conversion efficiency still remains as the main impediment for wider global acceptance of PV systems.

Concentrator PV systems use large sun-following reflecting or refracting optical elements to focus intense white light on to multi-junction (MJ) high efficiency cells of about 42% conversion efficiency and are highly attractive due to their high overall output. Currently there are several of such functioning solar plants with large traditional concentrators. The technology of MJ concentrator cells is highly emerging and these cells have already demonstrated an optimum conversion efficiency of over 45% and theoretical predictions of further higher values make it more attractive. However, traditional concentrator optical elements focus heat radiation also on to the cell and this drastically affects the efficiency. Hence special filters and cooling mechanisms are required.

2 Holographic Energy Couplers

Holographic Optical Elements (HOEs) are now effectively applied in a variety of scientific, industrial and Defense applications. High efficiency wavelength filters and dispersive elements, beam combiners and aim sight elements, beam shapers and diffusers, free space wavelength multiplexers and de-multiplexers are just a few examples to mention. Possibility of using HOEs as solar concentrators (Fig. 1) was proposed [1, 2] in the eighties and since then the technology gained increasing attention [3–6] and various module architectures [7] are under consideration. Traditional reflecting and refracting solar concentrator elements focus intense white light on to a MJ cell or MJ arrays, so that various spectral regions pass through the layers of junctions of different spectral sensitivity (Fig.2). On the other hand, Holographic Solar Concentrators (HSCs) are diffractive elements and hence have high spectral selectivity. Spectral response of an HSC under direct sunlight is shown in Fig. 3. Unlike traditional concentrators, HSCs can spectrally split, concentrate and properly direct the solar insolation to specific junctions of different spectral selectivity, resulting in higher and optimum conversion efficiency. HSCs can be designed in such a way that specific infrared wavelengths are diffracted away and the issue of overheating can also be effectively addressed. Thus, HSCs can function as sophisticated energy couplers. Continued improvements in MJ cell

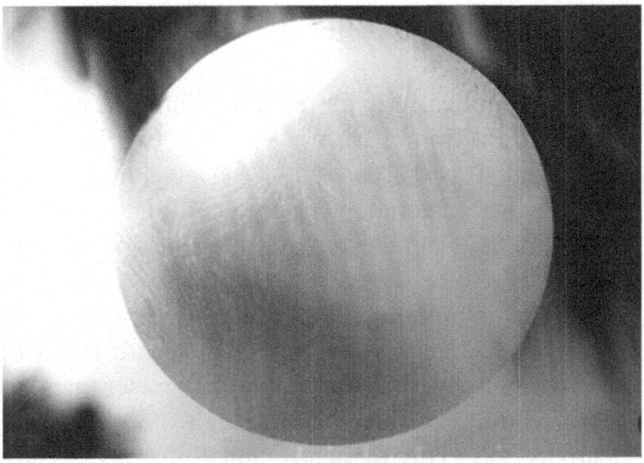
Fig. 1 A holographic solar element set against Sun

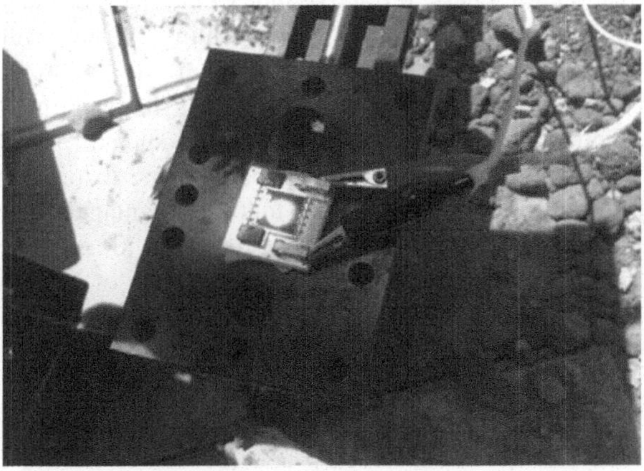
Fig. 2 A Multi-junction concentrator cell under white light illumination

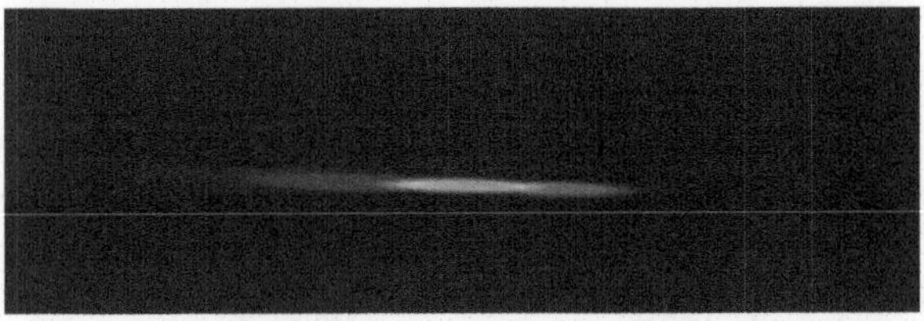
Fig. 3 Spectral response of a transmission type HSC

technology [8] with more number of spectrally selective junctions is a truly encouraging aspect that fuels active research in the development of HSCs and related concentrator modules. Along with this, availability of matured commercial grade photopolymer materials with near 100% diffraction efficiency offer lots of possibilities in recording optimum HSCs for next generation PV applications.

3 Design, Recording and Configuration

A standard optical scheme for recording a transmission HSC is shown in Fig. 4. Coherent light from a laser is split into two beams and both the beams are filtered and expanded to the desired size. Depending upon the spectral sensitivity of the photopolymer material, various recording wavelengths can be chosen. We have used commercial grade red sensitive photopolymer film of 15 μm thickness and homemade photopolymer plates of various thicknesses, coated with Polygrama green sensitive photopolymer emulsion. DPSS lasers of wavelength 639 and 532 nm respectively, were applied to record HSCs of various sizes and properties. Concentration ratio is an important parameter that decides the overall efficiency of a MJ concentrator cell. Along with high spectral selectivity, HOEs have high angular selectivity too. This demands Sun following architecture to effectively and continuously reap the solar insolation. Selection of transmission or reflecting geometry, material thickness, refractive index modulation, selection of substrate material, cost effective copying of the HCS, sealing of the emulsion, encapsulation etc. are essential practical aspects that are to be addressed to achieve commercially acceptable and viable designs. HOEs have high spectral selectivity and multi-function capability. Concentration and spread of each spectral region must match with the cell arrangement and parameters, to facilitate optimum energy coupling. Though photopolymers have lots of advantages over other holographic materials, they demand relatively longer exposure time. This imposes practical limits in recording large format photopolymer HOEs of extremely high concentration ability. However, availability of lasers with abundant power in the green region is a solution to the issue and there are matured photopolymer materials with relatively good response in the green region.

To have a practical HSC system that can yield a sizable electrical output, it is essential to design, develop and incorporate arrays of HOEs that, along with the multi-junction cells, form a holographic concentrator module. Figure 5 shows field

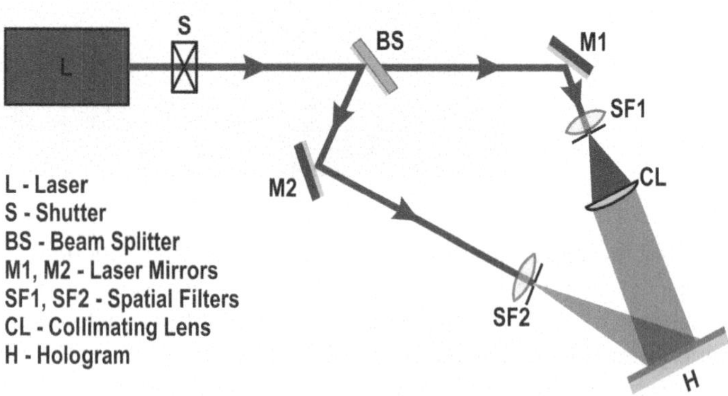

Fig. 4 Optical scheme for recording a transmission type HSC

Fig. 5 Field level spectral splitting and concentration response of one of the photopolymer HSCs

level response of one of the photopolymer HSCs we recorded. Advanced simulation and R&D on optimization of parameters to exactly match with ideal MJ cell arrays and viable module formation are further progressing.

4 Possibilities and Challenges

In sharp deviation from traditional concentrator system architecture, HSCs demand laterally arranged junctions (Fig. 6). Possibility of designing and recording HSCs of wide bandwidth with desired spectral response, higher energy coupling efficiency and lower price in comparison with other competing technologies etc. are added advantages of HSC modules. Possible collection of diffuse radiation component, frequency up conversion and passive breadboard type modules can also bring in revolutionary results. Compared to concentrator modules, one Sun modules demand larger collection area. As in any concentrator based system, HSCs also help to derive maximum energy yield from a minimum area. Weight of HSC based modules is much less than traditional concentrator modules. Possibility of mass copying of HSCs and lower cost of photopolymer materials, compared to silicon panels, are other added advantages. On the other side, per Watt price of MJ cells has to come down further in order to achieve development of commercially viable modules.

Though Sun tracking can improve the output yield, one Sun solar modules do not demand Sun tracking. Also, they are warranted for rated output for about 25 years. Hence, achieving stable long-term performance and freedom from Sun tracking are some of the challenges to be addressed with regard to HSC based module design and development.

Solar holography is emerging and getting matured as a subject of its own space and relevance [9, 10]. National Research Council, Institute for Microelectronics and Microsystems, Italy is one of the centers that conduct R&D in this avenue [11]. There are commercial modules [12] that use holographic planar concentrator films

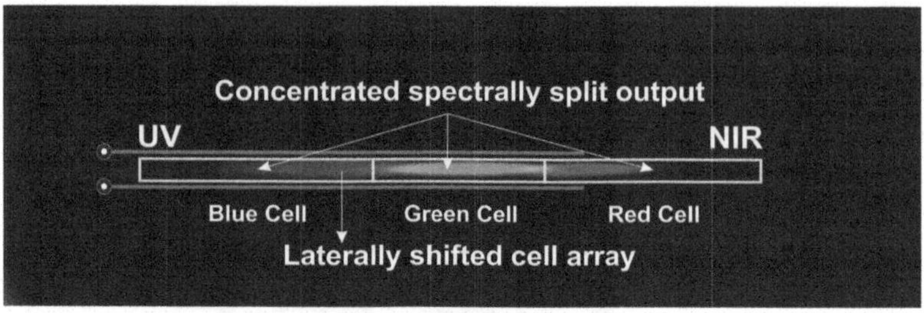

Fig. 6 Schematic of the cell arrangement

in conjunction with bifacial N-type silicon cells and here the architecture is different and the cells used are not multi-junction concentrator cells. Modules made of high efficiency MJ cells with three or more junctions, coupled with matching high efficiency HSCs, can yield exceptionally high energy output from unit area, that too for a lesser cost. However to address the above challenges, holography and semi-conductor technology companies/ laboratories must function hand in hand and in close coherence.

Acknowledgements Partial financial support from the Board of Research in Fusion Science and Technology (BRFST), Government of India, is gratefully acknowledged.

References

1. Ludman. J.E, "Holographic solar concentrator," Appl. Optics 21, 3057–3058 (1982).
2. Bloss, W.H., Griesinger, M., Reinhardt, E.R., "Dispersive concentrating systems based on transmission phase holograms for solar applications", Proc. of 16th IEEE photovoltaic specialist conference, San Diego, Sept, New York, CA, USA, 463, (1982).
3. Zhang, Y.W, Ih. C.S, Yan. H. F and M. J. Chang, "Photovoltaic concentrator using a holographic optical element," Appl. Optics, 27, 3556–3560 (1988).
4. Abhijit Ghosh, R. Ranajan, A.K. Nirala, and H.L. Yadav, Design and analysis of wavelength selective wide acceptance angle holographic concentrator for PV application, Proceedings of the 7th International Conference on Renewable Energy Sources 17, Malaysia (2013).
5. D. Chemisana. D, Collados. M.V, Quintanilla. M, and Atencia. J, "Holographic lenses for building integrated concentrating photovoltaics," Appl. Energ. 110, 227–235 (2013).
6. T. L. Shaji Sam, P. T. Ajith Kumar et al, "A multi-beam holographic light concentrator for solar applications," Proc. SPIE 6832, 68321E (2007).
7. Raymond K. Kostuk et. al, Holographic Applications in Solar-Energy-Conversion Processes, SPIE, Vol.: SL13, April 2016.
8. Thomas. N.D. Tibbits et. al, New Efficiency frontiers with wafer- bonded multi-junction solar cells, 29th European PV Solar Energy Conference, The Netherlands, 2014.
9. Juanita R. Riccobono and Jacques E. Ludman, The Art and Science of Holography - A Tribute to Emmett Leith and Yuri Denisyuk, Editor - H. John Caulfield, Chapter 18. 291, SPIE PRESS, The International Society for Optical Engineering (2003).
10. Jacques Ludman H. John Caulfield, Juanita Riccobono – Editors, Holography for the New Millennium, Chapter 7, 157, Springer, (2002).

11. G. Bianco et. al, Photopolymer-based volume holographic optical elements: design and possible applications, J. Eur. Opt. Soc.-Rapid 10, 15057 (2015).
12. And Prism Solar Technologies technical documentation, link: http://prismsolar.com/pdf/hb180ul.pdf, USA and US Patent No. US 20100186818-Holographic solar concentrator, 2009.

Status and Technology of Present Day Solar Cells

K.K. Ghosh

Abstract First efficient silicon solar cell was invented in 1954 at Bell Labs. Since then research and development of photovoltaic cells advance steadily through generations of much improved and better performing solar devices. First generation silicon solar cell is terminated into thin film second generation semiconductor cells. Organic and Quantum photo cells are penetrating in the market as third generation today to promise for even better optimal performance. The present article is a report of a survey of the past through present solar cell research and development in the photovoltaic industry. Relevant specifics of different theoretical principles of the cell actions and market viability considering the practical feasibilities are also discussed.

1 Introduction

The sun as a burning star sends huge light and heat energy in addition to other radiations for over 4 billion years as a result of nuclear fusion reaction inside. It is wonder to think of the awesome and unbelievable but yet the fact that if one would collect the solar light energy throughout the world in one minute that would supply the world need of electricity for one year; for collection of the same throughout a day would be sufficient to supply the need for 27 years and for a period of three day's storage of solar energy possibly be equivalent to the complete reserve of all fossil fuels in the earth. This strikingly huge possibility of practically unending supply of solar energy is the main motivation that led people to work and attempt to convert the solar energy into electricity. Alexandre-Edmond Becquerel a French physicist, in the year 1839 produced electricity directly from the sunlight. Solar cell research started effectively with this technological invention of Becquerel [1]. Since then solar photovoltaic research and development have passed through

K.K. Ghosh (✉)
Department of Electronics and Communication Engineering, Institute of Engineering and Management, Salt Lake, Sector V, Kolkata, West Bengal 700091, India
e-mail: kkghosh@iemcal.com

© Springer Nature Singapore Pte Ltd. 2017
I. Bhattacharya et al. (eds.), *Advances in Optical Science and Engineering*, Springer Proceedings in Physics 194, DOI 10.1007/978-981-10-3908-9_3

improvements in structures, materials and designing. These different phases of the photovoltaic research and development usher eras of different generations. First generation (1G) solar cells are mainly based upon bulk silicon wafers. The journey took off with the production of first silicon solar cell in the Bell laboratories in 1954 with 6% efficiency. These cells have efficiencies, though not high enough, still higher compared to the next generation cells. Structurally they are well stable and hardy. Disadvantages lie in the degradation of efficiency at high temperatures and in high production cost. Second generation cells are mainly based on thin films of semiconductors. These cells have even lower efficiencies; but they are light weight and require less material and lower production cost. The third generation solar cells are based upon novel use of altogether different types and combination of materials. These cells function through the exotic physical, chemical and opto-electronic properties of materials. This generic evolution is aimed at achieving continually better optimal performance with regard to production cost, efficiency, stability and finally the domestic and industry applicability. To understand the technology of solar cell, basic physics underlying the principle of action of such device is given at the beginning and that is followed by the evolution of generations of solar cells till it is reached at its present status.

2 Basic Physics

Basic principle of action of any solar cell remains almost the same. Three processes viz. (i) generation (ii) separation and (iii) collection of charge carriers are the fundamental steps to pass through in any solar device. The background physics is given below to explain the action of junction solar cell. It is basically a p-n junction diode. In absence of light only a feeble current due to the thermally generated minority carriers flow. This is the dark current. By optical radiation, minority carriers are enhanced in number due to the excitation of electrons from valence to the conduction band (Fig. 1) leaving holes behind in the valence band.

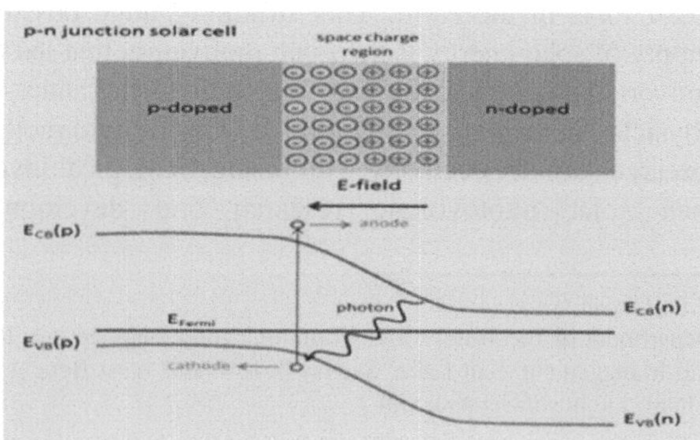

Fig. 1 Junction of p-n materials. Built-in field in the depletion region. Photogeneration of carriers

The electron-hole pair (EHP) forms an exciton with poor binding energy, called excitonic energy. Ambient thermal energy is enough to break this excitonic bond; the released and separated charge carriers are swept immediately in the built-in field of the depletion region such that the electrons are collected at the n-side while the holes are at the p-side. This collection and accumulation of photo-generated charge carriers continue till a balance of the photo-generated electric field with the opposing built-in field sets in. This happens at the point of open circuit voltage v_{oc}. This is the maximum voltage output from a single solar cell.

3 First Generation Solar Cell

Structure and performance of first generation (1G) solar cells mainly depend upon bulk silicon technology. Wafer based monocrystalline silicon (c-Si) p-n junction (Fig. 2) is formed by n-doping over the p-substrate. It is extremely difficult to extract the crystalline silicon from raw materials like SiO_2 though, however, the silica itself is nature abundant. High temperature and consequently huge energy consumption is the prime reason of very expensive technology in their production. The working principle follows the basic physics of photovoltaic effect. Recombination during the carrier transport of the generated carriers is minimised by surface passivation at the top n-surface emitter and by heavy p^+ doping at the rear end of the absorber p-type substrate. The c-Si being an indirect bandgap material it requires phonons to aid valence band electron transition to the conduction band and this make less probable the transition. Loss by heat and less performance efficiency result in. It also suffers from the efficiency instability at high working temperatures. Efficiency of the c-si solar cells typically falls within 16–20% while upper limit reported [2] for best laboratory based c-Si cells being 25%. The realization of drawbacks and demerits lying with this earliest version of solar cells soon prompted the photovoltaic research to make a different class of solar cells. This begins the era of second generation of solar cells.

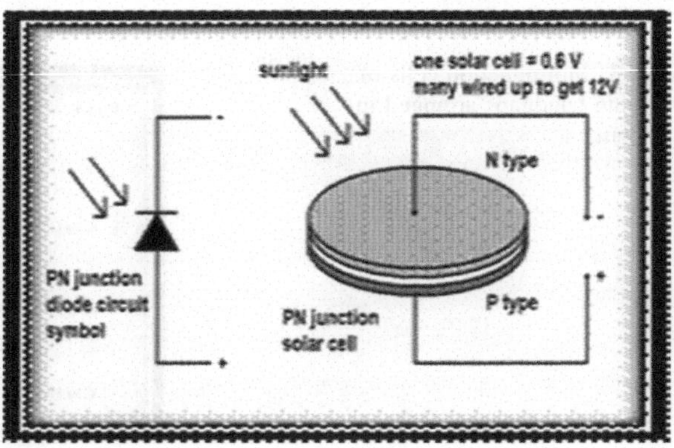

Fig. 2 Monocrystalline silicon junction solar cell

4 Second Generation Solar Cell

The second generation (2G) cells, for their structures and performance, deposition. Relatively lower temperature with less energy consumption and easier fabrication methods attracted industry and R&Ds primarily depend upon properties of thin film semiconductors. Thin films of amorphous silicon (a-Si), chalcogenide group of materials (sulphide, selenide, telluride) like CdTe, CIGS, CIGSe etc. are deposited on appropriate substrates. The fabrication of thin films is done by vacuum to innovate even better materials towards finding much improved performance efficiencies and lower cost per watt production. Laboratory based 2G cells reach efficiency to a level of 18.4% [3]. Multijunction cells of varying bandgaps are stacked one over another in tandem (Fig. 3) for broader coverage of the solar spectrum with enhanced efficiencies. Direct bandgap materials are chosen for these photovoltaic cells. Compared to the monocrystalline c-Si, these are cheaper than first generation but of lesser efficiencies. Because of slower rate of production and some sort of inherent instabilities in the cell functions owing to the bad contacts or weathering etc. this generation also had to leave space for the newer generation of devices to come up.

5 Third Generation Solar Cell

With a view to improve still better than the second generation in respect of performance efficiency, cost per watt, high production yield and to overcome any shortcomings in the way of achieving stability, acceptability and market penetration there is a meteoric rise in research work of solar cells with strong focus on finding and using novel materials with exotic photo-electrochemical based physical and chemical properties. These include nanomaterial based solar cells, plasmonic and polymer based solar cells, organic solar cell, organic based dye sensitized solar cells (DSSC) and concentration solar cells. The DSSCs are one of the most developed cells of present generation. Photosensitive organic dye coated transparent

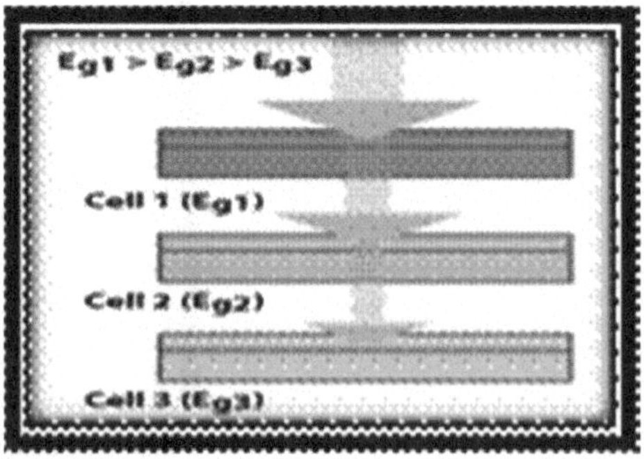

Fig. 3 Multijunction cells of varying bandgaps arranged in tandem

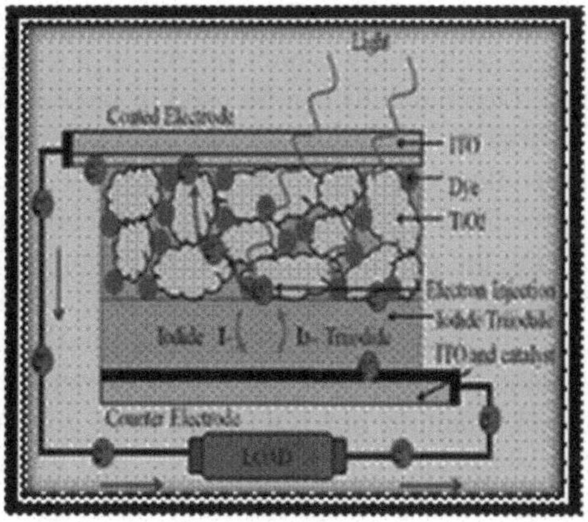

Fig. 4 Dye sensitized solar cell

conductive nanoparticles are embedded in a suitable electrolyte. Photoexcited electrons from the valence to the conduction band of the semiconducting medium find roadway (Fig. 4) through the electrolyte and accumulates at the anode and cathode producing the photovoltage.

These cells are grown on thin films with low production cost and easy to fabricate. Efficiency range is 10–15%. This low efficiency may be compromised against much lower cost per watt production. However attempt is being made to reach much better efficiency through the photo-electrochemical reaction of the dye coated nano-material TiO_2 with solar photons [4, 5]. Easy roll to roll production technology and flexibility attracted more research work on the DSSC devices. Organic photovoltaic (OPV) cells (Fig. 5) have also proved to be promising candidates in the photovoltaic industry. In OPV cells the organic materials, either in small molecular form or in long polymer chain (plastic solar cell), are used as the highly conductive absorbers. These cells work with broad absorption in the visible

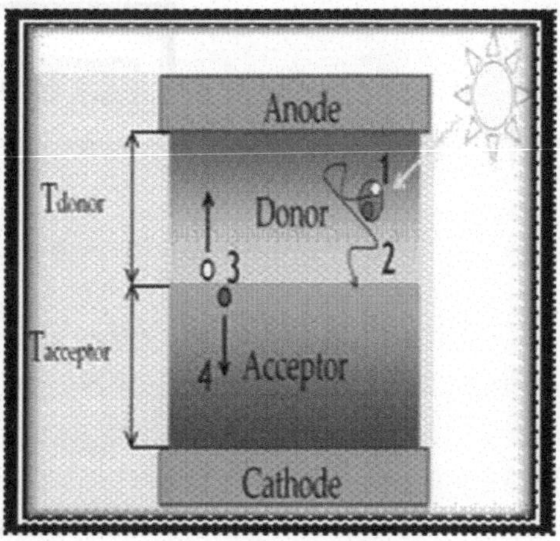

Fig. 5 Bilayer organic solar cell

and near infra-red regions. Fabrication is done by simple and quick process of roll to roll technology which results in a large scale production with less production cost. Organic cells can be used in large area with flexible module structure. Layers of organic materials in the structure act as donor and acceptor of electrons depending on their relative values of electron affinities and ionization energies. Accordingly, the charge carriers accumulate at the anode and cathode ends producing the photovoltage. Recently, conjugated molecule based donor and fullerene based acceptor are also being investigated in making the organic photovoltaic cells [6]. Efficiency of such cells is limited to 11% and reliability remains a problem. Another very prospective cell of present and future generation is the nanomaterial based quantum dot (QD) cell. The dots are made up normally of III–V, II–VI, IV–VI group of semiconductors. Not too complicated synthesis processes like the colloidal and pyrolytic produce these quantum cells reliably. The EHPs are trapped in the reorganized energy levels of the dots. This is due to the quantum confinement of charge carriers leading to the potential well formation in the conduction and valence bands (Fig. 6). Several such QDs are connected together by tunnel barriers and transport of charge carriers takes place by tunneling through these barriers. A very striking feature in these cells is the control of bandgap by control of size and shape of the quantum dots [7]. This band tuning very easily and efficiently makes a broader coverage of the solar spectrum compared to the tandem cells of varying alloy compositions arranged in varying bandgaps. Larger dots results in the red shift of the solar radiation while the smaller dots in the blue shift. In fact, the flexibility of using such cells lies in the smart use of the dots as up and down converters. Efficiency has not reached more than an approximated half of that of DSSCs. However, even against the best performing dye absorbers in DSSCs the dots in QD cells act as better absorbers by at least one order of magnitude. Large concentration of dots limits the photocurrent density and is prone to result in a higher rate of recombination. Future existence of QD cells depend on achieving a reasonably high optimized efficiency and a good cost effectiveness of the product.

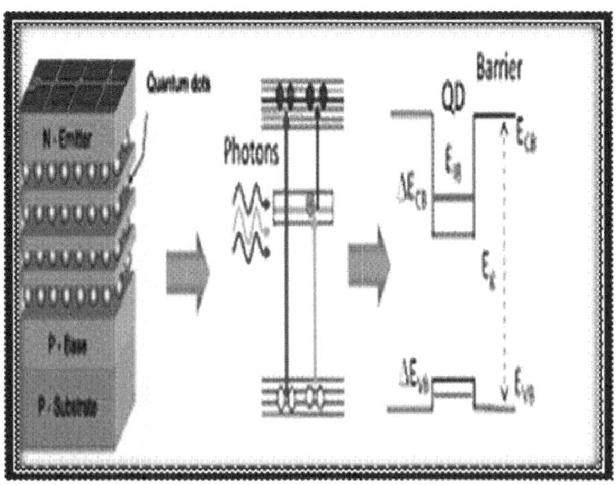

Fig. 6 Quantum dot solar cell

In recent years, perovskite based solar cells have emerged with a challenging offer of increasing performance efficiencies of the photovoltaic products [8]. Basics of physics and chemistry of the perovskite materials are still under studies. These are organic-inorganic hybrid of Pb or Sn-halide based materials. Such cells can be grown in thin films on flat substrates. Solar radiation hits the perovskite absorbing material which is sandwiched between two other layers, the electron transport medium/layer (ETM/ETL) and the hole transport layer (HTM/HTL). The interface field at the heterojunction separates the excitonic charges and helps charge accumulation at the anode metal contact and cathode substrate (Fig. 7).

Plasmonic and Chalcogenide like Kesterite solar cells are also being used in the photovoltaic industry of the present generation. Fundamental idea behind increasing the cell efficiency is to improve the light absorption. In plasmonic cells this is being tried by efficient light trapping by surface plasmons. Noble metal nanostructures e.g. Au, Ag etc. are embedded in the thin film of active semiconductor material grown on an appropriate substrate (Fig. 8). Surface plasmons of frequency identical with the radiation frequency resonate [9]. Light is scattered by plasmon-photon interaction and finally it is trapped in the active layer of semiconductor as it progresses through. This results in strong optical absorption. Efficiency rises to a level of 15% and production cost is justifiably low because of the low cost substrate material even like glass or plastic. Recently the metal-chalcogenide based thin film solar cells also provide efficient alternatives to the silicon based cells. Most notably, prospective potentialities of the CdTe, CIGS(Se) in efficiencies and cost per watt appear to be challenging contenders [10] to the present day solar cells and have already started to come to the market for being used in the industry. Long term degradation due to heat and moisture makes them very stable against weathering. Moderately high efficiency of 13–15%, tunable absorption edges and cost effectiveness are the factors in favour of their market acceptability. An architecture and working of CIGSe based cell (Fig. 9) is given here as an example. The window layer ZnO is highly doped. Due to very high absorption coefficient extremely thin film is enough to strongly absorb the light. Low temperature deposition of thin film of chalcogenide on glass results in very high performance and very low production

Fig. 7 Perovskite solar cell

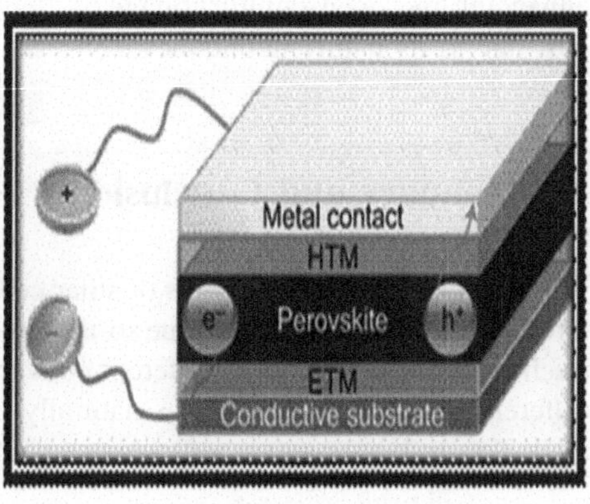

Fig. 8 Plasmonic solar cell

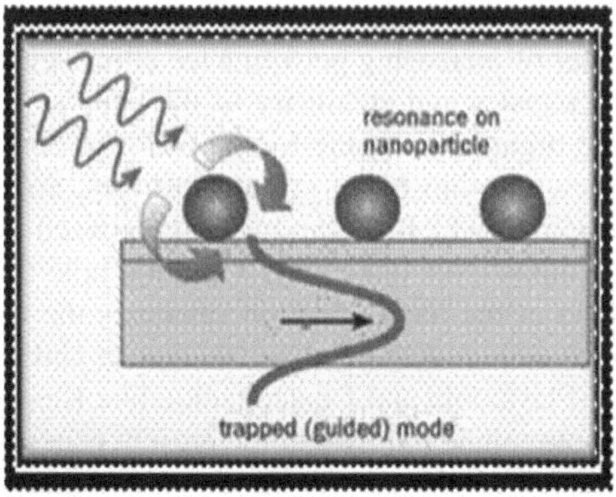

Fig. 9 Architecture and band diagram of CIGSe solar cell

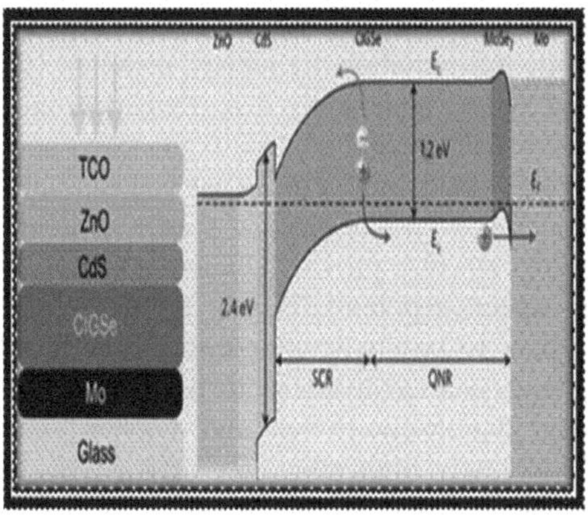

cost of such cells. Efficiency lie in the range 16–20% depending on the substrate. The sharpness of the band profile at the hetero-interface of CdS–CIGSe creates a strong built-in field to break the excitonic bond of EHP. The material component gallium in the structure increases the optical bandgap of the absorbing CIGSe enhancing the open circuit voltage V_{oc} and also helps minimizing the production cost due to its nature abundancy.

6 Summary and Conclusion

It is noticed that different types of solar cells of varying structures and technologies have been introduced from time to time with main objective of practical viability. Each category of them has different conversion efficiency, different output voltage, different output current and has naturally co-existing some merits and demerits as well. To survive, however, efficiency, energy requirement, over-all production cost, material availability and hygienic usability are the key issues to be considered judiciously. In this article, a gradual evolution of the scientific and technological developments of solar cells with their status at different generations is discussed. Number of improvements in the structure designing, material choosing and production technology to promote the solar cells for use in domestic and industrial sectors have been made. To replace the fossil fuels, two factors viz. increased efficiency and low production cost are to be considered with first importance. No generation of solar cells till today can claim 100% free from disadvantages. Each suffers simultaneously from some advantages and disadvantages. The 1G product till today credits for unbeatable conversion efficiency but falls behind in the production cost compared to the 2G and 3G products. While the cost per watt is low enough for these, the performance efficiencies are poor compared to 1G. Easy fabrication technology for 2G and 3G and low production cost raise hopes for their acceptability. However, on the top of all, the 1G silicon solar cell still captures the market with lion share. Strong R&D only can help the future human generation to see and enjoy the non-conventional solar energy at comfortable and at minimal cost economy.

Acknowledgements The author thankfully acknowledges Prof. Dr. S. Chakrabarti (Sr.), the Founder and Former Director of the Institute of Engineering & Management (IEM), Kolkata and presently the Honorable Chancellor of University of Engineering & Management (UEM), Kolkata for providing necessary support in preparing the article.

References

1. Solar Cell Research Activities at Plasma & Material Processing group-Adriana Creatore, Richard van de Sanden and Erwin Kessels Jurgen Palmans, Ilker Dogan, Samir Hanssen, Sjoerd Smit, Bas van de Loo, Bart Macco Stefan Bordihn, Harm Knoops, Diana Garcia Alonso, Kashish Sharma, Technische Universitiet Eindhoven- University of Technology
2. Zhao J, Wang A and Green MA, "24.5% efficiency Silicon PERT cell on MCZ substrates and 24.7% efficiency PERL cells on FZ substrates", Progress in Photovoltaics, Research and Applications, 7:471–4 (1999)
3. http://www.altenergy.org/renewables/solar.html
4. Li B, Wang L, Kang B, Wang P and Qui Y, "Review of Recent Progress in Solid State Dye Sensitized Silar Cells", Solar Energy Materials and Solar Cells, **90**, 549–573 (2006)
5. Jhan J, Sun P, Jiang S, Sun X and Lund T, "An investigation of the performance of Dye Sensitized Nano-crystalline Solar Cell with Anthocyanine Dye and Ruthenium Dye as the sensitizer", Roskilde University, Roskilde
6. Scharber M C, "On the efficiency limit of conjugated Polymer: Fullerene based Bulk Heterojunction Solar Cells", Advanced Materials, **28(10)**, 1994–2001 (2016); DOI: 10.1002/adma.2015049140
7. Kongkanand A, Tvrdy K, Takechi K, Kuno M and Kamat P V, "Quantum Dot Solar Cells. Tuning Photoresponse through Size and Shape control of CdSe-TiO_2 architecture, Jour. Am. Chem. Soc, **130 (12)**, 4007–4015 (2008)
8. Park N G, "Perovskite Solar Cells: an emerging photovoltaic technology", Materials Today, **18 (2)**, 65–72 (2015)

9. Jongmin K, Hongsik C, Changwoo N and Byungwoo P, "Surface-plasmon resonance for photoluminescence and solar cell applications", Electron. Mater. Lett., **8**, 351–364 (2012), doi: 10.1007/s13391-012-2117-8
10. Noufi R, "High Efficiency CdTe and CIGS Thin Film Solar Cells: Highlights of the Technologies Challenges", National Renewable Energy Laboratory, National Centre for Photovoltaics

Part II
Plenary and Invited Address

A Brief History of Aberrometry Applications in Ophthalmology and Vision Science

Mohana Kuppuswamy Parthasarathy and Vasudevan Lakshminarayanan

Abstract The measurement of wavefront aberrations of the human eye and applying techniques of adaptive optics to study the eye has exploded in the past two decades. Adapted from astronomy, adaptive optics has revolutionized the field of Ophthalmology and Vision Science since its successful introduction in 1991. It has been exploited for retinal imaging and improving vision correction. This paper reviews briefly the study of wavefront aberrations and its two major applications in vision.

1 Introduction

The human eye is not a perfect optical system. According to Helmholtz (quoted in [1]), "The eye has every possible defect that can be found in an optical instrument and even some which are peculiar to itself; but they are so counteracted that the inexactness of the image which results from their presence very little exceeds, under ordinary conditions of illumination, the limits which are set to the delicacy of sensation by the dimensions of the real cones" [1]. This summary of the optical system of the eye is as valid today as when it was first enunciated. The defects of the human eye can be classified as structural or functional. Structural defects are due to surfaces, media and alignment of various optical components. Functional defects occur due to the impact of these structural defects on retinal image quality. Because of these defects, a diffraction-limited eye would deviate the optical path of the light entering the light causing wavefront aberrations [2]; that is deviations from an ideal reference such as a plane wave or spherical wave. Even after correcting with spectacles or contact lens, normal eyes suffer from wavefront aberrations especially for large pupil sizes, leading to poor visual performance. In addition, the ophthalmologist examining the eye is limited by these aberrations and in

M.K. Parthasarathy (✉) · V. Lakshminarayanan
School of Optometry and Vision Science, University of Waterloo,
200, University Avenue West, Waterloo, ON N2L 3G1, Canada
e-mail: vengu@uwaterloo.ca

© Springer Nature Singapore Pte Ltd. 2017

I. Bhattacharya et al. (eds.), *Advances in Optical Science and Engineering*,
Springer Proceedings in Physics 194, DOI 10.1007/978-981-10-3908-9_4

instrumentation, they impose a limit on the resolution of the internal structures. Vision science has advanced to such an extent that patients with defective vision due to degraded optics can achieve the so called 'supernormal vision' through correction of the aberrated wavefront by means of adaptive optics technology [3].

Adaptive optics (AO) technology which is a closed loop wavefront control system, was originally developed in astronomy to correct the celestial images for dynamic wavefront errors caused by atmospheric turbulence. All AO system perform two general tasks, namely wavefront sensing and wavefront correction (Fig. 1). The shape of the incoming wavefront has to be rapidly determined to analyze the effect of the atmospheric variations. The incoming wavefront is analyzed by a wavefront sensor. The corrective optics consists of a deformable mirror and/or a tip/tilt mirror to compensate for the incoming wavefront shape. This system obviously requires a real time feedback control loop (see for example, Chap. 6 in the book [4] by Folcher et al.). A good review of adaptive optics is given by the classic book by Tyson [5]. This same technology can be used in correcting aberrations caused by the ocular imperfections [6]. AO was first proposed by Horace Babcock, (considered the founder of the field of AO) in 1953 for ground-based astronomical imaging but he was ahead of his time in terms of the technology required [7]. Full wavefront correction for white light imaging was not achieved until the first wavefront sensors and fast deformable mirrors were developed for the Real Time Atmospheric Compensation System at Itek Corporation in Boston in 1972–74 and this is the pioneering standard configuration for adaptive optical systems [8]. A modification of this basic system was installed at the Air Force Maui Optical Station in Hawaii in 1982 [9]. In 1989, the first diffraction-limited astronomical images obtained with AO were published by

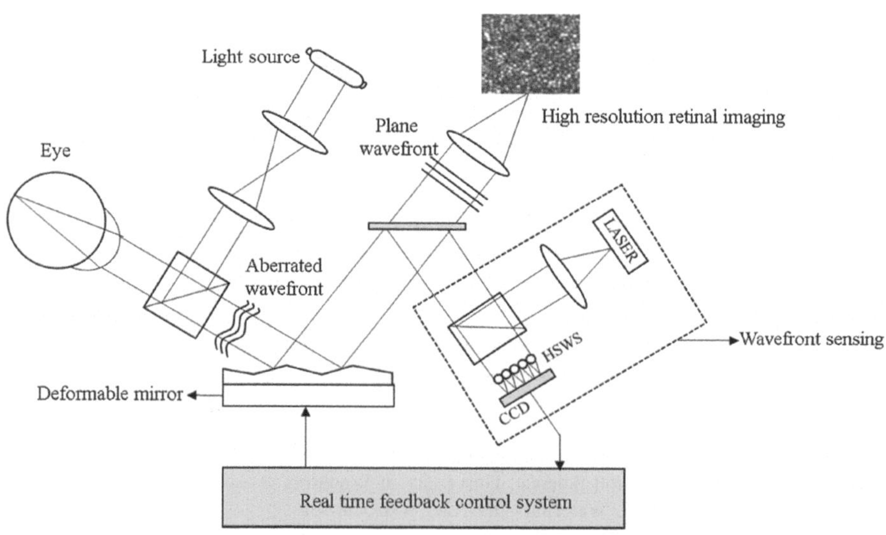

Fig. 1 Adaptive optics system fundus camera (adapted from [31])

Rousset et al. [10]. The history of adaptive optics, originally developed by DARPA and the United States Air Force makes for fascinating reading (see for example, Chaps. 1 and 2 in the book by Hardy (1998) or Duffner (2010) who are both pioneers in the field [9, 11]). The use of adaptive optics in astronomy is given in a recent review by Davies and Kasper [12].

The wavefront aberrations are mathematically described by Zernike polynomials, though other orthonormal polynomials can be used [13, 14]. These polynomials are characterized by two parameters, n, the radial parameter and m, the angular parameter (since we are dealing with circular pupils normally). Aberrations up to radial order 2 are called lower order aberrations and aberrations from radial order $n = 3$ and above are classified as higher order aberrations.

Clinically it should be noted that only the lower order aberrations, defocus (spherical correction), astigmatism (cylindrical correction) and tilt (prism correction) are corrected by spectacles and contact lenses. Higher order aberrations are not. Spectacles have been used to correct defocus since about the 13th century, and possibly earlier by the Chinese [15]. They were used to correct for astigmatism shortly after Thomas Young discovered this defect of the eye [16]. Clinically Herschel understood and applied a sensitive test for astigmatism and determination of the axis of astigmatism which are still used today [17]. However, for the past two centuries very little has been done for correction of higher order aberrations. The reasons for this include, the low order correction using spectacles improves vision to an acceptable level, measuring defocus and astigmatism is relatively easy compared to measuring higher order aberrations which were complex, cumbersome and time consuming. Recent technological advances using adaptive optics techniques have overcome these problems and adaptive optics has been transformed into a useful non-invasive tool to provide supernormal vision to patients with defective vision or to obtain highly resolved images of the retina that enables visualizing individual photoreceptors. Unlike in astronomy it is not particularly critical for real-time wavefront sensing to be very fast for applications in vision. In this short review, we provide a brief history on the study of wavefront aberrations and its applications in vision science for providing hyper vision and hyper resolution imaging of the retina.

2 Methods to Measure Ocular Aberrations

The first step in AO is to measure the ocular aberrations. The next step is to correct them in real time using AO. Several methods have been employed in the past to measure monochromatic aberrations dating back approximately 400 years. In 1619, Scheiner used a disc with two holes that produced two images on the retina [18]. At the beginning of the 20th century, Hartmann modified Scheiner's disc by drilling series of holes in the disc and using it to test the optics of telescope mirrors by tracing the light emerging from the disc [19]. This Hartmann screen test remained unchanged for nearly 70 years until there was a need to measure wavefront aberrations under low illumination conditions. In 1971, Roland Shack replaced the holes with an array of microlenses in order to maximize photon catch. The resultant light from the lenslets were focused onto a CCD array [20, 21] and this modification to Hartmann screen is now widely known as Hartmann-Shack sensor which was used clinically in Ophthalmology in 1991 [22, 23].

Several other methods were used in the past based on subjective ray tracing [24–26], as well as modified aberroscopes [27, 28] to measure aberrations of the eye. Early attempts to measure and correct monochromatic aberrations using an active mirror was initiated by Dreher along with Bille and Weinreb [29]. Dreher et al. at the University of California, San Diego, used an active optical system in the laser tomographic scanner (LTS) to partially compensate for ocular aberrations and provided cross-sectional retinal images of improved depth resolution. This further led to the development of HSWS-based optometer by Liang in the Bille's lab at Heidelberg in 1991 [23, 30]. Liang et al. could measure up to 4th order aberrations using their HSWS with 15 $\times$ 15 lenslets. Within a few years, David Williams and his group at the University of Rochester, built the first closed loop adaptive optics system for the eye, demonstrating the measurement and correction of higher order aberrations, achieving what is called supernormal vision and visualization of single cells in the human retina [3]. They improved the HSWS by increasing the sampling density of the wavefront slopes in the pupil and measured aberrations up to 10th order comprising 65 Zernike modes [31]. By increasing the number of actuators in the deformable mirror of their AO system, they along with Miller could correct the aberrations up to 5th order and obtain high resolution images of cone mosaic and improve the optical quality of the eye [3]. Instead of using deformable mirror, some have used liquid crystal spatial light modulators [32], phase plates [33] or customized contact lenses as an alternative [34]. Although there are many methods to measure ocular aberrations, HSWS that measures the wavefront emerging from the eye produced by reflecting a light spot on the fovea is considered the best method to precisely measure aberrations of the human eye [35] and is commonly used in clinical aberrometers. Fundus cameras, scanning laser ophthalmoscopy (SLO) and optical coherence tomography (OCT) have incorporated AO to achieve near diffraction-limited imaging system as AO improved the lateral and axial/depth resolution of the imaging system [36, 37].

3 Supernormal Vision

Higher the order of aberrations, the lower is its impact on vision, therefore, vision correction beyond sphero-cylinder was less compelling than AO application in retinal imaging, that's the reason why the vision correction modality had not changed in the past years before adaptive optics. As noted earlier after about 700 years of using fixed optics like lenses to correct vision, which is limited only to lower order aberrations, it was proposed by Smirnov in 1961 [26] that it is possible to measure and correct vision beyond defocus and astigmatism [38, 39]. He used subjective measurement of retinal misalignment of rays entering the different parts of the pupil to obtain the ocular aberrations up to 4th order, but it took many hours to get the slope of the light rays at several locations across the pupil [6, 39]. As noted in the previous section, after the development of HSWS based optometer in 1991, Liang, Williams and Miller successfully demonstrated the measurement and correction of ocular aberrations at the University of Rochester in 1997 [3]. They used their AO system with HSWS and 37 channel deformable mirror to correct not only lower order aberrations, but also spherical aberration, coma, and other irregular aberrations up to 5th order aberrations. Improvement in the optical quality was evident with the estimation the eye's modulation transfer function (MTF) and point spread function and measurement the contrast sensitivity. Eyes with adaptive correction could resolve high spatial frequency grating (50 cycles/s) with 40% contrast and had higher MTF with monochromatic light for a 6 mm pupil. Soon after this achievement in the research laboratories, wavefront-guided laser refractive surgery was introduced as a clinical treatment for refractive correction. In 1983, it was found that excimer laser was suitable for ablative photodecomposition that allows to remove corneal tissue in a sub-micron level precision [40]. This is called LASIK (Laser-Assisted In Situ Keratomileusis). This procedure combined with the assessment of wavefront aberrations led to the first wavefront guided refractive surgery on human eyes reported in 1999 by Mrochen et al. [41]. Here the wavefront errors are used with biophysical models of the cornea to determine how much tissue to remove to re-shape the cornea [42]. Mrochen et al. used Tscherning wavefront aberrometer to measure the aberrations [41]. The laser ablation pattern in this procedure, unlike the conventional LASIK, was based on the measurement of the total aberrations of the individual's eye, which reduced the common symptoms of glare, monocular diplopia after conventional surgery.

4 Retinal Imaging

Retinal imaging is an integral part of every ophthalmic examination for early diagnosis and follow up of retinal diseases. Retinal image quality has been a major concern since 1886 when the first retinal photograph was captured by Jackman and Webster [43]. Improving the resolution of the retinal images has always been the target for biomedical and optical engineers. The main limitation to improve resolution is diffraction in a small pupil and aberrations in a larger pupil. AO technology, has helped in improving the quality of the retinal images. As mentioned earlier Dreher et al. first applied AO in a LTS, which is a scanning laser microscope for studying the nerve fibre layer thickness and optic disc topography [29]. They could achieve two-fold improvement in depth resolution by using an active mirror with 13 actuators, which corrected mainly the measured astigmatism of the eye. A further improvement in depth resolution was shown by Liang et al. who incorporated HSWS in LTS [23]. Furthermore, AO was applied in a fundus camera in 1997 by Liang et al. and they obtained retinal images that could resolve the cone photoreceptors [3]. However, first direct images of the cone mosaic were obtained by Miller et al. in 1996 using a high magnification fundus camera and carefully correcting the lower order aberrations (defocus and astigmatism) [44]. Implementing AO in fundus camera was not difficult because it only involves introducing the sensor and deformable mirror within a simple camera optics (refer to Fig. 1) [25].

Invented by Robert Webb in 1980 [45], the SLO provides retinal images in real time by scanning a single focussed spot across the retina. SLO combined with a pinhole conjugate with the focal plane of the laser beam on the retina makes it confocal (cSLO), which blocks the light scatter from out-of-focus planes providing high contrast optical slices [46]. In 2002, Burns et al. used phase-correcting plates with cSLO to correct for higher order aberrations and could achieve 26% increase in contrast of the retinal blood vessels [47]. About the same time, Roorda et al. presented the first AO-cSLO, using HSWS to measure aberrations and a 37-channel deformable mirror to compensate for the measured aberrations, providing real-time, microscopic view of the human retina and even resolve the cells inside the blood vessels enabling the study of blood flow in the capillaries [36]. The resolution achieved with AO-cSLO was 2.5 μm lateral and 100 μm axial, compared with 5 μm lateral and 300 μm axial in conventional SLO. Yet another widely used retinal imager, optical coherence tomography (OCT) developed by Huang et al. [48] has been integrated with AO. The combination of enface coherence-gated camera and AO produced unprecedented axial and transverse resolution [49]. Soon after this, AO was integrated with ultrahigh resolution OCT [37] and spectral domain OCT [50].

With the advent of AO for retinal imaging and the photoreceptors being the first microstructure to be resolved by AO, researchers started studying the arrangement of cone mosaic [51], waveguide property of the photoreceptors [52], temporal variability in their reflectance [53], and so on. The neural cells namely ganglion cells, bipolar cells and photoreceptor nucleus are not resolved due to low light scatter but studies have used fluorophores to view them in animal studies [35]. First application of adaptive optics on a clinical patient was reported by Roorda in 2000 [54]. He observed disruptions in the cone mosaic of the patient who was diagnosed as cone-rod dystrophy. There is substantial literature on the study of AO images of retinal degenerative diseases like retinitis pigmentosa (RP), cone-rod dystrophy and stargardt disease and its associated genetic mutations [55]. Microstructures of the nerve fibers and lamina cribrosa of the optic disc provide insights into the disease state [56, 57]. A detailed review on clinical applications of AO can be seen in the publications by Carroll et al. [55] and Roorda et al. [35]. Using AO could possibly help diagnose early stages of eye diseases before it could be detected with standard fundus imaging tools, for example, in diseases like glaucoma where the structural change can be detected only after 25–35% of the ganglion cells are lost [58]. Makous et al. [59] used microperimetry equipped with AO to detect and map the microscotomas of a patient with deuteranopia. This combination of optics helped in reducing the spot size of the stimulus to confine to the size of the cone at 0.5° eccentricity. With AO images, the effects of experimental treatments on hereditary retinal degenerations like RP, can be monitored at cellular level and it makes it a sensitive test of disease progression [60]. The clinical applications of AO enabled ophthalmic imaging in a vast area of investigation.

5 Conclusion

Since the time Babcock, proposed the idea of adaptive optics, there have been dramatic advances in technology and applications. In fact, most ground based telescopes being built have adaptive optics incorporated in them Applications in ophthalmology and optometry has changed the outlook of vision correction and retinal imaging. Using wavefront optometers, it is possible to give a meaningful vision diagnosis and improving the optical quality of the eye would help in studying the neural limits of vision and early diagnosis of retinal disease conditions.

References

1. Charman, W.N. Optics of the human Eye, In: Visual Optics and Instrumentation, MacMillon, London, 1–14 (1991).
2. Helmholtz, H. Handuch der physiologischen optik, Leipzig: Leopold Voss, 137–47 (1867).
3. Liang, J., Williams, D.R. and Miller, D., J. Opt. Soc. Am. A 14, 2884–92 (1997).
4. Folcher, J.P., Carbillet, M., Ferrari, A. and Abelli, A. Adaptive Optics Feedback Control, In: New Concepts in Imaging: Optical and Statistical Models, D. Mary, C. Theys and C. Aime (eds), EAS Publications Series, 59, 93–130 (2013).
5. Tyson, R.K. Principles of Adaptive Optics, CRC Press, Boca Raton, Fl, (2010).
6. Porter, J., Queener, H., Lin, J., Thorn, K. and Awwal, A. A. Adaptive optics for vision science: principles, practices, design and applications, John Wiley & Sons. (2006).
7. Babcock, H.W., Publ. Astron. Soc. Pac. 65, 229–236 (1953).
8. Hardy, J. W., Feinleib, J. and Wyant, J. C., Digest of Technical Papers, Topical Meeting on Optical Propagation through Turbulence, sponsored by Opt. Soc. of America, Boulder, Colo., Paper ThB1, July (1974).
9. Hardy, J.W. Adaptive optics for astronomical Telescopes, Oxford University Press, Oxford, UK, (1998).
10. Rousset, G., Fontanella, J. C., Kern, P., Gigan, P. and Rigaut, F., Astronomy and Astrophysics, 230, L29–L32 (1990).
11. Duffner, R. The adaptive optics revolution: a history, University of New Mexico Press, Albuquerque, NM, (2010).
12. Davies, R. and Kasper, M., Annual Review of Astronomy and Astrophysics, 50: 305–351 (2012).
13. Lakshminarayanan, V. and Fleck, A., J. Modern Optics, 58, 545–561 (2011).
14. Lakshminarayanan, V. and Varadharajan, L.S. Special functions in optical science and engineering, SPIE Press, Bellingham, WA. (2015).
15. http://www.college-optometrists.org/en/college/museyeum/online_exhibitions/spectacles/invention.cfm, Last accessed 6 September 2016 5.26 pm.
16. Young, T., Bakerian Lecture, Philosophical Transactions of the Royal Society, 91, 23–88 (1801).

17. Enoch, J.M., Heitz, R.F. and Lakshminarayanan, V., Ophthalmic and Physiological Optics, 8, 349–350 (1988).
18. C. Scheiner, Oculus, sive fundamentum opticum, Innspruk, (1619).
19. Hartmann, J., Z. Instrumentenkd 20, 47 (1900).
20. Platt, B. C. and Shack, R., J. Refract. Surg. 17, S573–S577 (2001).
21. Schweigerling, J., Proc. SPIE, 9186, 91860U-1–91860U-8 (2014).
22. Goelz, S., Persoff, J.J., Bittner, G.D., Liang, J., Hsueh, C.T. and Bille, J.F., Proc. SPIE 1542, 502–511 (1991).
23. Liang, J., Grimm, B., Goelz, S. and Bille, J. F., International Society for Optics and Photonics. 543–554 (1991).
24. Tscherning, M., Z. Physiol Sinn, **6**, 456–471, (1894).
25. Ivanoff, A., C.R. Acad. Sci. 233, 170–172 (1946).
26. Smirnov, M.S., Biophysics, 6, 776–794 (1961).
27. Howland, H. and Howland, B., J. Opt. Soc. Am. 67, 1508–1518 (1977).
28. Walsh, G., Charman, W. N. and Howland, H., J. Opt. Soc. Am. A 1, 987–992 (1984).
29. Dreher, A., Bille, J.F. and Weinreb, R. N., Appl. Opt. 28, 804–808 (1989).
30. Liang, J., Grimm, B., Goelz, S. and Bille, J., J. Opt. Soc. Am. A 11, 1949–1957 (1994).
31. Liang, J. and Williams, D. R., J. Opt. Soc. Am. A, 14(11), 2873–2883 (1997).
32. Vargas-Martın, F., Prieto, P. M. and Artal, P., J. Opt. Soc. Am. A, 15(9), 2552–2562 (1998).
33. Navarro, R., Moreno-Barriuso, E., Bará, S. and Mancebo, T., Optics letters, 25(4), 236–238, (2000).
34. Dietze, H.H. and Cox, M.J., J Opt Soc Am A Opt Image Sci Vis. 21(4):473–85 (2004).
35. Roorda, A. and Duncan, J. L., Annual review of vision science, 1, 19–50 (2015).
36. Roorda, A., Romero-Borja, F., Donnelly III, W.J., Queener, H., Hebert, T.J. and Campbell, M.C.W., Optics Express, 10(9), 405–412 (2002).
37. Hermann, B., Fernández, E. J., Unterhuber, A., Sattmann, H., Fercher, A. F., Drexler, W., Prieto, P. M. and Artal, P., Optics letters 29, 18 2142–2144 (2004).
38. Le Gargasson, J. F., Glanc, M. and Léna, P., Comptes Rendus de l'Académie des Sciences-Series IV-Physics, 2(8), 1131–1138 (2001).
39. Roorda, A., Journal of Vision, 11(5), 6, 1–21 (2011).
40. Trokel, S. L., Srinivasan, R. and Braren, B., Am. J. Ophthal., 96(6), 710–715 (1983).
41. Mrochen, M., Kaemmerer, M. and Seiler, T., J. Refract. Surg., 16(2), 116–121 (2000).
42. Mrochen, M., Donitzky, C., Wüllner, C. and Löffler, J., Journal of Cataract & Refractive Surgery, 30(4), 775–785 (2004).
43. Jackman, W.M. and Webster, J.D., Philadelphia Photographer, 23, 340–341 (1886).
44. Miller, D. T., Williams, D. R., Morris, G. M. and Liang, J., Vision research, 36(8), 1067–1079 (1996).
45. Webb, R.H., Hughes, G.W. and Pomerantzeff, O., Appl. Opt., 19: 2991–2997 (1980).
46. Webb, R.H., Hughes, G.W. and Delori, F.C., Appl. Opt. 26, 1492–1499 (1987).
47. Burns, S. A., Marcos, S., Elsner, A. E. and Bara, S., Optics letters, 27(6), 400–402 (2002).
48. Huang, D., Swanson, E.A., Lin, C.P., Schuman, J.S., Stinson, W.G., Chang, W., Hee, M.R., Flotte, T., Gregory, K., Puliafito, C.A. and Fujimoto, J.G., Science. 254: 1178–1181, (1991).
49. Miller, D. T., Qu, J., Jonnal, R. S. and Thorn, K., Proc. SPIE 4956, 65–72 (2003).
50. Zhang, Y., Rha, J., Jonnal, R. and Miller, D., Optics Express, 13(12), 4792–4811 (2005).
51. Roorda, A. and Williams, D.R., Nature, 397, 520–2 (1999).
52. Roorda, A. and Williams, D.R., J Vis, 2, 404–12 (2002).
53. Pallikaris, A., Williams, D.R. and Hofer, H., Invest Ophthalmol Vis Sci, 44, 4580–92 (2003).
54. Roorda, A., J Refract Surg, 16, S602–S7 (2000).
55. Carroll, J., Dubis, A.M., Godara, P., Dubra, A. and Stepien, K.E., US Ophthalmic Review, 4 (2):78–83 (2011).
56. Takayama, K., Ooto, S., Hangai, M., Arakawa, N., Oshima, S., Shibata, N., Hanebuchi, M., Inoue, T. and Yoshimura, N., PloS one, 7(3), e33158 (2012).
57. Akagi, T., Hangai, M., Takayama, K., Nonaka, A., Ooto, S. and Yoshimura, N., Investigative ophthalmology & visual science, 53(7), 4111–4119 (2012).

58. Kerrigan–Baumrind, L. A., Quigley, H. A., Pease, M. E., Kerrigan, D. F. and Mitchell, R. S., Investigative ophthalmology & visual science, 41(3), 741–748 (2000).
59. Makous, W., Carroll, J., Wolfing, J. I., Lin, J., Christie, N. and Williams, D. R., Investigative ophthalmology & visual science, 47(9), 4160–4167 (2006).
60. Talcott, K.E., Ratnam, K., Sundquist, S.M., Lucero, A.S., Lujan, B.J., Tao, W., Porco, T.C., Roorda, A. and Duncan, J.L., Investigative ophthalmology & visual science, 52(5), 2219–2226 (2011).

Application of Phase-Shifted Fringe Projection Method with Linear Fiber Arrays Using Talbot Effect to Height Measurement of BGA

Motoharu Fujigaki, Takumi Hayashi and Yorinobu Murata

Abstract Recently, authors were proposed a small pitch fringe projection method using Talbot effect with an SLD (super luminescent diode) with inclining the appearing area of the fringe pattern. The phase of the projected fringe is shifted using a linear fiber array with four cores. In this paper, an application of these methods to height measurement of BGA (ball grid array) is performed.

1 Introduction

3D shape measurement systems that use contactless methods are required for quality inspection of metal molds and electronic parts in industrial fields. A fringe projection method with phase-shifting method offers the advantages of high precision and high speed. Some practical industrial inspecting machine employs the fringe projection method to measure the height of BGAs [1].

There are, however, several problems for shape measurement of a small object. One is that the pitch of a projected grating pattern should be produced smaller. A fringe projection method using Talbot effect is a useful method to produce a fringe pattern with small pitch. Several shape measurement methods using Talbot effect [2, 3] was studied [4–6].

M. Fujigaki (✉)
Human and Artificial Intelligent Systems, Graduate School of Engineering,
University of Fukui, Bunkyo 3-9-1, Fukui 910-8507, Japan
e-mail: fujigaki@u-fukui.ac.jp
URL: http://www.os.his.u-fukui.ac.jp

T. Hayashi
Graduate School of Systems Engineering,
Wakayama University, Sakaedani 930, Wakayama 640-8510, Japan

Y. Murata
Faculty of Systems Engineering,
Wakayama University, Sakaedani 930, Wakayama 640-8510, Japan

© Springer Nature Singapore Pte Ltd. 2017
I. Bhattacharya et al. (eds.), *Advances in Optical Science and Engineering*,
Springer Proceedings in Physics 194, DOI 10.1007/978-981-10-3908-9_5

Recently, authors were proposed a small pitch fringe projection method using Talbot effect with an SLD (super luminescent diode) with inclining the appearing area of the fringe pattern [7]. The phase of the projected fringe is shifted using a linear fiber array with four cores [8]. A technique to reduce noise with vibrating reference plate was proposed [9]. In this paper, an application of these methods to height measurement of BGA is shown.

2 Talbot Fringe Projected on Reference Plane and Object

2.1 Talbot Effect with Tilted Grating Plate

The Talbot effect is a well-known interference phenomenon of light [2, 3]. When collimated light passes through the grating plate, a fringe pattern of the same pitch as the grating plate appears at regular distances away from the grating plate. The fringe pattern can be projected without using a projection lens. The regular distance z_r called a Talbot length is obtained from (1),

$$z_r = \frac{d^2}{\lambda}, \tag{1}$$

where λ is the wavelength of the light source and d is the pitch of the grating plate.

Figure 1 shows the Talbot effect and phase shifting of the projected fringe in the case that the grating plate is tilted. When the angle of tilting from the optical axis is a, The Talbot length z_r is obtained from (2),

$$z_r = \frac{(d\cos\alpha)^2}{\lambda}. \tag{2}$$

The projected fringe patterns appear at $z = z_r, 2z_r \ldots nz_r$. The projected fringe pitch p is shown as (3),

$$p = d\cos\alpha. \tag{3}$$

The phase of the projected fringe patterns can be shifted by moving the light source.

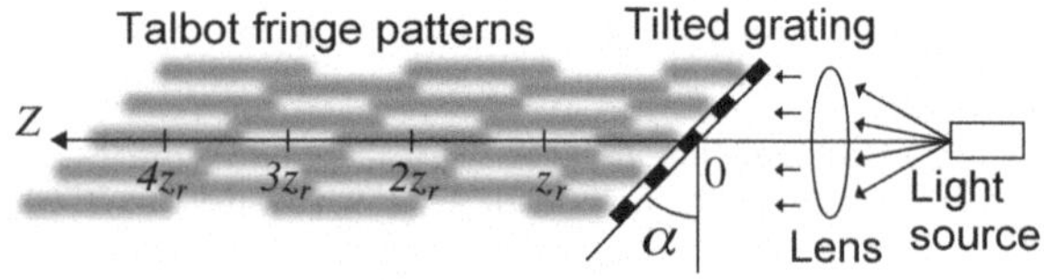

Fig. 1 Talbot fringe patterns produced with tilted grating

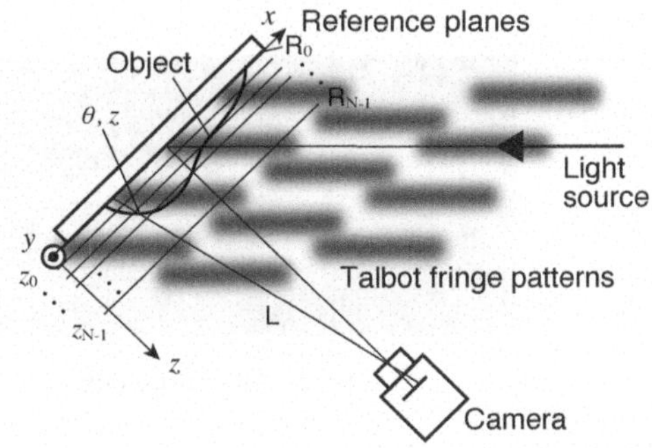

Fig. 2 Talbot fringe patterns projected on a reference plate and an object

2.2 *Phase-Height Table Produced with Whole-Space Tabulation Method*

Figure 2 shows an illustration of Talbot fringe patterns projected on a reference plate and an object. The appearance area of the Talbot fringe is also tilted with an angle of the tilted grating shown in Fig. 1. The reference plate is located on the optical axis with tilting with the same angle.

The reference plate is moving for the normal directions (z-direction) bit by bit. A phase-height table is produced from the phase of the projected fringe pattern and the z position pixel by pixel.

An object is located instead of the reference plate after producing the phase-height table. The phase map of the Talbot fringe pattern projected on the object is obtained with the phase-shifting method. The height distribution can be obtained from the phase map and the phase-height table using the whole-space tabulation method (WSTM).

3 Experiment

3.1 *Experimental Setup*

Figure 3 shows a developed light source. This light source has 4 super luminescent diodes (SLDs) inside a box. An SLD with 7 mW of power at a wavelength of 965 nm and a half-value width of the wavelength of 15 nm is used as a light source. A linear fiber array with 4 single mode optical fibers is used. Each light from SLD is emitted from each end of the optical fiber as shown in Fig. 3b. The 4 SLDs are turned on in sequence to perform phase-shifting. Changing lighting position gives phase-shifting of the projected Talbot fringe pattern. The phase-shifting amount can be adjusted with rotating the end of the optical fiber.

An arrangement and the photograph of an experimental setup is shown in Fig. 4a, b, respectively. A lens with the focal length of 100 mm is located in front

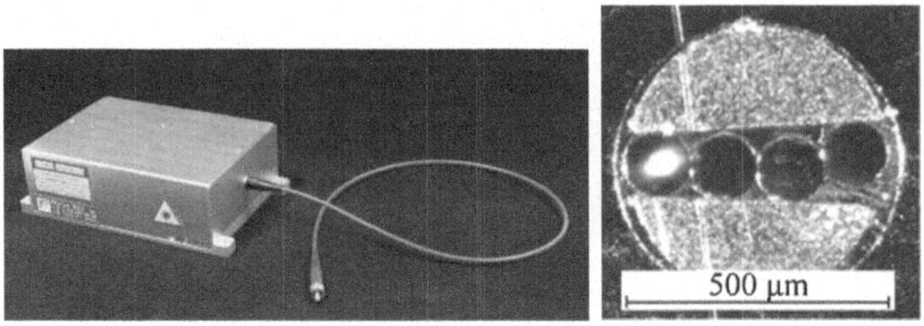

(a) Light source device (b) End of optical fiber

Fig. 3 Developed light source device

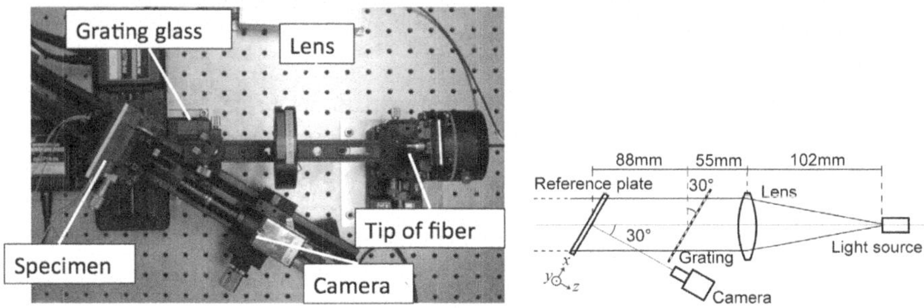

(a) Photograph (b) Arrangement

Fig. 4 Experimental setup

of the light source. A grating plate with 169.3 μm pitches is located on the optical axis and it is tilted with 30° from the optical axis. Talbot fringe patterns of 147 μm pitch are generated by the grating plate and it is projected on a reference plate tilted with 30° from the optical axis. A fringe pattern with around 170 μm pitch appears on the reference plane.

The reference plate has a scattered rough surface. The surface has some unevenness. Figure 5 shows an illustration of a proposed idea of smoothing of rough surface with vibrating reference plate. The unevenness causes much granulated noise in general. In this experiment, the amplitude of the vibration is 40 mm and the frequency of the vibration is 20 Hz.

3.2 Height Measurement of BGA

Figure 6 shows a specimen of BGA. The height of each ball is 200 μm approximately. The pitch of balls is 500 μm approximately. Figure 7 shows results of

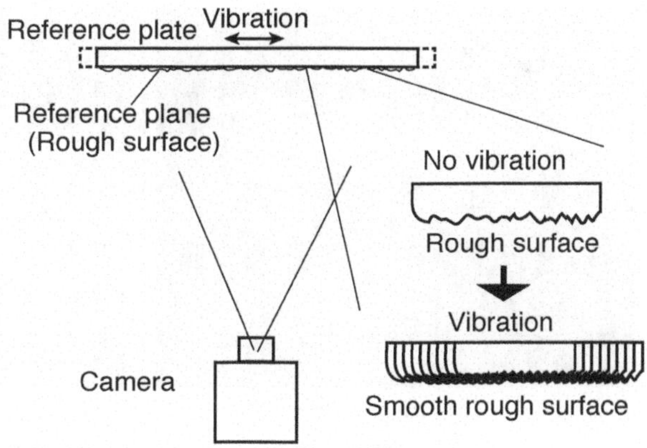

Fig. 5 Proposed idea of smoothing of rough surface with vibrating reference plate

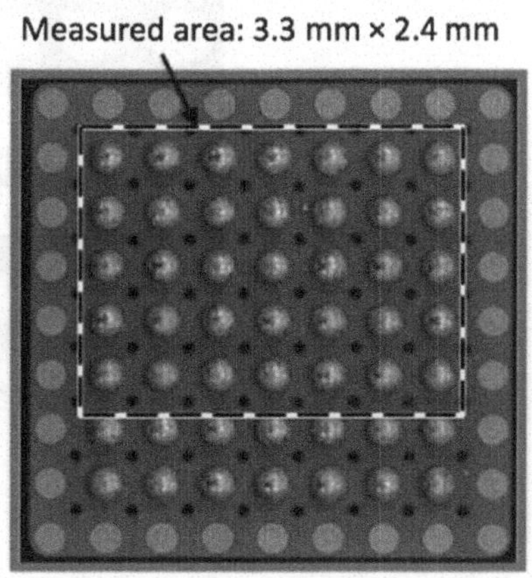

Fig. 6 Specimen of BGA

height measurement of BGA. Figure 7a shows one of phase-shifted fringe images when the exposure time is 3 ms. Figure 7b shows a phase map analyzed with 4-step phase-shifting method. Figure 7c shows a height map obtained from the phase map shown in Fig. 7b using phase-height tables. The phase-height tables are produced with the WSTM.

The measured height is compared with the heights obtained by a depth measuring microscope. Figure 8 shows height distributions obtained by both proposed method and the depth measuring microscope along a broken line shown in Fig. 7c. The difference near the top of the BGA is around 10 μm. This results show that the proposed measurement method is available for shape measurement of a small height object.

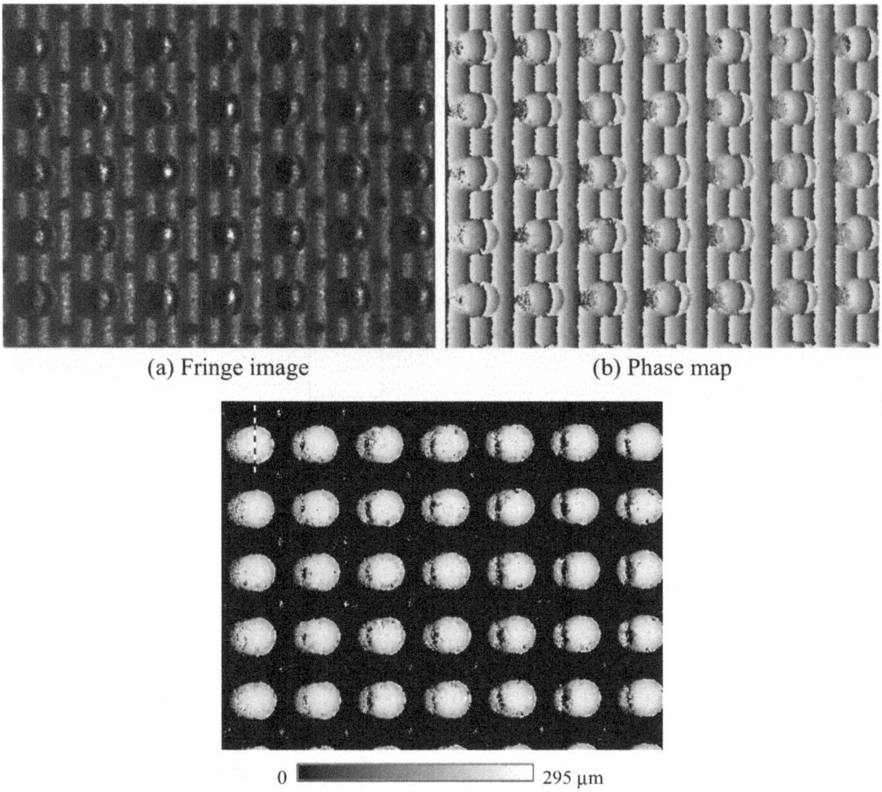

(a) Fringe image (b) Phase map

(c) Height map

Fig. 7 Results of height measurement of BGA

Fig. 8 Height distributions along a broken line shown in Fig. 7c

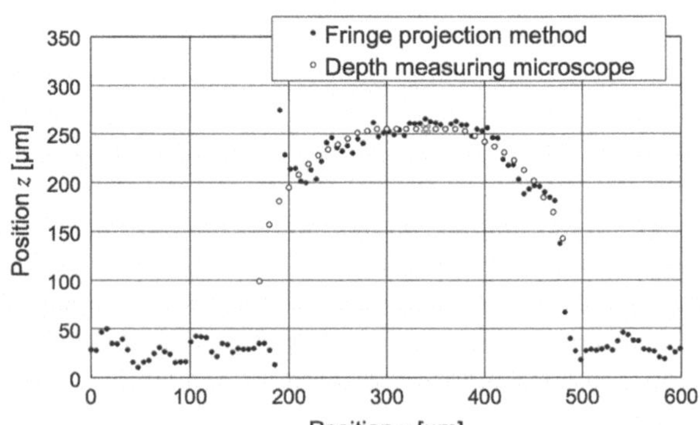

4 Conclusions

A small pitch fringe projection is realized with Talbot effect. Authors proposed a small pitch fringe projection method using Talbot effect with an SLD with inclining the appearing area of the fringe pattern. Authors also proposed a phase-shifting method using a linear fiber array and a method vibrating reference plate to reduce noise. In this paper, these techniques were applied to height measurement of BGA. A meaningful result of height measurement of BGA was obtained.

Acknowledgements The authors also would like to thank Mr. Tadashi Misaka in Yasunaga Corporation for supplying a specimen. The authors also would like to thank Mr. Takuya Sakano in University of Fukui for helping authors to evaluate measured results. This research had been supported by the Adaptable and Seamless Technology transfer Program (A-STEP), Japan Science and Technology Agency (JST).

References

1. Misaka, T., Murakami, M., Hikita, T. and Fujigaki, M., "Development of 100% appearance inspection sensor and applying to measurement of microscopic bumps of semiconductor component," Proceedings of 9th International Symposium on Advanced Science and Technology in Experimental Mechanics, #074 (2014).
2. Talbot, H. F., "Facts relating to optical science No. IV," London Edimburgh Philosophical Magazine and Journal of Science, 9(56), 401–407 (1836).
3. Rayleigh, L., "On copying diffraction-gratings, and on some phenomena connected therewith," Philosophical Magazine, 11(67), 196–205 (1881).
4. Rodriguez-Vera, R., Kerr, D. and Mendoza-Santoyo, F., "3-D Contouring of diffuse objects by Talbot-projected fringes," Journal of Modern Optics, 38(10), 1935–1945 (1991).
5. Oreb, B. F. and Dorsch, R. G., "Profilometry by phase-shifted Talbot images," Applied Optics, 33(34), 7955–7962 (1994).
6. Mehta, D. S., Dubey, S. K., Shakher, C., Takeda, M., "Two-wavelength Talbot effect and its application for three-dimensional step-height measurement," Applied Optics, 45(29), 7602 (2006).
7. Hayashi, T., Fujigaki, M. and Murata, Y., "Projection method of small pitch fringe pattern using Talbot effect with super luminescent diode for 3D shape measurement," Journal of JSEM, 15(Special Issue), s81–s86 (2015).
8. Hayashi, T., Murata, Y., Fujigaki, M., "3D shape measurement using incident small pitch fringe projection method with Talbot effect and phase shifting method with linear fiber arrays (in Japanese)," IEEJ Transactions on Electronics, Information and Systems, 136(8), 1063–1070 (2016).
9. Fujigaki, M., Hayashi, T., Murata, Y., "Reducing noise of shape measurement with fringe projection method using vibrated reference plate," Proceedings of 10th International Symposium on Advanced Science and Technology in Experimental Mechanics, #103 (2015).

Light Amplification in Photorefractive Ferroelectric Liquid Crystal Blends Containing Quarter-Thiophene Photoconductive Dopant

Takeo Sasaki, Shouta Morino and Khoa Van Le

Abstract Ferroelectric liquid crystal blends containing a photoconductive chiral dopant and an electron trap reagent exhibit a large photorefractivity and fast response. They can be utilized in dynamic amplification of moving optical signals. In the present study, the enhancement of working wavelength to a longer wavelength regions was investigated. A series of photoconductive chiral dopants was prepared and the photorefractive properties were examined.

1 Introduction

A ferroelectric liquid crystal (FLC) mixture composed of liquid crystal compounds and photoconductive chiral dopant exhibits a large photorefractivity. Blends of liquid crystalline compounds and photoconductive chiral compounds are used to achieve a high photorefractivity since it is difficult to obtain high transparency and high performance using a single compound. We have investigated the amplification of moving optical signals by photorefractive FLC blends [1]. The photorefractive effect forms a refractive index grating within a medium, based on a change in the refractive index of the medium resulted from photoinduced electric field and an electro-optic effect. The mechanism of the photorefractive effect in FLC blends containing photoconductive compounds is shown in Fig. 1.

The most characteristic phenomena of the photorefractive effect is the asymmetric energy exchange, in which the energy of one of the interfering laser beams transfer to the other. The asymmetric energy exchange can be utilized in optical signal amplification. The ferroelectric liquid crystals used for practical application are mixtures of several liquid crystalline compounds and chiral dopants. In order to obtain a photorefractive FLC, a photoconductive compound is also added to the FLC mixture. However, in the most case, the photoconductive compounds are not

T. Sasaki (✉) · S. Morino · K. Van Le
Department of Chemistry, Faculty of Science, Tokyo University of Science,
1-3 Kagurazaka, Shinjuku-ku, Tokyo 162-8601, Japan
e-mail: sasaki@rs.kagu.tus.ac.jp

© Springer Nature Singapore Pte Ltd. 2017

I. Bhattacharya et al. (eds.), *Advances in Optical Science and Engineering*,
Springer Proceedings in Physics 194, DOI 10.1007/978-981-10-3908-9_6

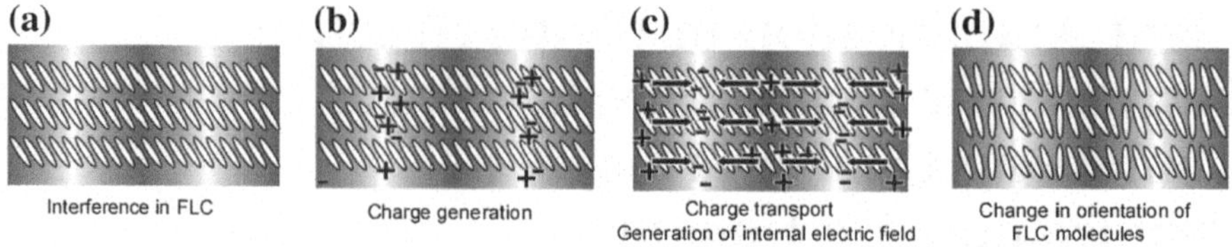

Fig. 1 Mechanism of the photorefractive effect in an FLC. **a** Two laser beams interfere in the SS state of themixture of a FLC and a photoconductive compound, **b** positive and negative charges are generated at the bright areas of the interference fringes, **c** while electrons are trapped at trap sites in the bright areas, positive charges (holes) migrate by diffusion or drift in the presence of an external electric field to generate an internal electric field between the bright and dark positions, and **d** a change in orientation of the spontaneous polarization vector is induced by the internal electric field

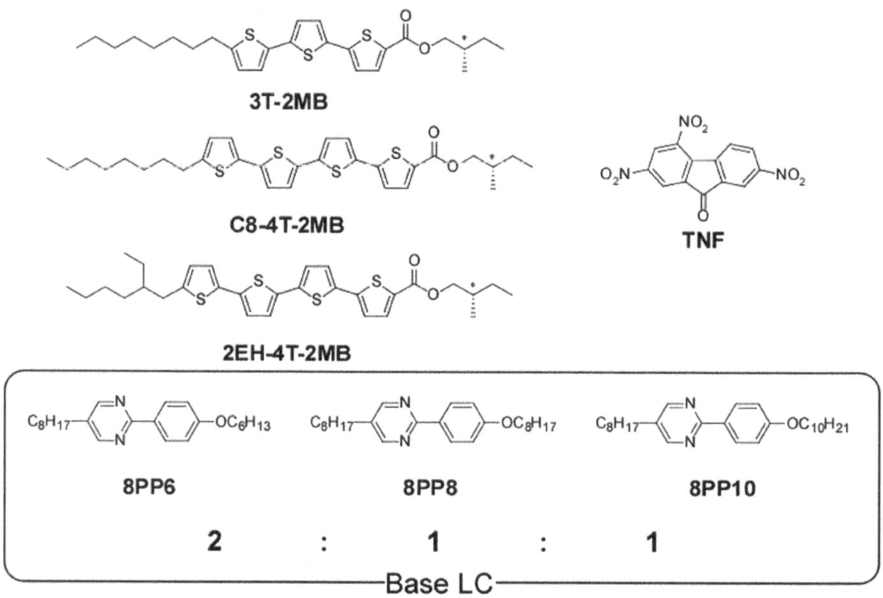

Fig. 2 Structures of the photoconductive chiral dopants smectic LCs, and the electron trap reagent (TNF)

liquid crystalline materials, addition of the photoconductive compound to the FLC mixture disturbs the alignment of FLC molecules. Thus the light scattering in the FLC medium increases. In order to avoid this problem, photoconductive compounds that also possess chiral structure were synthesized. One can obtain a photorefractive FLC just by mixing of the photoconductive chiral compound with a FLC mixture. We have reported that photorefractive FLC blends containing photoconductive chiral dopants exhibit a large photorefractivity and a fast response [2]. The photoconductive chiral dopants used in the previous study were terthiophene compounds (Fig. 2, 3T-2MB). The absorption of terthiophene is shorter than

500 nm so that 488 nm lasers were used to induce the photorefractive effect. In the present study, quarter-thiophenes (Fig. 2, C8-4T-2MB and 2EH-4T-2MB) were synthesized and mixed with a smectic liquid crystal to form a ferroelectric liquid crystals. The photorefractive properties in longer wavelength region were investigated.

2 Experimental

2.1 *Samples*

The structures of the compounds used in this study are shown in Fig. 2. A mixture of phenylpyrimidine type smectic liquid crystalline compounds, 8PP8, 8PP10 and 8PP6, was used as a host liquid crystal (base-LC). The mixing ratio of 8PP6, 8PP8 and 8PP10 was set to 2:1:1. Photoconductive chiral dopants possessing quaterthiophene chromophore were synthesized. The photoconductive chiral dopant was mixed with host liquid crystal and electron trap reagent TNF. The mixture of the base-LC, TNF and a photoconductive chiral dopant was injected into a glass-cell equipped with ITO electrode and polyimide alignment layer. The thickness of the FLC was adjusted to 10 μm.

2.2 *Measurement*

Phase transition temperatures were measured by differential scanning calorimetry (DSC; DSC822, Mettler) and by microscopic observations (FP-80, FP-82, Mettler; DM2700 polarizing microscope, Leica). Spontaneous polarization (Ps) was measured by the spontaneous polarization measurement system for ferroelectric liquid crystals (TS-LCM-6254C, Toyo Technica). The photorefractive effect was evaluated by two-beam coupling experiments. A linearly polarized 488 nm beam from a diode-pumped solid state laser (DPSS laser, Spectra Physics, Cyan) was divided in two by a beam splitter, and interfered within the sample film. The laser intensity was 2 mW for each beam and the diameter of each beam was 0.5 mm. The incident angles of the beams to the glass plane were 30° and 50°. The interference fringe interval was 1.87 μm. The measurement was conducted at room temperature (26 °C). An electric field (0–10 V/μm) was applied to the sample from a regulated DC power supply (Kenwood DW36-1). The transmitted beam intensity was monitored by photodiodes (ET-2040, Electro-Optics Technology, Inc.). The time required to form the refractive index grating in the FLC was ascertained based on the simplest single-carrier model of photorefractivity, in which the gain transient is exponential. The rising signal of the diffracted beam was fitted using a single exponential function:

$$\gamma(t)-1 = (\gamma-1)[1-\exp(-t/\tau)]^2, \quad (1)$$

$\gamma(t)$ represents the transmitted beam intensity at time t divided by the initial intensity ($\gamma(t) = I(t)/I_0$), and τ is the formation time. the two-beam coupling gain coefficient Γ was calculated assuming Bragg diffraction [1–5].

$$\Gamma = \frac{1}{D} ln\left(\frac{gm}{1+m-g}\right), \quad (2)$$

$D = L/\cos(\theta)$ is the interaction path for the signal beam (L = thickness of the sample, θ = propagation angle of the signal beam in the sample), g is the ratio of the signal beam intensities behind the sample with and without a pump beam, and m is the beam intensity ratio (pump/signal) in front of the sample.

3 Results and Discussions

Figure 3 shows the UV-visible absorption spectra of the photoconductive chiral dopants used in this study. The quarter-thiophene moiety absorbs longer wavelength than ter-thiophene moiety because of the expanded π-conjugation structure. It was confirmed that the quarter-thiophene dopants can be used for the writing beam wavelength of 532 nm.

The textures of the FLC blends containing C8-4T-2MB in 10-μm gap LC cell observed under polarizing microscope. C8-4T-2MB was soluble to the base-LC at the concentration of 4 wt%, however, precipitation of C8-4T-2MB crystals was observed at concentrations higher than 6 wt%. The size of the quarter-thiophene moiety is larger than that of the ter-thiophene moiety, so that the quarter-thiphene compound is more likely to crystalize. In order to prevent the quarter-thiophene dopants from crystallization in a LC medium, we made a modification to the molecular structure of the quarter-thiophene dopants. The flexible alkyl chain (tail unit), which is attached to the quarter-thiophene moiety, is necessary to achieve a

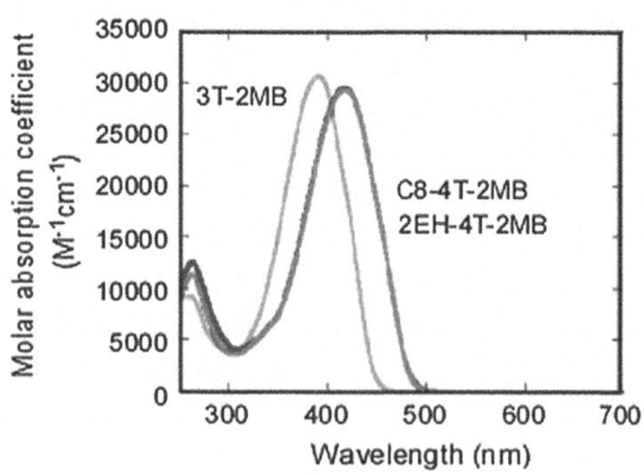

Fig. 3 UV-visible absorption spectra of the photoconductive chiral dopants measured in chloroform solution

high miscibility to a liquid crystalline medium. The tail unit is responsible for the lowering of the melting point and make the molecular structure more resemble to that of base LC. The quarter-thiphene dopant introduced with branched tail (2EH-4T-2MB) was synthesized.

The phase diagrams of the FLC blends mixed with quarter-thiophene chiral dopants are shown in Fig. 4. The temperature range of the SmC* phase was widened in the FLC blend with 2EH-4T-2MB. The textures of the FLC blend containing 2EH-4T-2MB in 10 μm-gap LC cells were observed under polarizing microscope.

No crystallization of the dopant was observed at the dopant concentration of 4 and 6 wt%. Thus, the miscibility of the quarter-thiophene dopant was improved by the introduction of branched tail structure. The transmittance of 633 nm laser through the FLC sample was investigated. Figure 5a shows the transmitted intensity of the laser beam through the FLC sample plotted as a function of the concentration of the quarter-thiophene dopants. The FLC sample containing 2EH-4T-2MB exhibited a higher transparency. The photoconductivities of the FLC samples were investigated as shown in Fig. 5b. No currents were observed in the FLC blends in the dark. When the 488 nm laser was irradiated to the sample, photocurrents were observed. The photocurrent of FLC blends with quarter-thiophene dopants were much larger than that of the FLC blend with ter-thiophene dopant. The photoconductivity was enhanced in the larger π conjugate system. No difference in photoconductivity was observed in FLC samples with C8-4T-2MB and 2EH-4T-2MB.

The photorefractive effects of the FLC blends were investigated by two-beam coupling method. Figure 6a shows the typical example of the asymmetric energy exchange observed in a FLC sample containing 2EH-4T-2MB. The asymmetric energy exchange was clearly observed at the writing beam wavelength of 532 nm. The photorefractive effect was observed only in the SmC* phase (ferroelectric phase). The result shows that the photorefractive effect of the FLC is caused by the response of the spontaneous polarization (Fig. 1). The magnitude of the gain

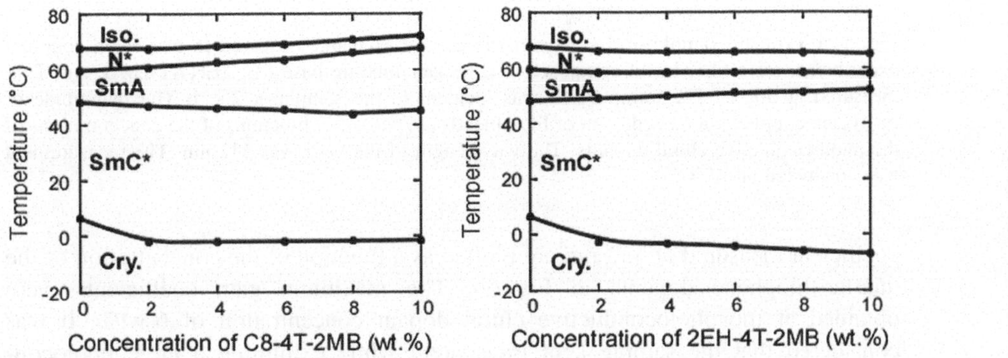

Fig. 4 Phase diagrams of base-LC and photoconductive chiral dopants

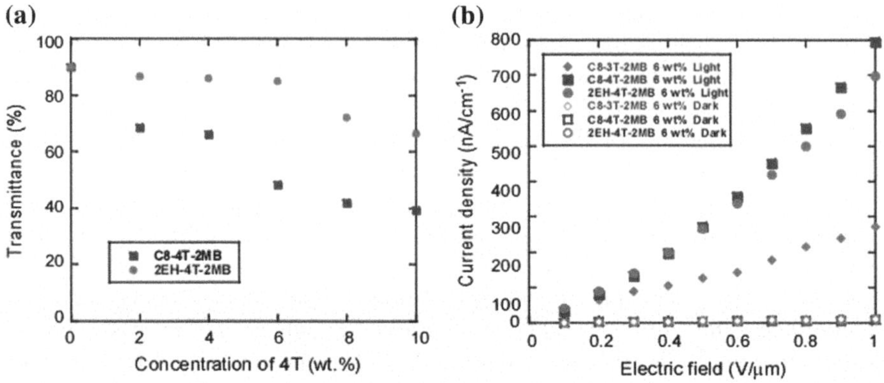

Fig. 5 **a** Transmitted beam intensities through FLC samples. The wavelength of the laser was 633 nm. The concentration of the photoconductive chiral dopant was set to 6 wt%. **b** The photoconductivities of the photorefractive FLC blends containing photoconductive chiral dopants. Current densities are plotted as a function of the applied electric field. The concentration of the photoconductive chiral dopants was kept at 6 wt%. A 488 nm laser (10 mW/cm^{-2}, 1 mm diameter) was used as the irradiation source

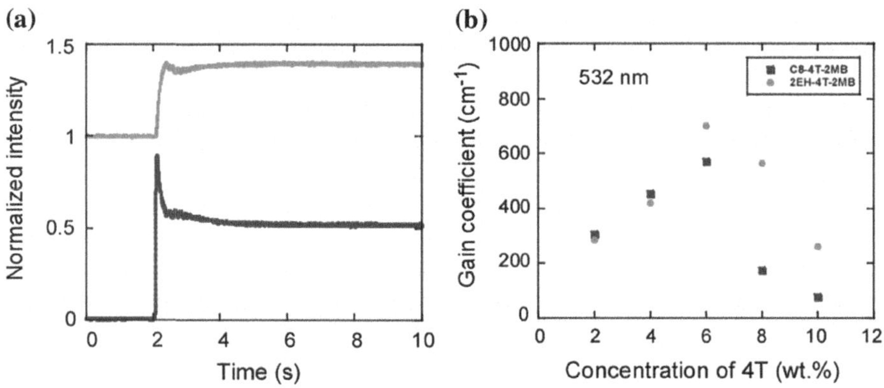

Fig. 6 **a** Typical example of the asymmetric energy exchange observed in two-beam coupling experiments for a FLC blend composed of a ternary mixture base LC, 2EH-4T-2MB, and TNF measured at 30 °C. The pump beam was incident to the sample at 2 s. **b** The magnitude of two-beam coupling gain coefficient of FLC blends are plotted as functions of the concentrations of the photoconductive chiral dopants. The wavelength of the laser was 532 nm. The measurement was conducted at 30 °C

coefficient measured at 532 nm are plotted as a function of the concentration of the quarter-thiophene dopants in Fig. 6b. The maximum gain coefficients were obtained at the photoconductive chiral dopant concentration of 6 wt%. It was considered that the sample kept transparent while exhibiting a large photoconductivity at this concentration. The gain coefficient was larger in the FLC blend with 2EH-4T-2MB than that with C8-4T-2MB because of the higher transparency in the 2EH-4T-2MB sample.

4 Conclusion

The photorefractive effects of FLC blends containing ter-thiophene and quarter-thiophene photoconductive chiral dopants were investigated. The low miscibility of the quarter-thiophene moiety into the liquid crystalline medium was improved by the introduction of branched chain into the tail unit of the quarter-thiophene chiral dopant. The FLC blend with branched tail quarter-thiophene compound exhibit a higher photoconductivity and longer absorption wavelength. The two-beam coupling experiment was conducted using a 532 nm laser. A large photorefractivity at 532 nm was observed in a FLC blend containing quarter-thiophene photoconductive chiral dopant.

References

1. Sasaki, T., Kajikawa, S. and Naka, Y. Farad. Discuss., 174, 203–218 (2014).
2. Sasaki, T. and Naka, Y. Opt. Rev., 21, 99–109 (2014).
3. Lynn, B., Blanche P.-A. and Peyghambarian, N. J. Polym. Sci. B Polym. Phys., 52, 193–231 (2014).
4. Kober, S., Salvador, M. and Meerholtz, K. Adv. Mater. 23, 4725–4763 (2011).
5. Ostroverkhova, O. and Moerner, W. E. Chem. Rev. 104, 3267–3314 (2004).
6. Skarp K. and Handschy, A. A. Mol. Cryst. Liq. Cryst. 165, 439–509 (1988).
7. Oswald P. and Pieranski P. Smectic and Columnar Liquid Crystals, Taylor & Francis: New York (2006)
8. Goonesekera, A., Wright D. and Moerner, W.E. Appl. Phys. Lett., 76, 3358 (2000).
9. Koukourakis, N., Abdelwahab, T., Li, M. Y., Höpfner, H., Lai, Y. W., Darakis, E., Brenner, C., Gerhardt, N. C. and Hofmann, M. R. Opt. Express 19, 22004–22023 (2011).
10. Sasaki, T., Miyazaki, D., Akaike, K., Ikegami, M. and Naka, Y., J. Mater. Chem., 21, 8678–8686 (2011).
11. Sasaki, T., Ikegami, M., Abe, T., Miyazaki, D., Kajikawa S. and Naka, Y. Appl. Phys. Lett. 102, 063306 (2013).

Photorefractive Optical Cryptography: A Personal Tour

Kehar Singh

Abstract Photorefractive crystals (PRCs) are used in a variety of applications including optical information processing where one such application has been in the area of image encryption. Due to their ability to record phase gratings, and to produce phase conjugate wave by non-linear wave-mixing, the PRCs have enriched the optical cryptographic techniques also. In this brief review, an attempt has been made to present glimpses of some of the techniques, with a focus on the research work carried out by the author's photonics group at IIT Delhi.

1 Introduction

The photorefractive (PR) materials have attracted a great deal of attention due to their unique characteristics such as near real-time recording capability and reusability [1–6]. As a result, the PR materials have been used widely, making possible an impressive array of applications. Investigations are continuing on the search for new materials, fabrication of the devices, and a variety of applications. Newer, versatile, stable, and low cost materials have been shown to exhibit the PR effect with impressive response times, and ease of processing for device applications. Special emphasis has been placed on the applications of PR materials in information photonics in which the devices use light as a carrier of information.

1.1 Optical Cryptography

Optical techniques offer many advantages for information security applications. Being inherently 2-D, optical systems can process and relay 2-D data in parallel, resulting in higher throughput rate compared to the electronic systems. Optics offers

K. Singh (✉)
The NorthCap University, Sec-23, Gurgaon 122017, India
e-mail: keharsiitd@gmail.com

© Springer Nature Singapore Pte Ltd. 2017

I. Bhattacharya et al. (eds.), *Advances in Optical Science and Engineering*,
Springer Proceedings in Physics 194, DOI 10.1007/978-981-10-3908-9_7

several dimensions such as phase, polarization, wavelength, and orbital angular momentum to encode data securely. These advantages, coupled with advancements in enabling technologies like P. R. crystals, spatial light modulators, and charge coupled devices have led to an increasing use of optical processing techniques for optical security applications [7–12]. A number of schemes have been introduced based on fractional Fourier-, Fresnel-, gyrator-, Hartley-, Hilbert-, Mellin-, Hadamard-, Gabor-, Hankel-, random sine and cosine, and wavelet transforms etc.

1.2 Photorefractive Cryptography

PRCs have been used mainly for producing phase conjugate of the encrypted signal by four-wave mixing, and for recording holograms in some of the techniques [12]. In the present article, an attempt has been made to present some glimpses of investigations carried out by the author's photonics group at IIT Delhi. The aspects covered are: using phase conjugate of encrypted signal, using PR crystal as a phase filter, and recording of holograms in a PR crystal.

2 Some Cryptographic Systems

2.1 Cryptosystem Using Phase Conjugate of the Encrypted Image

An encryption method proposed by Refregier and Javidi [13] uses a well-known 4-f coherent imaging system with two statistically-independent random phase masks (RPMs), one in the input plane and the other one in the Fourier plane. The encrypted image is recorded by a holographic method. The RPMs work as keys and the decryption is performed by propagating the encrypted image in reverse direction, and with a complex conjugate of the Fourier plane RPM. Unnikrishnan et al. [14] described a method in which the decryption is achieved by producing a phase conjugate of the encrypted image by four-wave mixing in a PRC (Figs. 1, 2), thus alleviating the need of the complex conjugate of the key. The investigations were extended to the fractional Fourier domain [15] so as to enlarge the key space by using the fractional FT order a as the key. An F.T. is a first-order FRT with $a = 1$.

2.2 Fully Phase-Image Encryption

Normally, the images that are encrypted are intensity images which can be detected by intensity detectors. It is advantageous to phase encode the images [16] and then encrypt them so that the decrypted image (also a phase image) can not be acquired without converting it into an intensity image. Nishchal et al. [17] described a fully-phase encryption system using FrFT in which the encrypted image is recorded in a barium titanate PRC. Decryption is achieved (Fig. 3) by using a phase

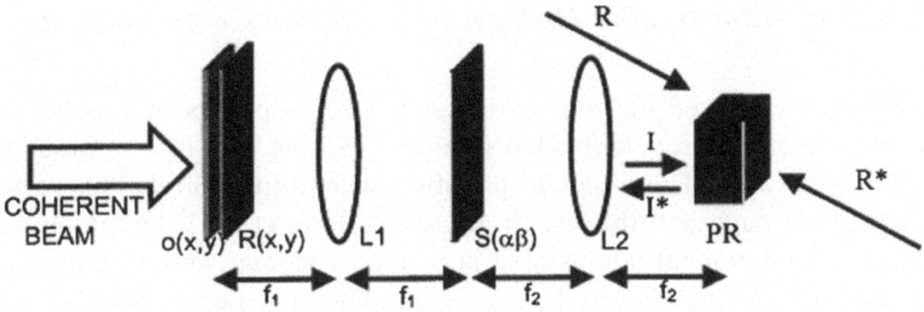

Fig. 1 Schematic illustrating the principle of cryptographic system; O (x, y)—original image, R (x, y)—first RPM, S (α, β)—second RPM, R and R*—counter propagating beams, I*—phase conjugate of encrypted image-bearing beam [14]

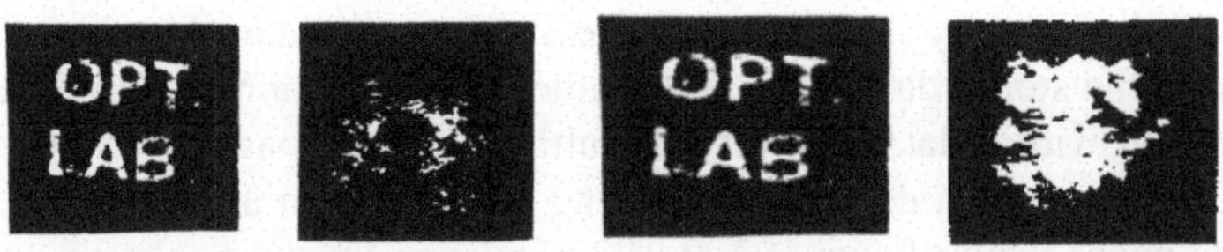

Fig. 2 Original image, encrypted image, decrypted image, and decryption with an incorrect key [14]

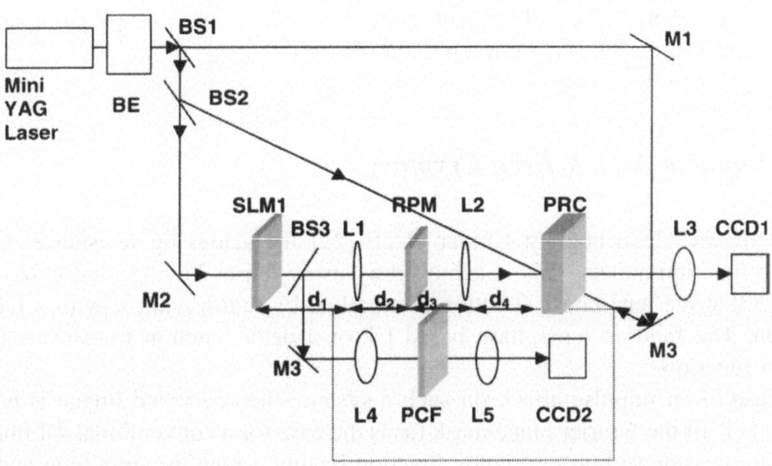

Fig. 3 Experimental set-up: *BE* beam expander, *BSs* beam splitters, *SLMs* spatial light modulators, *RPM* random phase mask, *PRC* PR crystal, *CCDs* charge-coupled devices, *Ls* lenses, *MSs* mirrors, and *PCF* phase-contrast filter [17]

conjugate beam generated by a PRC, and then converting the decrypted phase image into an intensity image by a phase-contrast filter using an electrically addressed SLM. Some other relevant publications [18–20] may also be noted.

2.3 Content Addressable Holographic Memory with Security

It is now well-recognized that one of the attractive properties of a volume holographic memory, is its content addressability, making possible a simultaneous search of the data base. This becomes possible due multiple correlations performed between a stored page and the search argument. John et al. [21] demonstrated a phase-image-based content addressable holographic storage system with security, using random phase encoding in the Fresnel domain. The phase- based data pages were recorded in an iron-doped lithium niobate PRC, using the technique of angular multiplexing, and encryption was done using an RPM. Only an authorized person having access to the completely correct key and its 3D position, can search through the data base. The shift invariance (sensitivity for lateral and longitudinal shift) of the RPM used was also analyzed. The use of RPM also ensures that the correlation-peak-intensity is proportional to the inner product between the search argument and the stored data pages. Experimental verification to prove the concept was carried out with a data bank of 16 multiplexed data pages. A study was also carried out on the search capability by using partial search argument.

The data pages were of size 192 $\times$ 192 pixels, and were displayed on an SLM (Jen Optik) of size 624 $\times$ 832 pixels with a pixel pitch of 32 μm $\times$ 32 μm. The PRC was of size 20 mm $\times$ 10 mm $\times$ 10 mm with the beams incident on the 20 mm $\times$ 10 mm face. The object and the reference beams had the mean angle of 23° between them. The subsequent reference beams for multiplexing, had an angular spacing of $\sim$0.08° between them (Fig. 4).

2.4 Impulse Attack-Free Cryptosystem

A system was described by Kumar et al. [22] for achieving resistance against impulse function attack. This is achieved by using lenses having randomized surfaces so that the lens phase function is modified by multiplying it with a random function. The random lens-phase based FT of a delta function transforms into a random function.

In case of an impulse attack on such a system, the encrypted image is not the inverse F.T. of the Fourier plane mask (as is the case for a conventional 4-f imaging based cryptosystem) but a complex function having a random amplitude and random phase distribution. It can be proved that the Fourier plane mask can not be retrieved with the variant form of the impulse function employing two similar looking images. Random phase variation is achieved by coating the lenses with

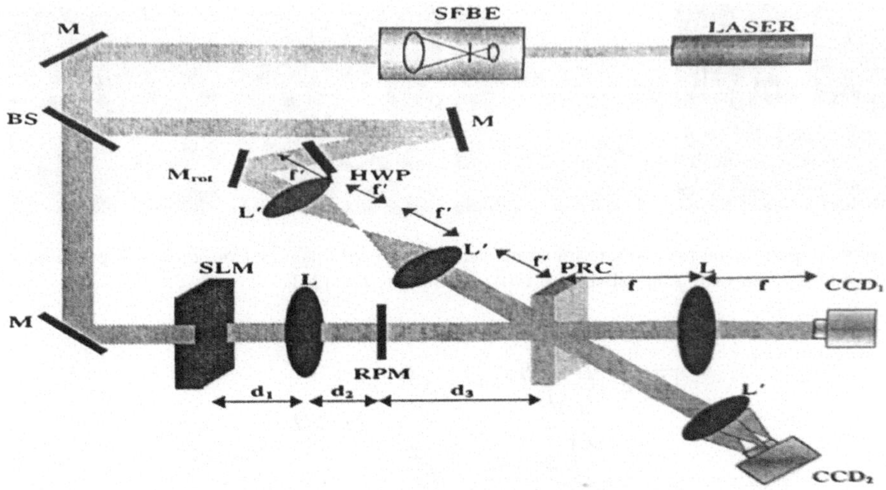

Fig. 4 Experimental set-up; *SFBE* spatial-filter beam-expander assembly, *Ms* mirrors. *BS* beam splitter, *SLM* spatial light modulator, *HWP* half wave plate, M_{rot} mirror mounted on a precision rotation stage, d_1, d_2, and d_3 distances to satisfy Fresnel transform condition [21]

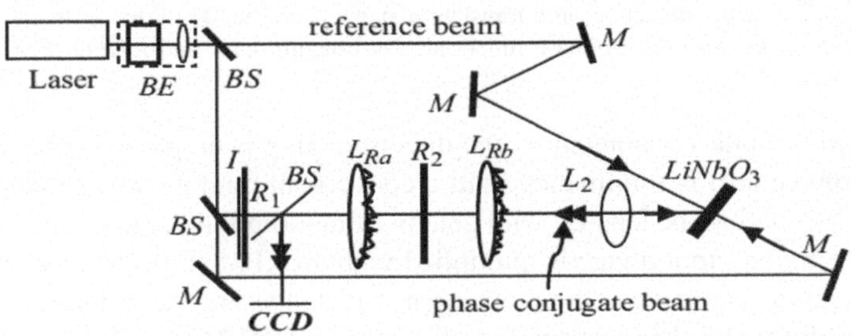

Fig. 5 Experimental set-up for the impulse attack-free scheme; *BE* beam expander, *BS* beam splitter; *M* mirror, R_1 and R_2 RPMs; L_{Ra} and L_{Rb} lenses with randomized phase functions, L_2 imaging lens, $LiNbO_3$-PR crystal [22]

gelatin with random thickness variations. Experimental results (Fig. 5) are shown for lenses with focal lengths 18 and 20 cm. Some other related work [23, 24] may also be noted in this connection.

2.5 Holographic Cryptosystem with Convergent Random Illumination

The cryptographic system (Fig. 6) [25] uses a convergent random wave front which improves the security of the system, only for the reference beam instead of both

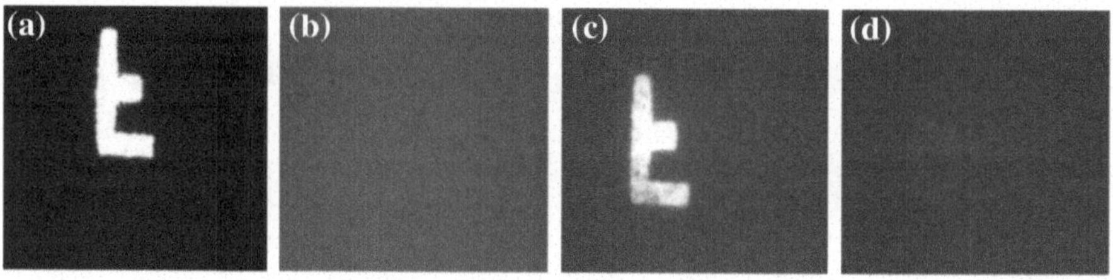

Fig. 6 Experimental results; **a** Original image, **b** Encrypted image, **c** Correctly decrypted image, and **d** Decrypted with a lateral shift of 0.01 mm in lens L_{Rb} [22]

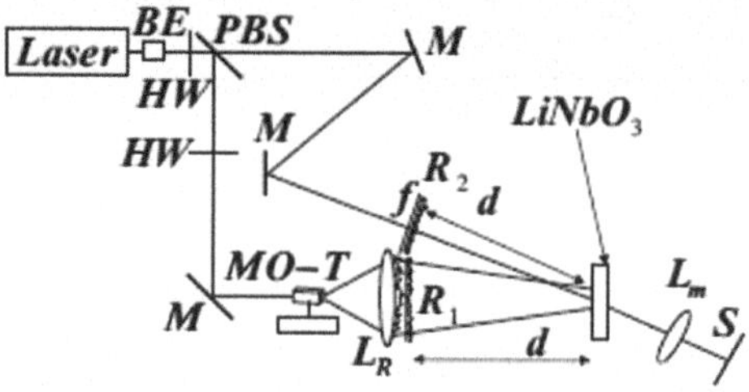

Fig. 7 Experimental set-up; *BE* beam expander, *PBS* polarizing beam splitter, *HWP* half wave plate, *MO-T* microscope objective on a translation stage, *S* screen, *M* mirror, *f* image, R_1 and R_2 RPMs, *LR* lens with randomized phase functions, *Lm* imaging lens, and $LiNbO_3$ PRC [25]

arms. Instead of phase conjugation, the direct imaging is employed. The RPM (R_1) in the reference arm is illuminated with a convergent random wave front which is generated by coating the lens L_R with gelatin of random thickness on the surface. In the signal arm, an input image is multiplied with an RPM (R_2).The system has two new encryption parameters (the convergence distance and randomized lens function) in addition to the propagation distance and RPMs in the reference arm (Fig. 7).

3 Conclusions

Due to the constraint of space, only four systems developed by the author's group have been very briefly described in the present article. During the last two decades or so, the area of optical image encryption has witnessed a steady growth as evidenced [7–12] by a large number of research papers. The subject is still growing with newer cryptosystems being described. There is also considerable effort to investigate the vulnerability of various schemes against various attacks such as the brute force-, known- plaintext-, chosen-plaintext-, and chosen-ciphertext attacks. Indeed the researchers have been able to prove that a number of schemes described in the literature are prone to attacks. Only the future will tell as to how many more systems would be proved weak against the attacks and what countermeasures can be taken to strengthen the system security.

References

1. Yeh, P., [Introduction to Photorefractive Nonlinear Optics] Wiley, N.Y. (1993).
2. Solymar, L., Webb D. J., and Grunnet-Jepsen A, [The Physics and Applications of Photorefractive Materials] Oxford Press UK (1996).
3. Yu, F.T.S. and Yin, S., (Eds). [Photorefractive Optics; Materials, Properties, and Applications] Academic Press, N.Y. (2000).
4. Günter, P. and Huignard, J-P. (Eds.), [Photorefractive Materials and Their Applications-1 Basic Effects] Springer-Verlag, Berlin (2006).
5. Günter, P. and Huignard, J-P. (Eds.), [Photorefractive Materials and Their Applications-2 Materials] Springer-Verlag, Berlin (2007).
6. Günter, P. and Huignard, J-P. (Eds.), [Photorefractive Materials and Their Applications-3 Applications] Springer-Verlag, Berlin (2007).
7. Singh, K., Unnikrishnan, G., and Nishchal, N.K., "Photorefractive optical processing for optical security", Proc. SPIE 4803, 205–219 (2002).
8. Javidi, B., Ed., [Optical and Digital Techniques for Information Security] Springer: N.Y. (2005).
9. Alfalou, A., and Brosseau,C., "Optical image compression and encryption methods". Adv. Opt. Photon. 1589–636 (2009).
10. Matoba, O., Nomura,T., Perez-Cabre, E., Millan, M.S., and Javidi, B., "Optical techniques for information security", Proc. IEEE. 97, 1128–1148 (2009).
11. Kumar, A., Singh, M., and Singh, K., "Speckle coding for optical and digital data security applications", in: [Advances in Speckle Metrology and Related Techniques], (Ed.) Kaufmann, G.H., Wiley-VCH: Weinheim, Chap. 6. pp. 239–299 (2001).
12. Kumar, P., Joseph, J., and Singh, K. "Double Random Phase Encoding Based Optical Systems Using Some Linear Canonical Transforms: Weaknesses and Countermeasures", in [Linear Canonical Transforms; Theory and Applications], Springer-Verlag, Healy, J.J. et al (Eds.) pp 367–396 (2016).
13. Refregier, P., and Javidi, B., "Optical image encryption based on input plane and Fourier plane random encoding" Opt. Lett. 20, 767–769.
14. Unnikrishnan, G., Joseph, J., and Singh, K. "Optical encryption system that uses phase conjugation in a photorefractive crystal", Appl. Opt. 37,8181–8186 (1998).
15. Unnikrishnan, G., Joseph, J., and Singh, K., "Optical encryption by double random phase encoding in the fractional Fourier domain", Opt. Lett. 25,887–889 (2000).
16. Towghi, N., Javidi, B., and Luo, Z., "Fully phase encrypted image processor" J. Opt. Soc. Am. A 36 (1999) 1915–1927.
17. Nishchal, N.K., Joseph, J., and Singh, K., "Fully phase encryption using fractional Fourier transform" Opt. Eng. 42 (2003) 1583–1588.
18. Nishchal, N.K., Joseph, J., and Singh, K., "Optical encryption using cascaded extended fractional Fourier transform" Opt. Mem. Neural Network 12, 39–145 (2003).
19. Nishchal, N.K., Joseph, and Singh, K., "Fully phase encrypted memory using cascaded extended fractional Fourier transform" Opt. Laser Eng. 42, 141–151 (2004).
20. Singh, M., Kumar, A., and Singh, K., "Secure optical system that uses fully phase-based encryption and lithium niobate crystal as phase contrast filter for decryption, Opt. Laser Technol. 40, 619–624 (2008).
21. John, R., Joseph, J., and Singh, K., "Phase-image-based content-addressable holographic data storage with security" J. Opt. A: Pure Appl. Opt. 7, 123–128 (2005).

22. Kumar, P., Kumar, A., Joseph, J., and Singh, K., "Impulse attack- free double-random phase encryption scheme with randomized lens-phase functions", Opt. Lett. 34, 331–333 (2009).
23. Kumar, P., Joseph, J., and Singh, K., "Impulse attack-free four random phase mask encryption based on a 4-f optical system", Appl. Opt. 48, 2356–2363 (2009).
24. Kumar, P., Joseph, J., and Singh, K., "Known-plaintext attack-free double random phase-amplitude optical encryption: vulnerability to impulse function attack" J. Opt. (IOP) 14,045401- 1/8 (2012).
25. Kumar, P., Joseph, J., and Singh, K., "Holographic encryption system in the Fresnel domain with convergent random illumination", Opt. Eng. 49,095803-1/6 (2010).

Interferometry: From Hooke till Date

Rajpal S. Sirohi

Abstract Coloured fringes seen in white light when the two glass plates in near contact enclose a very small angle were first observed by Boyle and independently by Hooke in the later part of 17th century and the explanation given using wave theory of light could be taken as the starting point of optical interferometry. Wave theory used by Boyle and Hooke, and in much refined form as proposed by Huygens in 1690 remained unaccepted until Young in 1801 demonstrated the interference between two waves by a very simple but ingenious way. Fizeau (Acad Sci 66:429, 1862 [1]) carried out experiments with a pair of plates using Na light and showed that the fringe pattern disappears when a certain distance separates the plates, essentially sowing the seeds for interference spectroscopy. Michelson in 1891 carried out measurement of visibility of fringes as a function of path difference between the two beams derived from various sources and showed that except for cadmium, other spectral lines showed the variation in visibility. Fizeau is also accredited to have suggested in 1868 that interferometry may be used for measuring stellar dimensions: the idea being taken forward by Michelson by inventing stellar interferometer. The manuscript presents the chronology of development in interferometry. Some current applications may also be highlighted.

1 Introduction

Interferometry makes use of the phenomenon of interference. When two or more waves of the same wavelength are superposed, irradiance varies in the region of superposition. The waves could be longitudinal or transverse and could have any wavelengths. In other words, the phenomenon of interference is observed with

R.S. Sirohi (✉)
Department of Physics, Tezpur University, Tezpur, Assam 78428, India
e-mail: rs_sirohi@yahoo.co.in

© Springer Nature Singapore Pte Ltd. 2017

I. Bhattacharya et al. (eds.), *Advances in Optical Science and Engineering*,
Springer Proceedings in Physics 194, DOI 10.1007/978-981-10-3908-9_8

sound waves and also with the light waves. The irradiance variation in the region of superposition shifts when the phase of one of the interfering waves changes. The variation repeats when the phase-shift is equal to 2π or a its multiples. In practice, this shift is introduced by the external variable that is being monitored.

2 Historical Perspective

Light, in times of Al Hazan and until the 17th century, was considered as stream of particles either emanating from the eye or the object. Newton provided explanation to the various kinds of colors that he observed. Somewhat different explanation of appearance of colors, particularly in thin air wedges, was provided by Robert Boyle (1664) and independently by Robert Hooke (1672) in the later half of the 17th century and their explanation based on somewhat akin to wave theory may be considered as the beginning of interferometry. Huygens in 1678 proposed the wave theory of light that could explain the known phenomena. However it was not accepted partly owing to the authority of Newton who was proponent of corpuscular theory of light and partly it could not explain rectilinear propagation of light. Young, while delivering the Bakerian Lecture in 1801, demonstrated the interference of light waves by a very simple but ingenious experiment, which has come to be known as Young's double-slit experiment. Since interference is a wave phenomenon, light was considered as a longitudinal wave motion. But Young did not attract support for the wave theory. His confidence in wave theory was further shaken when Malus in 1809 announced that the reflected light is polarized. During 1814–1818, Fresnel advanced the Huygens wave theory and wrote a brilliant memoir on diffraction in 1818 that contained the treatment of interference phenomenon. Fresnel also suggested arrangements to observe interference of light: these arrangements are known as Fresnel bi-mirror and Fresnel bi-prism. Lloyd described another way of observing interference by wavefront division in 1834, which is known as Lloyd's mirror arrangement. Complex phenomenon like appearance of a bright spot, Arago or Poisson spot, in the center of the shadows of circular objects could be explained using Fresnel theory of diffraction. Through careful experiments of reflection from good quality glass interfaces, Brewster in 1815 discovered what is today known as Brewster law. Further Arago and Fresnel conducted experiments with polarized light waves and showed that two orthogonally polarized waves do not interfere. These investigations led both Young and Fresnel to believe that light waves are transverse waves. In order for the transverse waves to propagate, Fresnel postulated the presence of an elastic medium pervading all matter—the *luminiferous aether*. He also showed that in a medium of refractive index μ moving with a velocity v, the aether should be carried along with a velocity $v(1 - 1/\mu^2)$.

2.1 Search for Aether

Hippolyte Fizeau in 1851 conducted an interference experiment in which one beam of light propagated along the flow of water and the other beam opposite the flow. He seemingly detected the *aether* drag effect but its magnitude was far smaller than expected. However, his result convinced the physicists to accept Fresnel's *aether* theory. Michelson and Morley in 1886, repeated Fizeau's experiment using a common path interferometer, which supported Fizeau result. Jamin and Hoek (1868) also repeated Fizeau's experiment. Fresnel's almost stationary theory of aether was accepted by almost all the 17th century physicists.

Maxwell in 1880 predicted that the earth motion through the aether should result in a change in the speed of light proportional to $(v/c)^2$, where v is the speed of the earth and c is the speed of light. He thought that the effect is too small to be measurable. However, Michelson in 1881 designed a two-crossed arms instrument and attempted to use it in Germany but with little success, as it could not support any of the existing aether drag theories. Later in 1887 he along with Morley set-up an interferometer with 11 m long arm that floated on mercury pool. Again they could not observe the expected shift of the fringe pattern. This failed experiment, known as Michelson-Morley experiment [2], rejected the Fresnel's stationary aether theory with partial dragging and confirmed the Stokes' hypothesis of complete aether dragging.

2.2 Variation of Fringe Visibility and Interference Spectroscopy

While experimenting with his interferometer using sodium lamp, Fizeau [1] found that the fringe pattern almost disappeared when the gap is increased such that 490 fringes passed and further increase in gap resulted in appearance of fringes reaching maximum contrast when 980 fringes passed. It suggested that the sodium light had two wavelengths—a doublet line. Michelson carried this work forward by plotting fringe visibility as a function of path difference for a number of sources using his interferometer in 1891. He found that the visibility exhibited maxima and minima for all the sources except the Cd red line, thereby showing that the radiation from these sources had multiple wavelength components. Ruben and Wood in 1911 extended this work to far infrared region (100–300 μm). However it was not until 1951 when Fellgett extracted the spectrum from the recorded interferogram as a function of path difference using Fourier transform theory and showed other advantages of the technique. Subsequent developments in Fourier Transform spectroscopy have made this technique preferable over the dispersive instruments in the whole optical region (IR-Vis-UV).

2.3 Physical Parameters Measurement

In 1846, Haidinger observed a fringe pattern at the focal plane of a lens when a plane parallel plate was illuminated normally by an extended source. The Haidinger fringes have been used to measure the wedge angle of thin plates. Jamin [3] invented a plates-based interferometer that was well suited to measure the refractive indices of gases and also to study the temperature dependence of refractive index of liquids and gases. The two waves were created by amplitude division and their paths were matched within the coherence length of the then available sources. This therefore required two identical thick plane parallel plates, which were illuminated by a collimated beam. The two beams generated by first plates were separated and recombined by the second plate.

In 1882 Michelson invented a new kind of refractometer, which came to be known as Michelson interferometer. The interferometer is used for the measurement of wavelength, wavelength difference and refractive indices of thin samples. The beams has to travel twice through the specimen, which results in an increase in sensitivity but at the same time may lead to erroneous results if the beam does not retrace its path. Mach [4] and Zehnder [5] independently described an interferometer, which came to be known as Mach-Zehnder interferometer in which the reference and test beams can be widely separated still retaining its almost equal path feature. This interferometer found many applications in aerodynamics and aerospace engineering. This had the feature to measure the refractive index of gases and also to study the temperature dependence of refractive index of liquids and gases. Rayleigh [6] developed an interferometer utilizing the Young's double slit geometry, which was earlier used by Fizeau and added some very convenient features for accurate measurement of fringe shifts. The interferometer was used to measure refractive indices of gases.

Sagnac [7] provided correct explanation of the fringe formation in a cyclic interferometer. This is a true common path interferometer and its variants have been used in optical testing. The interferometer is sensitive to the rate of rotation about an axis that passes through the interferometer loop. This interferometer finds application in gyros.

2.4 Optical Testing

Fizeau is accredited to have invented in 1862 what is now known as Fizeau interferometer, wherein the interference fringes are observed between a thin air-wedge illuminated by a broad source. This was a standard equipment in optical workshop for testing surfaces. Michelson interferometer was not suited for optical testing. Twyman and Green [8] modified the Michelson interferometer so that it could be used for testing optical components and it then became the workhorse of optical workshops. Linnik [9] modified the Michelson interferometer to examine surfaces under magnification thus expanding applications of interferometry to metallurgy. In the same year, he invented a simple interferometer for testing spherical wave. The point diffraction interferometer was revisited by Smartt and Strong [10], and Smartt and Steel [11]. Burch [12] provided the theory of scatter plate interferometer: another interferometer that could be used with low-coherence sources. Hariharan and Sen extended the Fizeau, and Rayleigh two-beam wavefront interferometer to three beams [13]. They described a cyclic interferometer for optical testing [14].

2.5 Optical Microscopy

Optical microscopy evolved to examine objects and surfaces under magnification. The microscopes were also equipped to make dimensional measurements. Microscopy had immense applications in medical examinations. However, the observation of transparent objects like the cells required staining. The theory of microscope imaging, known as Abbe's theory, was very well developed: imaging in a microscope was considered as two-step process involving diffraction at the object followed by diffraction at the objective. It was only around 1934 when Zernike devised a method to examine transparent/phase objects under microscope. The Zernike phase contrast was based on the simple fact that the directly transmitted and diffracted beams should be phase shifted by $\pi/2$ or odd integer multiple of $\pi/2$: the directly transmitted beam does then interfere with the diffracted beam resulting in positive or negative phase contrast. It also became necessary to incorporate an interferometer into a microscope objective or use Linnik like arrangement to examine surfaces under magnification. Dyson invented an interferometer microscope [15], which could be used to examine objects in transmission. Mirau objective [16] is a direct descendent of this arrangement. The microscope objectives could also be fitted with a tiny Michelson interferometer. Polarization based interferometer that forms part of a microscope was invented by Nomarskii [17]. This is known as Nomarskii microscope or Nomarskii differential interferential contrast (DIC) microscope.

3 Arrival of the Laser

Interferometry evolved as a technique for testing optical components and as a measurement tool. The interferometers were conceived and designed keeping the availability of sources and detectors. Arrival of lasers in 1960 relaxed several constraints on the interferometer design and new interferometers like Murty's plane parallel plate shear interferometer [18] came into existence. First one was the LUPI: the laser unequal path interferometer, allowing the testing of mirrors of long radii of curvatures.

With the availability of tabletop compute power and array detector, a new approach for interferogram evaluation was initiated in 1969. This approach has undergone continuous refinements and is a standard procedure for the evaluation of interferograms.

Initially the length measurement by counting the fringes electronically using Michelson type interferometer with mirrors replaced by corner cubes was carried out. Direction sensitivity was built-in by using two detectors in quadrature. Some interferometers used even four detectors to overcome the problem of irradiance variations. However, Hewlett-Packard came up with an AC fringe counting interferometer based on heterodyne in 1972 that revolutionized the dimensional measurements in the workshop environment. Barker and Hollenbach in 1965 used Michelson interferometer for impact studies and modified it in 1972 in which Doppler-shifted light was used for interferometry. This interferometer came to be known as VISAR: velocity interferometer system for any reflector. Currently it uses fiber-optics and has many good features. Holographic interferometry and speckle interferometry and its electronic version, ESPI, have found numerous applications. One of most exciting applications of interferometry is in the detection of gravitational waves.

4 Conclusion

The field of interferometry is an exciting field from both theory and applications. With the advent of lasers, and availability of tabletop compute power and array detectors, it has undergone immense changes in presentation of the final results and in the range of applications.

Acknowledgements To write the history of any subject is a very demanding task and it has not been easy to write the developments in the field of interferometry. I have used more space in the early part of the history and have very briefly gone over the developments after the advent of laser. I have drawn material from various sources, books [19–24], journals, magazines, Net etc. I would like to acknowledge all these media.

References

1. H. Fizeau, C.R. Acad. Sci. Paris, 66, p. 429, (1862)
2. Michelson-Morley: The Great Failure | The Scientist Magazine® www.the-scientist.com › *The Scientist › July 1987 Issue*
3. J. Jamin, C.R. Acad. Sci. Paris, 42, p. 482, (1856)
4. Ernst Mach. “Modifikation und Anwendung des Jamin Interferenz-Refraktometers”. Anz. Acad. Wiss. Wien math. Naturwiss. Klasse 28, p. 223–224, (1981)
5. Ludwig Zehnder, “Ein neuer Interferenzrefractor”, Z.Instrumentenkd. 11, p. 275–285, (1981)
6. Lord Rayleigh, Proc. Roy. Soc., 59, p. 198, (1896)
7. G. Sagnac, Comptes Rendus, 157:708 & 1410, (1913)

8. T. Twyman and A. Green, British Patent No. 103832, (1916)
9. Linnik, W. P. "A Simple Interferometer for the Investigation of Optical Systems". C.R. Acad. Sci. URSS 5: 210, (1933)
10. R. N. Smartt, and J. Strong, "Point-Diffraction Interferometer". Journal of the Optical Society of America 62: 737, (1972)
11. R. N. Smartt, and W. H. Steel "Theory and application of Point-Diffraction interferometers". Japanese Journal of Applied Physics, 14: 351–356, (1975)
12. J. M. Burch, Scatter Fringes of Equal Thickness, Nature (London) 171, 889–890 (1953)
13. P. Hariharan and D. Sen, "Three-beam interferometer", Journal of Scientific Instruments, 36 (2), 70–72, (1959)
14. P. Hariharan and D. Sen, "Cyclic Shearing Interferometer", Journal of Scientific Instruments, 7(10), 374–376, (1960)
15. J. Dyson, Proc. Roy. Soc., A., 204, p. 170, (1950) http://info.uibk.ac.at//c/c7/c704/museum/en/details/optics/fresnel.html
16. A. H. Mirau, "Interferometer," US Patent No. 2, 612,074 (1949)
17. G. Nomarski, "Microinterféromètre différentiel à ondes polarisées, " J. Phys. Radium 16, 9S–11S (1955)
18. M. V. R. K. Murty, "The Use of a Single Plane Parallel Plate as a Lateral Shearing Interferometer with a Visible Gas Laser Source", Applied Optics, 3(4), 531–534, (1964)
19. W. H. Steel, [Interferometry], Cambridge University Press, (1967)
20. M. Françon, [Optical Interferometry], Academic Press, (1966)
21. E. Hecht, [Optics], 4th Edition, Addison Wesley, (2002)
22. P. Hariharan, [Optical Interferometry], 2nd Edition, Academic Press, (2003)
23. J. C. Wyant, "Introduction to Interferometric Optical Testing (SC213)", Optical Science Center, University of Arizona, http://www.optics.arizona.edu
24. Rajpal S. Sirohi, [An Introduction to Optical Metrology], CRC Press, (2015)

Role of Light in Green Technology

Ajoy Chakraborty

Abstract Advancement of technology during past century has adversely affected the life-sustaining ecosystem of our planet by depleting natural resources and changing the global climate by carbon emission. This alarming situation has made the scientists aware of the necessity of mitigating or reversing the negative effects of human activities on environment. In the present discourse we propose to discuss various aspects of green or sustainable technology. Light or electromagnetic radiation plays a very important role in achieving green development of our economy and this forms the pivot of the present talk. In dealing with the uses of light and light-based technology on our ecosystem we must not overlook the dichotomy of light and dark. Darkness has also an important role to play in Green Technology. We propose to conclude by referring to the adverse effects of excessive, misdirected and obstructive artificial lighting, causing luminous pollution.

1 Introduction

The color green brings to mind the succulent green of grass, trees and forest. Green is veritably the color of Nature. Because of this association of green with Nature, it is often described as refreshing and tranquil color. In color psychology green is considered the color of balance, harmony and growth. Green relates to stability and endurance giving us persistence and strength to cope with adversity. In view of the above connotation of 'green', the technology that aims at the development of our society and economy without impairing ecological balance and that ensures sustenance is aptly called 'Green Technology'. Technological development during past century for maintaining our modern lifestyle has adversely affected the life-sustaining ecosystem of our planet. The accelerated consumption of fossil fuels and deforestation is monotonically increasing global temperature. We are exhausting our natural resources at a rapid pace and accelerating change in global

A. Chakraborty (✉)
Institute of Engineering and Management, Kolkata 700091, India
e-mail: akc_000@rediffmail.com

© Springer Nature Singapore Pte Ltd. 2017

I. Bhattacharya et al. (eds.), *Advances in Optical Science and Engineering*,
Springer Proceedings in Physics 194, DOI 10.1007/978-981-10-3908-9_9

climate that threatens our life-sustaining ecosystem and economy. Scientists have warned that if green-house gas emissions continue to increase we will soon pass threshold beyond which global warming would become catastrophic. Destroying environment any further can lead to a situation which can be irreversible. Thus the most urgent issue for present-day engineering and technology is sustainable development and reversal of negative effects of human activities on environment. During the last century we have acquired a huge amount of knowledge in various branches of technology. Mere knowledge however does not ensure sustainable development of our society and economy. What we need in addition is wisdom. In the context of green technology what we need may be called 'green wisdom'. Green technology is the technology endorsed by green wisdom. Green technology is intended to mitigate or reverse the adverse effects of human activities on the environment. It is developed and used in a way so that it does not disturb our environment and conserve natural resources. Green technology aims at sustainable development of our society and economy. Sustainable development is the development that meets the needs and aspirations of the society in ways that can continue indefinitely without damaging our environment or depleting natural resources. Thus, green technology or sustainable development ensures a better quality of life for everyone now and for generations to come.

1.1 The Sandestin Declaration

In 2003 approximately 65 scientists and engineers from industry, academic institutes and government met at the Sandestin resort in Florida for a conference on 'Green Engineering: Defining Principles.' By the end of the conference the scientists collectively agreed to a compiled set of nine principles now known as the 'Sandestin declaration.' These principles are listed below: (1) Conserve and improve natural ecosystem while protecting human health and well-being. (2) Engineer processes and products holistically, use systems analysis, and integrate environmental impact assessment tools. (3) Use life-cycle thinking in all engineering activities. (4) Ensure that all materials and energy inputs and outputs are as inherently safe and benign as possible. (5) Minimize depletion of natural resources. (6) Strive to prevent waste. (7) Develop and apply engineering solutions while being cognizant of local geography, aspiration and culture. (8) Create engineering solutions beyond current and dominant technologies, improve, innovate and invent new technologies to achieve sustainability. (9) Actively engage communities and stockholders in development of engineering solutions.

In this context we talk of three R's of sustainable development: Reduce, Reuse and Recycle. In order to ensure sustainable development we must try to reduce the use of natural resources as much as possible. We must also make efforts to use fewer materials so that we produce less waste. Besides, one must not throw away acceptable and reusable materials. Recycling of these materials is an important factor in conserving material resources.

2 Different Kinds of Green Technology

A technology endorsed by 'green wisdom' belongs to the field of Green Technology. Needless to mention that all branches of technology that take into account the imperatives dictated by green wisdom belong to green technology. Thus, green chemical technology is green counterpart of chemical technology; green electronics is the green counterpart of electronics and so on. The other kinds of green technologies include green energy technology, green information technology, green photonics, green lighting technology, green manufacturing technology, green building technology, green nanotechnology etc.

2.1 *Green Chemical Technology*

Green or sustainable chemical technology is based on a philosophy of chemical research and engineering that encourages the design of products and processes that minimize the use and generation of hazardous substances. Besides, it focuses on technological approaches to preventing pollution and reducing consumption of non-renewable resources. The most important green chemical reaction is photosynthesis. In this reaction carbohydrates and oxygen are synthesized from carbon dioxide and water in presence of sunlight. According to Genesis, light was created at the very beginning of the creation of the universe. It could not be otherwise because light happens to be the prerequisite for sustenance of life. Photosynthesis that sustains life is the first green technology on our planet.

2.2 *Biomimicry*

Biomimicry is a new discipline that studies Nature's design and processes and imitates them to solve human problems. One of the examples of biomimicry was the study of bird to enable human flight. Leonardo da Vinci (1452–1519) closely observed the anatomy and flight of birds and made various notes on the possibility of creating a heavier-than-air flying machine. Again, studying the structure of plant leaves enabled the scientists to invent a better solar cell. It is observed that plant leaves have wrinkles and folds. Imitating this structure of leaves, scientists of Princeton University, headed by Prof. Stephen Chou, were able to achieve major gains in light absorption and efficiency of the solar cells. They showed that the efficiency of solar cells can be increased considerably by using a nano-structured 'sandwich' of metal and plastic that collects and traps light. Prof. Stephen Chou and his team used nano-technology to overcome two main challenges that cause solar cell to lose energy—(1) light reflecting from the cell and (2) the inability of the cell to fully capture the light that enter the cell. The efficiency of solar cells has been considerably improved by using metal nano-particles in plasmonic solar cells and nano-particles of semiconducting materials in quantum dot solar cells.

2.3 Artificial Photosynthesis

Artificial photosynthesis is a chemical process that replicates the natural process of photosynthesis. Artificial photosynthesis is more versatile in its scope in sense that it is not just the formation of carbohydrates and oxygen from carbon dioxide and water using sunlight. Artificial photosynthesis system or photochemical cell that mimics what happens in plants could potentially create an endless supply of clean energy. Artificial photosynthesis is essentially leaf-mimicking. Photosynthesis is essentially two half-reactions of oxidation and reduction. In natural photosynthesis water molecules are photo-oxidized to release oxygen and protons. The second part of the reaction is a light-independent reaction that converts carbon dioxide into glucose. Researchers working on artificial photosynthesis are developing photo-catalysts that are able to perform both the reactions. By using photo-electrode in which gold nano-particles are loaded on an oxide semiconductor substrate, research at Hokkaido University, Japan, has worked to develop a method of artificial photosynthesis for successful synthesis of ammonia using visible light, water and atmospheric nitrogen.

2.4 Green Energy Technology

Green energy technology is very important branch of green technology. The global power sector is the biggest pollutant of our planet. It generates 40% of all global electricity from coal contributing most significant amount of green house gases that adversely affect global climate. According to the International Energy Agency, the power sector is responsible for 37% of carbon emission. Green energy technology therefore shares a big responsibility in producing clean energy for sustainable development. Photonic s or light-based technology plays important role in generating renewable energy without impairing our environment. The most promising source of green energy is the Sun. Green energy technology studies and develops optical systems for generating renewable energy. This includes solar cells and photovoltaic devices creating energy-efficient optical sources.

Solar or photovoltaic cells convert solar energy into electrical energy. These cells are based on photoelectric effect: ability of matter to emit electrons when irradiated by light. Silicon is a key ingredient in solar cells. As solar light hits silicon atoms they transfer their energy to dislodge electrons of the atoms that drive the current through the cell.

3 Lighting Technology

One of the outstanding inventions in lighting technology is LEDs of very high luminous efficacy. We may recall that the luminous efficiency of incandescent lamps is only about 12 lm/W and that of tungsten-halogen lamps is only about 18 lm/W. Thanks to the development of blue LEDs, the luminous efficacy of white LEDs with phosphorescent color mixing is as high as 260–300 lm/W.

Three Japanese scientists, Isamu Akasaki, Hirosi Amano, and Shuji Nakamura were awarded 2014 Nobel Prize in physics for their invention of blue LEDs. Some researchers at the University of Hiroshima in Japan have used silicon-based quantum-dots for a type of LED that promises to revolutionize lighting system. Japanese researchers have fabricated a hybrid organic/inorganic LED that produces white-blue electroluminescence using quantum dots. An organic LED is a device in which emissive electroluminescent layer is a film of organic compound that emits light in response to electric current.

3.1 Light Pollution

We hear a lot of discussion about air pollution, water pollution, sound pollution etc. Because of the artificial illumination of high level, particularly in urban areas during night, a new kind of pollution, known as light pollution, has become a matter of grave concern for biosphere of our planet. Light pollution, also known as photo-pollution or luminous pollution, is caused by excessive misdirected and obstructive artificial light. It is the result of outdoor lighting that is not properly shielded, allowing light to be directed into the eye and the night sky. As excessive sound causes sound pollution so also excess light causes light pollution. Light that shines into the eye is called ‘glare’ and light shining into the night sky above the horizon causes ‘sky glow’. The other major component of light pollution is ‘light trespass’. Urban sky glow is caused by brightening of night sky over inhabited areas, light trespass is light on areas where it is not intended or needed and glare is the excessive brightness that causes visual discomfort.

There are many adverse consequences of light pollution some of which are listed below: (1) It disrupts ecosystem. (2) It poses serious threat to nocturnal wild life. (3) It has negative impact on plant and animal physiology. (4) It can confuse the migratory patterns of animals and alter the competitive interaction of animals. (5) It can change predator –prey relationship. (6) Light pollution has devastating effects on many bird species. Birds that migrate at night navigate by moonlight and starlight. Artificial light can cause them wonder off towards the dangerous night-time landscapes of cities. (7) Every year millions of birds die colliding with needlessly illuminated building. Artificial light can cause them to migrate too early or too late and miss ideal climate conditions for nesting, foraging and other behavior. (8) Many insects are drawn to light. Artificial light can create fatal attraction and decrease insect population. (9) Declining insect populations have negative impact on species that rely on insects for food or pollination. (10) Some predators exploit this attraction to their benefit, affecting food webs in an unanticipated manner. (11) Artificial light near sea beach has devastating effects on sea turtles. These turtles live in sea water. When mature female turtles are ready to lay eggs they come to sandy beach and find a suitable dark place for the purpose. They dig shallow holes in the sandy beach with the help of their hind limbs and bury their eggs in these holes. After an incubation period of about eight weeks the eggs are hatched. However, the hatchlings stay under the sand until temperature and other conditions are right for the baby turtles to come out of the sand and move to the sea with the help of moonlight reflected from the sea. The presence of artificial lights from houses near beach, car parks and security light deceives baby turtles and disorients them. Consequently, they fail to reach the sea and move towards the artificially illuminated regions. Many of them die or get killed before they find the way to

4 Epilogue

From above considerations it is evident that light-based technologies have promising potentials to be green, smart, life-sustaining at the same time. Light, that played a vital role in originating life on earth, bears promise of ensuring sustenance of entire biosphere. This ethereal angel stands as a protector and savior of life in all its forms. Light has created many wonders that substantially contributed to green development of our technology. We are yet to explore various possibilities of photonic technologies for generating renewable energy, and for reducing energy consumption and green-house gas emission. 'Life throws challenges and every challenge comes with rainbows and lights to conquer it.' (Amit Ray, World Peace: The Voice of a Mountain Bird). If we ignore the figurative meanings of the words 'rainbows' and 'lights', the above quote may be interpreted to imply that light with all its variegations bears the promise of solving any human problem.

However, it is not the responsibility of the photonic technologists alone to keep our Mother Earth clean and life-sustaining. Herbert Marshal McLuhan, a Canadian Professor of English and an expert of mass communication, once said: 'There are no passengers on spaceship Earth. We are all crew.' Prof. McLuhan implied that all of us, irrespective of our nationality, religion and creed, are equally responsible for keeping our planet life-sustaining and healthy abode for all living creatures.

Pareto Optimality Between Far-Field Parameters of Lossless Phase-Only Filters

L.N. Hazra and S. Mukhopadhyay

Abstract Resolution capability of an optical imaging system can be enhanced by reducing width of central lobe of the point spread function (PSF) of the transverse intensity distribution on the far field plane. Attempts to achieve the same by pupil plane filtering, is usually accompanied by concomitant increase in side lobe intensity. The mutual exclusivity between these two objectives may be cast as a multi objective optimization problem that does not have a unique solution; rather a class of trade off solutions called Pareto optimal solutions may be generated. To achieve super resolution, lossless phase only filters with pre-specified lower limits for Strehl ratio are synthesized by using Particle Swarm Optimization technique. Practical validation of the theoretical results is also undertaken by realizing the phase filters on reflective, phase only liquid crystal on silicon spatial light modulator.

1 Introduction

The resolving power [1] of an optical imaging system is limited by several factors. The residual aberration of the imaging lens severely degrades the resolution capability. Even for an aberration-free objective, diffraction effects arising out of the finite aperture, produces a blurred spot instead of a point. The intensity distribution in that light spot is commonly known as intensity Point Spread Function (PSF). For a diffraction limited optical system with circular aperture, this PSF is called Airy pattern at the transverse focal plane. The Airy pattern consists of a central lobe surrounded by a number of bright and dark rings with gradually diminishing intensity.

L.N. Hazra (✉) · S. Mukhopadhyay (✉)
Department of Applied Optics & Photonics, University of Calcutta,
JD-2, Sector-3, Salt Lake, Kolkata 700106, India
e-mail: lakshminarayanhazra@gmail.com

S. Mukhopadhyay
e-mail: lnaphy@caluniv.ac.in

© Springer Nature Singapore Pte Ltd. 2017
I. Bhattacharya et al. (eds.), *Advances in Optical Science and Engineering*,
Springer Proceedings in Physics 194, DOI 10.1007/978-981-10-3908-9_10

Unlike the resolution enhancement techniques that involve extensive post-processing of data [2, 3], pupil plane filtering is a non-scanning, whole field imaging technique where the image is directly modified in real time without any post processing of data. The technique is commonly used for either suppression of side lobes and/or narrowing down the central lobe of the point spread function, as and when required. In most cases, the filters consist of annular zones with variation of either amplitude transmittance or complex amplitude transmittance among the annular zones. With the advent of diffractive optics technology, many investigators, notably Sales and Morris [4] explored the use of multiphase structures. In these structures the phase transmission of each zone can assume an arbitrary value in the interval $(0, 2\pi)$.

However, any attempt for narrowing the central lobe of the PSF is accompanied by enhanced intensity of the neighboring side lobes, of which the first side lobe plays a dominant role in degrading the quality of the image. The overall effect is reduction of contrast and poor Strehl Ratio in the final image. Simultaneous lowering of width of the central lobe and peak intensity in the side lobes of the PSF by pupil plane filtering is mutually exclusive, and, therefore search for optimal filters may be cast as a multi objective optimization (MO) problem. Instead of having a unique solution, an MO problem typically has a set of acceptable trade-off solutions called Pareto optimal solutions in the objective function space. The latter space is found to be divided into infeasible and sub-optimal regions, the separator between the two being called 'Pareto Front'. In case of a two objective problem the Pareto front reduces to a Pareto curve.

The concept of Pareto optimality was laid down by engineer/economist Vilfredo Pareto [5]. A solution belongs to Pareto set if there is no other solution that can improve one objective without degrading the other. In the problem under view, the width of the central lobe of the transverse PSF and the peak intensity in the side lobes are two conflicting objectives to be optimized. To solve this MO problem, we have adopted Particle Swarm Optimization (PSO) [6] technique. In a previous communication, Genetic Algorithm (GA) [7] was used by us as a tool to solve optimization tasks. GA approaches are inherently discrete in nature whereas PSO is inherently continuous and can be modified using sigmoid function for handling discrete variables. Both GA and PSO are qualitatively equivalent but in GA, the presence of three basic genetic operators namely, Selection, Crossover and Mutation increases computational intricacy. On the other hand, evaluation of only two equations for each variable in each iteration makes PSO computationally more efficient than GA. In this case, using PSO, optimal phase filters are generated and investigated for achieving resolution beyond diffraction limit at the transverse focal plane of an optical imaging system.

For experimental verification of the theoretically observed Pareto front, the optimal filters are realized on a programmable Reflective Liquid Crystal on Silicon Spatial Light Modulator (LCoS SLM). The Pareto optimality between width of the central lobe and the peak intensity in the side lobe in the transverse PSF is investigated experimentally. While formulating the far field parameters theoretically, we assumed a uniform plane beam. But in practice, the laser beam is

Gaussian instead of being a plane beam. As we move away from the center of the Gaussian beam along any azimuth, the intensity drops in a rapid rate. Therefore, the effect of increase in zone number of the optimum filters is not apparent.

2 Analytical Expression for the Transverse Intensity at the Far Field, Fitness Function and Overview of Pso

Figure 1 shows the exit pupil and the image plane in the image space of an axially symmetric imaging system. For an M zone filter on the exit pupil, the normalized transverse intensity distribution on the paraxial image plane of the axially symmetric imaging system may be expressed as

$$I_N(p) = 4\sum_{m=1}^{M}\sum_{n=1}^{M}[\cos k(\alpha_m - \alpha_n)\Im_m(p)\Im_n(p)] \tag{1}$$

where α_m is the phase over mth annular zone and $\Im_m(p)$ is given by

$$\begin{aligned}\Im_m(p) &= \left[\frac{r_m J_1(pr_m) - r_{m-1}J_1(pr_{m-1})}{p}\right] \\ &= \left[r_m^2\frac{J_1(pr_m)}{(pr_m)} - r_{m-1}^2\frac{J_1(pr_{m-1})}{(pr_{m-1})}\right]\end{aligned} \tag{2}$$

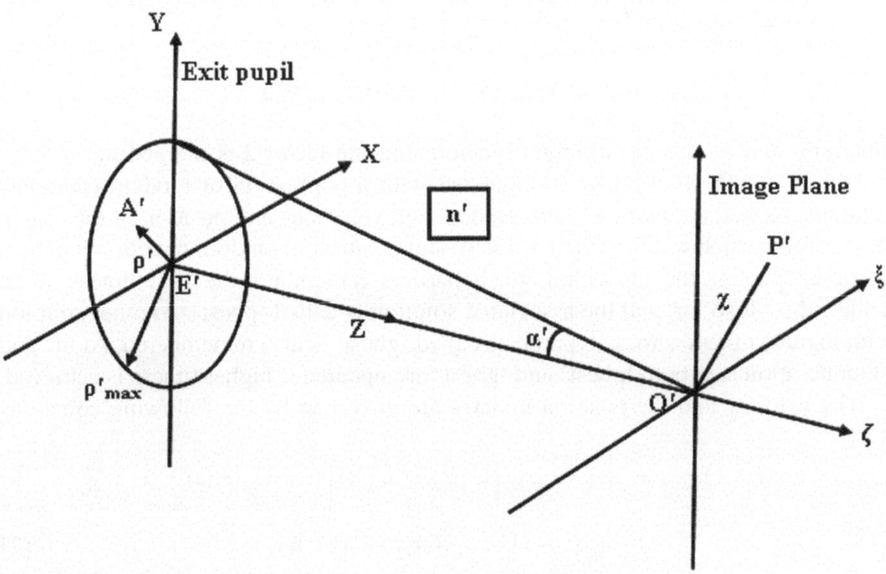

Fig. 1 Exit pupil and image plane of an axially symmetric imaging system

$J_N()$ is the Bessel function of the first kind, and of order N. r is the normalized radial distance of a point A′ on the exit pupil from the optical axis and is given by, $r = \rho'/\rho'_{max}$, where ρ′ is the radial distance of the point A′ from the center E′ of the exit pupil, and ρ'_{max} is the radius of the exit pupil and p is the reduced diffraction variable defined as

$$p = \frac{2\pi}{\lambda}(n' \sin\alpha')\,\chi \tag{3}$$

In the above, χ (= O′P′) is the geometrical distance of the point p' on the image from the center of the diffraction pattern, α' is the semi-angular aperture, $(n' \sin\alpha')$ is the image space numerical aperture and λ is the operating wavelength.

During the search procedure, the phase of innermost zone is set to zero and the radius of the innermost and outermost zone is set to 0 and 1 respectively. Any of the remaining zones are allowed to have any phase, out of some finite discrete phase levels in the range (0, 2π). The inner and outer radii of m-th zone of an M zone filter are r_{m-1} and r_m (m = 1,…,M) respectively. During stochastic search operation, r_m is allowed to take any value within the range (r_{m-1}, 1). The fitness function Φ in this case, is inversely related to the merit function ψ by

$$\Phi = \frac{1}{1+\psi} \tag{4}$$

ψ is defined as a linear combination of the squared difference of the target (superscripts T) and current (superscripts C) values of normalized Full Width Half Maximum (FWHM $(2\hat{p})$) of the transverse PSF, and β which is the peak intensity in its side lobes over a prespecified range.

$$\psi = \omega_1\left(\hat{p}^T - \hat{p}^C\right)^2 + \omega_2\left(\beta^T - \beta^C\right)^2 \tag{5}$$

where ω_1 and ω_2 are the weighting factors and the factor 2 is merged in ω_1.

In basic PSO, the system is initialized with a population of random candidate solutions termed as ‘particle’ with randomized velocities and position in the search space. The particles update their velocity and position in random direction which is a crucial part of the algorithm. Each particle remembers the best fitness value achieved by it so far, and the associated solution is called ‘pbest’. Another solution with highest fitness across all particles called ‘gbest’ is also remembered. At the end of an iteration step both ‘pbest’ and ‘gbest’ are updated if higher fitness is achieved.

The velocity and the position updates are governed by the following equations

$$v_{i,j}(t+1) = \omega v_{i,j}(t) + c_1 r_{1j}\left[x'_{i,j}(t) - x_{i,j}(t)\right] + c_2 r_{2j}[g_j(t) - x_{i,j}(t)] \tag{6}$$

$$x_{i,j}(t+1) = x_{i,j}(t) + v_{i,j}(t+1) \tag{7}$$

where $v_{i,j}(t)$ and $x_{i,j}(t)$ are the jth components of the velocity and the position vectors of the ith particle at time t, respectively. The term $x'_{i,j}(t)$ is the jth component of 'pbest' of ith particle and $g_j(t)$ is the jth component of 'gbest' of the swarm at time t, respectively. The time increases by unity for successive iteration steps. In the present problem, a 'particle' represents a phase filter comprising of phases and radii values. The associated intensity distribution is computed using (1). From this distribution transverse FWHM $2\hat{p}$ and side lobe intensity β are determined. These values are then used to evaluate the merit function as well as the fitness of that particular filter using (4) and (5). The velocity update takes values such that the particles converge towards their own best values as well as towards the swarm's best solution.

3 Numerical Results

Each search resulted in an optimal phase filter that provides a combination of $2\hat{p}$ & β values. Then the transverse FWHM $2\hat{p}$ is normalized by that of the Airy pattern. These sets of normalized transverse FWHM (Δ) and the corresponding side lobe intensity (β) define a curve called Pareto front which is concave in nature. The Pareto fronts for a 2 zone filter with available phase steps 8 at different pre specified lower limits of Strehl Ratio are shown in Fig. 2.

The advantage of generating a Pareto front is explained with an example as follows. To obtain a normalized transverse FWHM 0.8, the Strehl Ratio must be sacrificed; because the vertical dashed line does not intercept the Pareto fronts for the higher cut off values of Strehl Ratio 0.5 and 0.7. Possibility of obtaining FWHM 0.8 is increased as the Strehl Ratio is sacrificed to lower value; the vertical dashed line intercepts the Pareto fronts for lower cut off values of Strehl ratio 0.1 and 0.3. But in both cases one has to deal with a larger side lobe. Moreover, Pareto front provide a limiting value of the peak side lobe to achieve a target FWHM under a constrained value of Strehl Ratio. It has also been noted from our limited observations that increase in zone number with constant phase steps or vice versa does not have any major effect on the Pareto front.

4 Experimental Validation

The experimental arrangement is shown in Fig. 3. A series of simulated phase filters superposed on a lens function are displayed on the computer screen and so also on the SLM screen. Vertically polarized light is required for the SLM to work as a pure phase modulator. Some numerical intensity distribution and the corresponding experimentally captured intensity patterns for 2 and 3 zone filters are

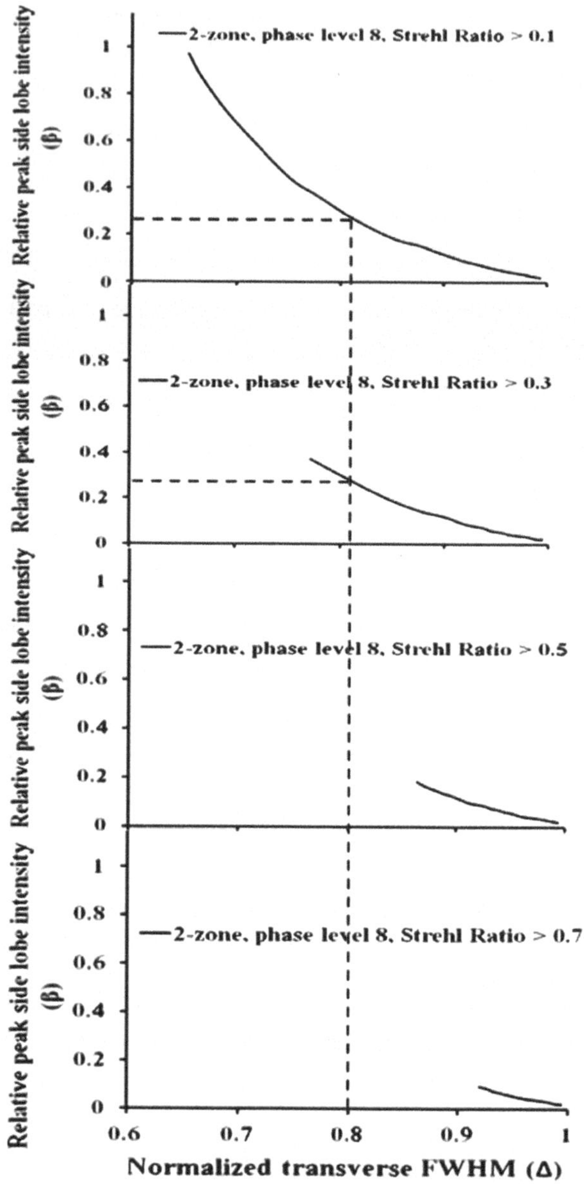

Fig. 2 Pareto front for a 2 zone filter at different lower cut off values of Strehl Ratio

shown in Fig. 4. Also the theoretically obtained and experimentally generated Pareto fronts for 2 and 3 zone filters are shown in Fig. 5.

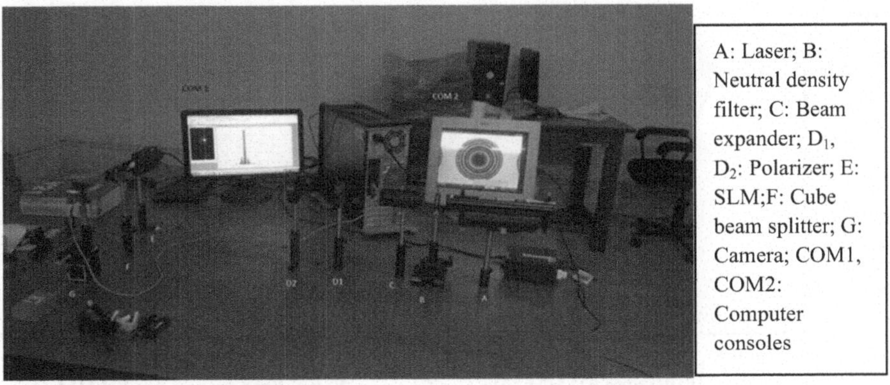

Fig. 3 Experimental arrangement

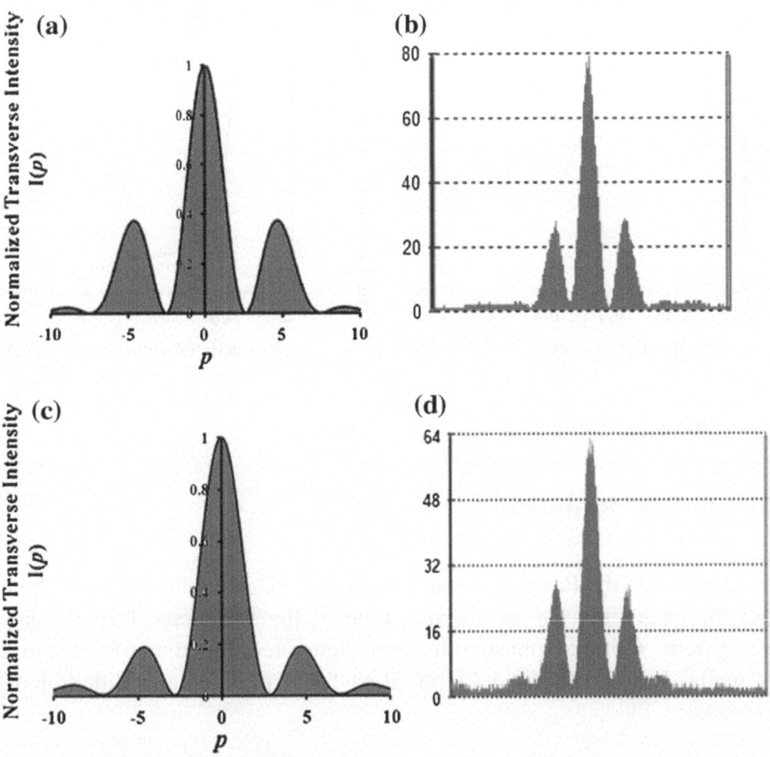

Fig. 4 Theoretically obtained and experimentally observed intensity distribution for 2 zone (**a–b**) and 3 zone filter (**c–d**) respectively

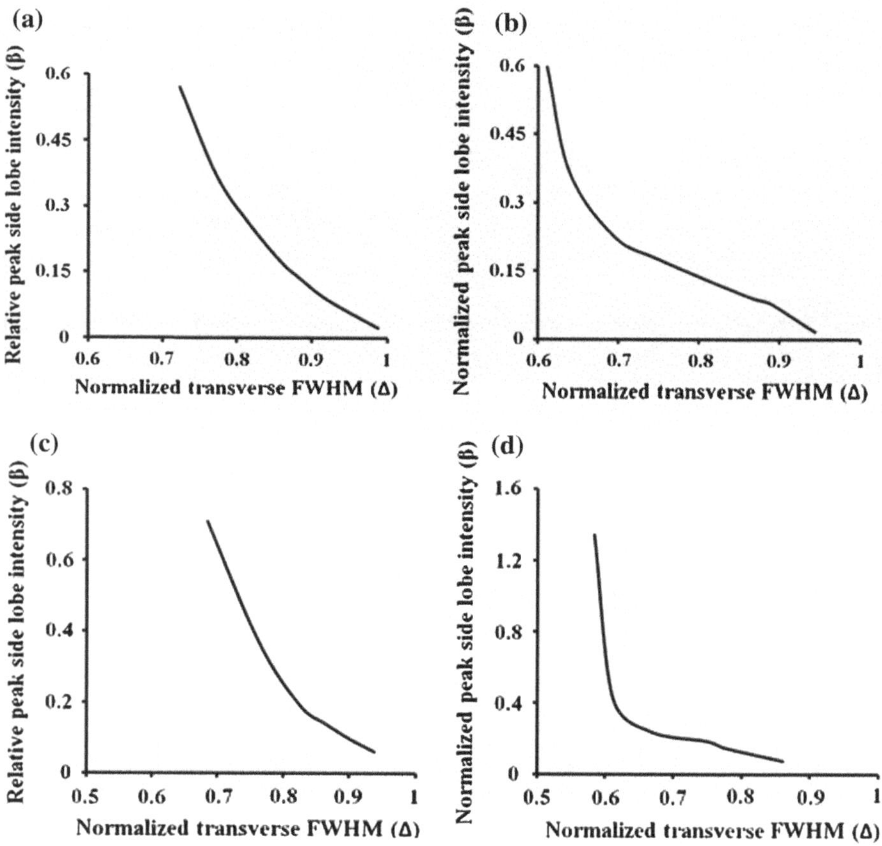

Fig. 5 Theoretically obtained and experimentally observed Pareto front for 2 zone (**a**–**b**) and 3 zone filter (**c**–**d**) respectively

5 Concluding Remarks

This report presents the Pareto optimal relationship between the width of the central lobe and the peak intensity in the side lobe of the transverse PSF in an optical imaging system. Optimal phase filters are generated based on PSO and subsequently realized on a reflective LCoS SLM for experimental validation of theoretical results.

Acknowledgements This work has been supported by the INSPIRE Fellow program for Somparna Mukhopadhyay under the scheme of "Assured Opportunity for Research Career" (AORC)/2010 [115] by the Department of Science & Technology (DST), India.

References

1. Jenkins F.A. and White H.E., [Fundamentals of Optics], 4th ed., McGraw-Hill (1976)
2. Paesler M. A. and Moyer P. J., [Near-Field optics: Theory, introduction, and applications], John Wiley, New York (1996)
3. Hell S. W. and Wichmann J., "Breaking the diffraction resolution limit by stimulated emission: stimulated-emission-depletion fluorescence microscopy," Opt. Lett. 19(11), 780–782 (1994)
4. Sales T. R. M. and Morris G. M., "Diffractive superresolution elements," J. Opt. Soc. Am. A 14(7), 1637–1646 (1997).
5. Pareto V., "Cours D'Economie Politique," Vol. I and Vol. II, F. Rouge, Lausanne (1886)
6. Kennedy J. and Eberhart R. C., "Particle swarm optimization," Proc. IEEE International Conference on Neural Networks IV, 1942–1948 (1995)
7. Reza N. and Hazra L. N., "Toraldo filters with concentric unequal annuli of fixed phase by evolutionary programming," J. Opt. Soc. Am. A 30(2), 189–195 (2013)

Can Photons Affect the Elastic Constants in Heavily Doped Nano Wires?

R. Paul, S. Ghatak, S. Das, M. Mitra, T. Datta and K.P. Ghatak

Abstract Effect of strong photo excitation on the elastic constant (EC) in extremely degenerate Nano-Wires (NW) forming Gaussian band tails has been investigated by deriving a fundamental carrier statistics formula using NWs of Heavily Doped (HD) n-InSb, n-InAs, $Hg_{1-x}Cd_xTe$ and $In_{1-x}Ga_xAs_yP_{1-y}$ $_y$ lattice matched to InP as examples. We observe that ΔC_{44} becomes invariant of the film thickness under the condition of relatively low values of the quantum thickness, indicating a very sharp fall at a particular value of the nano thickness manifesting the quantum size effect, in EC. The EC increases with decreasing light intensity, wavelength and alloy composition where the rate of change depends on the values of the band constants respectively. The EC can be experimentally determined by using the corresponding the experimental values of the thermo electric power.

R. Paul
Department of CSE, University of Engineering and Management,
Kolkata 700156, India

S. Ghatak
Department of BCA and M.Sc, Institute of Engineering and Management,
Kolkata 700091, India

S. Das
Department of CSE, Indian Institute of Engineering Science and Technology,
Shibpur, Kolkata 711103, India

M. Mitra
Department of ECE, Indian Institute of Engineering Science and Technology,
Shibpur, Kolkata 711103, India

T. Datta (✉)
Department of Basic Science and Humanities, Institute of Engineering and Management,
Kolkata 700091, India
e-mail: triparna.datta@iemcal.com

K.P. Ghatak
Department of ECE, University of Engineering and Management,
Kolkata 700156, India

© Springer Nature Singapore Pte Ltd. 2017
I. Bhattacharya et al. (eds.), *Advances in Optical Science and Engineering*,
Springer Proceedings in Physics 194, DOI 10.1007/978-981-10-3908-9_11

Since the inception of solid state physics, the importance of band structure in determining the physical properties of different materials under various physical conditions is well known. In this paper we wish to study the influence of energetic photon on the EC in extremely degenerate NW on the basis of the Hamiltonian and perturbation theory together with the heavy mathematical techniques of Quantum Mechanics in this context. With the advent of technologically important nano materials, the EC has been investigated under different conditions of reduced dimensions in the literature by the group of Ghatak et al. [1–12] and few others [13, 14]. In what follows, we study the EC under strong photon field in heavily doped nano wires taking the examples as stated in the abstract.

The electron energy (E) versus electron wave vector relation in Quantum Wires (QW) of heavily doped III–V and optoelectronics materials in the presence of energetic photons assumes the

$$(\hbar N_z/L_z)^2 + (\hbar N_y/L_y)^2 + (\hbar k_x)^2 = [M_e f_1/5] \tag{1}$$

where the notations mean as usual and f_1 is a complex function of electron energy, scattering potential and incident photon wavelength. From (1) the corresponding density of states per sub band can be expressed as

$$f_2 = \frac{g_v M_c}{5\pi\hbar} f_1' [\frac{M_c f_1}{5} - \{\hbar N_z/L_z)^2 + (\hbar N_y/L_y)^2\}]^{-1/2} \tag{2}$$

The linear electron density under the condition of extreme carrier degeneracy assumes the form

$$N_{1D} = \frac{2g_v}{\pi\hbar} \, Real\,part\,of \sum_{n_y=1}^{n_{y\max}} \sum_{n_z=1}^{n_{z\max}} [\frac{M_c f_1}{5} - \{\hbar N_z/L_z)^2 + (\hbar N_y/L_y)^2\}]^{1/2} \Big|_{E=E_{Fll}} \tag{3}$$

where E_{Fll} is the Fermi energy under the present constrained conditions.

The EC can be expressed as

$$\Delta C_{44} = \frac{-G_0^2}{9} Real\,part\,of [\frac{\partial N_{1D}}{\partial(E_{Fll} - E_{Sub-band})}] \tag{4}$$

Suggested relationship for experimentally determining ΔC_{44} from experimental values of corresponding Thermoelectric power (G)

$$\Delta C_{44} = \left[\frac{-N_{1D}(G_0)^2 |e| G}{(3\pi^2 k_B^2 T)} \right] \tag{5}$$

Therefore, by using (5) we can investigate ΔC_{44}.

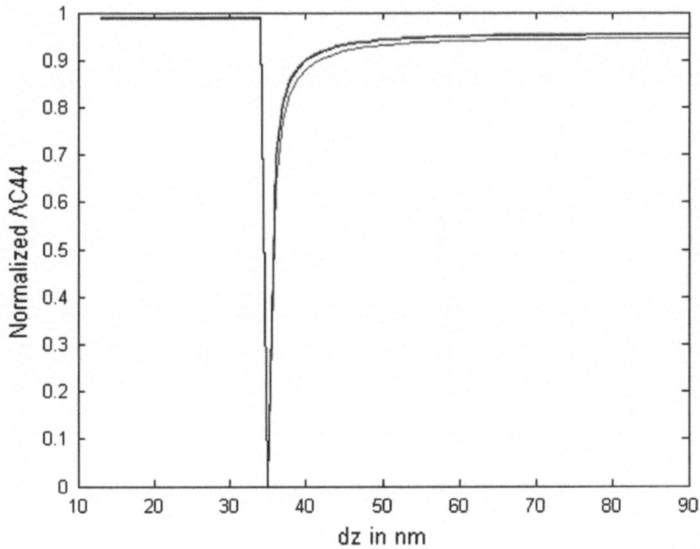

Fig. 1 The influence of quantum size effect on ΔC_{44} in the presence of intense photo excitation for all the highly degenerate Nano Wires as stated in the abstract

It appears from Fig. 1 that in HD Nano Wires the ΔC_{44} becomes invariant of the film thickness for the small value of film thickness, exhibits sharp fall at a particular value of thickness manifesting the quantum size effects in ΔC_{44} in Nano Wires and again increases rapidly with increasing thickness. Figure 2 shows the spiky oscillatory variation of ΔC_{44} with the electron statistics per unit length. The oscillatory variation occurs when the Fermi energy for the present system touches the edge of the sub-band energy. Figure 3 shows that the light intensity attenuates the ΔC_{44} which decreases as intensity increases although the amount of attenuation is different for different materials. Figure 4 shows that the influence of wave length on the ΔC_{44} decreases as with the increment of wave length of the external photo excitation for the whole range of wave lengths. Finally, from Fig. 5 we observed that with the decrement of alloy composition, the ΔC_{44} enhances.

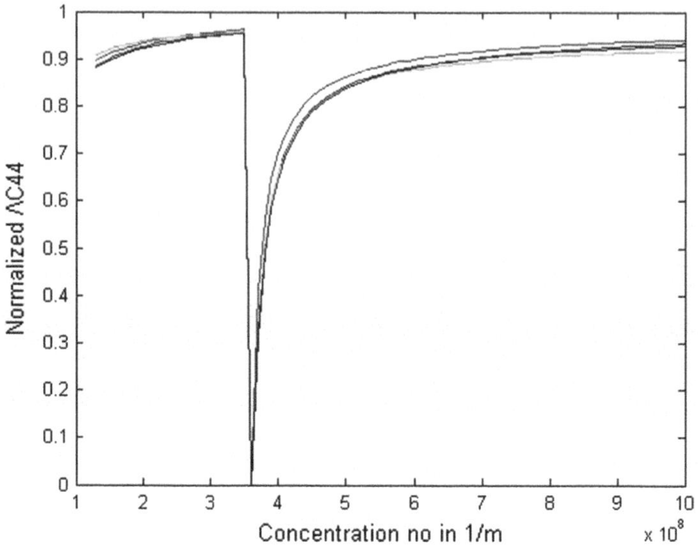

Fig. 2 The dependence of ΔC_{44} on carrier degeneracy in the presence of intense photo excitation for all the highly degenerate Nano Wires as stated in the abstract

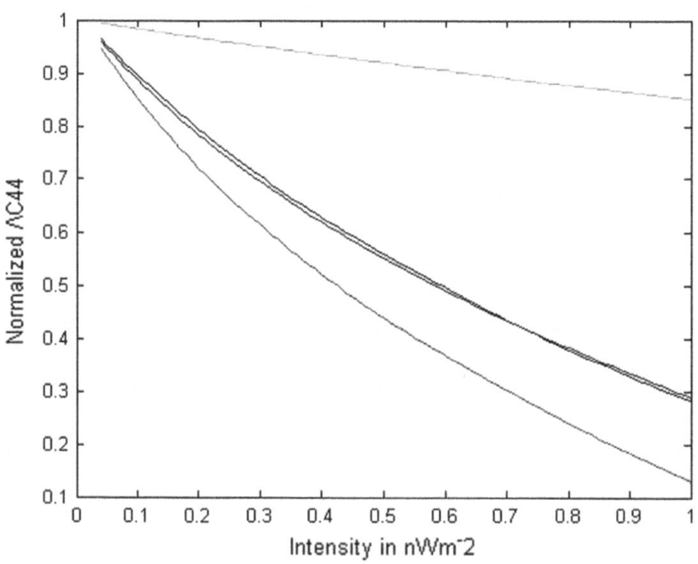

Fig. 3 The dependence of ΔC_{44} on intensity of photo excitation for all the highly degenerate Nano Wires as stated in the abstract

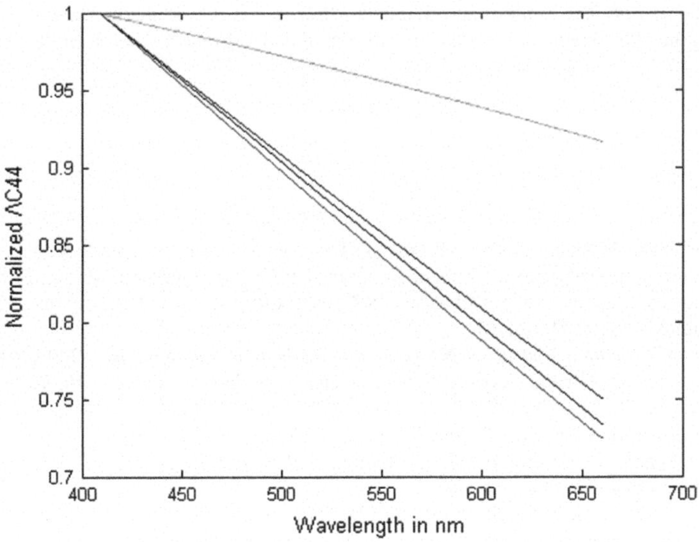

Fig. 4 The dependence of ΔC_{44} on wavelength of photo excitation for all the highly degenerate Nano Wires as stated in the abstract

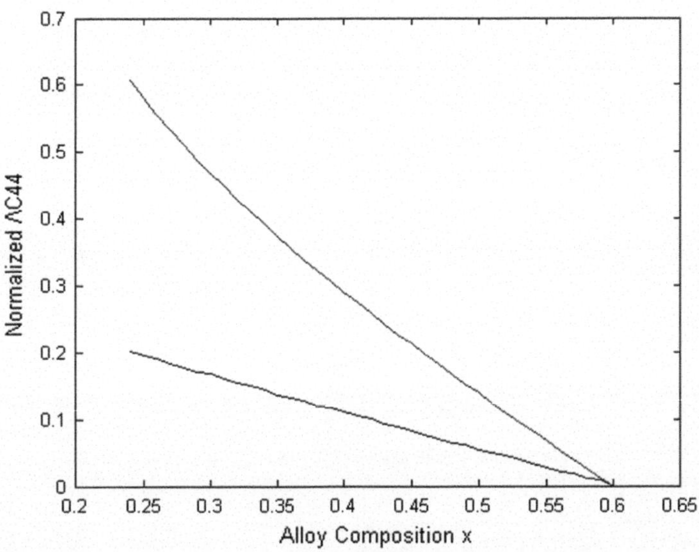

Fig. 5 The dependence of ΔC_{44} on wavelength of photo excitation for all the highly degenerate Nano Wires as stated in the abstract

Acknowledgements The authors are grateful to Prof. Dr. S. Chakrabarti, Director, Institute of Engineering and Management, Kolkata for inspiration and helpful discussion in the real sense of the term.

References

1. S Chakrabarti, M Chakraborty, KP Ghatak, Reviews in Theoretical Science, **4**, 10 (2016).
2. B Chatterjee, S Chakrabarti, SK Sen, M Mitra, KP Ghatak, Quantum Matter, **5**, 85 (2016).
3. M Mitra, M Chakraborty, S Debbarma, S Chakraborty, SK Sen, B Chatterjee, KP Ghatak, Quantum Matter, **5**, 58 (2016).
4. TN Sen, KP Ghatak, Journal of Nanoscience and Nanotechnology, **16**, 1229 (2016).
5. S Debbarma, KP Ghatak, Journal of Nanoscience and Nanotechnology, **16**, 1095 (2016).
6. B Chatterjee, S Chakrabarti, M Chakraborty, KP Ghatak, Reviews in Theoretical Science, **3**, 428 (2015); SM Adhikari, KP Ghatak, Quantum Matter, **4**, 599 (2015).
7. S Chakrabarti, B Chatterjee, S Debbarma, KP Ghatak, Journal of Nanoscience and Nanotechnology, **15**, 6460 (2015).
8. KP Ghatak, B Mitra, Physica Scripta **46** (2), 182 (1992).
9. B Nag, KP Ghatak, Journal of Physics and Chemistry of Solids **58** (3), 427, (1997).
10. KP Ghatak, B Nag, physica status solidi (b) **205** (2), 519, (1998).
11. KP Ghatak, JY Siddiqui, B Nag, Physics Letters A **282** (6), 428, (2001).
12. KP Ghatak, SN Biswas, MRS Proceedings **308**, 445, (1993).
13. R.W.Keyes, IBM jour. Res. Dev. **5**, 266 (1961).
14. A. K. Sreedhar, S. C. Gupta, phys. Rev **5B**, 1360 (1972).

Noble Metal Doped Optical Fiber for Specialty Light Source

Rik Chattopadhyay, Arindam Haldar, Mukul C. Paul and Shyamal K. Bhadra

Abstract We report noble metal doped optical fiber which is suitable for laser and helps enhancing fluorescence when it is doped with active elements. Silver nano-clusters (average diameter 1.5 nm) in the core of a standard step index fiber are doped using standard fiber fabrication method. These fibers show broad visible fluorescence in the wavelength range 420–700 nm under 405 nm excitation. This appears to occur due to long interaction length of pump light with the metal nano-clusters and quantum confinement effect. We observe enhanced fluorescence from rare earth ions (i.e. Tm^{3+} and Yb^{3+}) in presence of silver nano-clusters when the optimized length of the fiber is pumped by using 980 nm fiber-pig-tailed laser diode. The experimental results are explained with the help of analytical and quantum mechanical models. These fibers would be helpful to make efficient optical fiber based light sources mostly in the visible range.

1 Introduction

Conventional fiber based light source is made with the help of rare earth (RE) doped optical fiber where the core glass is doped with RE ions such as Tm^{3+}, Yb^{3+}, Er^{3+} etc. with limited fluorescence emission efficiency. The absorption and emission cross-sections of the RE ions in silica glass are insufficient to produce large optical emission band with available commercial pumping sources. Since

R. Chattopadhyay · A. Haldar · M.C. Paul · S.K. Bhadra
Fiber Optics and Photonics Division, CSIR-Central Glass and Ceramic Research Institute, 196 Raja S. C. Mullick Road, Kolkata 700032, India

A. Haldar
Optoelectronics Research Centre, University of Southampton, B53, Southampton, SO17 1BJ, UK

R. Chattopadhyay · S.K. Bhadra (✉)
Raman Centre for Atomic, Molecular and Optical Sciences, Indian Association for the Cultivation of Science (IACS), Kolkata 700032, India
e-mail: rcskb@iacs.res.in

© Springer Nature Singapore Pte Ltd. 2017

I. Bhattacharya et al. (eds.), *Advances in Optical Science and Engineering*, Springer Proceedings in Physics 194, DOI 10.1007/978-981-10-3908-9_12

noble metal-nano-clusters (MNCs) have interesting optical properties when they are doped in dielectric materials, we tried to dope some noble metals, especially Silver (Ag) nanoclusters (NCs) in the fiber core to fabricate a new family of fluorescent fiber which are capable of producing fluorescence emission without RE ions. On the other hand the characteristics fluorescence emission of RE doped fiber is enhanced in presence of MNCs. The main challenge in synthesis of MNCs in dielectric host is to maintain the desired structural stability. Without any scaffold the clusters tend to aggregate to larger particle, which is an irreversible process as it happens in fluid. It would be an added advantage if these Ag-NCs are synthesized in the core glass of an optical fiber to get niche properties. The immobility of the clusters will prevent aggregation of the NCs. Furthermore enhanced fluorescent output could be possible in the visible regime due to the long interaction length of propagating optical field with MNCs in fiber core. In order to understand the intricacies of the process we resort to some standard theoretical and numerical models that explain the experimental results quite satisfactorily.

2 Fluorescence from Ag NCs

The optical properties of MNCs differ considerably from those of the bulk metals due to confinement of conduction electrons [1]. The optical properties of particles with diameter (>10 nm) can be explained using classical Mie theory [2] or some finite element methods [3]. As the particle size reduces, the number of surface atoms increases dramatically, implying a change of the physical environment of the NCs [4]. In addition, systems with countable number of atoms, electrons are confined and the electronic densities of states become discrete, which bring discrete features to the dielectric spectra, namely, the quantum size effect. Both effects drive the dielectric characteristics of NCs to deviate from the bulk. The classical Drude model incorporates the surface effect on the conduction electrons. In case of MNCs we also have to consider the effect of bound electrons. The Lorentz model is widely used for this purpose. According to this model the dielectric function is given by:

$$\varepsilon(\omega) = 1 + \frac{\omega_p^2}{\omega_0^2 - \omega^2 - i\gamma\omega} \tag{1}$$

where ω_p is the plasmon resonance frequency, ω_0 is the Lorentz frequency, γ is the damping factor and ω is the frequency of incident light. In general the Lorentz parameters are constant for bulk metals but in case of MNCs the Density Functional Theory (DFT) shows that there exists certain frequency below which the intraband transition of electrons takes place and the property of the MNC can be approximated by the Lorentz model [5]. Hence the Lorentz parameters became a function

Table 1 Parameters used in (2)

Parameters	a_1	a_2	a_3	b_1	b_2
Values	1.95 nm^{-1}	0.18 eV^{-1}	0.192 nm eV	0.198 nm^{-1} eV^{-1}	0.62 nm

of the NC radius and the imaginary part of the dielectric function can be written as He and Zeng [5]:

$$\varepsilon_{Imag}(\omega r) = \frac{\omega_p^2\left[1 - e^{(a_1 r)^2}\right]^2\left(\gamma + \frac{a_3}{\hbar r}\right)\left[1 - e^{(r^3/b_2^3)}\right]\omega}{\left[\frac{1}{\hbar^2}\left(\frac{1}{a_2 + b_1 r}\right)^2 - \omega^2\right]^2 + \left[\left(\gamma + \frac{a_3}{r}\right)\left(1 - e^{(r^3/b_2^3)}\right)\omega\right]^2} \tag{2}$$

The parameters a's and b's are given in Table 1. We have used $\hbar\omega_{p,bulk} = 7.24\,\text{eV}$ and $\hbar\gamma_{bulk} = 0.0685\,\text{eV}$ [5].

After getting the imaginary part we can readily derive the real part of the permittivity using the Kramer-Kronig relation given by:

$$\varepsilon_{Real}(\omega r) = 1 + \frac{2}{\pi} P \int_0^{\infty} \frac{\varepsilon_{Imag}(r, \acute{\omega})\acute{\omega}}{\acute{\omega}^2 - \omega^2} d\acute{\omega} \tag{3}$$

where 'P' denotes principal value integral. Using (2) and (3) we can calculate the dielectric permittivity of MNCs. Figure 1 shows the dielectric permittivity of MNCs with different radius in vacuum.

When these MNCs are doped in a dielectric material, the effective permittivity of the composite material can be estimated by using Maxwell Garnett effective medium theory owing to the approximation that the MNCs are not connected to each other. In this case the effective permittivity is given by Koledintseva et al. [6]:

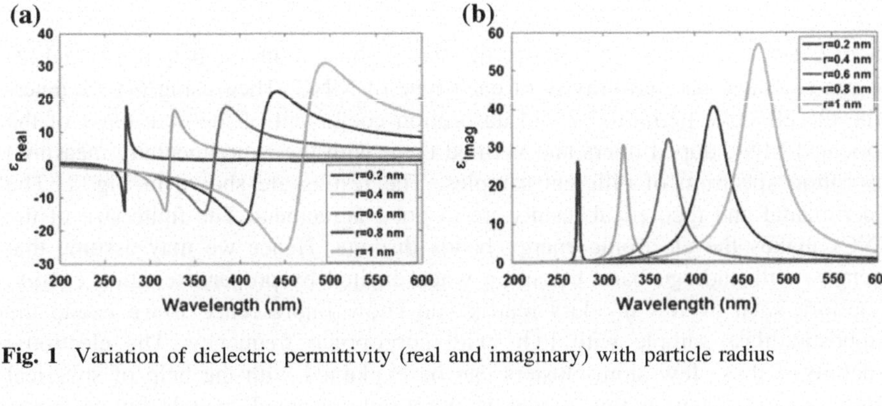

Fig. 1 Variation of dielectric permittivity (real and imaginary) with particle radius

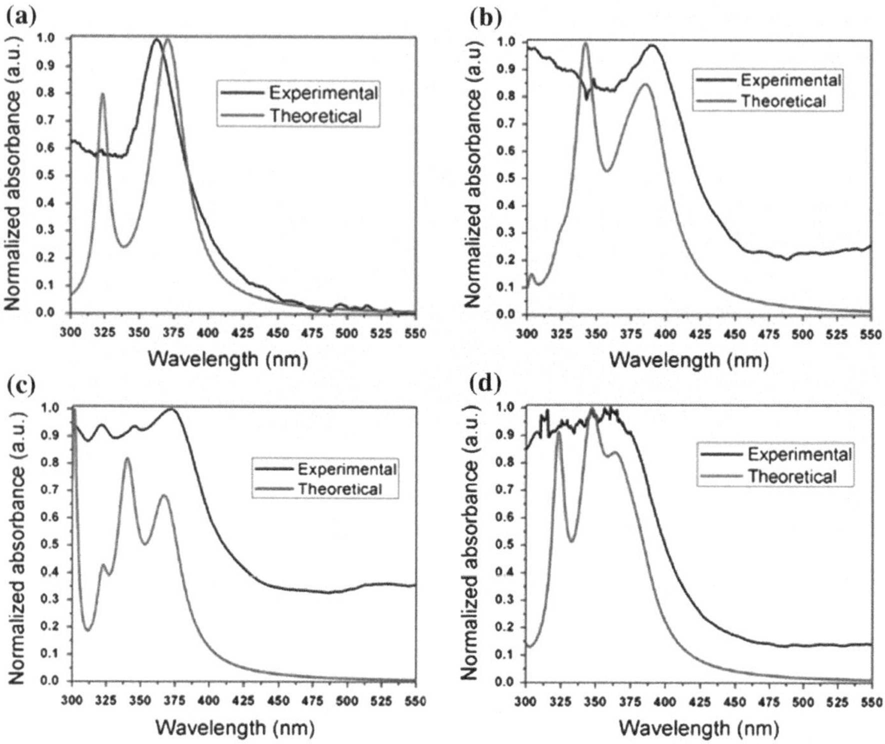

Fig. 2 Theoretical and experimental absorption spectra of fiber samples **a** AMNC-1, **b** AMNC-2, **c** AMNC-3 and **d** AMNC-4

$$\varepsilon_{eff} = \varepsilon_{host} + \frac{\sum_{i=1}^{n} f_i\left(\varepsilon_{i,metal} - \varepsilon_{host}\right)\frac{3\varepsilon_{host}}{\varepsilon_{i,metal} + 3\varepsilon_{host}}}{1 - \sum_{i=1}^{n} f_i \frac{\varepsilon_{i,metal} - \varepsilon_{host}}{\varepsilon_{i,metal} + 2\varepsilon_{host}}} \tag{4}$$

where f_i denotes volume fraction of i-th MNC in host. With the help of (2) and (3) we can calculate the permittivity of each type of MNC. Then using (4) we determine the effective permittivity and absorption coefficient of the core glass of the fabricated MNC doped fibers and verified them with the experimentally measured absorption spectrum of different samples. The results are shown in Fig. 2. The experimental and theoretical results are in good agreement. The finite size of the MNCs makes the electronic energy bands discrete. Hence we may assume that electrons will undergo band transition when excited by proper excitation energy. Therefore such MNCs in glass matrix can show fluorescence if we excite the composite glass sample with light with appropriate frequency. The electronic structure of these few atom clusters can be explained with the help of spherical jellium model. Though this model is proposed for alkali metals but it works satisfactorily in case of noble metals also, as they have single electron in the outer most orbit. According to this model the electrons in a MNC is distributed in different energy shells of a spherical harmonic potential well. Later Clemenger showed that for clusters having atoms ≥ 20 the potential is anharmonic and is given by Diez et al. [7]:

$$H = -\frac{p^2}{2m} + \frac{m\omega_0^2 q^2}{2} - Uh\omega_\Delta\left(l^2 - n\frac{(n+3)}{6}\right) \tag{5}$$

where p and q are the momentum and position operators of the single-electron respectively, l is the angular momentum, n is the shell number and U is the anharmonic distortion parameter which depends on the size of the cluster. The transition energy of the electrons can be written as Diez et al. [7]:

$$\Delta E = \frac{E_F r_s}{R}\left[1 - U\left(l_e^2 - l_g^2 - \frac{n+2}{3}\right)\right] \tag{6}$$

where l_e and l_g represent the angular momentum of excited and ground states respectively. The values taken for Ag are $r_s = 3.02$ atomic unit, $E_F = 5.49\,\text{eV}$. It is to be noted that there is no direct measurement technique of U. We choose the value by comparing the theoretical results with experimental observation. The experimentally measured emission spectrum and theoretically calculated transition probability for different emission lines of Ag NCs in our fabricated sample shows good agreement. The results are given in [8]. We observed that all calculated emission lines fall within the measured fluorescence band. This confirms the presence of Ag NCs in the fiber core glass and indicates that such fluorescence occur due to electronic band transition. We excited our fiber samples with a 405 nm pig-tailed laser diode of output power 50 mW.

3 Florescence Enhancement of RE by MNC

We have observed the fluorescence from Ag NCs doped optical fiber due to discontinuity in the energy bands. These discrete energy bands of the NCs serves another purpose in fluorescence light emission. We observed that when such clusters are codoped with conventional RE ions in the fiber core, the fluorescence emission from the RE increases in presence of MNCs. For large size MNPs such phenomena were observed and attributed as metal enhanced fluorescence (MEF) [9]. MEF explains the enhancement in fluorescence radiation with the help of radiative decay engineering. This theory states that larger MNPs produces localized electromagnetic (EM) field as they act as an dipole antenna. We calculated the local field enhancement factor for the Ag NCs present in our fabricated fiber samples. Since we codoped Ag NCs with Yb^{3+} and Tm^{3+} and excited the system with 975 nm laser diode to get resonant energy transfer (ET) between Yb^{3+} and Tm^{3+}, so we have to calculate the local field at 975 nm.

The calculated value comes out close to 1 indicating that MEF should not take place for such small NCs. It may happen that the MNCs may alter the ET process between Yb^{3+} and Tm^{3+}. So we measured the fluorescence energy at each emission line of the RE present in the system. The log-log plot of Yb^{3+} emission energy with that of Tm^{3+} is linear and the slope of the graphs represents number of photons involved in the ET process. We generated the graphs for our fabricated samples and found out that MNCs does not affect the ET process. Yet we observed an enhancement in the fluorescence of all the RE present in the system due to MNC, which is unconventional because in resonance ET process emission from one RE ion increases at the expense of other. Simultaneous enhancement was not observed. The fluorescence lifetime of the RE decreases in the presence of MNC. This property appears to be unusual. The results are shown in Fig. 3 [10]. Since MEF is inadequate to explain this enhancement mechanism therefore we apprehend the radiation coupling between MNC and RE is different. In next step we try to explain the presence of both MNCs and fluorophores together as a coupled system accordingly we have evaluated the dynamics of the coupled system from a quantum mechanical point of view. Subsequent studies show that this typical nature where MNCs act as an oscillating dipole in IR region is misleading. The MNC might be visualized as a single dipole if the electrons present in the system execute coherent oscillation. That may not be valid for such ultra-small metal particles.

We have already seen that the energy band of the electrons in MNC becomes discrete. These clusters behave as insulating cluster as the Kubo gap which is

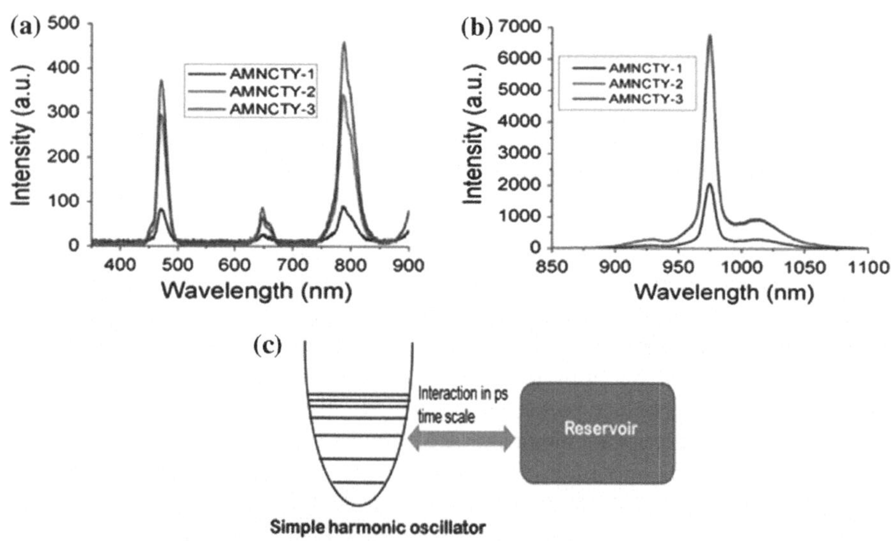

Fig. 3 Emission spectra of AMNCTY fibers **a** 350–850 nm and **b** 900–1100 nm. AMNCTY-1 sample does not contain Ag NCs [10] **c** the schematic of interaction between MNC and rare earth ion

greater than k_BT [11]. In the present case the electrons in MNC have closely spaced density of states (Kubo states) [11] with energy gap $\delta = 4E_F/3N$, where E_F is the Fermi energy and N is the number of valence electron of the MNC. The most probable cluster radius ($\sim$0.65 nm) as obtained from TEM analysis has $\delta \sim 0.13$ eV. The pump wavelength (975 nm) provides energy greater than this but less than that required for radiative transition [12]. Hence radiative transition of electrons is not possible. But electron can non-radiatively decay among various energy bands available in the valence band. So we may consider the MNC to be acting as a reservoir of many electronic oscillators having random frequencies and damping constant. From quantum mechanical point of view the rare earth ion can be visualized as a simple harmonic oscillator (SHO). The average damping time of electrons in Ag NCs is $\sim$16.6 fs [3] on the other hand the radiative decay time of Yb^{3+} form $^2F_{5/2}$ state is $\sim$924 µs. Therefore in the interaction picture of MNC and Yb^{3+}/Tm^{3+} we may consider that the rare earth stays in its corresponding excited state while interacting with the MNC. So in our model we assumed that the simple harmonic oscillator does not lose photon while the reservoir loses photon at some rate, γ. The schematic of the interaction picture is given in Fig. 3c.

The interaction of the system can be represented by a system Hamiltonian, where we consider the interaction is instantaneous and takes place at a resonant frequency. In our case the frequency corresponds to the pump signal at 975 nm. The evolution of the system is then represented by density matrix and the solution indicates that both the RE oscillation and the electron oscillation in MNC will decay at rate slower than the rate at which oscillation in the reservoir dies. Since MNCs act as reservoir and the oscillation in MNCs dies at within few fs, hence the RE oscillation will die faster than in the absence of MNC. This explains the lifetime decrease in presence of MNC.

This model explains the reduction in fluorescence lifetime of the Yb^{3+} and Tm^{3+} in presence of Ag NCs. Still, we observe enhanced fluorescence. This can be explained from the ground state absorbance of the AMNCTY samples. We observed that the absorption cross-section of all the rare earth increases in presence of Ag NCs. This enhanced absorption cross-section can be explained from this coupled model. Here we consider the MNC as a receiver and the rare earth ion acts as an antenna. The incident light is now strongly scattered by the rare earth ion and leads to an enhanced intensity in the position of Ag NC. Therefore larger dissipation of energy in MNC occurs. This does not lead to quantum quenching because the change in radiative and non-radiative decay rate of the molecule is very small as the distance between Ag NC and RE element is greater than the size of NCs [13]. This leads to enhanced absorption of the Yb^{3+} in presence of Ag NCs under coherent excitation and we observed enhanced fluorescence. Therefore we observed simultaneous fluorescence enhancement of Tm^{3+} and Yb^{3+} when Ag NCs are present in the system.

4 Fabrication and Material Analysis of Fabricated Core Glass of Ag-NCs and Ag-NCs Along with RE Doped Optical Fibers

Ag-NCs as well as Ag-NCs along with Tm^{3+} and Yb^{3+} doped alumino-silica glass optical preforms were fabricated using modified chemical vapour deposition process along with solution doping technique (MCVD-SD) [8, 10]. Different solutions were used having different concentration of the $AgNO_3$, $Tm(NO_3)_3$ and $Yb(NO_3)_3$ maintaining a constant concentration of $Al(NO_3)_3$ to create varying dopant concentration in the fabricated fibers. The compositions of the different fabricated preform core glasses are given in Table 2. Using these preforms, optical fibers of 125 μm diameter were fabricated using silica fiber draw tower at a drawing temp of 2050 °C. The MCVD set up and some fabricated Ag doped preform are shown in Fig. 4a, b. The refractive index profiles (RIP) of fabricated preforms were measured by optical perform analyser (PK 2600). From the RIPs, the numerical apertures of the fibers were calculated for 630 nm wavelength (Table 2).

To confirm the presence of Ag in the fiber core we have performed energy dispersive X-ray spectroscopy (EDS) of the fiber sample along its cross-sectional diameter. The result for AMNC-1 sample is shown in Fig. 4c. It reveals the presence of metallic Ag in the core glass. The core glass microstructures were analyzed by the transmission electron microscope (TEM), using thin film sampling technique (Fig. 5a–c). The images clearly show the presence of nano-particles into the core glass. The spot EDX analysis was also performed to get the structural information of the particles into the core glass (Fig. 5d). The electron diffraction (ED) was also performed to investigate the nature of the dopants into the core i.e. either amorphous or crystalline (Fig. 5d).

The XPS analysis was also performed using the same thin film sample to identify the exact oxidation state of Ag into the core glass (Fig. 5e). The peaks appeared in the XPS spectra confirms that the clusters present in the fiber are pure metallic clusters. These analyses show that Metallic Ag NCs are present in the core glass of

Table 2 Composition details of synthesis fibers

Sample id	Solution used	NA
AMNC-1	1.5(M) $AgNO_3$ − 1.0(M) $Al(NO_3)_3$	0.19
AMNC-2	1.75(M) $AgNO_3$ − 1.0(M) $Al(NO_3)_3$	0.15
AMNC-3	1.5(M) $AgNO_3$ − 1.0(M) $Al(NO_3)_3$	0.16
AMNC-4	1.0(M) $AgNO_3$ − 1.0(M) $Al(NO_3)_3$	0.17
AMNCTY-1	0.02(M) $Tm(NO_3)_3$ − 0.03(M) $Yb(NO_3)_3$ − 1.0(M) $Al(NO_3)_3$ − 0.0(M) $AgNO_3$	0.13
AMNCTY-2	0.02(M) $Tm(NO_3)_3$ − 0.03(M) $Yb(NO_3)_3$ − 1.0(M) $Al(NO_3)_3$ − 1.75(M) $AgNO_3$	0.15
AMNCTY-3	0.02(M) $Tm(NO_3)_3$ − 0.03(M) $Yb(NO_3)_3$ − 1.0(M) $Al(NO_3)_3$ − 1.5(M) $AgNO_3$	0.17

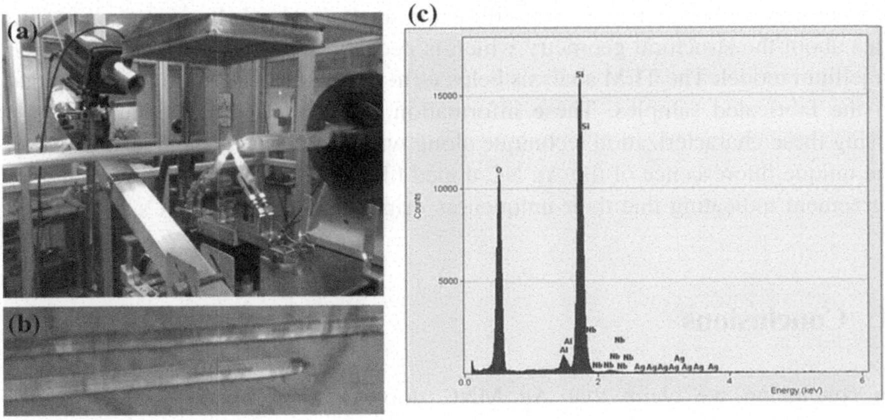

Fig. 4 **a** Sintering process during AMNC-2 preform fabrication showing yellow coloration at the tube wall and **b** Ag-NCs doped optical preforms. **c** EDS of AMNC-1 fiber along cross-sectional diameter

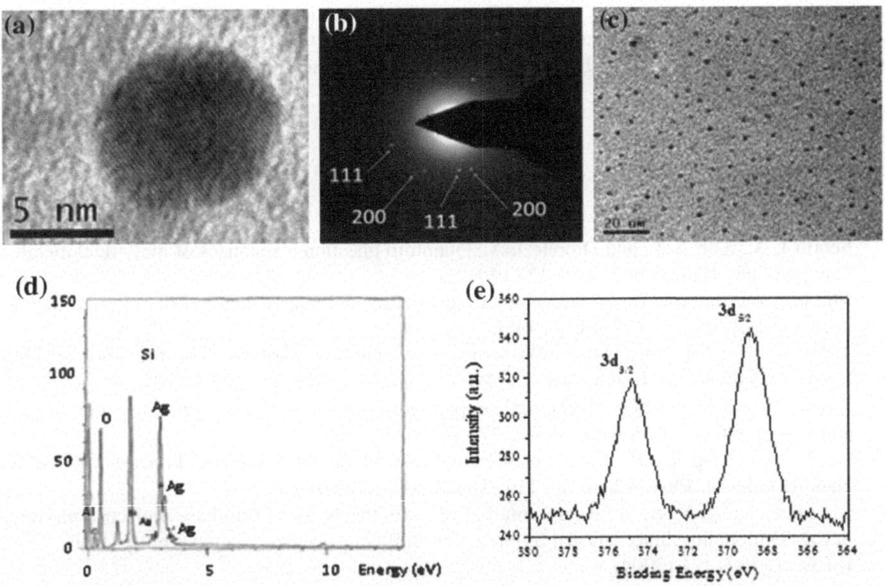

Fig. 5 **a** HR-TEM image of a large Ag particle obtained in one of the fabricated fiber sample **b** electron diffraction pattern along with crystallographic planes taken on the Ag particles of (**a**), **c** TEM image of Ag MNC doped fiber core glass, and **d** spot EDX spectra taken on the Ag particles at the core of Ag doped fiber [8], **e** XPS spectra of Ag particle of Ag MNC doped fiber, show two peaks at 374.8 and 368.8 eV for spin-orbit split peaks of Ag $3d_{3/2}$ and $3d_{5/2}$ representing the 'zero' oxidation state of Ag

our fabricated fibers. The electron diffraction analysis also helps us to generate an idea about the structural geometry which is needed while calculating shell number in jellium model. The TEM analysis helps us to create the NCs size distribution data in the fabricated samples. These information is vital for implementation of (4). Using these characterization technique along with our theoretical model we explain the unique fluorescence of the Ag NC doped fiber samples. The results are in good agreement indicating that their uniqueness originates from the MNCs.

5 Conclusions

In conclusion we claim that Ag MNC doped optical fiber possesses unique fluorescence property. When excited by adequate energy the MNC itself shows fluorescence. On the other hand if these fibers are codoped with RE ions and suitably pumped they show enhanced fluorescence of the RE ions without altering the ET process between the RE ions. These fibers can form a new family of fibers based low threshold light sources.

Acknowledgements Part of the work of RC & SKB is supported by CSIR Emeritus Scientist Scheme-21(1017)/15/EMR-II. Authors are indebted to Director, CSIR-CGRI, Kolkata for his support and encouragement and Director, IACS, Kolkata for his unstinted cooperation.

References

1. Scholl J. A., Koh A. L. and Dionne J. A., Quantum plasmon resonances of individual metallic nanoparticles, Nature, 483, 421–427 (2012).
2. Mie G., contributions to the optics of turbid media, particularly solution of colloidal metals, Ann. der Physik., 25, 377–445 (1908).
3. Kelly K. L. et. al., The Optical Properties of Metal Nanoparticles: The Influence of Size, Shape, and Dielectric Environment, J. Phys. Chem. B, 107, 668–677 (2003).
4. Baletto F. and Ferrando R., Structural properties of nanoclusters: Energetic, thermodynamic, and kinetic effects, Rev. Mod. Phys., 77, 371–423 (2005).
5. He Y. and Zeng T., First-Principles Study and Model of Dielectric Functions of Silver Nanoparticles, J. Phys. Chem. C, 114, 18023–18030 (2010).
6. Koledintseva M. Y. et. al., Representation of permittivity for multi-phase dielectric mixtures in FDTD modeling, International Symposium on Electromagnetic Compatibility (EMC 2004) IEEE,, 1, 309–314 (2004).
7. Diez I. et. al., Blue, green and red emissive silver nanoclusters formed in organic solvents, Angew. Chem. Int. Ed., 48, 2122–2125 (2009).
8. Halder A. et. al., Highly fluorescent silver nanoclusters in alumina-silica composite optical fiber, Appl. Phys. Lett., 106, 2, 011101 (2015).
9. Guzatov D. V., Plasmonic enhancement of molecular fluorescence near silver nanoparticles: Theory, modeling, and experiment, J. Phys. Chem. C, 116, 10723–10733 (2012).
10. Chattopadhyay R. et. al., Quantum sized Ag nanocluster assisted fluorescence enhancement in Tm^{3+}-Yb^{3+} doped optical fiber beyond plasmonics Appl. Phys. Lett., 107, 233107 (2015).

11. Morton S. M., Silverstein D. W., and Jensen L., Theoretical studies of plasmonics using electronic structure methods, Chem. Rev., 111, 3962–3994 (2011).
12. Maurizio C. et. al., Enhancement of the Er^{3+} luminescence in Er-doped silica by few-atom metal aggregates, Phys. Rev. B, 83, 195430 (2011).
13. Pandozii F. et. al., A spectroscopic analysis of blue and ultraviolet upconverted emissions from $Gd_3Ga_5O_{12}$:Tm^{3+}, Yb^{3+} nanocrystals, J. Phys. Chem. B, 109, 17400–17405 (2005).

Digital Holography for Recognition and Security of 3D Objects

Dhirendra Kumar and Naveen K. Nishchal

Abstract Three-dimensional (3D) object recognition has been achieved by comparing either holograms or the reconstructed images from respective holograms using conventional and nonlinear joint fractional transform correlators. For 3D information security, an encrypted image has been used as digital watermark. The encryption scheme is based on polarized light encoding and the photon counting technique. Multiple images may also be used as watermark. A multiple image encryption system has been proposed which is based on phase mask multiplexing and photon counting imaging.

1 Introduction

Digital holographic technique is used to record optical wavefront digitally and recreate the same wavefront optically or by numerical algorithms [1, 2]. The 3D object identification is an important area where digital holography is efficiently used [3–8]. Targets in 3D space cannot be identified by two-dimensional (2D) correlators because 2D correlators can't determine longitudinal distances accurately [8]. However, 2D correlators have been used to identify 3D objects using different perspectives of the object [8]. Many researchers have used joint fractional correlator (JFRTC) for 3D correlation [9–13]. The correlator's performance is improved by introducing nonlinearity in the joint power spectrum [11–13].

A fractional Fourier transform (FRT)-based scheme for 3D object recognition has been discussed [12, 13]. DH of different 3D objects are recorded numerically and reconstructed corresponding 3D images are compared with the reference object using JFRTC and nonlinear JFRTC [12]. These schemes are also verified by experimentally recorded digital hologram (DH). 3D correlation is also performed by comparing the holograms directly, since the holograms contain the complete

D. Kumar · N.K. Nishchal (✉)
Department of Physics, Indian Institute of Technology Patna, Bihta,
Patna, Bihar 801103, India
e-mail: nkn@iitp.ac.in

© Springer Nature Singapore Pte Ltd. 2017
I. Bhattacharya et al. (eds.), *Advances in Optical Science and Engineering*,
Springer Proceedings in Physics 194, DOI 10.1007/978-981-10-3908-9_13

information of the 3D object. A comparative study of the conventional and non-linear JFRTC for 3D recognition has been performed using both the methods of recognition of real 3D objects [13]. Various performance measure parameters have been computed to compare the strength of the proposed correlation schemes.

DHs may be used to store important data which needs to be secured. Hologram watermarking is an important technique to secure and prevent unauthorized copying of data [14]. In this technique, an image or a data is embedded in the digital media in a way which prohibits the image or a data from being accessed by unwarranted users and lets the image/data be read when needed. In many cases, encrypted images are used as watermark which offers additional level of security.

Over the years, researchers have developed various encryption schemes including double random phase encoding, phase truncation-based cryptosystem, interference-based image encryption, and multiple image encryption [14–21]. To add to the security of the encrypted image, photon counting imaging (PCI) technique has been clubbed with different encryption schemes [22–24]. An image encryption scheme has been discussed which is based on PCI technique and phase mask multiplexing for multiple image authentications [24]. The authentication scheme has also been used for securing 3D information contained in optically captured DH.

Another optical image security system utilizing the concept of polarized light encoding [25–30] and PCI technique has been briefly explained [30]. An input image is encoded using Stokes-Mueller formalism. The PCI technique is used to further encrypt the already encoded image. Attack analysis has also been carried out to check the robustness of the encryption scheme. The proposed encryption system has been used for 3D information security through hologram watermarking technique [31].

2 Three-Dimensional Correlation

2.1 3D Recognition with Digital Fresnel Hologram

The 3D correlation with the help of DH is realized because DH records different perspectives of a 3D object in different portions of a DH [4, 10]. The matching between the target 3D and given reference objects is performed by calculating correlation coefficient which is given by (1).

$$\mathrm{C(x,y)} = \Im^{-1}\left\{\Im^{\mu}\{\mathrm{H_{tar}(x,y;\ d;\ \alpha,\beta)}\}\{\Im^{\mu}\{\mathrm{H_{ref}(x,\ y;\ d;\ \alpha',\beta')}\}\}^{*}\right\} \quad (1)$$

$\Im^{\mu}$ and $\Im^{-1}$ denote fractional Fourier transform with order μ and inverse Fourier transform, respectively. $H_{\mathrm{tar}}(x,y;d;\alpha,\beta)$ and $H_{\mathrm{ref}}(x,y;d;\alpha',\beta')$, represent one perspective of the target and reference objects, respectively. $H(x,y;d;\alpha,\beta)$ denotes a region of the hologram which is used for recreating an angular view (α, β) of the 3D object and d represents the reconstruction distance.

2.2 *Joint Fractional Fourier Transform Correlators*

The conventional correlation scheme used for 3D correlation gives by Tripathi et al. [11],

$$C(x, y) = \Im^{-1}[\Im^{\mu}\{g(x, y)\} \times [\Im^{\mu}\{g_r(x, y)\}]^*] \tag{2}$$

Here, $g(x, y)$ and $g_r(x, y)$ represent images of target and reference objects. Nonconventional nonlinear JFRTC is an extension of JFRTC in which joint fractional power spectrum is nonlinearly transformed using power-law transformation [11]. A high correlation coefficient means that the reference image is similar to the input target image. n is nonlinearity factor and is one for JFRTC. Mathematically, nonlinear JFRTC is represented as,

$$C(x, y) = \Im^{-1}\left[|\Im^{\mu}\{g(x, y)\} \times [\Im^{\mu}\{g_r(x, y)\}]^*|^n \times \exp\{i \arg[\Im^{\mu}\{g(x, y)\}] - \arg[\Im^{\mu}\{g_r(x, y)\}]\}\right]. \tag{3}$$

3 Hologram Watermarking

The DH is secured by embedding an encrypted image into a host image. The encrypted image is used as a watermark and recorded DH is considered as a host image. The process of hologram watermarking is defined as Rajput et al. [24],

$$I_w(\xi, \eta) = I(\xi, \eta) + aE_p(\xi, \eta) \tag{4}$$

where, a is an arbitrary constant. The original as well as watermarked hologram reconstruct the same 3D object. The original watermark, which is known only to the actual owner, is retrieved from the watermarked DH using (5).

$$E_p(\xi, \eta) = \frac{I_w(\xi, \eta) - I(\xi, \eta)}{a} \tag{5}$$

Now, the decrypted retrieved watermark is verified using nonlinear JFRTC. The authenticity of the original hologram is verified using the correlation peaks obtained using the correlator.

3.1 Multiple Image Encryption Using Photon Counting Imaging and Phase Mask Multiplexing

A novel multiple image encryption technique has been proposed which utilizes modified GS algorithm for phase-retrieval and phase mask multiplexing technique [24]. The multiple images are converted into their corresponding phase-only images (POIs) with the help of the GS algorithm and then a single phase mask is synthesized by multiplexing these POIs. The individual keys are generated for successful retrieval of original images from the single phase mask. Further, FRT operation is performed on the single phase mask and then it is multiplied with a random phase mask to obtain complex valued function. To further strengthen the security of the encryption scheme, PCI technique is used. The output image obtained from PCI technique has sparse representation because it has limited number of photons as compared to the number of incident photons. Authenticity of the multiple images are checked by decrypting the encrypted image at first with proper decryption process and then the photon limited decrypted and original images are compared by calculating the correlation coefficient using nonlinear JFRTC.

3.2 Encryption Scheme Based on Polarized Light Encoding and Photon Counting Imaging

The encryption scheme uses polarized light encoding with the help of Stokes-Mueller formalism and then PCI technique has been applied to further encrypt the image [30]. Two different waves have been used to illuminate the image which is to be encrypted and an intensity image which acts as encryption key. The two light waves are combined to obtain a multiplexed stokes vector. The obtained multiplexed stokes vector, is further encoded by the pixilated polarizer with randomly distributed angles whose values lie between $-\pi$ and $+\pi$. The PCI technique has been utilized to further encrypt the polarized light encoded image. For the purpose of authentication verification, we used nonlinear optical correlator. The correlation between original input image and photon limited decrypted image can be obtained.

4 Results and Discussions

4.1 3D Object Recognition

Recognition of 3D object is demonstrated numerically as well as experimentally. Figure 1a shows the target image and Fig. 1b shows the experimentally recorded

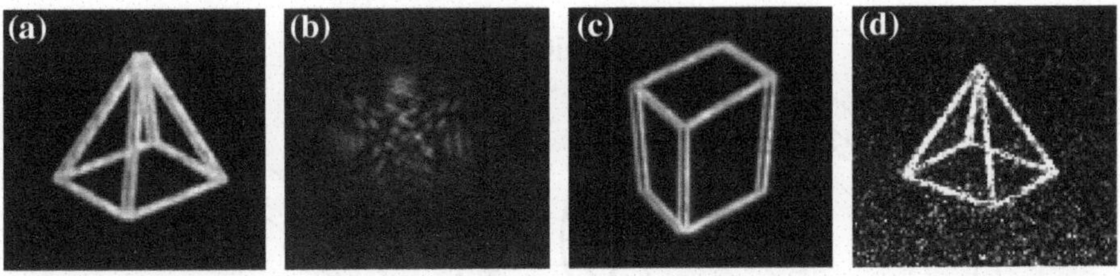

Fig. 1 **a** Pyramid, as the target object, **b** simulated DH for the pyramid, **c** rectangular prism, and **d** reconstructed pyramid [12]

DH. A rectangular prism, as shown in Fig. 1c, has been compared with the reconstructed pyramid image, shown in Fig. 1d. The JFRTC and nonlinear JFRTC have been used to obtain autocorrelation peaks between pyramid and the reconstructed pyramid. The cross-correlation between a rectangular prism and the reconstructed image of the pyramid is also obtained using JFRTC and nonlinear JFRTC. On comparing the correlation results, it is observed that nonlinear JFRTC discriminates between dissimilar objects better.

4.2 3D Correlation: A Comparative Study

The optical setup for recording DH of a real 3D object has been shown in Fig. 2. A He–Ne laser (make: Research Electro-Optics, USA) having 35 mW power and 632.8 nm wavelength has been used as a source of coherent light. Figure 3b, d show the recorded reference and target DHs, respectively. The numerically reconstructed 3D images have been shown in Fig. 3c, e. The two DHs and the two reconstructed 3D images have been compared with the help of JFRTC and nonlinear JFRTC. Performance measure parameters for the two correlation outputs have also been calculated to compare the conventional and nonlinear JFRTCs.

4.3 Hologram Watermarking

4.3.1 Phase Multiplexed Photon Limited Encrypted Image

Four different images each with size 256 × 256 pixels have been used for this study. The four gray-scale images have been shown in Fig. 3a–d. The multiplexed encrypted image, shown in Fig. 3e is obtained as a result of multiplexing of POIs obtained with the help of modified GS algorithm. The photon limited encrypted image has been shown in Fig. 3f. It is sparsely represented, therefore is indistinguishable. The auto-correlation peaks are obtained when the photon limited decrypted images obtained after using all correct keys, are compared with the

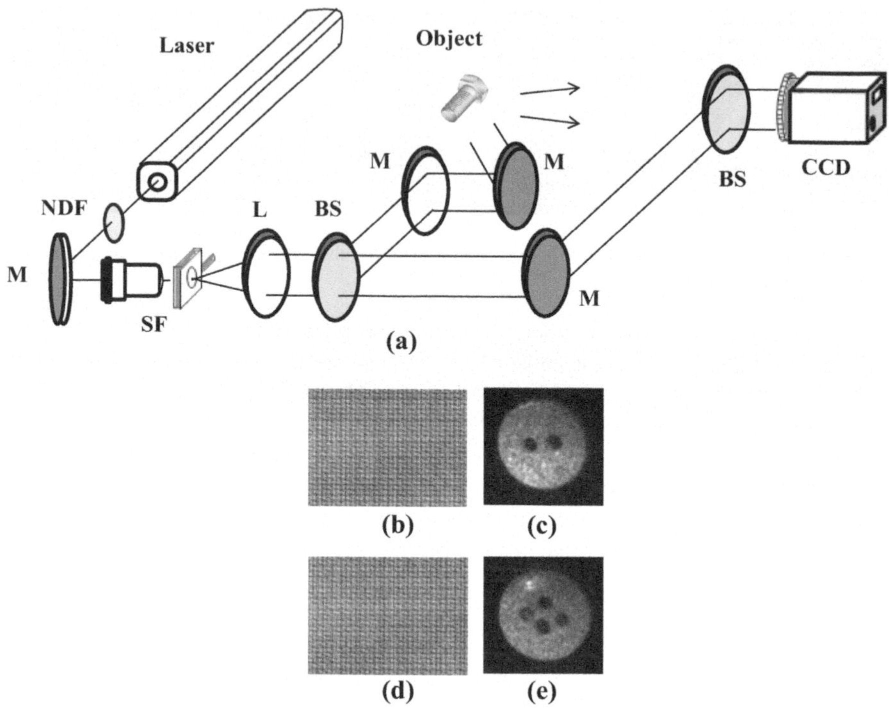

Fig. 2 **a** Experimental arrangement for recording DHs of 3D real objects. *M* Mirror, *BS* Beam Splitter, *L* Lens, *SF* Spatial Filter, and *NDF* Neutral Density Filter. **b** Reference hologram **c** reconstructed reference object **d** target hologram, and **e** reconstructed target object [13, 24]

Fig. 3 Simulation results: gray images of a **a** Baboon, **b** Cameraman, **c** Flower pot, **d** Capsicum **e** phase multiplexed encrypted image, and **f** encrypted image obtained with PCI technique [24]

corresponding original. Experimentally recorded DH is watermarked with the sparsely represented encrypted image using PCI technique containing the information of multiple images in the photon limited domain. The watermarked hologram has the information of multiple images. Reconstruction of original and watermarked DHs gives same image of the real 3D object. Authenticity of the retrieved watermark has also been checked by comparing the original and the photon limited decrypted images.

Fig. 4 Simulation results: **a** Cameraman, **b** intensity image, used as key, **c** polarized light encoded image, and **d** photon limited encrypted image, and **e** photon limited decrypted image [30]

4.3.2 Polarization-Encoded Photon Limited Encrypted Image

In Fig. 4a, a gray-scale image of a Cameraman having size 256 × 256 pixels has been used for encryption. Figure 4b, c, and d show the intensity image acting as encryption key, the polarization encoded image, and the sparsely represented encrypted image obtained on application of PCI technique, respectively. Figure 4e shows the decrypted image in photon limited domain.

Attack analysis has also been carried out to the verify robustness of the proposed encryption system. The photon limited encrypted image has been used for watermark. Reconstructed images from the original and the watermarked holograms are same, however the authenticity of the holograms can be checked with the help of retrieved watermark.

5 Conclusions

A comparative study of conventional JFRTC and nonlinear JFRTC has been done for real 3D object recognition. The study has been carried out by recording DHs numerically as well as experimentally. It is concluded from the study that the two slightly different 3D objects can be discriminate with nonlinear JFRTC better as compared to the conventional JFRTC. Security of the DHs is ensured by hologram watermarking with encrypted images. The encrypted images are obtained by multiple image phase multiplexing and PCI technique. Another encrypted image used for hologram watermarking is obtained by polarized light encoding and PCI technique.

Acknowledgements The financial assistance from the DRDO under Grant No. ERIP/ER/1200428/M/01/1473 is acknowledged. D. Kumar acknowledges the funding from the UGC under letter No. F.2-10/2011(SA-I).

References

1. Schnars, U. and Jueptner, W. P. O., "Digital recording and numerical reconstruction of holograms," Meas. Sci. Technol. 13, R85–R101 (2002).
2. Cuche, E., Marquet, P. and Depeursinge, C., "Spatial filtering for zero order and twin image elimination in digital off-axis holography," Appl. Opt. 39, 4070–4075 (2000).
3. Nelleri, A., Gopinathan, U., Joseph, J. and Singh, K., "Wavelet based three-dimensional object recognition using single off-axis digital Fresnel hologram," Proc. SPIE 5827, 30–37 (2005).
4. Javidi, B. and Tajahuerce, E., "Three-dimensional object recognition by use of digital holography," Opt. Lett. 25, 610–612 (2000).
5. Frauel, Y. and Javidi, B., "Neural network for three-dimensional object recognition based on digital holography," Opt. Lett. 26, 1478–1480 (2001).
6. Tajahuerce, E., Matoba, O. and Javidi, B., "Shift-invariant three-dimensional object recognition by means of digital holography," Appl. Opt. 40, 3877–3886 (2002).
7. Javidi, B. and Kim, D., "Three-dimensional-object recognition by use of single exposure on-axis digital holography," Opt. Lett. 30, 236–238 (2005).
8. Rosen, J., "Three-dimensional joint transform correlator," Appl. Opt. 37, 7538–7544 (1998).
9. Mendlovic, D., Ozaktas, H. M. and Lohmann, A. W., "Fractional correlation," Appl. Opt. 34, 303–309 (1995).
10. Rajput, S. K. and Nishchal, N. K., "Image encryption and authentication verification using fractional nonconventional joint transform correlator," Opt. Lasers Eng. 50, 1474–1483 (2012).
11. Tripathi, R., Pati, G. S. and Singh, K., "Nonlinear processing and fractional-order filtering in a joint fractional Fourier-transform correlator: performance evaluation in multiobject recognition," Appl. Opt. 40, 2844–2859 (2001).
12. Kumar, D. and Nishchal, N. K., "Three-dimensional object recognition using joint fractional Fourier transform correlators with the help of digital Fresnel holography," Optik 126, 2690–2695 (2015).
13. Kumar, D. and Nishchal, N. K., "Recognition of three-dimensional objects using joint fractional correlator and nonlinear joint fractional correlator with the help of digital Fresnel holography: a comparative study," Opt. Rev. 22, 256–263 (2015).
14. Nishchal, N. K. and Naughton, T. J., "Flexible optical encryption with multiple users and multiple security levels," Opt. Commun. 284, 735–739 (2011).
15. Chen, W., Javidi, B. and Chen, X., "Advances in optical security systems," Adv. Opt. Photon. 6, 120–155 (2014).
16. Situ, G. and Zhang, J., "Double random phase encoding in the Fresnel domain," Opt. Lett. 29, 1584–1586 (2004).
17. Hwang, H.-E., Chang, H. T. and Lie, W. N., "Multiple-image encryption and multiplexing using a modified Gerchberg-Saxton algorithm and phase modulation in Fresnel-transform domain," Opt. Lett. 34, 3917–3919 (2009).
18. Chen, W. and Chen, X., "Optical multiple-image encryption based on multiplane phase retrieval and interference," J. Opt. 13, 115401 (2011).
19. Deng, X. and Zhao, D., "Multiple-image encryption using phase retrieval algorithm and intermodulation in Fourier domain," Opt. Laser Technol. 44, 374–377 (2012).
20. Zalevsky, Z., Mendlovic, D. and Dorsch, R. G., "Gerchberg-Saxton algorithm applied in the fractional Fourier or the Fresnel domain," Opt. Lett. 21, 842–844 (1996).
21. Pérez-Cabré, E., Cho, M. and Javidi, B., "Information authentication using photon-counting double-random-phase encrypted images," Opt. Lett. 36, 22–24 (2011).
22. Perez-Cabre, E., Abril, H. C., Millan, M. S. and Javidi, B., "Photon-counting double-random-phase encoding for secure image verification and retrieval," J. Opt. 14, 94001 (2012).

23. Cho, M. and Javidi, B., "Three-dimensional photon counting double-random-phase encryption," Opt. Lett. 38, 3198–3201 (2013).
24. Rajput, S. K., Kumar, D. and Nishchal, N. K., "Photon counting imaging and phase mask multiplexing for multiple images authentication and digital hologram security", Appl. Opt. 54, 1657–1666 (2015).
25. Unikrishnan, G., Pohit, M. and Singh, K., "A polarization encoded optical system using ferroelectric spatial light modulator," Opt. Commun. 185, 25–31 (2000).
26. Tu, H.-Y., Cheng, C.-J. and Chen, M.-L., "Optical image encryption based on polarization encoding by liquid crystal spatial light modulator," J. Opt. A: Pure Appl. Opt. 6, 524–528 (2004).
27. Lin, C., Shen, X. and Xu, Q., "Optical image encoding based on digital holographic recording on polarization state of vector wave," Appl. Opt. 52, 6931–6939 (2013).
28. Alfalou, A. and Brosseau, C., "Dual encryption scheme of images using polarized light," Opt. Lett. 35, 2185–2187 (2010).
29. Dubreuil, M., Aflalou, A. and Brosseau, C., "Robustness against attacks of dual polarization encryption using Stokes-Mueller formalism," J. Opt. 14, 094004 (2012).
30. Rajput, S. K., Kumar, D., and Nishchal, N. K., "Photon counting imaging and polarized light encoding for secure image verification and hologram watermarking," J. Opt. 16, 125406 (2014).
31. Kishk, S. and Javidi, B., "Watermarking of three-dimensional objects by digital holography," Opt. Lett. 28, 167–169 (2003).

Recent Advances in Fiber Loop Ringdown Sensors

Tarun Kumar Gangopadhyay and Jijo V. Ittiarah

Abstract A review of recent trends in fiber loop ringdown spectroscopy (FLRDS) is presented. Various aspects of the chemical and physical sensing are discussed in detail. The comparison of time domain and frequency domain sensing is discussed. The pulsed laser source is used for time domain FLRDS and a sine modulated laser source is used for frequency domain method. In this chapter, sensing is demonstrated with the help of intensity modulated continuous laser in a fiber loop instead of a pulsed laser. Some experimental results are also presented from ringdown sensors.

1 Introduction

Cavity ringdown spectroscopy (CRDS) is a well-known method for gas sensing from the 1980s [1]. It is a highly sensitive technique to measure the mole fraction of gases in parts per trillions. This method utilises spectroscopic measurement where the wavelength absorption of gases, in turn, used to identify or quantify the molecules. The mirror cavity of CRDS creates multiple reflections of laser light through the gas molecules. The light exponentially decays after several reflections inside the cavity. The time required for the intensity decay depends on the absorption coefficient of gas molecules. Optical decay constant is the main parameter for sensing using this spectroscopic method. Finally, this decay time measurement helps in absorption based sensing of the chemicals and gases. Stewart et al. [2] suggested a new type of optical cavity made of optical fiber in 2001. Lehmann and Loock groups [3] further improved this idea by new versions of fiber loops in 2003. Thus, an alteration from traditional CRDS to the simplified and

---

T.K. Gangopadhyay (✉) · J.V. Ittiarah
Fiber Optics & Photonics Division, CSIR-Central Glass and Ceramic Research Institute, Kolkata, India
e-mail: tkg@cgcri.res.in
URL: http://www.cgcri.res.in

© Springer Nature Singapore Pte Ltd. 2017
I. Bhattacharya et al. (eds.), *Advances in Optical Science and Engineering*,
Springer Proceedings in Physics 194, DOI 10.1007/978-981-10-3908-9_14

convenient system using optical fiber is established. Fiber loop ringdown spectroscopy (FLRDS) is one of the generic methods for sensing chemical and physical parameters. In FLRDS, the bulk mirror cavity of CRDS is replaced with fiber loop cavity. The size of the sensing system reduced. The difficulty for alignment of mirrors is solved. Chemical composition can be identified using evanescent field sensors incorporated within the fiber loop [4]. Currently, the fiber loop scheme is also used for sensing of various physical parameters based on the sensor head configuration.

2 Theory

Fiber Loop ringdown spectroscopy is a quantitative measurement based on calculating the light attenuation of a high-finesse cavity. It is done by estimating the lifetime of light trapped inside the cavity. This lifetime is connected with scattering or absorption of the light in sensing medium. This measurement is independent of fluctuation in input light. The fiber loop ring down systems are broadly classified into time domain and frequency domain measurement schemes.

2.1 *Time Domain Fiber Loop Ring-Down Spectroscopy*

In time domain FLRD, after coupling of a laser pulse into a fiber loop, the intensity change of light can be modelled by [5],

$$\frac{\partial I}{\partial t} = -\frac{ITc_0}{nL} \tag{1}$$

where I indicates the light intensity, t is the time interval, c_0 is the speed of light in vacuum, n is the effective refractive index (RI) of the fiber cavity, L is the fiber loop length, and T is the total cavity losses which include fiber transmission loss, component insertion losses and so on. The intensity at a particular time can be derived as

$$I = I_0 e^{\frac{-c_0}{nL}Tt} \tag{2}$$

which implies FLRDS measures the light intensity decay rate instead of the absolute intensity change. This equation denotes an exponential decay behaviour of the light. The time required at which the intensity (I) decays to 1/e times of the initial intensity (I_0) in the absence of measurand is given by,

$$\tau_0 = \frac{nL}{c_0 T} \tag{3}$$

where, τ_0 is called ringdown time, which is the measurement parameter used in FLRDS system. For a given FLRDS, T is a constant, and is determined by the physical parameters such as the fiber transmission loss, coupler insertion loss, RI, and fiber length. The low losses inside the cavity give high ringdown time. When a sensing action occurs within the fiber loop, an additional optical loss T_1 will be introduced which results in a change in the ring-down time, τ_0 to τ. And τ is given by [5].

$$\tau = \frac{nL}{c_0(T + T_1)} \tag{4}$$

From (3) and (4),

$$T_1 = \frac{nL}{c_0}\left(\frac{1}{\tau} - \frac{1}{\tau_0}\right) \tag{5}$$

This equation provides the parameters of the quantity to be measured from the additional optical loss, T_1. A reference lookup table can be created based on this measurement for chemical or physical sensing. A schematic diagram of the standard time domain FLRD sensing scheme is shown in Fig. 1. In this figure, the simulation result of light intensity decay from a typical time resolved FLRDS system is presented.

Each of the separated spikes shows the intensity of the light coming out of the loop after each successive round trip. The time between two adjacent spikes is the round-trip time of the light inside the loop. The envelope follows a single exponential decay. A slower decay rate (longer ringdown time) means lower optical losses of the light in the loop, and vice versa.

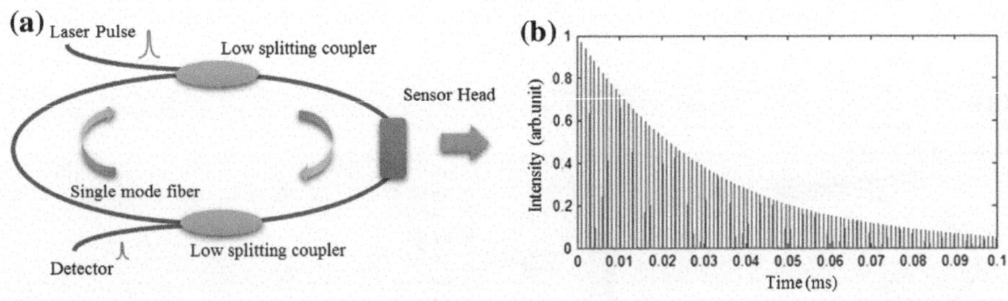

Fig. 1 Typical time-resolved FLRDS sensing scheme. **a** Schematic diagram, **b** plot of exponential decaying signals

2.2 *Frequency Domain Fiber Loop Ring-Down Spectroscopy*

Time-resolved FLRDS may not be a suitable method when the parameter is measured at high sampling rate. The measurement speed depends on the lifetime of light inside the cavity and pulse repetition rate which cannot be more than the life time. An alternative of time domain ringdown time is the frequency equivalent of decay constant. This frequency domain calculation method is named as Phase-Shift Fiber loop Ring-Down Spectroscopy (PS-FLRDS). A sinusoidally modulated light signal undergoes a phase shift that when traversing the cavity due to several parameters. Measurement can be done for this phase shift and was first used in 1980 by Herbelin et al. [1] to study the optical cavities. For single-exponential decay, the relationship between the ringdown time τ, modulation frequency ω and phase shift $\Delta\varphi$ is given by [5]

$$\tan(\Delta\varphi) = -\omega\tau \tag{6}$$

$$\Delta\varphi = \arctan(-\omega\tau) \tag{7}$$

$$T_1 = \frac{\omega \, n \, L}{c_0 \tan(\varphi - \varphi_0)} - T \tag{8}$$

From the calculation of additional optical loss, T_1, we can get the information of quantity to be measured by creating a lookup table. A schematic diagram of the standard PS-FLRDS sensing scheme is shown in Fig. 2. In this figure, the simulation result of phase shifting at the output port of PS-FLRDS is depicted.

The phase-shift method is suitable for flow measurement system. The PS-FLRDS is well-matched with very short cavity lifetime measurement. The high modulation frequency of the laser is implemented for measuring short cavity decay time. 10–100 MHz modulation frequency can be used for nanosecond range ringdown time calculation. Several measurements are required for calculating multiple exponential decays from PS-FLRDS, which is a major disadvantage of this system.

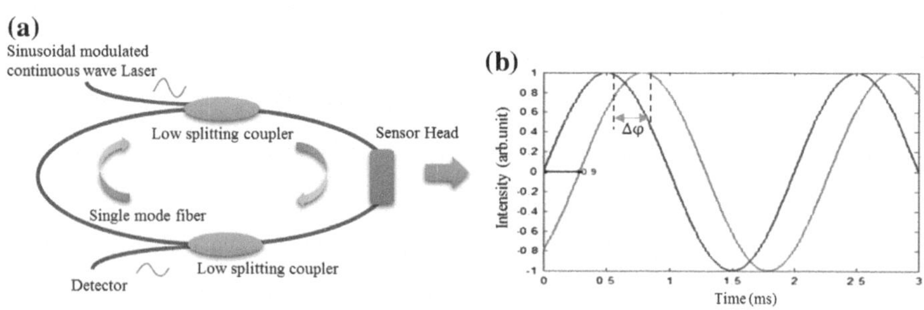

Fig. 2 Typical PS-FLRDS sensing scheme. **a** Schematic diagram, **b** phase shifting at the output

3 Recent Chemical Sensing Work on FLRDS

In this review, the main focus is given for chemical sensing by using different methods in last few years. The refractive index based evanescent field sensors are the prominent sensors used for chemical sensing along with FLRDS. The absorption of the spectral frequency by the chemicals are also playing a major role in identifying the chemical. The sensor head is the key component for introducing optical losses in the fiber loop and to facilitate interaction of the chemical with the light inside the cavity. Any intentional deformation in the optical fiber which enables light-chemical interaction can act as the sensor head for FLRDS. In 2011, Herath et al. [6] have proposed an idea about DNA and bacteria sensor based on FLRD-EF sensing scheme which is shown in Fig. 3a. In the 120 m fiber loop a partially etched single mode fiber of 24 cm is used as the sensor head. Three layers of coating is applied to the sensing head. The first layer is Poly-L-Lysine (PLL), the second is probe DNA and the third one is target DNA. It is demonstrated the detection of one bacteria and three DNA based on bulk RI and a label-free detection of DNA based on surface RI. The wavelength of diode laser is maintained at 1515.15 nm. The absorption of water and atmospheric constituents at the same wavelength is minimum. The ringdown signal is detected by the photodetector, then the pulse generator produces the pulses of duration equal to the ringdown time and it is in the tens of microsecond range. These pulses are used to trigger the laser. The laser produces spikes of width 20 ns for this pulse duration and the exponential decay signal is observed at the output. The variation of the ringdown time of each sample with respect to experimental time is shown in Fig. 3b.

In 2012, Wong et al. [7] demonstrated a RI sensor by integrating a photonic crystal fiber (PCF) Mach–Zehnder interferometer (MZI) into a cavity ring-down loop. The schematic diagram of the experimental setup and result are shown in Fig. 4. The cavity ring down fiber loop mainly consists of large length of single mode fiber, two 3 dB optical fiber couplers and an MZI based on PCF. A short

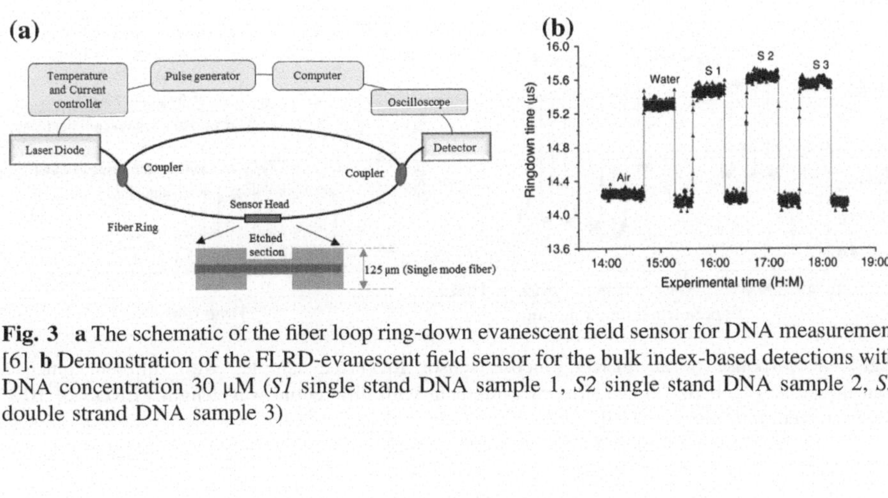

Fig. 3 **a** The schematic of the fiber loop ring-down evanescent field sensor for DNA measurement [6]. **b** Demonstration of the FLRD-evanescent field sensor for the bulk index-based detections with DNA concentration 30 μM (*S1* single stand DNA sample 1, *S2* single stand DNA sample 2, *S3* double strand DNA sample 3)

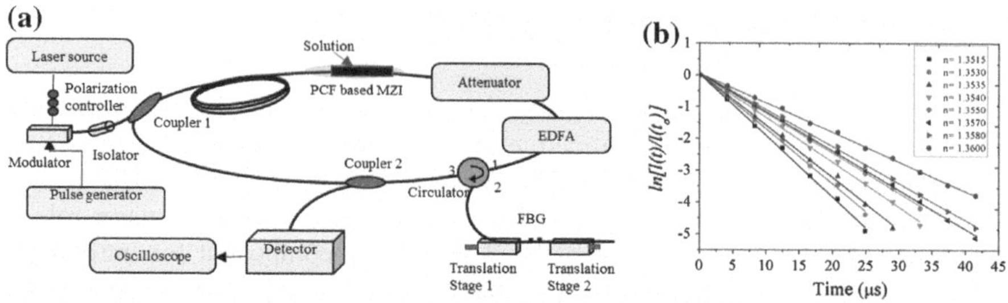

Fig. 4 **a** Schematic diagram of RI sensor using PCF based MZI ring down loop, **b** linear fit of ringdown values for different external refractive index [7]

section of PCF (LMA-10) is connected in between two single-mode fibers (SMF-28) and air holes at two splicing point is collapsed. Due to the coupling between core and cladding modes in the collapsed region, MZI is formed which can act as a sensor. In the effective RI from 1.3512 to 1.3600, a good linear response is obtained with the laser wavelength set as 1531.2 nm. For a corresponding minimum detectable effective RI of 7.8×10^{-5} RIU, the linear sensitivity of 11.7 μs^{-1} RIU^{-1} is obtained.

In 2012, Wang et al. [8] demonstrated a glucose sensor based on etched fiber sensor head and fiber loop scheme. The sensor head is 22 cm in length. The sensor head is an etched single mode fiber with immobilised glucose oxidase (GOD) on sensor surface. A fiber loop of 120 m is excited by a laser pulse of 1515.25 nm with a repetition rate of 100 Hz is used to obtain ring down time. The glucose solutions and synthetic urines in different concentrations ranging from 50 mg/dl to 10 g/dl are studied. The experimental result for glucose sensing is shown in Fig. 5a. The results are reproducible and had a quick response of around 0.1 s. Detection sensitivity is found to be 50 mg/dl.

In 2014, Zhang et al. [9] proposed a high-Q photonic Crystal Cavity based FLRDS system. The sensor head is designed with a point defect in the linear

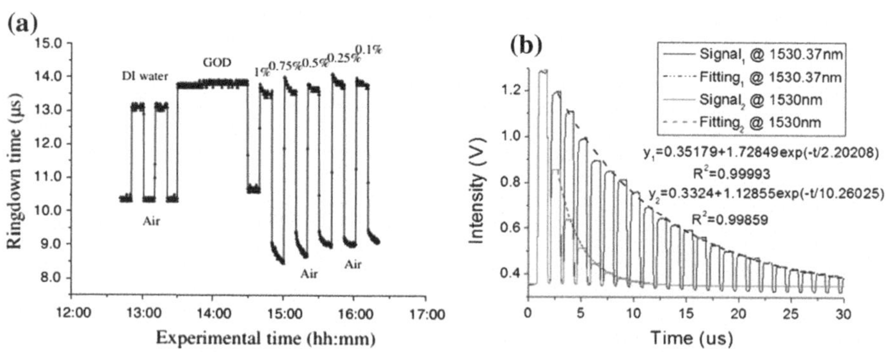

Fig. 5 **a** Response of an FLRD glucose sensor for urine samples with different glucose concentrations [8], **b** ring down time measurement with differential wavelength FLRDS of 10% ethyne mixed with nitrogen [10]

waveguide enclosed with one pair of air holes as reflectors. It creates a nanocavity between the reflectors. At a certain wavelength, this sensor head is extremely sensitive to the medium inserted into the surface of the sensor. A theoretical sensitivity of 20.34 ms/RIU is obtained for the RI range 1.33–1.40 with different wavelengths. A Q-factor of 605 is observed for the resonant cavity. In the same year, Zhao et al. [10] demonstrated a novel Ethyne sensor combined with active FLRDS. A dual wavelength differential absorption method is used in this system. A gas cell of 65 mm is used as sensor head which is made from a pair of collimators. A 900 ns pulse at a 10 kHz is alternatively emitted by two laser diodes with wavelengths 1530.37 and 1530 nm. This pulse is coupled to a fiber loop with delay line 252 m. An Erbium doped amplifier and FBG filters are employed inside the fiber loop to overcome extra losses by the cavity. Different concentration of Ethyne gas samples by mixing with nitrogen gas were tested. The results are shown in Fig. 5b. The ethyne gas concentration measurement shows a relative deviation less than ±0.4% of 0.1% with high stability over 24 h.

In 2015, Wu et al. [11] demonstrated a simple tapered fiber based RI sensor combined FLRDS. The tapered single mode fiber with a waist diameter of 28 μm and a taper length of 728 μm is used as sensor head. A fiber loop of 3000 m with a pair of couplers having splitting ratio 90:10 is used for making FLRDS. A laser pulse of 4.5 μs with a period of 150 μs is coupled to FLRDS for measuring the refractive index variation from 1.333 to 1.3737. The sensitivity of this simple system is up to −388.581 μs/RIU and achieved an RI detection limit below 2.57×10^{-5}.

4 Work at CGCRI

In this section, the fiber-loop sensor work done at our institute is explained in detail. A prototype of phase-shift FLRDS is made with single mode fiber loop of 23.5 m for chemical sensing. The schematic of the sensor is shown in Fig. 6.

In this PS-FLRDS system, the laser source of 1552 nm is selected for measurement. The continuous wave laser is modulated with a sine wave of frequency 8.3 MHz using an intensity modulator. The insertion loss of modulator is compensated by an Erbium doped fiber amplifier. A low splitting ratio coupler (99:1) is used for introducing light into the cavity. An isolator is used inside the loop to reduce the noise from back reflections. A side-polished fiber sensor is proposed for chemical sensing [12], which measures refractive index or wavelength absorption by the chemical. Another coupler is employed to couple out 1% of light from the cavity for phase measurement. A photodiode (InGaAs) is used to measure the sinusoidal signal from the cavity. The phase of modulated laser light from two detectors are compared using a LabVIEW program. The loss inside the cavity is responsible for a phase shift of the light at the output. This phase shift is used to calculate ringdown time and ultimately to measure the additional loss generated by the sensor head. The preliminary experimental results are given in Fig. 7. The phase shifting of the modulated wave is taken as a reference which is observed as

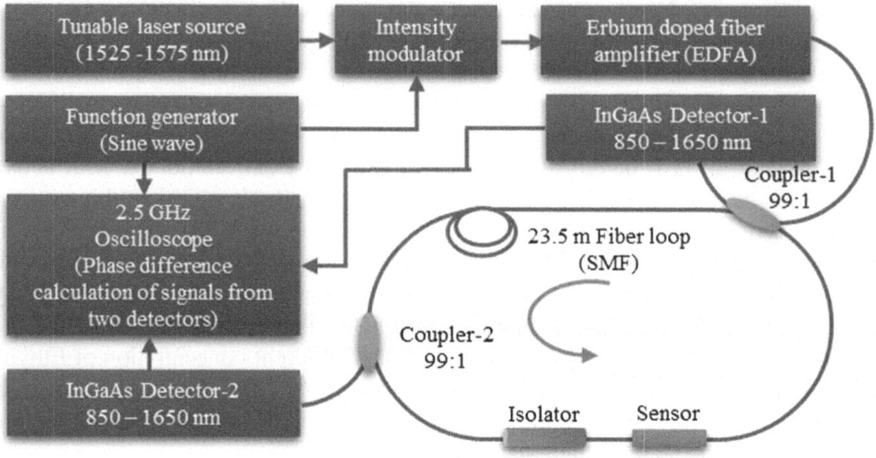

Fig. 6 Schematic diagram of PS-FLRDS with side-polished fiber sensor fabricated at CGCRI

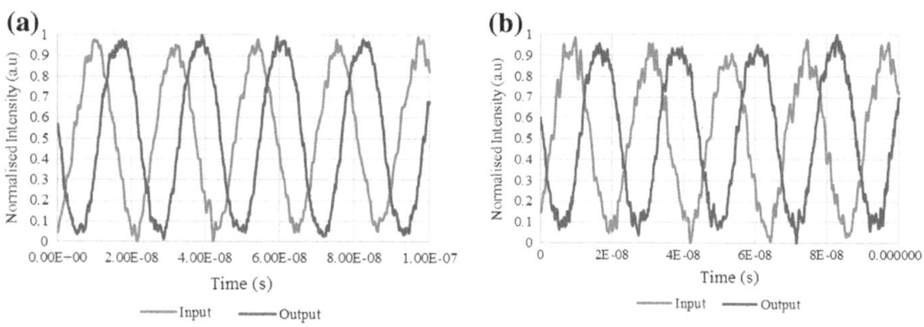

Fig. 7 **a** The plot showing a phase shift of 28.6857° before the measurand, **b** the plot showing a phase shift of 35.4157° after introducing a loss of 15 dB inside the fiber loop

28.6857°. An additional loss of 15 dB is inserted into the cavity using an attenuator for the system response study. A phase shift of 6.73° is observed for the loss parameter. This preliminary result gives a good indication that the PS-FLRDS system can be extended for chemical sensing based on evanescent field sensors.

5 Conclusion

A brief review on theoretical aspects of FLRDS system is done in this chapter. The main scheme of FLRDS using time domain and frequency domains systems are explained in detail. Some recent review on the fiber loop chemical sensor is explained. Different sensor heads for chemical sensing are pointed out in the review. The FLRDS setup proposed by the authors (at CGCRI) is also presented with phase-shift result. FLRDS is a well promising method for chemical sensing due to the simplicity and the accuracy of the method. Even though, some noise related problem is present in the system, which can be corrected by different methods like least mean square calculation and averaging samples. The main attraction of FLRDS system is that available sensors based on intensity measurements are also compatible with fiber loop systems. This combination avoids the major problem of intensity related errors in such sensors. FLRDS system will surely come up with lots of applications in the area of biomedical and chemical sensing applications.

Acknowledgements The research work supported by the sponsored project (No. GAP0141) from DST, Govt. of India. The author would like to acknowledge the support of the Director, CGCRI, Dr. Mukul Paul, Dr. M. Pal and Dr. Shyamal Das of FOPD, CGCRI, Kolkata for their help.

References

1. Herbelin, J. M., McKay, J. A., Kwok, M. A., "Sensitive measurement of photon lifetime and true reflectances in an optical cavity by a phase-shift method.," Appl. Opt. 19(1), 144–147, OSA (1980).
2. Stewart, G., Atherton, K., Yu, H., Culshaw, B., "An investigation of an optical fibre amplifier loop for intra-cavity and ring-down cavity loss measurements," Meas. Sci. Technol. 12(7), 843–849, IOP Publishing (2001).
3. Bescherer, K., Barnes, J. A., Loock, H., "Absorption measurements in liquid core waveguides using cavity ring-down spectroscopy.," Anal. Chem. 85(9), 4328–4334 (2013).
4. Gangopadhyay, T. K., Giorgini, A., Halder, A., Pal, M., Paul, M. C., Avino, S., Gagliardi, G., "Detection of chemicals using a novel fiber-optic sensor element built in fiber loop ring-resonators," Sensors Actuators B Chem. 206, 327–335, Elsevier B.V. (2015).
5. Waechter, H., Litman, J., Cheung, A. H., Barnes, J. a., Loock, H.-P., "Chemical sensing using fiber cavity ring-down spectroscopy.," Sensors (Basel). 10(3), 1716–1742 (2010).
6. Herath, C., Wang, C., Kaya, M., Chevalier, D., "Fiber loop ringdown DNA and bacteria sensors," J. Biomed. Opt. 16(5), 050501 (2011).
7. Wong, W. C., Zhou, W., Chan, C. C., Dong, X., Leong, K. C., "Cavity ringdown refractive index sensor using photonic crystal fiber interferometer," Sensors Actuators B Chem. 161(1), 108–113, Elsevier B.V. (2012).
8. Wang, C., Kaya, M., Wang, C., "Evanescent field-fiber loop ringdown glucose sensor.," J. Biomed. Opt. 17, 037004 (2012).
9. Zhang, Y. N., Zhao, Y., Wu, D., Wang, Q., "Fiber loop ring-down refractive index sensor based on high-Q photonic crystal cavity," IEEE Sens. J. 14(6), 1878–1885 (2014).
10. Zhao, Y., Chang, J., Ni, J., Wang, Q., Liu, T., Wang, C., Wang, Peng, G., "Novel gas sensor combined active fiber loop ring-down and dual wavelengths differential absorption method," Opt. Express 22(9), 11244 (2014).
11. Wu, D., Zhao, Y., Wang, Q., "SMF Taper Evanescent Field-Based RI Sensor Combined with Fiber Loop Ring Down Technology," IEEE Photonics Technol. Lett. 27(17), 1802–1805 (2015).
12. Ittiarah, J. V., Sidhik, S., Gangopadhyay, T. K., "Refractometry using evanescent field based fiber loop ringdown spectroscopy," Sensors Actuators A Phys. 223, 61–66, Elsevier B.V. (2015).

All-Optical Fiber-Cantilever Beam-Deflection Magnetometer: Detection of Low Magnetic Field and Magnetization Measurement

Partha Roy Chaudhuri and Somarpita Pradhan

Abstract All-optical fiber-cantilever beam deflection configuration using an optimized composition of cobalt doped nickel ferrite nanoparticles coated single-mode optical fiber cantilevers is demonstrated. Initially, a fiber-double-slit interferometer arrangement using coated fiber-cantilever-deflection is devised to detect the surrounding magnetic field by precisely measuring the changes in interference fringe-width. A theoretical platform is developed to model the fiber-cantilever deflection which in turn predicts magnetization value of the probe sample. In order to explore higher sensitivity, *etched* single-mode fiber cantilever is incorporated in double-slit arrangement and a marked improvement is achieved. Next, we refined the experiment by tracking the amplitude-modulation of propagating light through fiber-to-fiber coupling cantilever deflection-transmission arrangement which showed increased sensitivity further. In a series of experiments starting with single cantilever transmission technique, we ended up with a model of cascaded cantilevers to sense very low order (~ 1 *m*T) magnetic field. Developed theoretical model fairly predicts experimentally obtained results, in particular, the magnetization of the probe sample. We demonstrate that the scheme is capable of reproducing magnetization data obtained from high precision SQUID measurement. Finally, Sagnac loop assisted cascaded cantilever configuration is realized experimentally to establish the repeatability and more stable response of cascaded cantilever scheme. All our experimental configurations are all-optical with minimum system complexities. These results are new and will provide guideline and understanding towards designing fiber-optic low-magnetic field sensors $\sim m$T.

P.R. Chaudhuri (✉) · S. Pradhan
Indian Institute of Technology Kharagpur, Kharagpur, West Bengal 721302, India
e-mail: roycp@phy.iitkgp.ernet.in

S. Pradhan
e-mail: somarpita.lbc@gmail.com

© Springer Nature Singapore Pte Ltd. 2017
I. Bhattacharya et al. (eds.), *Advances in Optical Science and Engineering*,
Springer Proceedings in Physics 194, DOI 10.1007/978-981-10-3908-9_15

1 Introduction

The advancement in technology in the area of Photonics is very rapid compared to other technologies. Enormous application of photonic devices offers us an instance of the significance of this emerging discipline. Now-a-days, dielectric waveguide structures cover a large area in the field of fundamental research as well as in industry application. One outcome of these photonic waveguides in the form of optical fibers has a major role not only in communication but also in the development of different type of sensors. The fiber optic sensing offers some unique features which make it prominent among other sensing methods. Some of the essential motives for the recognition of optical fiber based sensor structures are small length, light weight, immunity to electromagnetic interference (EMI), capability of far field sensing, dielectric composition, higher sensitivity and multiplexing operation. Using fiber-optic sensors, one can measure different physical or chemical parameters for example voltage, magnetic field, temperature, humidity, strain, pH, rotation and so forth. Information about the measurand is conveyed in all optical fiber sensors by way of alternating polarization, intensity, frequency, phase or a combination of the above. Fiber optic sensors are often grouped into two basic classes referred to intrinsic and extrinsic sensors. In intrinsic fiber optic sensor, the fiber acts as active sensing area whereas in extrinsic fiber sensors, the fiber simply transmits light to and from the sensing medium.

Though the measurement of low magnetic field attracted huge research interest for decade, most of the reported measurement schemes either involve measurement of induced longitudinal strain caused by applied magnetic field [1] or require complex hardware or signal processing system [2]. A relatively new technology, cantilever-beam deflection, has attracted the research attention in recent years in the area of sensing and device applications [3–5]. Because of the high precision and scalability, the cantilever techniques are most often used in environmental monitoring, temperature, humidity, UV radiation sensing and also in medical, chemical, biochemical measurements. Consequent upon these features, we emphasized on devising a fiber-optic cantilever-beam deflection based sensor configuration which is capable of detecting low order magnetic field with minimum system complexity. The proposed scheme would be suitable for remote sensing, in particular at hazardous environment where electrical probes cannot be deployed. For designing such type of sensors, we need to at first identify a proper probe magnetic material. Spinel type oxide, Cobalt-doped nickel ferrite ($Ni_{0.97}Co_{0.03}Fe_2O_4$) was reported in literature for its exceptionally high magnetic properties [6] and was used by M. Sedlar et al. for devising a Mach-Zehnder interferometer type magnetic field sensor [7]. Consequent upon these findings, Cobalt-doped nickel ferrite ($Ni_{0.97}Co_{0.03}Fe_2O_4$) material was chosen as our probe magnetic material. In the following sections, in brief we describe our preparation methodology of probe magnetic material, end characterization results and chronology of developments of magnetic field sensing experimental set-ups based on cantilever deflection.

2 Probe Material Preparation Method and Characterisation

Among different routes, we adopted sol-gel method for sample preparation due to its capability of producing high purity products, low processing temperature and low cost. The material was prepared using nitrate salts of iron [$Fe(NO_3)_3 \cdot 9H_2O$], nickel [$Ni(NO_3)_3 \cdot 6H_2O$] and cobalt [$Co(NO_3)_2 \cdot 6H_2O$] [8]. Firstly, solution of iron nitrate in ethylene glycol was prepared at temperature 70 °C and kept in an ultrasonic cleaner. Then, nickel and cobalt nitrates were added in preheated 2-methoxyethanol (2-MOE) maintaining the stoichiometry. This solution was then mixed with previously prepared iron nitrate solution and stirred at 70 °C until gelation. Finally, the prepared sample was dried and grinded properly and annealed at 600 °C for 3 h and various characterization processes were performed. The crystalline structure of the probe material ($Ni_{0.97}Co_{0.03}Fe_2O_4$) was investigated by X-ray diffraction (XRD) analysis using CuK_α source (wavelength 1.54 Å) at room temperature. The lattice constant was estimated to be ~8.35 Å. We also evaluated coercive field (H_C) value of the magnetic probe material and reported [9] value was 175.50 Oe.

3 Fiber-Cantilever Deflection Setups and Magnetic Field Sensing

3.1 Interferometric Configuration

As an initial step, we attempted a fiber double-slit interferometer experiment where two exit fibers were placed adjacent to each other by making cantilevers of desired length to act as twin-source of interference (see Fig. 1).

3.1.1 Configuration Details and Result

Light from a HeNe laser was coupled into one input of a 3 dB coupler (at 632.8 nm) which acts as a beam-splitter and splits light equally into two output

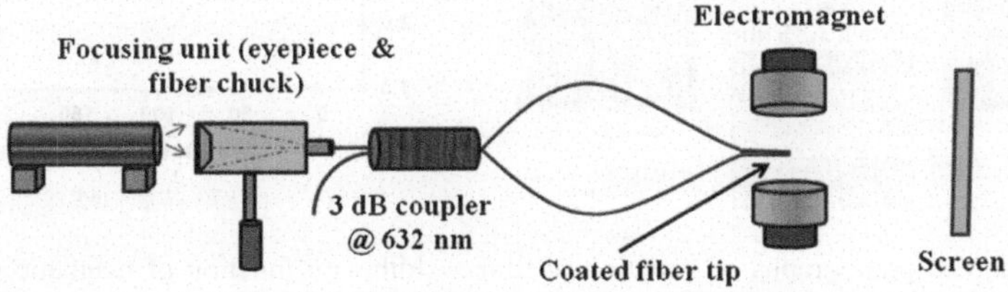

Fig. 1 Schematic of interferometric configuration [9]

ports. One of these ports was coated with the cobalt-doped nickel-ferrite particles and placed parallel with other one making a cantilever of length 2.0 cm. Among different coating methods, we adopted a gluing technique of attaching magnetic nanoparticles over the fiber surface that showed high response [10] in low field.

In presence of magnetic field, due to induced magnetization and consequent bending of coated fiber, separation between two exits fiber ports (twin-source) increases. This leads to dynamic change in the fringe-width (β) at various fields as observed at a screen placed at a large distance as given by

$$\beta = \frac{\lambda D}{d} \tag{1}$$

Here, D denotes the distance from the sources to the screen and the separation of two interfering sources is expressed by d. Experimental result in terms of fringe with variation is shown in Fig. 2(left).

3.1.2 Etching Single-Mode Fiber as Cantilever

As the etched fibers are recognized for their flexible nature and anticipated for higher response, we next introduced etched fiber cantilever in the setup. Single-mode fibers (S405-XP) were etched [11] by using HF 40% pure (MERCK) solution at room temperature (300 K). The etching rate was estimated to be 2.27 μm/min. An optimized etched fiber with diameter 50 μm ± 5 μm coated with probe magnetic material having the same coated fiber length of 1.2 (±0.1) cm (as used for normal fiber case) was incorporated in the interferometric

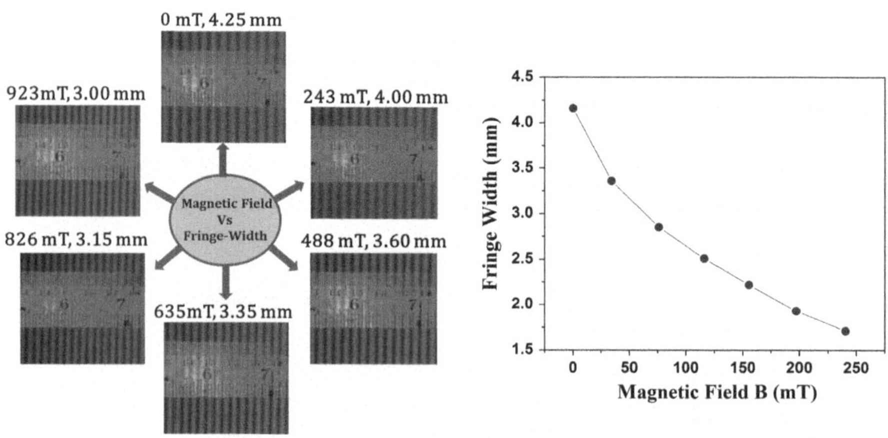

Fig. 2 Recorded photographs of variation of fringe-width as a function of magnetic field using normal coated single-mode fiber (S405-XP) (*left*) and measured fringe-width as a function of magnetic field using etched coated fiber (*right*) [9]

configuration. As expected, sensitivity was seen to improve noticeably. Fringe-width variation results for etched fiber cantilevers (coated fiber thickness 0.12 mm) are shown in Fig. 2(right) which nicely depicts the enhanced sensitivity obtained in this modified scheme.

3.1.3 Model of Fiber Cantilever Deflection: Determination of Magnetization of Probe Material

Magnetization (*M*) of the sample for a constant magnetic field (*B*) can be estimated using the basic torque (τ_m) relation [12]:

$$\tau_m = V_m M \times B \tag{2}$$

Here, volume of magnetic particles used is denoted by V_m. For fiber cantilever, torque experienced due to the external magnetic field is balanced by the internal bending moment (*EI*/*R*) of the fiber. *E* and *I* represent the elastic modulus and the geometrical moment of inertia of the cantilever substrate respectively. By equating these two, for the case of distributed torque acting over the coated length ($b - a$) of the fiber [see Fig. 3(left)], deflection (Δ) can be expressed as [13],

$$\Delta = \frac{V_m BM}{6EI}\left[\frac{(b-z)^3}{(b-a)} - 3(b+a)(b-z) + 2b(b+a) - a^2\right] \tag{3}$$

So, deflection of the fiber end with cantilever length *b* ($z = b$ point) will be given by [9]

$$\Delta = \frac{V_m BM}{6EI}\left[2b(b+a) - a^2\right] \tag{4}$$

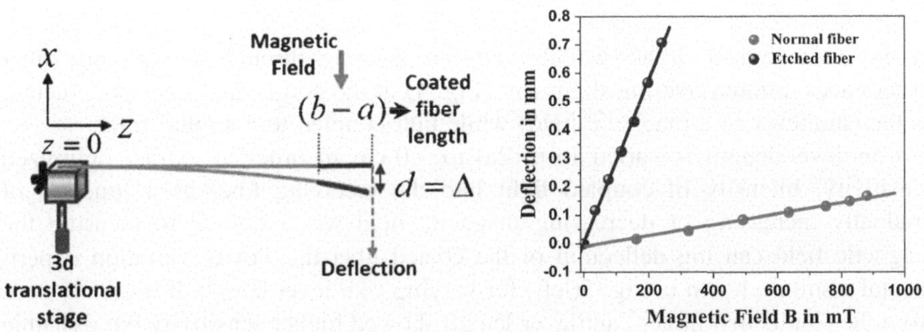

Fig. 3 Schematic of bending of the fiber due to application of magnetic field with specified co-ordinate (*left*) and variation of estimated deflection for normal and etched coated fiber obtained from interferometric configuration (*right*) [9]

Table 1 Magnetization values of probe material obtained from SQUID measurement and interferometric configuration [9]

Magnetic field (Gauss)	Magnetization M (emu/cc) from SQUID data $\times 10^{-3}$	Magnetization M (emu/cc) from experiment $\times 10^{-3}$
1614	347	582
3230	752	801
4915	1130	1005
6355	1420	1018
7510	1920	1026
8260	2070	1061
8810	2150	1115

We calculated deflection of normal and etched fiber tips under calibrated magnetic field using (1) and these are shown in Fig. 3(right). Improved sensitivity of cantilever deflection of the etched fiber is evident from the plot. These calculated deflection values for normal coated fiber cantilever with 0.32 mm thick coating were incorporated in (4) to extract magnetization of the magnetic material (listed in Table 1). In the table, thc magnetization data obtained from SQUID measurement is also quoted to appreciate the performance of this experiment which fairly predicts the order of magnitude of the magnetization value under different field strength.

3.2 Fiber-to-Fiber Cantilever Transmission Arrangement

In this configuration, magnetic field was measured by modulating the amplitude of propagating light through misalignment zone between two cascaded optical fibers (butt-coupled) one being the coated deflection cantilever.

3.2.1 Single Cantilever Transmission Configuration

In this arrangement, light (@632.8 nm) propagating through single-mode fiber experiences a transverse misalignment (Fig. 4) at the end coated fiber port (acting as the cantilever in a magnetic field) while butt-coupled to a similar fiber.

Cantilever length is varied from 2.0 to 3.0 cm in order to extract optimized sensitivity. Intensity of coupled light into the receiving fiber as a function of gradually increasing or decreasing magnetic field was recorded to measure the magnetic field causing deflection of the coated fiber-tip. Power variation experimental result is shown in Fig. 5(left) for varying cantilever length. It is clearly seen from the results that larger cantilever length showed higher sensitivity but dynamic range of operation reduces. Thus optimization of cantilever configuration owes much to given range. We then reconfigured the set-up by replacing the fiber cantilever with etched coated single mode fiber with fiber diameter 50 μm ± 5 μm.

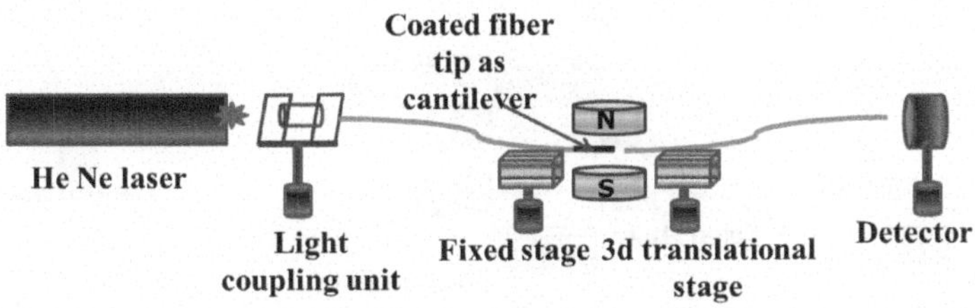

Fig. 4 Schematic of single cantilever transmission configuration [9]

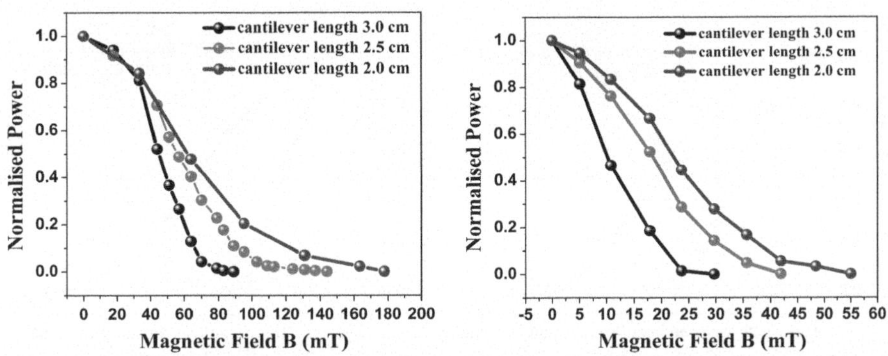

Fig. 5 Transmitted power variation with magnetic field using single-mode fiber (S405-XP) (*left*) and etched fiber (*right*) [9]

Keeping the coating length same (1.2 (±0.1) cm), response of the etched port was recorded for different cantilever lengths. Modified system also establishes enhanced sensitivity of the bigger cantilever length (3.0 cm). We observed reversible and repeatable response for both the cases of normal and etched fiber cantilever. Improved response of the modified scheme is shown in Fig. 5(right).

3.2.2 Double Cantilever Transmission Configuration

Based on the former work, to realize better sensitivity without chemical etching process, we then reconfigured the single cantilever model to double cantilever transmission scheme. By the term "double cantilever", we describe a configuration that involves two successive transverse misalignment zones (see Fig. 6) unlike the single cantilever scheme.

We maintained the coated fiber length as 1.2 ± 0.1 cm and cantilever length (3.0 cm) throughout the experiment as these are the optimized values obtained from the single cantilever arrangement. In this double cantilever set-up, one transverse off-set (between fiber 1 and 2) followed by another (between fiber 2 and 3) result in rapid decrement in transmitted power which makes the configuration more sensitive to weak magnetic field compared to single cantilever scheme.

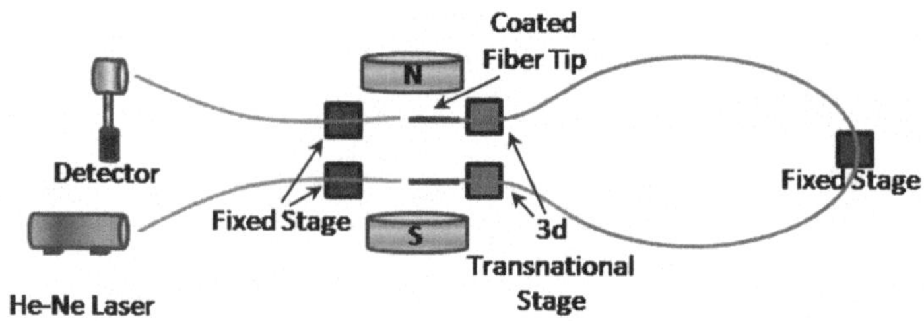

Fig. 6 Schematic of double cantilever transmission configuration

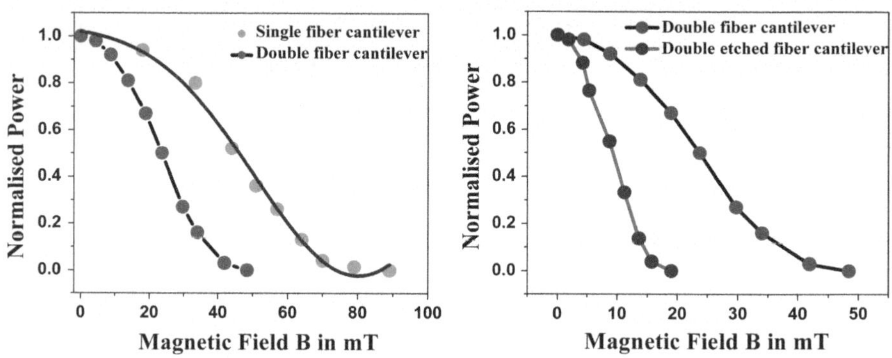

Fig. 7 Comparison between response of single and double cantilever configuration (*left*) in a calibrated magnetic field and response of double cantilever transmission configuration using both normal and etched coated fiber cantilevers (*right*)

We recorded the light intensity coupled into the last receiving fiber as a function of varying magnetic field. The experimentally obtained data are plotted in Fig. 7 (left). We also placed the results obtained from single cantilever setup to distinguish the improved response of double cantilever [14]. Response of the etched fiber (60 μm ± 5 μm) cantilevers is also plotted in this figure (Fig. 7(right)). This scheme is capable of detecting magnetic field as low as 1 *m*T.

3.2.3 Theoretical Model of Fiber-to-Fiber Transmission: Estimation of Deflection

Considering the normalized Gaussian fundamental modes of two single-mode fibers that have transverse offset d_1 as

$$\psi_1(x,y) = \sqrt{\frac{2}{\pi}}\frac{1}{w_1}e^{-\frac{(x^2+y^2)}{w_1^2}} \quad \text{and} \quad \psi_2(x,y) = \sqrt{\frac{2}{\pi}}\frac{1}{w_2}e^{-\frac{\left[(x-d_1)^2+y^2\right]}{w_2^2}} \tag{5}$$

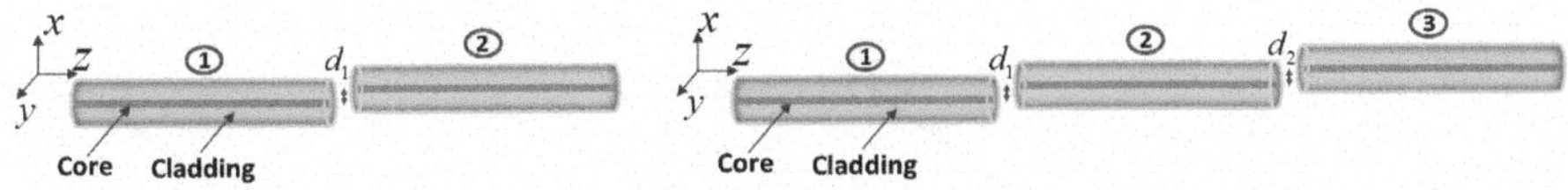

Fig. 8 Schematic of transverse misalignment between two (*left*) or three (*right*) optical fibers

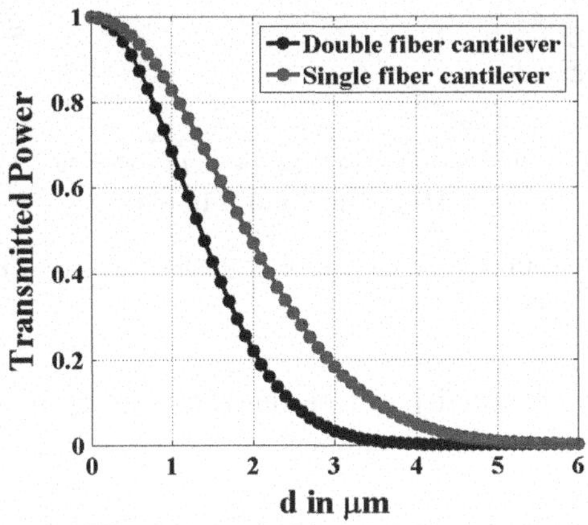

Fig. 9 Transmitted power variation profile with increasing transverse misalignment for both single and double fiber cantilever configuration using S405-XP single mode fibers

in the direction of x (as shown in Fig. 8) and propagating along z, the fractional transmitted power to the second receiving fiber from the first is obtained from the overlap integral [15]

$$T_{12} = \left| \int_{-\infty}^{+\infty} \int_{-\infty}^{+\infty} \psi_1 \psi_2^* dxdy \right|^2 = \left(\frac{2w_1 w_2}{w_1^2 + w_2^2} \right)^2 e^{-\frac{2d_1^2}{w_1^2 + w_2^2}} \tag{6}$$

Here, w_1 and w_2 are the spot sizes of fundamental modes of fiber 1 and fiber 2 respectively.

Equation (6) can be adopted for three fiber-system incorporating two successive transverse misalignments of the fibers:

$$T_{13} == \left(\frac{4w_1 w_2^2 w_3}{(w_1^2 + w_2^2)(w_2^2 + w_3^2)} \right)^2 e^{-2\left[\frac{d_1^2}{w_1^2 + w_2^2} + \frac{d_2^2}{w_2^2 + w_3^2}\right]} \tag{7}$$

For three identical optical fibers imposing the condition $d_1 = d_2 = d$, transmitted power to third fiber will be:

$$T = e^{-\frac{2d^2}{w^2}} \tag{8}$$

The transmitted power variation profile for the both single and double misalignment cases involving identical single mode fibers (S405-XP) having mode field diameters 4.6 µm ± 0.5 µm at 632 nm wavelength is plotted in Fig. 9.

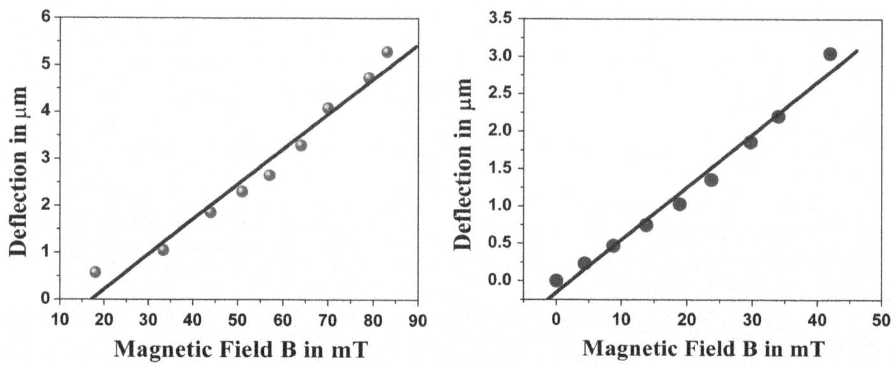

Fig. 10 Deflection values at different magnetic field for both single (*left*) and double (*right*) cantilever transmission

Sensitivity can be tuned further by realizing the cascaded model. For n such cascaded misalignments,

$$T = \mathrm{e}^{-\frac{nd^2}{w^2}} \tag{9}$$

Equation (9) clearly indicates the effect of cascading in terms of sensitivity.

3.2.4 Determination of Magnetization of Probe Material

Deflection (Δ) of the coated fiber-tip due to the applied magnetic field was then calculated by fitting the experimental results with the theoretical model and depicted in Fig. 10 for both single and double cantilever cases for cantilever length 3.0 cm. Estimated deflections values were used to calculate the magnetization (M) of the probe sample. For our case of single and double cantilever configurations with 0.47 mm and 0.56 mm of coated fiber thickness respectively, magnetization values are listed in Table 2. We have used the value of E = 70 GPa and $I = \pi D^4/64$, where D is the diameter of the fiber (substrate material).

3.3 *Sagnac Loop Assisted Cascaded Cantilever Configuration*

We next implemented a Sagnac loop assisted cascaded cantilever configuration in order to realize two exactly identical cantilever paths for two opposite (forward and

Table 2 Magnetization of probe material obtained from both single and double cantilever transmission configuration

Single cantilever		Double cantilever	
Magnetic field (Gauss)	Magnetization M (emu/cc) from experiment $\times\ 10^{-3}$	Magnetic field (Gauss)	Magnetization M (emu/cc) from experiment $\times\ 10^{-3}$
180	32.3	43	36.9
334	32.4	88	37.4
440	43.0	138	38.0
510	46.0	189	38.2
570	48.1	237	40.1
640	52.4	297	44.0
700	59.4	340	45.5
790	61.0	419	57.1

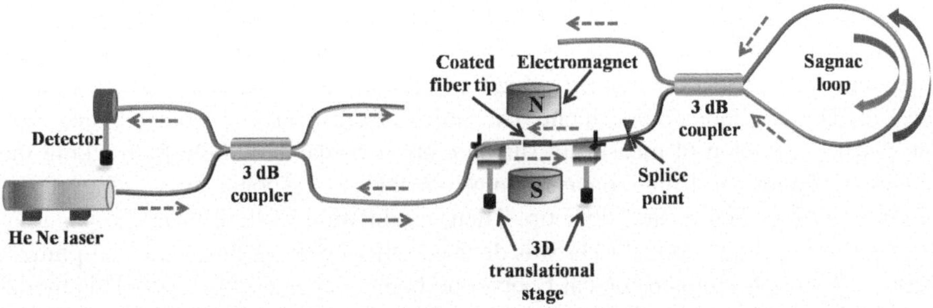

Fig. 11 Schematic of Sagnac loop assisted cascaded cantilever configuration

backward) propagating light. In this configuration, laser light was coupled to one input of a 3 dB coupler while the other input was used to detect transmitted light as shown in Fig. 11.

One output port of the coupler acts as the transmitting/receiving end of the cantilever stage with a Sagnac loop mirror using a second 3 dB coupler. The cantilever (coated fiber length 1.2 ± 0.1 cm) is formed at the input port of the Sagnac loop made by splicing the two output ports of a 3 dB coupler. Due to the fiber-mirror characteristics of Sagnac loop, light traveling through the misalignment zone (caused by external magnetic field) traverses the identical path twice leading to double the response as detected from the backward propagating light through the first 3 dB coupler. The response of this scheme as intensity variation with applied magnetic field is depicted in Fig. 12(left). Then, we reconfigured our set-up using etched coated fiber cantilevers and their response in terms of power variation is depicted in Fig. 12(right).

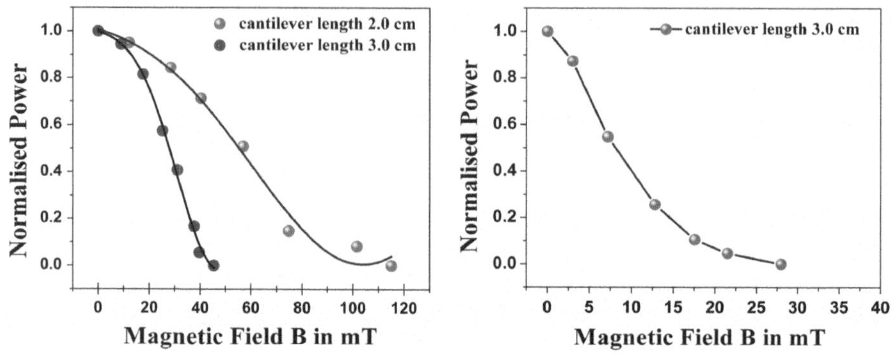

Fig. 12 Response of sagnac loop assisted cascaded cantilever configuration using normal (*left*) and etched (*right*) fibers

4 Conclusion

In this article, we detailed our experimental research aimed at devising a weak magnetic field detection system which is all-optical, and particularly useful in electrically hazardous environment. The series of experimental studies began with interferometric setup of measuring surrounding low-magnetic field by tracking the dynamic change in fringe pattern. Improvement in response was achieved by incorporating etched coated fiber tip. Then, modulation of light-beam propagating through single-mode optical fiber was devised utilizing the transverse misalignment between two butt-coupled optical fibers, one being a coated cantilever. This model was reconfigured as double cantilever transmission arrangement with a visible improvement in sensitivity. A theoretical platform is provided which nicely interprets the experimental data. Magnetization of the probe sample is also calculated using our theory and validated by quoting SQUID measurement data. Our approach is capable of detecting magnetic field $\sim 1\ m$T or even lower without any system complexity or involving any complex circuit. Sagnac loop assisted double-pass identical cantilever scheme is demonstrated for higher sensitivity. These results could be helpful for developing low magnetic field detection schemes.

Acknowledgements The authors sincerely acknowledge BRNS, Govt. of India for the financial support to carry out this research.

References

1. Dai, Y., Yang, M., Xu, G., and Yuan, Y., "Magnetic field sensor based on fiber Bragg grating with a spiral microgroove ablated by femtosecond laser," Opt. Expr. 21, 17386–17391 (2013).
2. Lenz, J., and Edelstein, A. S., "Magnetic sensors and their applications," IEEE Sens. 6, 631–649 (2006).

3. Koch, R. *et al.*, "Perpendicular magnetic fields in cantilever beam magnetometry," J. Appl. Phys. 96, 2773–2778 (2004).
4. Duc, T. C., *et. al.*, "Piezoresistive cantilever beam for force sensing in two dimensions," IEEE Sens. 7, 96–104 (2007).
5. Kumar, A., *et. al.*, "Cantilever based MEMS pressure sensor using different piezoelectric materials: a comparative study," IJEDR 2, 4022–4026 (2014).
6. Dung, N. K., and Tuan, N. H., "The effect of cobalt substitution on structure and magnetic properties of nickel ferrite," VNU Journal of Science, Mathematics-Physics 25, 153–159 (2009).
7. Sedlar, M. *et al*, "Optical fiber magnetic field sensors using magnetostrictive jackets," Sens. Actu. A 84, 297–302 (2000).
8. Sedlar, M. *et al,* "Preparation of cobalt nickel ferrite films on optical fibers by dip-coating", Ceram. Int. 21, 21–27 (1995).
9. Pradhan, S., and Chaudhuri, P. R., "Experimental demonstration of all-optical weak magnetic field detection using beam-deflection of single-mode fiber coated with cobalt-doped nickel ferrite nanoparticles," Appl. Opt. 54, 6269–6276 (2015).
10. Yang, M., Dai, J., Zhou, C., and Jiang, D., "Optical fiber magnetic field sensors with TbDyFe magnetostrictive thin films as sensing materials," Opt. Expr. 17, 20777–20782 (2009).
11. Kumar, A.*et al.*, "Optimize etching based single mode fiber optic temperature sensor" Int. J. Eng. Res. Appl. 4, 4–8 (2014).
12. Weber, M., Koch, R., and Rieder, K. H., "UHV Cantilever Beam Technique for Quantitative Measurements of Magnetization, Magnetostriction, and Intrinsic Stress of Ultrathin Magnetic Films," Phys. Rev. Lett. 73, 1166–1169 (1994).
13. Adhikari, R.*et al.*, "The cantilever beam magnetometer: a simple tool for characterization" Am. J. Phys. 80, 225–231 (2012).
14. Pradhan S. and Chaudhuri P. R., "Performance of Low Magnetic Field detection using Double Cantilever Fiber-Beam Deflection-Transmission Configuration," JSAP-OSA Joint Symposia, Sept. 13–16, 2015.
15. Ghatak. A., and Thyagarajan, K., "Introduction to Fiber Optics," Cambridge University Press, Reprint 2011.

Dynamics of Dissipative Solitons in Active Silicon Waveguides

Samudra Roy

Abstract Dissipative soliton is a fascinating nonlinear stable structure, which arises as a result of double balance between gain and loss as well as dispersion (diffraction) and nonlinearity. Formation of the dissipative soliton is critical and may leads to some chaotic dynamics if some external perturbations are imposed. In this work, we mainly study the nonlinear temporal dynamics of such waves inside active Si based waveguides where the pulse is governed by the complex Ginzburg-Landau equation.

1 Introduction

Nonlinear silicon photonics as a field is exploring due to novel nonlinear properties of silicon (Si) and its compound elements [1]. Silicon-on-insulator (SOI) technology is growing up rapidly because of the ability of Si to amplify, generate, process and sense signals [1]. The high refractive index and nonlinear coefficient of Si provide addition advantage like modulation of nonlinear waves. The situation becomes intriguing when active medium is considered that leads to rich nonlinear dynamics. In this work, we focus on the generation and dynamics of a special kind of stable temporal structure called *dissipative solitons* (*DS*). DSs are generated in a non-conservative system far from equilibrium [2]. Here in addition to the balance between dispersion and nonlinearity, we require the balance between gain and loss in order to keep DS alive. However DS reacts abruptly in presence of perturbations (like free carrier interaction, higher order dispersion etc.) and may leads to some chaotic dynamics. In this work we try to study complex dynamics of a perturbed DS in active medium.

S. Roy (✉)
Department of Physics, IIT Kharagpur, Kharagpur, West Bengal 721302, India
e-mail: samudra.roy@iitkgp.ernet.in

© Springer Nature Singapore Pte Ltd. 2017

I. Bhattacharya et al. (eds.), *Advances in Optical Science and Engineering*,
Springer Proceedings in Physics 194, DOI 10.1007/978-981-10-3908-9_16

2 Theory

The propagation of an optical pulse with envelope $u(z,t)$ and carrier frequency ω_0 in the proposed Si-based active nano photonic waveguide, is governed by the complex dimensionless Ginzburg-Landau (GL) equation [3],

$$\begin{aligned} &i\frac{\partial u}{\partial \xi} - \frac{1}{2}\mathrm{sgn}(\beta_2)\frac{\partial^2 u}{\partial \tau^2} - i\left(g_0 + g_2\frac{\partial^2}{\partial \tau^2}\right)u - i\delta_3\frac{\partial^3 u}{\partial \tau^3} \\ &+ (1+iK)|u|^2 u + \left(\frac{i}{2} - \mu\right)\phi_c u = 0, \end{aligned} \tag{1}$$

which is coupled with an ordinary differential equation accounting for the two photon absorption (2PA)-induced free-carrier (FC) dynamics $\frac{d\phi_c}{d\tau} = \theta|u|^4 - \tau_c\phi_c$ [3]. Here g_0 and g_2 are gain coefficient and gain-dispersion coefficient respectively. The effect of third order dispersion (TOD) is included in the equation by normalized TOD coefficient δ_3. Two photon absorption (TPA) loss and free carrier dispersion are also included by the parameters K and μ. Note that, the free carrier density (ϕ_c) is related to a source term involving θ which contain TPA coefficient and free carrier recombination time (t_r) is rescaled by the input pulse width (t_0) as $\tau_c = t_r^{-1}t_0$.

DSs are formed under proper dynamical equilibrium conditions, and are associated with certain discrete parameters of the GL equation that satisfy the energy balance condition. Stable solution (in absence of higher order dispersion and free carrier effect) for DSs can be found as [3],

$$u(\xi,\tau) = u_0[\mathrm{sech}(\eta\tau)]^{(1+ia)}e^{i\Gamma\xi}. \tag{2}$$

where the four parameters u_0, η, a, and Γ satisfy the following algebraic relations,

$$|u_0|^2 = \frac{(g_0-\alpha)}{K}\left[1 - \frac{\frac{1}{2}\mathrm{sgn}(\beta_2)a + g_2}{(g_2(a^2-1) - \mathrm{sgn}(\beta_2)a)}\right], \tag{3a}$$

$$\eta^2 = \frac{(g_0-\alpha)}{(g_2(a^2-1) - \mathrm{sgn}(\beta_2)a)}, \tag{3b}$$

$$\Gamma = \frac{\eta^2}{2}\left[\mathrm{sgn}(\beta_2)(a^2-1) + 4ag_2\right], \tag{3c}$$

$$a = \frac{H - \sqrt{H^2 + 2\delta^2}}{\delta}. \tag{3d}$$

where, $H = -\left(\frac{3}{2}\mathrm{sgn}(\beta_2) + 3g_2K\right)$ and $\delta = -(2g_2 - \mathrm{sgn}(\beta_2)K)$. The exact values of u_0, η, a and Γ are fixed by the physical parameters of the system: g, g_2, K and α which generate the stable DSs.

3 Results

In this section we show the dynamics of DS, without and with perturbation by solving the complex GL equation numerically. In Fig. 1a, b the evolutions of the exact DS and Kerr soliton are shown in temporal domain. Here we try to show the difference between the dynamics of conventional and dissipative soliton. From the figure it is clear that, if we launch an exact DS whose parameters satisfy (3a) we get stable pulse dynamics where pulse shape remain unchanged. On the other hand, an input Kerr soliton (having $\mathrm{sech}(\tau)$ form) exhibits a chaotic dynamics.

Next, we elaborate our study by adopting the perturbing effect in DS dynamics. TOD is the prime perturbation that should be included in the GL equation to capture the full picture regarding the pulse dynamics. DSs are very sensitive to the perturbation and react abruptly. Under the inclusion of TOD a breathing effect of the stable DS is observed where shape of the soliton varies periodically in time (and frequency) domain over the distance as shown in Fig. 2a. Though TOD initiates the breathing, the period of oscillation is mainly dominated by 2PA coefficient *K*. The numerical simulation confirms that, for a fixed TOD coefficient, the period of breathing increases linearly with 2PA coefficient. Apart from the breathing TOD initiate a phase matched radiation beyond the zero dispersion point as shown in Fig. 2b. Under a phase matched radiation, energy is transferred from soliton to a linear wave having same phase (with soliton). The phenomenon of phase matched radiation is well studied in the context of supercontinuum generation [4]. The

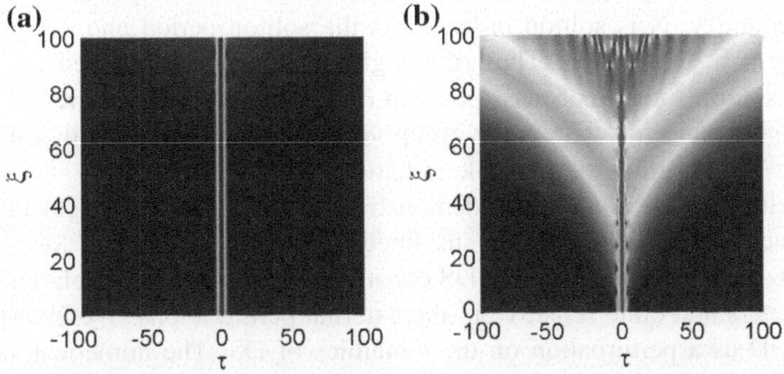

Fig. 1 **a** Time domain structure of the exact solution of GL equation as input pulse over the propagation distance. **b** Time domain structure of unchirped Sech pulse over the propagation distance. Both the figures are plotted for the medium characteristics parameters: $K = 0.05, g_0 = 0.05, g_2 = 0.0001$. The chaotic dynamics is evident in plot (**b**)

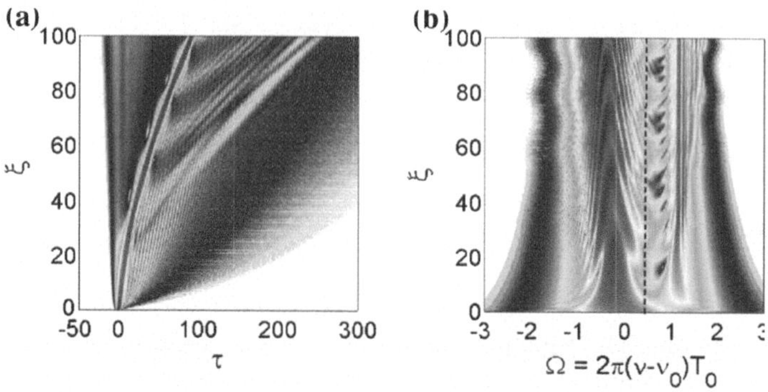

Fig. 2 **a** Time and **b** frequency domain structure of the sDS when TOD is present ($\delta_3 = 1$). Both the figures are plotted for the medium parameters: $K = 0.01, g_0 = 0.01, g_2 = 0.01$ and $\alpha = 0$. The vertical *dotted line* represents the position of zero GVD point. The multiple phase match radiation is observed beyond the zero GVD point

dispersive radiation generally falls in normal dispersion regime and plays a significant role in the generation of blue spectral component [4]. Different aspects of the DWs had already been reported in articles. The previous study is mainly confined for standard optical soliton [5]. In this work, however, we extend the study for more complicated DSs. The numerical simulation has already predicted the existence of DS mediated DWs in the form of radiation.

As shown in Fig. 2, instead of one station DW (which normally the case in Kerr soliton) we observe multiple radiations which corresponds to each breathing. The phase matching equation for oscillating DS can be calculated as,

$$\frac{\text{sgn}(\beta_2)}{2}\Omega^2 + \delta_3\Omega^3 - \tau_g\Omega = \frac{2\pi n}{\xi_0} + 2\Gamma \tag{4}$$

Here $\Omega = 2\pi(v_d - v_s)T_0$ is detuning frequency of the dispersive waves from soliton frequency. N is soliton order, ξ_0 is the soliton period and n is an integer number ($n = 0, \pm1, \pm2\ldots$). The soliton group delay τ_g is defined as a rate of change in soliton's centre of mass position (in the moving frame of reference) with propagation distance ξ (or τ_g is the group velocity mismatch). The modified phase matched (4) can predict the multiple radiations.

In conclusions, we numerically demonstrate the dynamics of the DS in active Si based waveguide. The DS can be sustained under dual balance of loss- gain and dispersion–nonlinearity. The exact DS consists of the complicated relation between the parameters and quite sensitive to the external perturbations. We investigate the role of TOD as a perturbation on the dynamics of DS. The numerical simulation confirms that the TOD parameter leads to the multiple phase matched radiations. Further we try to establish a modified phase matched equation that predicts the multiple radiations.

References

1. J. Leuthold, C. Koos and W. Freude, Nat. Photonics **4**, 535 (2010).
2. N. Akhmediev and A. Ankiewicz, "Dissipative Soliton: Lecture Notes in Physics", (Springer, Berlin, 2005).
3. S. Roy, A. Marini, and F. Biancalana, Phys. Rev. A **87**, 065803 (2013).
4. G. P. Agrawal, "Nonlinear Fiber Optics", 5th ed. (Academic, 2013).
5. S. Roy, S. K. Bhadra, and G. P. Agrawal, Phys. Rev. A**79**, 023824 (2009).

Part III
Green Photonics: Applied Photovoltaics, Application of Solar Energy, Organic Photovoltaics

Wavelength Response Analysis and Optimization of Photopolymer Volume Holographic Elements for Solar Energy Applications

T.L. Shaji Sam and P.T. Ajith Kumar

Abstract Holograms as spectrum splitting concentrator elements are achieving high relevance in improving overall efficiency of photovoltaic systems. Photopolymer holographic elements have the added advantages of near hundred percent diffraction efficiency, easy recording and handling, lower cost etc. Along with high diffraction efficiency, achieving optimum spectral response is of critical importance here. This paper reports recording of volume holographic elements of 91% optimum diffraction efficiency at 639 nm and the spectral and angular response studies conducted.

1 Introduction

Volume Holographic Solar Elements (VHSE) as spectrum splitting concentrators has many advantages over traditional concentrators [1]. In Conventional concentrator system, the unwanted spectral components are concentrated on to the photovoltaic element, which will drastically reduce the conversion efficiency. Concentrator solar cells deliver maximum efficiency and output if and only if intense cool solar radiation is concentrated on to them and coupled in a proper manner. One added advantage of holographic solar elements is the possibility of eliminating unwanted radiation foot print by diffracting it away and concentrating spectral range which is matched with the solar cell [2, 3]. The use of doped silicon as a semiconductor material utilizes only a small fraction of the solar spectrum,

T.L. Shaji Sam (✉) · P.T. Ajith Kumar (✉)
Light Logics Holography and Optics, Crescent Hill, Thiruvallom,
Thiruvananthapuram, Kerala 695027, India
e-mail: shajisam@gmail.com

P.T. Ajith Kumar
e-mail: ajithpt@gmail.com

T.L. Shaji Sam
Optical Image Processing Division and Holography Studio, Centre for Development of Imaging Technology (C-DIT), Thiruvananthapuram, Kerala 695027, India

© Springer Nature Singapore Pte Ltd. 2017
I. Bhattacharya et al. (eds.), *Advances in Optical Science and Engineering*,
Springer Proceedings in Physics 194, DOI 10.1007/978-981-10-3908-9_17

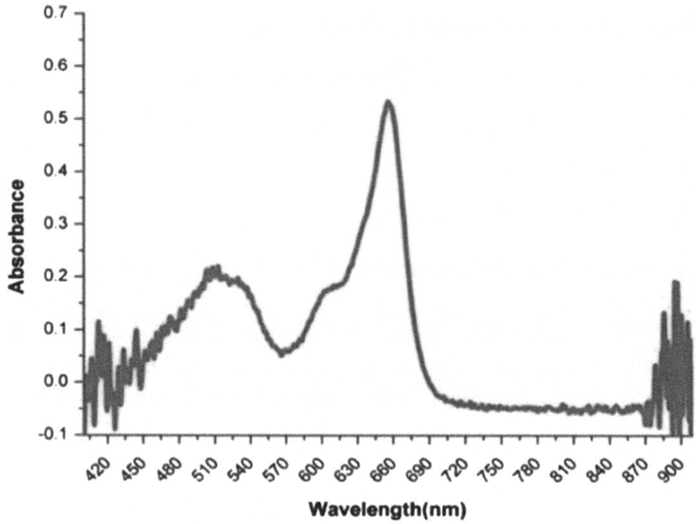

Fig. 1 Absorption spectrum of photopolymer

due to the defined band gap of the material. This limits the performance of single junction cells. To increase the performance and power output of a cell, multiple sub cells connected in series are to be used. The sub cells are made of different band gap materials that have different absorption regions within the solar spectrum. This results in absorption of a much larger portion of the solar spectrum, leading to higher performance. Hence Multi-junction cells offer a better solution to the absorption of energy from a wider spectral range than that of conventional PV cells. Holographic coupling of solar radiation onto spectrally matched multi-junction (MJ) photovoltaic cells can result in dramatic increase in the energy yield. The aim of the present study is to design an off axis holographic element for multi-junction solar cell. Diffraction efficiency, spectral selectivity and angular response are core general parameters that have to be analyzed in optimizing such designs.

The efficiency and the spectral selection of holographic solar elements depend on the properties of the recording media as well as the recording geometry. Silver halide, Dichromated Gelatin (DCG), photoresists and photopolymers are generally used to record holographic optical elements. DCG and photopolymer (PP) materials have high diffraction efficiency; hence these materials are suitable for recording solar elements for PV applications [2, 4]. Photopolymers have significant advantages over other materials, making them suitable for applications such as data storage, HOEs, fiber optic components, variable data security holograms and multifunction elements for various applications. Real-time photopolymer materials eliminate issues related to wet processing in comparison with the conventional recording materials. The aim of the present study is to optimize a holographic element for solar application in a professional grade photopolymer. The material exhibits fairly good refractive index modulation (~ 0.03), high diffraction efficiency and stability, hence is good medium for recording VHSEs. The absorption spectrum of photopolymer before holographic exposure was measured with a spectrometer and is shown in Fig. 1. The design and experimental aspects of VHSEs recorded in this material are discussed in the subsequent section.

2 Theoritical Analysis of Solar Elements

To analyze diffraction efficiency of thick phase wavelength selective holographic solar elements, we use Coupled Wave Theory [5, 6]. This theory provides theoretical equation for diffraction efficiency by making the assumption that refractive index modulation is sinusoidal and also the thickness of the recorded VHSE remains unaltered during the recording and post recording processes. Since Coupled Wave Theory gives good results for thick hologram, care has been taken to make sure that parameters satisfy the criteria for thick grating. Diffraction efficiency $\boldsymbol{\eta}$ is the ratio of the diffracted intensity to the incident intensity and also associated with refractive index modulation $\mathbf{n_1}$. The expression for diffraction efficiency from Coupled Wave Theory is written in the form

$$\eta = \sin^2 \frac{\left(\xi^2 + v^2\right)^{\frac{1}{2}}}{\left(1 + \frac{\xi^2}{v^2}\right)}$$

where

$$\xi = \frac{\delta 2\pi n d \sin \theta}{\lambda}$$

$$\upsilon = \frac{\pi n_1 d}{\lambda \cos \theta}$$

Here $\mathbf{n_1}$ is the depth of refractive index modulation, $\mathbf{n}$ is the average refractive index of the recording medium, $\mathbf{d}$ is the emulation thickness, $\boldsymbol{\lambda}$ is the wavelength of the reconstructed light and $\boldsymbol{\delta}$ is the angular deviation with respect to Bragg's angle $\boldsymbol{\theta}$. When $\boldsymbol{\delta} = \mathbf{0}$ i.e. the reconstruction is done at Bragg angle condition, diffraction efficiency can be written as

$$\eta = \sin^2 \left(\frac{\pi n_1 d}{\lambda \cos \theta}\right)$$

From the above formula $\mathbf{d}$, $\mathbf{n_1}$ and $\boldsymbol{\Lambda}$ (fringe spacing) can be optimized to have higher diffraction efficiency for a VHSE at a particular wavelength which can be used for spectral selective concentrators. Since the photopolymer material available has a thickness of 16 μm, wavelength response was analyzed for that thickness.

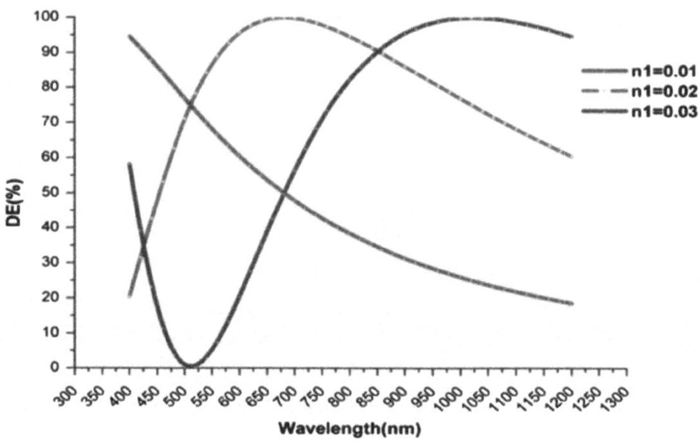

Fig. 2 Variation of diffraction efficiency with wavelength for different values of refractive index modulation for a Bragg angle of 20° and d = 16 μm

Maximum refractive index modulation for the material is 0.03. Wavelength response of the VHSEs, based on Kogelnik's Coupled Wave Theory, was simulated by using *LightGrate* software, developed by Light Logics, as part of a funded project from the BRFST*. Bragg angle was selected as 20° so that the spatial frequency ensures a broad band operation. The wavelength response was calculated for different refractive index modulation for Bragg angle 20° is shown in Fig. 2.

3 Optical Recording and Spectral Analysis

Our recording geometry was based on the interference of an off-axis spherical wave with a plane reference wave. Optical arrangement for recording the off-axis VHSEs is shown as Fig. 3. A red DPSS laser of 639 nm was used as the source. Reference beam and the object beam were expanded by using a spatial filter with 20X microscope objective. The reference beam is made parallel by a converging lens L1. The reference beam was set normal to the recording medium and the diverging beam subtended an angle of 40° with this, at the center of the recording medium. The entire setup was arranged in a specially designed vibration isolation table. A set of transmission solar elements were recorded in the material. Exposure time was varied with different beam intensity, to optimize the diffraction efficiency. Optimum beam strength identified on the PP plate was about 2 mW/cm^2. Thus a set of optical elements each of area 60 cm^2 was recorded and is shown in Fig. 4. The VHSEs were fixed by illuminating with a Mercury vapor lamp, for about 30 min. Diffraction efficiencies at the recording wavelength (639 nm) were measured and the best efficiency obtained was 91%.

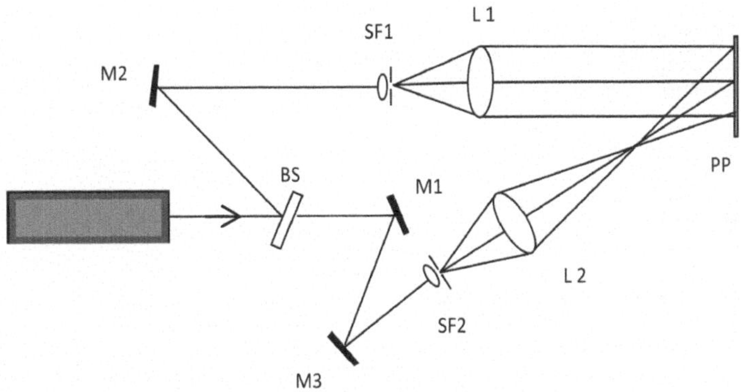

Fig. 3 Optical arrangements for recording VHSE in photopolymer

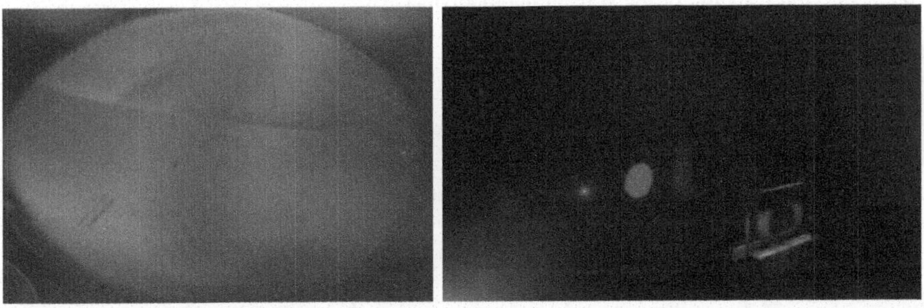

Fig. 4 Holographic solar element in photopolymer and its reconstruction with 639 nm

Spectral response of the optimized elements was measured at different laser wavelengths of 639, 633, 532 and 442 nm and diffraction efficiencies were calculated. The hologram was first oriented normal to the collimated beam from the laser sources. The diffracted light was focused to an off-axis point according to the recording geometry. The intensities of diffracted and undiffracted light were measured with a calibrated power meter. The experimental variation of diffraction efficiency with wavelength is shown in Fig. 5.

Angular response of optimized VHSE in photopolymer element was studied at the recording wavelength. To measure angular response, the elements were placed in a rotational stage. Initially the incident beam was set normal to the hologram. Rotational stage is rotated at regular interval of 0.5° and diffraction efficiency was measured at each case for 639 nm. Figure 6 shows the variation of efficiency of volume a holographic element as a function of the reconstruction angle (in air) with a spatial frequency of 1080 ln/mm. The recorded volume holographic solar element in 16 μm thick photopolymer showed a narrow angular selectivity and hence the associated system has to be Sun following.

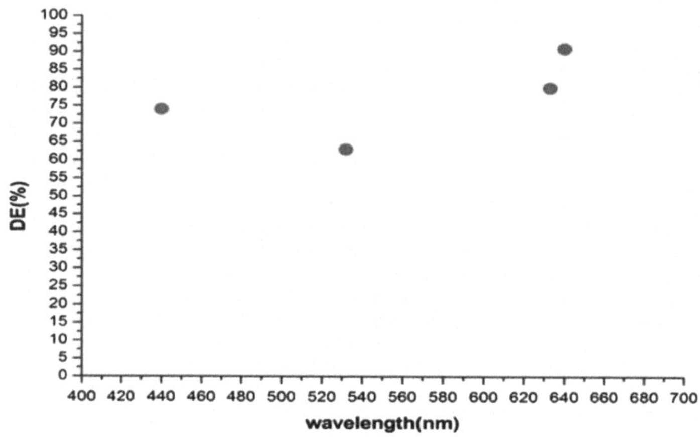

Fig. 5 Spectral response of the HSE in photopolymer

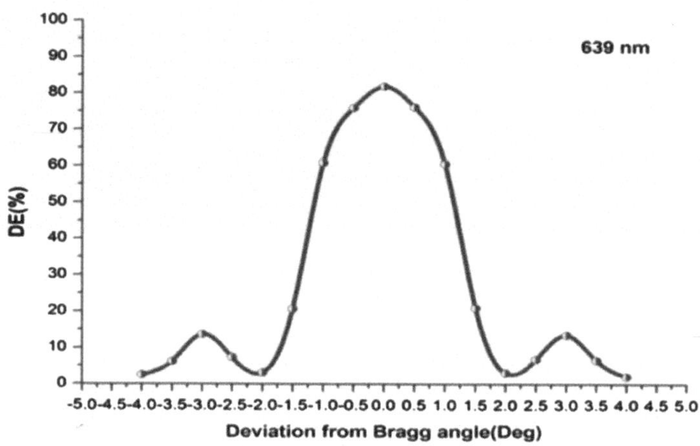

Fig. 6 Angular response of HSE at 639 nm

4 Conclusion

We have recorded and optimized a photopolymer VHSE of 9 cm diameter and fairly high diffraction efficiency in the whole visible region of the solar spectrum. Scaling up of this will result in VHSEs that can function in conjunction with commercial grade multi-junction solar cells. Further studies on bandwidth and scaling up are progressing. This preliminary work was done as part of our efforts to develop and optimize VHSEs that yield a concentration ratio 500× on InGaP/InGaAs/Ge multi-junction cell of 1 cm^2 area.

Acknowledgements Partial financial support from the Board of Research in Fusion Science and Technology (BRFST), Department of Atomic Energy, Government of India, is gratefully acknowledged.

References

1. Jacques E. Ludman, "Holographic solar concentrator", Appl. Optics, Vol 21, 3057–3058 (1982)
2. Shaji Sam T.L and Ajith Kumar P.T "Design and optimization of photopolymer based solar concentrator", AIP. Proc 248, 1391 (2011)
3. T. L. Shaji Sam, P. T. Ajith Kumar et al, "A multi-beam holographic light concentrator for solar applications," Proc. SPIE 6832, 68321E (2007)
4. HodaAkbari, IzabelaNaydenova, and Suzanne Martin, "Using acrylamide-based photopolymers for fabrication of holographic optical elements in solar energy applications," Appl. Opt. 53, 1343–1353 (2014)
5. H. Kogelnik, "Coupled wave theory for thick hologram gratings" The Bell system technical journal. 48, 9, 2909–2945 (1969)
6. Raymond K. Kostuk and Glenn Rosenberg, "Analysis and design of holographic solar concentrator" *Proc. of SPIE*, 7043, 704301 (2008)

Investigating the Growth-Time Dependent Comparative Performance of Vapour-Liquid-Solid (VLS) Grown p-CuO/n-Si Thin Film Hetero-Junction Solar Cells

Jenifar Sultana, Somdatta Paul, Anupam Karmakar and Sanatan Chattopadhyay

Abstract This work investigates the impact of growth time on the performance of p-CuO/n-Si hetero-junction solar cells, where CuO thin films are grown by employing vapour liquid solid method, using CBD synthesized CuO powder. Studies are performed in terms of the film morphology, chemical composition, crystallite structure, optical properties and energy conversion efficiency. The potential for photovoltaic applications of such films are investigated by measuring the p-CuO/n-Si diode current-voltage characteristics and by extracting its relevant photovoltaic parameters, including open circuit photo-generated voltage, short circuit current density and energy conversion efficiency.

1 Introduction

The optical properties, energy band gap and charge carrier diffusion length [1] are the primary parameters, on the basis of which the semiconductor materials are preferred for solar cell applications. Copper oxide (CuO) is the transition metal oxide with fascinating physical and chemical properties and hence has drawn

J. Sultana · S. Paul
Centre for Research in Nanoscience and Nanotechnology (CRNN), University of Calcutta, Kolkata 700098, India
e-mail: sultanajenifar@gmail.com

S. Paul
e-mail: paul.somdatta@gmail.com

A. Karmakar · S. Chattopadhyay (✉)
Department of Electronic Science, University of Calcutta, Kolkata, West Bengal 700009, India
e-mail: scelc@caluniv.ac.in

A. Karmakar
e-mail: akelc@caluniv.ac.in

© Springer Nature Singapore Pte Ltd. 2017
I. Bhattacharya et al. (eds.), *Advances in Optical Science and Engineering*, Springer Proceedings in Physics 194, DOI 10.1007/978-981-10-3908-9_18

considerable attention in recent days in various domains. The reason behind the superiority of CuO for photo-detection and optical switching applications in the visible or near-infrared region is its relatively smaller band gap [2]. Reports are available on the development of CuO based solar cells and photovoltaic's by exploiting its efficient solar absorbance and low thermal emittance properties [3].

The CuO thin film can be obtained by employing several methods such as thermal oxidation, electro-deposition, chemical conversion, chemical brightening, spraying and molecular beam epitaxy technique [2]. The combination of CBD and VLS growth techniques for the deposition of CuO ultrathin films from CuO nano-powder provides a relatively simple and cost-effective technological route for developing device quality films. In the current work, CuO thin films are grown by vapour liquid solid (VLS) process using the indigenously synthesized CuO powder by CBD method. The impact of variation of growth time on the quality of VLS grown films and hence, on the electrical and photovoltaic characteristics of p-CuO/n-Si hetero-junction is investigated.

The p-CuO films are physically characterized by employing Field Emission Scanning Electron Microscopy (FESEM), Energy Dispersive X-ray Spectroscopy (EDS) and X-Ray Diffraction (XRD) study. Optical properties of the grown films are investigated by using Spectroscopic Ellipsometer (SE). Electrical performance of the p-CuO/n-Si hetero-junction diode is investigated by measuring its current-voltage (I-V) characteristics and thereby extracting the values of open circuit photo-generated voltage (V_{oc}), short circuit current density (J_{sc}) and power conversion efficiency (η) from the relevant electrical measurements in both dark and illuminated conditions.

2 Experimental Details

The CuO thin films are grown on Si (100) substrate by employing VLS process. Initially the CuO powder is synthesized by CBD process, where 0.1 M of $Cu(NO_3)_2 \cdot 3H_2O$ and 1 M of NaOH solution are added in 100 ml of DI water to form the reaction mixture. The above solution is heated at 95 °C under constant stirring until CuO precipitates and is finally filtered out and dried up. For the growth of CuO thin film, the Si wafers are subjected to the standard RCA-I and RCA-II cleaning, followed by a dip in 10% HF solution for native oxide removal. Initially the cleaned Si wafers are coated with a thin film ($\sim$7 nm) of gold (Au) by DC sputtering which acts as the metal catalyst required for VLS growth. The growth is performed at 600 °C in a 2-zone furnace. Argon (Ar) is used as the carrier gas at a deposition pressure of 183.9 Torr.

Two different samples are prepared by varying the growth times for 30 min and 40 min. The back contact is taken by evaporating Au and the top CuO film is covered by an ITO layer of thickness 100 nm and then Al dots of radius 10 μm is deposited to take electrical contact by the W-probe of radius 9.5 μm. Elemental as well as compositional analysis of the CuO powder is estimated by EDAX

(JEOL-JSM 7600F) and the surface morphology and thickness of the grown samples are characterized by using field-emission scanning electron microscopy (FESEM) (Zeiss Auriga 39–63) and spectroscopic ellipsometric (SENTECH SE850) techniques. The analysis of structural and crystallographic orientation of the grown samples is conducted by employing X-ray diffraction (XRD) method. The current-voltage characterizations of p-CuO/n-Si hetero-junction are performed by using Keithley 4200-SCS, under both the dark and illuminated condition. The photovoltaic parameters are measured under white light illumination of 0.037 mW/cm^2 input power.

3 Results and Discussion

3.1 *Structural and Material Characterization of p-CuO Thin Film Grown on n-Si*

Figure 1a–c shows the SEM micrographs of CuO powder, synthesized by CBD method, and the as-deposited CuO thin film on n-Si substrate of growth time 30 and 40 min, respectively. The CBD grown CuO powder is observed to attain a spindle like shape as shown in Fig. 1a, however, Fig. 1b, c confirms the formation of a uniform CuO film without any visible individual grains for both the growth times. The average thickness of the film is measured to be ~20 and ~32 nm from SEM images. It is interesting to note that only the thickness of the CuO films increases with increase in reaction time, however, the morphology remains the same. Figure 2a shows EDAX spectra of the CBD grown CuO powder measured by employing JEOL-JSM 7600F FESEM system. It is apparent from the elemental plot that the ratio of copper and oxygen is ~1:1, thereby, indicates the formation of a proper stoichiometric CuO film. The EDAX pattern also exhibits some elemental trace of carbon (C), as carbon tape was used for holding the powder during the measurement.

To investigate the crystallographic and structural orientation of the grown samples, the X-ray diffraction measurements of the grown powder and the thin films

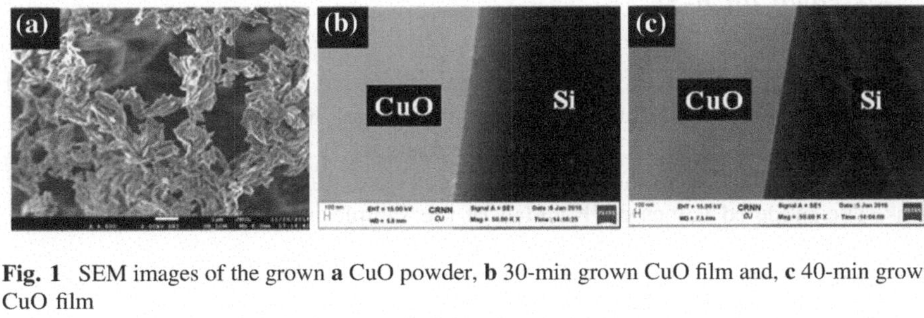

Fig. 1 SEM images of the grown **a** CuO powder, **b** 30-min grown CuO film and, **c** 40-min grown CuO film

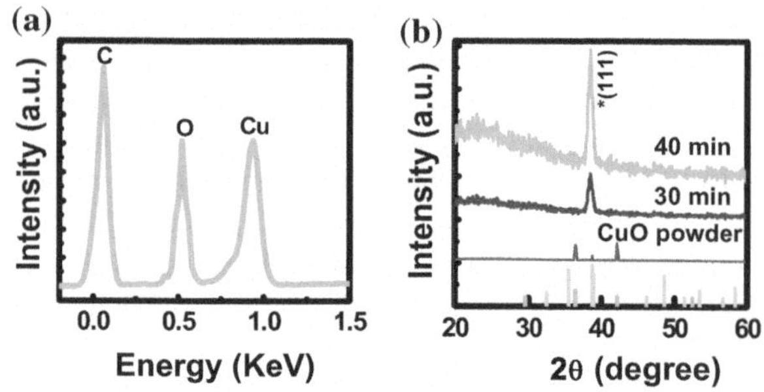

Fig. 2 **a** EDAX spectra of the grown CuO powder and **b** XRD spectra of the grown powder and the films [Ref. JCPDS (80-1917, *green* for CuO) and (05-066, *red* for Cu_2O)] (color figure online)

are performed and plotted in Fig. 2b. XRD pattern of the prepared powder exhibits a polycrystalline nature, however, the XRD pattern of both the CuO films exhibits single crystalline nature. Peak positions of the plot of the grown CuO powder corresponds to both monoclinic cupric oxide (CuO) and orthorhombic cuprous oxide (Cu_2O) phases which have been verified by JCPDS card No. 80-1917 and 05-0667. However, the single crystalline nature of the CuO thin films may be attributed to the fact that at a growth temperature of 600 °C, the orthorhombic Cu_2O phase changes to CuO phase. It is interesting to note from Fig. 2b that for the longer reaction time, diffraction peaks ascribed to CuO phase is enhanced. The average crystallite size and inter-planar (d) spacing of the grown powder are estimated to be 11.92 and 0.23 nm, respectively and those for the thin film are measured to be 23, 0.5 nm and 35.32, 0.38 nm for the films with growth time 30 and 40 min respectively. This increase in average crystallite size and inter-planar spacing in the grown films as compared to the grown powder is attributed to the atomic diffusion and change of the crystallographic phase from Cu_2O to CuO at a growth temperature of 600 °C. Also the increase in intensities and decrease of widths (FWHM) of the X-ray peaks for increasing growth time from 30 to 40 min confirms the improvement of crystalline quality of the VLS grown films. The micro-strain (ε) has been calculated and is obtained to be 0.09 for the powder and 0.16, 0.13 for the films of growth time 30 and 40 min respectively.

3.2 Optical Characterization of the p-CuO Thin Films Grown on n-Si

The spectroscopic ellipsometric (SE) measurement has been performed on the 30 and 40 min grown samples in the wavelength range of 300–800 nm and the plots of relative amplitude change (Ψ) and phase change (Δ) are shown in Fig. 3. Such results are simulated to fit the experimental data to extract the thickness of the CuO films. The modeled values of Ψ and Δ of the sample is also plotted in Fig. 3.

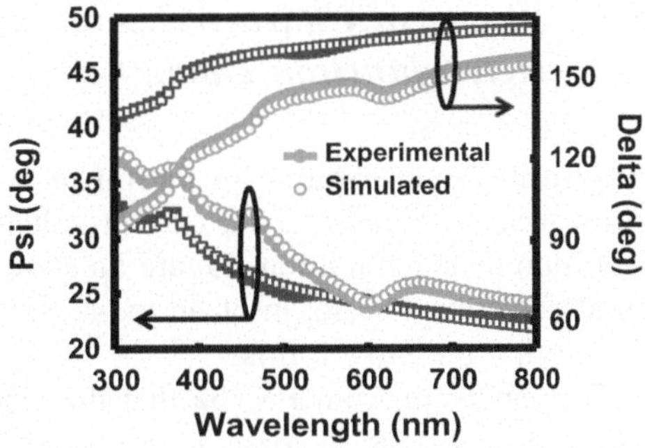

Fig. 3 Measured (*symbols*) and model fit (*lines*) ellipsometer parameters, Ψ and Δ, for the CuO film on Si substrate in the range of 300–800 nm (*blue* for 30 min grown sample and *red* for 40 min grown sample) (color figure online)

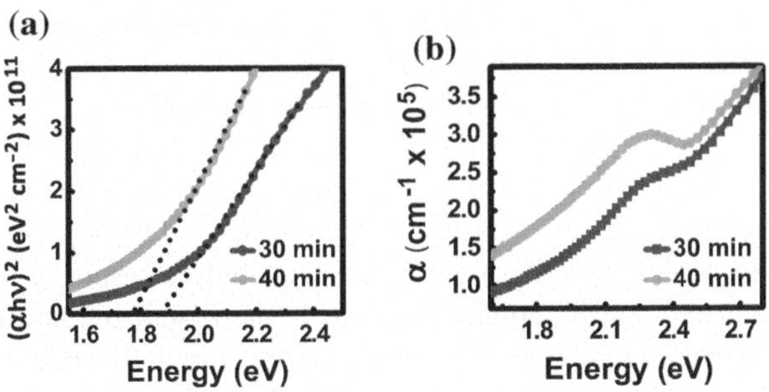

Fig. 4 **a** Plots of the extracted absorption coefficient with incident photon energy for the samples grown at different time duration and; **b** Tauc's plots for the samples to extract band gap

It is apparent from such plots that the best-fit data match well with the measured data, thereby, suggesting the model to be extensive over a wide range of wavelength. The thicknesses of the grown CuO films are obtained to be 20.38 and 31.79 nm for the growth time of 30 and 40 min respectively, which corroborate with the values obtained from tilted SEM images. The absorption coefficients of the films grown for different time durations are extracted from the measured extinction coefficient of spectroscopic ellipsometer (SE) data. It is apparent from the plots of Fig. 4a that the absorption coefficient for the 40 min grown sample is higher than the 30 min grown sample which may be attributed to the largest crystallite size. Thus, the amount of reflection from the surface of 40 min grown sample decreases, and hence, enhances the effective absorption. The band gaps of the grown thin films are also estimated from its absorption spectra by using Tauc's plots, as shown in Fig. 4b. The values of direct band gap of CuO films which are obtained to be 1.87 and 1.77 eV, respectively, for the samples grown for 30 and 40 min of growth time. The refractive indices are determined by using Moss relation, $n^2 = \sqrt{k/E_g}$ where k is a constant with a value of 108 eV [4]. The values of refractive indices are obtained to be 2.75 and 2.8 for the films with growth time 30 and 40 min, respectively. Thus the high refractive index of the 40 min grown film also dictates that such films are suitable for anti-reflecting coating.

3.3 Electrical Characterization of p-CuO/n-Si Heterojunction Diode

Electrical characterization of the p-CuO/n-Si thin film hetero-junction diodes are performed by measuring its current-voltage (I-V) characteristics under both dark and illuminated condition. Figure 5a shows the comparative I-V plots under dark condition (linear scale; inset: in log-scale) of the hetero-junction diodes fabricated by varying the growth time.

It is apparent from Fig. 5a that the dark current under forward bias condition increases with the increase in growth time from 30 to 40 min which may be attributed to the increase in thickness of the film. The respective values of the rectification factor, ideality factor and the barrier height are calculated from the I-V plots and are obtained to be 3.3, 5.54; 3.86, 1.44 and 0.638, 0.15 for 30 and 40 min grown samples respectively. Thus the rectification factor increases whereas ideality factor and barrier height decreases with increase in growth time.

Figure 5b shows the J-V plots under illumination (inset shows the area marked by circle) of the hetero-junction diodes fabricated by varying the growth time. The negative photocurrent at 0 V bias is due to the photo generated carriers which overcome the heterojunction barrier height resulting in the reverse photo-generated current. As the applied voltage is increased, the contribution of bias current dominates and the negative photocurrent gradually decreases and approaches to zero. With further increase in applied voltage the heterojunction is forward biased and the electrical carriers dominate the current transport, reducing the effect of photo-carrier generated current. The 40-min grown sample exhibits superior performance in comparison to the 30-min grown sample in terms of rectification factor, ideality factor and photovoltaic parameters, which is attributed to the superior crystalline quality and enhanced absorption. The power conversion efficiency of the 30 and 40-min grown samples are obtained to be 0.29 and 4.04% respectively, at a power input of 0.037 mW/cm^2.

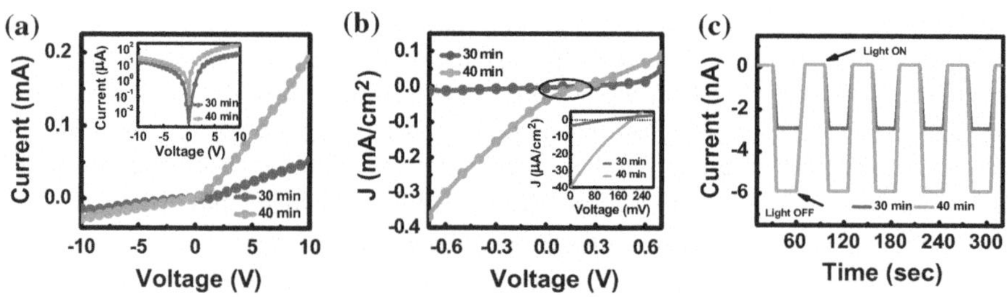

Fig. 5 **a** Comparative I-V plots under dark condition (linear scale; *inset*: in log-scale), **b** J-V plots under illumination (*inset* shows the area marked by *circle*) of the hetero-junction diodes fabricated for varying growth time and; **c** plot of light 'ON'/'OFF' switching characteristics of the hetero-junction photodiodes at zero-bias (0 V)

The time-dependent photocurrent response of the sample is measured periodically by switching ‘ON’ and ‘OFF’ the light source for a cycle of 60 s each and the relevant rise- and fall-time, are shown in Fig. 5c. From Fig. 5c the rise and fall time are obtained to be 3.02, 2.94 s and 3.8, 3.2 s respectively for 30 and 40 min grown samples. The light ‘ON’/‘OFF’ characteristics indicate that such CuO thin film/Si hetero-junction can also be used for photo-detection. This comparatively slow response of the 30 min grown samples can be attributed to the fact that the built-in potential prevents the transport of carriers causing delay in response time.

4 Conclusions

CuO thin films have been grown by VLS technique using the CuO powder synthesized by employing CBD technique. The thickness, morphology, crystalline quality, optical and photovoltaic properties of the grown CuO thin films have been systematically studied with reaction time. It is observed that the reaction time has a significant influence on the thickness, crystallite size, optical and electrical properties of the films. The crystalline nature and the optical properties of the films have been verified by employing XRD and SE analysis respectively. It is observed that a good stoichiometric CuO film with a thickness of 31.79 nm has been possible to grow by 40 min of growth time using VLS technique, which has been subsequently used for the fabrication of p-CuO/n-Si heterojunction diodes and is observed to deliver the superior optical and electrical properties in comparison to the 30 min grown sample. Thus, the VLS grown 31.79 nm film of p-CuO on n-Si substrate provides a promising low-cost alternative technological route for the development of superior performance photovoltaic devices.

Acknowledgements Ms. S. Paul like to acknowledge University Grants Commission (UGC), India, for providing National Fellowship to pursue her research work. The authors also like to acknowledge Center for Research in Nanoscience & Nanotechnology (C.U.) and Department of Electronic Science (C.U) for providing necessary infrastructure to conduct this work.

References

1. N. Serin, T. Serin, S. Horzum and C.Yasemin, “Annealing effects on the properties of copper oxide thin films prepared by chemical deposition”, Semicond. Sci. Technol. 20, 398–401 (2005).
2. J. Sultana, A. Das, A. Das, N. R. Saha, A. Karmakar, and S. Chattopadhyay. “Characterization of nano-powder grown ultra-thin film p-CuO/n-Si hetero-junctions by employing vapour-liquid-solid method for photovoltaic applications”, Thin Solid Films, 612, 331–336 (2016).

3. Paul, S., Das, A., Sultana, J., Karmakar, A., Chattopadhyay, S. and Bhattacharyya, A., "Perfomance investigation of n-ZnO nanowire/p-CuO thin film heterojunction solar cell grown by chemical bath deposition and vapour liquid solid technique," Proc. IEEE 978-1-4673-9513-7/15 (2015).
4. Yunus. A., Yıldırım. M. A., Ateş. A. and Yıldırım. M., "The relationship between refractive index-energy gap and the film thickness effect on the characteristic parameters of CdSe thin films," Optics Communications 284, 2307–2311 (2011).

Electrical Characterization of n-ZnO NW/p-CuO Thin Film Hetero-Junction Solar Cell Grown by Chemical Bath Deposition and Vapor Liquid Solid Technique with Varying Reaction Time

Somdatta Paul, Jenifar Sultana, Aritra Banerjee, Pintu Singha, Anupam Karmakar and Sanatan Chattopadhyay

Abstract The current work reports the fabrication of p-CuO/n-ZnO heterojunction grown on n-Silicon (100) substrate by varying the growth time of ZnO nanowires. The p-CuO thin film has been grown by vapor-liquid-solid (VLS) technique and the n-ZnO nanowires by employing chemical bath deposition (CBD) method. The surface morphology, crystalline structure and optoelectronic properties of the heterojunction are investigated by employing SEM, XRD and spectroscopic ellipsometry respectively. The ZnO NWs grown for 40 min on the CuO film, indicates best performance in terms of crystal quality, band gap, refractive index and solar performance parameters like short circuit current density, open circuit voltage and efficiency.

S. Paul · J. Sultana
Centre for Research in Nanoscience and Nanotechnology (CRNN), University of Calcutta, Kolkata, West Bengal 700098, India
e-mail: paul.somdatta@gmail.com

J. Sultana
e-mail: sultanajenifar@gmail.com

A. Banerjee · P. Singha
Department of Physics, University of Calcutta, Kolkata, West Bengal 700009, India
e-mail: arbphy@caluniv.ac.in

P. Singha
e-mail: pintusingha7@gmail.com

A. Karmakar · S. Chattopadhyay (✉)
Department of Electronic Science, University of Calcutta, Kolkata
West Bengal 700009, India
e-mail: c_sanatan@yahoo.com

A. Karmakar
e-mail: akarmakar.cu@gmail.com

© Springer Nature Singapore Pte Ltd. 2017
I. Bhattacharya et al. (eds.), *Advances in Optical Science and Engineering*,
Springer Proceedings in Physics 194, DOI 10.1007/978-981-10-3908-9_19

1 Introduction

In recent years, photovoltaics has been a technological symbol to protract energy supply system. Solar panels consisting of a number of solar cells are used to obtain power generation. Recently, there has been a considerable development in the manufacturing of solar cells and photovoltaic arrays due to the growing demand of renewable energy sources.

Copper oxide (CuO) has emerged as a potential material due to its natural p-type conductivity, having low band gap in the range of 1.5–2.1 eV, which makes it possible to absorb light throughout the visible spectrum [1]. On the other hand, Zinc oxide (ZnO) is one of the most interesting material due to its wide direct band gap of 3.37 eV, excellent chemical and thermal stability and large excitonic binding energy of 60 meV [2]. Among the different variety of ZnO nanostructures, ZnO nanowires has gain much interest due to its application in various electronic devices. Various chemical and physical deposition processes are used for the growth of ZnO nanowires [3]. Out of these, chemical bath deposition method is most preferred due to its low cost, low temperature and environment friendly growth mechanism [4].

The surface morphology of the grown CuO film and ZnO nanowires are characterized by using field emission scanning electron microscopy (FESEM) (Zeiss Auriga 39–63). The crystallographic orientation of the grown samples is observed by employing X-Ray diffractometer (PANanalytical X'Pert Pro X-ray Diffractometer). Optical properties of the grown heterojunction are investigated by the spectroscopic ellipsometer (SENTECH SE850). The electrical characterizations are carried out using Keithley 4200 SCS parameter analyzer. The current work focuses on the fabrication of p-CuO/n-ZnO NW heterojunctions by varying the growth time of the ZnO NWs for 30, 40 and 60 min and then optimization of the deposition time for the best performance of the solar cell in terms of its efficiency.

2 Experimental Procedure

Vapor-liquid-solid (VLS) technique has been employed for the growth of CuO thin film on an n-Si (100) substrate. Prior to the growth procedure, the wafer has been cleaned by following the standard RCA-I and RCA-II cleaning and then the wafer is dipped in 10% HF solution for the oxide removal. After cleaning, sputtering technique has been employed to gold coat the wafer with a thickness of $\sim$7 nm. The CuO thin film is grown from CuO powder, in a 2-zone furnace where the powder is kept in the higher temperature zone and the Au-coated sample in the relatively lower temperature zone of 600 °C for 30 min. Argon gas has been used as a carrier at a deposition pressure of 183.9 Torr. Following the same procedure three CuO thin film samples has been prepared.

The ZnO NWs have been grown by employing chemical bath deposition (CBD) technique on the CuO film coated samples for 30, 40 and 60 min. For the growth of ZnO nanowires an equi-molar aqueous solution of 0.05 M of Zinc Nitrate Hexahydrate ($Zn(NO_3) \cdot 6H_2O$) and Hexamethlenetetramine ($C_6H_{12}N_4$) are prepared by using 50 ml DI water and then mixed together in a beaker which produces transparent solution. The CuO thin film coated samples are dipped vertically in the solution and stirred at 250 rpm. With increasing time and temperature, the transparent solution turns slightly whitish due to the precipitation and deposition of ZnO on CuO film coated n-Si substrates. After the completion of respective deposition time, each of the samples are taken out from the solution and rinsed under running DI water, followed by drying up in N_2 ambient. The contacts are taken by coating the CuO film and the ZnO NWs with ITO layer having a thickness of 100 nm followed by Ag dots of diameter 50 μm, for creating the electrical contact by the tungsten probe of diameter 9.5 μm.

3 Results and Discussion

Figure 1 shows the FESEM micrographs of ZnO nanowires grown on CuO thin film for 30, 40 and 60 min respectively. The average thickness of the CuO film is ~20 nm as shown in Fig. 1a. It is evident from Fig. 1b that after completion of 30 min of deposition time the ZnO nanowires have grown however do not cover up the entire CuO film. With increasing reaction time (from 30 to 40 min), the density of the ZnO nanowires increases (Fig. 1c). However, further increasing the reaction time (from 40 to 60 min), the nanowires coalesce and finally form clusters (Fig. 1d).

The XRD pattern of the ZnO NW's/CuO film heterojunction for different deposition time are plotted in Fig. 2. Peak position of the plots confirm the growth of single crystalline CuO with [111] plane and polycrystalline ZnO with [100], [002] and [101] planes, which are verified from the JCPDS card no. 80-1917 and 36-1451 respectively. The average crystallite size (D) and micro-strain (ε), estimated by using Scherrer formula, $D = \frac{0.9\lambda}{\beta \cos \theta}$ and $\varepsilon = \frac{\beta \cos \theta}{4}$ where λ is the wavelength used for X-ray radiation and β is the full-width-at-half-maximum (FWHM) of the peaks at the diffracting angle θ.

The single crystalline CuO with [111] plane is having a crystallite size of 15.47 nm. However for the ZnO NWs [100] plane is dominating, having the crystallite size of 28.02, 32.97 and 29.6 nm and micro-strain of 0.0031, 0.0026 and 0.0029 for 30, 40 and 60 min respectively. So, it is apparent from the above values that the heterojunction formed for 40 min deposition time is having the highest

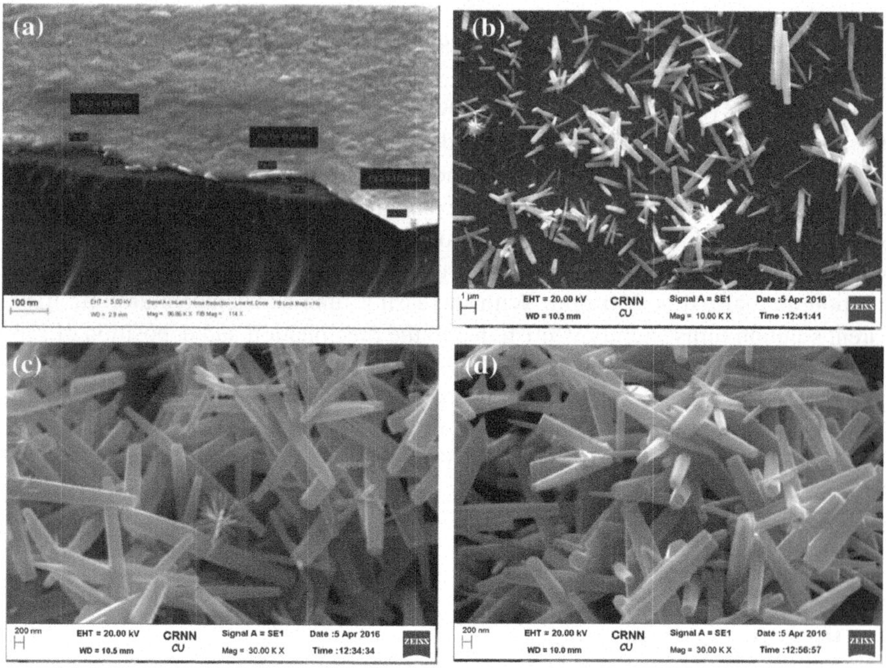

Fig. 1 SEM images of **a** CuO film and ZnO nanowires grown on CuO film for **b** 30 min, **c** 40 min, **d** 60 min

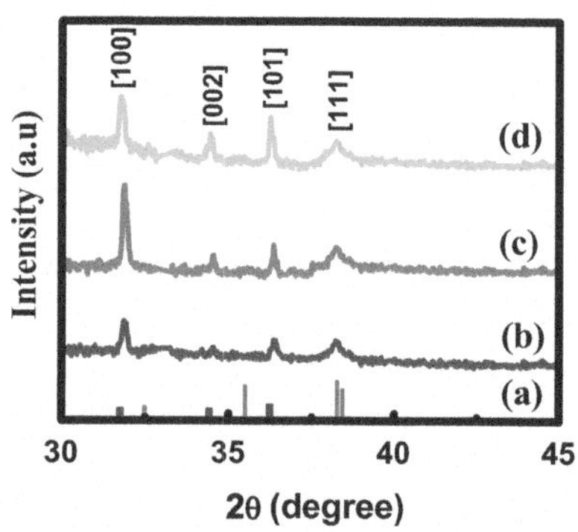

Fig. 2 XRD spectra of: **a** ZnO from JCPDS (36-1451, *blue*) and CuO from JCPDS (80-1917, *red*), ZnO NW's/CuO film **b** 30 min, **c** 40 min, **d** 60 min (colour figure online)

crystallite size, which exhibits superior light absorption in comparison to the other heterojunctions fabricated in the current work.

The bandgap is extracted using Tauc's relation $\alpha hv = A(hv - Eg)^n$ in Fig. 3a and it is obtained to be 3.28, 3.33 and 3.41 for 30, 40 and 60 min deposition time respectively.

Figure 3b shows the variation of refractive indices and it has been plotted with deposition time. The refractive indices are calculated by using Moss relation and

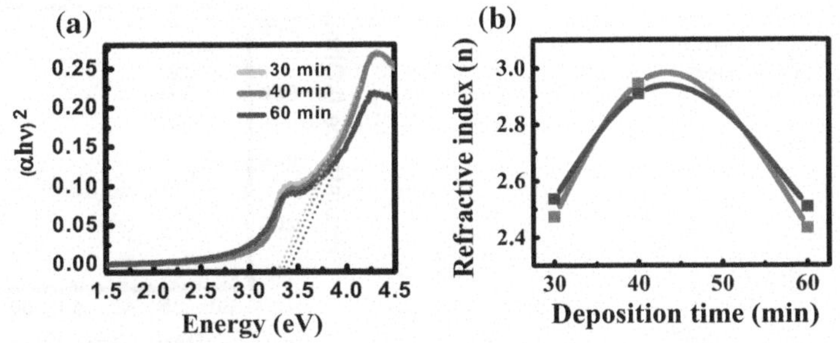

Fig. 3 **a** Tauc's plot and **b** variation of refractive indices by Moss relation (*red*) and Herve & Vandamee relation (*blue*) of the heterojunction with deposition time (colour figure online)

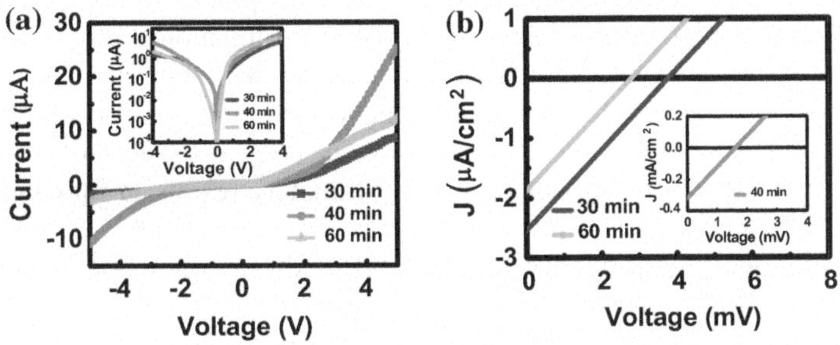

Fig. 4 **a** Current-voltage (I-V) characteristics of **a** p-CuO/n-ZnO heterojunctions in *dark*. *Inset* shows the plot in log scale. **b** Current density-voltage (J-V) characteristics of p-CuO/n-ZnO under UV illumination

Herve and Vandamee relation [5]: $n^2 = \sqrt{k/E_g}$ and $n = \sqrt{1 + \left(\frac{A}{E_{g+B}}\right)}$ respectively. The values of the refractive indices are obtained to be 2.54, 2.92 and 2.48 for 30, 40 and 60 min respectively. Such high value of refractive index for the 40 min grown heterojunction can be used for anti reflection coating.

Figure 4a and b shows the current—voltage and current density—voltage plots of n-ZnO/p-CuO hetero-junction on Si substrate at room temperature under dark and illuminated condition, respectively. The heterojunction exhibits a diode like characteristics in dark condition and a photocurrent is observed under white light illumination of 0.037 mW/cm^2. It is apparent from Fig. 4b, that with increasing the deposition time from 30 to 40 min, the photo-generated current increases which may be attributed to the increase in density of the ZnO nanowires. However, with further increase in deposition time from 40 to 60 min, the photo generated current decreases. This decrease in the photo generated current is a replication of the enhancement of the series resistance, due to the cluster formation of ZnO nanowires for such high deposition time as apparent from Fig. 1c.

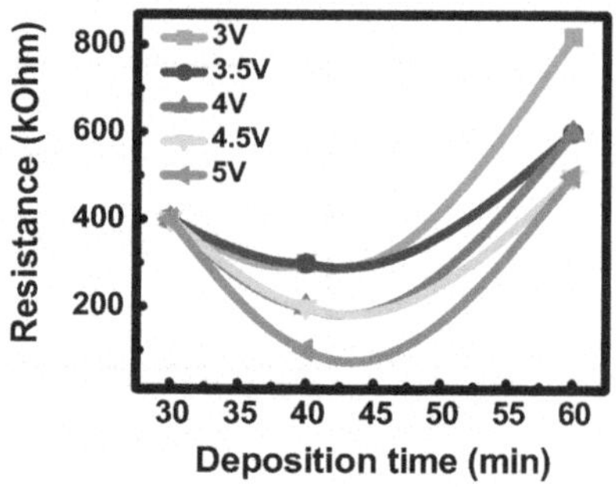

Fig. 5 Variation of resistance with deposition time at different voltage

To determine the series resistance values, the inverse slope of I-V characteristics ($R_i = dV_i/dI_i$) is considered, under illuminated condition, and its variation with deposition time at different applied forward voltage is plotted in Fig. 5. The highest series resistance for the 60 min deposition time heterojunction is responsible for the lowest photo-generated current which is consistent with Fig. 4b. The maximum photocurrent is observed for the heterojunction having the deposition time of 40 min, which can be attributed to the lowest series resistance of 100 kΩ and highest carrier collection probability which ultimately results in maximum power conversion efficiency.

The photovoltaic performance of the fabricated heterojunction is analysed in terms of open circuit voltage, short circuit current density and efficiency. The fill factor is calculated as: $FF = \frac{J_{m*}V_m}{J_{sc*}V_{oc}}$ where J_m and V_m are the current and voltage respectively at maximum power of the J–V curve. The fill factors are obtained to be 0.16, 0.24 and 0.23 for 30, 40 and 60 min respectively. The conversion efficiency is calculated from: $\eta(\%) = \frac{FF*J_{sc*}V_{oc}}{P_{in}}$ which is obtained to be 0.003, 0.37 and 0.002% for 30, 40 and 60 min deposition time, at an input power incident of 0.037 mW/cm^2.

4 Conclusions

The n-ZnO NWs/p-CuO film has been grown by employing the combined CBD and VLS techniques. The optical and photovoltaic properties of the grown samples are investigated. The open circuit voltage, short circuit current density and efficiency are obtained to be 0.0018 V, 0.32 mA/cm^2 and 0.37% respectively for the heterojunction having 40 min deposition time of ZnO NWs. Hence the preliminary study indicates that the n-ZnO NWs grown for 40 min on the previously grown p-CuO film is much superior as compared to the other samples and so it can be chosen as a promising candidate for future photovoltaic applications.

Acknowledgements Ms. S. Paul likes to thank the University Grants Commission (UGC), India, for providing National Fellowship to pursue her research work. The authors also like to acknowledge Center for Research in Nanoscience and Nanotechnology (C.U.), COE, TEQIP Phase-II and Department of Electronic Science (C.U) for providing necessary infrastructure to conduct this work.

References

1. Huang, J., Morshed, M.M., Zuo, Z., and Liu, J., "Distributed Bragg reflector assisted low-threshold ZnO nanowire random laser diode," Appl. Phys. Lett., 104, 131107 1–3 (2014).
2. Joshi, S., Mudigere, M., Krishnamurthy, L. and Shekar. G. L., "Growth and morphological studies of NiO/CuO/ZnO based nanostructured thin films for photovoltaic applications," Chemical Papers, 68, 1584–1592 (2014).
3. Paul. S., Das. A., Sultana. J., Karmakar. A., Bhattacharyya. A., and Chattopadhyay, S., "Growth performance investigation of n-ZnO nanowire/p-CuO thin film heterojunction solar cell grown by chemical bath deposition and vapor liquid solid technique," Proc. IEEE 978-1-4673-9513-7/15 (2015).
4. Das. A., Palit. M., Paul. S., Chowdhury. B.N., Dutta. H.S., Karmakar. A. and Chattopadhyay. S., "Investigation of the electrical switching and rectification characteristics of a single standalone n-type ZnO-nanowire/p-Si junction diode," Appl. Phys. Lett., 105, 1–4 (2014).
5. Chand. P., Gaur. A., Kumar. A. and Gaur, U. K., "Effect of NaOH molar concentration on morphology, optical and ferroelectric properties of hydrothermally grown CuO nanoplates," Materials Science in Semiconductor Processing, 38, 72–80 (2015).

Analytical Modeling of Vertically Oriented Standalone Si-Nanowire Metal-Oxide-Semiconductor Capacitors for Wavelength Selective Near-Infrared Sensing Applications

Subhrajit Sikdar, Basudev Nag Chowdhury and Sanatan Chattopadhyay

Abstract In the current work, photogeneration in a vertically oriented standalone Si-nanowire metal-oxide-semiconductor (MOS) capacitor has been investigated by developing a theoretical model that incorporates phase-breaking phenomena such as electron-photon interaction. For this purpose, a set of quantum field equations associated with second quantization electron and photon field operators have been solved through non-equilibrium Green's function (NEGF) formalism by achieving self-consistency with Poisson's equation. The obtained photocapacitance profile shows sharp peaks at specific wavelengths, detected by their relevant diameter-voltage combinations in the near-infrared region. Such peaks are observed to be shifted towards shorter wavelengths with increasing the applied voltage. Thus, the proposed device scheme can contribute significantly to wavelength selective photosensing applications with high selectivity.

1 Introduction

In recent times, immense applications of semiconductor nanostructure based devices in the field of optical sensing/imaging have led to significant progress in fabricating such devices for detecting ultraviolet to infrared spectrum with high sensitivity [1, 2]. Moreover, the detection of near-infrared region (720–2500 nm) is

---

S. Sikdar · S. Chattopadhyay (✉)
Department of Electronic Science, University of Calcutta, Kolkata, India
e-mail: scelc@caluniv.ac.in

S. Sikdar
e-mail: subhrajit.sikdar@yahoo.com

B.N. Chowdhury
Centre for Research in Nanoscience and Nanotechnology (CRNN),
University of Calcutta, Kolkata, India
e-mail: basudevn@gmail.com

© Springer Nature Singapore Pte Ltd. 2017
I. Bhattacharya et al. (eds.), *Advances in Optical Science and Engineering*,
Springer Proceedings in Physics 194, DOI 10.1007/978-981-10-3908-9_20

essential for applications in the domain of astronomy [3], agriculture [4], remote sensing [5], material science [6] and medical imaging [7]. Theoretical investigation/modeling of such photosensors is essential for developing a comprehensive understanding of the inherent optical and electrical transport properties [8]. Incidentally, Si based nano-devices are always in the central attraction of research since ~97% of the semiconductor industry is dominated by it and MOS capacitor/metal-semiconductor-metal (MSM) architectures are explored as the potential candidates for optoelectronic applications [9, 10]. The commercially available near-infrared sensing instruments offer the full-width-at-half-maximum (FWHM) ~10 to 15 nm whereas the spectral resolution is reported to be ~4 nm [11, 12] thereby indicates a significant overlap between each spectral bands. Such unavoidable overlap results a loss of spectral information due to aliasing. Further, in such instruments [13], wavelengths in the ultraviolet region are detected by utilizing photo-multiplier tube (PMT) whereas Indium-Gallium-Arsenide (InGaAs) is used for infrared detection which results a noisy behavior while detecting wavelengths in the near-infrared region. Therefore, in the current work, a theoretical model has been developed to study the photogeneration phenomena in vertically oriented standalone Si-nanowire MOS capacitor for near-infrared sensing applications. A set of coupled quantum field equations associated with second quantization electron and photon field operators has been solved in non-equilibrium Green's function (NEGF) framework by achieving self-consistency through a simultaneous quantum-electrostatic solver. Finally, a design window of nanowire diameter and applied voltage is represented to selectively detect the wavelengths in the near infrared region.

2 Mathematical Modeling

In the theoretical model, generation of electron-hole pair is considered as scattering of electrons due to interaction with the incident photons where the incoming electron with the effective mass of hole deviated from its trajectory with effective mass of an electron after interacting with the photon, as shown in Fig. 1. This process is represented as the annihilation of incoming electron (with effective mass of hole) and photon, and subsequently creation of the outgoing electron (with effective mass of electron). Here, valence band is considered as the 'source' or 'reservoir' of electrons whereas the conduction band supposed to be the 'device'.

The transition of a fermion (electron) from the valence band to the conduction band after interacting with a boson (photon) is described by following equations [14]:

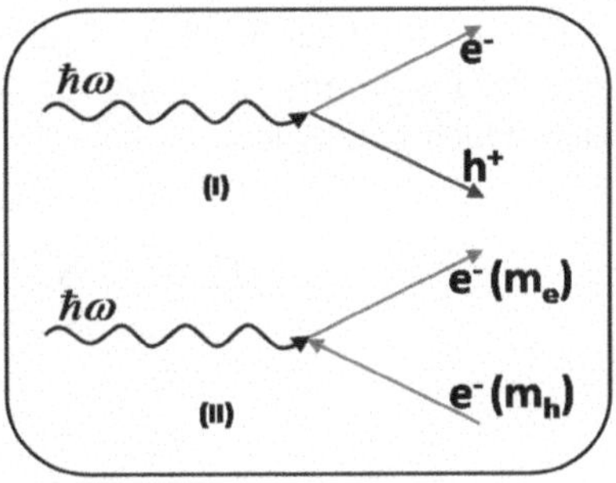

Fig. 1 Generation of electron-hole pair after interacting with a photon (**I**). Equivalent picture of electron-photon interaction considered in theoretical model (**II**)

$$i\hbar\frac{d}{dt}C_i^c = H_{ISO}^c C_i^c + \sum_{r,\alpha}(\tau_{ir}^{\alpha-pht}C_r^v b_\alpha^{pht} + \tau_{ir}^{\alpha-pht*}C_r^v b_\alpha^{pht+}) \tag{1a}$$

$$i\hbar\frac{d}{dt}C_s^v = H_{ISO}^v C_s^v + \sum_{j,\alpha}(\tau_{sj}^{\alpha-pht}C_j^c b_\alpha^{pht} + \tau_{sj}^{\alpha-pht*}C_j^c b_\alpha^{pht+}) \tag{1b}$$

$$i\hbar\frac{d}{dt}b_\alpha^{pht} = \hbar\omega_\alpha^{pht} b_\alpha^{pht} + \sum_{j,r}(\tau_{jr}^{\alpha-pht*}C_j^c C_r^{v+} + \tau_{jr}^{\alpha-pht*}C_j^{c+}C_r^v) \tag{1c}$$

where C_i^c and C_s^v are the i-th and s-th sub-band electron annihilation operator in the conduction and valence band respectively and b_α^{pht} is the annihilation operator for the photon in α mode with angular frequency of ω_α. All the b's and C's follow the Bose-Einstein (BE) commutation and Fermi-Dirac (FD) anti-commutation relations. H_{ISO}^C and H_{ISO}^v are considered as the Hamiltonian for isolated state in conduction and valence band respectively. Solving (1a)–(1c), using relevant Green's function (G^D) to calculate the two-time correlation function $n_{ik}^c(t,t') = \langle C_k^{c+}(t')C_i^c(t)\rangle$ for the filled state of electrons in the conduction band which after Fourier transform in energy domain can be expressed in matrix form as,

$$n_{ik}^c(E) = [G^D(E)]\left[\sum_{sc}^{In-pht}(E)\right][G^{D+}(E)] \tag{2}$$

where, n_{ik}^c represents photogenerated electron density and $\sum_{sc}^{In-pht}$ is the photon in-scattering function.

The relevant Green's function (G^D) of a quantum device is given by,

$$G^D(E) = \delta(E - E_{sub}) = (1/2\pi)\int_{-\infty}^{\infty} e^{i/\hbar(E-E_{sub})t}\,dt. \tag{3}$$

Finally, the total number of photogenerated carriers per second (n_{1D}^{ph}) is calculated by integrating (2) using (3) over the entire energy range in the conduction band which results to,

$$n_{1D}^{ph} = \sum_{allsub-bands} [(1/2\pi) \sum_{sc}^{In-pht} (E_{sub})]/\hbar]. \tag{4}$$

3 Results and Discussions

Schematic of the vertically oriented standalone Si-nanowire MOS capacitor is shown in Fig. 2 where the Si nanowire is assumed to be grown on [100] direction on a Si substrate. The silicon-dioxide (SiO_2) layer of 2 nm thickness is deposited on top of the nanowire, followed by a metal electrode. Such a system is assumed to be operated at room temperature (300 K).

When a positive voltage (V) is applied on the metal electrode, the conduction band bends more in the downward direction as shown in Fig. 3, forming a quantum well at the oxide/semiconductor interface. Electrons accumulated in such a quantum well are confined in all possible directions due to structural and electrical quantization. Consequently, troughs and crests are observed in the one dimensional (1-D) carrier concentration profile in dark condition which signifies the existence of discrete energy states due to such confinement in the conduction band regime. Further, illuminating the MOS capacitor with light of appropriate wavelength generates electron-hole pairs throughout the length of the nanowire however; the applied voltage attracts photoelectrons near the oxide/semiconductor interface while pushing back holes towards the substrate. Thus, the applied electric field reduces

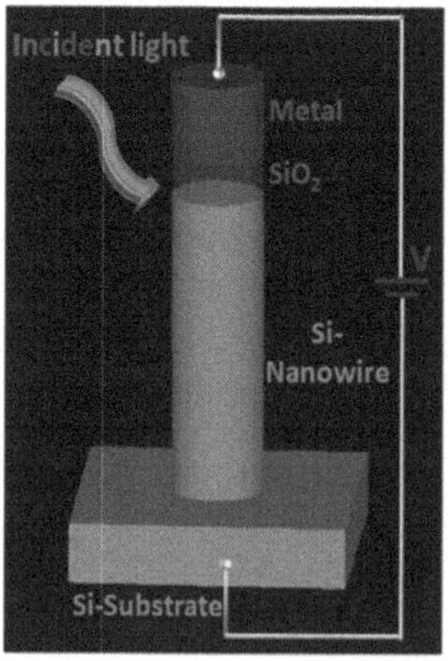

Fig. 2 Vertically aligned standalone MOS capacitor

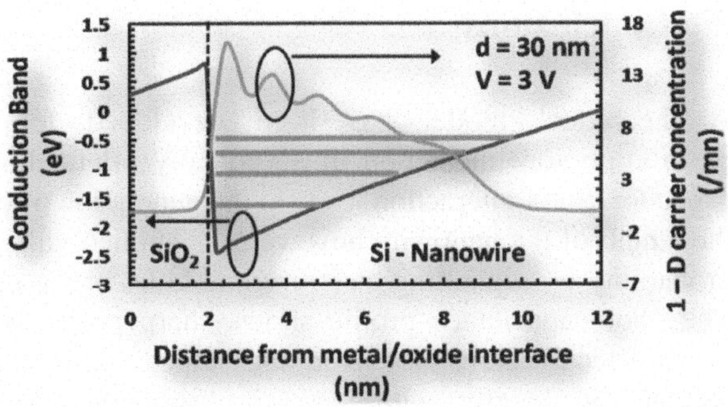

Fig. 3 Conduction band potential and 1-D carrier concentration with distance from metal/oxide interface

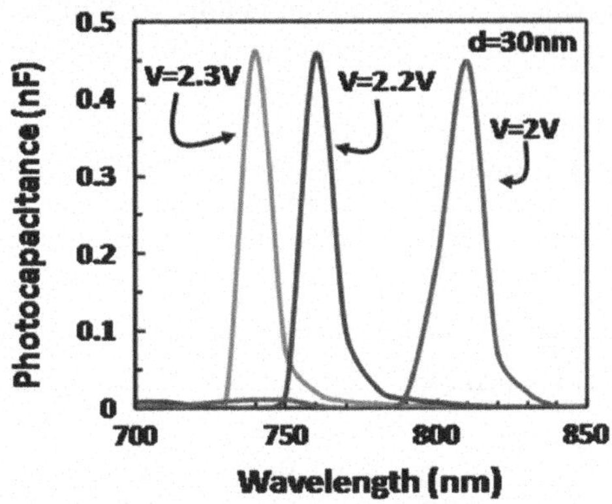

Fig. 4 Plot of photocapacitance profile with incident light wavelength for a nanowire comprising 30 nm diameter with varying voltage

the probability of recombination at the oxide/semiconductor interface by separating photogenerated electrons from holes. Such photogenerated charge carriers in addition to the charge carriers in dark condition are integrated over the extension of 1-D carrier concentration profile and dividing it by the relevant voltage provides the photocapacitance. The distribution of such photocapacitance under the illumination of 30 mW source power for different diameter-voltage combinations has been shown in Fig. 4.

The photocapacitance profile shows a sharp peak at 810 nm wavelength for a combination of 30 nm diameter and 2 V applied voltage. When the applied voltage is increased to 2.2 V the photocapacitance peak is observed to be shifted towards the shorter wavelength region (blue shift) and results a sharp peak at 760 nm. Similarly, on application of 2.3 V such a profile is shifted more towards blue and results a peak at 740 nm. Such a shifting of photocapacitance peaks occurs with increasing the voltage since the energy gap between two interacting sub-bands is increased. Corresponding FWHMs for these three spectral bands, calculated using Gaussian distribution function obtained as 14.35, 10.56 and 9.90 nm respectively. This situation signifies the ability of the proposed device to detect wavelengths in the near-infrared region with high selectivity by their relevant diameter-voltage combinations.

4 Conclusions

A theoretical model has been developed to investigate the impact of electron-photon interaction in a vertically oriented standalone Si-nanowire MOS capacitor. Such interaction leads to the generation of electron-hole pairs throughout the length of the nanowire however, the applied voltage reduces the recombination probability by separating electrons and holes at the oxide/semiconductor interface. The photocapacitance profile shows sharp peaks at specific wavelengths in the near-infrared region, moreover, such a peak is observed to be shifted towards the shorter wavelengths with increasing applied voltage. Most importantly, wavelengths in the near-infrared region can be detected by their relevant diameter-voltage combinations with high selectivity. Thus, photo-detection in the proposed device scheme is independent of any variations in the illumination condition.

Acknowledgements The authors would like to acknowledge the Centre of Excellence (COE) for the Systems Biology and Biomedical Engineering, University of Calcutta, funded by the World Bank through TEQIP Phase II for providing the necessary infrastructural support.

References

1. Dai, X., Zhang, S., Wang, Z., Adamo, G., Liu, H., Huang, Y., Couteau, C., Soci, C., "GaAs/AlGaAs nanowire photodetector," Nano letters, 14(5), 2688–2693, (2014).
2. Fu, X.-W., Liao, Z.-M., Zhou, Y.-B., Wu, H.-C., Bie, Y.-Q., Xu, J., Yu, D.-P., " Graphene/ZnO nanowire/graphene vertical structure based fast-response ultraviolet photodetector," Applied Physics Letters, 100(22), 223114, (2012).
3. Tokunaga, A., Simons, D., Vacca, W., "The Mauna Kea observatories near-infrared filter set. II. Specifications for a new JHKL'M'filter set for infrared astronomy," Publications of the Astronomical Society of the Pacific, 114(792), 180, (2002).
4. Liu, D., Sun, D-W., Zeng, X-A., "Recent advances in wavelength selection techniques for hyperspectral image processing in the food industry," Food and Bioprocess Technology, 7(2), 307–323, (2014).
5. Gao, B-C., "NDWI—A normalized difference water index for remote sensing of vegetation liquid water from space," Remote sensing of environment, 58(3), 257–266, (1996).
6. Tessler, N., Medvedev, V., Kazes, M., Kan, S., Banin, U., "Efficient near-infrared polymer nanocrystal light-emitting diodes," Science, 295(5559), 1506–1508, (2002).
7. Weissleder, R., Tung, C H., Mahmood, U., Bogdanov, A., "In vivo imaging of tumors with protease-activated near-infrared fluorescent probes," Nature biotechnology, 17(4), 375–378, (1999).
8. Aeberhard, U., "Theory and simulation of photogeneration and transport in Si-SiOx superlattice absorbers," Nanoscale research letters, 6(1), 1–10, (2011).
9. Tsai, D S., Lin, C A., Lien, W C., Chang, H C., Wang, Y L., He, J H., "Ultra-high-responsivity broadband detection of Si metal–semiconductor–metal schottky photodetectors improved by ZnO nanorod arrays," ACS nano, 5(10), 7748–7753, (2011).
10. Bae, J., Kim, H., Zhang, X.-M., Dang, C. H., Zhang, Y., Choi, Y. J., Nurmikko, A., Wang, Z. L., "Si nanowire metal–insulator–semiconductor photodetectors as efficient light harvesters," Nanotechnology, 21(9), 095502, (2010).

11. Lambrechts, A., Gonzalez, P., Geelen, B., Soussan, P., Tack, K., Jayapala, M., "A CMOS-compatible, integrated approach to hyper-and multispectral imaging," Electron Devices Meeting (IEDM), 1–10, (2014).
12. Geelen, B., Tack, N., Lambrechts, A., "A compact snapshot multispectral imager with a monolithically integrated per-pixel filter mosaic," Spie Moems-Mems, International Society for Optics and Photonics, 89740L-89740L-8, (2014).
13. www.perkinelmer.com.
14. Datta, S., [Quantum transport: atom to transistor], Cambridge University Press, (2005).

Computing Spectral Response of GaAs Solar Cell with Quasi-Fermi Level Approximation

Somshuddha Datta, Tamalika Chakraborty and Arpan Deyasi

Abstract Spectral response of GaAs solar cell is analytically computed incorporating the existence of quasi-Fermi level generated after photon shower. Analytical computation is carried out for simultaneous solution of continuity equations and diffusion equations, and contribution of drift is also taken into account. Results are obtained by varying structural parameters within lower photon energy range to obtain higher response magnitude.

1 Introduction

Renewable energy becomes the subject of research in this age due to present energy crisis, and solar cell already exhibits [1, 2] as one of the promising alternative of conventional energy sources with lower cost and moderate efficiency. Estimation of conversion efficiency (CE) and spectral response (SR) are the two prime factors to measure its performance, and hence accurate computation with realistic physical consideration plays important role for its performance analysis. Chegaar [3] calculated the effect of irradiance intensity on short circuit current, open circuit voltage, saturation current, ideality factor, figure of merit, fill factor, series resistance and shunt resistance. Ghitas [4] presented the solar irradiance variation and radiation intensity at different time of the day as a function of wavelength.

Chander [5] experimentally studied the spectral response of a mono-crystalline Si solar cell and also the quantum efficiency to verify the bandgap of the material. Spectral response of organic solar cell is also measured in recent past [6] and charge

S. Datta (✉) · T. Chakraborty · A. Deyasi
Department of Electronics and Communication Engineering,
RCC Institute of Information Technology, Kolkata, West Bengal 700015, India
e-mail: somshuddha95@gmail.com

T. Chakraborty
e-mail: itamalikachakraborty95@gmail.com

A. Deyasi
e-mail: deyasi_arpan@yahoo.co.in

© Springer Nature Singapore Pte Ltd. 2017
I. Bhattacharya et al. (eds.), *Advances in Optical Science and Engineering*,
Springer Proceedings in Physics 194, DOI 10.1007/978-981-10-3908-9_21

transport efficiencies are calculated form experimental result. For heterojunction solar cell, spectral response is calculated [7] considering the excess minority carrier concentration at the base region and also at the top of solar cell. Effect of cell temperature and maximum power point tracking on efficiency are highlighted [8] for future design with enhanced performance.

In the present paper, detailed analytical calculation is made for SR computation of p-n junction solar cell with the consideration of excess carrier generation, i.e., generation of quasi-Fermi level, which is absent in previous literatures [9, 10]. As per the physical insight, quasi-Fermi level formation due to high shower of excess photons is a realistic phenomenon, and thus modelling of SR incorporating this effect adds novelty in this regard.

2 Mathematical Modeling

Spectral response of p-n junction solar cell can be derived considering the solution of electron and hole concentration equations in depletion region at steady state incorporating the effect of recombination rate. Equations are given by

$$\frac{\partial^2(\Delta p)}{\partial z^2} + \frac{\Delta p}{L_p} - \frac{G_p}{D_p} = 0 \tag{1}$$

$$\frac{\partial^2(\Delta p)}{\partial z^2} + \frac{\Delta p}{L_p} - \frac{G_p}{D_p} = 0 \tag{2}$$

where $G_{(n/p)}$ is the generation rate of excess electron, given by

$$G_{(n)} = \alpha\varphi(1 - R)\exp[-\alpha z] \tag{3}$$

Substituting (3) in (1) and (2) under low injection condition

$$D_n \frac{\partial^2 n_p}{\partial z^2} + \alpha\varphi(1 - R)\exp[-\alpha z] - \frac{n_p - n_{p0}}{\tau_n} = 0 \tag{4}$$

$$D_p \frac{\partial^2 p_n}{\partial z^2} + \alpha\varphi(1 - R)\exp[-\alpha z] - \frac{p_n - p_{n0}}{\tau_p} = 0 \tag{5}$$

General solution of (4) is

$$n_p = n_{p0} + C_1 \cosh(z/L_\mathrm{n}) + C_2 \sinh(z/L_\mathrm{n}) - \frac{\alpha\varphi(1 - R)\tau_n}{\alpha^2 L_n^2 - 1}\exp[-\alpha z] \tag{6}$$

and of (5) is

$$p_n = p_{n0} + C_3 \cosh(z/L_p) + C_4 \sinh(z/L_p) - \frac{\alpha\varphi(1-R)\tau_p}{\alpha^2 L_p^2 - 1}\exp[-\alpha z]. \tag{7}$$

Surface recombination velocities are given by

$$D_n \frac{d(\Delta n)}{dz} = S_n(\Delta n) \tag{8.1}$$

$$D_p \frac{d(\Delta p)}{dz} = S_p(\Delta p) \tag{8.2}$$

where S_n and S_p are the surface recombination velocities.

Boundary conditions are given by

$$\frac{dn_p}{dz}_{z=0} = \frac{S_n}{D_n}[n_p(0) - n_{p0}] \tag{9.1}$$

$$\frac{dp_n}{dz}_{z=0} = \frac{S_p}{D_p}[p_n(0) - p_{n0}] \tag{9.2}$$

As per the best knowledge of authors, Researchers calculated the current density of solar cell so far by assuming that excess carrier generation is negligible [11–13] at depletion edge, which mathematically can be represented as

$$p_n - p_{n0} \approx 0 \qquad n_p - n_{p0} \approx 0.$$

But considerable amount of excess carriers are generated at the edges of depletion region. In that case, quasi-Fermi levels are generated. Therefore, introducing quasi-Fermi level conditions and using (8.1), (8.2), (9.1) and (9.2), one may obtain

$$n_p = n_{p0} \exp\left[\frac{qV}{k_B T}\right] \tag{10}$$

$$p_n = p_{n0} \exp\left[\frac{qV}{k_B T}\right] \tag{11}$$

We obtain

$$n_p = \frac{T\exp\{-\alpha(x_j + w_D)\}[\sinh(\frac{H-x}{L_n}) + \frac{1}{K_n}\cosh(\frac{H-x}{L_n})] + T\exp(-\alpha H)[1 - \frac{1}{K_n}\alpha L_n]\sinh(\frac{x-x_j-w_D}{L_n})}{\sinh(\frac{H'}{L_n}) + \frac{1}{K_n}\cosh(\frac{H'}{L_n})} - T\exp[-\alpha x] + n_{p0} + \frac{n_{p0}[\exp(\frac{qv}{kt}) - 1][\sinh(\frac{H-x}{L_n}) + \frac{1}{K_n}\cosh(\frac{H-x}{L_n})]}{\sinh(\frac{H'}{L_n}) + \frac{1}{K_n}\cosh(\frac{H'}{L_n})} \tag{12}$$

$$p_n = \frac{T[1+\frac{\alpha L_P}{K_P}][\sinh(\frac{x_j}{L_P})cosh(\frac{x}{L_P}) - \sinh(\frac{x}{L_P})\cosh(\frac{x_j}{L_P})] + T\exp(-\alpha x_j)\,[\sinh(\frac{x}{L_P}) + \frac{1}{K_P}cosh(\frac{x}{L_P})]}{[\sinh(\frac{x_j}{L_P}) + \frac{1}{K_P}\cosh(\frac{x_j}{L_n})]}$$
$$- T\exp[-\alpha x] + p_{n0} + \frac{p_{n0}[\exp(\frac{qv}{kt}) - 1][\sinh(\frac{x}{L_P}) + \frac{1}{K_P}cosh(\frac{x}{L_P})]}{[\sinh(\frac{x_j}{L_P}) + \frac{1}{K_P}\cosh(\frac{x_j}{L_n})]} \tag{13}$$

where

$$T = \frac{\alpha\varphi(1-R)\tau_n}{\alpha^2 L_n^2 - 1} \tag{14.1}$$

$$K_{n/p} = \frac{S_{n/p}L_{n/p}}{D_{n/p}} \tag{14.2}$$

W_D is the width of the depletion region, H is the width of the device, H $' = H - z - W_D$.

Using diffusion equations, current densities can be obtained as

$$J_n = -\frac{qD_n}{L_n}T\exp\{-\alpha(x_j+w_D)\}L_n\alpha - \left[\frac{K_n[\cosh(\frac{H'}{L_n}) - \exp(-\alpha H')] + \sinh(\frac{H'}{L_n}) + \alpha L_n\exp(-\alpha H')}{K_n\sinh(\frac{H'}{L_n}) + \cosh(\frac{H'}{L_n})}\right]$$
$$+ \frac{qD_n}{L_n}n_{p0}\left[\exp(\frac{qv}{kT}) - 1\right]\left[\frac{\{K_n\cosh(\frac{H'}{L_n}) + \sinh(\frac{H'}{L_n})\}}{K_n\sinh(\frac{H'}{L_n}) + \cosh(\frac{H'}{L_n})}\right] \tag{15}$$

$$J_p = -\frac{qD_p}{L_p}T\exp\{-\alpha(x_j+w_D)\}L_p\alpha - \left[\frac{K_p[\cosh(\frac{H'}{L_p}) - \exp(-\alpha H')] + \sinh(\frac{H'}{L_p}) + \alpha L_p\exp(-\alpha H')}{K_p\sinh(\frac{H'}{L_p}) + \cosh(\frac{H'}{L_p})}\right]$$
$$+ \frac{qD_p}{L_p}p_{n0}\left[\exp(\frac{qv}{kT}) - 1\right]\left[\frac{\{K_p\cosh(\frac{H'}{L_p}) + \sinh(\frac{H'}{L_p})\}}{K_p\sinh(\frac{H'}{L_p}) + \cosh(\frac{H'}{L_p})}\right] \tag{16}$$

Without introducing quasi-Fermi level conditions, current densities may be put into the following form

$$J_n = -\frac{qD_n}{L_n}T\exp\{-\alpha(x_j+w_D)\}L_n\alpha$$
$$- \left[\frac{K_n[\cosh(\frac{H'}{L_n}) - \exp(-\alpha H')] + \sinh(\frac{H'}{L_n}) + \alpha L_n\exp(-\alpha H')}{K_n\sinh(\frac{H'}{L_n}) + \cosh(\frac{H'}{L_n})}\right] \tag{17}$$

$$J_p = -\frac{qD_p}{L_p} T \exp\{-\alpha(x_j + w_D)\} L_p \alpha - \left[\frac{K_p[\cosh(\frac{H'}{L_p}) - \exp(-\alpha H')] + \sinh(\frac{H'}{L_p}) + \alpha L_p \exp(-\alpha H')}{K_p \sinh(\frac{H'}{L_p}) + \cosh(\frac{H'}{L_p})}\right] \quad (18)$$

Spectral response is given by Sze and Ng [10]

$$S_R = \frac{J_n + J_P + J_{dr}}{q\varphi(1 - R)}. \quad (19)$$

3 Results and Discussions

Using the (15), spectral response of GaAs p-n junction solar cell is computed and plotted as a function of incident photon energy. It should be noted down first that with increase of incident energy, spectral response monotonically decreases. Figures 1 and 2 show the variation of spectral response with photon energy for different doping concentrations of either side of p-n junction. For electron concentration, result is showed in Fig. 1, whereas Fig. 2 exhibits for hole concentration

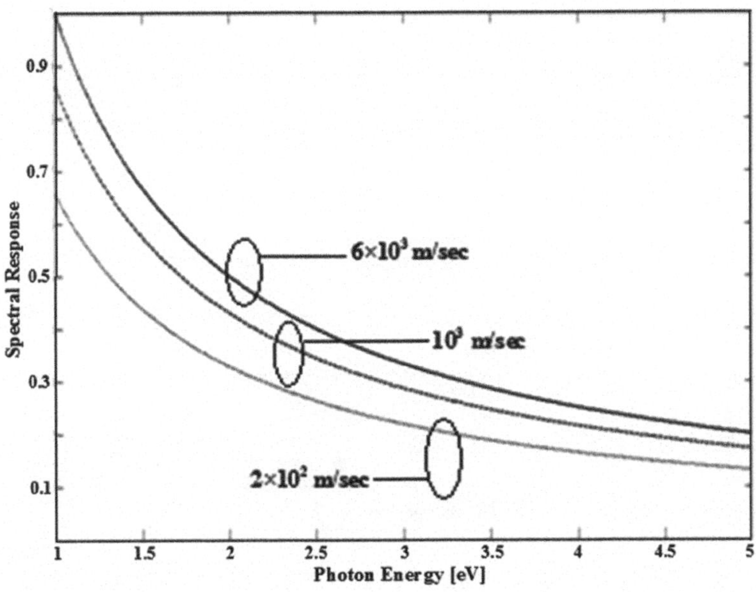

Fig. 1 Spectral response variation with incident photon energy for different electron concentrations

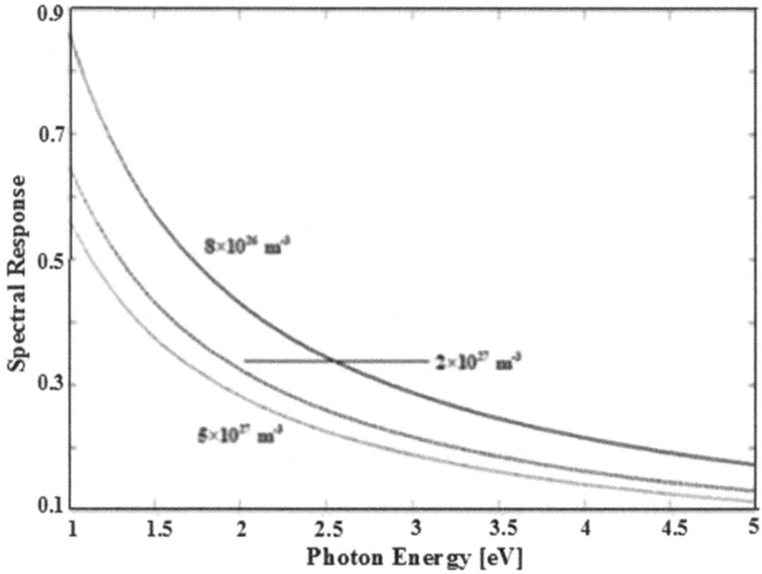

Fig. 2 Spectral response variation with incident photon energy for different hole concentrations

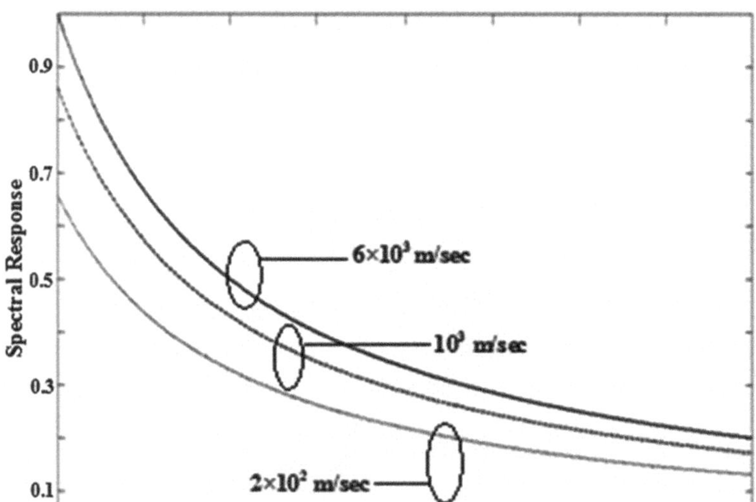

Fig. 3 Spectral response variation with incident photon energy for different surface recombination velocities

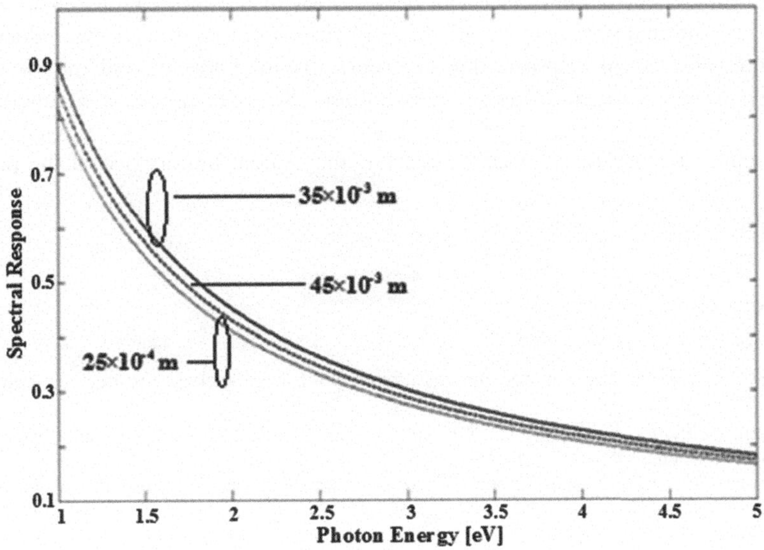

Fig. 4 Spectral response variation with incident photon energy for different width of the structure

variation. Results suggest that a variation of doping can tune the spectral response of the device prior to fabrication. Also it may be noted that the concentration gradient has significant effect on spectral response for lower photon energy. In both the plots, it is observed that SR decreases monotonically with increasing photon energy, and also with increasing concentration. Hence moderate doping concentration is required for optimum performance.

Figure 3 exhibits SR profile for different surface recombination velocities (S), whereas Fig. 4 shows it for width of the structure. It is observed that higher $S_{(n/p)}$ within the practical limit provides higher SR, whereas by increasing width of the cell, SR decreases. The variation is monotonic for both the cases, whereas for the ideal condition (neglecting space-charge effect), the variation is quite different [4]. This difference is due to the consideration of quasi-Fermi level approximation as exhibited in the mathematical modelling section, whereas in all the previous calculations as far known to authors, this excess carrier generation factor is neglected.

4 Conclusion

Spectral response of GaAs p-n junction solar cell is analytically computed and plotted as a function of incident photon energy for different internal parameters (carrier concentrations in either side of junction, depletion region width, surface recombination velocity). Simulated findings reveal that spectral response, and thus quantum efficiency of the device can be set prior to the fabrication if knowledge about a few internal parameters is given. Higher incident photon energy reduces the magnitude of spectral response due to excess loss of photons, and thus lower to moderate range is essential as input condition. Since response is a function of electron and hole currents, hence it also controls the fill factor of the device, which is the figure of merit of solar cell. Here lies the critical importance of the present paper.

References

1. Debije. M, "Renewable Energy: Better Luminescent Solar Panels in Prospect", Nature, 519, 298–299 (2015).
2. Galad. M, Spanik. P, "Design of Photovoltaic Solar Cell Model for Stand-Alone Renewable System", Electro, 285–288 (2014).
3. Chegaar. M, Petit. P, Hamzaoui. A, Aillerie. M, Namoda. A, Herguth. A, "Effect of Illumination Intensity on Solar Cells Parameters", Energy Procedia, 36, 722–729 (2013).
4. Ghitas. A. E, "Studying the Effect of Spectral Variations Intensity of the Incident Solar Radiation on the Si Solar Cells Performance", NRIAG Journal of Astronomy and Geophysics, 1, 165–171 (2012).
5. Chander. S, Purohit. A, Nehra. A, Nehra. S. P, Dhaka. M. S, "A Study on Spectral Response and External Quantum Efficiency of Mono-Crystalline Silicon, International Journal of Renewable Energy Research, 5, 4144 (2015).
6. Kumar. S, Iyer. S. S. K, "Variations in Spectral Response Behaviour in Single Layer Organic Solar Cells with Active Layer Thickness and Bias", 42nd Photovoltaic Specialist Conference, 1–6 (2015).
7. Biswas. S, Biswas. A. K, Chatterjee. A, Sinha. A, "An Analytical Study of a GaAs-Si n/p Heterojunction Solar Cell and Suggestion for A Structure for Improved Performance", International Journal of Applied Engineering Research, 11, 1855–1858 (2016).
8. Dinçer. F, Meral. M. E, "Critical Factors that Affecting Efficiency of Solar Cells", Smart Grid and Renewable Energy, 1, 47–50 (2010).
9. Hovel. H. J, "Photovoltaic Materials and Devices for Terrestrial Solar Energy Applications", Solar Energy Materials, 2, 277–312 (1980).
10. Sze. S. M, Ng. K. K, "Physics of Semiconductor Devices", Wiley, 3rd ed. (2007).
11. Chapin D. M, Fuller C. S, Pearson G. L, "A New Silicon p-n Junction Photocell for Converting Solar radiation into Electrical Power", Journal of Applied Physics, 25, 676 (1954).
12. Reynolds. D. C, Leies. G, Antes L. L, Marburger. R. E, "Photovoltaic Effect in cadmium Sulphide", Physical review, 96, 533 (1954).
13. Henry. C. H, "Limiting Efficiency of Ideal Single and Multiple Energy Gap Terrestrial Solar Cells", Journal of Applied Physics, 51, 4494 (1980).

Optimal Designing of Higher Efficiency Chalcogenide Thin Film Solar Cell

Koushik Sarkar, Sudipta Banerjee, A.K. Chakraborty and K.K. Ghosh

Abstract Efficiency and performance of a solar cell greatly depends on physical processes on junction in addition to device material characteristics and size. In the present work, attempt is made to explore a better designing of solar cell with some trade-offs and compromise amongst different parameters to obtain maximum power output and better efficiency. Chalcogenide based CdSTe thin film solar cell is studied for the purpose. The proposed structure takes care of dealing against relatively low band gap of window material and its lattice mismatch with the absorber. Efficiency as high as 26.6% and power output to the level of (0.266 W/10 cm^2) has been observed in our numerical simulation with PC1D. The result thus convincingly proves the superiority of our designed structure over conventional structures.

1 Introduction

Due to high absorption coefficient, CdTe is a promising candidate [1, 2] for thin film solar cell absorber and CdS plays a chosen partnership in forming the heterojunction window. Low cost fabrication and processing gives an added advantage to use CdTe cell. Back surface scattering loss is checked by using a high band gap barrier material ZnTe [1]. It helps to reduce carrier recombination creating a schottky barrier [3]. Enhanced output voltage and increased power conversion efficiency as depicted in the following graphs motivates our investigation to greater detail of works. In practical issue, as sheets of solar cells be very large thus limit the

---

K. Sarkar (✉)
Department of ECE, SVIST, Kolkata, West Bengal 700145, India
e-mail: kousraj2009@gmail.com

S. Banerjee
Department of ECE, NSHM Knowledge Campus, Durgapur, West Bengal 713212, India

A.K. Chakraborty · K.K. Ghosh
Institute of Engineering & Management, Electronics Complex, Salt Lake, Sector–V, Kolkata, West Bengal 700091, India
e-mail: kkghosh@iemcal.com

© Springer Nature Singapore Pte Ltd. 2017
I. Bhattacharya et al. (eds.), *Advances in Optical Science and Engineering*,
Springer Proceedings in Physics 194, DOI 10.1007/978-981-10-3908-9_22

performance due to increase of résistance of the sheets [6]. So reduction of the upper sheet layer has been a key to get success in the field of solar panels [6]. It has been studied that SnO_2 acts as a low resistive electrode that can trap light with the help of its large band gap and low absorption capability. It has been found that is easy on the pocket too [5].

2 Simulation and Results

The performance of the solar diode is reflected from the plots and the table based on simulated data output from the simulator as shown in Fig. 2 and Table 1.

It is observed from the Fig. 2 and Table 1 that the maximum V_{oc} is 1.16 V and remains unaltered for all the film thicknesses while the short circuit current increases with the thicknesses. Also it is noticed that the output power and conversion efficiency significantly enhanced with the film thicknesses of the absorber material. Table 1 is furnished with the data accrued from the simulator on the basis of one of our proposed model with long diffusion length, band gap (1.7 eV), mole fraction (x = 0.75) and carrier life time (0.09893 μs) for CdS_xTe_{1-x} layer. The thickness and doping concentration of CdTe layer is reserved 1.5 μm and 10^{18} cm^{-3} accordingly.

Following Fig. 1, the simulated structure is differently taken. Here we investigated the performance of the CdTe layer thickness considering the diffusion length (2.06 μm), carrier life time (4.997×10^{-3} μs), diffusivity (8.495 cm^2/s) [4]. The results are reflected through Table 2 and Fig. 3. The investigation showed the maximum efficiency at 1.5 μm thickness of the CdTe layer. The parameters of CdSTe layer was unchanged.

Restraining all the other layer's parameters same, we then investigated variation of doping concentrations of the CdTe layer (1.5 μm) to find the best possible cell efficiency. The Table 3 and Fig. 4 show the results. For doping concentration 1×10^{19} cm^{-3}, we got the highest efficiency. Finally we measured the effect of variable band gap on cell efficiency according to the value of mole fraction of CdS_xTe_{1-x}. It has been found that the band gap 1.7 eV for x = 0.75, is the best option to reach the highest efficiency for this diode model [4]. Figure 5 shows the results of the simulation using PC1D.

Table 1 Results of the first simulation of CdSTe solar cell

CdSTe (μm)	V_{oc} (v)	I_{sc} (A)	FF (%)	η (%)
2.2	**1.166**	**2.407**	**92.55**	**25.98**
2.1	1.164	2.369	92.58	25.53
2.0	1.163	2.351	92.53	25.30
1.9	1.161	2.331	92.56	25.05

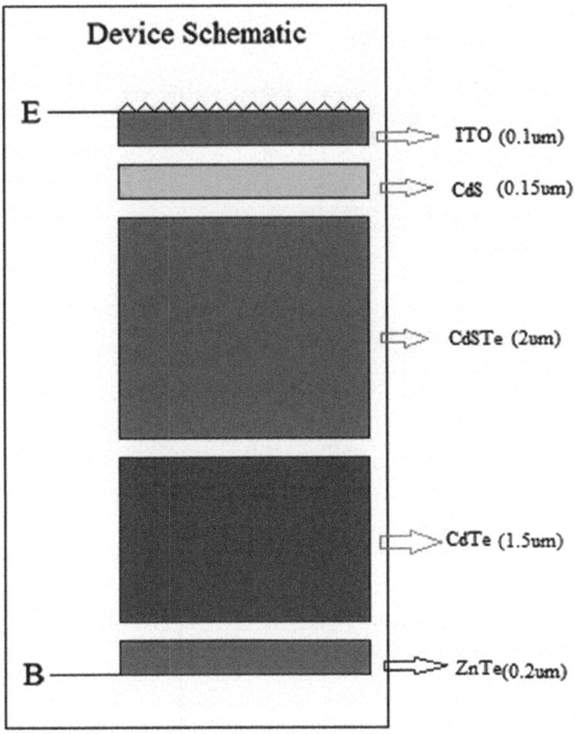

Fig. 1 PC1D Simulation Structure

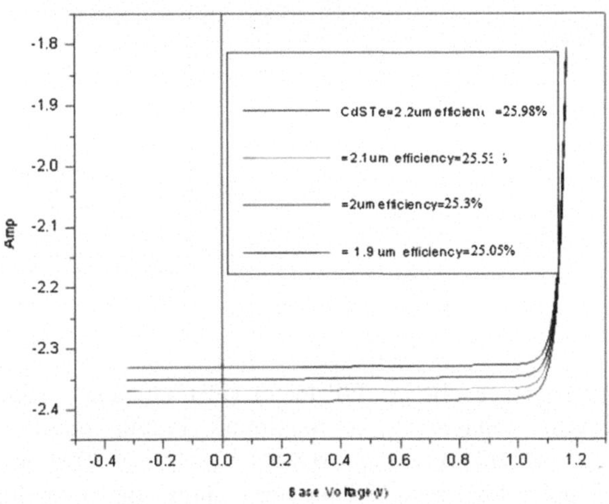

Fig. 2 V-I characteristics of the simulated solar diode

Table 4 enlighten us about the significant parameters we have taken for the diode model to get the best possible efficiency (26.63%). The I-V characteristic curve for the highest efficiency that is shown in the Fig. 6, describes the most

Table 2 Thickness vs efficiency for CdTe layer

Thickness (μm)	Efficiency (%)
2.5	25.88
3	25.8
5	25.59
8	25.52
1.4	25.97
1.3	25.97
1.2	25.95
1.1	25.91
1	25.86
0.8	25.73
0.6	25.5
1.5	25.98
1.6	25.98
1.7	25.98
2.6	25.97
0.1	25.45

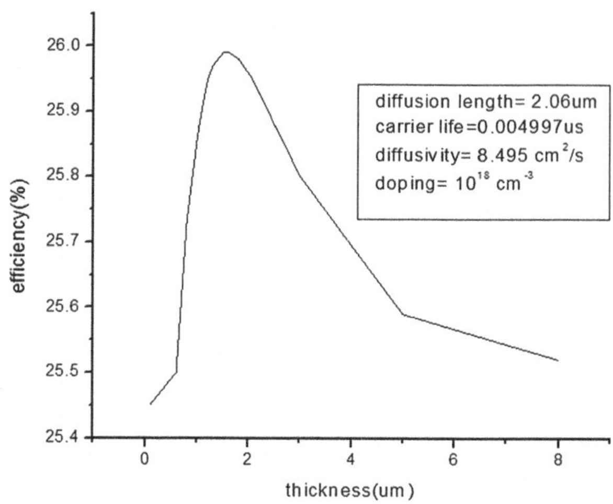

Fig. 3 Thickness vs efficiency curve for CdTe layer

important results i.e. Fill factor (FF), efficiency, (Short circuit current (I_{sc}), Open circuit voltage (V_{oc}), maximum output power ($I_m \times V_m$). $I_{sc} = 0.3416$ A, $V_{oc} = 0.8904$ V, $V_m = 0.8084$ V, $I_m = 0.3294$ A, $\eta = 26.63\%$, FF = 96.4%. Figure 6 I-V characteristic curve shows the solar diode for the highest efficiency (26.63%) configuration.

Table 3 Doping concentration vs efficiency for CdTe

Doping (cm^{-3})	Efficiency (%)
1.00E+13	17.13
1.00E+14	17.21
1.00E+15	17.9
1.00E+16	21.09
1.00E+17	24.81
1.00E+18	25.98
1.00E+20	24.96
1.00E+21	22.18
1.00E+22	20.85
1.00E+19	26.63

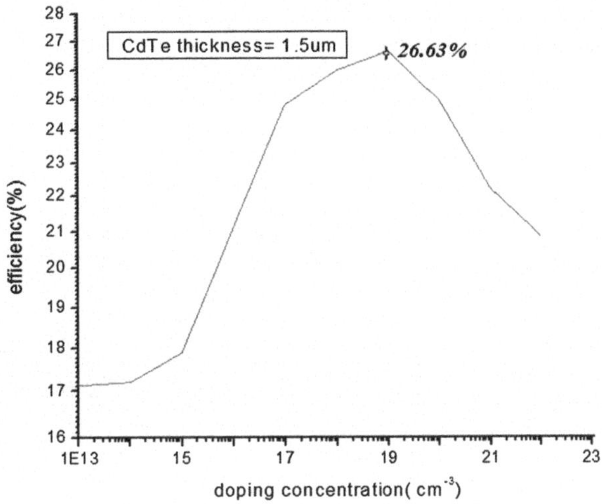

Fig. 4 Doping concentration vs efficiency curve: CdTe

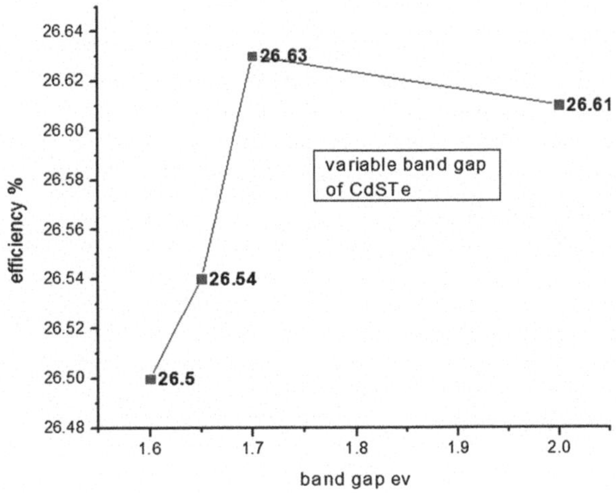

Fig. 5 Band gap vs efficiency curve: CdSTe

Table 4 The significant parameters for the diode mode

Material	Thickness (μm)	Band gap (eV)	Doping (cm^{-3}) concentration	Dielectric constant
SnO_2 (ITO)	0.1	3.6	10^{17}	9
CdS	0.15	2.4	10^{17}	10
CdS_xTe_{1-x}	1.9–2.2 (2.2 best)	1.5–2.4	10^{17}	9.4–10
CdTe	0.18 (1.5 best)	1.5	10^{13}–10^{22} (10^{19} best)	9.4
ZnTe (back contact)	0.2	2.26	10^{18}	9.67

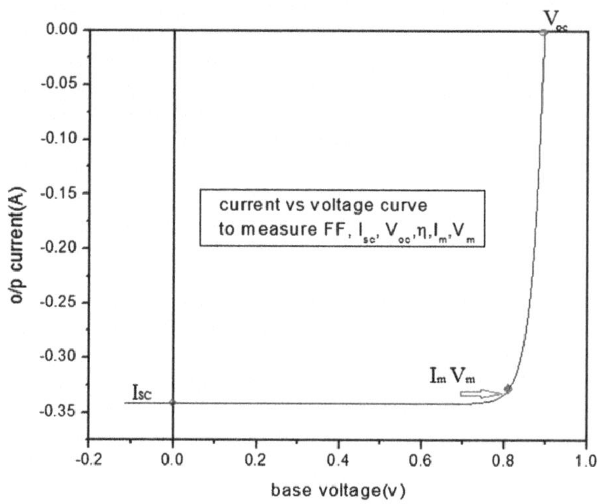

Fig. 6 I-V characteristic curve of the solar diode for the highest efficiency (26.63%) configuration

3 Conclusion

The proposed structured is a promising candidate to offer high output and high efficiency solar diode with very good fill factor (96.4%). Also it is covering the broad spectrum [ITO (330–350 nm), CdS (500–520 nm), CdSTe (500–800 nm), CdTe (800–850 nm)] of sunlight. Enough scope is still there to invade upon searching even better solar chalcogenide thin film solar diode of higher quality performance.

References

1. Design of a high efficiency ultrathin CdS/CdTe solar cell using back surface field and backside distributed Bragg reflector, S. Khosroabadi and S.H. Keshmiri- Opt. Express 22, A921(2014).
2. Achievements and Challenges of CdS/CdTe Solar Cells, Zhou Fang et al., Hindawi Publishing Corporation International Journal of Photoenergy, Volume 2011, Article ID 297350, 8 pages.
3. Optimization of Thickness of Znte Thin Film AsBack Contact for Cdte Thin Film Solar Cells, Shailaja Jeetendra et al., International Journal of Engineering Research & Technology (IJERT), ISSN: 2278-0181, Vol. 3 Issue 5, May – 2014.
4. Advantages of transparent conducting oxide thin films with controlled permittivity for thin film photovoltaic solar cells, T.A. Gessert et al., Thin Solid Films 519 (2011) 7146–7148.
5. Transparent Conducting Oxides for Photovoltaics, Elvira Fortunato et al., MRS BULLETIN • VOLUME 32 • MARCH 2007.
6. Properties of transparent conducting oxides formed from CdO and ZnO alloyed with SnO_2 and In_2O, X. Wu, T. J. Coutts et al., J. Vac. Sci. Technol. A 15(3), May/Jun 1997.

Part IV
Fibre and Integrated Optics, Photonic Crystals, Silicon Photonic Devices

Highly Non-linear Simple Designed Solid Core Photonic Crystal Fiber

Moutusi De, Rahul Kumar Gangwar and Vinod Kumar Singh

Abstract A highly non-linear solid core photonic crystal fiber (SCPCF) is designed in the present work. Three hexagonal air hole rings in cladding region and four very small air holes are present in a symmetric manner in the core region. By using full vectorial finite element method (FVFEM) with the perfectly matched layer, we study numerically the effective area of modal pattern as well as the nonlinear coefficient of this proposed SCPCF. For this proposed fiber a small modal effective area 5.58 μm^2 and a high nonlinear coefficient 21.38 W^{-1} km^{-1} are obtained at communication wavelength 1.55 μm for the small air holes in the core with diameter 0.15 μm. This type of SCPCF is useful for different nonlinear applications.

1 Introduction

The PCF is an optical fiber whose cladding comprises a 2D periodic array of air holes [1]. Numerous forms of PCF exist, and many schemes can be envisaged with which to categorize them. The classification chosen here is based on the material which forms the core region and through which the majority of the optical guided mode field travels [2]. A fiber which contains solid material within the guiding core will be termed a solid core PCF [3, 4]. The SCPCF can guide light by a modified form of total internal reflection (TIR) if an appropriate average of the refractive index of the cladding is below the refractive index of the core. The microscopic array of air hole within the cladding of PCFs running down the fiber length that offer design flexibility in tuning effective area as well as nonlinearity. Highly nonlinear PCFs on the other hand, are suitable for various novel applications including wavelength conversion, optical parametric amplification, super continuum generation etc. [5].

M. De (✉) · R.K. Gangwar · V.K. Singh
Optical Fiber Laboratory, Department of Applied Physics, Indian Institute of Technology (Indian School of Mines), Dhanbad, Jharkhand 826004, India
e-mail: demoutusi@gmail.com

© Springer Nature Singapore Pte Ltd. 2017

I. Bhattacharya et al. (eds.), *Advances in Optical Science and Engineering*, Springer Proceedings in Physics 194, DOI 10.1007/978-981-10-3908-9_23

In this paper, we propose a newly designed solid core photonic crystal fiber with a nonlinear coefficient more than 21 W^{-1} km^{-1} at the communication wavelength 1.55 μm also this SCPCF has small modal effective area 5.68 μm^2 at this wavelength. We use FVFEM based COMSOL software for this simulation [6].

2 Proposed SCPCF

We demonstrate a solid core PCF which has four symmetrically placed very small air holes near the center and inside the core region. Figure 1a shows the proposed design for SCPCF using three hexagonal air hole rings in the cladding region and Fig. 1b shows the electric field distribution in the proposed fiber. The air-holes in the cladding region have diameter d and the air-hole pitch i.e. the separation between the two air-hole centers is Λ. These two parameters are related by the relation, air filling fraction, $d/\Lambda = 0.8$. For this air filling fraction the confinement loss will be very small in three ring hexagonal SCPCF [7, 8]. In this fiber core four very small air holes having diameter d1 are placed symmetrically near the center of the fiber. The pitch of these very small air-hole is $\Lambda 1$. Two different values of d1 are considered for the fixed $\Lambda 1$ value and for these values the effective area and nonlinear coefficient are calculated over a long wavelength range (0.8–1.65 μm). Then the measured effective areas and nonlinear coefficients are relatively studied for the two d1 values. This type of micron air holes can be manufactured by a combined FIB (Focused Ion Beam) milling and Scanning Electron Microscopy (SEM) technique [9].

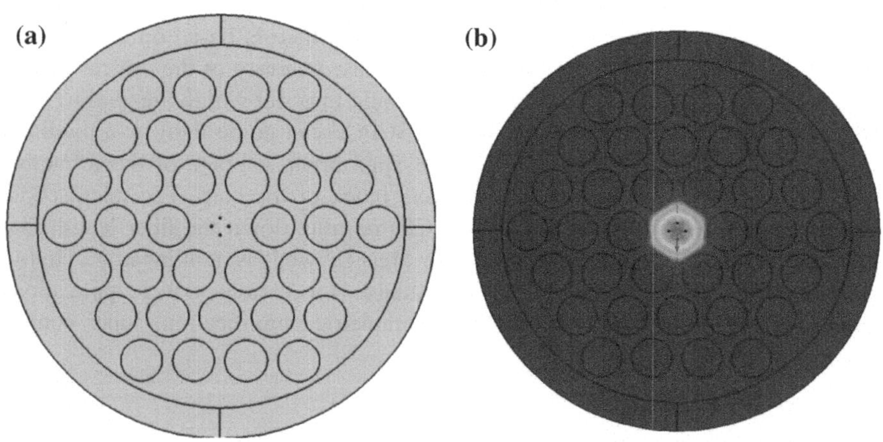

Fig. 1 **a** Proposed SCPCF image, **b** Electric field distribution inside the proposed SCPCF (using comsol software)

3 Results and Discussions

We consider the SCPCF which is made of silica. Refractive index of silica and air holes are 1.45 and 1 respectively. The modal effective area (A_{eff}) and nonlinear coefficient (Υ) of the PCF can be calculated by using the equation as given bellow,

$$A_{eff} = \frac{\left(\int_{-\infty}^{+\infty} |E|^2 dxdy\right)^2}{\int_{-\infty}^{+\infty} |E|^4 dxdy}\ \mu\text{m}^2 \tag{1}$$

$$\Upsilon = \frac{2\pi n_2}{\lambda A_{eff}}\ \text{W}^{-1}\ \text{km}^{-1}. \tag{2}$$

where $n_2 = 3.0 \times 10^{-20}$ m^2 w^{-1} is the nonlinear index of silica and λ is the operating wavelength [10–13].

In this paper the d and p are kept constant throughout the simulation. The very small air hole diameter (d1) is varied, keeping their pitch ($\Lambda 1$) fixed. The d1 taken in two studies are 0.10 and 0.15 μm. At first, keeping the air hole diameter (d) and pitch (p) constant modal effective areas and nonlinear coefficients are calculated using (1) and (2) for d1 = 0.10 μm with $\Lambda 1$ = 0.40 μm. Same thing is repeated for d1 = 0.15 μm keeping other parameters unchanged. It is found that for both the cases effective area increase with increasing wavelength. Also, effective area increase when the small air hole diameter (d1) decreases as shown in Fig. 2. It is also

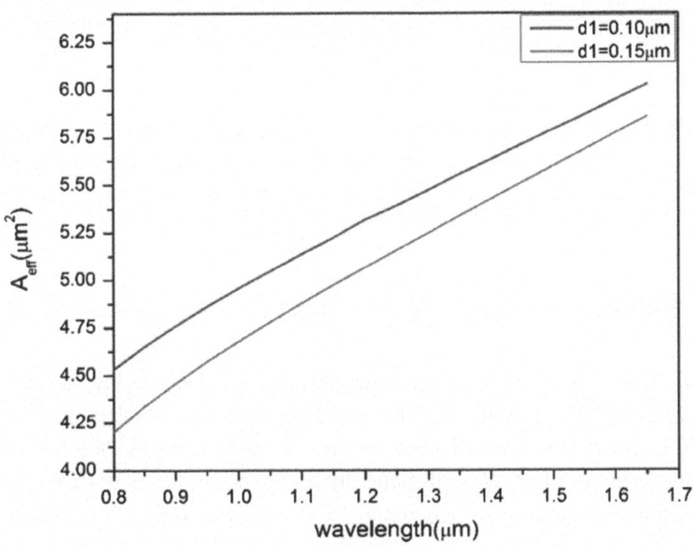

Fig. 2 Effective area curve with varying wavelength for d/Λ = 0.8 and $\Lambda 1$ = 0.4 μm

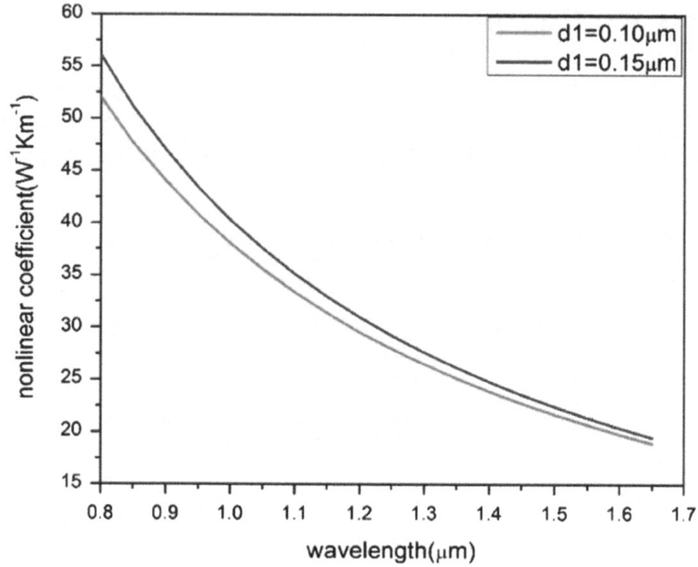

Fig. 3 Non-linear coefficient curve with varying wavelength for d/ʌ = 0.8 and ʌ1 = 0.4 μm

Table 1 Wavelength versus effective area and nonlinear coefficient representation

Wavelength (λ) (μm)	For dl = 0.10 μm		For dl = 0.15 μm	
	A_{eff} (μm^2)	ϒ (W^{-1} km^{-1})	A_{eff} (μm^2)	ϒ (W^{-1} km^{-1})
1.30	5.47	26.49	5.24	27.63
1.55	5.86	20.73	5.68	21.38

found that for both the cases, nonlinear coefficient decreases with the increasing wavelength. Also the nonlinear coefficient increases with increasing small air hole diameter as shown in Fig. 3. The simulation results are represented in the Table 1.

4 Conclusion

The modal effective area and nonlinear property of the proposed designed solid core PCF are studied with suitable structural parameters, i.e. air filling fraction (d/ʌ), small air hole pitch (ʌ1) by full vector finite element method. We have found the effective area and nonlinear coefficient are inversely related with each other i.e. nonlinear coefficient increases when effective area decreases and vice versa. The nonlinear coefficient is maximum 21.38 W^{-1} km^{-1} at 1.55 μm for dl = 0.15 μm, ʌ1 = 0.4 μm, and d/ʌ = 0.8 μm for the proposed SCPCF. This type of SCPCF can be used in the different nonlinear optical applications.

References

1. J.C. Knight, T.A. Birks, P.St.J. Russell, D.M. Atkin, "All-silica single-mode optical fiber with photonic crystal cladding", Opt. Lett. 21 (19) 1547 (1996).
2. F. Benabid and P.J. Roberts, "Linear and nonlinear optical properties of hollow core photonic crystal fiber", Journal of Modern Optics, Vol. 58, No. 2, pp. 87–124 (2011).
3. J. C. Knight, T. A. Birks, P. St. J. Russell, and D. M. Atkin, "All silica single-mode photonic crystal fiber," Opt. Lett., vol. 21, no. 19, pp. 1547–1549 (1996).
4. T. L. Wu and C. H. Chao, "A novel ultraflattened dispersion photonic crystal fiber," IEEE Photon. Technol. Lett., vol. 17, no. 1, pp. 67–69 (2005).
5. https://www.comsol.co.in/comsol-multiphysics.
6. D. Ferrarini, L. Vincetti, M. Zoboli, A. Cucinotta, and S. Selleri, "Leakage properties of photonic crystal fibers", Optics Express, vol. 10, pp. 1314–1319 (2002).
7. Rahul Kumar Gangwar, Vanita Bhardwaj and Vinod Kumar Singh, "Highly nonlinear solid core photonic crystal fiber with one nano hole", AIP Conf. Proc. 1675, 030077 (2015).
8. B. C. Gibson, S. T. Huntington, S. Rubanov, P. Olivero, K. Digweed-Lyytikäinen, J. Canning and J. D. Love, "Exposure and characterization of nano-structured hole arrays in tapered photonic crystal fibers using a combined FIB/SEM technique", Optics Express, Vol. 13, No. 22, 2005.
9. S.S. Mishra, Vinod Kumar Singh, "Study of non-linear properties of hollow core photonic crystal fiber", Optik, 122, 687–690 (2011).
10. S. M. Abdur Razzak, Student Member, IEEE, and Yoshinori Namihira, "Proposal for Highly Nonlinear Dispersion-Flattened Octagonal Photonic Crystal Fibers", IEEE photonics technology letters, vol. 20, no. 4 (2008).
11. S. S. Mishra and Vinod K. Singh, "Study of Fundamental Propagation Properties of Solid Core Holey Photonic Crystal Fiber in Telecommunication window", Chinese Journal of Physics, vol. 48, no. 5, pp. 592–606, 2010.
12. S. S. Mishra and Vinod K. Singh, "Study of Dispersion properties of Hollow-core Photonic Crystal Fiber by Finite Element Method" Optoelectronics and Advanced Materials-Rapid Communications, vol. 3, no. 9, pp. 874–878, 2009.
13. Jonathan C. Knight, "Photonic crystal fibres", Nature, 424, 847–851 (2003).

Butterworth Filter Design Using Ternary Photonic Crystal Structure Under Polarized Incidence of E.M. Wave

Meenakshi Banerjee, Romi Dey, Arpan Deyasi, Sandip Dey and Antara Das

Abstract Transmittivity of ternary photonic crystal (SiO_2/air/TiO_2) is numerically computed using transfer matrix method for bandpass filter application at 1.55 μm. Structural parameters are varied to analyze the ripple in passband width, and result is computed for both the polarization conditions. Result is also compared with that obtained for normal incidence. Number of layers and incidence angle are also varied to observe the variation of transmittivity. Simulated finding indicates better Butterworth response for s-polarized wave that that obtained for p-polarization condition.

1 Introduction

Photonic crystal (PhC) is the alternative periodic arrangement of metal/dielectric layers which allows propagation of e.m wave for specified spectral range [1], whereas other spectra are rejected. Thus the device can be specified as optical bandpass filter [2], which is realizable due to the formation of electromagnetic bandgap [3]. This bandgap formation is possible in one, two or three dimensions. Researchers already considered it as an innovative building block for the

M. Banerjee (✉) · R. Dey · A. Deyasi · S. Dey · A. Das
Department of Electronics and Communication Engineering, RCC Institute of Information Technology, Kolkata, West Bengal 700015, India
e-mail: meenakshiriabanerjee@gmail.com

R. Dey
e-mail: romidey070@gmail.com

A. Deyasi
e-mail: deyasi_arpan@yahoo.co.in

S. Dey
e-mail: sandipdeyarmstrong@gmail.com

A. Das
e-mail: dasantara21@gmail.com

© Springer Nature Singapore Pte Ltd. 2017

I. Bhattacharya et al. (eds.), *Advances in Optical Science and Engineering*, Springer Proceedings in Physics 194, DOI 10.1007/978-981-10-3908-9_24

next-generation optical communication [4]. To add flexibility in the design, it has already been studied that proper choice of structural parameters and incident angle can tune the position of passband and corresponding bandwidth, thus the filter becomes tunable [5]. Researches on this aspect are already available in different literatures of repute [6–9], but works on ternary structures are rarely available because of complex mathematical formulation, and corresponding algorithm.

Banerjee [10] suggested the use of ternary PhC as refractometric sensing elements, and proposed that it may work better than binary PhC. Gharaati and co-workers [11] calculated width of bandgap for ternary photonic crystal structures. Arafa [12] made comparative analysis between dielectric photonic crystal and metallic photonic crystal for transmission characteristics. Aghajamali et al. [13] calculated the effect of incidence angle on lossy dielectric based PhC structures. Gharaati [14] later calculated photonic bandgap of 2D ternary PhC. But tenability of ternary PhC based optical filter is yet to be studied in details, as far the knowledge of the authors. In this paper, Butterworth characteristics of ternary PhC based optical filter is computed for both type of polarized incidence, and comparative study has been carried out.

2 Mathematical Modeling

For s-polarized incident wave at angle θ1, interface reflectivities are given by

$$r_{12} = -r_{21} = \frac{n_1 \cos(\theta_1) - n_2 \cos(\theta_2)}{n_1 \cos(\theta_1) + n_2 \cos(\theta_2)} \tag{1}$$

Similarly for p-polarized wave,

$$r_{12} = -r_{21} = \frac{n_1 \cos(\theta_2) - n_2 \cos(\theta_1)}{n_1 \cos(\theta_2) + n_2 \cos(\theta_1)} \tag{2}$$

From the wave equations, transfer matrix corresponding to the interface can be obtained as

$$M_{1,2}^T = \frac{1}{t} \begin{pmatrix} 1 & r_{21,12} \\ r_{21,12} & 1 \end{pmatrix} \tag{3}$$

Considering the phase factor of the field propagating through uniform medium, propagation matrix is given as

$$P_{1,2} = \begin{pmatrix} \exp[jk_{1,2}d_{1,2}] & 0 \\ 0 & -\exp[jk_{1,2}d_{1,2}] \end{pmatrix} \tag{4}$$

where di is the propagation length in i layer, and k is the wavevector in that layer. Thus, transfer matrix for the elementary cell is

$$M = M_1^T P_1 M_2^T P_2 \tag{5}$$

For a perfectly periodic medium composed of N such elementary cells, the total transfer matrix for such a structure is

$$M_{tot} = M^N \tag{6}$$

Transmittivity is given by

$$T = \frac{1}{M_{11}^2(tot)} \tag{7}$$

3 Results and Discussions

Figure 1 shows the transmittivity profile for the p-polarized (TM wave, Fig. 1a) and s-polarized (TE wave, Fig. 1b) for the SiO_2/air/TiO_2 structure with different thicknesses of SiO_2 material. Comparative study reveals that for TE wave, ripple at the desired passband is less than that obtained for TM wave propagation. This speaks in favor of TE wave propagation inside ternary PhC structure. But for TM wave propagation, length of notch in transmittance scale is higher which speaks about better noise rejection property.

In Fig. 2, comparative study is made by varying the width of TiO_2 layer. The observations for this case are exactly same as obtained in Fig. 1. But in this case, it is seen that several passband appears in vicinity of the desired spectrum which gives

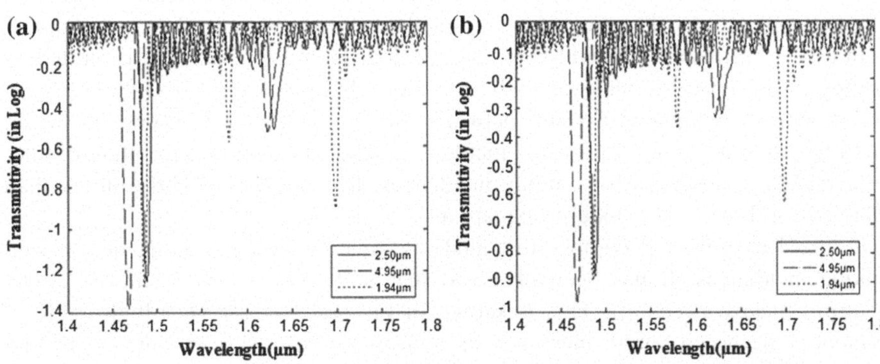

Fig. 1 Transmittivity profile with wavelength under oblique incidence for SiO_2-air-TiO_2 composition with different thicknesses of SiO_2 layer for **a** TM wave; **b** TE wave

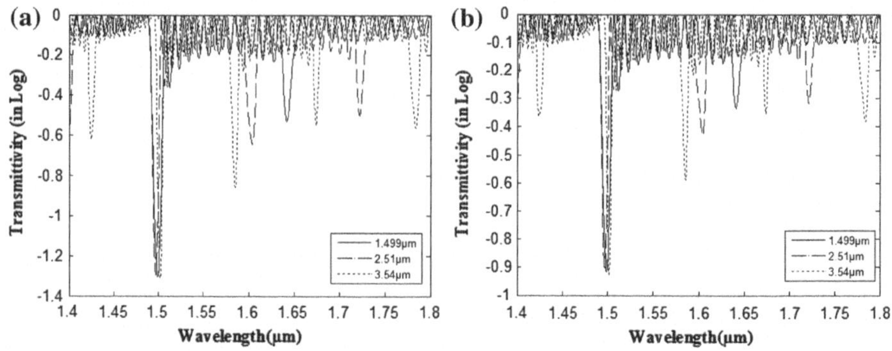

Fig. 2 Transmittivity profile with wavelength under oblique incidence for SiO_2-air-TiO_2 composition with different thicknesses of TiO_2 layer for **a** TM wave; **b** TE wave

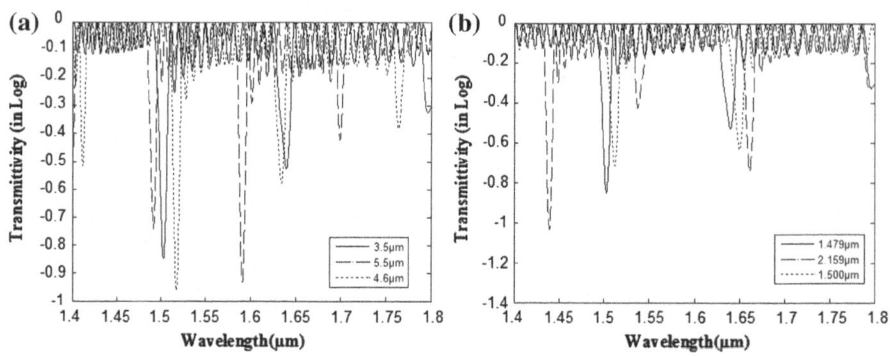

Fig. 3 Transmittivity profile with wavelength under normal incidence for SiO_2-air-TiO_2 composition with different thicknesses of **a** SiO_2 layer; **b** TiO_2 layer

the advantage of shifting of operating region near the low attenuation zone for different applications. This provides the added advantage of multiplexing technique.

In Fig. 3, results are plotted for normal incidences. In Fig. 3a, result is plotted by varying SiO_2 layer dimensions, whereas Fig. 3b represents for different TiO_2 thicknesses. A comparative study between Fig. 3a with Fig. 1 shows that better guard length is observed for normal incidences, whereas asymmetric nature of filter performance is observed for oblique incidences. But ripple is slightly higher than obtained for TM or TE mode propagation.

Figure 4 exhibits the comparative analysis for transmittance under normal and oblique incidences. It has been observed that with increase of incidence angle, spectrum makes a redshift. This is shown in Fig. 4a. But for very high angle of incidence, the notch length increases, as seen in Fig. 4b. This is quite interesting from application point of view as small incidence angle detoriates the filter performance w.r.t normal incidence, whereas very high angle enhances the filter

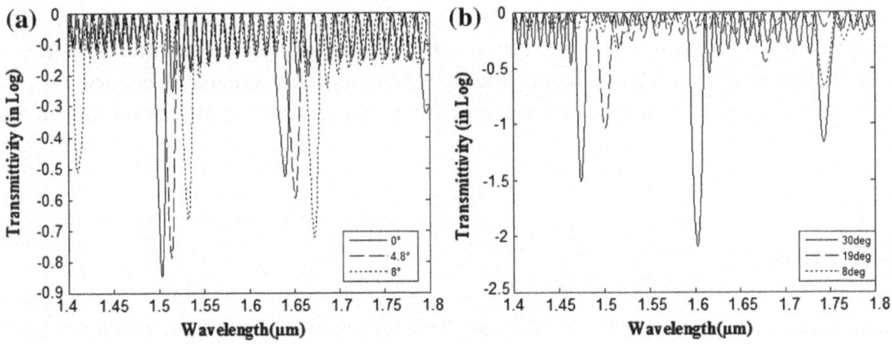

Fig. 4 Transmittivity with wavelength **a** comparative analysis between normal and oblique incidences; **b** for different angle of incidences

Fig. 5 Transmittiivty with wavelength for different number of layers

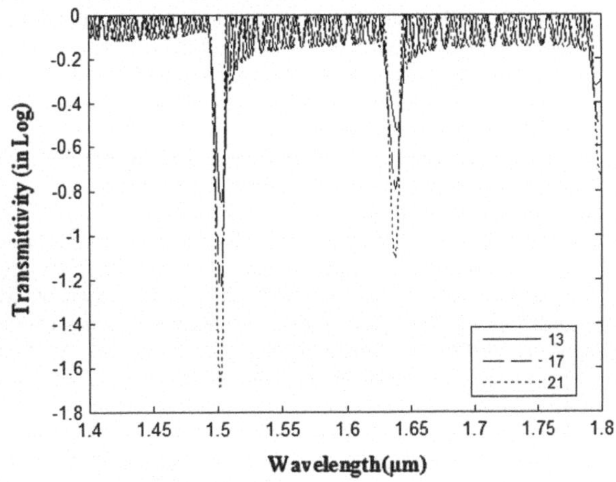

performance. By increasing number of layer along quantized direction, it has been observed that filter performance is enhanced keeping the magnitude of bandwidth constant, as shown in Fig. 5. In this case, since magnitude of ripple is also unaffected, so it may be concluded that better Butterworth response is obtained if higher number of layers are grown.

4 Conclusion

Comparative study between p and s-polarized wave incidence on ternary PhC structure reveals the fact that s-polarized wave incidence (TE mode propagation) gives better Butterworth response in the optical communication zone owing to lower ripples, but has the slight disadvantage of lower transmittance value at either side of passband which may lead to the interference. Choice of layer dimensions play major role in tuning the passband in the desired region. Comparative study is carried out with normal incidence. Incidence angle has a greater influence on the transmittivity, and by increasing number of layers, quality of Butterworth performance is enriched.

References

1. Yablonovitch. E., Gmitter. T. J., "Photonic Band Structure: The Face-Centered-Cubic Case", Physical Review Letters, 63, 1950–1953 (1987).
2. Mao. D, Ouyang. Z, Wang. J. C, "A Photonic-Crystal Polarizer Integrated with the Functions of Narrow Bandpass and Narrow Transmission Angle Filtering", Applied Physics B, 90, 127–131 (2008)
3. Xu. X, Chen. H, Xiong. Z, Jin. A, Gu. C, Cheng. B, Zhang. D, "Fabrication of Photonic Crystals on Several Kinds of Semiconductor Materials by using Focused-Ion Beam Method", Thin Solid Films, 515, 8297–8300 (2007)
4. Hansryd. J, Andrekson. P. A, Westlund. M, Li. J, Hedekvist. P. O, "Fiber-based Optical Parametric Amplifiers and their Applications," IEEE Journal of Selected Topics on Quantum Electronics, 8, 506–520 (2002)
5. Arafa. S, Bouchemat. M, Bouchemat. T, Benmerkhi. A., "Design of Optical Filter Based on Photonic Crystal Squared Silicon Nitride Microcavity", Nanoscience and Nanotechnology, 6 (1A), 97–104 (2016)
6. Jugessur. A. S, Pottier. P, De La Rue. R. M, "Engineering the Filter Response of Photonic Crystal Microcavity Filters", Optics Express, 12, 1304–1312 (2004)
7. Mahmoud. M. Y, Bassou. G, Taalbi. A, Chekroun. Z. M., "Optical Channel Drop Filters based on Photonic Crystal Ring Resonators", Optics Communications, 285, 368–372 (2012)
8. Maity. A, Chottopadhyay. B, Banerjee. U, Deyasi. A, "Novel Band-Pass Filter Design using Photonic Multiple Quantum Well Structure with p-Polarized Incident Wave at 1550 nm", Journal of Electron Devices, 17, 1400–1405 (2013)
9. Mandal. S, Bose. C, Bose. M. K, "A Generalized Design of One Dimensional Photonic Crystal based Optical Filter with Lossy Materials", Optical and Quantum Electronics, 48, 200 (20160
10. Banerjee. A, "Enhanced Refractometric Optical Sensing by using One-Dimensional Ternary Photonic Crystals", Progress in Electromagnetics Research, 89, 11–22 (2009)
11. Gharaati. A, Serajfard. S. A, "Investigation of a Ternary 1D Photonic Crystal Band Gap Width", International Conference On Photonics 1–5 (2010)
12. Arafa H. A, Ismaeel. M, Abdel-Rahman. E, "Comparative Study of the One Dimensional Dielectric and Metallic Photonic Crystals", Optics and Photonics Journal, 2, 105–112 (2012)
13. Aghajamali. A, Akbarimoosavi. M, Barati. M, "Properties of the Band Gaps in 1D Ternary Lossy Photonic Crystal Containing Double-Negative Materials", Advances in Optical Technologies, 2014, 1–7 (2014)
14. Gharaati. A, Mohamadebrahimi. L, Roozitalab. Z, "Photonic Band Gap in Negative Ternary Refractive IndicesDavis, A. R., Bush, C., Harvey, J. C. and Foley, M. F., "Fresnel lenses in rear projection displays," SID Int. Symp. Digest Tech. Papers 32(1), 934–937 (2001)

Numerical Simulation of Millimeter Wave Generation in a Digital Fiber Optic Link

V. Charan Teja and M. Ganesh Madhan

Abstract Millimeter wave (MMW) generation using optical techniques has gained popularity in recent days. In this work, a 1 Gbps conventional single mode fiber optic link is utilized to remotely generate a 60 GHz MMW signal by heterodyning principle. Two independent distributed feedback lasers (DFB) operating at 193.5 and 194.1 THz, in the transmitter side, are used to generate a 60 GHz MMW signal at the remote receiver. One of the laser is externally modulated by 1 Gbps data and another operating in CW mode, are present in the transmitter side. The important aspect of this approach is generation of 60 GHz RF signal and transmission of 1 Gbps NRZ data, simultaneously in an optical fiber link. The 60 GHz RF signal and digital data are extracted by using a narrow band pass filter and low pass filter respectively. The performance of the data reception and RF generation are investigated by varying the link distance from 10 to 50 km.

1 Introduction

Wireless networks need to provide data rates in Gbps in order to satisfy the newly emerging applications [1]. Use of millimeter-wave (MMW) carrier frequencies has been considered as a solution for high data rate transmission, as different radio frequency bands has already been occupied. The main issue of using MMW for wireless communication system is high atmospheric attenuation. The microwave-photonics based system such as radio over fiber (RoF) suggests utilizing optical fiber as a solution for delivery of the MMW signal at the required locations [2]. Optical methods of millimeter wave generation has gained popularity owing to wide deployment of RoF schemes [3]. In RoF systems, the RF/microwave modulated signal is generated at the central station and transmitted through optical fiber to the base stations (BSs). There are various techniques [4] developed for remote

V. Charan Teja (✉) · M. Ganesh Madhan
Department of Electronics Engineering, Madras Institute of Technology Campus, Anna University, Chennai, Tamil Nadu 600044, India
e-mail: charanteja0705@gmail.com

© Springer Nature Singapore Pte Ltd. 2017

I. Bhattacharya et al. (eds.), *Advances in Optical Science and Engineering*,
Springer Proceedings in Physics 194, DOI 10.1007/978-981-10-3908-9_25

transfer of MMW band signals. Most common and simple one is the direct intensity modulation scheme, in which a MMW carrier source directly modulates a high speed laser and then the MMW signal can be recovered at remote photodiode by direct detection. Alternatively, externally modulation of a continuous wave laser source using Mach-Zehnder modulator (MZM) or Electro Absorption Modulator (EAM) can also be utilized. Another scheme uses an Intermediate frequency (IF) band signal, which is transported over optical fiber instead of RF band signal, and IF-to-RF up conversion is accomplished at the BS level. Optical heterodyning technique is a scheme where two or more optical signals are simultaneously transmitted and are heterodyned in the receiver. An electrical beat note is then generated at the output of the photodetector with a frequency corresponding to the wavelength spacing of the two optical waves [5–9]. Kuri and Kitayama [5] experimentally demonstrated a 25-km-long fiber-optic transmission and the optical heterodyne detection of a 59.6-GHz radio-on-fiber signal with 155.52-Mb/s differential-phase-shift-keying formatted data. Madjar and Berceli [6] given a review on microwave generation by optical techniques. Kassa et al. [7] simulated by using electrical modeling of semiconductor laser diode for heterodyne RoF system. Hofstetter et al. [8] generated optical MMW using heterodyne method with dispersion effects. The recent work of Ramin et al. [2], examines the effect of intensity noise on the remote generation of MMW. However the effect of fiber length and dispersion effect on the received digital data has not been investigated. In this paper, we analyze the performance of MMW signal generation in a 1 Gbps data link using optical heterodyning principle incorporating fiber dispersion. The quality of received data interms of Bit error rate [9], Optical signal to noise ratio [9] and Q factor [9] along with the generated RF power is investigated under different link lengths as well as input signal powers.

2 Remote Heterodyning Scheme

(a) System blocks and operation

The overall block diagram for 60 GHz MMW generation in a 1 Gbps Single mode fiber (SMF) link is given in Fig. 1. The two DFB laser diodes are modelled using single mode rate equations. The optical power of laser diode 1 is modulated by 1 Gbps Non return to zero (NRZ) data using MZM modulator. The MZM output and CW optical power of laser diode 2 are combined and transmitted through a single mode fiber. The 60 GHz MMW signal is generated by optical heterodyne at the photodetector. The power divider divides the output of photo detector into two equal powers. By using band pass filter, the 60 GHz signal is generated, whereas a low pass filter is used to retrieve the digital data. This approach is similar to the report of Ramin et al. [2].

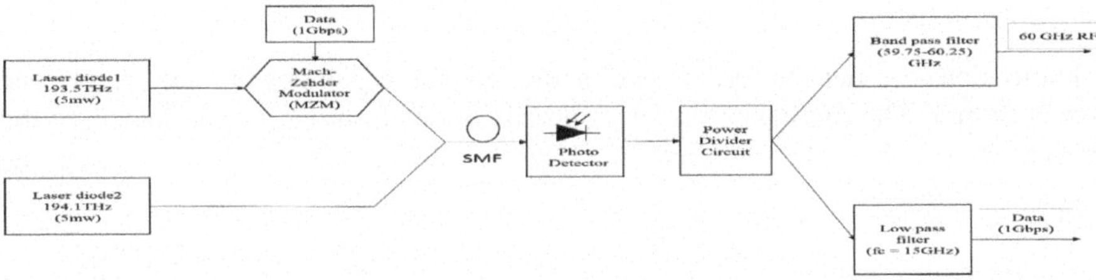

Fig. 1 A schematic of remote heterodyning based 60 GHz generation by using two distributive feedback (DFB) laser diodes at transmitter

(b) **Laser model**

For a DFB laser diode, three rate equations [10] are used to obtain the laser characteristics. They are given by (1)–(3). The solution for these equations can be used for determining the frequency chirp and output power.

$$\frac{dN(t)}{dt} = \frac{I(t)}{qv_a} - \frac{N(t)}{\tau_n} - v_0 a_0 \frac{N(t) - N_0}{1 + \varepsilon_c S(t)} \mathrm{s(t)} \tag{1}$$

$$\frac{dS(t)}{dt} = \left(\Gamma a_0 v_g \frac{N(t) - N_0}{1 + \varepsilon_c S(t)} - \frac{1}{\tau_p} \right) \mathrm{s(t)} + \frac{\beta \Gamma N(t)}{\tau_n} \tag{2}$$

$$\frac{d\emptyset_m}{dt} = \frac{\alpha}{2} \left(\Gamma v_g a_0 (N(t) - N_0) - \frac{1}{\tau_p} \right) \tag{3}$$

The time variations of the optical power output is given by,

$$p(t) = \frac{s(t) V_a \eta_0 h\upsilon}{2 \Gamma \tau_p} \tag{4}$$

where hv denotes the photon energy and η_0 denotes the total differential quantum efficiency.

(c) **Fiber**

The single mode fiber is modelled using an impulse response [11] as shown

$$\mathrm{H(f)} = (4\pi\beta_2 L)^{\frac{1}{2}} \exp(-\mathrm{j} f^2 / 2\beta_2 \mathrm{L}) \tag{5}$$

where L is length of fiber (in km) and β_2 is second order dispersion parameter. The fiber output is obtained by convolving the optical pulse from the laser and the fiber response. The output power from fiber gives the dispersion effect [12].

(d) **Photo detector**

A photo detector detects and converts the optical input power into an electric current output. The mixing of the two optical signals from laser1 and laser2, at the photodiode provides the required 60 GHz RF signal. The photo current at the detector is given as

$$\begin{aligned} I(t) &= R|E(t)|^2 \\ &= R\left[\frac{E_1(1+\cos(\Delta\emptyset(t)))}{2} + E_2^2 + 2E_1E_2\cos\left(\frac{\Delta\emptyset(t)}{2}\right) \times \cos\left(2\pi f_d t - \frac{\Delta\emptyset(t)}{2}\right)\right] \end{aligned} \quad (6)$$

where

$$\Delta\emptyset(t) = \frac{\pi}{V_\pi}\left(\frac{V_\pi}{2} + v(t)\right) \quad (7)$$

In the above equation,

R—Responsivity of photodiode
$v(t)$—Data signa,
V_π—extinction voltage of MZM,

$$f_d = f_2 - f_1$$

3 Simulation Results

The rate (1)–(3) are solved numerically using an ODE solver in Matlab for a 15 mA step current input. The parameters of DFB laser diode rate equations are given [13].

The first laser diode with wavelength 1549 nm generates a power of 5 mW for a drive current of 15 mA, whereas the second laser diode generates the same power at 1545.9 nm (194.1 THz). The characteristics are shown in Fig. 2(1a–b). Figure 2 (1c) illustrates the power versus wavelength, in which a power of 7.05 dBm is shown at both wavelengths of laser diodes. NRZ data of 1 Gbps is provided as a drive voltage to the Mach-Zehnder modulator (MZM) given in Fig. 2(1d). The Mach-Zehnder modulator (MZM) provides a maximum output of 4.6 mw as plotted in Fig. 2(1e).

The output power of Laser diode1 with 193.5 THz, which is modulated with given data along with a CW signal at 194.1 THz is given as input to fiber. The Group Velocity Dispersion (GVD) [14] is the main effect in 1550 nm SMF link and dispersion coefficient of 17 ps/nm/km is fixed for this study. From the (5), the

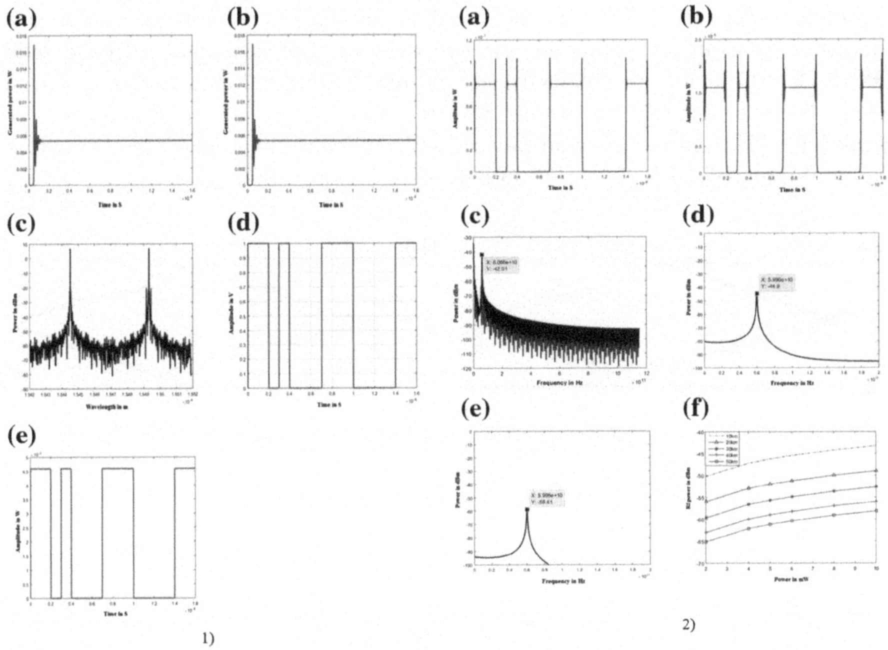

Fig. 2 **1(a)** Optical power of laser diode1 after transient switch on, (**b**) Optical power of laser diode2 after transient switch on, (**c**) Optical spectrum of two laser diodes, (**d**) Data given to MZM modulator, (**e**) Output power of MZM. **2**(**a**) Fiber output for 10 km, (**b**) Fiber output for 50 km, (**c**) power spectrum of photo detector output, (**d**) Power spectrum of band pass filter output for 10 km fiber length, (**e**) Power spectrum of band pass filter output for 50 km fiber length, (**f**) Power in dBm for different input power values for different lengths of fiber

dispersion β_2 (ps^2/km) isn't negligible for short pulses, which causes the temporal pulse broadening. Figure 2.2a–b shows the output power from the fiber, which illustrates the effect of GVD for 10 and 50 km fiber lengths. The photo detector provides an output current with difference frequency of two optical signals which is coupled from the output of fiber. The frequency of the output current of photo diode is shown in Fig. 2(2c). The power spectrum is obtained by using Fast Fourier Technique (FFT) with 16,384 samples which is given in Fig. 2(2c) in which the power at 60 GHz is 37 dBm, which means difference between 60 GHz signal and its harmonics is −42 dBm.

The 60 GHz signal is retrieved by using a microwave band pass filter. The generated signals are −44.9 and −58.61 dBm for 10 and 50 km respectively and is shown in Fig. 2(2d–e). The simulation is repeated for different link distance and input signal power and the results are plotted in Fig. 2(2f). When the link distance increases, the RF power decreases for same input signal power, moreover the RF power increases as input signal power increases as expected. However this analysis is carried out up to 10 mW, as SBS effect may become dominant beyond this value [15].

Similarly, Low pass filter seperates the data and the bit error rate performance is calculated by using eye diagram, the eye opening, optical signal to noise ratio (OSNR) and Q factor are obtained from the received digital data and is shown in Fig. 3. The Eye diagram for 10 and 50 km are plotted in Fig. 3a–b and also eye opening for different input signal powers with different link distance are shown in Fig. 3c. Similarly, BER, Q factor, OSNR also shown in Fig. 3d–f.

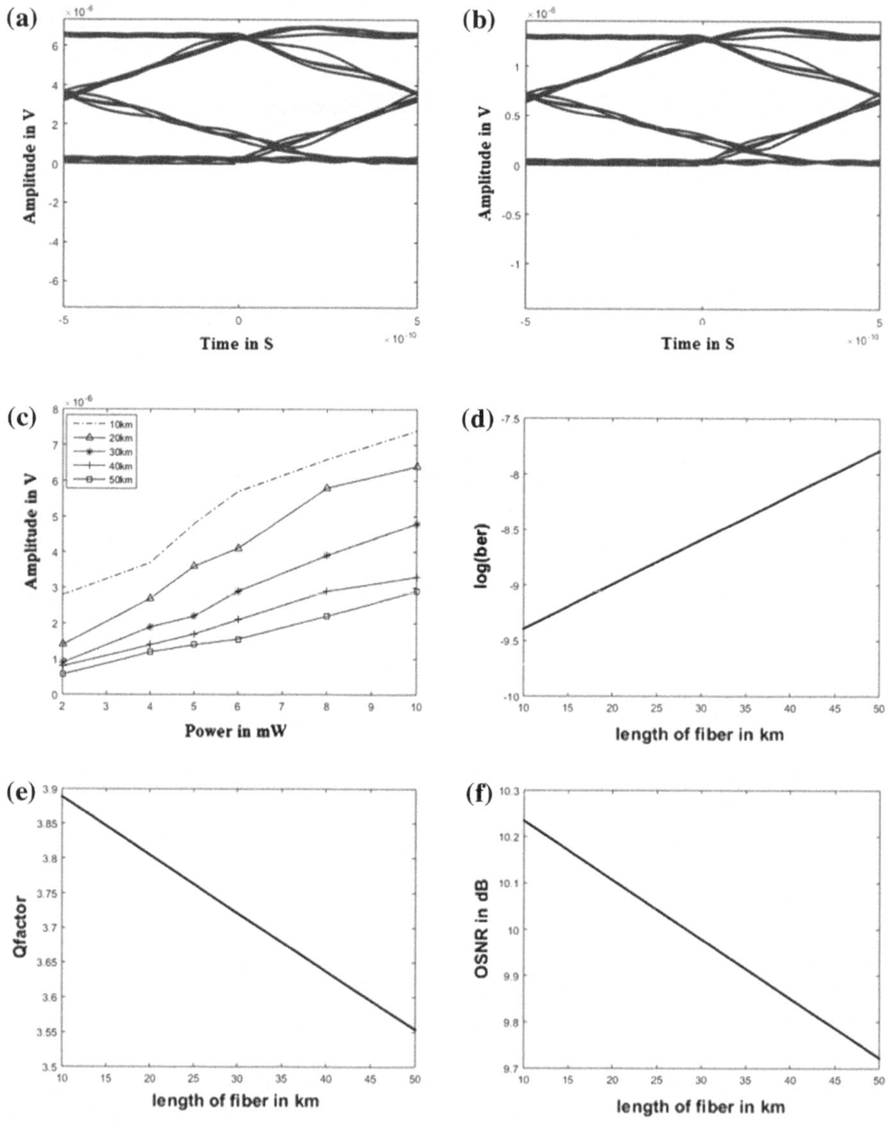

Fig. 3 **a** Eye diagram for 10 km. **b** Eye diagram for 50 km. **c** Eye opening versus input power for different length. **d** Bit error rate versus length of fiber. **e** Qfactor versus length of fiber. **f** OSNR versus length of fiber

These figures illustrates that as the link distance increases, BER increases and also a linear decrease in Q factor and OSNR are observed. This coincides well with the theory [16]. According to IEEE 802.3Z (1998) standard, where 1000BASE-ZX is used for Gigabit Ethernet transmission using 1550 nm wavelength to achieve distance of up to 70 km over single mode fiber with BER < 10^{-9}.

4 Conclusion

This paper investigates the effect of fiber length on the simultaneous MMW generation at 60 GHz and 1 Gbps data transmission. The effect of input signal power of laser diodes and fiber link on the generated 60 GHz RF signal is evaluated. The digital link performance is determined by eye opening, BER, and Q factor under this scheme. The standard BER for 1 Gbps is obtained at 20 km of fiber length with RF power −49 dBm

References

1. Braun R.P., Grosskopf G., von Helmolt C.H., Kruger K., Kruger U., Rohde D. and Schmidt F., 'Optical microwave generation and transmission experiments in the 12 and 60 GHz-region for wireless communications', IEEE Transactions on Microwave Theory and Techniques, Vol. 46, No. 4, pp. 320–330 (1998).
2. Ramin K., Hamza Hallak E., Julien P, and Beatrice C., 'Impact of Amplitude Noise in Millimeter-Wave Radio-Over-Fiber Systems', Journal of Lightwave Technology, Vol. 33, No. 13, pp. 2913–2919 (2015).
3. Radio over Fiber Transmission by Sub Carrier Multiplexing, Available from: <http://dspace.bracu.ac.bd/bitstream/handle/10361/697/RADIO%20OVER%20FIBER%20TRANSMISSION.pdf>
4. Techniques for generating 60 GHz signal, Available from: <http://www.ijcait.com/IJCAIT/1210R.pdf>
5. Toshiaki Kuri., Ken-ichi Kitayama., 'Optical Heterodyne Detection of Millimeter-Wave-Band Radio-on-Fiber Signals With a Remote Dual-Mode Local Light Source', IEEE Transactions on Microwave Theory and Techniques, Vol. 49, No. 10 (2001).
6. Asher Madjar., Tibor Berceli., 'Microwave Generation by Optical Techniques-A Review', Proceedings for 36th European Microwave Conference.
7. Wosen-Eshetu Kassa, Anne-Laure Billabert, Salim Faci, and Catherine Algani, 'Electrical Modeling of Semiconductor Laser Diode for Heterodyne RoF System Simulation', IEEE Journal of Quantum Electronics,Vol. 49, No. 10 (2013).
8. Rolf Hofstetter, Harald Schmuck, and Rolf Heidemann, 'Dispersion Effects in Optical Millimeter-Wave Systems Using Self-Heterodyne Method for Transport and Generation', IEEE Transactions on Microwave Theory and Techniques, Vol. 43, No. 9 (1995).
9. Performances: eye diagram, osnr, Qfactor, ber, Available from: < http://optiwave.com/wp-content/uploads/2015/10/TC-Optical-Signal-to-Noise-Ratio-OSNR.pdf?51da48>
10. Binh L.N, Optical Fiber Communication Systems with MATLAB and Simulink Models, CRC Press, New York (2015).

11. Vuorinen K., Gaffiot F. and Jacquemod G., 'Modeling Single-Mode Lasers and Standard Single-Mode Fibers using a Hardware Description Language', IEEE Photonics Technology Letters, Vol. 9, No. 6, pp. 824–826 (1997).
12. Fiber optic III: attenuation and dispersion, Available from: < www1.ceit.es/asignaturas/comuopticas/pdf/chapter4.pdf>
13. Koussalya Balasubramanian., M.Ganesh Madhan.,'Simulation of Thermal Effects in Laser Diode and its Impact on High Speed Fiber Optic Link', Journal of High Speed Networks 17 (2010).
14. Fiber optic: Group Velocity Dispersion, Available from: < http://www.mitr.p.lodz.pl/evu/lectures/Abramczyk3.pdf>
15. M. Baskaran., M. GaneshMadhan.,'A novel approach for simultaneous millimeter wave generation and high bit rate data transmission for Radio over Fiber (RoF) systems' Elsevier (2014).
16. Vishal Sharma., Amarpal Singh., Ajay K. Sharma., 'Analysis of the Impact of P-Ratio on BER, Q-Factor and OSNR of Radio over Fiber (RoF) System' Optics and Photonics Journal (2011).

Enhanced of Photonic Bandgaps in One-Dimensional Plasma Photonic Crystal with Defect

G.N. Pandey, Anil Kumar Shukla, Khem. B. Thapa and J.P. Pandey

Abstract The paper presents the transmittance characteristics of electromagnetic (EM) waves in one-dimensional photonic crystal with the insertion of defect layer within the regular structure of plasma photonic crystal. The Plasma Photonic Crystal (PPC) consists of alternate layers of thin micro-plasma with dielectric material in one-dimensional periodic structure. The reflectance and transmittance of considered structure are calculated using transfer matrix method. From the study, it is found that a number of photonic band gap increases by introducing a defect layer inside the regular structure of plasma photonic crystal.

1 Introduction

Photonic Crystals are the periodic layered structures of different type of dielectric materials with contrast refractive indices on the scale of wavelength radiation. Photonic crystals exhibit many unique features and have been gaining attention in the field of solid state and optical physics for making opto-electronic devices [1, 2]. First time, Keskinen et al. [3] and Hojo and Mase [4, 5] have studied photonic band gaps in photonic crystals using dusty plasma and discharged micro-plasma respectively, and

G.N. Pandey (✉)
Department of Physics, Amity Institute of Applied Sciences, Amity University, Noida, Uttar Pradesh, India
e-mail: gnpandey2009@gmail.com

A.K. Shukla
Amity Institute of Telecom Engineering and Management, Amity University, Noida, Uttar Pradesh, India
e-mail: akshukla2@amity.edu

Khem.B. Thapa
Department of Physics, B.B.A. University, (A Central University), Lucknow, Uttar Pradesh, India

J.P. Pandey
Department of Physics, M.L.K. P.G. College, Balrampur, Uttar Pradesh, India

© Springer Nature Singapore Pte Ltd. 2017

I. Bhattacharya et al. (eds.), *Advances in Optical Science and Engineering*, Springer Proceedings in Physics 194, DOI 10.1007/978-981-10-3908-9_26

termed as ***Plasma Photonic Crystal*** (PPC). H. Hojo and N. Uchida, group of JAPAN, have been studied theoretically the dispersion relation by solving Maxwell's wave equations and found that band gap will increase with increase of plasma width and plasma density [6]. The band structure and transmittance properties are theoretically studied using conventional transfer matrix method [7, 8]. The other optical characteristics properties: group velocity, effective group and phase index of 1-D ternary PPCs are theoretically studied [9].

G.N. Pandey et al. have already studied the reflection properties of plasma dielectric photonic crystal for making omnidirectional reflector with effect of variation of plasma width and plasma density [10–15]. This study very clearly reveals that the width of 100% reflection bands increases for both TE- and TM-modes as the density of plasma increases and hence ODR band also increases. The broad band omni-directional reflector can be obtained by increasing of plasma density for the given thickness of the plasma layers [16, 17].

In this paper, we present the transmittance characteristics of EM waves in one-dimensional photonic crystal with the insertion of defect layer within the regular structure of plasma photonic crystal. The reflectance and transmittance of considered structure are calculated using transfer matrix method. From the study, it is found that a number of photonic band gap increases by introducing a defect layer inside the regular structure of plasma photonic crystal.

2 Mathematical Modeling

The simple and conventional method of mathematical modeling is the transfer matrix method which is also known as the translational matrix method.

2.1 Transfer Matrix Techniques for 1-D Photonic Crystals

The simplest way to study the propagation of electromagnetic wave in 1-D photonic crystal is by using the transfer matrix technique or transfer matrix method (TMM) (Figs. 1 and 2).

We consider a periodic arrangement of multilayer structure with refractive indices n_1 and n_2, each with thicknesses d_1 and d_2 respectively. Let the plane waves

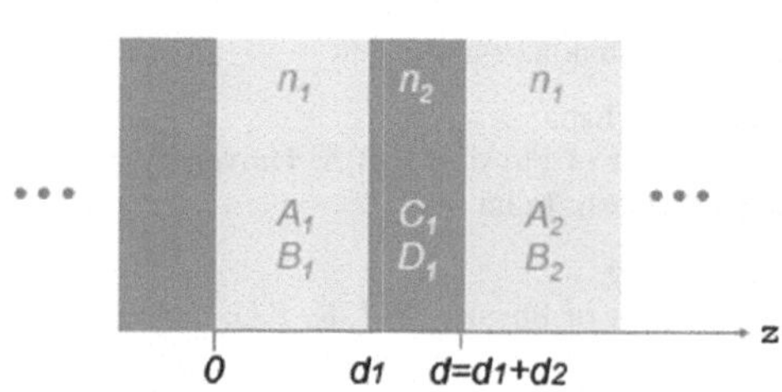

Fig. 1 Periodic arrangements of multilayer structure with refractive indices n_1 and n_2

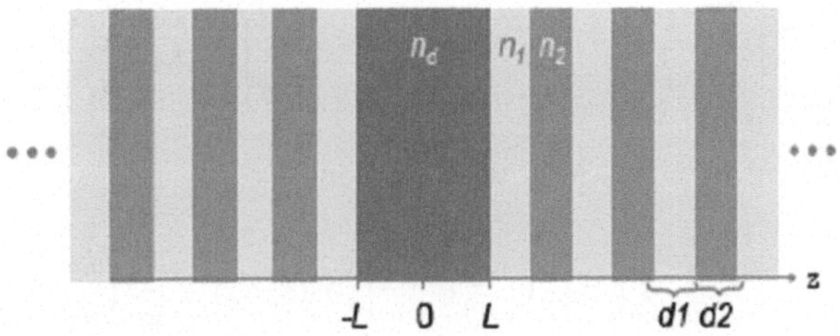

Fig. 2 A defect layer with index n_d surrounded with infinite layers in the left and right

is traveling to the right and to the left, say, for layer with index n_1 the amplitudes are A_1 and B_1 respectively and for layer with index n_2 the amplitudes are C_1 and D_1 respectively. Hence electric field in term of wave vector along z-axis for layer with index n_1 the solution

$$E(z) = A_1 e^{ik_1 z} + B_1 e^{-ik_1 z}$$

Similarly the electric field in term of wave vector along z-axis for layer with index n_2 the solution

$$E(z) = C_1 e^{ik_2(z-d_1)} + D_1 e^{-ik_2(z-d_1)}$$

The parameter k_1 and k_2 will be called the wave vectors, and can be expressed as $k_1 = \omega n_1$ and $k_2 = \omega n_2$. At the interface between layers ($z = d_1$), the solution and its derivative should be continuous. This gives a relation between plane waves amplitudes,

$$\begin{bmatrix} C_1 \\ D_1 \end{bmatrix} = \mathrm{M}_{12} \begin{bmatrix} A_1 \\ B_1 \end{bmatrix}$$

with

$$\mathrm{M}_{12} \begin{bmatrix} \frac{1}{2}(1 + \frac{k_1}{k_2})e^{ik_1 d_1} & \frac{1}{2}(1 - \frac{k_1}{k_2})e^{-ik_1 d_1} \\ \frac{1}{2}(1 - \frac{k_1}{k_2})e^{ik_1 d_1} & \frac{1}{2}(1 + \frac{k_1}{k_2})e^{-ik_1 d_1} \end{bmatrix}$$

and also at z = d, the interface between layer with index n_2 and n_1, continuity of the plane waves and its derivative gives,

$$\begin{bmatrix} A_2 \\ B_2 \end{bmatrix} = \mathrm{M}_{21} \begin{bmatrix} C_1 \\ D_1 \end{bmatrix}$$

where the matrix M_{21} is the same as M_{12} but with interchanging the indices. Concluding from the two matrix equations above,

$$\begin{bmatrix} A_2 \\ B_2 \end{bmatrix} = \mathrm{M} \begin{bmatrix} A_1 \\ B_1 \end{bmatrix}$$

where M = $M_{21}M_{12}$ with elements.

$$\begin{aligned}
\mathrm{M}(1,1) &= e^{ik_1 d_1}\left[\cos(k_2 d_2) + \frac{1}{2} i\left(\frac{k_2}{k_1} + \frac{k_1}{k_2}\right)\sin(k_2 d_2)\right] \\
\mathrm{M}(1,2) &= e^{-ik_1 d_1}\left[\frac{1}{2} i\left(\frac{k_2}{k_1} + \frac{k_1}{k_2}\right)\sin(k_2 d_2)\right] \\
\mathrm{M}(2,1) &= e^{ik_1 d_1}\left[-\frac{1}{2} i\left(\frac{k_2}{k_1} - \frac{k_1}{k_2}\right)\sin(k_2 d_2)\right] \\
&= \overline{\mathrm{M}(1,2)} \\
\mathrm{M}(2,2) &= e^{-ik_1 d_1}\left[\cos(k_2 d_2) - \frac{1}{2} i\left(\frac{k_2}{k_1} + \frac{k_1}{k_2}\right)\sin(k_2 d_2)\right] \\
&= \overline{\mathrm{M}(1,1)}
\end{aligned}$$

The matrix M shall be called as the Transfer Matrix of one unit cell of the structure. Using the transfer matrix technique, it can be shown easily that the relation between the plane waves amplitudes in the left and right exterior is as follows;

$$\begin{bmatrix} t \\ 0 \end{bmatrix} = \mathcal{M} \begin{bmatrix} 1 \\ r \end{bmatrix}$$

where t and r is the transmittance and reflectance of the wave.

A periodic layer medium which consist of the N unit cell and bounded by homogenous media for which the reflection and transmission coefficient are given by

$$\mathrm{r} = \left(\frac{\mathrm{b_0}}{\mathrm{a_0}}\right)_{\mathrm{b_N}=0} = \frac{\mathrm{M_{21}}}{\mathrm{M_{11}}}$$

$$\mathrm{t} = \left(\frac{\mathrm{a_N}}{\mathrm{a_0}}\right)_{\mathrm{b_N}=0} = \frac{1}{\mathrm{M_{11}}}$$

Transmittance is defined as;

$$\mathrm{T} = |\mathrm{t}|^2$$

Now if we plot a graph between normalized frequency versus transmittance then it gives the complete photonic band gap.

3 Results and Discussions

In this section, we consider the transmittance of the defect of PPCs. The defect PPCs is created when a layer of defect of plasma inserted in periodic alternate layered structure of air and SiO_2. For calculating the transmittance of the proposed periodic structures, we have taken $n_1 = 1$ (air), $n_2 = 1.46$ (SiO_2); thicknesses of air and SiO_2 are a = 0.125 mm and b = 0.086 respectively. The thickness of the defect layer of micro plasma is 0.5 mm. Using all these parameters; we have calculated the transmittance of the considered pure air/SiO_2 periodic structure, and symmetry and asymmetry plasma defect structure. Figure 3 shows the transmittance versus normalized frequency for different structures: (a) pure periodic structure i.e. $(air/SiO_2)^N$, (b) asymmetry defect periodic structure i.e. $(air/SiO_2)^N$/plasma/$(air/SiO_2)^N$ and (c) symmetry defect periodic structure i.e. $(SiO_2/air)^N$/plasma/$(air/SiO_2)^N$.

In Fig. 3a shows the transmittance of the simple photonic structure of $(air/SiO_2)^{10}$ without the defect, we obtained a large reflection photonic band gap approximately 2 multiples of normalized frequency when transmittance is zero.

Now we calculate the transmittance versus normalized frequency for micro-plasma as defect in air/SiO_2 asymmetry periodic structure and symmetry periodic structure. The Fig. 3b shows the transmittance of a defect in the asymmetry air/SiO_2 periodic structure, we predict that the band gap is enhanced and containing two transmission peaks in symmetry. It means the defect of the

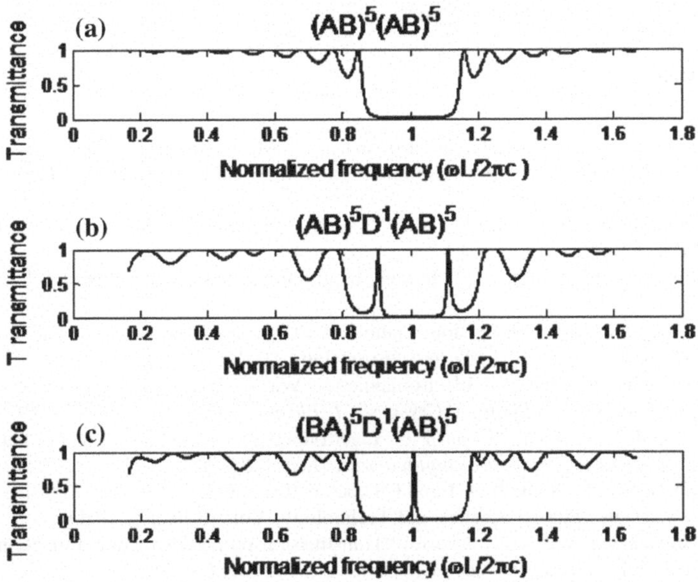

Fig. 3 Transmittance versus normalized frequency for **a** pure $(air/SiO_2)^{10}$ **b** asymmetric $(air/SiO_2)^5$/plasma/$(air/SiO_2)^5$ and **c** symmetric $(SiO_2/air)^5$/plasma/$(air/SiO_2)^5$

micro-plasma in the asymmetry air/SiO_2 is found the symmetry transmission around the edges of the enlarged band gap. Secondary transmission peaks of such micro-plasma defect in the asymmetry structure are also found symmetry. Such asymmetry structures with micro-plasma defect are may be used in optical switch. The transmittance of micro-plasmadefect in symmetry periodic structure of the air/SiO_2 is obtained asymmetry transmission peak with enlarged band gap as shown in the Fig. 3c. Such asymmetry structure with micro-plasma defect may be used as a sensor.

4 Conclusion

From the above discussion, we found the even number of peak in defect of micro-plasma in asymmetry periodic structure of air/SiO_2 and enlarged photonic band gap of defect micro-plasma in asymmetry periodic structure of air/SiO_2. Such defect asymmetry structures are may be used in optical switch. The defect of micro-plasma in asymmetry periodic structure of air/SiO_2 is obtained odd number of peak and also enlarged photonic band gap of defect micro-plasma in symmetry periodic structure of air/SiO_2. Such structure may be used in sensor.

Acknowledgment One of the authors Dr. G. N. Pandey is thankful to Dr. Ashok K Chauhan, Founder President, Amity University, Uttar Pradesh, for his interest in research and constant encouragement.

References

1. M. Scalora, J. P. Dowling, C. M. Bowden, and M. J. Bloemer, “Optical Limiting and Switching of Ultrashort Pulses in Nonlinear Photonic Band Gap Materials”, *Phys. Rev. Lett.*, **73**, 1368 (1994).
2. J. D. Joannopoulos, R. D. Meade and J. N. Winn, “*Photonic Crystals*”, Princeton University Press, Princeton (1995).
3. M. J. Kiskinen and R. Fernsler, “Photonic band gaps in dusty plasma crystals”, *Appl. Phys. Lett.*, 77, 1925 (2000).
4. H. Hojo and A. Mase, “Dispersion relation of Electromagnetic Waves in one-dimensional Plasma Photonic Crystals”, *J. Plasma Fusion Res.,* 80, 89, (2004).
5. H. Hojo and A. Mase, “Electromagnetic—Wave Transmittance Charecteristics in one-dimensional Plasma Photonic Crystals”, *J. Plasma Fusion Res.*, 8, 477–479, (2009).
6. H. Hojo and N. Uchida, “Control of Electromagnetic waves by 2-D plasma Photonic Crystals”, *plasma and industrial applications, IEEE,* (2005).
7. Laxmi Shiveshwari, “Some New Band Characteristics in One-Dimensional Plasma Dielectric Photonic Crystals” Plasma Science and Technology, Vol. 13, No. 4, (2011).
8. Laxmi Shiveshwari and S. K. Awasthi “Transmission properties of one dimentional ternary plasma photonic crystal” Physics of Plasmas, 22, 092129, (2015).

9. S. Prasad, V. Singh and A. K.Singh, "Model Propagation characteristics of EM waves in ternary one dimensional plasma photonic crystals, Optik, Vol 121, No 16, pp 1520–1528, (2010).
10. G. N. Pandey, Khem. B. Thapa, and S. P. Ojha, "Reflection Properties of One—Dimensional Magnetic Photonic Crystals", Journal of Ovonic Research (U S A), Vol. 6, 87–92, (2010).
11. G. N. Pandey, K. B. Thapa and S. P. Ojha, "Ominidirectional Reflection Band in One Dimensional Plasma Photonic Crystal", *Optik*—International Journal for Light and Electron Optics, Volume 124, Issue 18, pp 3396–340 (2013).
12. G. N. Pandey and S. P. Ojha, " Band Structure, Group velocity, Effective group index and Effective phase index of one Dimensional Plasma Photonic Crystal", Optik—International Journal for Light and Electron Optics, Volume 124, Issue 18, Pages 3514–3519,(2013).
13. G. N. Pandey,J. P. Pandey, A., K., Mishraand S. P. Ojha, "Three Dimensional Reflectance Properties of Plasma Dielectric Photonic Crystal" Proceedings of the International Conference on "Condensed Matter & Applied Physics" (ICC—2015)" AIP Conference Proceedings, Volume 1728, pp 020312-1–020312-4, (2016).
14. G. N. Pandey, Narendra Kumar, Khem B. Thapaand S. P. Ojha, "Reflectance Properties of One-Dimensional Metal-Dielectric Ternary Photonic Crystal" Proceedings of the International Conference on "Condensed Matter & Applied Physics" (ICC—2015) AIP Conference Proceedings Volume 1728, 020310-1–020310-3, (2016).
15. G. N. Pandey, J. P. Pandey, U. K. Pandey, BhagyashreeSancheti and S. P. Ojha, "Three Dimensional Reflectance Properties of Superconductor- Dielectric Photonic Crystal" Proceedings of the International Conference on "Condensed Matter & Applied Physics" (ICC—2015) AIP Conference Proceedings, 1728, pp 020306-1–020306-4, (2016).
16. Khem, B. Thapa, S. K. Singh, S. K. Srivastava, Om Prakash, G. N. Pandey, U. N. Singh and S. P. Ojha, "Reflection properties, anomalous group velocity and negative refraction index in one-dimensional Plasma Photonic Crystal" Proceedings of "Seventh International Conference on Optoelectronics, Fiber Optics and Photonics (PHOTONICS 2004)", 9–11 December, pp OMD-P-360 (2004).
17. S. K. Srivastava, G. N. Pandey, Khem B. Thapa, S. K. Awasthi and S. P. Ojha, " Design of Omnidirectional Optical Reflector using Plasma-Dielectric Photonic Crystal" Proceedings of "National workshop on Optoelectronic Material and Devices (AOMD-2007)", 27–29 Dec, pp 305–311, (2007).

Design of All-Optical Logical AND Gate Based on Photonic Crystal Using T-Shaped Waveguide

Tanvi Sachdev, Anil Kumar Shukla and G.N. Pandey

Abstract In this paper design of all optical logical AND gate is proposed based on 2-D photonic crystal. The structure proposed consists of Si rods placed in air background with T-junction using line defect and point defect in the photonic crystal. Finite difference time domain (FDTD) and plane-wave expansion (PWE) methods are used for the simulation and analysis of the performance of the proposed gate. The size of the logic gate is only about 16 μm × 16 μm and it works in the 1550 nm wavelength. The contrast ratio obtained for this structure is 11 dB. A good forbidden band gap in the normalized frequency range of $0.277883 \leq (1/\lambda) \leq 0.417455$ for Transverse Electric (TE) mode is obtained for this structure. Therefore, band gap width is 0.139572. The simplicity and the small sized T-shaped structure along with high contrast ratio make this device suitable candidate for integration in optical communication.

1 Introduction

Logic gates form the basis of the digital system. With the use of these gates various sequential and combinational circuits are achieved. Various switching operations along with operations at the nodes of the network such as data encoding, pattern matching, data decoding, etc. are performed with the use of these logic gates. Optical signal processing is the essential need in this era of high data speed along

T. Sachdev (✉) · A.K. Shukla
Amity Institute of Telecom Engineering and Management, Amity University, Noida, Uttar Pradesh, India
e-mail: tanvisachdeva1791@gmail.com

A.K. Shukla
e-mail: akshukla2@amity.edu

G.N. Pandey
Department of Physics, Amity Institute of Applied Science, Amity University, Noida, Uttar Pradesh, India
e-mail: gnpandey2009@gmail.com

© Springer Nature Singapore Pte Ltd. 2017

I. Bhattacharya et al. (eds.), *Advances in Optical Science and Engineering*, Springer Proceedings in Physics 194, DOI 10.1007/978-981-10-3908-9_27

with enhanced capacity in the telecommunication systems. Optical logic gates have rigorous relevance in the fields of optical switching.

Photonic crystals are materials which are artificially created by modulating the refractive index (RI) of the material in 1-D, 2-D or 3-D. This modulation of RI is done in comparison with the operating wavelength. With this varying RI, photonic crystal possess a property of trapping the light in the crystal itself and propagating the flow of light forcefully around spiky bends [1]. And sometimes the propagation of the light is also avoided by appropriate reflections from the crystal [2]. They generally possess frequency range of light in which light propagation is forbidden through the structure known as the photonic band gap of the crystal [1–9].

Technology based on photonic crystal (PhC) has attracted many researchers and scientists for recognition of optical equipments for the next generation Photonic Integrated Circuits (PICs). The knowledge can be employed in micrometer size instead of millimeter and centimeter dimensions [2]. Phc are believed to be used as the basic building block for creating future generation ultra fast optical integrated circuits [10]. Many advantages are offered by the Phc such as quick response time, low power consumption, compact size and fast recovery etc.

In the presented paper, a design is proposed on all-optical AND logic gate based on two-dimensional photonic crystal.

To understand the flow of light through the crystal, the simulation has been done using the PWE and FDTD methods [1, 4, 5]. The appropriateness of proposed design is determined by the contrast ratio of logic gate. It is ratio of power of logic "1" to logic "0". Important feature of the proposed gate, as compared to the previously designed gates, are its high contrast ratio which is 11 dB, and its small footprint size, which is in the order of 16 μm × 16 μm. Additionally, the band of its operating wavelength is centred at 1550 nm, which makes it suitable candidate for application in optical communication.

This paper is arranged as mentioned. Section 2 elaborates the design of the all-optical logic AND gate, which includes the detail of band structure of the PhC used in the model. This is followed by, Sect. 2.1 which explains the optimization of the point defect ('r_e'), followed by, Sect. 3, the description of simulation results which are used to establish the working principle of the optical logic gate. Finally, in Sect. 4, a conclusion is drawn on the presented work.

2 Structure Design and Band Diagram

In the paper, we present all-optical logical AND gate structure using the T-shaped waveguide. In this structure, there are 15 rods in the x direction and y direction each, thus forming a square lattice.

To analyze the behavior of the wave, dispersion diagram i.e. known as band diagram is studied to have the idea of band gap [3]. The analysis of the band gap is done using the plane wave expansion method (PWE) as shown in Fig. 2, which gives the propagation modes of the 16 μm × 16 μm photonic crystal lattice [11].

The band diagram gives a good forbidden band gap in the normalized frequency range of $0.277883 \leq (1/\lambda) \leq 0.417455$ for Transverse Electric (TE) mode. Band gap width is 0.139572. The schematic structure of the all-optical AND gate is shown in Fig. 1. The structure is designed using a 2-D 15 × 15 array of cylindrical Si rods. These rods are placed in the air background. The silicon rods have a radius of 0.2a. 'a' refers to the lattice constant which depicts the distance between two neighbour rods in the lattice.

The proposed structure operates as an AND gate with different input combinations and we get suitable power levels at the output port. The structure so presented is formed with line defect and point defect. Line defect is formed by removing dielectric rods which is treated as linear waveguide [8]. Point defect is formed by increasing the size of rod or removing the rod [1, 5]. To create the point defect optimization of a rod '*re*' is done in the lattice. The optimization of the point defect is so done that the fundamental working of the logical AND gate is achieved. Low power levels are defined as logic-0 and high power levels are defined as logic-1. When the input at both the input ports is low then a logic-0 output is obtained. For other combinations of inputs the optimization of the dielectric rod '*re*' is done. The optimized radius of the rod '*re*' is $0.3a$ and this optimised value maintains the output power value near to the desired power. The radius of '*re*' is so chosen that it absorbs some portion of light and prohibits the propagation of light through the waveguide giving a low level output. Thus at '*re*' a cavity is formed where absorption of light happens. If the obtained power at the output port is 50% below than the input power it is considered as logic-0 and if the obtained power at the output port is 50% above than the input power then it is taken as logic-1. For the proposed structure the power levels achieved at the output for different input combinations satisfy the working of the logical AND gate.

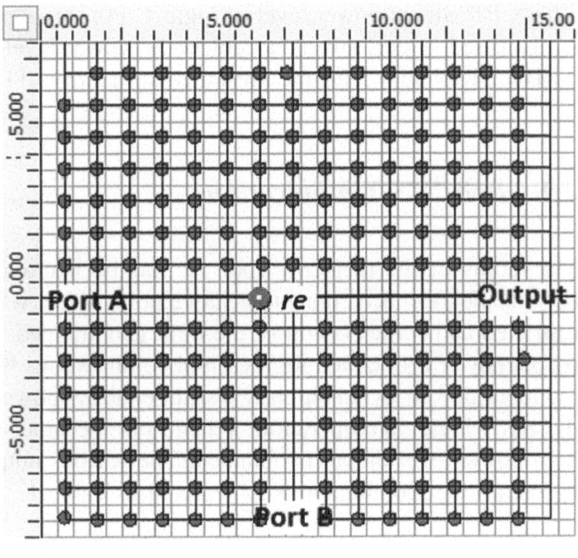

Fig. 1 Schematic structure of a two-dimensional silicon PhC AND gate

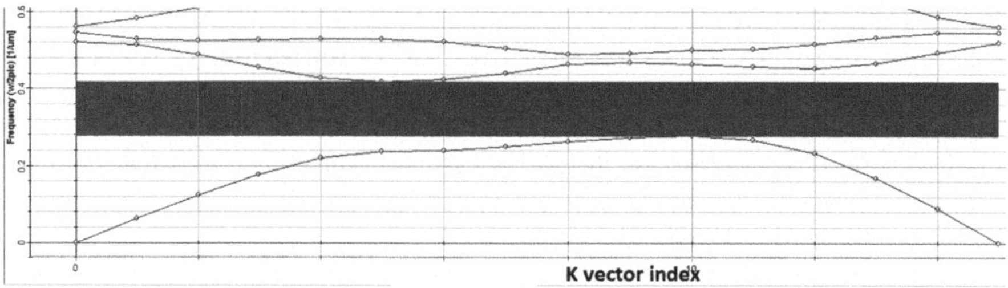

Fig. 2 Band diagram of 16 μm × 16 μm square lattice 2-D photonic crystals

Table 1 Truth table of the proposed AND gate along with obtained output power levels

Input port A	Input port B	Output port C	Output power from observation point (W/m)
0	0	0	0
0	1	0	0.08
1	0	0	0.19
1	1	1	0.44

Contrast ratio (CR) of logic gate is an aspect that determines the appropriateness of the design. It evaluates the performance of the device. Contrast ratio is calculated by taking the ratio of the signal power level for logic-1 with the signal power level of logic-0. This ratio is mathematically expressed as:

$$CR = 10 \log \frac{\text{P1}}{\text{P0}} \tag{1}$$

where, P1: Signal power level of logic-1, P0: signal power level of logic-0 and CR is expressed in decibels as it depicts the power level of on to off power ratio.

The CR obtained for the presented structure is 11 dB.

2.1 *Rod 'Re' Optimization*

The radius of the dielectric rod '*re*' has been modified from 0.2*a* to 0.3*a*. This optimization has been done to achieve point defect where the flow of light has been restricted and some portion of the light gets trapped. With this phenomenon a low output power is obtained for the input combinations '01' and '10'. So, if any one of the input power level is high, we still get a low power level output, which satisfies the AND logic gate operation. When both the input combinations are high then a high power level output is obtained. The various output power levels achieved for different input combinations is depicted in Table 1.

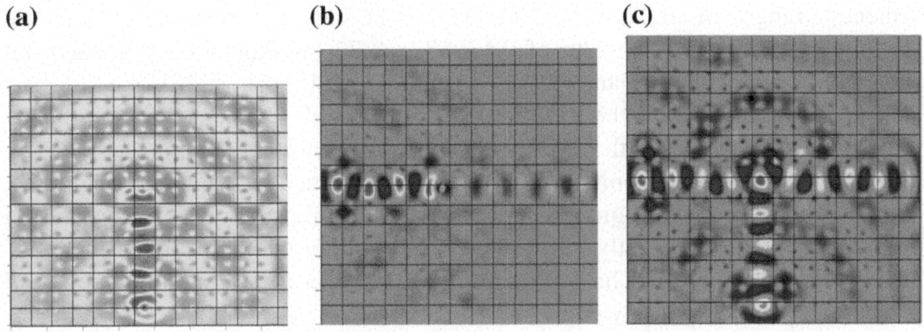

Fig. 3 Simulation results of the presented AND gate for the input combinations **a** 01, **b** 10, and **c** 11 along with their optical field patterns

3 Simulation Results

The simulation is carried out by using finite-difference time-domain (FDTD) method for the different combinations of the inputs with TE polarization [12]. Figure 3 displays the simulation results where the optical field patterns for the various input combinations can be observed. No output is observed when there is no light at the input ports. When the input is launched at port B and there is no input at port A(01), output power obtained is 0.08 W/m. As this value is very less than the launched input power so this considered to be a low value or logic-0. Likewise when there is no input launched at the input port B and a light is launched at the input port A(10), the obtained output power is 0.19 W/m. As this value is also very less as compared to the launched input at port A, so this value is also considered to be low value or logic-0. While a high output power level of 0.44 W/m is obtained at the output when the light is launched at both the input ports (11). The output power achieved for this input combination is high and is in accordance to the launched input powers, so this value is taken as a high value or logic-1.

The 1550 nm wavelength which is a standard optical communication carrier, is chosen for the wave propagation analysis of the structure. An elliptical rod, is placed at the intersection of the T-shaped waveguide plays the key role for the proper working of the proposed structure.

4 Conclusion

A structure of all optical AND logic gates based on 2D photonic crystal using silicon dielectric rods in air is proposed. This structure behaves as logic AND gate and efficiently acts as logic switch. The structure has been simulated and analyzed using FDTD and PWE methods on the Optiwave software. The band diagram obtained for this structure gives a good forbidden band gap in the normalized frequency range of $0.277883 \leq (1/\lambda) \leq 0.417455$ for Transverse Electric (TE) mode with band gap width of 0.139572. A 32 bit simulation is carried out using numerical FDTD and PWE methods with the transverse electric (TE) polarization to verify the truth table of All-optical AND gate. A Continuous input signal with wavelength of 1.55 μm is launched to this structure is partially reflected and partially transmitted. The structure is formed by point and line defect forming a T-shaped waveguide. The AND gate encompasses two inputs and an output. The output is logically "1" if and only if both of the inputs are at logic "1" and exhibits '0' otherwise. The working of the gate is verified by its simulations and by measuring the output powers.

The main advantages of this structure are the size of logic gate is small (about 15 μm × 15 μm) and it operates at 1.55 μm which is primary third optical window. Also, the obtained contrast ratio is 11 dB. This structure shows the possibility of designing other logic functions. It can be used for various logic circuits, flip flop, decoders and many other photonic integrated circuits.

References

1. Aghajamali, B. Javanmardi, M. Barati and C. J. Wu, "Defect modes properties in periodic lossy multilayer containing negative index materials with symmetric and asymmetric geometries structures," *Optik,* vol. 125, no. 2, pp. 839–843, 2014.
2. K. Sakoda, *Optical Properties of Photonic Crystals*, Springer, Berlin (2001).
3. M. J. Kiskinenand R. Fernsler, "Photonic band gaps in dusty plasma crystals", Appl. Phys. Lett., 77, 1925 (2000).
4. Mortaza Noshad, Amin Abbasi, Reza Ranjbar, Reza Kheradmand" Novel All-Optical Logic Gates Based on Photonic Crystal Structure" Journal of Physics: ConferenceSeries 350 (2012) 012007.
5. Q. Zhu and Y. Zhang, "Defect modes and wavelength tuning of one-dimensional photonic crystal with lithium niobate," *Optik*, vol. 120, no. 4, pp. 195–198, 2009.
6. J. D. Joannopoulos, S. G. Johnson, J. N. Winn and R. D. Meade, *Photonic Crystals: Molding the Flow of Light*, 2nd ed., Princeton University Press, 2008.
7. V. Dinesh Kumar, T. Srinivas, A. Selvarajan, "Investigation of ringresonators in photonic crystal circuits," Photonics Nanostruct. Fundamentals Applic., 2, pp. 199–206.
8. H. Takeda and K. Yoshino 2003 "Tunable light propagation in Y-shaped waveguides in two-dimensional photonic crystals utilizing liquid crystals as linear defects," Phys. Rev. B, vol. 67, p. 073106.
9. Igor A. Sukhoivanov, Igor V. Guryev, *Physics and Practical Modeling*, Springer Series in Optical Sciences, ISSN 0342-4111 e-ISSN 1556-1534.
10. G. N. Pandey, Khem. B. Thapa, and S. P. Ojha, "Reflection Properties of One—Dimensional Magnetic Photonic Crystals", Journal of Ovonic Research (U S A), Vol. 6, 87–92, (2010).
11. J. H Yuan, Y. Y. Lu, "Photonic bandgap calculations with Dirichlet-to-Neumann maps," J. Opt. Soc. Am. 23, pp. 3217–3222.
12. S.D. Gedney, *Introduction to finite-difference time-domain(FDTD) method for electromagnetics*, Morgan and Claypool (2010).

Part V
Lasers, Interferometry, Imaging, Devices

Surface Plasmon Resonance Based Differential Phase Analysis Using Mach-Zehnder Interferometric Set-up

Jayeta Banerjee, Mahua Bera and Mina Ray

Abstract A novel scheme of experimental surface plasmon resonance (SPR) based differential phase analysis using Mach-Zehnder interferometric set-up has been demonstrated. Two glass prisms have been used in the two arms of the interferometric set-up. SPR modulated interferometric fringes can be observed with the use of thin Al coating on the hypotenuse surface of the glass prism, placed in one of the arms of interferometric set-up. Moreover, the change in fringe pattern can also be interpreted in terms of the direct difference of phase between Al-coated and uncoated glass prisms. SPR based direct difference of phase between test and reference beams can be measured using this technique which can be utilized for nanoplasmonic sensing.

1 Introduction

Surface plasmons are the free charge oscillations that occur at the interface of metal and dielectric having real dielectric permittivity of opposite signs. Surface plasmon resonance (SPR) measurement technique has been used extensively for nanoplasmonic sensing [1]. SPR measurement using phase detection based interferometric technique can improve the resolution of the device over conventional intensity measurement [2]. SPR based phase imaging [3] by differential phase measurement [4] between the p-and s-polarized beam has been proposed for the detection of RI change of the order of 10^{-5} RIU. Huang et al. have reviewed quite a number of SPR based phase measurement techniques [5]. In most of the literature, some kind of interferometric set-up [6] is involved which requires critical alignment. In our previous work, we have already demonstrated SPR modulated radially sheared interference fringes [7] by using Birefringent lens (BL) and SPR based moiré pattern generation [8] using a BL and a Wollaston Prism under dual shearing

J. Banerjee · M. Bera · M. Ray (✉)
Department of Applied Optics and Photonics, University of Calcutta, JD-2, Salt lake, Sector-III, Kolkata, West Bengal 700106, India
e-mail: mraphy@caluniv.ac.in

© Springer Nature Singapore Pte Ltd. 2017

I. Bhattacharya et al. (eds.), *Advances in Optical Science and Engineering*, Springer Proceedings in Physics 194, DOI 10.1007/978-981-10-3908-9_28

(lateral as well as radial) environment. Differential phase measurement is more robust technique than conventional phase jump shift measurement, which has also been studied theoretically by us [9].

Adjustable two window Mach-Zehnder interferometer [10] has been already reported for analyzing phase profile based on 4f set-up. Both reflectance and phase can be demonstrated based on Fabry-Perot interferometer [11] in spectral interrogation. Metal thickness can also be measured from interference fringe pattern using Mach-Zehnder interferometer based on phase difference due to SPR modulation [12]. For bio-sensing application differential phase analysis [13] is advantageous as it increases the sensitivity up to the order of 10^{-8} RIU. In this paper, we have experimentally demonstrated the differential phase measurement using Mach-Zehnder interferometric set-up and the effect of SPR on the fringe pattern obtained.

2 Description of Experimental Set-up

The schematic diagram of the experimental set-up is demonstrated in Fig. 1. The beam expander and collimator are used for beam expansion and collimation of the beam coming out of the He–Ne laser of working wavelength 632.8 nm. The collimated beam passed through a polarizer followed by a quarter wave plate (QWP). In order to get circular polarized beam, transmission axis of polarizer is kept at 45° and fast axis of QWP is kept vertical. An aperture is used to limit the diameter of the beam. The beams coming out of the polarizing beam splitter (PBS) are orthogonally polarized with respect to each other. As shown in Fig. 1 two beams are incident almost normally on the right-angled fused silica glass prisms. Two reflected beams from two prisms are subsequently passed through the second PBS. The two beams interfere after the analyzer and form straight fringes. The emergent beam is focused by a lens and captured by a charge coupled device (CCD).

The refractive index of Fused silica glass prism and Al-metal layer can be calculated using Sellmeier's dispersion equation [14] and Drude formula [15] respectively. The resonance parameters such as reflectance and phase response of a

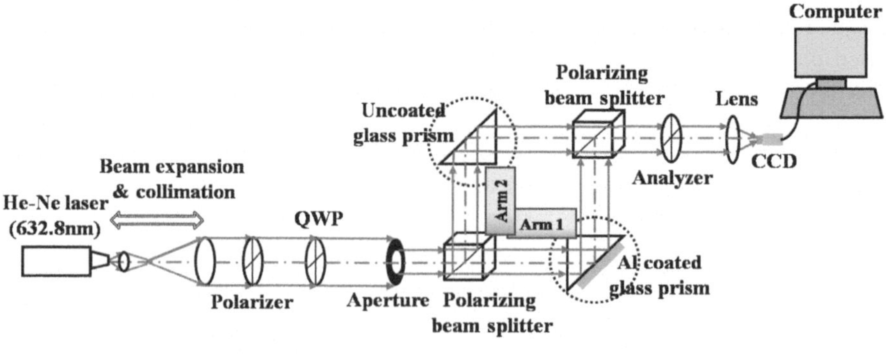

Fig. 1 Experimental set-up for the evaluation of SPR based differential phase

three-layer Kretschmann configuration can be calculated using Fresnel equation and by characteristic transfer matrix (CTM) method [16, 17]. Selection of optimized metal thickness and resonance angle is very important for visualizing SPR effect on interference fringe pattern [18]. After analyzing the resonance parameters, 40 nm thicknesses Al film is coated on the hypotenuse face of equilateral right angled glass prisms. The beam is incident on the Al-coated hypotenuse surface of the prism at SPR angle, 44.5°. Thus the beam is incident almost normally at the front surface (base) of the prism.

3 Mathematical Background

Mach-Zehnder interferometer with PBS has been analyzed using tensor algebra. The circular polarized beam is represented as

$$In = \begin{pmatrix} 1 & 0 \\ 0 & -i \end{pmatrix} \times \begin{pmatrix} 1 \\ 1 \end{pmatrix} = \begin{pmatrix} 1 \\ -i \end{pmatrix} \quad (1)$$

where linear polarizer is at 45° and fast axis of QWP is kept vertical.

The PBS transmits horizontal polarization and reflects vertical polarization. The general matrix representation for PBS is

$$PBS = \begin{pmatrix} t_H & ir_H & 0 & 0 \\ ir_H & t_H & 0 & 0 \\ 0 & 0 & t_V & ir_V \\ 0 & 0 & ir_V & t_V \end{pmatrix} \quad (2)$$

Here, PBS^t has values of $t_H = 1$ and PBS^r has values of $r_V = 1$.

TIR can be expressed by the following matrix

$$p_{TIR} = \begin{pmatrix} r_{p_{TIR}} exp\left(i\phi_{p_{TIR}}\right) & 0 \\ 0 & r_{s_{TIR}} exp\left(i\phi_{s_{TIR}}\right) \end{pmatrix} \quad (3)$$

Similarly, SPR can be expressed as

$$p_{SPR} = \begin{pmatrix} r_{p_{SPR}} exp\left(i\phi_{p_{SPR}}\right) & 0 \\ 0 & r_{s_{SPR}} exp\left(i\phi_{s_{SPR}}\right) \end{pmatrix} \quad (4)$$

The matrix representation for the beam coming out of arm 1 is

$$O_1 = PBS^r P_{TIR} PBS^t In \quad (5)$$

The matrix for the beam coming out of arm 2 is

$$O_2 = PBS^t P_{TIR} PBS^r In \tag{6}$$

Matrix multiplications are done by making matrices of same order using Kroneker function.

The intensity distribution in the interference pattern is given by

$$I_{out} = |O_1 + O_2|^2 \tag{7}$$

4 Experimental Results and Discussions

Straight line fringes are produced by interference of two beams reflected from two uncoated prism satisfying only TIR condition (greater than critical angle, 43.5°) as shown in Fig. 2a. The two uncoated prisms have also been placed on the calibrated rotating table here in such a way that the incident/reflected beam at the hypotenuse surface is at 44.5° angle for the comparison with the results of Al-coated prism. SPR induced phase change is visible in Fig. 2b when the Al-coated prism is placed in one arm of the interferometric set-up. In our case, fringes get modulated due to phase change when one uncoated prism is replaced by Al-coated prism satisfying resonance conditions. The "**STEP**" slope-like fringes with sharp bend has been

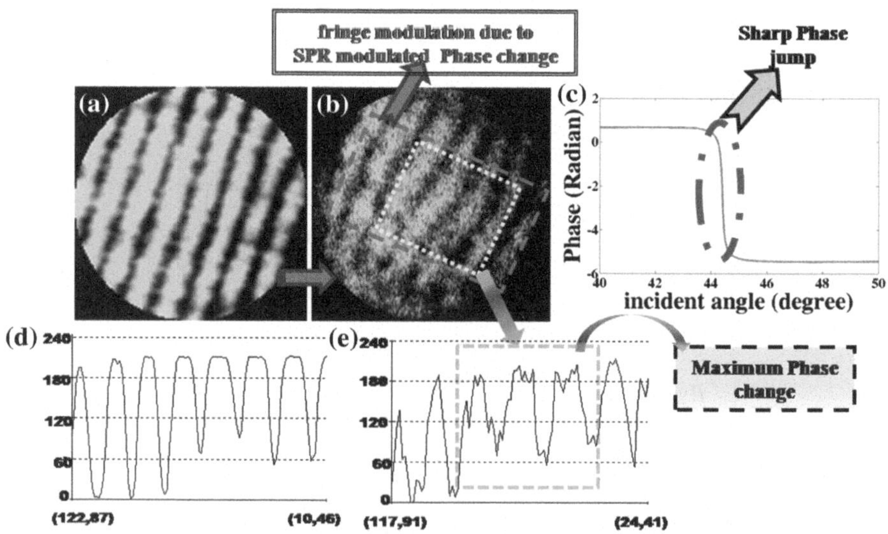

Fig. 2 Fringe pattern for **a** both uncoated prisms, **b** Al-coated and uncoated prisms in two opposite arms, **c** simulated phase response in angular interrogation, **d**, **e** line profiles of patterns shown in **a**, **b**

observed. As one arm has uncoated prism like previous case, and another arm has Al coated prism, thus the change in the fringes has been occurred due to the coated prism at resonance condition with optimum metal thickness. So we can conclude that step like phase jump in the interferometric fringes are due to SPR [19]. As the two interfering beams contain information of phase for both beams (one reflected from uncoated prism and another from coated prism), resultant fringe pattern gives information regarding differential phase between those two beams coming from the two arms. The orientation of the fringe pattern is kept same in both cases. Here we have placed the imaging plane of the CCD parallel to the surface of the analyzer and also to the PBS in order to compare both the results. The distance between PBS and analyzer and also between analyzer and CCD are kept same in order to get same sized image. We have ensured the same position of the uncoated and Al-coated prism in calibrated rotating mounts. So the modification in the fringe pattern has been observed only due to SPR.

Theoretically phase response has been studied for three layer Kretschmann structure with air as analyte in angular interrogation. It can be seen from Fig. 2c that there is a step like phase jump at resonance angle, which is also indicated in Fig. 2b by red colored rectangular region. Line profiles of fringes in Fig. 2a, b are shown in Fig. 2d, e respectively. Modification in line profile in Fig. 2e is due to change of the phase due to SPR. Yellow region in line profile indicates maximum phase change as already discussed theoretically.

5 Conclusions

Differential phase analysis based on SPR phenomena has been studied using Mach-Zehnder interferometer. Interferometric straight fringe is modulated by SPR and a sharp jump is prominent in straight fringes. This is only because of the phase change introduced by SPR phenomena. The direct difference in phase between the reference and unknown sample can also be observed by the interferometric fringe pattern if we place two Al-coated prisms in the two arms with test and reference sample adjacent to the metal layer.

Acknowledgements The author, Jayeta Banerjee would like to acknowledge Department of Science and Technology, Government of India for financial support under Women Scientist Scheme A.

References

1. Homola, J., "Surface plasmon resonance sensors for detection of chemical and biological species", Chemical Review **108**, 462–493 (2008).
2. Nelson, S.G., Johnston, K.S. and Yee, S.S., "High sensitivity surface plasmon resonance sensor based on phase detection", Sens Actuators B **35–36**, 187–191 (1996).

3. Notcovich, G.A., Zhuk, V. and Lipson, G.S., "Surface plasmon resonance phase imaging", Appl. Phys. Lett., **76**, 1665–1667 (2000).
4. Ho, H.P. and Lam, W.W., "Application of differential phase measurement technique to surface plasmon resonance sensors", Sens Actuators B, **96**, 554–559 (2003).
5. Huang, H.Y., Ho, P.H., Wu, Y.S. and Kong, K.S., "Detecting phase shifts in surface plasmon resonance: A review", Advances in Optical Technologies Article ID **471957,** 1–12, doi:10.1155/2012/471957 (2012).
6. Kashif, M., Bakar, A.A.A., Arsad, N. and Shaari, S., "Development of phase detection schemes based on surface plasmon resonance using interferometry", Sensors **14**, 15914–15938 (2014).
7. Bera, M., Banerjee, J. and Ray, M., "Experimental surface plasmon resonance modulated radially sheared interference imaging using a birefringent lens", Appl. Phys. Lett. **104**, 251104-1-5 (2014).
8. Bera, M., Banerjee, J. and Ray, M., "Moiré pattern generation by dual shearing and its modulation by surface plasmon resonance", Opt. Lett. **40**, 1857–1860 (2015).
9. Bera, M., Banerjee, J. and Ray, M., "Resonance parameters based analysis for metallic thickness optimization of a bimetallic plasmonic structure", Journal of Modern Optics **61**, 182–196 (2014).
10. García, D.I.S., Arellano, N.I.T., García, A.M., Álvarez, J.A.R., Zurita, G.R. and Pérez, A.M., "Adjustable-window grating interferometer based on a Mach-Zehnder configuration for phase profile measurements of transparent samples", Optical Engineering **51**(**5**), 055601-1-7 (2012).
11. Hadjar, Y., Renault, M., Blaize, S., Bruyant, A., Vincent, R. and Hmima, A., "Compact interferometer transducer based on surface plasmon phase resonance", J. Opt. Soc. Am. A **32** (**5**), 771–777 (2015).
12. Liu, C., Liu, O. and Hu, X., "SPR phase detection for measuring the thickness of thin metal films", Optics Express 22(**7**), 7574–7580 (2014).
13. Wu, S.Y., Ho, H.P., Law, W.C., Lin, C. and Kong, S.K., "Highly sensitive differential phase-sensitive surface plasmon resonance biosensor based on the Mach–Zehnder configuration", Optics Letters **29**(**20**), 2378–2380 (2004).
14. The Practical Application of Light, MELLES GRIOT, Vol (Catalog) X, Barloworld Scientific, p. 4.8.
15. Jha, R. and Sharma, A., "High performance sensor based on surface plasmon resonance with chalcogenide prism and aluminum for detection in infrared", Opt. Lett., **34**(**6**), 749–751 (2009).
16. Born, M., and Wolf, E. Principles of Optics, 7th expanded edition (Cambridge University Press, Cambridge, 1999).
17. Abeles, F. Recherches sur la propagation des ondes electromagnetiques sinusoidales dans les milieux stratifies, Application aux couches minces. Ann Phys (Paris) **5**, 596–640 (1950).
18. Bera, M., Banerjee, J. and Ray, M., "Surface plasmon resonance mediated fringe modulation using a birefringent lens creating radial shearing environment", Journal of the Optical Society of America B, **32**(**5**), 961–970 (2015).
19. Nikitin, P.I., Beloglazov, A.A., Kochergin, V.E., Valeiko, M.V. and Ksenevich T.I.,"Surface plasmon resonance interferometry for biological and chemical sensing", Sens Actuators B **54**, 43–50 (1999).

Modeling of a Transversely Pumped Aprotic Liquid Laser

A.K. Varshney, Avinash C. Verma, Gaurav Singhal, Mainuddin and R.K. Tyagi

Abstract The present paper discusses laser kinetic modeling of diode pumped flowing medium aprotic liquid laser containing Nd3+ ions in POCl3 solvent host. Transverse pumping has been examined, which is ideal from point of view of laser scalability to higher power levels. In the computations, spectroscopic parameters of aprotic solution having an Nd3+ ion concentration of 0.3 M have been considered and numerical calculations were performed for analyzing the effect of various parameters including; active ion concentration, gain medium length, absorption depth, output coupler transmission and input pump power on laser output and efficiency. The results indicate that nearly 1.5 kW laser output may be obtained for an input pump power of 5.0 kW with a slope efficiency of nearly 31% having a threshold pump power requirement of 300 W for a given cell length of 15 cm. Hence, a set of optimized parameters have been obtained for designing an efficient transversely diode pumped kW level flowing liquid laser.

1 Introduction

Lasers which may be scaled up to large power levels have significant potential for both civil and military applications. Various lasers ranging from solid-state, fiber, chemical and gas lasers have been developed towards this end. However each has its respective merits and demerits. Rare earth doped glass and crystal based solid-state lasers [1] have been extensively used in variety of applications. However, these lasers suffer from thermal issues at higher power levels. Fiber lasers [2] are another promising candidate owing to their large surface to volume ratio providing easier thermal management. However, nonlinear scattering and thermal

A.K. Varshney (✉) · A.C. Verma · G. Singhal · R.K. Tyagi
Laser Science and Technology Centre, Metcalfe House, 110054 New Delhi, India
e-mail: agvarshney@yahoo.com

Mainuddin
Department of Electronics & Communication Engineering,
Jamia Millia Islamia, 110025 New Delhi, India

© Springer Nature Singapore Pte Ltd. 2017
I. Bhattacharya et al. (eds.), *Advances in Optical Science and Engineering*,
Springer Proceedings in Physics 194, DOI 10.1007/978-981-10-3908-9_29

fracture in fiber at multi-kilowatt level presently restrict their applications. Chemical lasers such as Chemical Oxygen Iodine Laser (COIL) [3] is one the most potent laser as far as power scalability aspect is concerned, however few logistic constraints and overall system volume are issues that need to be resolved.

Another approach in the direction of developing a compact scalable laser with large power output is the use of liquid medium. Liquid lasing medium such as the ones based on inorganic solvents may be employed, filled in a laser cavity of given size and shape with heat removal being affected by circulating the medium through the heat exchanger. In this context, Lawrence Livermore National Laboratory successfully demonstrated high average power dye laser [4]. However, inorganic solvent based liquid medium laser were not extensively pursued in initial phase of laser development [5, 6] due to broadband pumping yielding low efficiency and high thermal load. In recent years with the development of high power diode lasers, liquid lasers have again attracted attention of researchers towards development of scaled up liquid lasers as selective excitation using diode laser may greatly reduce thermal load leading to easier thermal management [7–9].

These liquid lasers employing inorganic aprotic liquid solvent [7] with Lewis acid impregnated with rare earth ions such as Nd^{3+}, Nd^{3+}: phosphate glass micro-balls or thin sheets dispersed in organic match liquid [9] are being studied using transverse laser diode pumping for scaled up laser development.

Keeping in view the potential of this laser and also to examine its scalability to high power levels, a laser kinetic model applicable to transversely pumped flowing inorganic aprotic liquid laser has been developed. The liquid lasing media consisting of phosphorus oxychloride ($POCl_3$) host solvent containing rare earth ions Nd^{3+} with Lewis acid $SnCl_4$ is preferred as the solution since it yields high efficiency and exhibits relatively good stability against decomposition. In the computations, spectroscopic parameters of aprotic solution having an Nd^{3+} concentration of 0.3 M have been considered and numerical calculations were performed for analyzing the effect of various parameters viz., active ion concentration, gain medium length, absorption depth, output coupler transmission, input pump power etc. on to the laser output and efficiency of the liquid laser. The simulation has also been carried out separately to check the validity of the model and found in good agreement with the few results reported in patent US-6600766B1-2003. A set of optimized parameters is obtained for designing an efficient transversely diode pumped kW level flowing liquid laser.

2 Description of the Model

The energy level diagram of Nd^{3+} in $POCl_3$ with Lewis acid $SnCl_4$ depicting the pump excitation and laser wavelengths along with involved energy levels is shown in Fig. 1. The solution having an Nd^{3+} concentration of 0.3 M is shown to be resonantly pumped at 863 nm corresponding to ${}^4F_{3/2}$–${}^4I_{9/2}$ transition using diode laser excitation and then laser emission at 1054 nm is achieved corresponding to

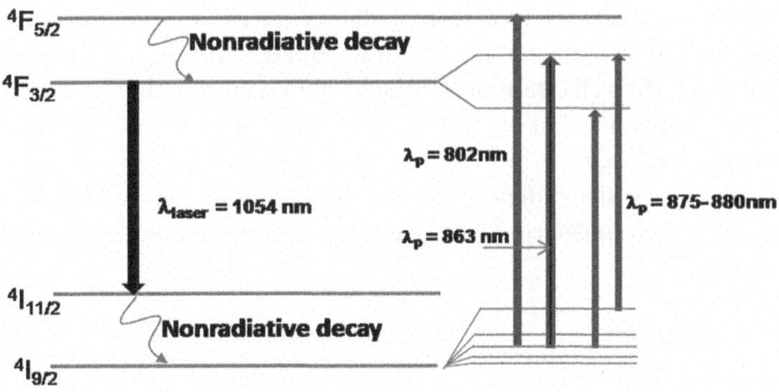

Fig. 1 Energy level diagram of Nd^{3+}: $POCl_3$ liquid laser

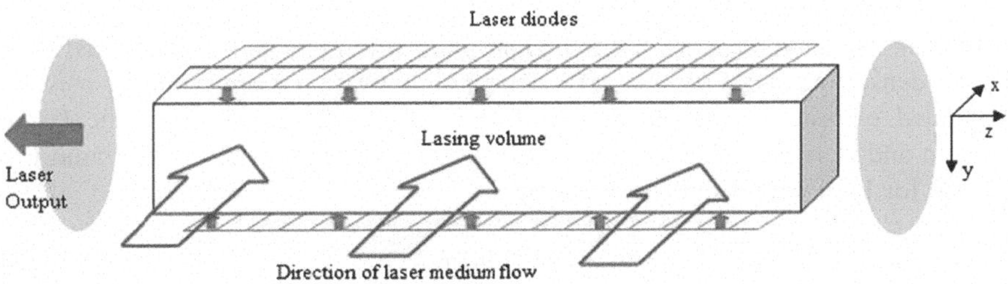

Fig. 2 Schematic of transversely pumped Nd^{3+}: $POCl_3$ flowing liquid laser

$^4F_{3/2}$–$^4I_{11/2}$ transition. The solution has peak absorption at 802 nm wavelength corresponding to transition in the $^4F_{5/2}$ pump band above the laser level $^4F_{3/2}$.

In our analysis we have considered 863 nm wavelength for resonant pumping as it reduces the total quantum defect between the pump and laser emission wavelengths, thus reducing the thermal effects induced by optical pumping. As per literature, among the peak absorption lines, the absorption coefficient is $\sim$1.8 cm^{-1} at wavelength of 863 nm and stimulated emission cross-section is nearly 6×10^{-20} cm^2 with fluorescence lifetime being $\sim$330 μs at emission wavelength of 1054 nm. The lower absorption cross-section at 863 nm is compensated with possible higher rare earth ionic concentration in the liquid. It also provides more uniform pumping of the lasing liquid medium as an accrued benefit.

A scalable version of liquid laser would essentially require a circulating lasing medium along with transverse pumping in a direction orthogonal to the lasing direction. Hence, in such a configuration the direction of circulation, pumping and lasing may be mutually perpendicular. The pump light from laser diode arrays as shown in Fig. 2 is closely coupled to the liquid medium which is transversely pumped from opposite sides. It is also prudent to mention that pump beam variation along the x axis is considered to be Gaussian being the fast axis of laser diodes and in the *z-axis* is taken as uniform being the slow axis for laser diodes. The pump radiation absorption direction is the *y-axis*. Figure 2 also shows the basic schematic depicting all these directions in a rectangular cell with a cross-section of 10 mm 10 mm and a cell length (*l*) of 150 mm filled with liquid lasing medium. A two-dimensional discretization of the liquid gain medium along *x-axis* and *y-axis* is made in such a manner that each divided volume element has dimension of Δx. Δy. *l* where Δx, Δy < < *l*.

The standard space dependent laser rate equations in steady state condition in each volume element considering homogeneous pumping along z-direction may be written as below,

$$\frac{dn(x,y)}{dt} = p(x,y) - \frac{n(x,y)}{\tau} - \frac{c\sigma n(x,y)r(x,y)}{n_l} = 0 \tag{1}$$

$$\frac{dR}{dt} = \frac{c\sigma}{n_l}\int n(x,y)r(x,y)dV - \frac{c\delta R}{2n_l l} = 0 \tag{2}$$

In these equations, *n*(*x*, *y*) and *r*(*x*, *y*) are population inversion density and laser photon density respectively. *p*(*x*, *y*) is the pumping rate per unit volume. σ is the stimulated emission cross- section, τ is the upper state life time of the lasing medium and n_l is the refractive index at lasing wavelength. The cavity round trip loss (δ) for laser wavelength is given by

$$\delta = 2\alpha_l l - \ln(R_1 R_2) \tag{3}$$

where R_1 and R_2 are the mirror reflectivity and α_l is the absorption coefficient at laser wavelength. *p*(*x*, *y*) is the pumping rate for single traverse of the pump beam through the gain medium and is expressed as,

$$p(x,y) = \frac{1}{V_p}\frac{P(x,y)e^{-\left(\frac{2x^2}{\omega_{po}^2}\right)}}{h\nu_p}\left(1 - e^{-\alpha_p \Delta y}\right) \tag{4}$$

where $V_p = \Delta x.\ \Delta y.\ l$ is the pump volume of each element and *P*(*x*, *y*) is the total pump power.

Using the developed algorithm in MATLAB, these equations are solved for the population distribution *n*(*x*, *y*) in each volume element. Finally, the laser output power in each volume element using Rigrod model is computed using the following expression,

$$P_l(x,y) = I_s \Delta x \Delta y T\left[\frac{-2g(x,y)l}{\ln(1-S-T)} - 1\right] \tag{5}$$

where, I_s is the saturation intensity and *g*(*x*, *y*) is the small signal gain in each element. $T = (1 - R_1)$ is the output coupler transmission and S is the scattering losses.

3 Simulation Results and Analysis

The simulations have been carried out to evaluate pump power distribution, the total extractable power by determining the gain of the medium for given transmission and active ion concentration. The effect of variation in mirror transmission and active ion concentration on threshold power and output power has also been studied.

The foremost requirement before carrying out computations of each of the above stated parameters is to confirm grid independence of the results. This is because the selection of grid numbers along x-y direction may influence the overall simulation results. Therefore, the dependence of the optical efficiency, one of the most critical parameters, on grid numbers was tested by varying the number of grids in x-y direction as shown in Fig. 3. The results stabilize for grid no. of 30 $\times$ 30 beyond which there is practically no variation in the results obtained. Further increase in grid no. would essentially make the solution time intensive without significant enhancement in accuracy.

The determination of pump power distribution is critical in ascertaining whether the depth of the medium is sufficient to absorb the overall pump power. Figure 4 shows the computed variation of pump intensity for both sided transverse pumping at various locations in the bulk of the lasing medium for a total irradiating pump power of 5 kW. Here the effects of diffused reflection leading to pumping equalization have been neglected and only direct irradiation of the lasing medium through laser diodes is taken into account. It is evident that better uniformity of pumping along y-direction is achieved at pump wavelength of 863 nm as compared to the case of 802 nm pump. The ratio of minimum to peak intensity variation along y-direction is $\sim$0.7 for the former and $\sim$0.2 in case of latter. The residual power exiting the medium volume for cell size of 10 mm in y-direction is nearly zero for both cases. Hence, the absorption depth is required to be chosen such that an

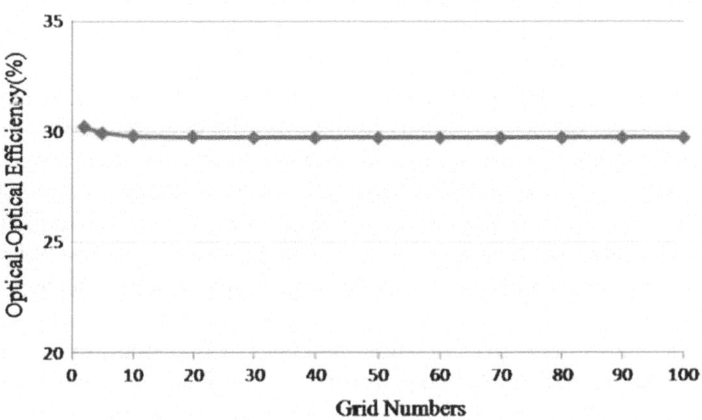

Fig. 3 Dependence of optical efficiency on grid numbers

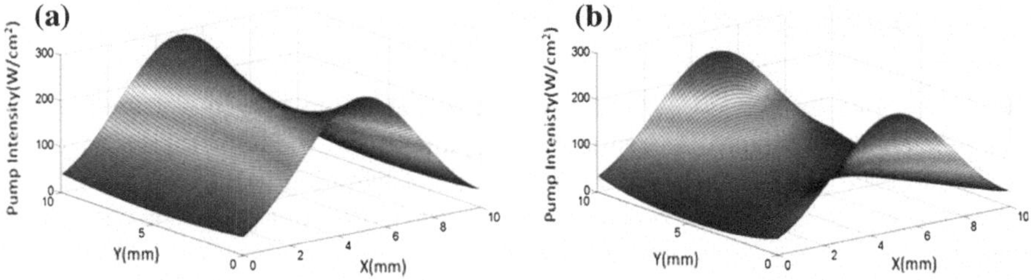

Fig. 4 Pump power distribution along x and y direction at wavelength **a** 863 nm **b** 802 nm

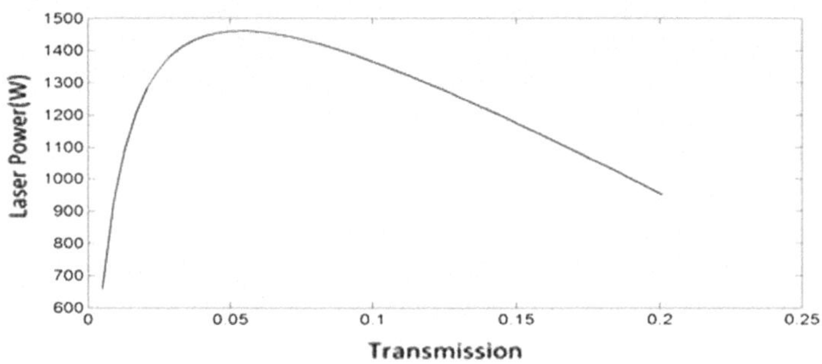

Fig. 5 Laser power versus output coupler transmission

optimum trade- off between pumping uniformity and absorption efficiency is achieved. Typically in the present case the absorption efficiency is 99.5% for 802 nm and 86% for 863 nm. However, as mentioned the benefits in terms of thermal load and pump uniformity are significant in case of 863 nm.

Subsequently, it is essential to compute the total expected power for the specified input parameters. Figure (5) depicts the variation of total expected multimode power corresponding to the pumping profile of Fig. 4 (a) for different coupler transmission. It is clear that maximum power of nearly 1.5 kW is achievable at output coupler transmission of 0.05 (5%).

The other aspect that is required to be analyzed is the slope efficiency of the medium. The slope efficiency actually varies with the coupler transmission which primarily controls the overall loss provided to the medium. Hence, considering the optimal transmission value of 5% the slope efficiency curve is represented in Fig. 6.

Hence, it can be inferred that for a typical case of Nd species activation of 0.3 M, the laser power increases continuously with pump power. The threshold power for given coupler transmission is nearly 300 W with a typical slope efficiency of nearly 31%.

Further, it is also imperative to study the effect of variation in pump power on optical conversion efficiency for an active medium of given concentration, the plot is shown in Fig. 7. It is seen that optical to optical conversion efficiency of 25–30% is achievable for pump power in excess of 3 kW for present medium configuration.

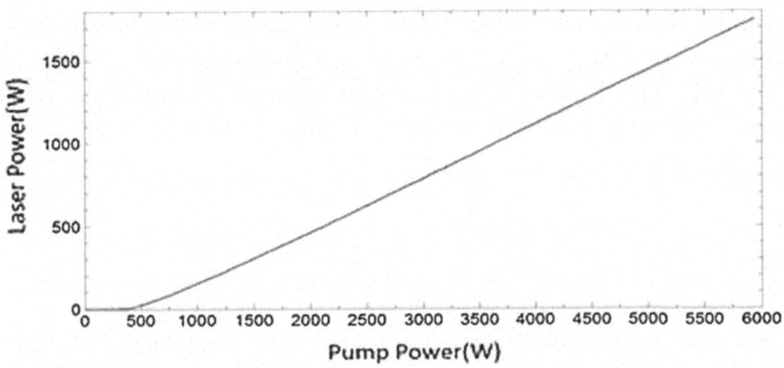

Fig. 6 Laser power versus pump power

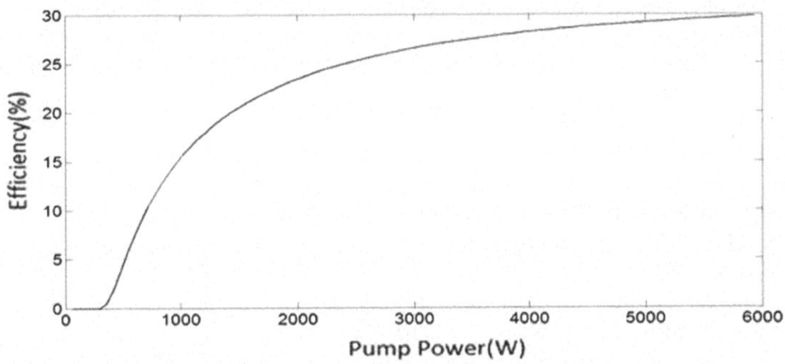

Fig. 7 Slope efficiency curve

In the region of pump power <3 kW the conversion efficiency is lower on account of gain of the medium being comparable to the cumulative medium loss along with necessity for achieving the laser threshold. It can also be inferred from the results that medium activated with 0.3 M Nd^{3+} ions is sufficient to handle the incident pump power in the range of 3–6 kW.

4 Conclusion

The transversely pumped flowing aprotic liquid laser is a promising candidate from point of view of scalability to higher power levels. In the paper, the modeling of transversely pumped Nd^{3+}: $POCl_3$ liquid laser has been carried out. The output characteristics of aprotic liquid laser are analyzed under the influence of various parameters like Nd^{3+} ion concentration, output coupler transmission and input pump power. The simulated results have been validated against few results presented in US patent. The exhaustive numerical studies have resulted in indentifying a set of near optimized parameters for designing an efficient transversely diode pumped kW class flowing liquid laser. The results indicate that nearly 1.5 kW laser output power may be obtained for an input pump power of 5.0 kW with a slope efficiency of nearly 31% having a threshold pump power requirement of 300 W for a cell of dimension 10 mm × 10 mm × 150 mm.

References

1. Bowman, S. R., "High power diode-pumped solid state laser," Opt. Eng. 52(2), 021012 (2012).
2. Zervas, M. N. and Codemard, C. A., "High power fiber lasers: a review," IEEE Journal of Selected Topics in Quantum Electronics 20(5), 1 (2014).
3. Singhal, G., et. al., "Test bed for a high throughput supersonic chemical oxygen iodine laser," Quantum Electronics 41(5), 430 (2011).
4. Bass, I.L., Bonnno, R.E., Hackel, R.P., and Hammond, P.R., "High average power dye laser at Lawrence Livermore National Laboratory," Applied Optics 31(33), 6993 (1992).
5. Lempicki, A., Heller, A., "Characteristics of Nd^{3+}; $SeOCl_2$ liquid laser," Applied Physics Letters 9, 108 (1966).
6. Schimitschek, E.J., "Laser emission of a neodymium salt dissolved in $POCl_3$," Journal of Applied Physics 39, 6120 (1968).
7. Ault, E.R., Comaskey, B.J., Kulko, T.C., "High average power laser using a transverse flowing liquid host," United States Patent-US 6600766 B1, (2003).
8. Jones-Bey, H.A., "Flowing neodymium offers improved heated management," Laser Focus World 15(3), 18 (2004).
9. Luo, G., Chaoqi, H., Bo, P., Dianyuan, F., "A new fluid state laser system realizes laser output," Proc. of SPIE 7843, 78431I-1 (2010).

Differential Phase Imaging of Evanescent Wave in Total Internal Reflection for Determining Refractive Index

Tania Das and Srinjini Roy

Abstract A non-interferometric differential phase measuring technique is used to determine the refractive index of a medium. In total internal reflection (TIR), phase change is a function of the refractive indices of a pair of media involved. Also, the said phase change is different for p-polarized and s-polarized light components. Difference between the phases of these two polarized light components leads to the refractive index of the sample. The theory of the technique is discussed and a simple experimental method is demonstrated to determine the phase difference of the totally internally reflected p and s components.

1 Introduction

Measurement of RI is important because it gives different information about bio-specimens, especially the cell morphology. As an example, computed tomography is used to measure the three dimensional refractive index of breast cancer cell where optical tweezers provide a non-contact method for reorienting the cell [1]. Refractive index measurements based on light scattering is works on the assumption that microstructures of varying geometries and refractive indices create unique diffraction patterns. Using the far field diffraction pattern, the quantitative phase profile of a living cell has been evaluated [2]. Different methods such as SPR technique, interferometry etc., exist for determination of the localized refractive

T. Das (✉)
Department of Electronics and Communication Engineering, Heritage Institute of Technology, Kolkata, West Bengal 700107, India
e-mail: taniadas.physics@gmail.com

S. Roy
Department of Applied Optics and Photonics, University of Calcutta, JD-2, Sector III, Kolkata, West Bengal 700106, India
e-mail: srinjini.ray@gmail.com

© Springer Nature Singapore Pte Ltd. 2017
I. Bhattacharya et al. (eds.), *Advances in Optical Science and Engineering*, Springer Proceedings in Physics 194, DOI 10.1007/978-981-10-3908-9_30

index. Thickness independent measurement of refractive index is important to measure refractive index of a non uniform sample. Although interferometric techniques yield high accuracy, the technique is sensitive towards noise [3], requires a complex setup with the additional criterion that the sample must be optically flat. Differential phase measurement is an efficient technique for determining the refractive index with high efficiency [4]. In the present work we propose an in-line optical configuration to measure the differential phase by amplitude addition and subtraction through the use of a polarizer. Here a simple set up is used to image a liquid sample applied on a prism surface from where the incident light beam suffers total internal reflection. The decaying evanescent light field which extends only a few hundreds of nanometers into the sample interacts with the sample. Hence, the measurement is independent of the sample thickness as long as it is greater than the depth of the evanescent field depth. In such cases the phase change of the totally internally reflected (TIR) wave is indicative of the refractive index of the sample [5] and independent of the sample thickness.

2 Theory

Fresnel's theory explains that if light is incident from the higher refractive index side of a dielectric interface with refractive indices n_1 and n_2 ($n_1 > n_2$) at an angle θ_1 which is greater than the critical angle for the interface as shown in Fig. 1, then reflectivity r_s and r_p and the absolute phase's φ_s and φ_p associated with TM and TE components respectively of the total internally reflected light field, written as [6],

$$r_s = \frac{n_1 \cos\theta_1 - i\sqrt{n_1^2 \sin^2\theta_1 - n_2^2}}{n_1 \cos\theta_1 + i\sqrt{n_1^2 \sin^2\theta_1 - n_2^2}} \tag{1}$$

$$r_p = \frac{in_1\sqrt{n_1^2 \sin^2\theta_1 - n_2^2} - n_2^2\cos^2\theta_1}{in_1\sqrt{n_1^2 \sin^2\theta_1 - n_2^2} + n_2^2\cos^2\theta_1} \tag{2}$$

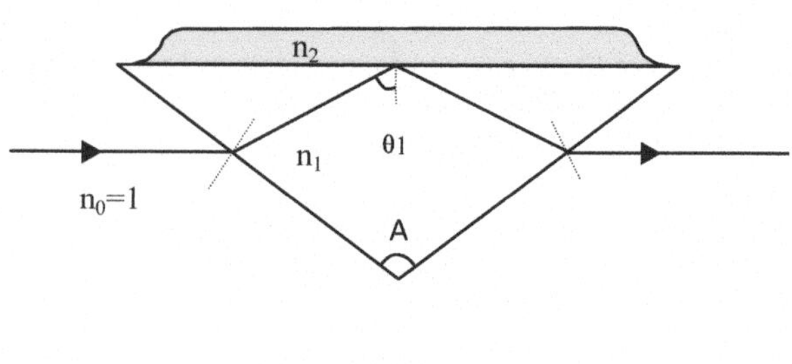

Fig. 1 Diagrammatical representation of total internal reflection utilized to evaluate n_2

$$\varphi_p = -1/2 \tan^{-1} \frac{n_1\sqrt{n_1^2 \sin^2\theta_1 - n_2^2}}{n_2^2 \cos\theta_1}$$
$$\varphi_s = -1/2 \tan^{-1} \frac{\sqrt{n_1^2 \sin^2\theta_1 - n_2^2}}{n_1 \cos\theta_1} \quad (3)$$

It is our target to measure the differential phase ϕ_{diff} given by,

$$\varphi_{diff} = \varphi_p - \varphi_s \quad (4)$$

The penetration depth of evanescent wave is expressed as

$$d = (\lambda/4\pi)\left(n_1^2 \sin^2\theta_1 - n_2^2\right)^{-1/2} \quad (5)$$

Equations (4) shows that phase difference between TE and TM polarized light is directly related to the incident angle and refractive indices of involving medium. If total internal reflection is implemented in a prism of prism angle A as shown in Fig. 1, then θ_1 can be expressed as $\theta_1 = [90 - A/2 + \{\sin^{-1}(\sin A/2)/n_1\}]$. The phase difference is also a function of the sample refractive index n_2. Figure 3a gives the variation of phase difference with respect to change of refractive index. Figure 3b represents change of penetration depth of evanescent wave with respect to refractive index. Both are used as calibration curves to find out refractive index or its variation over the sample. Figure 2 represents in line experimental set up. An expanded and collimated beam from He–Ne laser (λ = 632.8 nm) is passed through a circular polarizer to convert linearly polarized light to circularly polarized light. Thereafter, a linear polarizer oriented at 45° angles is placed for ensuring presence of equal amount of s-polarized and p-polarized light in beam. This beam is incident on the prism (BK7, RI = 1.521) at such an angle such that it satisfies total internal reflection condition at the interface of prism and sample. The output beam from totally internally reflected from the sample carries information of sample refractive index. At the output the two linear polarizer are used. The first polarizer combines the p and s components as desired and the second polarizer is suitably oriented so that the CCD operates in the linear region. If the first polarizer is oriented at 45° and 135° then the recorded intensity on the CCD is given by,

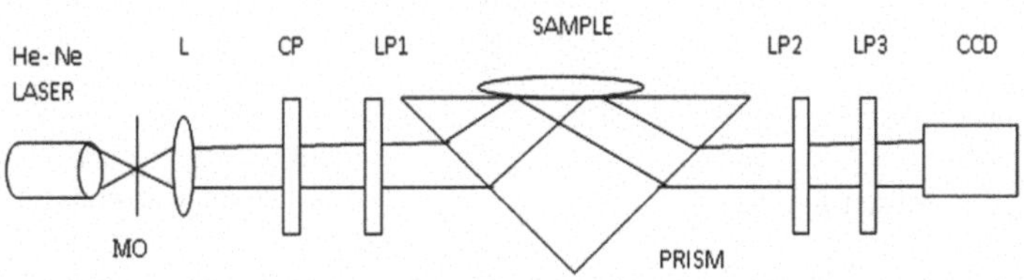

Fig. 2 Experimental setup. *MO* microscope objective, *L* Lens, *CP* Circular Polarizer, *LP1* linear polarizer with 45° angle, *LP2, LP3* Linear Polarizer, *CCD* Charge coupled Device (CCD)

$$I_{45} = r_s^2 + r_p^2 + 2r_s r_p \cos \varphi_{diff} \tag{6}$$

$$I_{135} = r_s^2 + r_p^2 - 2r_s r_p \cos \varphi_{diff} \tag{7}$$

where r_p and r_s are the amplitudes associated with the p and the s components respectively, and ϕ_{diff} is the phase difference between these components.

Subtracting (7) from (6) the phase difference can be evaluated as,

$$\varphi_{diff} = \cos^{-1} \frac{I_{45} - I_{135}}{4r_s r_p} \tag{8}$$

r_s and r_p also can be represented as function of intensity at polarizing angle 0° and 90°.

Where

$$r_s = \sqrt{I_{0^\circ}} \quad \text{and} \quad r_p = \sqrt{I_{90^\circ}} \tag{9}$$

3 Results

From Fig. 3, it is noticed that the phase range lies for measured refractive indices from 1.1 to 1.35 for a 120° prism and is used for samples expected to have refractive indices in this range. Figure 4a represents the four frames of intensity variation for polarizing angle θ = 0°, 45°, 90° and 135° recorded on the CCD for a drop of water serving as the sample. Using these four frames and the calculations mentioning in above theory, the unwrapped phase obtained is shown in Fig. 5a, and it is observed that the phase difference is −6.702 (in degrees) for water and Fig. 5b

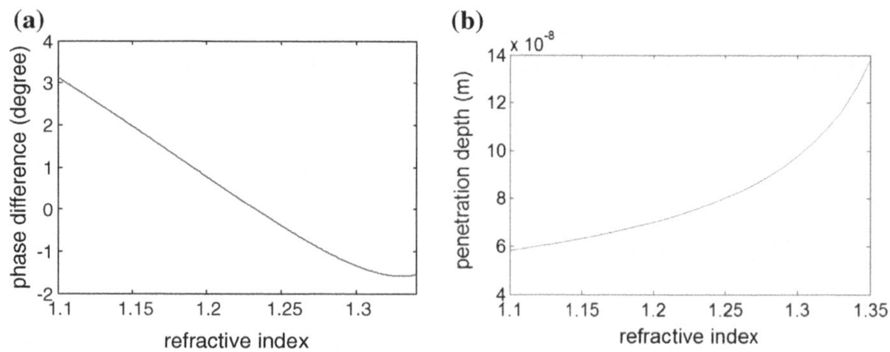

Fig. 3 **a** Plot for phase difference with respect to refractive indices for prism angle 120° **b** plot for penetration depth associated with evanescent wave with respect to refractive indices for prism angle 120°

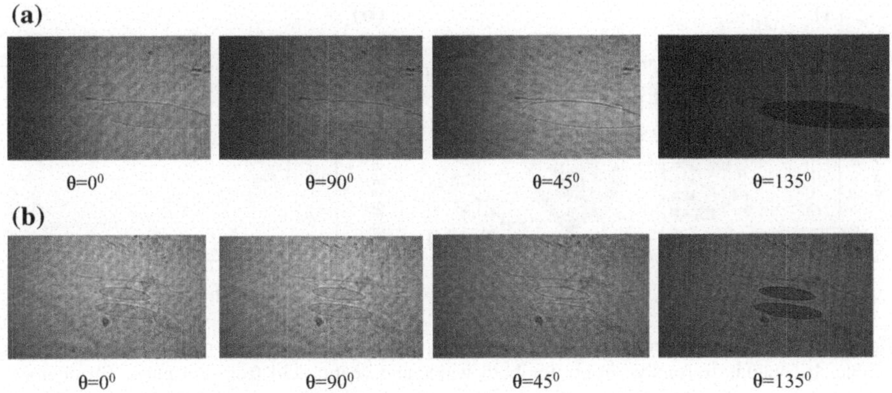

Fig. 4 Four frames of intensity variation for polarizing angle $\theta = 0°$, 45°, 90° and 135° are recorded on the CCD **a** water as sample and **b** water and acetic acid as sample

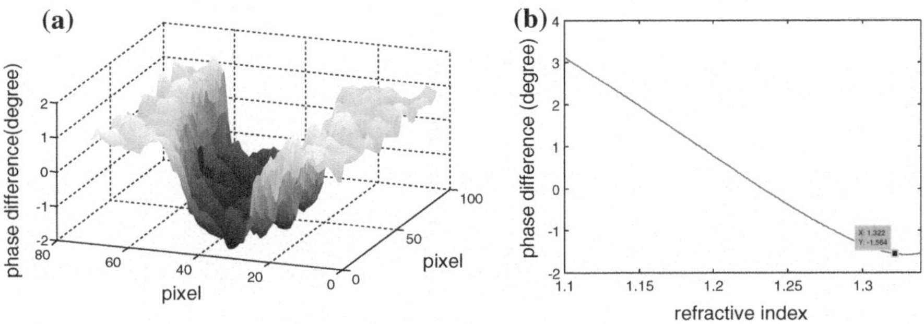

Fig. 5 **a** Plot of the unwrapped phase profile of water as sample. **b** Evaluate refractive index for water from the phase (in degrees) versus refractive index calibration curve and the value is 1.32

represents the corresponding refractive index is found to be 1.32 for water. A non uniform sample of two liquids, water and acetic acid placing side by side on the prism surface of prism angle 120° are served as sample to show the spatial variations of refractive index and following the procedure as discussed previously, the captured frames are shown in Fig. 4b and the phase was obtained as shown in Fig. 6a, b, corresponding to refractive indices of 1.32 and 1.30 for water and acetic acid and is confirmed by the results obtained when these two liquids were applied separately on the prism surface.

Figure 7a shows cross sectional variation of refractive index and Fig. 7b represents different penetration depth for two different liquids along cross section, here water and acetic acid. Phase difference φ_{diff} is indicative of the refractive index of the sample. The sensitivity (S) of refractive index in this measuring technique is expressed as

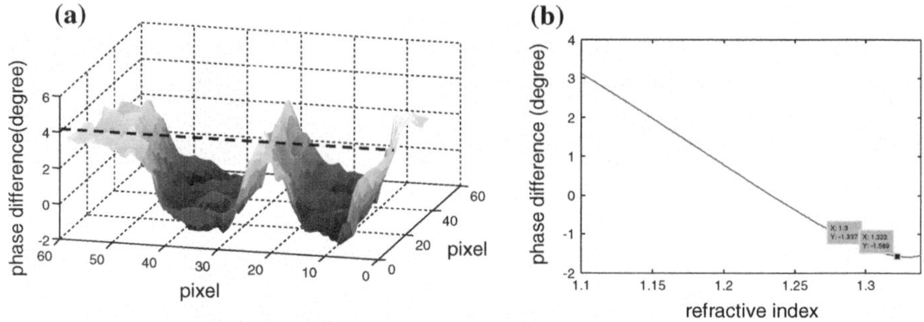

Fig. 6 **a** Plot of the unwrapped phase profile of water and acetic acid placed side by side as sample. **b** Evaluate refractive index for both water and acetic acid from the phase (in degrees) versus refractive index calibration curve and the values are 1.32 for water and 1.30 for acetic acid

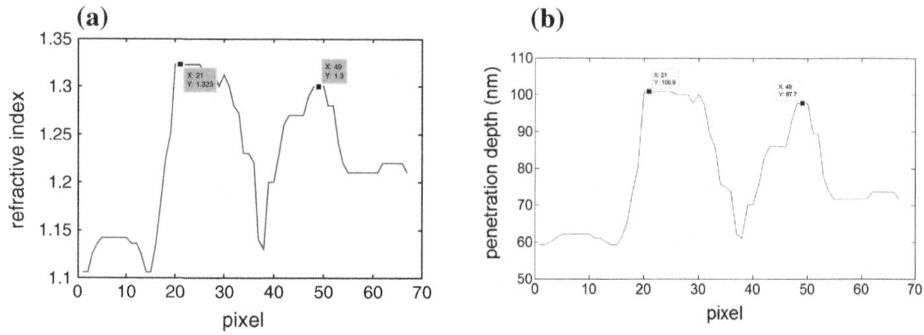

Fig. 7 **a** Plot of refractive index along a y-axis direction as represented by a *dotted line* in Fig. 6a **b** plot of penetration depth along y-axis direction as shown by a *dotted line* in Fig. 6a

$$S = \frac{d\phi_{diff}}{dn_2} = \frac{4n_2^3 \cos^2\theta_1 ab^2 + n_2^4 \cos^2\theta_1 b^2 (4n_2^3 \cos^2\theta_1 - 2n_1^2 n_2) - 2n_1^2 n_2^2 a^2 \cos^2\theta_1}{a^2 b^2} \tag{10}$$

where

$$a = n_2^4 \cos^2\theta_1 + n_1^4 \sin^2\theta_1 - n_1^2 n_2^2 \quad \textit{and} \quad b = n_1^2 - n_2^2$$

Equations (10) shows that sensitivity of measurement is a function of the glass refractive index n_1 and the angle of the prism (which determines θ_1) as well as the sample refractive index n_2 and penetration depth also function of angle of prism, refractive index of prism, sample refractive index and wavelengths. Experimental summary is given in Table 1.

Table 1 Experimental summary of prism angle

Prism angle A	Sample	Change of phase in degrees	Refractive index (RI) (λ = 632.8 nm)	Penetration depth (in nm) (From 5)	Sensitivity $\frac{d\varphi_{diff}}{dn_2}$ (from 10)
A = 120^0	Water	−1.56	1.323	100.9	−0.069
	Water	−1.56	1.323	100.9	−0.045
	Acetic acid	−1.33	1.30	97.7	

4 Conclusion and Discussion

This technique is used to determine spatial variation of refractive index in for nonuniform samples. In the present work we have considered only liquid samples but it is possible to measure refractive index variations in solid samples, in which case a thin layer of a high index liquid (α-bromonapthelene) needs to be applied between the prism and the solid sample and using this technique, the results for refractive index variations in various solid structures will be reported in future. The drawback of this technique is that the prism angle determines the range of measurable refractive indices. To increase the refractive index range it is required to increase refractive index of prism. If we use an equilateral prism the measurable refractive index range is shifted to 1.47–1.48 but the major advantage of this inline experiment is its simplicity and that the differential phase can be derived from intensity measurements.

References

1. Toshiki Yasokawa et al, 'A Method for Measuring the Three-Dimensional Refractive-Index Distribution of Single Cells Using Proximal Two-Beam Optical Tweezers and a Phase-Shifting Mach—Zehnder Interferometer', *Optical Review*, Volume 14, Issue 4, pp 161–164, 2007.
2. J Beuthan, et al, 'The spatial variation of the refractive index in biological cells', *Phys. Med. Biol.* 41,369–382,1996.
3. Hee Joo Choi, Hwan Hong Lim, Han Seb Moon, Tae Bong Eom, Jung Jin Ju, and Myoungsik Cha, 'Measurement of refractive index and thickness of transparent plate by dual-wavelength interference', *Opt Express, OSA, Vol.* 18, No. 9, 2010.
4. S. Patskovsky, M. Meunier, and A. V. Kabashin, 'Phase-sensitive silicon-based total internal reflection sensor', Optics Express, Vol. 15, No. 19,2007.
5. Muhammad Kashif, Ahmad Ashrif A. Bakar*, Norhana Arsad and Sahbudin Shaari, 'Development of Phase Detection Schemes Based on Surface Plasmon Resonance Using Interferometry', *Sensors, 14*(9), 15,914–15,938, 2014.
6. William M. Ash III and Myung K Kim, 'Digital Holography of total internal reflection', *Opt Express, OSA*, Vol. 16, No. 13,2008.

Gray to Binary Code Converter Using Ti-Indiffused Lithium Niobate Based Mach-Zehnder Interferometer

Harsh Kumar, Laxman Kumar, Vijay Janyani, Buryy Oleh, Ubizskii Serhij and Ghanshyam Singh

Abstract In this paper, we propose a 3-bit gray to binary code converter using Ti-indiffused Lithium Niobate based on Mach-Zehnder Interferometer (MZI) waveguiding structure. All optical XOR logic gate is the basic building block for this code converter. At the output port, we get binary output in form of optical signal and also their complements which can be used in further computations and reduce the required hardware. The proposed device offers high switching speed and tolerable losses. The extinction ratio achieved is around 20 dB. The proposed combinational circuit can be useful in optical logic devices.

1 Introduction

All optical logic circuits is an evolving topic of research in the field of optical communication and a large number of devices have been proposed. Code converters are the essential circuits for secure optical digital communication. Binary to gray, gray to binary, gray to BCD code converters have already been simulated using different technologies. A 4-bit digital encoder using semiconductor optical amplifier (SOA) based MZI is simulated on Opti-system [1]. All optical gray to binary code converters purposed with the help of semiconductor optical amplifier-assisted Sagnac switch which needs pump laser [2]. The binary to gray to binary converter is designed using the principle of all optical switching by Kerr non-lincar prism [3].

In this paper, electric gray to optical binary code converter using Ti-indiffused Lithium Niobate (Ti-LN) based Mach-Zehnder Interferometer (MZI) is proposed.

H. Kumar (✉) · L. Kumar · V. Janyani · G. Singh
Department of Electronics and Communication Engineering,
Malaviya National Institute of Technology, Jaipur, India
e-mail: 2015pwc5305@mnit.ac.in

B. Oleh · U. Serhij
Institute of Telecommunication Radioelectronics and Electronic Engineering,
Lviv Polytechnic National University, Lviv, Ukraine

© Springer Nature Singapore Pte Ltd. 2017

I. Bhattacharya et al. (eds.), *Advances in Optical Science and Engineering*,
Springer Proceedings in Physics 194, DOI 10.1007/978-981-10-3908-9_31

LN is used as substrate material because it shows very good electro-optic (EO), acousto-optic (AO) and piezoelectric effects. In the EO-effect, the progression in refractive index is incited by applying an electric field while in the AO-effect, the change is brought on by interaction between acoustic and optical waves in the crystal. Different dopants like magnesium oxide, titanium, etc. are utilized to form channel waveguides with LN [4]. The EO–switches and modulators are fabricated with either Ti-indiffusion or annealed proton exchange process. The MZI structure has been used to design the XOR logic gate, which is the elemental building block for the proposed gray to binary code converter [5].

1.1 Working Principle of MZI

MZI structure can be used to design an EO-switch, consisting two interferometric arms and the switching action takes place based on the relative phase change between them [6, 7]. The relative phase shift and the output power at port 1 and 2 can be calculated as:

$$P_{out1} = P_{in}\sin^2\frac{\Delta\emptyset}{2} \quad (1)$$

$$P_{out2} = P_{in}\cos^2\frac{\Delta\emptyset}{2} \quad (2)$$

$$\Delta\emptyset = \frac{\pi}{V_\pi}V \quad (3)$$

where, V_π is the electrode voltage corresponding to phase difference, $\Delta\emptyset = \pi$ and P_{in} is input power. When switching voltage V is zero switch operates in cross state and when switching voltage is equal to V_π i.e. 6.75 V, switch will work in bar state [5]. The schematic diagram of MZI is shown in Fig. 1.

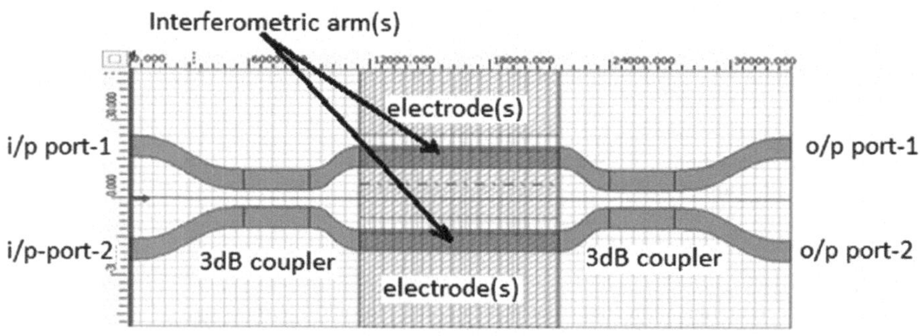

Fig. 1 Schematic view of Mach-Zehnder Interferometer

Table 1 Truth table for gray to binary conversion

Gray code			Binary code		
G2	G1	G0	B2	B1	B0
0	0	0	0	0	0
0	0	1	0	0	1
0	1	0	0	1	1
0	1	1	0	1	0
1	0	0	1	1	1
1	0	1	1	1	0
1	1	0	1	0	0
1	1	1	1	0	1

2 Gray to Binary Code Conversion

The 3-bit gray code can be converted into 3-bit binary code using the truth table given in Table 1. A two input and a three input XOR logic gate is used in this conversion. Two input all optical XOR gate can be implemented by cascading two MZI switches [5]. The gray code input is provided as the electrode voltage of MZI. The output binary code bits B2, B1 and B0 are represented in terms of gray code bits G2, G1 and G0 as given below:

$$B2 = G2 \tag{4}$$

$$B1 = G2 \oplus G1 \tag{5}$$

$$B0 = G2 \oplus G1 \oplus G0 \tag{6}$$

where, the symbol $\oplus$ denotes the XOR logic operation.

3 Implementation of Code Converter

The electric gray to optical binary code converter is designed using Ti-indiffused LN waveguide. The layout of design is shown in Fig. 2 that consists of 6 MZI switches. It has 3 ports where optical source is connected and 6 output ports in which port 1, 4 and 5 corresponding to binary output B0, B1 and B2 and port 2, 3 and 6 give complementary binary output. Gray code input is given as electrode voltage. Logic 1 of gray code input is equal to the switching voltage of MZI i.e. 6.75 V [6, 7].

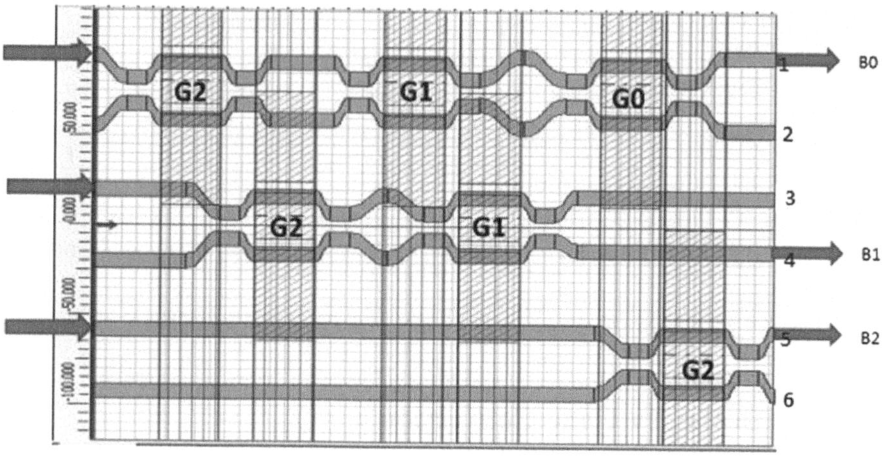

Fig. 2 The layout of 3-bit gray to binary converter

4 Result and Discussion

The layout is simulated using beam propagation method and all the combinations of truth table given in Table 1 are verified as shown in Fig. 3. The length of device is around 10 cm. Here the electrode voltages represent the gray code input with V = 0 V for logic 0 and V = 6.75 V for logic 1.

The extinction ratio (ER) and insertion loss (IL) are calculated using the following relations

$$ER = 10 \log_{10} \frac{P_{OH}}{P_{OL}} \tag{7}$$

$$IL = 10 \log_{10} \frac{P_{in}}{P_{out}} \tag{8}$$

where, P_{OH} and P_{OL} are the output power corresponding to the binary 1 and binary 0 respectively. Table 2, depicts the calculated values for the extinction ratio (ER) and insertion loss (IL) of the proposed convertor logic.

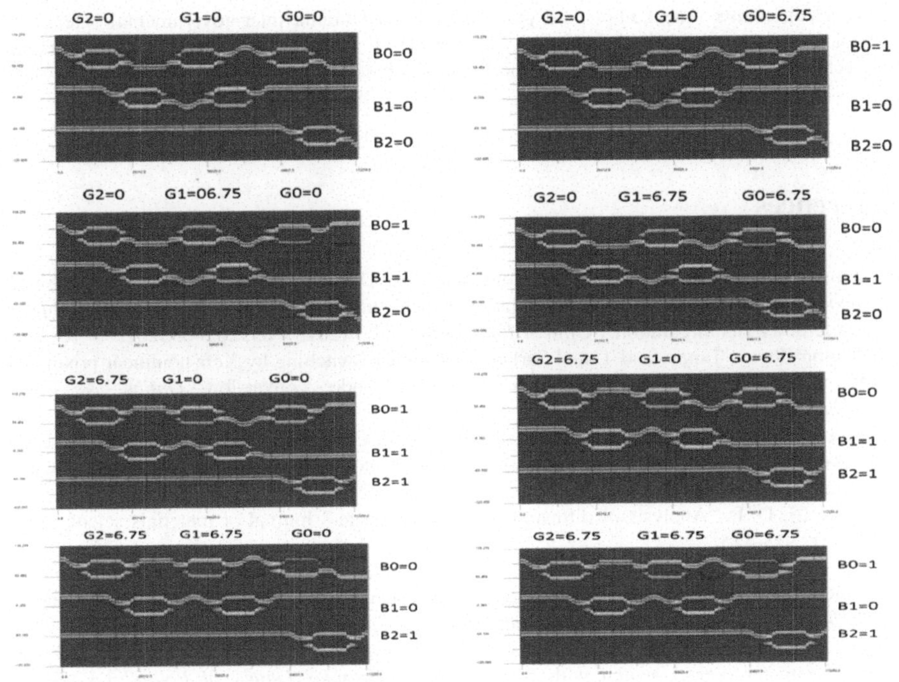

Fig. 3 Simulation result waveforms

Table 2 The extinction ratio at output ports

Output port	1	2	3	4	5	6
Extinction ratio (dB)	25.5	19.6	22	23.4	25.1	22
Insertion loss (dB)	0.7	0.4	0.02	0.08	0.05	0.02

5 Conclusion

In this paper, the structuring of an electric gray to optical binary code converter circuit is done utilizing the conventional MZI structure based on Ti-LN waveguides. This device can further be used to obtain many other optical logic gates and circuit operations. Gray code input is applied as electrode voltage and is converted into binary code output which is optical signal. The extinction ratio at all output ports is calculated and the minimum obtained is around 20 dB. The insertion loss is found to be 0.02 dB at port 3 and 6 and 0.7 dB at port 1. In future, the device can be further optimized to reduce the length of converter.

Acknowledgements The authors are grateful to India–Ukraine inter-governmental science & technology cooperation programme between the MNIT Jaipur (India) and the Lviv National Polytechnique Institute, Lviv (Ukraine) for technical support. Project sanction no: INT/RUS/UKR/P-15/2015.

References

1. Mehra, Rekha, Shikha Jaiswal, and Hemant Kumar Dixit. "Ultrafast all-optical 4-bit digital encoders using differential phase modulation in semiconductor optical amplifier-Mach-Zehnder interferometer configuration." *Optical Engineering* 52.3 (2013): 035202–035202.
2. Chattopadhyay, Tanay, and Tamal Sarkar. "All-optical switching by Kerr nonlinear prism and its application to of binary-to-gray-to-binary code conversion." *Optics & Laser Technology* 44.6 (2012): 1722–1728.
3. Gayen, Dilip Kumar, et al. "All-optical binary to Gray code and Gray to binary code conversion scheme with the help of semiconductor optical amplifier-assisted Sagnac switch." *IET circuits, devices & systems* 5.2 (2011): 123–131.
4. Birnie III, D. P. "Analysis of diffusion in lithium niobate." Journal of materials science 28.2 (1993): 302–315.
5. Kumar, Ajay, Santosh Kumar, and Sanjeev Kumar Raghuwanshi. "Implementation of XOR/XNOR and AND logic gates by using Mach–Zehnder interferometers." *Optik-International Journal for Light and Electron Optics* 125.19 (2014): 5764–5767.
6. Singh, G., V. Janyani, and R. P. Yadav. "Modeling of a 2 × 2 electro-optic Mach-Zehnder Interferometer optical switch with s-bend arms." *Photonics Letters of Poland* 3.3 (2011): pp-119.
7. Singh, Ghanshyam, Vijay Janyani, and R. P. Yadav. "Modeling of a high performance Mach-Zehnder interferometer all optical switch." *Opt. Appl* 42.3 (2012): 613–625.

Fingerprint Detection and Analysis Using Talbot Interferometry

Jitendra Dhanotia, Litesh Bopche, Vimal Bhatia and Shashi Prakash

Abstract In the present communication, a full field technique for mapping latent fingerprint using Talbot interferometry has been proposed. Collimated light from He–Ne laser illuminates a specimen comprising of a fingerprint implanted onto a reflecting surface. Reflected light carries information regarding the depth and orientation of furrows and ridges in the fingerprint. The topological information of the fingerprint is retrieved using phase shifting interferometric technique. The slope of the phase provides the information regarding variation in ridges of latent fingerprint. The proposed technique does not require any kind of chemical or physical treatment.

1 Introduction

Fingerprint is a unique and distinguishable pattern of curves, which makes its detection one of the biometric parameter of importance in forensic science [1]. A fingerprint impression is obtained on a surface due to friction of ridge skin with the surface. The fingerprint is composed of ridge lines and valleys. The ridge lines are mainly parallel but bifurcations as well as terminations also appear. The pattern in a fingerprint can be basically classified as arch, loop and whorl. Wide ranges of techniques have been developed for detection and analysis of finger prints. The important things of concern in choosing a technique are the time taken in analyzing the results, simplicity of the setup and portability. Based on the time consumption, the techniques can be classified as based on the physical and chemical processing of a sample and those based on real-time analysis. The techniques based on physical

J. Dhanotia · V. Bhatia
Signals and Software Group, Discipline of Electrical Engineering,
Indian Institute of Technology Indore, Indore, Madhya Pradesh 453446, India

L. Bopche · S. Prakash (✉)
Photonics Laboratory, Institute of Engineering & Technology,
Devi Ahilya University, Indore, Madhya Pradesh 452001, India
e-mail: sprakash_davv@rediffmail.com

© Springer Nature Singapore Pte Ltd. 2017

I. Bhattacharya et al. (eds.), *Advances in Optical Science and Engineering*,
Springer Proceedings in Physics 194, DOI 10.1007/978-981-10-3908-9_32

and chemical processing of a sample, are time consuming and use hazardous chemicals. Also they are relatively older techniques. Real-time techniques are comparatively safer and faster. Optical Coherence Tomography (OCT) is one such technique. OCT methods have been reported [2–10] to recognize genuine fingerprints by providing in-depth information of the finger. Huang et al. [2] presented fringe projection based technique to obtain 3D fingerprint with color texture information in real time by projecting color sinusoidal fringe patterns onto a finger surface. Dubey et al. [9, 10] described full field swept source OCT based technique to detect latent fingerprint. The authors used super-luminescent diode and acousto-optic tunable filter as wavelength tuning device for non-mechanical scanning. Interferometric systems including the OCT based systems have relatively high sensitivity to the environmental perturbations and are difficult to align at locations outside the laboratory. Also, relatively high cost of tunable laser source increases the overall system cost.

In present communication, we report application of Talbot interferometer (TI) for fingerprint detection and mapping. To increase the sensitivity and reliability in detection and identification process, a four-step phase shifting technique has been incorporated. The phase profiling of the latent fingerprint can be achieved by visualization of the 2D and 3D phase plots obtained by mapping the phase profile of fingerprints. The technique is full field and incorporates amplitude and phase profiling of the latent fingerprints. Amplitude profiling can be attained by analyzing the light reflected from the specimen. The phase profiling can be achieved by visualization of the 2D and 3D phase plots. The experimental arrangement is simple, inexpensive and flexible.

2 Basic Theory

Figure 1 shows the schematic of experimental setup based on TI. TI involve two gratings with a separation equals to Talbot distance. When a plane wave is incident upon a diffraction grating, its image is repeated at regular distances away from the grating plane and is given by $d = 2p^2/\lambda$, where p is the period of the grating and λ is the wavelength of the light used. The regular distance is called the Talbot length, and the repeated images formed at regular intervals are called as self images or Talbot images. Various applications of the TI have been reported in the literature [11, 12].

The intensity distribution, formed at a point (x, y) on the CCD chip, is a co-sinusoidal function which can be expressed as:

$$I(x,y) = A(x,y) + B(x,y)\cos[\psi(x,y) + \delta] \tag{1}$$

where, $A(x,y)$ is average intensity, $B(x,y)$ is intensity modulation, $\psi(x,y)$ is the desired phase information and δ is the known phase shift. In the experimental setup as shown in Fig. 1, the reflected beam from the specimen, carrying information

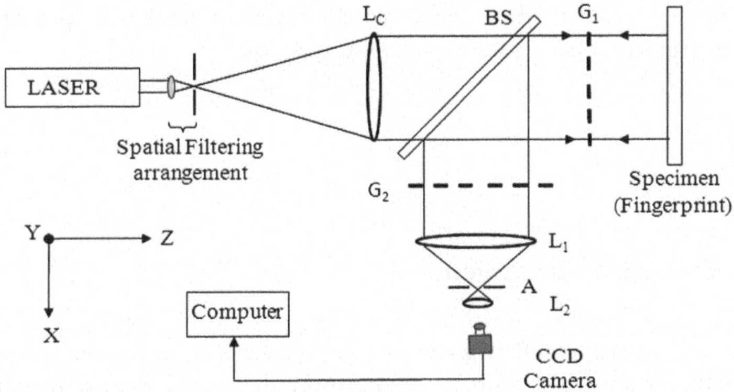

Fig. 1 Schematic of experimental arrangement used for fingerprint detection using Talbot interferometer

regarding the details in the finger print is analysed using the TI. The TI comprises of a pair of gratings of period p. One of the two gratings is shifted in four-steps such that the phase difference of $\pi/2$ results in each successive recording. Intensity profile of the grabbed fringe patterns is given by [12]

$$I_1(x,y) = A(x,y) + B(x,y)\cos[\psi(x,y)] \tag{2}$$

$$I_2(x,y) = A(x,y) + B(x,y)\cos\left[\psi(x,y) + \frac{1}{2}\pi\right] \tag{3}$$

$$I_3(x,y) = A(x,y) + B(x,y)\cos[\psi(x,y) + \pi] \tag{4}$$

$$I_4(x,y) = A(x,y) + B(x,y)\cos\left[\psi(x,y) + \frac{3}{2}\pi\right] \tag{5}$$

Phase at each point of the specimen can be determined from the above equations using the following relation

$$\psi(x,y) = \arctan\left[\frac{I_4 - I_2}{I_1 - I_3}\right] \tag{6}$$

The phase $\psi(x,y)$ is calculated using phase shifting algorithm and plotted against pixel position.

The deflection angle $\Phi(x,y)$ can be written as,

$$\Phi(x,y) = \frac{p}{2\pi d}\psi(x,y); \tag{7}$$

where p is the period of the grating. The deflection angle $\Phi(x,y)$ is related to $\psi(x,y)$ and is effectively the measure of the slope of the rays in the direction perpendicular to the grating lines. The relative surface height $h(x, y)$ or depth of ridges can then be determined from the phase data by integrating [13]

$$h(x, y) = \frac{\lambda}{4\pi} \int \Phi(x, y) dx dy \tag{8}$$

3 Experimental Arrangement

Schematic of experimental arrangement for fingerprint detection using Talbot interferometry is shown in Fig. 1. Light from the laser of power 15 mW and wavelength 632 nm is spatially filtered and illuminates collimating lens L_C to obtain a collimated beam. The collimated beam after transmission through a beam splitter BS and grating G_1, is incident on the specimen comprising of a fingerprint implanted on a plane reflecting surface. The collimated beam after transmission through the BS is incident on the specimen comprising of a fingerprint implanted on a plane reflecting surface. The back reflected light from the specimen has been diverted using beam splitter, towards a set of gratings G_1 and G_2, each of pitch 0.08 mm and separated by a Talbot distance. The first self imaging plane is at 20 mm away from the first grating while the second grating was kept at fourth self imaging plane which was at 80 mm. The second grating was kept at fourth Talbot distance because it is not possible to keep the second grating at first Talbot plane due to small distance. The choice of grating pitch is according to their availability. The spatial filtering arrangement comprising of lens L1, L2 and aperture A2 was used in the setup. Focal lengths of lenses L_1, L_2 are 250 mm and 40 mm respectively. The aperture A of variable size was used. One of the gratings was mounted on a precision translation stage. The grating has been shifted, in-plane, in four steps. Four subsequent patterns have been recorded using a CCD camera, and the phase shifting program in MATLAB environment was used to obtain 2D and 3D phase maps.

4 Results and Discussion

A plane reflecting surface with fingerprint implanted on it has been used as a specimen. When a light beam passes through or is reflected from the specimen under study, the beam acquires a phase change. This phase change exhibits the parameter of interest, in this case, the depth. The interferogram corresponding to this fingerprint is recorded and stored in computer memory for analysis. The image size has been chosen to be 800 × 1000 pixels. Figure 2a shows directly recorded fringe pattern without gratings of the right hand thumb. Figure 2b to Fig. 2e shows the recorded interference patterns with the grating G_1 shifted in steps of 0, $\pi/2$, π, $3\pi/2$, and 2π, with respect to grating G_2, respectively.

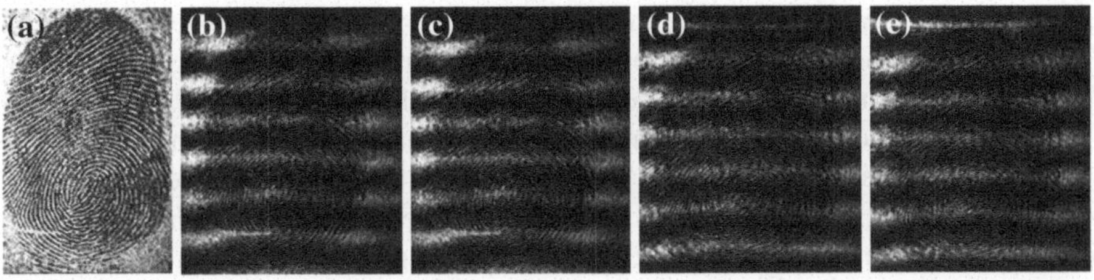

Fig. 2 **a** shows the directly recorded fingerprint, **b–e** show the fringe patterns recorded subsequently after a phase shift of $\pi/2$, π, $3\pi/2$, and 2π with respect to the directly recorded patterns

For determination of phase, the computation has been done in MATLAB environment. Phase shifting algorithm has been developed to evaluate the phase values at each point of the interferogram. Phase shifting interferometry (PSI) electronically records a series of interferograms while the reference phase of the interferometer is changed. The wavefront phase is encoded in variation in the intensity pattern of the recorded interferograms and a simple point by point calculation recovers the phase. Once the phase is determined across the interference field, the corresponding measurement variable can be measured. To introduce a phase shift, a suitable optical element is used in TI. In PSI, one of the two gratings is subjected to in-plane translation in direction perpendicular to the grating lines. Calibrated micrometer screw based translation stage and piezo-electric transducer (PZT) are used for introducing phase shifts in steps. (PZT) is a very precise device in which piezo-electric crystal expands or contract with an externally applied voltage. During introduction of a phase shift, all physical parameters must necessarily remain fixed. In PSI is the lack of registration between the frames and noise due to speckles (in case of laser source is used) are presents. It seems that both of these are adding to the loss of resolution in the unwrapped phase map. These can be reduced by using an appropriate filtering scheme. Figure 3a shows the slope of the phase map for a small portion (300 pixels) which provides the information regarding inter-ridge distance. The small portion of the image is chosen to display the intensity variation in clearer

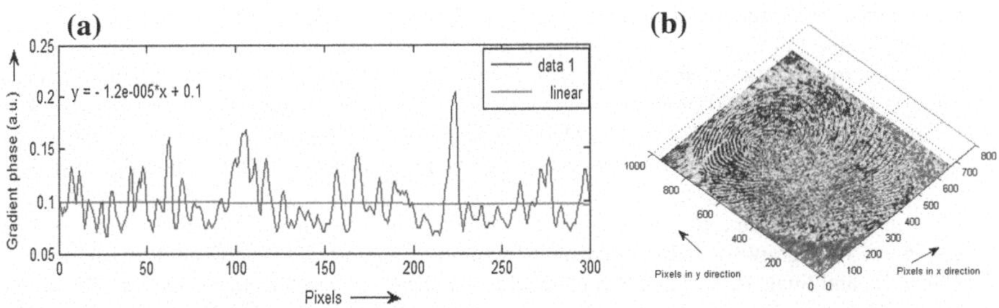

Fig. 3 **a** 2D phase map obtained using PSI technique. **b** Reconstructed 3D phase map obtained by applying PSI on the fringe patterns obtained with the specimen carrying a fingerprint

fashion. 3D reconstructed phase map for the fingerprint has also been plotted in Fig. 3b. The slope of the phase provides the information regarding variation in ridges of latent fingerprint. It has been estimated to be as 0.449 mm.

This inter ridge spacing calculated using 2D phase map is comparable to the value as reported in literature [14]. Thus phase map can be used for recognizing as well as matching fingerprints with finer details. The slope data obtained has been integrated using least square method to obtain the ridge height. Using the formula in (8) the depth information has been determined. The average depth has been determined to be 0.224 mm.

5 Conclusions

In this paper TI has been experimentally demonstrated for fingerprint detection and mapping. The results obtained using TI match well with the techniques reported till date. Local and global variations in the ridge lines and valleys have been successfully mapped. Highlight of the technique has been its extreme simplicity, low cost and ability for the technology to be transferred to the locations outside the laboratory. 2D and 3D phase map of grabbed fingerprint images provide finer details about the depth profile of fingerprint. Also, the technique does not require any chemical or physical processing. Hence, repeated tests may be performed without any degradation in the sample under test. The technique is simple, full field, low cost and fast.

References

1. Menzel, E. R., “Detection of latent fingerprints by laser-excited luminescence,” Analytical Chemistry 61(8), 557–561(1989).
2. Huang, S., Zhang, Z., Zhao, Y., Dai, J., Chen, C., Xu, Y., Zhang, E. and Xie, L., “3D fingerprint imaging system based on full-field fringe projection profilometry,” Optics and Lasers in Engineering 52, 123–130 (2014).
3. Cheng, Y. and Larin, K. V., “Artificial fingerprint recognition by using optical coherence tomography with autocorrelation analysis,” Applied Optics 45(36), 9238–9245 (2006).
4. Cheng, Y. and Larin, K. V., “In Vivo Two- and Three-Dimensional Imaging of Artificial and Real Fingerprints With Optical Coherence Tomography,” IEEE Photonics Technology Letters 19(20), 1634–1636 (2007).
5. Chang, S., Cheng, Y., Larin, K. V., Mao, Y., Sherif, S. and Flueraru, C., “Optical coherence tomography used for security and fingerprint-sensing applications,” IET Image Processing, 2 (1), 48–58 (2008).
6. Bossen, A., Lehmann, R. and Meier, C., “Internal fingerprint identification with optical coherence tomography,” IEEE Photonics Technology Letters 22(7), 507–509(2010).
7. Gabai, H. and Shaked, N. T., “Dual-channel low-coherence interferometry and its application to quantitative phase imaging of fingerprints,” Optics Express 20(24), 26906–26912 (2012).

8. Auksorius, E. and Boccara, A. C., "Fingerprint imaging from the inside of a finger with full-field optical coherence tomography," Biomedical Optics Express 6(11), 4465–4471 (2015).
9. Dubey, S. K., Anna, T., Shakher, C. and Mehta, D. S., "Fingerprint detection using full-field swept-source optical coherence tomography," Applied Physics Letters 91, 1811061–1811063 (2007).
10. Dubey, S. K., Mehta, D. S., Anand, A. and Shakher, C., "Simultaneous topography and tomography of latent fingerprints using full-field swept source optical coherence tomography," J. Opt. A: Pure Appl. Opt. 10(1), 015307 (2008).
11. Dhanotia, J. and Prakash, S., "Automated collimation testing by incorporating Fourier transform in Talbot interferometry," Applied Optics 50(10), 1446–1452, (2011).
12. Prakash, S., Rana, S., Prakash S. and Sasaki, O., "Automated collimation testing using a temporal phase shifting technique in Talbot interferometry," Applied Optics 47(31), 5938–5943, (2008).
13. Caber, P. J., "Interferometric profiler for rough surfaces," Applied Optics 32(19), 3438–3441 (1993).
14. Hastings, R., "Ridge enhancement in fingerprint images using oriented diffusion," Digital Image Computing Techniques and Applications, 245–252 (2007).

Birefringence Analysis Using Mach-Zehnder Interferometer

Santa Sircar and Ipsita Chakraborty

Abstract The theory for interferometric analysis of an arbitrarily oriented birefringent sample is developed. The object is placed in one of the arms of a Mach-Zehnder Interferometer. Interferograms are recorded for vertically and horizontally polarized light passing through the sample. The spatial phase information for these two cases is suitably combined through an algorithm developed for this purpose to yield the complete spatial distribution of retardance and direction of birefringence over the sample plane. The proposed method is able to measure spatial variations of birefringence from 0 to π and is targeted to measurement of low birefringence as in biological specimens. Simulation results presented validate the proposed theory. Experimental results are forthcoming.

1 Introduction

For a birefringent specimen, analysis of birefringence essentially means finding the phase difference between the orthogonal exit beams along the principal directions of the sample as well as locating the direction of the optic axis with respect to a reference coordinate. These parameters are often spatially varying. Properties of birefringence in optical crystals are used to control light in various applications. In telecommunication industry, birefringence finds its application in optical isolators, circulators and optical interleavers, brightness enhancer in illumination systems is among non imaging applications [1]. Analysis of birefringence also finds application in the field of clinical medicine [2], ophthalmology [3], biomedical devices, polarimetry and characterization of non-linear optical devices and electro-optic materials. Interferometric analysis of birefringence has gradually gained importance

S. Sircar (✉) · I. Chakraborty
Department of Applied Optics and Photonics, University of Calcutta,
Kolkata, West Bengal 700098, India
e-mail: santa_sircar@hotmail.com

I. Chakraborty
e-mail: ipsitachak@rediffmail.com

© Springer Nature Singapore Pte Ltd. 2017

I. Bhattacharya et al. (eds.), *Advances in Optical Science and Engineering*,
Springer Proceedings in Physics 194, DOI 10.1007/978-981-10-3908-9_33

due to the wide spread use of birefringent optical materials in various areas of science and technology including the important area of nonlinear optics. In general, the polariscope and the Mueller matrix polarimetry are used for evaluation of birefringence. A new method has been proposed which utilizes phase shifting interferometry which yields greater resolution of measurement. In the present technique, a phase shifting Mach-Zehnder interferometer is utilized for the purpose.

2 Experimental Set Up

A collimated laser beam is passed through a Cube Beam Splitter (CBS) where it is amplitude divided as shown in Fig. 1. The transmitted beam is then linearly polarized by a polarizer P1 set at 45°. It is then passed through a spatially varying birefringence sample S. The resultant beam is again passed through a CBS after passing through a polarizer P2. The amplitude split reflected beam coming from the first CBS after reflection from the mirrors M1 and M2 is linearly polarized by the polarizer P3. Thus the interferogram recorded at the output port for two orthogonal positions of the polarizer has maximum fringe contrast after passing through the quarter wave plate and polarizer P4.

3 Theory

When a beam of light linearly polarized at 45° is transmitted through an arbitrarily oriented birefringent sample, there exist in general a phase difference $\Delta = \delta x \sim \delta y$ between the absolute phases transmitted along the reference x and y directions in the laboratory frame.

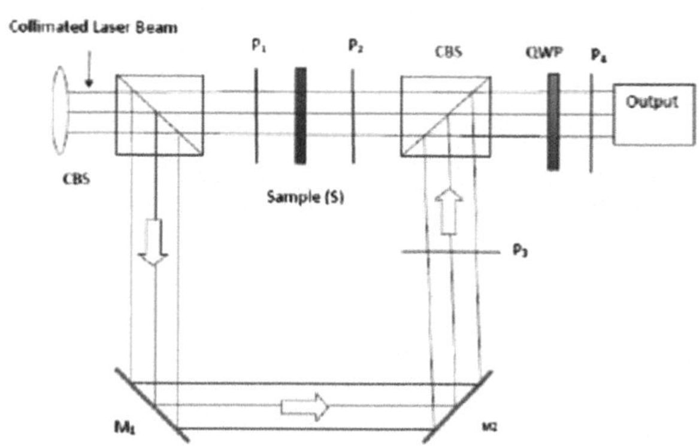

Fig. 1 Experimental setup

According to the Jones formulation, the Jones matrix of the sample is given by,

$$W(\delta,\varphi) = \begin{pmatrix} \cos\left(\frac{\delta}{2}\right) + i\sin\left(\frac{\delta}{2}\right)\cos(2\varphi) & i\sin\left(\frac{\delta}{2}\right)\sin(2\varphi) \\ i\sin\left(\frac{\delta}{2}\right)\sin(2\varphi) & \cos\left(\frac{\delta}{2}\right) - i\sin\left(\frac{\delta}{2}\right)\cos(2\varphi) \end{pmatrix} \tag{1}$$

Where, δ is the local value of birefringence of the sample and φ is the orientation of the fast axis with respect to the reference abscissa

The light transmitted from the sample is given as a Jones vector,

$$\mu = W(\delta,\varphi)P(\theta)$$

where, $W(\delta,\varphi)$ is the Jones matrix of the sample of spatial distribution of retardance $\delta(x,y)$ and the orientation of the fast axis φ(x, y) and P(θ) is the Jones vector of a linear polarizer P1 as shown in Fig. 1.

Assuming that the light beam incident on the sample is linearly polarized along a direction θ to the positive direction of the reference abscissa, the beam exiting from the sample is represented by,

$$\mu = \begin{pmatrix} \cos\left(\frac{\delta}{2}\right) + i\sin\left(\frac{\delta}{2}\right)\cos(2\varphi) & i\sin\left(\frac{\delta}{2}\right)\sin(2\varphi) \\ i\sin\left(\frac{\delta}{2}\right)\sin(2\varphi) & \cos\left(\frac{\delta}{2}\right) - i\sin\left(\frac{\delta}{2}\right)\cos(2\varphi) \end{pmatrix} \begin{pmatrix} \cos(\theta) \\ \sin(\theta) \end{pmatrix} \tag{2}$$

$$\mu = \begin{pmatrix} \cos\left(\frac{\delta}{2}\right)\cos(\theta) + i\sin\left(\frac{\delta}{2}\right)\cos(2\varphi-\theta) \\ \cos\left(\frac{\delta}{2}\right)\sin(\theta) + i\sin\left(\frac{\delta}{2}\right)\sin(2\varphi-\theta) \end{pmatrix} \tag{3}$$

which may be expressed $\begin{pmatrix} a_x e^{i\delta_x} \\ a_y e^{i\delta_y} \end{pmatrix}$ as where,

$$\tan\delta_x = \frac{\tan\left(\frac{\delta}{2}\right)\cos(2\varphi-\theta)}{\cos(\theta)} \tag{4a}$$

$$\tan\delta_y = \frac{\tan\left(\frac{\delta}{2}\right)\sin(2\varphi-\theta)}{\sin(\theta)} \tag{4b}$$

From the above equations, the values of δ and φ are calculated as follows;

$$\phi = \frac{1}{2}\tan^{-1}\left(\frac{\tan\delta_x + \tan\delta_y}{\tan\delta_x - \tan\delta_y}\right) \tag{5}$$

$$\delta = \frac{1}{2}\tan^{-1}\left(\frac{\tan\delta_x - \tan\delta_y}{2\cos(2\varphi)}\right) \tag{6}$$

4 Simulated Results

For establishing the validity of the proposed algorithm, a spatially varying birefringent object has been simulated as shown in Fig. 2.

With this simulated sample, δx and δy has been calculated from 4a and 4b and used in 5 and 6 to calculate φ and δ. The results for φ and δ are given in Figs. 3 and 4 and are seen to match exactly with that of the sample.

Fig. 2 Simulated sample

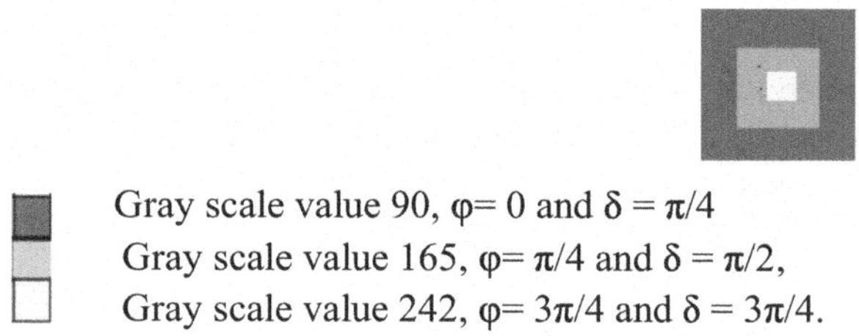

Fig. 3 3D profile of φ

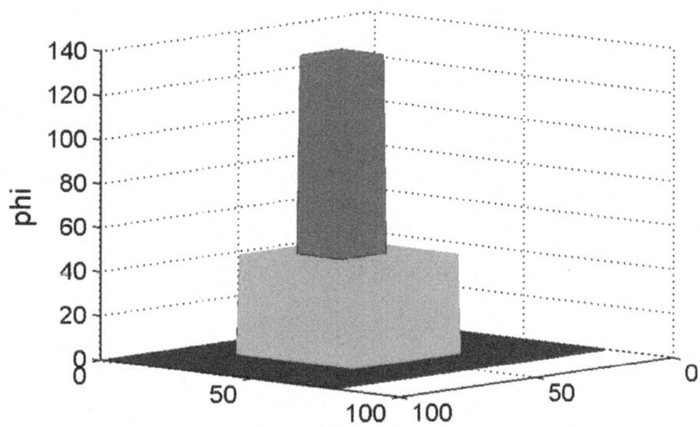

Fig. 4 3D profile of δ

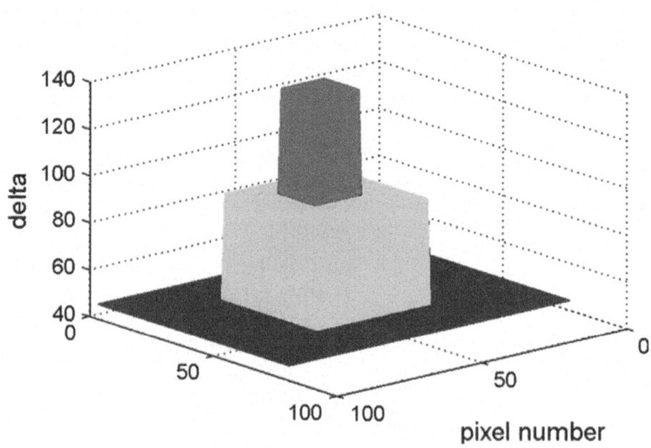

5 Discussion

To eliminate the effects of unequal phase introduced in the output due to the beam splitting interfaces and other optical surfaces, this process should be repeated without the sample. The resultant phase thus obtained is the difference of the above two processes. The measurement procedure discussed above is restricted to phase measurements from 0 to pi, therefore it is more appropriate for birefringence measurements of biological samples.

References

1. Hassler, R. A., Gregory, G. G. and Freniere, E. R., Proc. SPIE 4769 (July 2002).
2. Iftimia, N., Brugge, W. R. and Hammer, D. X., [Advances in Optical Imaging for Clinical Medicine], John Wiley & sons (2010).
3. Lim, Y. et al., "Birefringence measurement of cornea and anterior segment by office-based polarization-sensitive optical coherence tomography," Biomed. Opt. Express 2(8), 2392–2402 (2011).

Part VI
Optical Communication and Networks

Fuzzy Based Relay Selection for Secondary Transmission in Cooperative Cognitive Radio Networks

Jyoti Sekhar Banerjee, Arpita Chakraborty and Abir Chattopadhyay

Abstract Cooperative communication plays the vital role in cognitive radio network where intermediate nodes are employed as relays. But it is really tough to select the desired or so called the best relay in a multiple-relay cognitive radio system in order to improve the performance of the secondary network while ensuring the quality-of-service (QoS) of the primary network. In this paper we propose a new fuzzy logic-based decision-making procedure for relay selection unlike to many existing works where Signal-to-Interference-plus-Noise Ratio (SINR) is considered as the only parameter for relay selection. The underlying decision criterion considers SINR with some other important parameter like Relative Link Quality (RLQ) of the relay node from destination & Reliability of the relay node. To find out the best relay using our proposed scheme, we have conducted an extensive simulation study. The simulation results reveal the impact of different parameters on selection of Best relay.

1 Introduction

Cognitive Radio (CR), the recent buzz word [1–3] in wireless communication is empowered with the solution to overcome the spectrum scarcity problem by adopting dynamic spectrum access in which secondary user (SU) is allowed to access the licensed spectrum simultaneously with the primary user (PU) under certain threshold value i.e. interference temperature. In this paper, we focus on the formulation of best relay selection criteria in cooperative cognitive radio networks [4–7]. Our major contributions can be summarized as follows. Firstly, we have proposed a unique fuzzy logic-based relay selection scheme [7] which considers not

J.S. Banerjee (✉) · A. Chakraborty
Department of ECE, Bengal Institute of Technology, Kolkata, West Bengal 700150, India
e-mail: tojyoti2001@yahoo.co.in

A. Chattopadhyay
Department of ECE, University of Engineering & Management,
Kolkata, West Bengal 700156, India

© Springer Nature Singapore Pte Ltd. 2017
I. Bhattacharya et al. (eds.), *Advances in Optical Science and Engineering*,
Springer Proceedings in Physics 194, DOI 10.1007/978-981-10-3908-9_34

only the achievable highest received SINR at secondary destination (SD) of the relay node but few other important parameters like relative link quality and reliability. Second, we derive an exact form of reliability of the relay node for secondary transmissions while guaranteeing the quality-of-service (QoS) of the primary transmissions. The paper is organized as follows. The system model and data transmission process are defined in Sect. 2. Section 3 provides our proposed FLBRS scheme for cooperative cognitive transmission. Section 4 presents the simulation results and some discussions. Finally, the conclusion is presented in Sect. 5.

2 System Model

Here we consider an underlay cognitive relay [8] network where both primary transmitter (PT) and secondary transmitter (ST) transmit their data message simultaneously to the primary destination (PD) and secondary destination (SD) respectively using the same frequency band. In this paper the authors have designed a secondary network model with a single Source (ST) and a Destination node (SD) along with K relaying nodes, those are assumed to be Decode and Forward (DF) types (Shown in Fig. 1). Here, the concept of aggregate channel model has been employed that considers both path loss and slow Rayleigh fading (i.e. channel coefficients are stationary during one time slot) [8] in order to simulate the wireless environment. In the underlay approach of this paper, time is divided into number of slots $(t_f | f \text{ is any integer} > 0)$ and each one is further divided into two sub-slots $\langle t_f(\iota) | \iota \in (1,2) \rangle$ namely *observation interval* and *relaying interval*

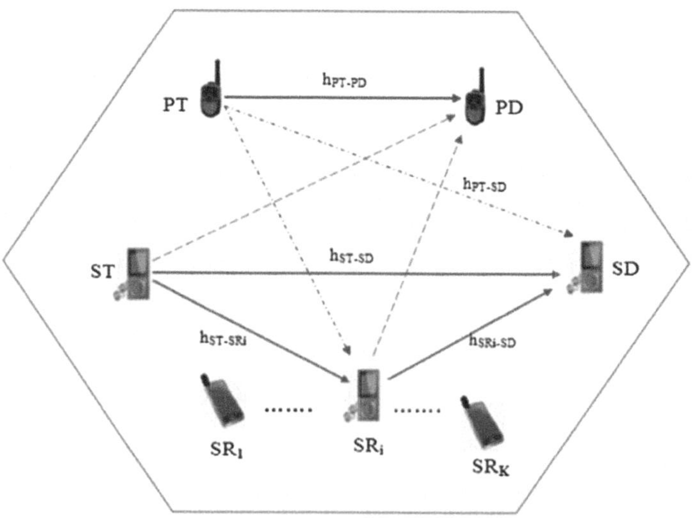

Fig. 1 Basic relay model

that correspond to two transmission phases. In the observation interval, all CR users listen to the primary user signal, and in the relaying interval each CR user acts as a relay. The fading coefficient is assumed fixed during one slot and varies independently with each time slot (Fig. 2).

In observation interval/the first sub-slot $\langle t_f(\iota = 1)\rangle$: ST transmits its data $(x_S\langle t_f(\iota = 1)\rangle)$ of data rate R_{ST} to both secondary Relays (SRs) and secondary destination (SD) with power P_{ST} along with the PT, who transmits its data $(x_P\langle t_f(\iota = 1)\rangle)$ of data rate R_{PT} with power P_{PT}. Hence the received signal at the secondary relays (SRs) and SD becomes:

$$\begin{aligned} Y_{SR_{i\in\Re}}\langle t_f(\iota = 1)\rangle = {} & \sqrt{P_{ST}}h_{ST,SR_i}\langle t_f(\iota = 1)\rangle x_S\langle t_f(\iota = 1)\rangle \\ & + \sqrt{P_{PT}}h_{PT,SR_i}\langle t_f(\iota = 1)\rangle x_P\langle t_f(\iota = 1)\rangle \\ & + n_{SD}\langle t_f(\iota = 1)\rangle \end{aligned} \quad (1)$$

$$\begin{aligned} Y_{SD}\langle t_f(\iota = 1)\rangle = {} & \sqrt{P_{ST}}h_{ST,SD}\langle t_f(\iota = 1)\rangle x_S\langle t_f(\iota = 1)\rangle \\ & + \sqrt{P_{PT}}h_{PT,SD}\langle t_f(\iota = 1)\rangle x_P\langle t_f(\iota = 1)\rangle \\ & + n_{SD}\langle t_f(\iota = 1)\rangle \end{aligned} \quad (2)$$

where $h_{ST,SD}$ and $h_{PT,SD}$ are fading coefficients of the channel from ST to SD and from PT to SD respectively and $n_{SD}\langle t_f(\iota = 1)\rangle$ is an additive white Gaussian noise. As in this paper we study the impact of best relay selection strategy, we have considered only the maximum average power transmitted by the ST. After receiving the signal from ST, all the SRs make an effort to decode the received message. Only the SRs who successfully decode the received message form a group named Decoding set and is represented as below:

$$\Phi_D\langle t_f(\iota = 1)\rangle = \{\varphi | \varphi \in \phi \cup \varphi_K, \quad k = 1, 2\ldots, 2^K - 1\}$$

where ϕ represents null set and φ_K represents a non empty sub set of 'K' SRs.

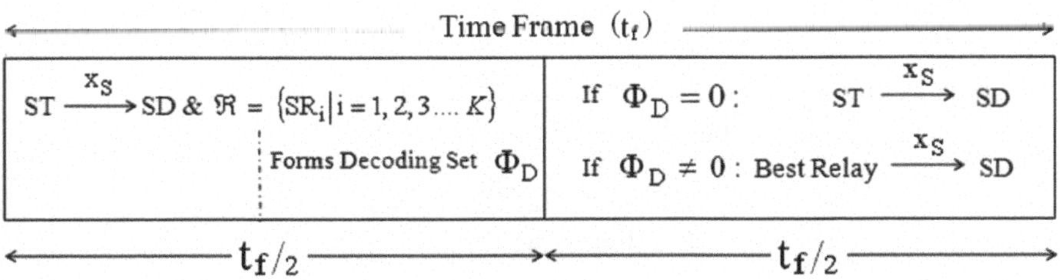

Fig. 2 Time slots of secondary transmission

In the relaying interval/second sub-slot $\langle t_f(\iota = 2)\rangle$: In this phase if $\{\varphi \neq \phi \text{ and } \varphi = \varphi_k\}$ i.e.—the Best Relay (SR_{BEST}) gets selected from the non empty Decoding set (φ_k) and transmits its correctly decoded message to the SD. In this paper we have proposed a new Fuzzy Logic based Best Relay Selection scheme (FLBRS). Hence the received signal at SD is as follows:

$$\begin{aligned} Y_{SD}\langle t_f(\iota = 2)\rangle|_{\varphi=\varphi_k} = {} & \sqrt{P_{ST}} h_{SR_{BEST},SD}\langle t_f(\iota = 2)\rangle x_S\langle t_f(\iota = 2)\rangle \\ & + \sqrt{P_{PT}} h_{PT,SD}\langle t_f(\iota = 2)\rangle x_P\langle t_f(\iota = 2)\rangle \\ & + n_{SD}\langle t_f(\iota = 2)\rangle, \quad SR_{BEST} \in \varphi_K \end{aligned} \tag{3}$$

On the other hand, if $\varphi = \phi$, i.e.—not a single SR is able to decode the message correctly and consequently ST has to re-transmit the original message to the SD over the direct link. Finally, SD executes MRC (Maximum ratio Combining) on the two copies of the received signal.

3 Proposed FLBRS Scheme

So far we have studied different Best Relay selection schemes which are based on received SINR at the SD [8], where as in practical scenario few other parameters may be considered for Best Relay selection in order to get better outage performance. Here in this very paper we have employed Fuzzy Logic for selecting the Best Relay (FLBRS) from the Decoding set (φ_k).

3.1 Signal-to-Interference-Plus-Noise Ratio ($SINR_{SD}$)

It is a very popular metric for relay selection and is calculated at the destination by performing MRC on the two copies of the received signal transmitted by ST and SRi node respectively. Hence the obtained SINR at SD is:

$$\begin{aligned} SINR_{SD}(SR_i)\Big|_{i\in\varphi_k} = {} & \frac{P_{ST}\left|h_{ST,SD}\langle t_f(\iota = 1)\rangle\right|^2}{P_{PT}\left|h_{PT,SD}\langle t_f(\iota = 1)\rangle\right|^2 + N_0} \\ & + \frac{P_{ST}\left|h_{SR_{BEST},SD}\langle t_f(\iota = 2)\rangle\right|^2}{P_{PT}\left|h_{PT,SD}\langle t_f(\iota = 2)\rangle\right|^2 + N_0}, \quad SR_{BEST} \in \varphi_k \end{aligned} \tag{4}$$

3.2 Relative Link Quality (RLQ)

It describes the selected Relay node's channel quality relative to the other relay nodes of the decoding set (φ_k).

$$\gamma_{SR_i,SD}\Big|_{i\in\varphi_k} = \frac{H_{SR_{SR_i,SD}}}{\frac{1}{K}\sum_{i=1}^{K} H_{SR_{SR_i,SD}}} = \frac{\left|h_{SR_{SR_i,SD}}\right|^2}{\frac{1}{K}\sum_{i=1}^{K}\left|h_{SR_{SR_i,SD}}\right|^2} \quad (5)$$

where $h_{SR_{i,SD}}$ is the channel impulse response between nodes SR_i and SD.

3.3 Reliability

Reliability basically expresses a relay node's performance to transmit a signal to the destination successfully. As the correct copy of the signal reaches to the destination, an ACK (Acknowledgement) is sent to both the ST and the corresponding SRi indicating successful transmission, and similarly NACK is sent to indicate failure transmission. Let during the time interval Δt (Where $\Delta t = t_1 - t_2$) total numbers of ACK and NACK received by the relay node SRi are $N^{ACK}_{SR_i|\Delta t}$ and $N^{NACK}_{SR_i|\Delta t}$ respectively.

Rate of successful transmission $(\lambda_S(SR_i, \Delta t))$ of SRi node may be defined as:

$$\lambda_S(SR_i, \Delta t)\big|_{i\in\varphi_K} = \frac{N^{ACK}_{SR_i|\Delta t}}{N^{ACK}_{SR_i|\Delta t} + N^{NACK}_{SR_i|\Delta t}} \quad (6)$$

Now let's assume, during Δt time probable number of transmissions carried by the SRi node be denoted by $\xi_{(SR_i,\Delta t)|\in\varphi_K}$. ∴ Average number of successful transmissions carried by the SRi relay node thus becomes

$$\overline{\Gamma}_S(SR_i, \Delta t)\big|_{i\in\varphi_K} = \lambda_S(SR_i, \Delta t)\big|_{i\in\varphi_K} \times \xi_{(SR_i,\Delta t)|\in\varphi_K} \quad (7)$$

Thus we may calculate the Reliability (β_{SR_i}) of the particular node SRi which is nothing but the average number of successful transmissions carried by the SRi relay node over a longer period of time say T $\left(\text{Where } T = t \times \Delta t|_{\text{tisanyintegerand} \neq 0}\right)$ as considered in this paper.

$$\beta_{SR_i} = \sum_{j=1}^{T} \overline{\Gamma}_S(SR_i, \Delta t_j) = \sum_{j=1}^{T} \lambda_S(SR_i, \Delta t_j)\big|_{i \in \varphi_K} \times \xi_{(SR_i, \Delta t)| \in \varphi_K} \quad (8)$$

3.4 Relay Selection Criteria Proposed by FLBRS

Next, considering the above three inputs, Fuzzy Logic Based Relay Selection (FLBRS) criteria may be written as:

$$\begin{aligned} \mu_{D_q}(W) &= f\left\{T\left(\mu_{A_m}(X), \mu_{B_n}(Y), \mu_{C_p}(Z)\right)\right\} \\ &= \arg\text{Max}\left[\text{Min}\left\{\mu_{A_m}(X) t \mu_{B_n}(Y) t \mu_{C_p}(Z)\right\}\right], \quad m, n, p, q \in (1, 2, 3) \end{aligned} \quad (9)$$

where X, Y, Z are input variables and the corresponding Output variable is W and are described as follows:

X $SINR_{Sd} \in \{A_1, A_2, A_3\} \in$ {Low, Medium, High},
Y Relative Link Quality (RLQ) $\in \{B_1, B_2, B_3\} \in$ {Weak, Moderate, Strong},
Z Reliability $\in \{C_1, C_2, C_3\} \in$ {Risky, Acceptable, Desired},
W Relay $\in \{D_1, D_2, D_3\} \in$ {Rejected, Good, Best}

After applying inputs to the Fuzzy Inference System (FIS) (shown in the Fig. 4.), which is considered to be Mamdani here, the output fuzzy set (Relay) is obtained as

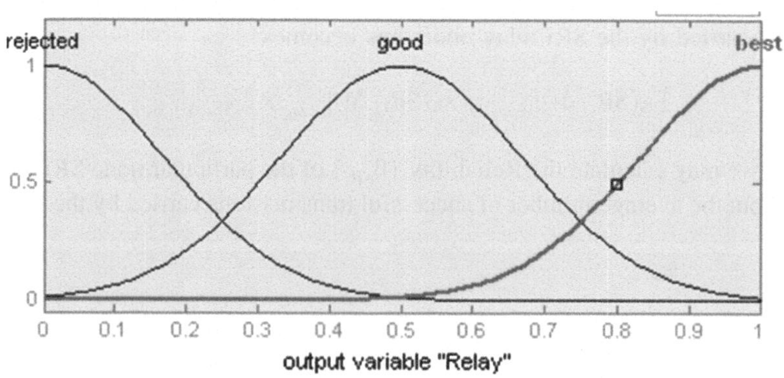

Fig. 3 Output membership function of relay

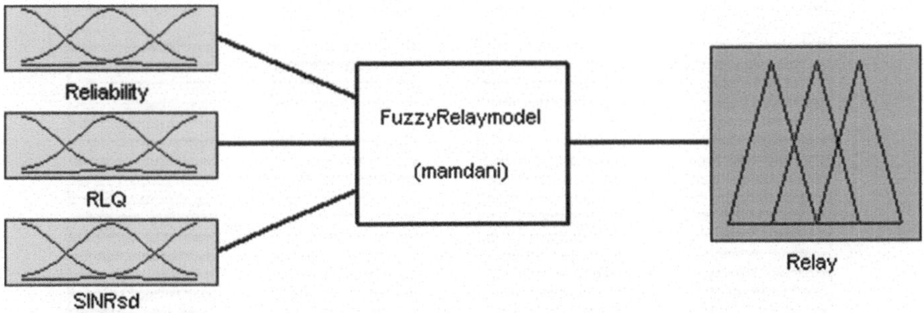

Fig. 4 Model of the proposed FLBRS system (Mamdani type)

shown in the Fig. 3. But to obtain the fuzzy system output, according to the inputs applied to the system, we have defined the 27 fuzzy rules. The output membership function of the Relay gets selected employing any of the 27 rules and thus Best Relay is selected whose output membership function $\left(\mu_{BEST}\left(SR_i|_{i\in\varphi_K}\right)\right)$ of the group 'Best' is with the highest value.

$$\begin{aligned} \therefore \text{Best Relay } (SR_{BEST}) &= \arg \underset{i}{\mathrm{Max}} \left\{\mu_{BEST}\left(SR_i|_{i\in\varphi_K}\right)\right\} \\ &= \arg \mathrm{Max} \left[\mathrm{Min}\left\{\mu_{A_m}(SINR_{Sd}) t \mu_{B_n}(RLQ) t\, \mu_{C_p}(\text{Reliability})\right\}\right], \\ &\quad m,\, n,\, p \in (1, 2, 3) \end{aligned} \tag{10}$$

4 Results and Discussion

This section provides simulation results to verify the analytical results. Here Fuzzy logic is used because it is a multi-valued logic and many input parameters can be considered to take the decision. Mamdani type fuzzy rule based system (FRBS) provides a natural framework to include expert knowledge in the form of linguistic rules. The simulation results are shown in Fig. 5a, b. Figure 5a describes the situation of Best Relay as the three input criteria are high enough. Figure 5b describes the surface view of Mamdani-type FIS.

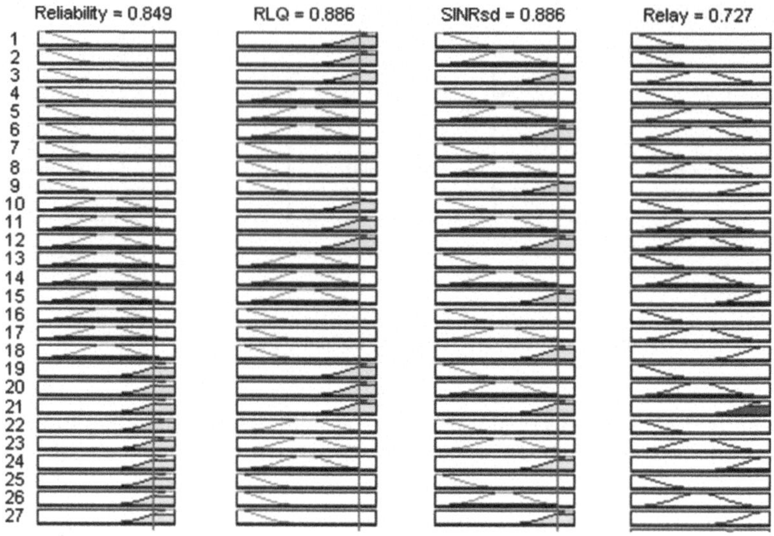

(a) Rules decide the Relay type to be 'Best'

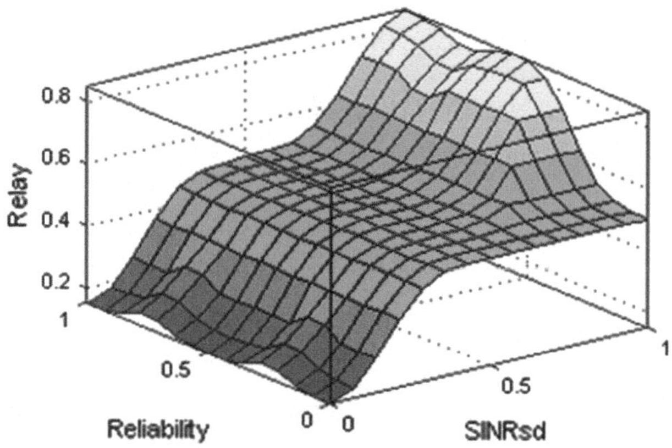

(b) Surface view of Mamdani-type FIS

Fig. 5 The fuzzy inference rules based on the Mamdani fuzzy inference system (FIS) and its output

5 Conclusion

A fuzzy logic based Best Relay selection technique is proposed here which will help to take wise decision during secondary data transmission phase in cognitive networks. The method considers three input parameters and thus is a practicable solution for secondary data transmission. The simulation software programs for the proposed system are neither complex nor consume much time to respond. Hence, it can be easily embedded into application programs and can be implemented in real systems.

References

1. Akyildiz, I. F., Lee, W.-Y., Vuran, M. C., & Mohanty, S., "Next Generation/ Dynamic Spectrum Access/ Cognitive Radio Wireless Networks: A Survey". Computer Networks, 50 (13), 2127–2159 (2006).
2. Bletsas, A., et. al., "A simple cooperative diversity method based on network path selection" . IEEE J. Select. Areas in Commun., 24(3), 659–672 (2006).
3. Banerjee, J. S., et. al., [Fundamentals of Software Defined Radio and Cooperative Spectrum Sensing: A Step Ahead of Cognitive Radio Networks], In N. Kaabouch, & W. Hu (Eds.) Handbook of Research on Software-Defined and Cognitive Radio Technologies for Dynamic Spectrum Management, Information Science Reference, Hershey, Pennsylvania, USA, 499–543 (2015).
4. Banerjee, J.S., et. al., [Modeling of Software Defined Radio Architecture & Cognitive Radio, the Next Generation Dynamic and Smart Spectrum Access Technology], In M.H. Rehmani & Y. Faheem (Ed.), Cognitive Radio Sensor Networks: Applications, Architectures, and Challenges, Information Science Reference, Hershey, Pennsylvania, USA, 127–158 (2014).
5. Banerjee, J.S., et. al., [Architecture of Cognitive Radio Networks], In N. Meghanathan & Y.B. Reddy (Ed.), Cognitive Radio Technology Applications for Wireless and Mobile Ad Hoc Networks, Information Science Reference, Hershey, Pennsylvania, USA, 125–152 (2013).
6. Banerjee, J.S., et. al., "A Comparative Study on Cognitive Radio Implementation Issues". International Journal of Computer Applications, 45(15), No.15, 44–51(2012).
7. Banerjee, J. S. & Chakraborty, A., "An Advance Q Learning (AQL) Approach for Path Planning and Obstacle Avoidance of a Mobile Robot". International Journal of Intelligent Mechatronics and Robotics, 3(1), 53–73 (2013).
8. Zou, Y., et. al.,"An adaptive cooperation diversity scheme with best-relay selection in cognitive radio networks". Signal Processing, IEEE Transactions on, 58(10), 5438–5445 (2010).

Optical Generation of Subcarrier Multiplexed (SCM) Millimeter Wave Signal and Its Application in Radio Over Fiber System

Madhumita Bhattacharya

Abstract Millimeter-wave radio over fiber is an attractive scheme which is widely used in broadband wireless communication system. In this paper, a scheme for the optical generation of subcarrier multiplexed (SCM) millimeter wave signal is proposed. In SCM, each RF carrier is modulated by the baseband signal. A group of baseband modulated subcarriers modulate the intensity of an optical carrier. The components required for the scheme consists of an optical pulse source, a WDM demultiplexer, two laser diodes, a photodiode and optical couplers. Two lightwaves are selected from the optical pulse source whose frequency difference is equal to the frequency of the desired mm wave signal. When two lightwaves, one intensity modulated by the SCM signals and the other CW lightwave, are combined and fed to the input of a wideband photodiode, the subcarrier multiplexed millimeter wave signal is generated. The two lightwaves are made phase coherent through optical injection locking of laser diodes. The novelty of the scheme is that no intermodulation distortions (IMD) are produced at the output. The generated mm wave modulated by the SCM signal can be radiated by an antenna over a base station. The phase noise of the generated mm wave signal is also very low in this scheme. No reference low noise mm-wave source is required in this scheme.

1 Introduction

Millimeter wave signals span over the frequency range 30–300 GHz of the electromagnetic spectrum. It finds potential use in wireless broadband communication system. This band of frequency is less congested and can be utilized for wireless cellular mobile communication services. Electronic generation of mm wave signals contains noise. Optical generation and optical transport of mm-wave signal is widely used [1–5]. The 60 GHz mm wave signal is used in fiber optic mm-wave

M. Bhattacharya (✉)
Department of Physics, Gushkara Mahavidyalaya, Gushkara, West Bengal 713128, India
e-mail: mita_bhatt@yahoo.co.in

© Springer Nature Singapore Pte Ltd. 2017

I. Bhattacharya et al. (eds.), *Advances in Optical Science and Engineering*, Springer Proceedings in Physics 194, DOI 10.1007/978-981-10-3908-9_35

radio. In a mm wave radio over fiber (ROF) system, mm wave modulated optical carriers is transported through low loss optical fiber from the central station to base stations. The modulated optical wave is detected at the base stations by a wideband photodiode and the mm wave is radiated over a pico cell. The frequency band around 60 GHz is used for short range communication since it falls in the absorption band of oxygen and thus the band can be reused. It is useful for short range low power communication.

2 System Description

In this paper, we propose a scheme for the optical generation of subcarrier multiplexed 60 GHz mm wave signal which finds application in ROF system. The components required for realization of the proposed scheme is commercially available. It consists of an optical pulse source, a WDM demultiplexer, two laser diodes, an optical combiner and a wideband photodiode. The schematic circuit diagram of the proposed SCM mm wave generator is shown in Fig. 1. The modelocked laser diode is used as the optical pulse source, typical repetition frequency of which can be 10 GHz and the central wavelength can be chosen to be 1.55 μm. An optical WDM demultiplexer is used to separate the optical pulse into different CW lightwaves separated by 10 GHz around the central frequency of the optical pulse. An arrayed waveguide grating (AWG) can be used as the demultiplexer. Two CW lightwaves, having frequencies f_1 and f_2 are so chosen that the frequency separation between them is 60 GHz, which is the desired frequency of the mm wave signal to be generated. Two laser diodes, LD1 and LD2 are so chosen so that their free running oscillation frequencies are close to f_1 and f_2 respectively. We assume the laser diodes are injection locked to the input lightwaves [2] and are thus made phase coherent and get power amplified. LD1 is intensity modulated by bias current modulation. A group of microwave subcarriers, each carrying different information of a specific channel is combined in the microwave domain and applied

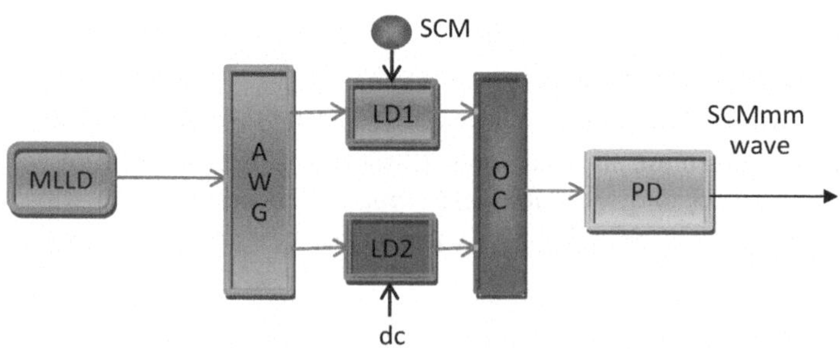

Fig. 1 The schematic circuit diagram of the proposed SCM mm wave generator. *MLLD* mode-locked laser diode; *AWG* arrayed waveguide grating; *LD1* and *LD2* laser diodes; *SCM* microwave subcarriers; *OC* 2X1 optical coupler; *PD* wideband photodiode

to bias of the laser diode. The SCM modulated lightwave and CW lightwave from LD2 are superimposed and are incident on the wideband photodiode (PD). At the output of the photodiode we get the desired mm-wave signal modulated by SCM signals. The SCM signal which directly modulated the optical intensity of the lightwave from LDI gets transferred to the difference frequency signal having frequency ($f_2 - f_1$). This system produces no IMD which is a unique feature of this ROF communication system.

3 Analysis

A group of microwave subcarriers is multiplexed in the microwave domain and the combined wave is applied to the laser diode, LD1 through a bias tee. The laser diode, LD1 is intensity modulated by the microwave subcarrier multiplexed signal. Direct current modulation of the laser diode produces intensity modulation as well as frequency chirp. This frequency chirp can be strongly reduced using optical injection locking [6].

Under optical injection-locked condition, the output lightwaves from the laser diodes, LD1 and LD2 can be expressed as

$$E_1(t) = |E_1(t)| \exp[j(\omega_1 t + \theta_1)] \quad (1)$$

and

$$E_2(t) = |E_2| \exp[j(\omega_2 t + \theta_2)] \quad (2)$$

respectively. $|E_1(t)|^2$ is the intensity of the modulated optical wave. Mathematically, it can be expressed as $|E_1(t)|^2 = P_{01}(1 + \sum_{i=1}^{n} m_i \cos \Omega_i t)$. P_{01} is the free-running power of LD1. m_i is the intensity modulation index of the i-th microwave subcarrier having angular frequency Ω_i. n is the number of microwave subcarriers. $E_2(t)$ is the output CW lightwave from LD2. The lightwaves from the laser diodes, LD1 and LD2 are combined in the 2X1 optical coupler and the combined wave is incident on the photodiode. The composite wave entering the photodiode is given by

$$Y(t) = \frac{1}{\sqrt{2}}[|E_1(t)| \exp j(\omega_1 t + \theta_1) + |E_2| \exp j(\omega_2 t + \theta_2)] \quad (3)$$

After photo detection, the photocurrent can be calculated to be

$$I_{pd} = \frac{\eta}{2}\left[|E_1(t)|^2 + |E_2|^2 + 2|E_1(t)||E_2| \cos\{(\omega_2 - \omega_1)t + (\theta_2 - \theta_1)\}\right] \quad (4)$$

η is the responsivity of the photodiode. From (4) we can observe the generated mm wave signal having angular frequency $(\omega_2 - \omega_1)$ is amplitude modulated by the microwave SCM signal. In mathematical calculation, we assume that the microwave subcarrier frequencies are much less than the frequency of the generated mm wave signal. A high pass filter following the photodiode will filter out the desired SCM modulated mm wave signal. This SCM modulated mm wave signal is amplified at the base station and radiated by the base station antenna over a pico cell.

4 Conclusion

A scheme for the optical generation of SCM modulated mm wave signal is proposed. The components required for the implementation of the proposed scheme is commercially available. It can be used in mm wave radio over fiber system. The noise of the generated mm wave signal is small and the output intermodulation distortion is much less.

References

1. J. Li, T. Ning, L. Pei and C. Qi, "Scheme for high capacity 60 GHz Radio-over-Fiber Transmission System", J. of Opt. Commun. Netw., 1(4), 324–330 (2009).
2. M. Bhattacharya, B. Sarkar and T. Chattopadhyay, "Subcarrier-multiplexed fiber-optic mm-wave radio using sideband injection locking of semiconductor lasers", Proceedings of International conference ICOCN 2002, Singapore, 11–14 Nov (2002).
3. A.K. Vyas, Dr. N. Agarwal, "Radio over Fiber: Future Technology of Communication", IJETTCS, 1(2), 233–237(2012).
4. J. Yao, "Microwave Photonics", IEEE J. Lightwave Technology, 27(3), 314–335 (2009).
5. J. Yu, Z. Jia, T. Wang and G. Chang, "A novel radio-over-fiber configuration using optical phase modulator to generate an optical mm-wave and centralized lightwave for uplink version", IEEE Photonics Technology Letters, 19(3), 140–142, (2007).
6. M. Bhattacharya, R. Mukherjee and T. Chattopadhyay, "Reduction of frequency chirping in semiconductor lasers through injection locking", IETE Journal of Research, 45 (5–6), 1999.

Design and Performance Analysis of Traffic Rerouting Based Congestion Control Technique in Optical WDM Network

Manoj Kr. Dutta

Abstract In the present paper a traffic rerouting based architectural model for contention resolution in optical wavelength division multiplexing (WDM) network has been proposed. The proposed switching architecture is consisted of two different types of Erlang traffic processing models namely Node B and Node C. Traffic from source is divided into two unequal parts at the router and are diverted to the above mentioned nodes. There is a channel shifter between the two nodes which can shift a number of available output channels from one node to another. A control signal looks after the blocking probability and gives necessary instructions to the channel shifter. If one node is heavily loaded while the other one is less then to reduce the congestion and traffic loss the channel shifter shifts some free output channels from lightly loaded node to the heavily node and vice versa. This model may be used for optimum utilization of given number of available channels in a WDM network and to achieve minimum blocking probability. Performance of the proposed architectural model has been verified by appropriate mathematical models and simulations.

1 Introduction

Bandwidth requirement is increasing day by day because of different applications like e-commerce, social networking sites, telemedicine etc. Optical fiber offers huge raw bandwidth which can satisfy the ever increasing bandwidth requirement. WDM is a very useful technology to explore the available raw bandwidth for high speed communication. WDM technology along with wavelength re-routing, enables the optical backbone network to provide a tremendous speed of 100 Gbps or more. Though very efficient but contention may create unwanted problems in WDM application. Contention may occur in WDM network when more than one signal tries to reserve same wavelength channel at the output. Contention in a WDM network can be

M.Kr. Dutta (✉)
Department of Physics, Birla Institute of Technology, Off-Campus Deoghar, Ratanpur, Jasidih, Jharkhand 814142, India
e-mail: mkdutta13@gmail.com

© Springer Nature Singapore Pte Ltd. 2017

I. Bhattacharya et al. (eds.), *Advances in Optical Science and Engineering*, Springer Proceedings in Physics 194, DOI 10.1007/978-981-10-3908-9_36

resolved by using different conventional technologies viz. space domain, time domain and wavelength domain [1–4]. But all these contention resolution policies require extra hardware which increases the network complexity and cost.

A different congestion control technique for WDM network is used in this paper. The proposed architectural model consisted of two different Erlang traffic processors. Poisson arrival process of traffic is considered at the intermediate router. The router then reroutes the traffic to the processing nodes. One control signal will maintain the correlation between the router and the channel shifter. It will reallocate the available output wavelengths from one node to another and vice versa. Incoming traffic flow to the node is random but measurable. An empirical formula has been developed by which the required number of channels to be shifted can be calculated and required action will be taken by the channel shifter to keep blocking probability low. Appropriate analytical model has been developed to simulate the performance of the proposed switching architecture [5–11].

2 Modeling of Switching Architecture

To evaluate the performance of the switch architecture it is assumed that the maximum incoming traffic at the router is ‘ρ’. The function of the router is to divide the incoming traffic into two parts and to send ‘kρ’ traffic to node B and ‘(1 − k)ρ’ traffic to the node C. The parameter ‘k’ is a variable quantity and the value ‘k’ will vary from 0 to 1. The value of ‘k’ is a user defined quantity. The user will set a proper ‘k’ value to obtain a required overall blocking probability. Figure 1 shows

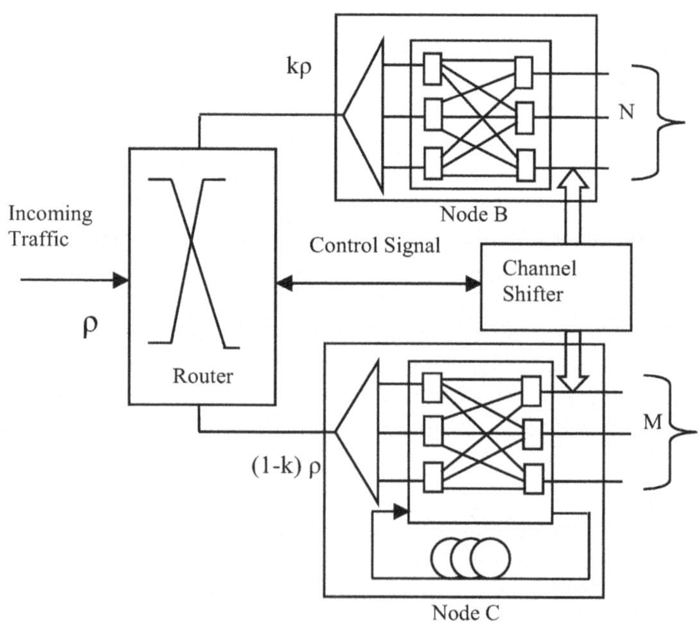

Fig. 1 Node architecture

Fig. 2 Flow chart

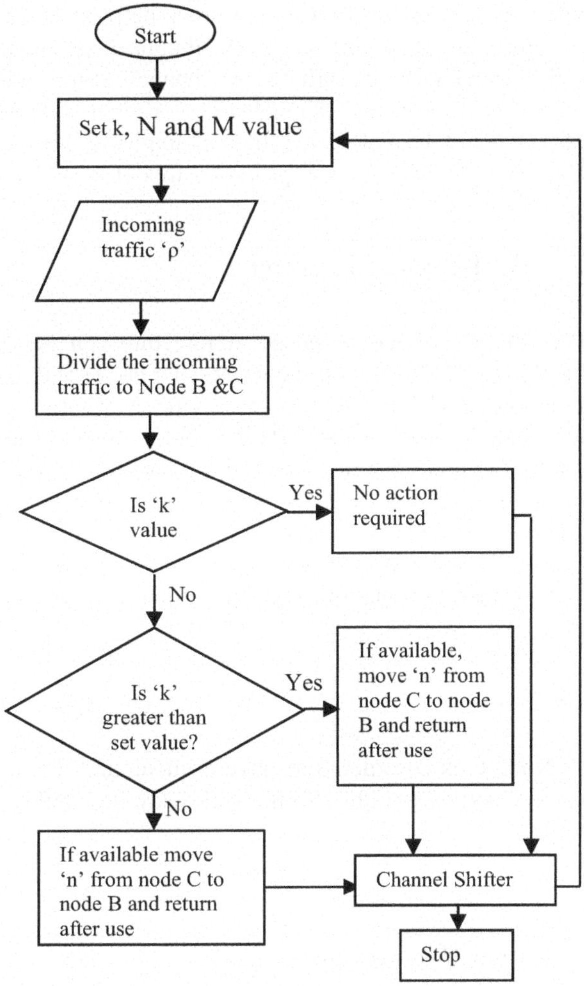

the proposed switching architecture. Here it is assumed that processor B has N and processor C has M number of output wavelength channels. These number of assigned output wavelength channels to Node B and C can significantly maintain the required value of 'k' which in turn restrict the dropping probability. It is essential to keep the value of 'k' unchanged otherwise performance of the system may be degraded. When 'k' value changes then the corresponding blocking probability of node B changes which may in turn starts creation of congestion in the node. The value of desired blocking probability will depend on 'k'. The value of 'k' is set initially by the user. The value of 'k' will be continuously monitored by the router. If the set value of 'k' is changed then channel shifter will be instructed by the router to take necessary action as depicted in Fig. 1. Channel shifter is capable to shift a number of free available output channels from one node to another and vice versa if required to maintain blocking probability low. The channel shifter will be directed by three different types of control signals from the router viz. no need of any channel shifting, shift 'n' no channels from node B to node C or vice versa, return back the shifted channels to its original node when action is over. The value of 'n' could be calculated by an empirical formula where, n = [a(1.6^{k}) − N]. Complete flowchart of the proposed model is shown in Fig. 2.

3 Mathematical Model

Performance of the proposed architecture can be evaluated by determining the blocking probabilities of two processing nodes and that of the overall model. In the proposed model the B node is having N, C node is having M no of available wavelengths respectively. Erlang B traffic model is applied to node B. In this traffic model the node has no intermediate queue to keep data for a certain time so an incoming call will either be processed or rejected at this node. In the proposed architecture random incoming traffic is considered. The router will route 'ρk' traffic to node B and rest to node C.

Considering standard Erlang B traffic formula blocking probability of Node B is,

$$P_B = \frac{\frac{(k\rho)^N}{N!}}{\sum_{i=0}^{N} \frac{(k\rho)^N}{i!}} \tag{1}$$

Node C is considered to have a queue and this is used for holding the unprocessed incoming traffic. So the call delay probability node C is given by

$$P_c\,[Call\ Delayed] = \frac{\rho^M}{\rho^M + M!\left(1 - \frac{\rho}{M}\right)\sum_{i=0}^{M} \frac{\rho^i}{i!}} \tag{2}$$

For proposed network the modified blocking probability for node C is

$$P_C = \frac{\{(1-k)\rho\}^M}{\{(1-k)\rho\}^M + M!\left(1 - \frac{\{(1-k)\rho\}}{M}\right)\sum_{i=0}^{M} \frac{\{(1-k)\rho\}^i}{i!}} * \exp \frac{-[M - \{(1-k)\rho\}]}{H} t \tag{3}$$

where, t is the delay time and H is the average duration of the call.

The above equations have been used to find out the call delay and blocking probability of the incoming traffic.

The grade of service (GOS) of the proposed network can be calculated by,

$$GOS = \frac{\rho^M H_M^{N-1}}{\sum_{l=1}^{M} \rho^l H_l^{N-1}} \tag{4}$$

4 Simulation and Results

The simulation results show that the implementation of channel shifter improves the performance of the switch architecture. Figure 3 depicts the variation of blocking performance with incoming traffic for node B, node C and overall switch before the establishment of channel shifter. Figure 4 shows the same parameter after the establishment channel shifter. Both the figure shows that the blocking probability offered by the node C is best but the comparative study reveals that the channel shifter improves blocking probability capacity for all types of nodes and the overall performance of the switch. For incoming traffic load of 5 Erlang the blocking probability of node C is around 0.225 in Fig. 3 but same value is around 0.19 in case of Fig. 4. This result shows the improvement of the blocking probability of the switch architecture. The channel shifter readjusts the number of available output wavelength channels between node B and C so as to get the optimum utilization of the available resource. Figure 5 depicts the grade of service for the proposed architecture. GOS initially increases with increasing traffic and after a certain input traffic it remains almost constant. This nature of the graph may be due to the fact that initially as the incoming traffic is increasing then the number of available wavelength channels of each node are being occupied to process the traffic so the grade of service is increasing and reaches up to the serviceable capacity of the corresponding processing node and the GOS graph becomes saturated and constant, but when the incoming traffic goes beyond the serviceable capacity of the processor then the excess traffic cannot be processed so it is dropped or in the hold condition in the queue.

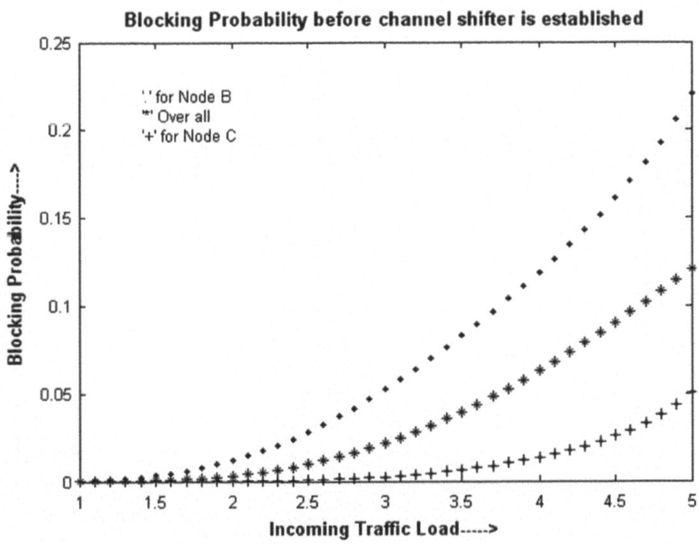

Fig. 3 Blocking probability before establishment of channel shifter

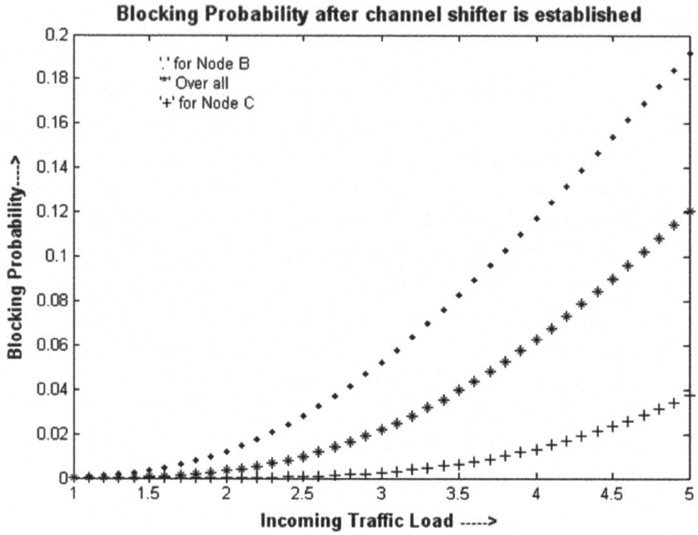

Fig. 4 Blocking probability after establishment of channel shifter

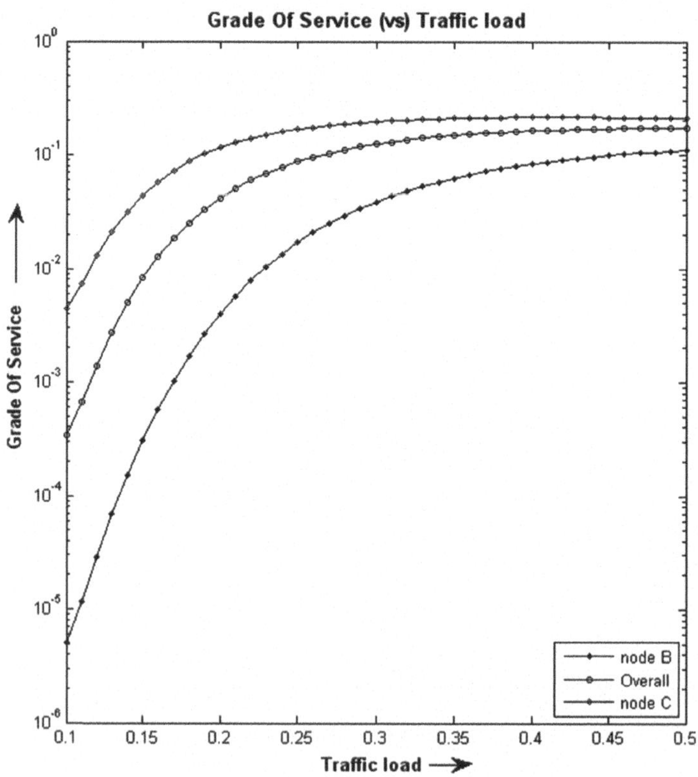

Fig. 5 Grade of service analysis

5 Conclusions

An architectural model for contention resolution in WDM network is proposed in this research work. The model is consisted of two different types of processors namely node B and node C where node B follows Erlang B traffic formula and node C follows Erlang C traffic formula. A channel shifter along with the controls signal looks after the traffic condition in terms of blocking probability of the switch. The channel shifter is capable of shifting a number of free output available channels from heavily loaded node to the lightly loaded node and vice versa. The simulation result obtained in MATLAB environment shows a significant improvement in the blocking performance of the proposed model. The model may be used for optimum utilization of a given number of available channels in a WDM network and to achieve minimum blocking probability.

Acknowledgements The author would like to thank all the authorities of Birla Institute of Technology, Mesra and Off-campus Deoghar for providing all kinds of supports and granting institute seed money scheme for carrying out this work.

References

1. Qiwu W., Xianwei Z., Jianping W., Zhizhong Y. and Lin L., “Multicast routing and wavelength assignment with delay constraint in WDM networks with sparse wavelength conversions”, Springer Science+Business Media, LLC (2009).
2. I. Chlamtac, A. Ganz, and G. Karm., “Light path communications: An approach to high bandwidth optical WAN’s”, IEEE Transaction on Communications, vol. 49, no. 7, pp. 1171–1182, July (1992).
3. X. Chu, B. Li, and C. Imrich., “Wavelength converter placement under different rwa algorithms in wavelength-routed all-optical networks”, IEEE Transaction on Communications, vol. 51, no. 4, pp. 607–617, Apr. (2003).
4. Sho Shimizu, Yutaka Arakawa, Naoaki Yamanaka., “A wavelength assignment scheme for WDM Networks with Limited Range Wavelength Converters”, IEEE ICC (2006).
5. Li-Wei Chen, Eytan Modiano, “Efficient Routing and Wavelength Assignment for Reconfigurable WDM Networks with Wavelength Converters”, IEEE INFOCOM (2003).
6. Kundu. R, Chaubey. V.K. “Analysis of optical WDM Network Topologies with Application of LRWC under Symmetric Erlang –C Traffic”, Springer Science+Business Media B.V. (2008).
7. Ramaswami. R, Sasaki. G.H, “Multi wavelength optical networks with limited wavelength conversion”, Proc. INFOCOM’97, vol. 2, pp. 489–498. April (1997).
8. Subramaniam S., Azizoglu M., Somani A., “On optical converter placement in wavelength-routed networks”, III/ACM Trans, Networking, vol. 7, no. 5, pp. 754–766, Oct (1999).
9. M.K. Dutta, V.K. Chaubey, “Optical Network Traffic Control Algorithm under Variable Loop Delay: A Simulation Approach”, Int. J. Communication, Network and System Sciences, 2009, 7, 651–655.

10. M. De Leenheer, C. Develder, F. De Truck, B. Dhoedt, P. Demeester, "Erlang Reduced Load Model for Optical Burst Switched Grids", Third International Conference on Networking and Services. ICNS 2007, Ghent University, pp. 177.
11. Y.L. Hsue, M.S. Rogge, S. Yamamoto, L.G. Kazovsky, "A highly flexible and efficient passive optical network employing dynamic wavelength allocation", IEEE J. Lightwave Technol. 23 (2005) 277–286.

Outage Analysis of an Inter-relay Assisted Free Space Optical Communication System

Himanshu Khanna, Mona Aggarwal and Swaran Ahuja

Abstract In this work, we analyze the outage performance of a one way inter-relay assisted free space optical link. We assume the absence of direct link or line of sight path between the source and destination nodes. The analytical closed form expressions for the outage probability have been derived and are illustrated by numerical plots. We attempt to show that the absence of line of sight path between the source and the destination doesn't lead to significant performance degradation. Moreover, we conclude that inter-relaying not only provides an additional degree of freedom, but can also further help to increase the distance for transmission between the source and the destination nodes, while keeping the total transmitted power within permissible levels.

1 Introduction

The free space optical communication (FSO) has emerged as one of the key research areas because of its several potential benefits such as higher bandwidth, cost effective, secure and flexible solutions for the last mile problems [1]. However, the performance of FSO systems is affected significantly by the path losses, atmospheric conditions between transmitter and receiver, and pointing errors. In order to mitigate the turbulence induced fading, relay assisted communication is employed in FSO systems [2]. The relay assisted FSO system can be implemented as a multi-hop serial relaying or a parallel relaying system [2, 3]. Moreover, the relaying can either be active in which all relays cooperatively listen and transmit [4], or it can be selective where a subset of relays is selected from the available relays based on the state of FSO network [5, 6].

If relays are inter-connected in FSO systems, they increase the diversity order and enhance the performance of FSO networks [7, 8]. However, the system models presented in the literature consider a direct link between source and destination

H. Khanna (✉) · M. Aggarwal · S. Ahuja
Department of EECE, The NorthCap University, Gurgaon, Haryana, India
e-mail: khanna.h@rediffmail.com

© Springer Nature Singapore Pte Ltd. 2017

I. Bhattacharya et al. (eds.), *Advances in Optical Science and Engineering*,
Springer Proceedings in Physics 194, DOI 10.1007/978-981-10-3908-9_37

nodes, which in a practical scenario may not be feasible for long distances. We investigate the outage performance of the system and derive closed form expression for the outage probability. We also analyze the outage performance of the system under consideration, in the absence of pointing errors, and compare the same with non-interconnected (NIR) and dual-hop serial relaying techniques, in the presence and absence of channel state information (CSI), which is illustrated by numerical plots.

Rest of the paper is organized as follows: In Sect. 2, the system model is presented. In Sect. 3, we derive analytical expressions for outage probability, for the system model under consideration. The numerical plots and conclusions are given in Sects. 4 and 5, respectively.

2 System Model

We consider a binary pulse position (PPM), intensity modulated FSO communication system that employs a direct detection scheme. The source (S) transmits data to destination (D) with the help of two relays (R_1, R_2), as shown in Fig. 1. We assume all channels to experience independent but not necessarily identically distributed (i.n.i.d.) fading. The transmission of data is in the half duplex mode. The decode and forward (DF) relaying protocol employed here, decodes the incoming signal at the relay and re-transmits the decoded version of signal on correct reception of cyclic redundancy check (CRC) bits, to the destination node. The channel coefficients h_{xy} for $\{xy\} \in \{SR_1, SR_2, R_{12}, R_{1D}, R_{2D}\}$, where SR_i is the link between S and relay i, R_{12} is the link between R_1 and R_2, and R_{iD} is the link between relay i and D, for $i \in \{1, 2\}$, are modeled taking into consideration the effect of atmospheric turbulence, path losses, and pointing error impairments. The atmospheric turbulence is characterized by gamma-gamma fading statistics with its probability density function (pdf) given by [9]. The atmospheric path loss is modeled using the exponential Beers-Lambert Law, and is given as $h_l = \exp(-\sigma L)$, where σ is the attenuation coefficient, and L is the link length in m. Further, the pointing error losses are considered to be Rayleigh distributed and given as [10]. The pdf of the channel h_{xy}, considering the influence of all the three channel impairments discussed above is given by [10, (20)]. The pdf $f_{\gamma_{xy}}(\gamma)$ of the instantaneous SNR, Υ_{xy}, can then be derived as [10].

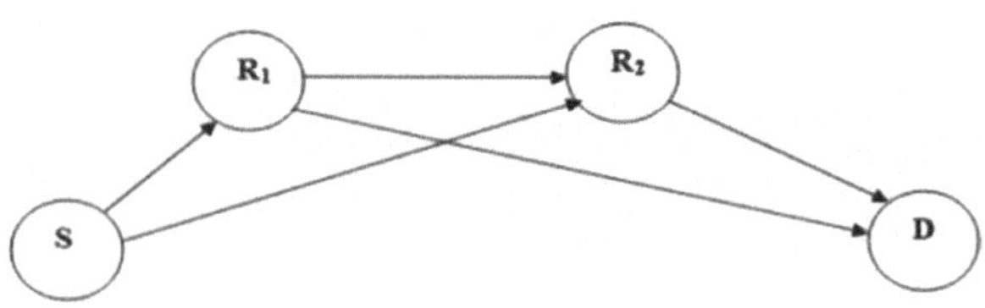

Fig. 1 Two relay FSO communication system with a unidirectional interconnected relay link

3 Outage Analysis

3.1 Outage Analysis for Inter-connected Relaying FSO System in the Absence of CSI, but with Considerable Pointing Error Impairments

Firstly, we consider a case that the pointing errors are present and the CSI is not available at relay and destination nodes. Further, an all active relaying strategy is considered for the analysis. We define γ_{th} as the threshold value for the SNR, post reception and decoding. If the instantaneous SNR, γ, over any link falls below γ_{th}, then that link is said to be in outage. We now define outage events corresponding to four cases for the given system model described earlier in this section as follows.

Case—1: When both relays R_1 and R_2 decode correctly, i.e. $\gamma_{R_1} > \gamma_{th}$ and $\gamma_{R_2} > \gamma_{th}$, the outage event is

$$P(\gamma_{R_{1D}} < \gamma_{th})P(\gamma_{R_{2D}} < \gamma_{th})$$

Case—2: When R_1 decodes correctly but R_2 does not, i.e $\gamma_{R_1} > \gamma_{th}$ and $\gamma_{R_2} > \gamma_{th}$, the outage event is

$$P(\gamma_{R_{12}} < \gamma_{th})P(\gamma_{SR_2} < \gamma_{th})P(\gamma_{R_{1D}} < \gamma_{th})$$

Case—3: When R_2 decodes correctly but R_1 does not, i.e. $\gamma_{R_2} > \gamma_{th}$ and $\gamma_{R_1} > \gamma_{th}$, the outage event is

$$P(\gamma_{SR_1} < \gamma_{th})P(\gamma_{R_{2D}} < \gamma_{th})$$

Case—4: When both relays R_1 and R_2 don't decode correctly, i.e. $\gamma_{R_1} > \gamma_{th}$ and $\gamma_{R_2} > \gamma_{th}$, the outage event is

$$P(\gamma_{SR_1} < \gamma_{th})P(\gamma_{SR_2} < \gamma_{th}).$$

Assuming all the links to be statistically independent, the overall outage probability for these four disjoint cases can be written as

$$\begin{aligned} P_{out} = {} & P(\gamma_{SR_1} > \gamma_{th})P(\gamma_{R_{1D}} < \gamma_{th})P(\gamma_{R_{2D}} < \gamma_{th})\left[P(\gamma_{SR_1} > \gamma_{th})P(\gamma_{R_{12}} > \gamma_{th}) + P(\gamma_{SR_2} > \gamma_{th})\right] \\ & + P(\gamma_{SR_1} > \gamma_{th})P(\gamma_{SR_2} < \gamma_{th})P(\gamma_{R_{12}} < \gamma_{th})P(\gamma_{R_{1D}} < \gamma_{th}) \\ & + P(\gamma_{SR_1} < \gamma_{th})P(\gamma_{SR_2} > \gamma_{th})P(\gamma_{R_{2D}} < \gamma_{th}) + P(\gamma_{SR_1} < \gamma_{th})P(\gamma_{SR_2} < \gamma_{th}), \end{aligned} \tag{1}$$

where the probability of outage for a given link, xy, can be written as

$$P(\gamma_{xy} \leq \gamma_{th}) = \int_{0}^{\gamma_{th}} f_{\gamma_{xy}}(\gamma)d\gamma,$$

and is given as [11]

$$\begin{aligned} F_{\gamma_{xy}}(\gamma_{th}) &= P(\gamma_{xy} \leq \gamma_{th}), \\ F_{\gamma_{xy}}(\gamma_{th}) &= \frac{\xi_{xy}^2}{\Gamma(\alpha_{xy})\Gamma(\beta_{xy})} G_{2,4}^{3,1}\left(\alpha_{xy}\beta_{xy}\sqrt{\frac{\gamma_{th}}{\Omega_{xy}}} \,\middle|\, \begin{matrix} 1, \xi_{xy}^2+1 \\ \xi_{xy}^2, \alpha_{xy}, \beta_{xy}, 0 \end{matrix}\right), \end{aligned} \tag{2}$$

where $F_{\gamma_{xy}}(\gamma_{th})$ is the cumulative distribution function (CDF) of a single link xy, $G_{p,q}$ is Meijer's G-function, α_{xy} and β_{xy} are the atmospheric turbulence parameters, ξ_{xy} is the pointing error parameter, and Ω_{xy} denotes the average electrical SNR for link xy and is defined as $\Omega_{xy} = (\eta \mathrm{E}[h_{xy}])^2/N_o$ with E denoting the expectation operator. Moreover, it may be further noted that $P(\gamma_{xy} > \gamma_{th}) = 1 - P(\gamma_{xy} \leq \gamma_{th}) = 1 - F_{\gamma_{xy}}(\gamma_{th})$. Therefore, the outage probability, P_{out}, can be re-written as

$$\begin{aligned} P_{out} = {} & (1 - F\gamma_{SR_1}(\gamma_{th}))F\gamma_{R_{1D}}(\gamma_{th})F\gamma_{R_{2D}}(\gamma_{th})[(1 - F\gamma_{SR_1}(\gamma_{th}))(1 - F\gamma_{R_{12}}(\gamma_{th})) \\ & + (1 - F\gamma_{SR_2}(\gamma_{th}))] + (1 - F\gamma_{SR_1}(\gamma_{th}))F\gamma_{SR_2}(\gamma_{th})F\gamma_{R_{12}}(\gamma_{th})F\gamma_{R_{1D}}(\gamma_{th}) \\ & + F\gamma_{SR_1}(\gamma_{th})(1 - F\gamma_{SR_2}(\gamma_{th}))F\gamma_{R_{2D}}(\gamma_{th}) + F\gamma_{SR_1}(\gamma_{th})F\gamma_{SR_2}(\gamma_{th}). \end{aligned} \tag{3}$$

By substituting $F_{\gamma_{xy}}(\gamma_{th})$ in (3), the closed form expression for outage probability can then be obtained in terms of Meijer's G-function.

3.2 *Outage Performance for a Non-interconnected Relaying FSO System with No Pointing Errors*

For no or negligible pointing error impairments, the outage probability of the link xy, defined in terms of cdf of the gamma-gamma distribution, is given as [5]

$$p_{xy}^{(N)} = P(\gamma_{xy} \leq \gamma_{th}) = \frac{1}{\Gamma(\alpha_{xy})\Gamma(\beta_{xy})} G_{1,3}^{2,1}\left(\frac{\alpha_{xy}\beta_{xy}N}{G_{xy}P_M} \,\middle|\, \begin{matrix} 1 \\ \alpha_{xy}, \beta_{xy}, 0 \end{matrix}\right), \tag{4}$$

where G_{xy} is the gain of the link xy and given as [2], P_M is the power margin given as $P_M = \frac{RT_bP_t}{\sqrt{N_o\gamma_{th}}}$, with R being the responsivity of the photodetector, T_b is the bit duration, P_t is the total transmitted power, N is the number of active links, and N_o is the AWGN.

3.2.1 When CSI Is Available

When CSI is available at relay and destination nodes, selective relaying protocols can be employed for transmitting data along the unique strongest path [5, 6, 8]. In this case, the outage probability of an NIR-FSO system is given as [8]

$$\begin{aligned} P_{out} &= p_{SR_1D}^{(N)} \times p_{SR_2D}^{(N)} \\ &= (p_{SR_1}^{(N)} + p_{R_{1D}}^{(N)} - p_{SR_1}^{(N)} p_{R_{1D}}^{(N)})(p_{SR_2}^{(N)} + p_{R_{2D}}^{(N)} - p_{SR_2}^{(N)} p_{R_{2D}}^{(N)}), \end{aligned} \tag{5}$$

where $p_{SR_1D}^{(N)}$ and $p_{SR_2D}^{(N)}$ are the outage probabilities for paths S-R_1-D and S-R_2-D, respectively, $p_{SR_1}^{(N)}$ and $p_{SR_2}^{(N)}$ are the outage probabilities for S to R_1 and S to R_2 links, $p_{R_{1D}}^{(N)}$ and $p_{R_{2D}}^{(N)}$ are the outage probabilities for R_1 to D and R_2 to D links, respectively, and are defined by (4).

3.3 *Outage Performance of Inter-connected Relaying FSO System with No Pointing Errors*

We now analyze the outage performance of inter-connected relaying FSO system considering negligible pointing error impairments, and when CSI is available and when it is unavailable at the relay and destination nodes.

3.3.1 When CSI Is Not Available

When no CSI is available at the relay and destination nodes, the outage probability can be found by substituting (4) in (3), after replacing $F_{\gamma_{xy}}(\gamma_{th})$ with $p_{xy}^{(N)}$ in (3).

3.3.2 When CSI Is Available

When CSI is available at relay and destination nodes, the selective relaying protocol can be used for transmitting the data along the strongest path, as mentioned earlier. In this case, the outage probability of the system is given as [8]

P_{out} = (P_{out} for NIR-FSO system with CSI) × Q, where $Q = 1 - Q_{11}Q_{12}Q_{13}$, and $Q_{11} = 1 - p_{R_{12}}^{(3)}$, $Q_{12} = \frac{(1-p_{SR_1}^{(3)})p_{R_{1D}}^{(2)}}{p_{SR_1D}^{(2)}}$, $Q_{13} = \frac{(1-p_{R_{2D}}^{(3)})p_{SR_2}^{(2)}}{p_{SR_2D}^{(2)}}$, respectively, with $p_{R_{12}}^{(N)}$ denoting the outage probability of the link R_{12}, and is given by (4).

4 Numerical Results

In this section, we present the simulation results for the outage performance of the system. We assume the values of C_n^2 for moderate and strong turbulence as 3×10^{-14} and $1 \times 10^{-13}\ m^{-2/3}$, respectively. The wavelength of the optical signal, λ, is assumed to be equal to 1550 nm, the attenuation coefficient $\sigma = 0.43$ dB/km, $R = 0.625$ A/W, and T_b is in the order of nanoseconds.

Figure 2 considers the independent and identically distributed (i.i.d.) characteristics for the channels, where the distances $d_{SR_1} = d_{R_{12}} = d_{R_{2D}} = d_{SR_2} = d_{R_{1D}} = 1000$ m, and where d_{SR_1} is the distance between S and R_1, $d_{R_{12}}$ is the distance between R_1 and R_2, $d_{R_{2D}}$ is the distance between R_2 and D, d_{SR_2} is distance between S and R_2, and $d_{R_{1D}}$ is the distance between R_1 and D, respectively. Figure 2 illustrates a plot of outage probability as a function of average SNR per hop Ω (dB) for various turbulence conditions with pointing errors $\xi_1 = 1.2$ and $\xi_2 = 4.0$. As per the expectation, the outage performance deteriorates with increase in turbulence strength and pointing errors.

For Fig. 3, we consider the i.n.i.d. fading characteristics for the optical links, with $d_{SR_1} = d_{R_{12}} = d_{R_{2D}} = 1000$ m, and $d_{SR_2} = d_{R_{1D}} = 2500$ m. It can be seen that the inter-connected relaying technique outperforms the non-interconnected (parallel) relaying and serial relaying techniques, for the given values of distances between the nodes. Moreover, inter-connected relaying without CSI performs better than the one where CSI is available, for higher values of SNR. It is due to the use of DF relaying protocol which leads to better noise rejection at higher SNRs.

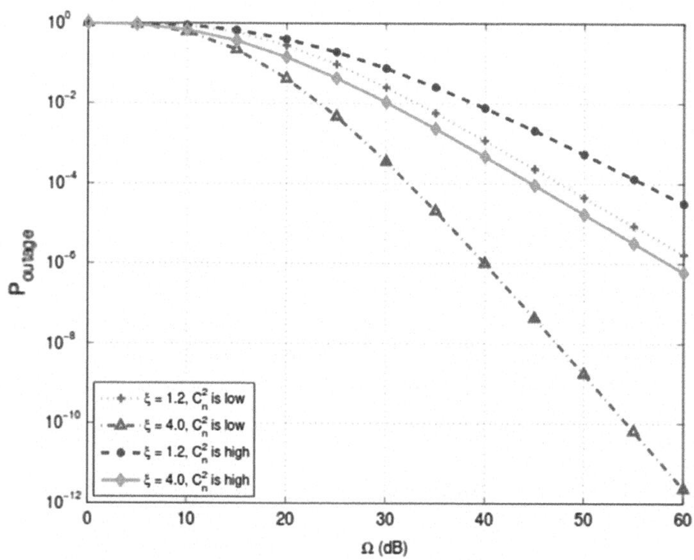

Fig. 2 Outage probability versus avg. SNR (dB) of a relayed FSO system, $\gamma_{th} = 5$ dB

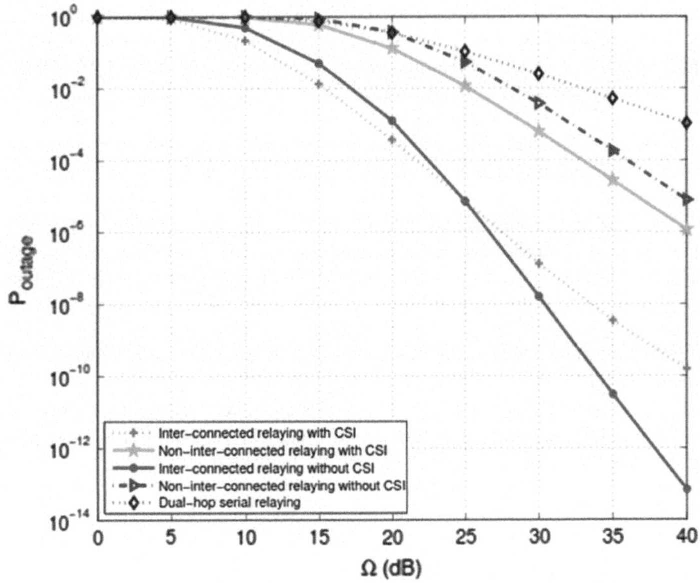

Fig. 3 Outage probability versus avg. SNR (dB) of a relayed FSO system with no pointing errors, $\gamma_{th} = 5$ dB

5 Conclusion

In this paper, we investigated the outage performance of a one way inter-connected dual relay FSO communication system. The closed form expression for outage probability were obtained and the effect of various system parameters on it, was studied. We observed that the absence of direct link between source and destination nodes does not significantly degrades the performance of the system, due to the use of inter-relay connections.

References

1. Chan, V., "Free-space optical communications," IEEE/OSA Journal of Lightwave Technology 24(12), 4750–4762 (2006).
2. Safari, M. and Uysal, M., "Relay-assisted free-space optical communication," *IEEE Transactions on Wireless Communications* 7(12), 5441–5449 (2008).
3. Aghajanzadeh, S. and Uysal, M., "Outage performance and DMT analysis of DF parallel relaying in FSO IM/DD communications," in [*Vehicular Technology Conference, Quebec, Canada*], *Proc. IEEE*, 1–5 (2012).
4. Karimi, M. and Nasiri-Kenari, M., "Free-space optical communications via optical amplify-and-forward relaying," *IEEE/OSA Journal of Lightwave Technology* 29(2), 242–248 (2011).

5. Chatzidiamantis, N. D., Michalopoulos, D. S., Kriezis, E. E., Karagiannidis, G. K., and Schober, R., Relay selection protocols for relay-assisted free-space optical systems, *IEEE/OSA Journal of Optical Communication and Networking* 5(1), 4790–4807 (2013).
6. Rjeily, C. A. and Haddad, S., "Cooperative FSO systems: Performance analysis and optimal power allocation," *IEEE/OSA Journal of Lightwave Technology* 29(7), 1058–1065 (2011).
7. Rjeily, C. A. and Haddad, S., "Inter-relay cooperation: A new paradigm for enhanced relay-assisted FSO communications," *IEEE Transactions on Communications* 62(6), 1970–1982 (2014).
8. Rjeily, C. A., "All-active and selective FSO relaying: Do we need inter-relay cooperation?," *IEEE/OSA Journal of Lightwave Technology* 32(10), 1899–1906 (2014).
9. Al-habash, M., Andrews, L., and Philips, R., "Mathematical model for the irradiance pdf of a laser beam propagating through turbulent media," *Optical Engineering* 40(8), 1554–1562 (2001).
10. Gappmair, W., "Further results on the capacity of free-space optical channels in turbulent atmosphere," *IET Communications* 5(9), 1262–1267 (2011).
11. Aggarwal, M., Garg, P., and Puri, P., "Analysis of subcarrier intensity modulation based optical wireless DF relaying over turbulence channels with path loss and pointing error impairments," *IET Communications* 8(17), 3170–3178 (2014).

Part VII
Optical and Digital Data and Image Processing

Single Public Key Dependent Attack on Optical Asymmetric Cryptosystem

Areeba Fatima and Naveen K. Nishchal

Abstract This work aims to study the vulnerability of the phase-truncated Fourier transform (PTFT) based cryptosystem towards single key attack analysis. The existing attack algorithms aimed at retrieving the input information from the PTFT cryptosystems require the knowledge of both the encryption keys. In this work, we aim to retrieve the plaintext through phase retrieval algorithm that uses only one of the keys as its constraints. The proposed method is supported by simulation results.

1 Introduction

Optical cryptosystems are being extensively studied due to their benefits they offer in the form of fast parallel processing and large degrees of freedom serving as the security keys [1]. In literature, various schemes have been proposed to secure information optically [1–10]. The broad areas into which these cryptosystems fall are symmetric and asymmetric cryptosystems. The asymmetric cryptosystems fair better than the symmetric ones because the key that is used during the encryption is not used for decryption, thus making the system nonlinear. This nonlinearity helps in enhancing the security and makes the system less vulnerable to various attacks. One such asymmetric cryptosystem namely, the phase-truncated Fourier transform (PTFT) was proposed that uses two random phase keys for encryption [2]. In this scheme, the decryption keys are generated during the encryption procedure and are different from those used for encryption. Thus, being a nonlinear system, the security of the cryptosystem is enhanced. However, later it was shown that the PTFT scheme is vulnerable to the specific attack [3, 4]. In this attack, the attacker knows the ciphertext and the encryption keys. The plaintext is retrieved through the iterative phase retrieval algorithm using the known quantities as the constraints.

A. Fatima (✉) · N.K. Nishchal
Department of Physics, Indian Institute of Technology Patna,
Bihta, Patna, Bihar 801103, India
e-mail: areeba@iitp.ac.in

© Springer Nature Singapore Pte Ltd. 2017
I. Bhattacharya et al. (eds.), *Advances in Optical Science and Engineering*,
Springer Proceedings in Physics 194, DOI 10.1007/978-981-10-3908-9_38

Many variants of the PTFT schemes have been introduced to overcome the specific attacks [6–10].

The cryptanalysis forms an integral part of cryptography. Along with establishing newer and more robust cryptosystem, there is a constant effort to study attack schemes that can break the security of the various proposed schemes. In this regard, some of the attack schemes that are relevant are the known plaintext attack, chosen plaintext attack, ciphertext only attack and the specific attack [7]. The known plaintext attack and the chosen plaintext attack aim at unauthorized retrieval of the keys. These keys can be further used for retrieving other plaintexts. For the PTFT scheme, the known plaintext attack is not much significant because the keys generated in this technique are object dependent and hence if retrieved, will not be used any further because for different plaintexts, the keys would be different. The important attack scheme for this encryption system is the specific attack, wherein the ciphertext and the encryption keys are used to retrieve the information of the plaintext. Various specific attack schemes have been proposed to retrieve the input information [3, 4]. It is of interest to study whether the input information can be retrieved or not when the number of parameters known to the attacker is minimized. The idea is to reduce the available information to the attacker, yet the constraints should be enough to help the iterative phase retrieval algorithm to converge.

The attack schemes are usually based on iterative phase retrieval algorithms. The modified GS algorithm has been the most common of all the algorithms to be used for studying attack schemes [6]. Later on, Fienup's algorithm for quicker convergence namely the error reduction algorithm (ERA) and the hybrid input output algorithm (HIOA) were introduced for more efficient phase retrieval [11–14].

2 Principle

2.1 Introduction to PTFT

In this section, the PTFT cryptosystem is briefly described [2]. The input image field is combined with the random phase key (RPK) R_1, and is Fourier transformed. The resulting output is phase-truncated to get the amplitude $G(u, v)$ and in the process the phase is reserved to be used as the decryption key K_1. If $I(x, y)$ is the input image field, then this first step of the procedure can be mathematically written as:

$$G(u,v) = |FT(I(x,y) \times R_1)| \tag{1}$$

$$K_1 = AT\{FT(I(x,y) \times R_1)\} \tag{2}$$

Here AT{.} denotes amplitude truncation.

The amplitude $G(u, v)$ is again combined with another RPK, R_2 and then subjected to the Fourier transform. The resulting output is again phase-truncated to obtain $E(m, n)$, which serves as the ciphertext. The phase is reserved and is used as the second key K_2.

$$E(m,n) = |FT(G(u,v) \times R_2)| \tag{3}$$

$$K_2 = AT\{FT(G(u,v) \times R_2)\} \tag{4}$$

In the PTFT scheme, the encryption keys differ from the decryption keys and hence a nonlinearity has been introduced which helps in resisting various attacks. However, it was shown that when the encryption keys are used as the public keys, the security is compromised and the plaintext can be retrieved. This work led to other studies which formulated newer and better specific attacks on the PTFT. The common thing between the existing attacks is that they require the knowledge of the encryption keys to carry out the phase retrieval. In another approach, the crypt-analysis of the PTFT scheme can involve the situation wherein only one of the keys is available to the attacker. This work aims to study the formulation of a phase retrieval algorithm which uses only one of the keys to retrieve the plaintext.

2.2 Attack Algorithm

The phase retrieval algorithm consists of two stages. In the first stage, the amplitude $G(u, v)$ is evaluated using Fienup's method of HIOA.

In the next section, an overview of the ERA and the HIOA is discussed. Though ERA is an upgrade over the GS algorithm, its limitation in our study is discussed. Consequently, HIOA has been used in the first stage of the proposed attack scheme.

In the second stage, the retrieved amplitude $G(u, v)$ and the first encryption key R_1 are used to retrieve the plaintext.

2.3 Error Reduction Algorithm (ERA)

The error reduction algorithm, proposed by Fienup, is an upgrade over the GS algorithm [13]. Initially, a random estimate of the phase is multiplied with the known Fourier domain intensity and the obtained quantity is inverse Fourier transformed. The output $G'_k(u,v)$ is then made to comply with the object-domain constraints, which are usually the non-negativity of the object. Thus, for the kth iteration, the estimate of the object can be written as:

$$G_{k+1}(u,v) = \begin{cases} G'_k(u,v) & u,v \notin D \\ 0 & u,v \in D \end{cases} \tag{5}$$

Here, D denotes the set of points at which G_k violates the object-domain constraints. Apart from the non-negativity, another object-domain constraint can be that the expanse of the object should be restrained to the support of the object. In our study, the object to be retrieved is complex. Hence, the ERA is not the best algorithm to be used, as the non-negativity condition cannot be applied. In that case, the ERA would require only the support constraint.

2.4 Hybrid Input Output Algorithm (HIOA)

The HIOA is a further upgrade on the ERA [14]. Here, a random estimate of the phase is multiplied with the known Fourier domain intensity and the obtained quantity is inverse Fourier transformed. The output $H'_k(u, v)$ is then conformed to the constraints. In this method, the input from a previous iteration is modified to form the next input. Mathematically, it is stated as:

$$H_{k+1}(u,v) = \begin{cases} H_k(u,v) & u, v \notin D \\ H_k(u,v) - \beta H'_k(u,v) & u, v \in D \end{cases} \quad (6)$$

Here, β is a constant that lies between 0 and 1. D denotes the set of points at which H_k violates the object-domain constraints. This method is known for faster convergence and is used in the first stage of the attack scheme.

3 Results and Discussion

The numerical simulation has been carried out on MATLAB platform (R2009a). The input image shown in Fig. 1a, is zero padded to a binary image of size 256×256 pixels. This binary image is subjected to the PTFT encryption scheme,

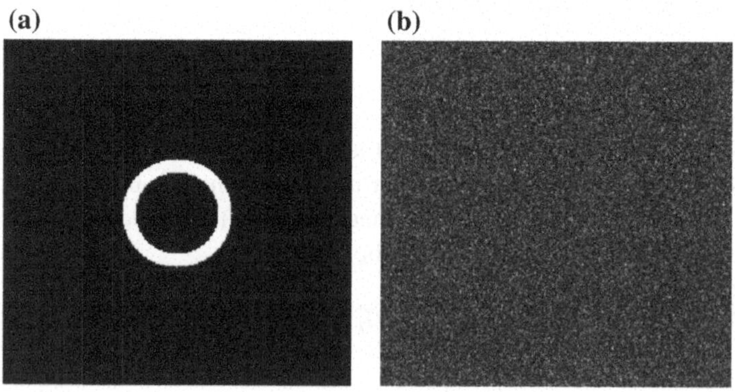

Fig. 1 **a** Input image used for PTFT encryption **b** ciphertext as a result of PTFT encryption

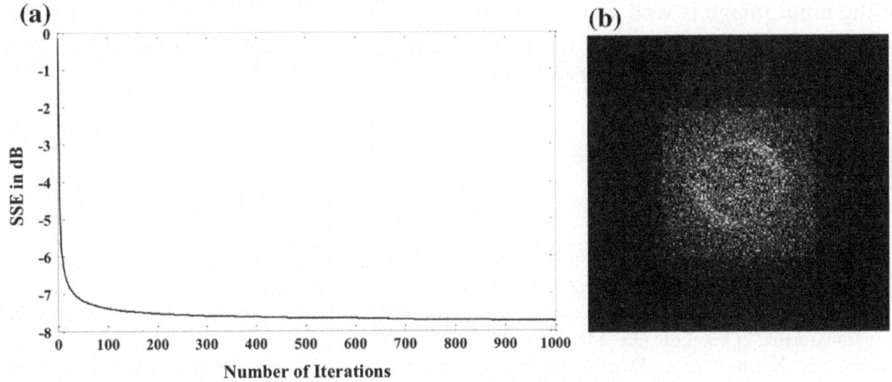

Fig. 2 **a** Plot of the SSE corresponding to the iteration number, **b** retrieved image at the end of the second stage

using random phase masks as the encryption keys. Figure 1b shows the ciphertext, which is assumed to be available with the attacker.

The ciphertext is used as the input to the first stage of the proposed attack scheme, which consists of the HIOA. This stage evaluates the estimate of the intermediate function $G(u, v)$. Next, this function is used in the second stage wherein the first encryption key is used as the constraint to evaluate the plaintext. A cost function, namely the discrete sum squared error (SSE), is used to study the convergence of this stage of the scheme [14].

$$SSE = 10\log_{10}\left\{\frac{\sum_u \sum_v \{|G'_i(u,v)| - G'(u,v)\}^2}{\sum_u \sum_v \{G'(u,v)\}^2}\right\} \tag{7}$$

Here, the subscript i denotes the iteration number. Figure 2a denotes the plot of the evaluated SSE with respect to the iteration number. The retrieved image at the end of the second stage is shown in Fig. 2b. It can be seen that the contour of the input image is well visible and hence the proposed scheme successfully retrieves the input information.

4 Conclusion

To summarize, we have proposed an iterative phase retrieval algorithm that reduces the number of parameters required to break the PTFT encryption scheme. In all the existing specific attack schemes, both the encryption keys are necessary for successful attack. In this study, we have proposed an algorithm that uses only the first encryption key for successful attack. The simulation carried out shows the contour of the input image is well visible and this supports our proposition. Hence it can be concluded that the encrypted information can be retrieved even in the absence of knowledge of one of the encryption keys.

References

1. Refregier, P. and Javidi, B., "Optical image encryption based on input plane encoding and Fourier plane random encoding," Opt. Lett. 20, 767–769 (1995).
2. Qin, W. and Peng, X., "Asymmetric cryptosystem based on phase-truncated Fourier transforms," Opt. Lett. 35, 118–120 (2010).
3. Wang, X. and Zhao D., "A special attack on the asymmetric cryptosystem based on phase-truncated Fourier transforms," Opt. Commun. 285, 1078–1081 (2012).
4. Wang, X., Chen, Y., Dai, C., and Zhao, D., "Discussion and a new attack of the optical asymmetric cryptosystem based on phase-truncated Fourier transform," Appl. Opt. 53, 208–213 (2014).
5. Kumar, P., Joseph, J., and Singh, K., "Vulnerability of the security enhanced double random phase-amplitude encryption scheme to point spread function attack," Opt. Lasers Eng. 50, 1196–1201 (2012).
6. Rajput, S. K. and Nishchal, N. K., "Asymmetric color cryptosystem using polarization selective diffractive optical element and structured phase mask," Appl. Opt. 51, 5377–5386 (2012).
7. Rajput, S. K. and Nishchal, N. K., "Known-plaintext attack on asymmetric cryptosystem," Proc. SPIE 8855, 88550U (2013).
8. Rajput, S. K. and Nishchal, N. K., "An optical encryption and authentication scheme using asymmetric keys," J. Opt. Soc. Am. A 31, 1233–1238 (2014).
9. Mehra, I. and Nishchal, N. K., "Image fusion using wavelet transform and its application to asymmetric cryptosystem and hiding," Opt. Express 22, 5474–5482 (2014).
10. Fatima, A., Mehra, I., and Nishchal, N. K., "Optical asymmetric cryptosystem using equal modulus decomposition and multiple diffractive imaging," J. Opt. 18 (2016) 085701.
11. Gerchberg, R. W., "A practical algorithm for the determination of phase from image and diffraction plane pictures," Optik 35, 237–246 (1972).
12. Fienup, J. R., "Phase retrieval algorithm: a comparison," Appl. Opt. 21, 2758–2769 (1982).
13. Guo, C., Liu, S., and Sheridan, J. T., "Iterative phase retrieval algorithms. Part I: Optimization," Appl. Opt. 54(15), 4698–4708 (2015).
14. Guo, C., Liu, S., and Sheridan, J. T., "Iterative phase retrieval algorithms. Part II: Attacking optical encryption systems," Appl. Opt. 54(15), 4709–4719 (2015).

Weighted Laplacian Energy Based Image Fusion

Pradeep Shankhwar, A.K. Singh and B.S. Chauhan

Abstract Laplacian Pyramid (LAP) based image fusion has been implemented in FPGA. A concept of weighted laplacian energy based fusion has been applied. Eighty percent of pixel weight from high Laplacian energy and twenty percent pixel weight from low Laplacian energy gives better results than traditional high Laplacian energy based fusion algorithm. Entropy as a measure of quality of fused image has been considered. This approach is simulated in PC environment using MATLAB and has also been implemented in Xilinx's Virtex 5 SX FPGA.

1 Introduction

There is always possible to capture different details by taking multiple shots with same camera or same scene by two different cameras. Storing multiple images of same location is a burden and cost more resources even if they have different details in them. Similarly, it is cumbersome to display videos of two different camera simultaneously. Fusion is a better way to harness complementary information obtained from different camera. There are numerous methods available for image fusion viz. weighted average, high pass filtering, Principal Component analysis (PCA), Laplacian Pyramid and Discrete Wavelet Transform (DWT) and contrast pyramid etc. These image fusion algorithms have been tried in past to fuse the images. Laplacian Pyramid based technique has been selected due to low computation cost and relatively better fusion result [1]. A novel scheme of weighted Laplacian energy based fusion is proposed in this paper.

The paper is divided into five sections. Section 2 describes traditional Laplacian image fusion, Sect. 3 gives implementation details and Sect. 4 describes the approach of weighted Laplacian fusion. Results and discussions are mentioned in Sect. 5.

P. Shankhwar (✉) · A.K. Singh · B.S. Chauhan
Instrument Research & Development Establishment (IRDE), Dehradun, India
e-mail: pradeep.irde@gmail.com

© Springer Nature Singapore Pte Ltd. 2017

I. Bhattacharya et al. (eds.), *Advances in Optical Science and Engineering*,
Springer Proceedings in Physics 194, DOI 10.1007/978-981-10-3908-9_39

2 Laplacian Image Fusion

Laplacian pyramid based image representation is firstly introduced by Burt and Adelson in 1983 [2]. The Laplacian pyramid is derived from the Gaussian pyramid which is a multi-resolution image representation. The Gaussian pyramid is obtained through a recursive reduction (low-pass filtering and decimation) of the image data set. LAP fusion algorithm implementation has following stages for each video channel [3, 4].

(a) Gaussian filtering and down sampling
(b) Up scaling and subtractions to get Laplacian pyramid (LAP) image
(c) Fuse each level of pyramid to get fused image pyramid
(d) Up sample the lower level pyramid and add to higher pyramid
(e) Do above steps recursively depending upon no of pyramid level

The reduce operation will be achieved from (1) (Fig. 1).

The Gaussian pyramid is expanded by (2) to match the size of underlying level. Finally, the expanded image data set is subtracted with the same size image of Gaussian pyramid to get the Laplacian pyramid Subtract Operation as (3).

Let G_0 be the original input image, $G_1(i,j)$ is the 1st level Gaussian pyramid image, i = 0,…, R_i and j = 0,…,C_i. (R_i, C_i) is the image size of l level Gaussian pyramid, w(m, n) is the Gaussian template [4]. The reduce operation is defined as

$$G_1(i,j) = \sum_{m=-1}^{1} \sum_{n=-1}^{1} w(m,n)\, G_{l-1}(2i+m, 2j+n) \tag{1}$$

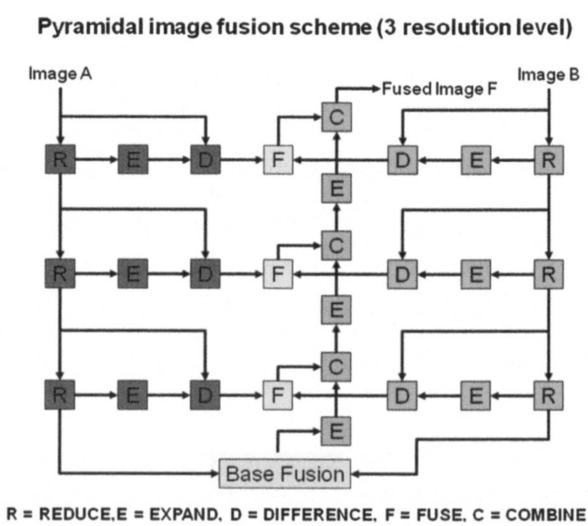

Fig. 1 Laplacian image fusion pipeline

For $k = 0, 1,\ldots,l$

$$G_{1,k}(i,j) = \left\{ 4 \sum_{m=-1}^{1} \sum_{n=-1}^{1} w(m,n)\, G_{l,k-1}\left(\frac{i+m}{2}, \frac{j+n}{2}\right) \right. \tag{2}$$

$G_{1,k}(i,j)$ is the expanding operation results of $G_{l,k-1}(i,j)$, where summation terms are taken to be null for non integer values of $(i + m)/2$ and $(j + n)/2$ [4].

Let $L_1(i, j)$ be the l level Laplacian pyramid image, N is the total decomposing level, it is obtained by [4],

$$\begin{cases} L_N = G_N \\ L_l = G_l - G_{l+1,1} \; (0 \le l < N) \end{cases} \tag{3}$$

Different fusion strategies can be chosen on different Laplacian pyramid level. Average method (4) will be adopted for the top level of Laplacian pyramid which represent the low frequency information of original image. $G_N I_1$ and $G_N I_2$ are the top level of N level Laplacian pyramid decomposed from input image I_1 and I_2 respectively. $G_N F$ is the top level of N level of fusion Laplacian pyramid.

$$G_N F(I,j) = (G_N I_1(i,j) + G_N I_2(i,j))/2, \quad 0 \le i < R_N, \;\; 0 \le j < C_N \tag{4}$$

Regional energy (RE) or Laplacian energy (LE) fusion rules are selected for choosing the high frequency information at different level of Laplacian pyramid. The region energy of each coefficient of every sub-band is calculated using (5) [4].

$$LE_1(ij) = \sum_{-m}^{m} \sum_{-n}^{n} w(m,n) \cdot |I(i+m, j+n)| \tag{5}$$

where

$$m = 1, n = 1, w = \frac{1}{9}\begin{pmatrix} 1 & 1 & 1 \\ 1 & 1 & 1 \\ 1 & 1 & 1 \end{pmatrix}$$

For $1 \le l \le N$, $L_l I_1$ and $L_l I_2$ are the l level decomposing Laplacian pyramid of input image I_1 and I_2 respectively. $L_l F$ is the l level fusion Laplacian pyramid. Comparing the value of LE_{l1I1} and LE_{L1I2} the maximal one's pixel value is chosen as the corresponding pixel value of $L_l F$ [4].

$$L_l F(i,j) = \begin{cases} L_l I_1(i,j) & \textit{if } LE_{l1I1}(i,j) \ge LE_{L1I2}(i,j) \\ L_l I_2(i,j) & \textit{otherwise} \end{cases} \tag{6}$$

The fused image will be constructed by the fusion Laplacian pyramid, G_0F is corresponding to the final result [4].

$$G_{l-1}F(i,j) = l_{l-1}F(i,j) + EXPAND(G_lF(i,j)) \quad 1 \le l \le N \tag{7}$$

3 Implementation

Figure 2 shows that the image of each channel is processed in parallel during decomposing process. Considering the processing and operation character of this algorithm, different modules are designed to implement this algorithm including controlling module, decomposing module, fusion module and reconstruction module [4].

The two decomposing modules will process the input data to get the Laplacian pyramid and perform synchronously. Fusion module will execute the fusion strategies in all levels of the Laplacian pyramid. Reconstruction module, which is part of fusion module, rebuilds the fused image with the fused Laplacian pyramid by expanding and adding operation. Images of each channel are processed in parallel with decomposing process [4].

FPGA implementation of Laplacian pyramid involves buffering of Laplacian pyramid stages and it is possible by introducing a pair of static RAM (SRAM) devices at each pyramid level. So, if number of pyramid levels are 3, 3 pairs of SRAM are needed for single channel implementation. Total number of SRAM implementation will be 12 which is expensive affair in terms of cost and interface design. Block RAM (BRAM) based implementation has been done to realize this algorithm. Total delay in computation can be calculated in number of lines. For each stage, convolving function is involved, and for this function three lines must be ready before starting the convolution. So total delay in terms number of lines is calculated as follows.

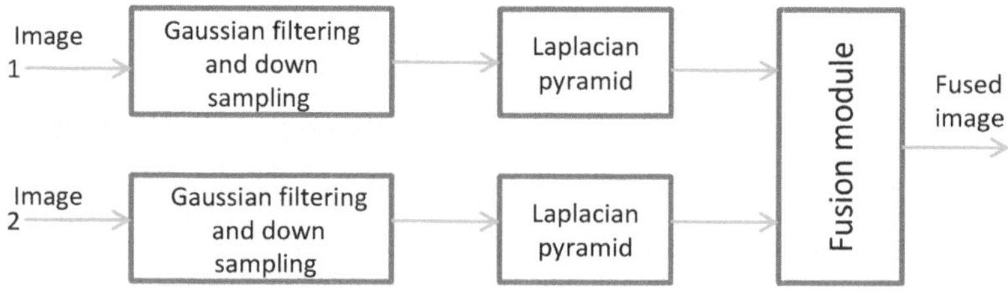

Fig. 2 Image decomposition pipeline for fusion process

(a) 3 lines delay in buffering for R function
(b) 3 Lines delay for E and D functions
(c) so 6 lines delay in one stage pyramid and total delay will be minimum 18 lines for three stage pyramid implementation.

While reconstruction of fused image, Fusion and combine functions will be needed twice and each with 3 lines delay. After construction of fused image, total delay is around 24 lines. Single clock of 13.5 MHz is considered. Entire 3 stage pyramid fusion pipeline can be filled in 24 lines of delay. Fused pixel output will start coming after this delay.

Image Fusion Rules

Low frequency coefficients of fusion image are calculated with average operation using (4), for high frequency coefficients, the implementation of (5) and (6) are to be done using Laplacian energy comparison method [4].

Design of Reconstruction Operation

When fusion by Laplacian pyramid is done at each level, the reconstruction operation is achieved through expand and adding operations to get higher level of image. Because there is fix delay time for expanding operation delay, module is used to guarantee the time request. The expanding results of $G_N F$ is added with $L_l F$ to obtain the fusion results. If the MSB bit of $L_l F(i,j)$ is 1, this means that $L_l F(i,j)$ is a negative value and the adding operation will be replaced by subtracting operation. In this case, if the calculation is a negative value, the corresponding pixel value will be set as 0 [4].

4 Weighted Laplacian Fusion

Using traditional method of image fusion, results are not containing complete information. It is found that contrast of fused image is very good in case of high Laplacian energy fusion. When weighted Laplacian energy based approach is applied, fused image gets enriched in terms of information. e.g., Weighted energy Fusion for 80% Laplacian energy is defined as:

$$\begin{cases} L_l F(i,j) = L_l I_1(i,j) * 0.8 + L_l I_2(i,j) * 0.2 & \text{if } LE_{l1I1}(i,j) \geq LE_{LlI2}(i,j) \\ \qquad L_l I_1(i,j) * 0.2 + L_l I_2(i,j) * 0.8 & \text{otherwise} \end{cases} \tag{8}$$

Table 1 Entropy of fused images

%HE/LE	Set1	Set2	Set3	Set4	Set5
HLE	6.790463	6.904101	7.044823	6.686868	6.417245
90_10	7.188268	7.172002	***7.288515***	6.978414	6.651329
80_20	***7.225088***	***7.189417***	***7.282732***	***6.999864***	***6.701603***
70_30	7.209094	7.174774	7.249606	6.989514	6.688864
60_40	7.166959	7.141097	7.195538	6.962057	6.652486

There are quantitative methods which can measure the quality of fusion without reference images. These could be Standard deviation, entropy, cross entropy and spatial frequency etc. It is found that entropy is a very good measure which signifies the information contents in an image. Results are evaluated based on entropy (He) computation which is defined as [2].

$$He = -\sum_{i=0}^{L} h_{I_f}(i) \log_2 h_{I_f}(i)$$

where L is gray levels, $h_{I_f}(i)$ is the histogram of an image $I_f(x, y)$. Interesting fact is that when part of darker pixel is added, information represented by low gray values is also highlighted and overall information increases so entropy also increases. In general, a combination of 80% pixel weight from higher LE and 20% pixel weight of lower regions gives highest entropy. In other combination also entropy of fused image always remains above the high Laplacian energy based fusion. If fusion is seen from application point of view, high contrast image will be appreciated, i.e. in tracking application. If information contents and target detection are main interests, weighted energy based fusion will be more useful. To establish this fact, five set of registered images are considered for experimentation. Entropy of each fused image is calculated which is shown in Table 1.

5 Results and Discussions

There are two input images, one of CCD camera and another of thermal camera. Both shows different spectral contents of same scene. This picture has been taken using co-aligned cameras and image registration was done before using them for fusion. In default rule of high Laplacian energy (HLE) based fusion, some CCD contents are absent; when part of low Laplacian energy (LE) has been added, the

Fig. 3 CCD image

Fig. 4 TI image

fused image got enriched. However some drop in contrast has been observed (Figs. 3 and 4).

Fusion of one set of images with varying combination of Laplacian energy is shown in Figs. 5, 6, 7 and 8 (marked with red circles) for visual appreciation.

Fig. 5 High laplacian energy fusion

Fig. 6 80_20 weighted energy fusion

Fig. 7 70_30 weighted energy fusion

Fig. 8 60_40 weighted energy fusion

6 Conclusion

Multi-resolution pyramid fusion approach gives better result and both sensors' features are highlighted in the fused imagery. Different fusion approaches at different pyramid level have been applied and it was found that weighted Laplacian energy based fusion gives better results to capture both the sensors perfectly.

A suitable combination of high Laplacian and low Laplacian energy is suggested and demonstrated in results as well. Various method of fusion in pyramidal domain will remain an open area of research for further work in image fusion.

References

1. Chunling Zhao, Quanxin Ding, and Jianxun Li, "The performance analysis of image fusion algorithm", Intl. Sym. on Computational Intelligence and Design, 83–88, (2008)
2. PETER J. BURT, Member, IEEE and Edward H. Adelson, "The Laplacian Pyramid as a Compact Image Code", IEEE Transaction on Comm., Vol. Com-31, No. 4, 532–540, Apr 1983
3. Wencheng Wang and and Faliang Chang, "A Multi-focus Image Fusion Method Based on Laplacian Pyramid", Journal of Computers, Vol. 6, No. 12, 2559–2566, Dec 2011
4. Yajun Song*, Kun Gao, Guoqiang Ni, Rong Lu, "Implementation of real-time Laplacian pyramid image fusion processing based on FPGA", Proc. of SPIE Vol. 6833 683316-(1–8), (2007)

Radon Transforms and Chaotic Mask Based Image Encryption for Information Security

Avinash Kumar Jha, Sajan Ambadiyil and Himanshu Shekhar

Abstract Recent advances in image encryption techniques are capable of protecting the digital images which are being communicated over various transmission media from leakage. Images related to medical or military applications, corporate video conference, etc. need reliable and secure transmission, which can be achieved by encryption. Here we propose an efficient optical image encryption technique using Radon Transforms and Chaotic phase mask.

1 Introduction

Recent advances in image encryption techniques are capable of protecting the digital images which are being communicated over various transmission media from leakage. Images related to medical or military applications, corporate video conference, etc. need reliable and secure transmission, which can be achieved by encryption. With the help of efficient optical encryption and decryption technique, one can fulfill the requirements of security needs of digital images. Multiple image encryption systems based on optical means have been proposed by various research groups' earlier [1–10]. For real time applications, optics and optoelectronics techniques are very useful as these are accurate, fast computing and support parallelism. These methods provide parameter such as wavelength, phase, polarization, etc. which can be used to hide information more securely in various types of images [11]. Li et al. proposed propose a new method for color image encryption by wavelength multiplexing on the basis of two-dimensional (2-D) generalization of fractional Hartley transform [12]. Nishchal et al. proposed and implemented a phase-encrypted memory system they utilized the cascaded extended fractional Fourier transform (FRT) [13]. Madan Singh et al. proposed an encryption method based on double random phase encoding and

A.K. Jha (✉) · H. Shekhar
Hindustan Institute of Technology & Science, Chennai, India
e-mail: ak.vit2006@gmail.com

S. Ambadiyil
Center for Development of Imaging Technology, Trivandrum, India

© Springer Nature Singapore Pte Ltd. 2017
I. Bhattacharya et al. (eds.), *Advances in Optical Science and Engineering*,
Springer Proceedings in Physics 194, DOI 10.1007/978-981-10-3908-9_40

decoding system for two-dimensional gray scale image [14, 15]. Li et al. proposed a double-image encryption algorithm. This algorithm can encrypt two input images into a single encrypted output image. The method exploits the amplitude of gyrator transform with two different groups of angles to decrypt the images [16, 17]. Narendra Singh and Aloka Sinha proposed a novel method for image encryption, utilizing gyrator transform and chaos theory [18].

Here we propose an efficient optical image encryption Technique Using Radon Transforms and Chaos function. The proposed technique uses Radon Transform and Chaotic Random Phase Mask (CRPM). The technique is highly robust and has great immunity to unauthorized decryption. The original and decrypted image are highly correlated. It is possible to implement this encryption using optical imaging technique which makes it more relevant for radio over fiber communication systems.

1.1 Radon Transform

The radon transform is represented by integral of a function over straight lines. It is utilized predominantly in the field of Medical Imaging, Ground Penetrating Radar (GPR), electron microscopy, hyperbolic partial differential equation etc.

The mathematical model of radon transform in for a two dimensional function can be represented as follows.

Considering f as a mathematical function bound by a large disc in Eucildian plane $\mathbf{R}^2$.

Then it's radon transform can be defined as function $R_{f,}$ which is defined on the space of lines L in $\mathbf{R}^2$ as given in (1).

$$Rf(L) = \int_L f(x)d\sigma(x) \tag{1}$$

the integration is done w.r.t. the arc length measure $d\sigma$ on L. L can be parameterized as follows;

$$L(x(t), y(t)) = (t \sin\alpha, +s\cos\alpha) + (-t\cos\alpha + \sin\alpha)$$

here the distance between L and origin is denoted by s and α signifies the angle it makes with the x axis. Thus (α, s) represents coordinates on the space of all lines in $\mathbf{R}^2$, and in terms of these coordinates Radon transform can be expressed as

$$Rf(\alpha, s) = \int_{-\alpha}^{\alpha} f(x)(t), y(t)dt = \int_{-\infty}^{\infty} f(t(\sin\alpha, -\cos\alpha) + s(\cos\alpha, \sin\alpha))dt \tag{3}$$

1.2 Chaotic Phase Mask

Chaotic functions describe the behavior of certain dynamic systems i.e., systems whose state evolves with time and which may exhibit dynamics that are highly sensitive to initial conditions. Due to this sensitivity, behavior of chaotic systems appears to be random. For certain chaotic systems if the initial conditions are known, their future dynamics can be completely predicted. This is defined as deterministic chaos. In our encryption technique a random phase mask has been generated using logistic map as chaotic map. It is a discrete 1-D function as mentioned in the equation

$$x_{n+1} = rx_n(1 - x_n) \tag{2}$$

here x_n is a positive number between 0 and 1 which represents the population at year *n* and *r* represents the rate of growth i.e.; combined rate of reproduction and starvation in the population.

1.3 Cryptographic Enhancement

For an m × n image two sequence of random number of length m and n corresponding to each row and column of the image are generated. At first each pixel of the image is replaced by the r'th pixel from the right of the original pixel, where r is the random number corresponding to that particular row. Same action is performed with respect to column. During decryption same action is performed in the reverse direction yielding the correct image. The procedure is explained in Fig. 1. The sequence of random number acts as a set of key, and this whole operation adds additional security feature to the encryption.

2 Proposed Technique

The proposed technique utilizes radon transforms and double chaotic random phase mask. Let f(x, y) denotes the original image to be encrypted. The block diagram in Fig. 2a gives an overview of the encryption process.

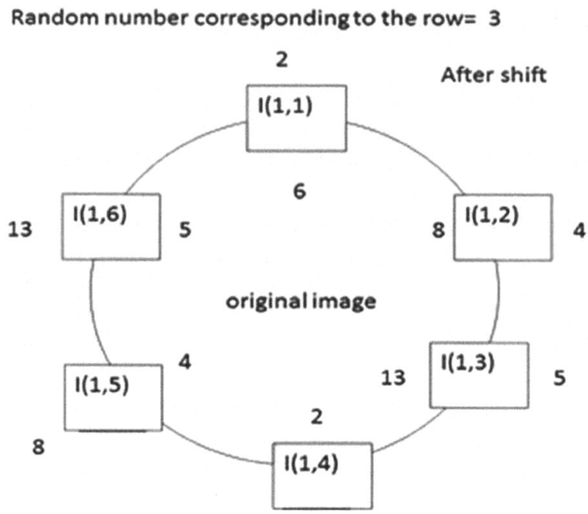

Fig. 1 Cryptographic enhancement

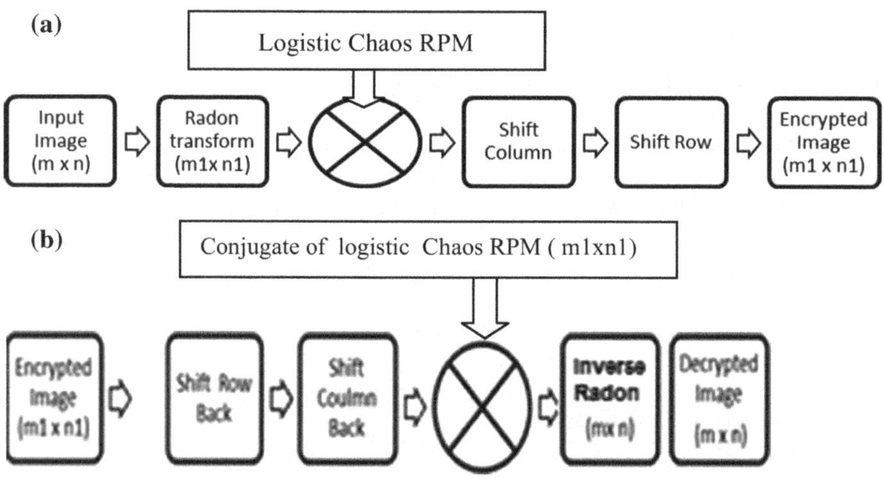

Fig. 2 **a** Block diagram for encryption, **b** block diagram for decryption

2.1 Encryption Process

A gray scaled image of size 256 × 256 is used as input. Radon transform is applied on the input image. The transformation generates an intensity image of different size (367 × 451) with respect to the input image. The number of rows of the transformed image depends on the number of intensity level present in the image while the number of column depends on the total number of angle for which RT is taken, which acts as a security feature. The transformed image is multiplied with the random phase mask represented by exp(ipiC(x)) here C(x) is the logistic map function. The row column shift operation is performed on this output to generate the encrypted image. The encryption procedure is shown in Fig. 2a.

2.2 Decryption Process

For decryption the pixels are shifted to their original coordinates by applying row column shift operation in the reverse direction, and then this image is multiplied with the conjugate of the random phase mask. Now inverse radon transform applied for the same set angle as of radon transform yields the original input image. Figure 2b illustrates the decryption process.

3 Result and Analysis

The above figure shows the results obtained by the proposed method. To find the robustness of the proposed system MSE error analysis, histogram analysis, correlation of pixel distribution among input, output and un-authenticated decryption image were performed. The original and decrypted image were found to have high correlation coefficient and the method is highly immune to un-authorized decryption.

As shown in Table 1, different value of chaotic mask seed were applied to an image encrypted with the seed value 3.8. It is clearly evident from the graph and table that the MSE value for correct key is negligible. 48.5199, however for wrong

Table 1 MSE w.r.t variation in rate of the logistic function

Rate of logistic map	MSE for lena
3.76	253565.5016
3.78	261938.1427
3.79	262371.1103
3.799	265253.1471
3.7999	264224.9905
3.8	**48.5199**
3.8001	269381.7596
3.801	285538.7191
3.81	276406.204
3.82	355639.5019

key such as 3.7999 the MSE value is extremely high (264224.9905) implying failed decryption as shown in Fig. 3h.

The proposed technique results into a correlation co-efficient of 0.9932 between the input and output and decrypted image (Fig. 4).

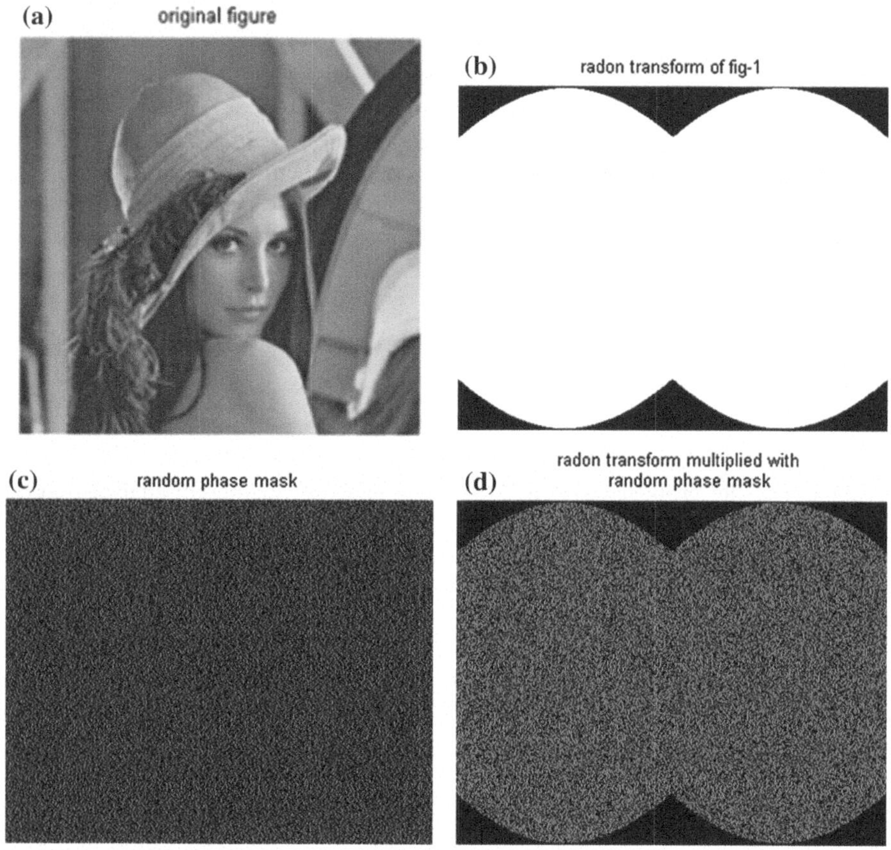

Fig. 3 **a** Input image, **b** radon transform of image, **c** random phase mask, **d** radon transformed image multiplied with phase mask, **e** encrypted image after row column shift, **f** decrypted image with correct procedure, **g** image decrypted without shifting the row-column back, **h** image decrypted with wrong conjugate mask

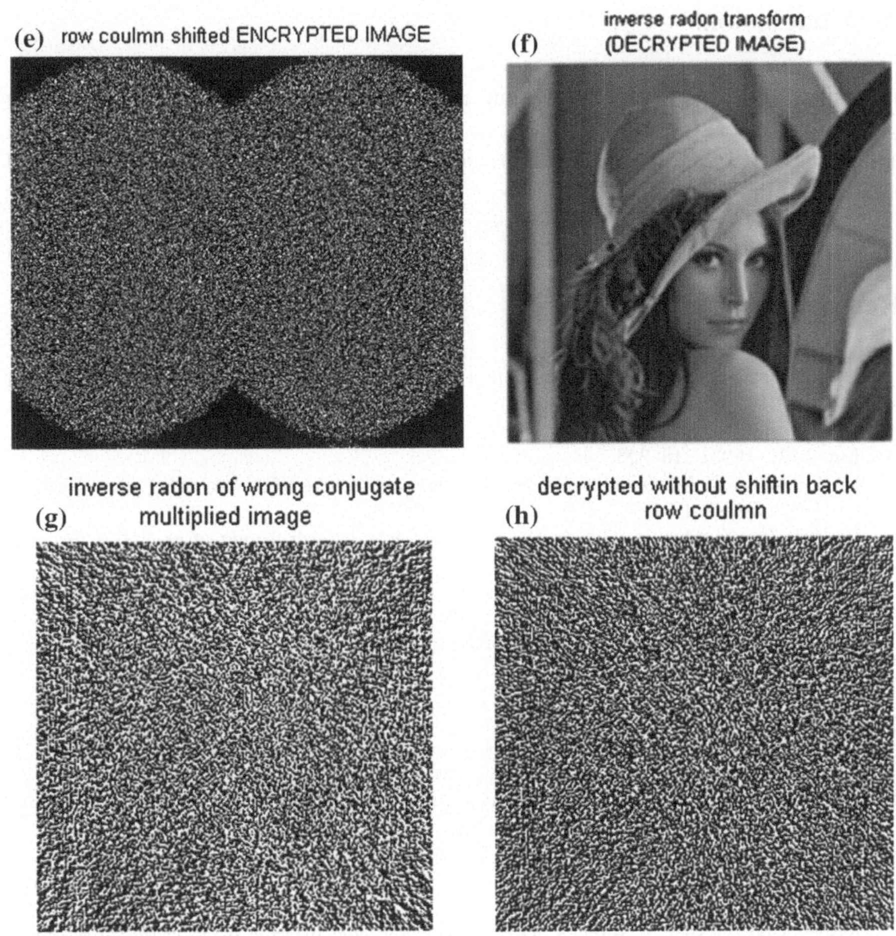

Fig. 3 (continued)

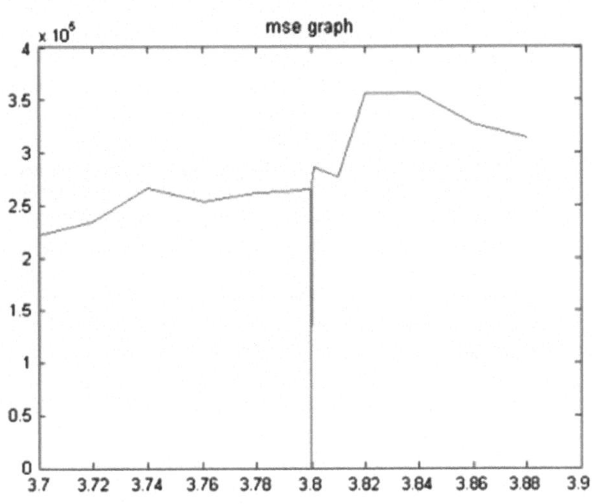

Fig. 4 MSE w.r.t variation in rate of the logistic function

4 Conclusions

A highly efficient image encryption and decryption Technique Using Radon Transforms and Chaos function has been proposed. The technique is highly robust and has great immunity to unauthorized decryption. The original and decrypted image are highly correlated.

References

1. H.K.L. Chang, J.L. Liu, A linear quad tree compression scheme for image encryption, Signal Process. 10 (4) (1997) 279–290.
2. 769 J. Scharinger, Fast encryption of image data using chaotic Kolmogrov flow, J. Electronic Eng 7 (2) (1998) 318–325.
3. N. Bourbakis, C. Alexopoulos, Picture data encryption using SCAN pattern, Pattern Recogn. 25 (1992) 567–581.
4. Fridrich Jiri, Symmetric ciphers based on two dimensional chaotic maps, Int. J. Bifurcat Chaos 8 (6) (1998) 1259–1284.
5. Refregier, B Javidi, Optical image encryption based on input plane and fourier plane random encoding, Opt. Lett. 20 (1995) 767.
6. H. Cheng, X.B. Li, Partial encryption of compressed image and videos, IEEE Trans. Signal Process. 48 (8) (2000) 2439–2451.
7. J.C. Yen, J.I. Guo, An efficient hierarchical chaotic image encryption algorithm and its VLSI realization, IEE Proc. Vis. Image Process. 147 (2000) 167–175.
8. J.C. Yen, J.I. Guo, A new image encryption algorithm and its VLSI architecture, in: Proceedings of the IEEE workshop signal processing systems, 1999, pp. 430–437.
9. C.C. Chang, M.S. Hwang, T.S. Chen, A new encryption algorithm for image cryptosystems, J. Syst. Software 58 (2001) 83–91. 749–761.
10. J.C. Yen, J.I. Guo, A new chaotic key based design for image encryption and decryption, Proceedings of the IEEE International Symposium Circuits and Systems, vol. 4, 2000, pp. 49–52.
11. Nishchal NK, Joseph J, Singh K. Fully phase-encrypted memory using cascaded extended fractional Fourier transform. Opt Lasers Eng 2004;42(2):141–51.
12. X. Li, D. Zhao, Optical color image encryption with re defined fractional Hartley transform, Opt. Int. J. Light Electron. Opt. (2009), doi:10.1016/j.ijleo.2008.10.008.
13. Naveen Kumar Nishchal, Joby Joseph, Kehar Singh, "Fully phase-encrypted memory using cascaded extended fractional Fourier transform", Optics and Lasers in Engineering 42 (2004) 141–151.
14. Madan Singh a, Arvind Kumar b,_, Kehar Singh, "Optical security system using jigsaw transforms of the second random phase mask and the encrypted image in a double random phase encoding system", Optics and Lasers in Engineering 46 (2008) 763–768.
15. M. Singh, et al., Encryption and decryption using a sandwich phase diffuser made by using two speckle patterns and placed in the Fourier plane: Simulation results, Opt. Int. J. Light Electron. Opt. (2008), doi:10.1016/j.ijleo.2008.03.025.
16. Yong-Ying Wang_, Yu-Rong Wang, Yong Wang, Hui-Juan Li, Wen-Jia Sun, "Optical image encryption based on binary Fourier transform computer-generated hologram and pixel scrambling technology", Optics and Lasers in Engineering 45 (2007) 761–765.
17. Huijuan Li, Yurong Wang, "Double-image encryption based on iterative gyrator transform", Optics Communications 281 (2008) 5745–5749.
18. Narendra Singh, Aloka Sinha, "Gyrator transform-based optical image encryption, using chaos", Optics and Lasers in Engineering 47(2009) 539–546.

Pose Invariant Face Recognition Technique Based on Eigen Space Approach Using Dual Registration Techniques After Masking

Tumpa Dey and Dibyendu Ghoshal

Abstract A novel method is proposed to solve different pose related problems related to face images in recognition system. The method removes the background of the image using masking. Subsequently, both training and testing images are registered by manual landmark detection and modeling the mapping process using affine transformation. The proposed method is found to solve the complications during scaling and rotation. Another registration method based on log-polar transformation is then proposed. Application of this method is found to improve arbitrary rotation angles and scale change. Lastly, log-polar images are projected into eigen space. These eigenface images are classified with the help of Euclidean distance. In the simulation based experimentation, IRIS face database is used. Recognition rate applying the proposed method is found to be 89.65%.

1 Introduction

Face recognition is carried out mainly for two primary purposes viz. verification which is one to one matching and another is the identification that means matching a particular image among many such images [1–6]. During the past several years, the evolution of recognition techniques has been applied in the age verifications, illumination variation, and pose variation. Here, variations of pose and scaling of images are treated as potential challenges for correct classification and identification. A pose invariant face recognition technique is applied where two image registration techniques are used to get the approximate alignment used in an image. The main characteristics of image registration are the geometric alignment of a set of images. The set consisting of two or more images has been taken from a single scene at different times, sensed by different sensors, and different points of view. In

T. Dey
Department of IT, Women's College, Agartala, Tripura 799004, India

D. Ghoshal (✉)
Department of ECE and EIE, NIT Agartala, Jirania, Agartala, Tripura 799055, India
e-mail: tukumw@gmail.com

© Springer Nature Singapore Pte Ltd. 2017

I. Bhattacharya et al. (eds.), *Advances in Optical Science and Engineering*,
Springer Proceedings in Physics 194, DOI 10.1007/978-981-10-3908-9_41

the proposed study, a simulated masking technique is applied to crop the background of input image followed by image registration using affine transformation and log-polar transformation for better alignment. In the last phase, again eigenface analysis is applied on face images of various poses and scaling collected from IRIS Database. Although a good many studies on face recognition is reported [7–16] in various journals and conference proceedings, no study based on the current proposed methodology is found in any published or on-line literature.

2 System Overview

In this paper a face recognition technique is proposed to find improved performance. Every face image is first cropped using masking, followed by the registration to each image. Then, polar transformation is done on images into two sets, i.e. training and testing. Here masking technique has been applied to crop the background without changing the pixel values inside the mask. A registration by feature point mapping with affine transformation in the masked images is applied to get more accurate aligned images and this is followed by log-polar transformation. The eigenspace is computed on the log-polar transformed images. Figure 1 shows the system overview.

2.1 *Masking*

A profile picture may have any unwanted complex background which creates distorted output. Therefore, masking method is used to subtract the background. In the present study, region of interest from the face image is determined first and masking is done to crop the background. Masking process sets the pixel values in an image to zero from some other values in the background of the image. There are two types of masking. First method uses an image as a mask where some of the pixel intensity values are zero, and others pixel values are non-zero. For example,

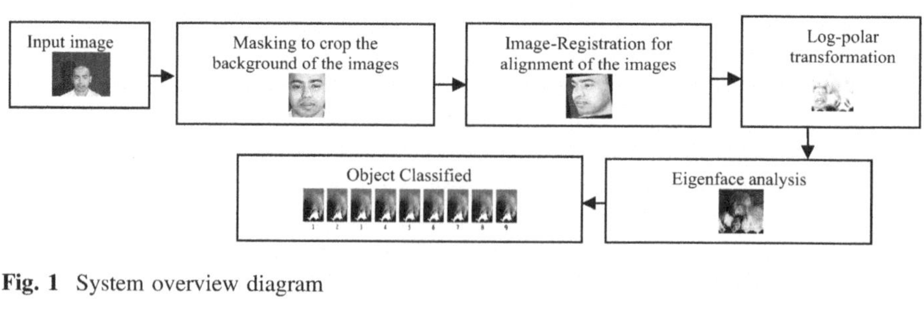

Fig. 1 System overview diagram

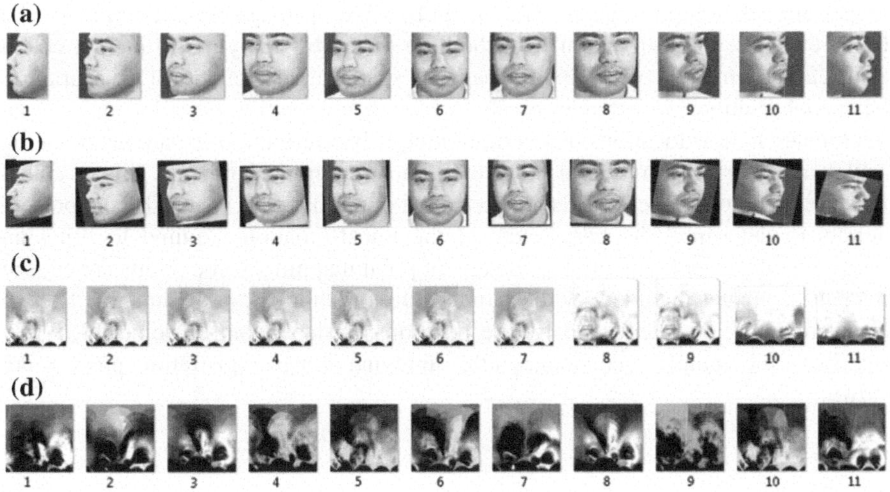

Fig. 2 Steps showing sample images from IRIS face database after **a** masking, **b** registration, **c** log-polar transformation, **d** eigenfaces

create a mask image. The second method selects a region of interest as the mask. In Fig. 2a, some simulated outputs of face images are displayed which have been obtained by applying masking during pre-processing step.

2.2 Masking Algorithm in the Proposed Approach

Input: I is an image of size K × L.
Output: M is an image of size A × B after masking of the image.
Step 1: J is an image of size of C × D which is chosen as region of interest of facial part including eyes, nose and lips.
Step 2: The outside area of J is burned.
Step 3: Part of the image that exists inside the mask and zero outside the mask only left i.e. M = A × B.
Step 4: Cropped image is displayed.

2.3 Registration

Image registration [7–12] is an alignment process of two or more images of the same scene and having same dimension. First, one image is taken as the base image or reference image and other images are compared to the base image and the set of images are called input images. The main objective of image registration is to take the input image into alignment with the base image by applying a function called spatial transformation of the input image. A spatial transformation [10] works on the spatial relationship between pixels in an image, mapping pixel locations in an input image to new locations in an output image. It determines the parameters of the spatial transformation required to take the images into alignment. This is called image registration process. A technique called point mapping is used by the present authors to determine the parameters of the transformation required to bring an image into alignment with another image. In point mapping, some points are chosen in a pair of images that identify the same feature or landmark in the images. Then, a spatial mapping is determined by the positions of these control points [8, 9]. In Fig. 2b, some sample face images after applying image registration process are shown.

2.4 Registration Algorithm

Input: I is an image of size M $\times$ N.
Output: K is an image of size S $\times$ T after image registration.
Step 1: Control points in image I are selected.
Step 2: Spatial transformation structure of I is created after selecting control points.
Step 3: Spatial transformation is done and the aligned image K = S $\times$ T dimension is obtained.

2.5 Log-Polar Transformation

The main reason behind the log-polar transformation is to bring images into close alignment even in the presence of large scale changes for large geometric transformations, as well as arbitrary rotations and translations. Log-polar coordinates are a two dimensional coordinate system where point are identified by two numbers, one is the logarithm of the distance between two points from the center of the input image, and another is an angle between the lines present in the image. Log-polar coordinates are connected to polar coordinates and normally used to describe domains in the plane with some sort of rotational symmetry. The Cartesian coordinates are not mapped one-to-one onto pixels in the Log-Polar coordinate space. An average of the surrounding pixels requires to be calculated to get the log polar transformed image. Log-polar coordinates convert the scale and rotational differences into vertical and horizontal offsets which can be calculated by using the method in [12–15].

2.6 Algorithm: Log-Polar Transformation

Input: An image with A × B size is taken in Cartesian coordinate space.
Output: An image with $C^p \times C^p$ size in Log-polar coordinate space is obtained.
Step 1: The centre (m, n) and radius (R) from the input image of size A × B is found out. Centre of the circle is

$$a = \lfloor A/2 \rfloor, \quad b = \lfloor B/2 \rfloor \tag{1}$$

Step 2: The polar images are calculated.
Let pixel in the input image (s_i, t_i) will have the pixel at location (r_1, Φ) in the polar image, where

$$r_1 = \sqrt{(s-a)^2 + (t-b)^2} \quad 0 \le r_1 \le R \tag{2}$$

$$\Phi = \tan^{-1}\left(\frac{t-b}{s-a}\right) \quad 0 \le \Phi \le 360^0 \tag{3}$$

Step 3: The log-polar transformation is carried out. Log-polar transform is written as (l, Φ), where $l = \log_e^{r_1}$
Step 4: Output image with size is $C^p \times C^p$, where $p = \log_C^R$ is obtained.

2.7 Eigenspace Analysis for Face Recognition

Eigenspace approach is used in this paper to recognize face images. Here steps are given [16–19].

Step 1: Set s comprises m number of face images. Every face image is converted into a vector of size n.

$$s = \{\Gamma_1, \Gamma_2, \Gamma_3 \ldots, \Gamma_m\} \tag{4}$$

Step 2: Then the mean image is calculated as Ψ.

$$\Psi = \frac{1}{m}\sum_{n=1}^{m} \Gamma_n \tag{5}$$

Step 3: The difference between the input image and the mean image i.e. Φ

$$\Phi_i = \Gamma_i - \Psi \text{ is calculated} \tag{6}$$

Step 4: u_n is set of orthonormal vectors. M is the number of orthonormal vectors. kth vector is u_k.
Here the eigenvectors u_k and eigenvalues λ_k of the covariance matrix C are calculated.

$$\lambda_k = \frac{1}{M}\sum_{n=1}^{M}\left(u_k^T \Phi_n\right)^2 \tag{7}$$

Step 5: The covariance matrix **C** is calculated as,

$$\begin{aligned} C &= \frac{1}{M}\sum_{n=1}^{M}\Phi_n\Phi_n^T \\ &= AA^T \end{aligned} \tag{8}$$

$$A = \{\Phi_1, \Phi_2, \varPhi_3, \ldots \Phi_n\} \tag{9}$$

$$L_{mn} = \Phi_m^T \Phi_n \tag{10}$$

when eigenvectors v_l of the L matrix is calculated then eigenfaces u_l is obtained as,

$$u_l = \sum_{k=1}^{M} V_{lk}\Phi_k \quad l = 1, \ldots M \tag{11}$$

3 Experimental Results and Discussions

In the proposed method, experiments are carried out on IRIS Database. A thorough system performance is analyzed, which covers 7 (seven) different conditions of human face recognition like variation of size, pose, illumination conditions, etc. The experiment is carried out by using 90 images from IRIS database.

3.1 IRIS (Imaging, Robotics, and Intelligent Systems) Thermal/Visible Database

This database contains Visual/Thermal face images. It has unregistered thermal and visible face images under variable illuminations, expressions, and poses. There are 176–250 images of persons, 11 rotations per illumination, expression of a particular person. Expressions are surprise, laughing, angry (varying poses) and each expression have 11 thermal and 11 visual images. There are different types of illumination condition, i.e. left light on, right light on, both lights on, dark room, left and right lights off, glass off-bright, glass-off-off, glass-off medium, and glass-off-dark, with varying poses. Sample images with different poses, their corresponding masking, registered and log-polar transformed images are shown in Fig. 2.

3.2 Eigenspace Analysis with IRIS Face Database

In the present study, the complete sets of log-polar transformed images are divided into two sets as training and testing, respectively. There are 90 images from IRIS database taken. Nine images per persons have been taken. The analysis is started by taking one image each from each set (i.e. training and testing sets) from a single class (i.e. person) of IRIS. In eigenface analysis first, a comparison is done between the input face image and mean image and after getting the difference, it is multiplied with each eigenvector of the matrix. These values represent a weight and would be saved in a vector (matrix). The minimum Euclidean distance is calculated for each test image. Each person is symbolized as P1, P2 and so on in Table 1. From Table 1, it can be seen that in few cases, 95% recognition rate is achieved although in some other cases the rate is found to be less. This is due to the illumination variance and error in pose estimation during image acquisition. In the case of IRIS database, it is observed that the recognition rate increases proportionately with the number of training images.

Table 1 Recognition rates for IRIS database

No of training images of each class (P_1, P_2...P_{10})	No of testing images randomly taken from each class (P_1, P_2...P_{10})	Average % of recognition rate
1+⋯+1	1+⋯+1	80
2+⋯+2	2+⋯+2	80
3+⋯+3	3+⋯+3	86.66
4+⋯+4	4+⋯+4	90
5+⋯+5	5+⋯+5	92
6+⋯+6	6+⋯+6	93.33
7+⋯+7	7+⋯+7	94.28
8+⋯+8	8+⋯+8	95
9+⋯+9	9+⋯+9	95.55
Average percentage		89.65

3.3 Conclusion and Future Work

In this paper, the efficacy of eigenface method is established followed by distance vector calculation for the classification of images. During the process, masking algorithm to eliminate the unwanted background objects is applied. Then the registration technique is applied for alignment. The variances of transformed input images are calculated into log-polar form to minimize the effect of rotational, positional variances. The proposed method has been demonstrated on IRIS Database and it contains face images with extreme pose variations. The average recognition rates achieved with various numbers of training and testing images from IRIS database is 89.65%. It has also been seen that the proposed method gives better recognition rate than the principal component analysis (PCA) with the face databases mentioned above. In future, attempts may be made to develop an adaptive multilevel pre-processing technique followed by a recognition algorithm to achieve high recognition rate irrespective of pose, illumination or rotational variances.

References

1. Nastar, C. and Mitschke, M., "Real time face recognition using feature combination," Proc. 3rd IEEE International Conference on Automatic Face and Gesture Recognition. Nara, Japan, 312–317 (1998).
2. Brunelli, R. and Poggio, T., "Face Recognition: Features versus Templates," IEEE Trans. Pattern Analysis and Machine Intelligence. 15(10), 1042–1052 (1993).
3. Chellappa, R., Wilson, C. L., and Sirohey, S., "Human and machine recognition of faces: A survey," Proc. IEEE, 83, 705–740 (1995).
4. Hsieh, C.K. and Chen, Y.C., "Kernel-based pose invariant face recognition," Proc. IEEE Int. Conf. Multimedia and Expo, 987–990 (2007).
5. Jafri, Rabia. and Arabnia, Hamid R., "A Survey of Face Recognition Techniques," Journal of Information Processing Systems. 5 (2), 41–67 (2009).
6. Ding, X. Q., Fang, C., "Discussions on some problems in face recognition," Proceedings. Advances in Biometric Person Authentication 3338, Lecture Notes Computer Science. Springer, 47–56 (2004).
7. Zitova, B. and Flusser, J., "Image registration methods: a survey," Image Vis Compute, 21 (11), 977–1000 (2003).
8. Allney, S., Morandi C., "Digital image registration using projections," IEEE Trans. on Pattern Analysis and Machine Intelligence. PAMI-8, 222–233 (1986).
9. Lee, D. J., Kpile, T. F., Mitra, S., "Digital registration techniques for sequential fundus images," International Society for Optics and Photonics, 293–300 (1988).
10. Fitzpatrick, J., Michael, D., Hill, L.G., Calvin, R. Maurer Jr., [Image registration], 447–513 (2000).
11. Tistarelli, M., Grosso, E., "Active vision-based face recognition: issues, applications and techniques in Face Recognition," Springer. Berlin Heidelberg. 262–286 (1998).
12. Brown, L. G., "A survey of image registration techniques," ACM Computing Surveys. 24(4), 325–376 (1992).
13. Zokai, S. and Wolberg, G. "Image registration using log-polar mappings for recovery of large-scale similarity and projective transformations," IEEE Trans Image Process, 14(10), 1422–1434 (2005).

14. Matungka, R., Zheng, Y.F., Ewing, R.L., "2D invariant object recognition using log-polar transform," Proc. World Congress on Intelligent Control and Automation, 223–228 (2008).
15. Matungka, R., Zheng, Y.F., Ewing, R.L., "Image registration using adaptive polar transform," IEEE Trans Image Process. 2009.
16. Turk, M. and Pentland, A., "Eigenfaces for recognition," Journal of Cognitive Neuro-science. 3(1), 71–86 (1991).
17. Turk, M., Pentland, A., "Face recognition using eigenfaces," Proc. IEEE Computer Society Conference on Computer Vision and Pattern Recognition, 586–591 (1991).
18. Martínez, A. M., and Kak, A. C., "PCA versus LDA," IEEE Trans. on Pattern Analysis and Machine Intelligence. 23(2), 228–233 (2001).
19. Serrano, S., "Eigenface Tutorial," <http://www.pages.drexel.edu/EigenfaceTutorial.htm>.

Part VIII
Opto-Electronic Devices, Terahertz Technology

Hybrid VLC-RF System for Real Time Health Care Applications

Anshul Vats, Mona Aggarwal, Swaran Ahuja and Sharda Vashisth

Abstract Now a days, modern hospitals are adopting electronic health (E-Health) over traditional health care services. The E-Health services are provided by a radiation free, high capacity, high transmission rate and radiation free communication. In this paper, a hybrid radio frequency-wireless visible light communication (VLC-RF) system gives a newest solution to change the traditional wireless communication systems in health organizations. This paper provides an all-purpose system model where the biomedical information from patient are transmitted using VLC uplink to a decode and forward relay which further radiocasts that information to the laboratory. Finally a feasible demonstration which extracts the carotid signals data from patient is implemented in the laboratory.

1 Introduction

Wireless technologies are used in medical area networks to increase flexibility and ease for medical staff and patient, but it has been reported that many frequencies of radio waves induce strong electric field intensity, which is prone to interfere with electronic medical equipments, which is a risk towards the accuracy of medical equipments [1]. This paper gives a inventive design and implementation of a medical healthcare data system using the rising wireless visible light communication (VLC) technology in combination with existing radio frequency communication network. The hybrid VLC-RF based medical healthcare system model can be used to give data services and keeping track of patients in radio frequency restricted health organization areas. This system delivers services such as telemedicine, client-patient portal, diagnostic reports, home care, E-medical documents, health database, old records management, etc. [2, 3]. Networks using VLC technology can be implemented using light emitting diodes (LED), to transmit medical and healthcare information [4]. In this paper, we propose a hybrid VLC-RF setup for the

A. Vats (✉) · M. Aggarwal · S. Ahuja · S. Vashisth
North Cap University, Sector 23 A, Gurgaon, India
e-mail: nalurajput2@gmail.com

© Springer Nature Singapore Pte Ltd. 2017

I. Bhattacharya et al. (eds.), *Advances in Optical Science and Engineering*,
Springer Proceedings in Physics 194, DOI 10.1007/978-981-10-3908-9_42

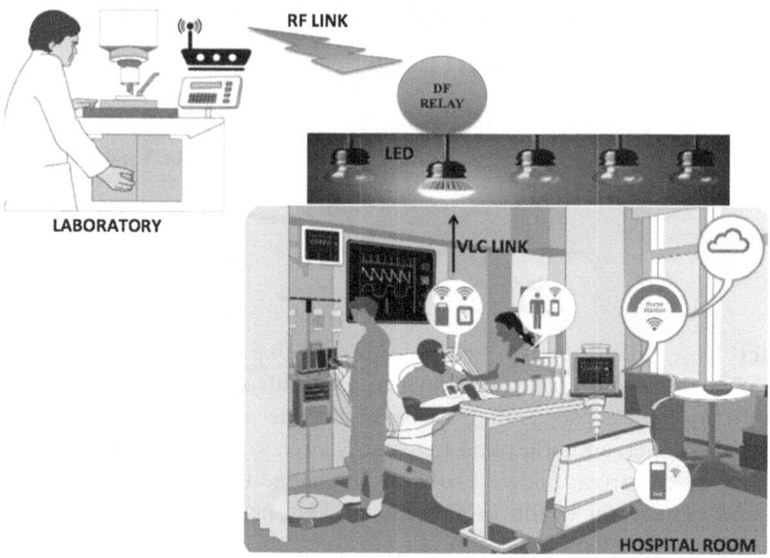

Fig. 1 Medical health care system model using hybrid VLC-RF communication link

hospital database management and applications in indoor environment. Hybrid VLC-RF setup implementation in the hospitals is a novel idea, which is demonstrated by transmitting real bio medial signals in the laboratory. In this paper we considered an radio frequency influenced environment in which EM waves cannot pass through the walls of the room to keep the indoor environment free from RF radiations [4]. Biomedical instruments generate signals which are received at a relay node and this node is incorporated with the optical transceiver. The relaying node decodes the signal and forward it to some distant Laboratory or research lab via RF link as shown in Fig. 1.

2 System Model

Hybrid visible light communication-radio frequency (VLC-RF) system gives a way out to change the traditional wireless transmission systems in healthcare organizations. This model gives a system where the medical signals from the patient are transferred using VLC link to decode and forward relay which further broadcasts that data to the laboratory. The smart devices like mobile phone, laptops, tabs or any other instrument able to transmit information are presumed to be with optical photo detector interfaces. These devices transmit the E-Health information in the form of optical signal towards the relay. Hence, the downlink becomes a hybrid VLC-RF link. The coverage in the room or any laboratory or location is given by number of LEDs. In this system any time any relay can be selected according to the signal strength but for reducing the complexity of the system we have opted a single relay system. The received optical signal from the selected relay is the best and strongest signal [4]. We analyse the hybrid VLC-RF link performance by simulation using Matlab and results in an expression for the outage probability with decode and forward relay in the hybrid VLC-RF link.

2.1 Visible Light Communication Link Model

The source to relay (S-R) link is modelled as VLC link with the PDF of the instantaneous channel coefficient h_{SR}

$$P_{h_{SR}}(x) = \alpha x^{-\beta} \tag{1}$$

Since γ_{SR} is related to squared of h_{RD}, as $\gamma = \frac{E_s}{N_0} h_1^2$ therefore the probability distribution function (PDF) of instantaneous SNR (γ_{SR}) is derived and given as

$$f_{\gamma_{SR}}(\gamma) = \frac{\alpha}{2} \times \frac{\gamma^{-\frac{\beta+1}{2}}}{\bar{\gamma}^{\frac{1-\beta}{2}}} \tag{2}$$

where $\bar{\gamma} = \frac{E_s}{N_0}$, γ is the average electrical SNR with $\alpha = 2.381 \times 10^9$ and $\beta = 1.161$ [4]. By integrating (2) we can get the corresponding cumulative distribution function (CDF) of the link shown below

$$F_{\gamma_{SR}}(\gamma) = \int_0^{\gamma} \frac{\alpha}{2} \times \frac{\gamma^{-\frac{\beta+1}{2}}}{\bar{\gamma}^{\frac{1-\beta}{2}}} d\gamma = \frac{\alpha}{1-\beta} \times \frac{\gamma^{\frac{1-\beta}{2}}}{\bar{\gamma}} \tag{3}$$

2.2 Radio Frequency Link Model

In recent years, the generalized-K (K_G) fading model for RF communication models, is gaining more popularity [5], which is the derived from the lognormal distribution using Gamma distribution models [6] and, then, incorporating with the Nakagami distribution model. In this paper, we analyze the performance of RF downlink using K_G distribution. The PDF of instantaneous SNR of generalised K distribution is given by (4)

$$f_{\gamma_{RD}}(\gamma) = \frac{2(E_{RD})^{\frac{\beta+2}{2}}}{\Gamma(m_{RD})\Gamma(k_{RD})}(\gamma)^{\frac{\beta-1}{2}}K_{\alpha}(2\sqrt{E_{RD}\gamma}) \quad (4)$$

where $k_{\alpha}(\cdot)$ is the bessel function with order α, $\alpha = k - m$ and $\beta = k + m - 1$, k and m are the distribution shaping parameters and $\Gamma(\cdot)$ is Gamma function. $E(\cdot)$ is the mean of the shadowing fading determined by propagation loss. The instantaneous signal-to-noise (SNR) per symbol is defined as $\gamma_{RD} \rightarrow X^2 \frac{E_S}{N_O}$, where E_S is energy signal and N_O AWGN noise signal. Using the instantaneous SNR, the corresponding average SNR $\gamma_{RD} = \Omega\left(\frac{E_S}{N_O}\right)$, where Ω is the mean [7]. For simplicity, K_G distribution is converted into corresponding Meijer G function as shown below

$$f_{\gamma_{RD}}(h) = \frac{1}{\Gamma(m_{RD})\Gamma(k_{RD})\gamma_{RD}} G_{1,3}^{2,1}\left(E_x\gamma_{RD} \middle| \begin{matrix} 1 \\ \alpha_{RD}, \beta_{RD}, 0 \end{matrix}\right). \quad (5)$$

3 Outage Probability

In this section outage probability of the hybrid VLC-RF link is analyzed using DF relaying protocol. Decode and forward relay enhance the overall performance of wireless optical link in indoor environment (in this case). The DF relay decodes the signal, process it and further re-encodes it and finally forwards it to the distant node. The term outage probability is defined as—when the instantaneous SNR goes below a predefined threshold or critical SNR, which is a alarmed value of SNR, below which the hybrid link performance is un-satisfactory [8]. Using DF relaying protocol, the equivalent instantaneous SNR at the single LED relay link γ_{DF}, can be written as

$$\gamma_{DF} = \min(\Gamma_{SR}, \Gamma_{RD}) \quad (6)$$

Therefore the CDF ($F_{DF}(\gamma)$) of γ_{DF} derived as shown

$$F_{DF}(\gamma) = Pr(\gamma_{DF} \leq \gamma) = Pr(\min(\Gamma_{SR}, \Gamma_{RD}) \leq \gamma) \quad (7)$$

where Γ_{SR} and Γ_{RD} are instantaneous SNRs of the source to relay (S-R) and relay to destination (R-D) links respectively. Further we can rewrite (7) as

$$\begin{aligned} F_{DF}(\gamma) = {} & Pr(\Gamma_{SR} \leq \gamma, \Gamma_{RD} \geq \Gamma_{SR}) + Pr(\Gamma_{RD} \leq \gamma, \Gamma_{SR} \geq \Gamma_{RD}) \\ & + Pr(\Gamma_{SR} \leq \gamma, \Gamma_{RD} \leq \gamma) \end{aligned} \quad (8)$$

Since Γ_{SR} and Γ_{RD} are not dependent, (8) is modulated as

$$\begin{aligned} F_{DF}(\gamma) &= Pr(\Gamma_{SR}<\gamma)+Pr(\Gamma_{RD}<\gamma)-Pr(\Gamma_{SR}<\gamma)Pr(\Gamma_{RD}<\gamma) \\ &= F_{\Gamma_{SR}}(\gamma)+F_{\Gamma_{RD}}(\gamma)-F_{\Gamma_{SR}}(\gamma)F_{\Gamma_{RD}}(\gamma) \end{aligned} \tag{9}$$

where $F_{\Gamma_{SR}}(\gamma)$ and $F_{\Gamma_{RD}}(\gamma)$ are the CDF's of Γ_{SR} and Γ_{RD} respectively. Now substituting (3) and (5) in (9), we obtain the P_{out} of the DF relayed hybrid VLC-RF link given as

$$\begin{aligned} P_{out} = &\frac{\alpha}{1-\beta}\times\frac{\gamma_{SR}^{\frac{1-\beta}{2}}}{\bar{\gamma}_{SR}}+\frac{1}{\Gamma(m_{RD})\Gamma(k_{RD})}G_{1,3}^{2,1}\left(E\gamma_{RD}\Big|\begin{matrix}1\\ \alpha,\beta,0\end{matrix}\right) \\ &-\frac{\alpha}{1-\beta}\times\frac{\gamma_{SR}^{\left(\frac{1-\beta}{2}\right)}}{\bar{\gamma}_{SR}}\times\frac{1}{\Gamma(m_{RD})\Gamma(k_{RD})}G_{1,3}^{2,1}\left(E\gamma_{RD}\Big|\begin{matrix}1\\ \alpha,\beta,0\end{matrix}\right). \end{aligned} \tag{10}$$

4 Numerical Result

The hybrid VLC-RF E-health care system model has been studied and analyzed. Its performance is measured in terms of outage probability (P_{OUT}) with various parameters. Figure 2 shows the P_{OUT} varying with avg. electrical SNR for different

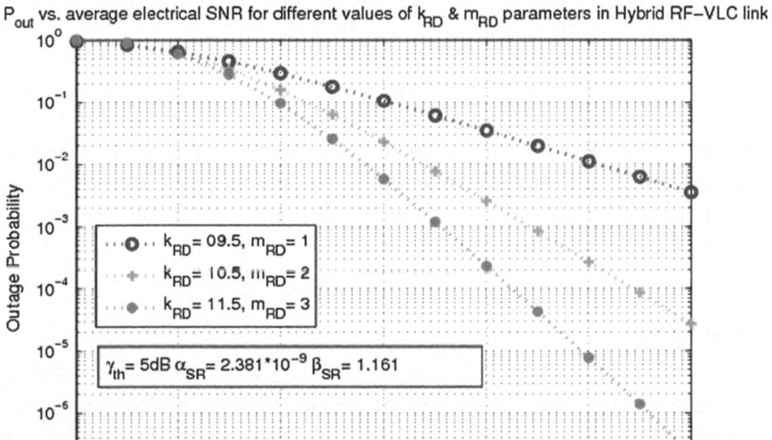

Fig. 2 Outage probability versus average SNR for varying RF hop parameters in hybrid VLC-RF link

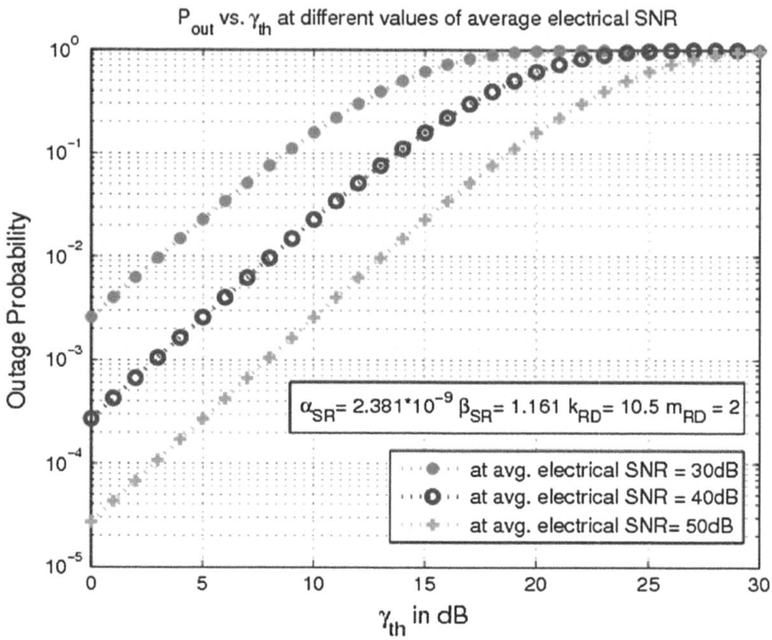

Fig. 3 Outage probability versus threshold SNR of the hybrid VLC-RF link at different values of average SNR

values of the shaping parameters in RF hop. We can conclude from the figure that slight change in the shaping parameters, may lead to increase in the system performance. For example, at 10^{-5} SNR of the hybrid link is approx. 49 dB at $k_{RD} = 11.5$ and $m_{RD} = 3$. Whereas, for the same P_{OUT} SNR is 60 dB at $k_{RD} = 10.5$ and $m_{RD} = 2$. Now, for a fixed SNR value, with increase in the threshold SNR will tends to enhance the P_{OUT} of the hybrid VLC-RF link. This can be noted from Fig. 3, that at SNR 30 dB, as γ_{th} increases, P_{out} approaches to 1. Hence by choosing an optimal Threshold SNR we can get better results for the same model.

5 Conclusion

The proposed hybrid system for the indoor communication via light which is used to communicate fast, the biomedical information from patient through VLC uplink to a decode and forward relay which further radio casts that information to the laboratory. From the results we can see that as the system performance is reaching an ideal SNR of the system if the modelled hybrid link threshold SNR is kept optimum. Also, after analysing the RF hop link, we conclude that by increasing the shape parameters of the distribution, we can get better overall performance of the system model. Therefore, overall performance of the system is enhanced significantly, and the e-health information can be communicate with better rates and cost effectively.

References

1. Van Der Togt, R., van Lieshout, E. J., Hensbroek, R., Beinat, E., Binnekade and J. P., Bakker,, "Electromagnetic interference from radio frequency identification inducing potentially hazardous incidents in critical care medical equipment," Jama, 24 (299), 2884–2890 (2008).
2. M. J., Ball and J., Lillis, "E-health: transforming the physician/patient relationship," International Journal of Medical Informatics, 1 (61), 1–10 (2001).
3. Ding, Wenbo, Fang Yang, Hui Yang, Jintao Wang, Xiaofei Wang, Xun Zhang, and Jian Song. "A hybrid power line and visible light communication system for indoor hospital applications", Computers in Industry, 68(C), 170–178 (2015).
4. S. H., Hussain, M. M., Abdallah and K. A. Qaraqe, "Hybrid Radio-Visible Light Downlink Performance in RF Sensitive Indoor Environments," 6th International Symposium on Communications, Control and Signal Processing (ISCCSP), 81–84 (2014).
5. P., Bithas, N., Sagias, P., Mathiopoulos, G., Karagiannidis, and A. Rontogiannis, "On the performance analysis of digital communications over generalized-k fading channels," Communications Letters IEEE, 5 (10), 353–355 (2006).
6. P., Shankar, "Outage probabilities in shadowed fading channels using a compound statistical model Communications," IEEE Proceedings, 6 (152), 828–832 (2005).
7. Petros S. Bithas, Nikos C. Sagias, P. Takis Mathiopoulos, George K. Karagiannidist, and Athanasios Rontogiannis, "Digital Communications over Generalized-K Fading Channels," 2nd International Symposium on Wireless Communication Systems, 684–687 (2005).
8. N. C., Beaulieu and J., Hu, "A closed-form expression for the outage probability of decode-and-forward relaying in dissimilar rayleigh fading channels," Communications Letters, IEEE, 12 (10), 813–815 (2006).

Arc-Induced Long Period Gratings: Analysis of the Fabrication Parameters on the Surrounding Refractive Index Sensitivity

Rajeev Ranjan, Flavio Esposito, Stefania Campopiano and Agostino Iadicicco

Abstract This paper reports on recent numerical and experimental studies about the effects of the fabrication parameters on the Long Period Gratings (LPGs) sensitivity to the surrounding refractive index. The LPGs are fabricated by Electric Arc-Discharge (EAD) technique and the fabrication parameters are optimized by acting on the arc duration, power and electrodes gap. This permits to fabricate several LPGs in standard fiber with period ranging from 400 to 500 μm gratings where the coupling is with LP_{05} and LP_{06} cladding modes. Furthermore, we have investigated the surrounding refractive index (SRI) sensitivity of the LPGs as function of the grating period and compared the experimental results with numerical analysis based on coupled mode theory. As we show here, we achieve a good agreement between numerical and experimental results.

1 Introduction

Since first long period grating (LPG) based device in 1994 [1], LPGs have attracted scientific community [2, 3] because of their appealing spectral features suitable for communication and sensing application. LPGs rely on coupling light from the fundamental core mode (LP_{01}) of a single mode fiber into the m-th cladding modes (LP_{0m}), thus producing one or more attenuation bands, at specific wavelengths, in the fiber transmission spectrum. This coupling mechanism allows for LPGs sensitivity to external parameters such as strain, bending, temperature and surrounding

R. Ranjan · F. Esposito · S. Campopiano (✉) · A. Iadicicco (✉)
Department of Engineering, Centro Direzionale di Napoli Isola C4, University of Naples "Parthenope", 80143 Naples, Italy
e-mail: campopiano@uniparthenope.it

A. Iadicicco
e-mail: iadicicco@uniparthenope.it

© Springer Nature Singapore Pte Ltd. 2017

I. Bhattacharya et al. (eds.), *Advances in Optical Science and Engineering*, Springer Proceedings in Physics 194, DOI 10.1007/978-981-10-3908-9_43

medium refractive index (SRI) [4, 5]. Indeed, the evanescent field of the cladding modes goes beyond the cladding perimeter, thus making LPGs sensitive to the optical properties of materials encompassing to fiber (that is the SRI). This important characteristic has led to study the LPGs as candidate toward chemical sensing. As of now, there have been no general systematic experimental investigations reported on the chemical sensitivity of LPGs as function of their period and the numerical justification of the results.

Production of LPGs is always a challenge in terms of cost, repeatability and performance whereas their design is challenging for the control of bandwidth, central wavelength etc. Among different LPGs fabrication technique, EAD procedure [6] is such that allows for easy control of several features. It's simple, flexible, low cost, permits the control of bandwidth and central wavelength [7, 8]. Furthermore with EAD we are able to inscribe LPGs in almost all kind of fibers [1, 6]. In this paper we present an assessment of the EAD based LPGs fabrication by acting on the electrodes gap enabling the fabrication of gratings with low period. Then we present a comparison between experimental and numerical investigation of the SRI sensitivity of high order cladding modes as function of the grating period. We have inscribed several LPGs in standard Ge-doped SMF supplied by OZ-Optics. Grating period Λ was changed from 400 to 500 μm, in order to have LP_{05} and LP_{06} attenuation bands always visible in the wavelength range of 1200–1650 nm. Furthermore, we investigated the sensitivity of such LPGs with SRI changes by placing them in different liquids with refractive index varying from 1.33 to 1.44. We report here the wavelength shifts of LP_{05} and LP_{06} bands due to a change in refractive index from air to water (1.00–1.33) for several LPGs written with different grating periods. A numerical model for the simulation of LPGs spectra when it imposes to different refractive index was also developed following the coupled Mode Theory (CMT) based approach proposed by Anemogiannis et al. [9, 10], in order to validate the experimental results.

2 Experimental Details

The schematic of the experimental set up for LPG fabrication is plotted in the Fig. 1a. For fabrication of LPG with EAD technique we have used two electrodes based on commercial fiber optics fusion splicer (i.e. Sumitomo Type-39). A short section of uncoated optical fiber (OZ-Optics standard single mode fiber with Ge-doped core diameter of 8.2 μm and pure-silica cladding diameter of 125 μm, Mode Field Diameter (1.55 μm of 10.4 ± 0.8 μm, Numerical Aperture of 0.14) is placed in the middle portion between the two electrodes of the splicer. One end of the optical fiber is fixed in the V-grooves placed on a computer controlled motorized translator (Thorlabs, LTS 300 mm) whereas the other end of the optical fiber is kept under axial tension with a weight of 12 g attached to it. The insertion of Fig. 1a shows a microscope image of the fiber after EAD treatment, where it can be clearly seen the geometrical modification induced by the EAD. While this technique is simple and economical, it does not easily permit to decrease the grating period.

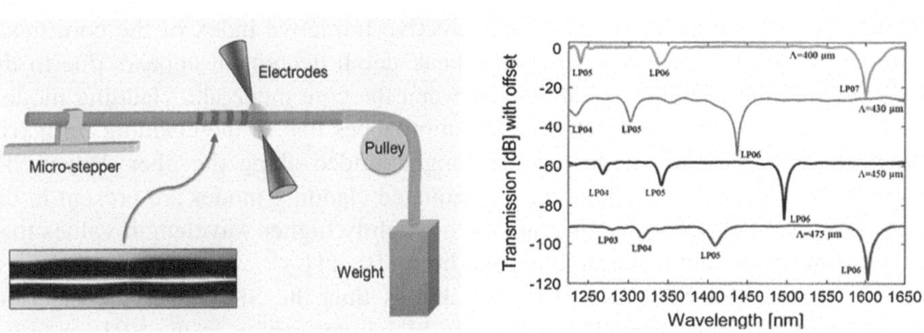

(a) LPG Fabrication Set-up (b) Transmission Spectra

Fig. 1 **a** LPG fabrication set-up: insert is the modulation in the fiber after EAD treatment. **b** Transmission spectra of LPGs with different period Λ = 400, 430, 450, and 475 μm

To overcome this limitation, in order to excite attenuation bands related to the higher order cladding modes we act on the gap between the electrodes of the splicer. Furthermore, in order to apply controlled arc discharge, we have customized the input parameters of splicer viz. arc time duration and arc discharge current. For the presented LPGs, the typical range of the electric arc current was 10–15 step (proprietary unit of the splicer) while the time duration was 400–800 ms. Electric arc discharge plays a vital role in localization of the tapering of the transversal size of the core and cladding regions along the fiber and changes of the silica refractive index due to the stress relaxation induced by local hot spots [7]. The transmission spectrum of the grating has been monitored and recorded after each discharge using Optical Spectrum analyzer (Yokogawa AQ6370B) with the resolution of 0.1 nm with the Super Luminescent Light Emitting Diode (SLED) as a source of the range of 1200–1650 nm.

We fabricated several LPGs in OZ-Optics fiber with different grating periods. In Fig. 1b we report the transmission spectra of the LPGs with period of 400, 430, 450 and 475 μm written by using electric arc current of 12 step and duration of 600 ms. For the period of 400 μm the resonant band LP_{05} is at 1240 nm and LP_{06} one is at 1340 nm, for 430 μm LP_{05} is at 1303 nm and LP_{06} is at 1437 nm, for 450 μm, LP_{05} is at 1342 nm and LP_{06} is at 1497 nm and for 475 μm, LP_{05} is at 1413 nm and LP_{06} is at 1605 nm. It can be seen that, in agreement with the theory of LPG [6], the resonant wavelength increases with the period.

3 Results and Discussion

One of the properties of LPGs is that the resonant wavelengths of each attenuation bands are high sensitive to the changes in the SRI. In particular, if the refractive index of surrounding medium is lower than that of the cladding, an increase of the SRI induces a blue shift of each resonant wavelength as well as a slightly decrease of the attenuation band depth in the transmission spectra. The former is due to a decrease of the difference between the effective refractive index of the core mode and of the cladding modes whereas the peak depth decrement appears due to the decreasing of the coupling coefficient between the core mode and cladding modes. As the value of external refractive index approaches that of the cladding refractive index, distinct cladding modes are no longer guided along the fiber. When SRI exceeds the cladding refractive index, attenuated cladding modes are present in the cladding. The attenuation bands reappear at slightly higher wavelength values than for the case where air is the ambient medium [10, 11].

The most useful SRI range for the LPG is thus the SRI lower the cladding refractive index. Here, the LPGs show the higher sensitivity to the SRI when the refractive index of the surrounding medium approach the cladding region one. Moreover, it is well know that the sensitivity to SRI increase for the higher order cladding modes.

The resonant wavelength dependence on SRI is expressed by the following equation [8]:

$$\frac{d\lambda_r}{dn_{ex}} = \frac{d\lambda_r}{dn_{cl}}\frac{dn_{cl}}{dn_{ex}}$$

where λ_r is resonant wavelength, n_{ex} is the surrounding medium refractive index, n_{cl} is effective refractive index of the cladding mode.

We have carried out experiments to investigate the sensitivity to surrounding refractive index changes for the attenuation bands related to LP_{05} and LP_{06} cladding modes in several LPGs with period ranging from 400 to 500 μm. The aim is to identify the relationship of the SRI sensitivity versus the grating period as design criteria for high performance sensors.

To this aim medium with different refractive indexes ranging from 1 (air) to 1.34 (slightly higher than water refractive index) were used in the experiments. Shift of the attenuation bands was analyzed in order to calculate the sensitivity around the water refractive index (around 1.33). Figure 2 shows the wavelength shifts of the

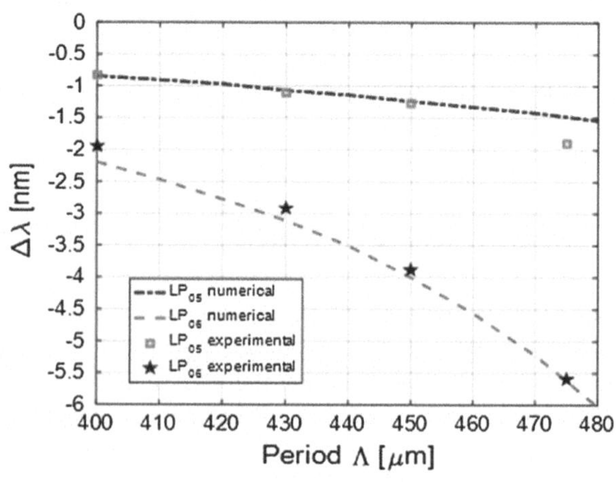

Fig. 2 Sensitivity of LP_{05} and LP_{06} in water versus the grating period

LP_{05} (square markers) and LP_{06} (star markers) attenuation bands when the SRI changes from air to water versus the grating period. Both curves show a sublinear behavior showing that the sensitivity increases with grating period.

We also develop numerical tools in MATLAB environment using CMT [9, 10] and simulated the effect of SRI on the LPG with the different period. Simulated period of gratings are 400, 430, 450, and 475 μm. The gratings with same period were tested with the SRI in the range of 1.00–1.33. In Fig. 2 the dotted plot is experimental data plot whereas stars and square points are numerical data plot. Obtained results are plotted together with numerical outputs, showing close agreement.

4 Conclusions

In this paper we present a numerical and experimental evaluation of the sensitivity of higher order modes of LPG written by electric arc discharge. We successfully demonstrate the theoretical behavior of attenuation bands with numerical simulation which exhibit a blue shift with SRI increasing, with the magnitude of shift increasing with the order of the cladding mode. The numerical results are in very good agreement with the experiments. Moreover, as shown, the magnitude of shift also increases with grating period Λ with a sublinear behavior. We successfully report and co-relate it with the CMT and justify our results with numerical analysis.

Acknowledgements This work was supported by the University of Naples "Parthenope" through the "Bando di sostegno alla ricerca individuale per triennio 2015–2017, Annualità 2015".

References

1. M. J. P. H. Poole CD, "Two-mode fibre spatial-mode converter using periodic core deformation", Electron Lett., vol. 30, pp. 1437–1438, (1994).
2. Vengsarkar AM, Lemaire PJ, Judkins JB, Bhatia V, Erdogan T, Sipe JE. Long-period fiber gratings as band-rejection filters. J. Lightwave Tech. 14:58–65, (1996)
3. Gaspar Rego, "Arc-Induced Long Period Fiber Gratings," Journal of Sensors, vol. 2016, Article ID 3598634, 14 pages, (2016).
4. Ranjan, R., Esposito, F., Iadicicco, A., Stăncălie, A., Sporea, D., & Campopiano, S, "Comparative Study of Long-Period Gratings Written in Standard and Fluorine-Doped Fibers by Electric Arc Discharge". IEEE Sensors Journal, 16(11), 4265–4273, (2016).
5. L. Coelho, D. Viegas, J. L. Santos, and J. M. M. M. Almeida, "Characterization of zinc oxide coated optical fiber long period gratings with improved refractive index sensing properties," Sensors and Actuators B: Chemical, vol. 223, pp. 45–51, (2016).
6. R. Ranjan, F. Esposito, A. Iadicicco, A. Stăncălie, D. Sporea, S. Campopiano, "Long period gratings written in fluorine-doped fibers by electric arc discharge technique." in Proc. Sixth European Workshop on Optical Fibre Sensors (EWOFS'2016), pp. 991622–991622. International Society for Optics and Photonics, (2016).

7. M. Smietana, W. J. Bock, P. Mikulic, J. Chen "Increasing sensitivity of arc-induced long-period gratings - Pushing the fabrication technique toward its limits", Meas. Sci. Technol. 22, 015201 (2011).
8. S.Y. Tan, Y.T. Yong, S.C Lee and F.A. Rahman, "Review on an arc induced long period-grating fiber grating and its sensor applications," Journal of Electromagnetic Waves and Applications, vol 29, no. 6, pp 703–826, (2015).
9. E. Anemogiannis, E. N. Glytsis, K. Gaylord, "Transmission characteristics of long-period fiber gratings having arbitrary azimuthal/radial refractive index variations". J. Lightwave Technol. 21, 218–227 (2003).
10. A. Cusano, A. Iadicicco, P. Pilla, L. Contessa, S. Campopiano, A. Cutolo, M. Giordano and G. Guerra, "Coated long-period fiber gratings as high-sensitivity optochemical sensors," Journal of Lightwave Technology, vol. 24, no. 4, pp. 1776–1786, (2006).
11. Stegall DB, Erdogan T.," Leaky cladding mode propagation in long-period fiber grating devices". IEEE Photon. Technol. Lett. 11:343–345 (1999).

Design of Data Transmission System for Application in Terahertz Frequency

Sanjana Meshram, Manoj Dongre, Kshitij Mittholiya and Saurabh Bhardwaj

Abstract The terahertz recurrence correspondence frameworks have developed because of expanding interest for higher transfer speed and information rates. We have designed Personal Computer (PC) to PC data transmission system for application in terahertz frequency. In this system, data is transfer from Personal computer (PC) through serial port RS232 using HyperTerminal software. The intensity modulator modulates the optical carrier signal by data signal which is send from the Personal computer (PC-1). The wavelength of the laser is ranges from (1527 to 1565 nm) and optical output power ranges from 3.5 to 40 mW. The modulated signal is given to the photodiode which convert optical signal into electrical signal and electrical output of photodiode is amplified using Pre-amplifier. This amplified signal logic is converted to RS232 logic and then given to another Personal computer (PC-2). The aim of this system is to transfer data from PC to PC using optical components which can operate at higher frequency. We believe that this system can be integrated with the terahertz generation and detection system.

1 Introduction

Recently, there has been an increasing growth in the application of terahertz (THz) waves for broadband transmission communications [2]. The frequency range from 0.1 to 10 THz with the corresponding wavelength ranges from 3 to 0.03 mm,

S. Meshram (✉) · M. Dongre
Department of Electronics and Telecommunication Engineering,
Ramrao Adik Institute of Technology, Mumbai, India
e-mail: sanjanameshram64@gmail.com

M. Dongre
e-mail: mmdongre04@gmail.com

K. Mittholiya · S. Bhardwaj
Department of Photonics, Society for Applied Microwave Electronics Engineering
and Research (SAMEER), Mumbai, India
e-mail: kmittholiya@gmail.com

© Springer Nature Singapore Pte Ltd. 2017
I. Bhattacharya et al. (eds.), *Advances in Optical Science and Engineering*,
Springer Proceedings in Physics 194, DOI 10.1007/978-981-10-3908-9_44

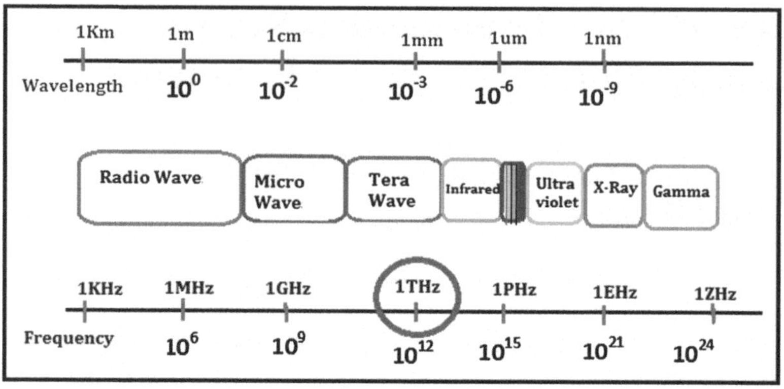

Fig. 1 The electromagnetic spectrum

this frequency range is located between the microwave and infrared in the electromagnetic spectrum which is called as Terahertz region (is shown in Fig. 1). The photonic technologies for generating and modulating terahertz waves are the most efficient and easiest ways because of its inherent broadband nature, and allow us to deliver photonic terahertz waves over long distance using optic-fibers [3].

We propose the system that have application in the terahertz wireless communication setup [1]. This system is used to transfer data from one computer to another using optical components i.e. intensity modulator and photodiode (optical to electrical converter). In this system, the optical carrier signal generated from the two tunable lasers is modulated using intensity modulator driven by the data signal from the computer. This signal is transferred through photodiode and amplifier to the oscilloscope which display the output waveform and to the personal computer (PC-2) which receives the data. This system is further used in the terahertz wireless communication systems which transfer data from one computer to another. In the terahertz communication system instead of using Photodiode it requires source (UTC-PD) and detector (SBD) to transfer terahertz wave except that, all the modulation and demodulation part of the this communication system is included in our data transmission system.

Section 2 describes the system configuration. i.e. describes data flow of the data transmission system and explain working of each component in details. Section 3 describes the application of this system in terahertz frequency. Experimental results are shown in Sect. 4. Conclusion and future work of the system discusses in Sect. 5.

2 System Configuration

In the data transmission system, we can transfer data from one computer to another through optical components. The hardware components used in this PC to PC data transmission system are as follows:

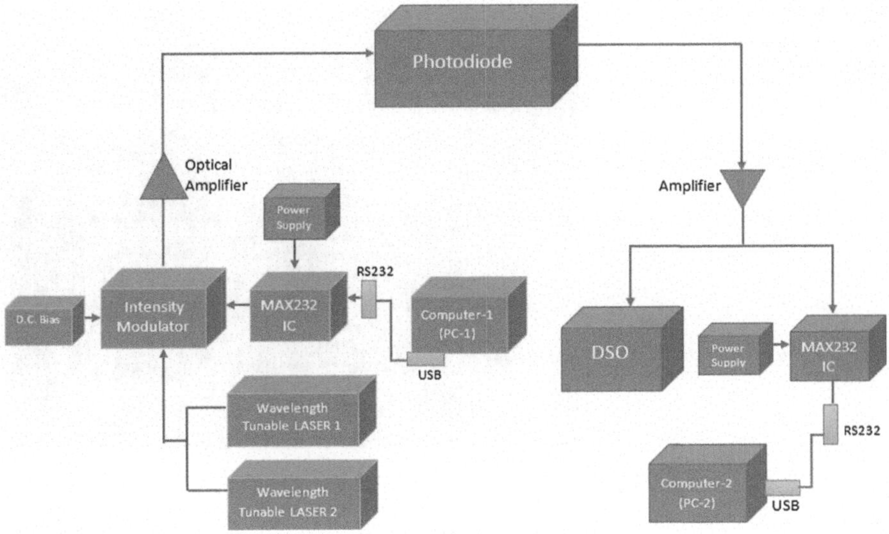

Fig. 2 Block diagram of the data transmission system

- Universal Serial Bus (USB) to RS-232 Serial cable
- Oscilloscope
- Personal Computer (PC)
- Intensity Modulator
- Max232 IC
- Wavelength Tunable Laser
- Photodiode
- Amplifier

The block diagram of the data transmission system through optical components is shown in Fig. 2.

In this system, the HyperTerminal software and USB to RS232 cable is used to send the ASCII character (i.e. the letters and alphanumeric characters) from the Computer (PC-1). The output of USB to RS232 cable is in RS232 logic having a voltage level of ±13 V. The data from USB to RS232 Cable is feed to the MAX232 IC which converts RS232 logic to TTL logic. This conversion of the RS232 logic level into TTL logic level is required because this signal further feed to the input of the Intensity Modulator and the required voltage for this modulator should not be exceed 5 V.

The Transmitter pin of USB to RS232 cable i.e. pin no. 3 is connected to the pin no. 8 of MAX232 IC which converts the RS232 logic to TTL logic and TTL output from pin no. 9 of MAX232 IC is connected to intensity modulator RF input. Figure 3 shows the circuit implemented for RS232 to TTL logic conversion. The

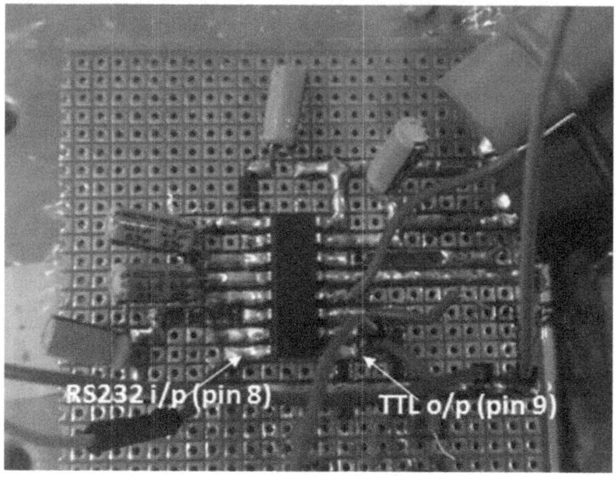

Fig. 3 RS232 to TTL converter implemented on PCB

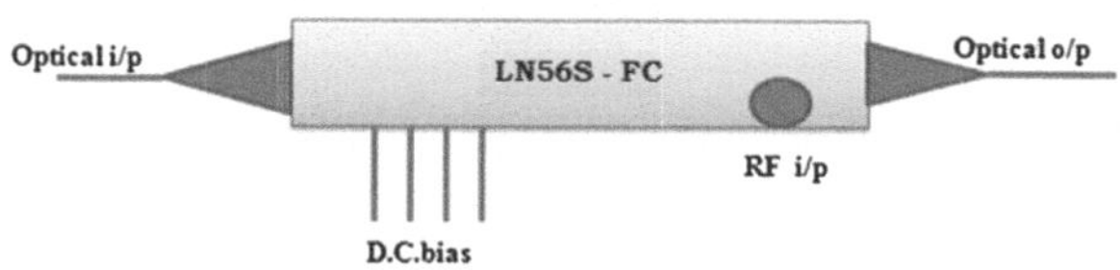

Fig. 4 LN56S-FC intensity modulator

optical signal is generated by using two laser with wavelength tunability of 1527–1565 nm and optical output power ranges from 3.5 to 40 mW. The intensity modulator have two inputs, one for optical carrier signal from the laser and another to provide RF modulating signal from computer. The optical carrier signal is modulated using the intensity modulator with the modulating data signal from the computer.

The diagram of intensity modulator is shown in Fig. 4. The operation of the intensity modulator is based on the Mach Zehnder interferometer, Phase of optical signal changes in proportion to applied voltage in one arm of the modulator. The output signal of the intensity modulator is optical signal, so to convert this optical signal into electrical signal Photodiode is used. As the maximum output voltage of photodiode is 137 mV which is not sufficient to drive MAX232 IC so it is amplified using a amplifier. The output of the pre-amplifier is connected to the oscilloscope which will display the output of the corresponding bit pattern waveform of the ASCII character which is send from the computer.

The amplifier is shown in Fig. 5. This pre amplified signal is now converted to RS232 logic by MAX232 IC which is given to receiver pin (Pin no. 2) of RS232 to USB cable of the Personal Computer-2 (Receiver PC) which will receive the data send from Personal Computer-1 (Transmitter PC).

Fig. 5 Pre-amplifier

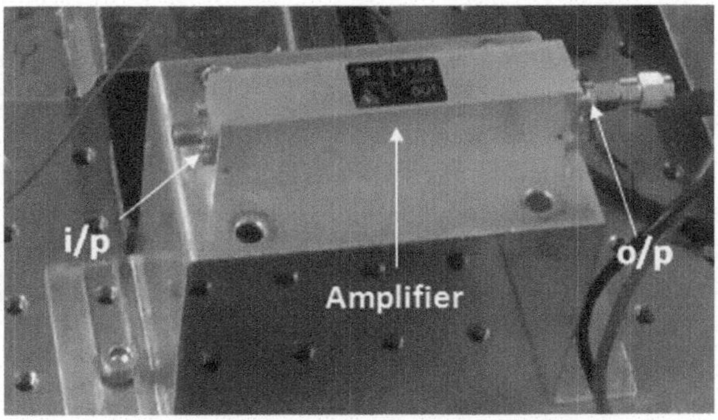

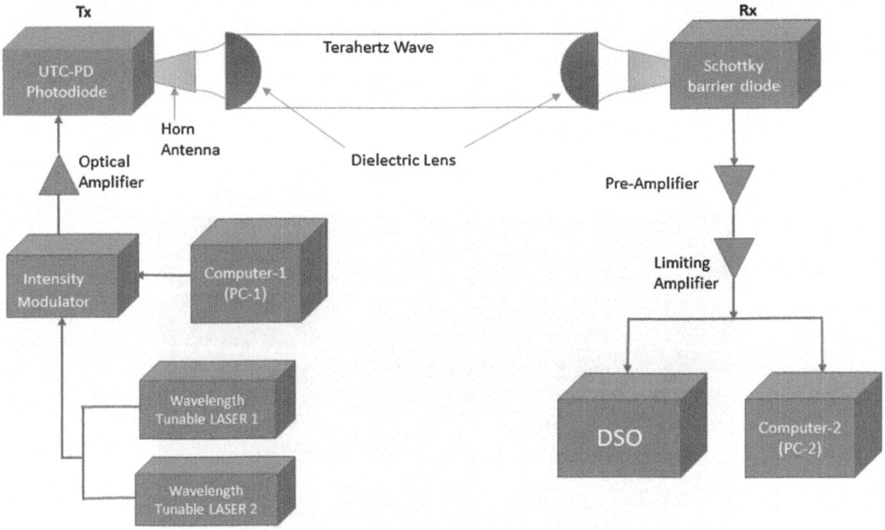

Fig. 6 PC to PC communication through Terahertz frequency

3 Application in Terahertz Frequency

This scheme of the PC to PC data transmission system using optical components will be further used in the terahertz wireless communication system. To implement this data transmission system into terahertz systems in the future, the photodiode can be replace by the source UTC-PD (Uni-Travelling Carrier Photodiode) and detector SBD (Schottky barrier diode) except this all the modulation and demodulation process is same as in this data transmission system.

The block diagram of PC to PC communication through Terahertz frequency is shown in Fig. 6, THz-wave signal is generated by heterodyning the two wavelengths of light from the wavelength-tunable light sources. Heterodyned signal is modulated by Data signal using Intensity modulator and the output is converted to high frequency signal by the UTC-PD module which act as a THz source, and it is radiated to the free space by means of a horn antenna with a gain of 25 dBi [4]. The transmitted terahertz wave is very much collimated by Teflon lens. At receiver side, high frequency signal is detected using Schottky Barrier Diode detector (SBD) then amplified using pre-amplifier. The output of Pre-Amplifier is displayed on the oscilloscope and given to the Personal Computer-2 on the receiver side which will receive the data.

4 Experimental Results

1. ASCII character output on oscilloscope (Fig. 7)
2. ASCII character transferred from PC-1 to PC-2 (Figs. 8 and 9).

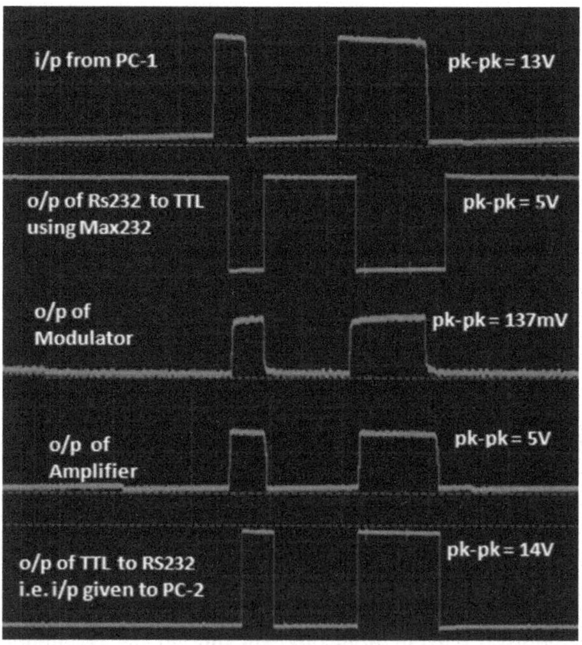

Fig. 7 Output of ASCII character displayed on oscilloscope at different stages

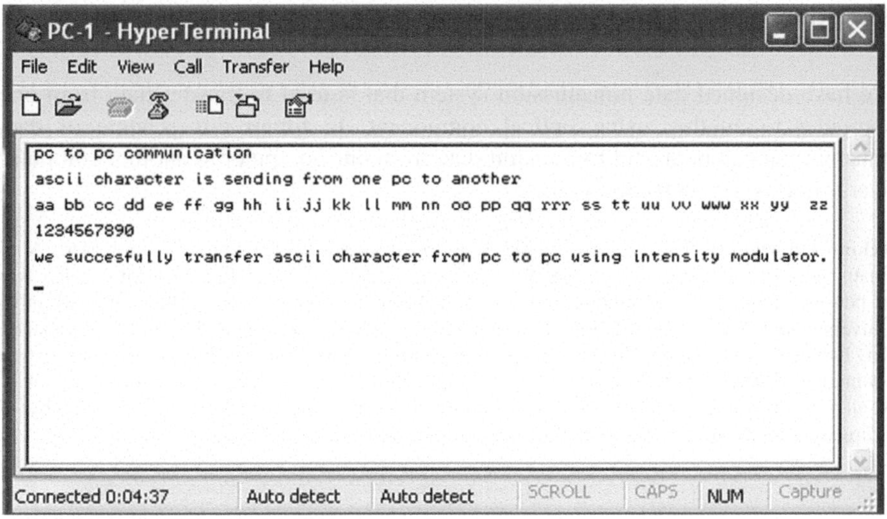

Fig. 8 Dialog box of ASCII character sent from PC-1

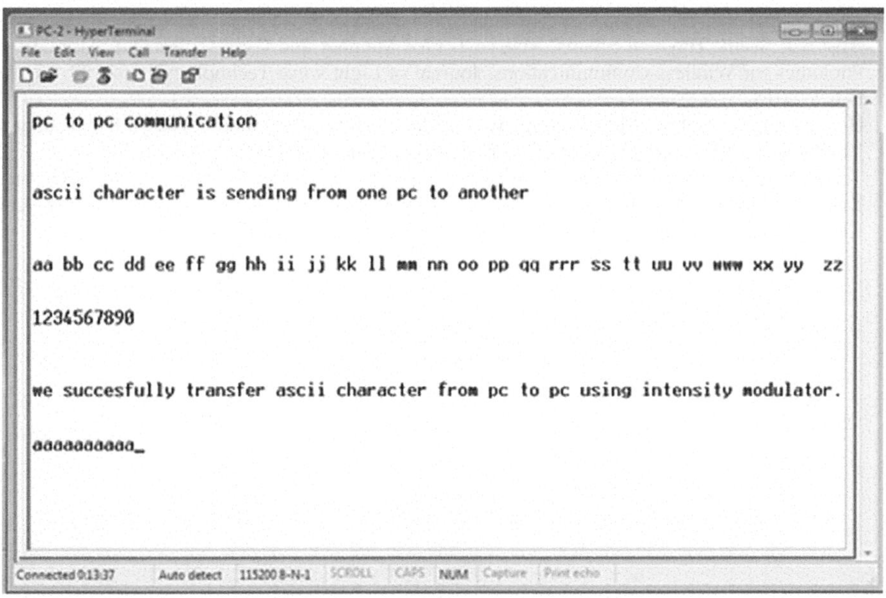

Fig. 9 Dialog box of ASCII character received on PC-2

5 Conclusion

We have designed data transmission system that is used to transfer data from one computer to another using optical component. In future this scheme of data transmission can be useful to transmit data from one computer to another computer over wireless Terahertz frequency.

Acknowledgements The authors would like to express their gratitude towards Dr. Anuj Bhatnagar for being a guiding angel for this work. Also we would like to acknowledge two immensely cooperative institutions 'Ramrao Adik Institute of Technology', affiliated to Mumbai University and 'Society for Applied Microwave Electronics Engineering & Research', Mumbai (R & D Lab of Department of Information Technology, Ministry of Communications and Information Technology, Govt. of India) for providing with an inclusive, encouraging environment. On behalf of all the authors, I would like to acknowledge our friends and families who have directly or indirectly joined hands to make this application development successful.

References

1. Basil A.M, Kshitij Mittholiya, Archana Hegde, N.Shinde, Dr Anuj Bhatnagar, "Short distance terahertz communication demonstration using photonics components", CP-11.12, DOC:7377, NLS-24, 2015.
2. Alwyn J. Seeds, Haymen Shams, Martyn J. Fice Member and Cyril C. Renaud, "TeraHertz Photonics for Wireless Communications, Journal of Light wave Technology", Vol. 33, No. 3, Feb 1, 2015.
3. Ho-Jin Song, Katsuhiro Ajito, Atsushi Wakatsuki, Yoshifumi Muramoto, Naoya Kukutsu, and Yuichi Kado, "Terahertz Wireless Communication Link at 300 GHz", 2010 IEEE.
4. Tadao Nagatsuma, Ho-Jin Song, Yoshihide Fujimoto, Kazumasa Miyake, Akihiko Hi-rata, Katsuhiro Ajito2, Atsushi Wakatsuki3, Tomofumi Furuta3, Naoya Kukutsu2, and Yuichi Kado, "Giga-bit Wireless Link Using 300–400 GHz Bands", IEEE 2010.

Gesture Based Audio/Video Player

Indrajeet Vadgama, Yash Khot, Yash Thaker, Pranali Jouras and Yogita Mane

Abstract In this research paper we wish to develop a Windows-based application for live motion gesture recognition using web-camera as input built using Java, and using this input to control a video/audio player (VLC media player). A combination of user motion detection and gesture recognition will be used. This application uses the webcam to detect gestures made by the user and perform basic operations such as play/pause, volume up, volume down, next, previous.

1 Introduction

In this research paper we propose a system to control VLC media player without physical interaction with the computer. Image is captured and verified with our system in which image pre-processing and other techniques are used for detection of gesture. Image is captured and provided as the input to the application via a camera of minimum 1 MP quality for good results. Processing will take place in the system after providing the input as image. Here the input gesture will be recognized on the basis of finger count. The desired action will be performed. Errors may arise due to invalid gestures or quick movement of hand resulting in the system not recognizing the gesture. Invalid gestures will result in no action being performed similarly quick movement of hand will be ignored to avoid accidental gestures. System uses frames from an input video stream, performs morphological filtering upon it, RGB/HSV filtering is performed to detect hand glove, contours and defects are detected and gesture is recognised and compared with number of outstretched fingers, based on the number of outstretched fingers operations are performed in VLC media player. Communication with VLC media player is done via TCP Connection. RGB/HSV filtering techniques used have a high rate of detection,

I. Vadgama (✉) · Y. Khot · Y. Thaker · P. Jouras · Y. Mane
Department of IT, Universal College of Engineering, Thane, Maharashtra 401212, India
e-mail: vadgamaindrajeet@gmail.com

© Springer Nature Singapore Pte Ltd. 2017

I. Bhattacharya et al. (eds.), *Advances in Optical Science and Engineering*,
Springer Proceedings in Physics 194, DOI 10.1007/978-981-10-3908-9_45

also the performance is high. The implementation of this system renders easy media viewing experience for users. Enables access to the computer (media player) from a distance. Also it acts as a help for differently abled individuals to operate the media player without physical interaction. No additional hardware cost is required for laptops, for desktops web-cameras are available at a very low cost. This is achieved by Java code using OpenCV libraries and using the inbuilt web-camera in laptops or external web-cameras for desktops. Open Source Computer Vision Library (OpenCV), is a software library. OpenCV was built to provide a common framework for computer vision applications. It has Java interfaces that are used by us and it supports all OS's.

Roadmap: We first take you through the survey of existing systems, their advantages and disadvantages. Next we take you through our proposed system. Further we specify the methodology used in our system, and the algorithms used by our system. Then we define the performance of our system and also the minimum hardware requirements. Then we conclude and discuss future scope. We end our paper with references.

2 Existing Systems

Many applications have been developed that implement gesture control. These applications are present for gaming, controlling media players, etc.

Xu et al. [1] in IEEE 2012 developed a system that recognizes hand gestures. MEMS 3-axes accelerometers send signals which act as input. Bluetooth transmission is used to transmit the detected hand motion which is in 3 perpendicular directions, this is detected by 3 accelerometers. Segmentation algorithm was implemented and gestures are recognized by comparing gestures stored in the system.

Li et al. [2] in IEEE 2011 used hand moments to retrieve information present on the internet. Appropriate function is executed when the user gestures to the system and the system in turn reports the action to user via speech. This system uses face recognition to identify unique user as requirements differ as per user. Hand and face recognition is done with PCA method. Recognised gesture is compared to stored gesture template to perform action.

Thomas [3] in VSRD-IJEECE 2011, compared the results obtained by various hand detection techniques. Edges method, pixel by pixel comparison and orientation histogram were the methods used. Various static hand gesture images were stored in image database which was used. These images relative to American sign languages.

Agrawal and Rautaray [4] in IJCA 2010 designed a system to control the VLC media player using hand gestures. The K nearest neighbour algorithm was implemented to recognize gestures. Lucas Kanade Pyramidical Optical Flow algorithm detects hand from video.

Krishna Chaitanya and Janardhan Rao [5] in 2014 used skin detection algorithm to control windows media player. Advantage of proposed system: Our system uses RGB/HSV filtering which has higher detection rate than above methods, also proposed system uses gloves for better recognition of hand.
Patil and Varalakshmi [6] in September 2014, used a three axis controller together with a PIC16F8779 micro controller and MATLAB to control a MP3 player.
Chappidi and Thunuguntla [7] in June–July 2012, used the MEMS Accelerometer MMA7660 IC with an ARM7 micro controller to control a java music player.
Stenger et al. [8] in 2006, recovered the three-dimensional hand motion given an image sequence.
Loi and Fujimura [9] in 2004, used an active sensor to recognize hand gestures from a sequence of real-time images.

3 Proposed System

We have proposed a system solely based on java, as java is platform independent. Our system uses web-cameras which are acquitted in almost all laptops and are easily available for desktops. The input through a web-camera is accepted and recognized by our system to perform the desired action required by the user. As shown in Fig. 1 first image is captured and then morphological filtering is done over this image, after this step RGB/HSV filtering is done and later contour detection and defect detection is performed. Gesture is recognised and corresponding command is sent to VLC media player using TCP Connection.

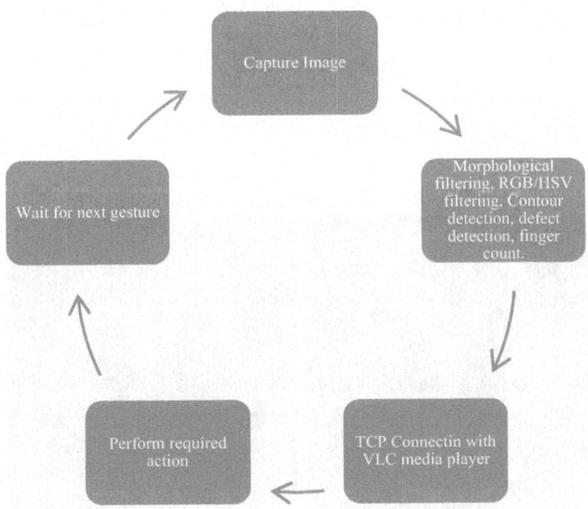

Fig. 1 System architecture

4 Methodology

4.1 Proposed System Follows the Following Algorithm

Step 1 Web-camera provides a constant feed of inputs; this input is read one frame at a time to recognize the gesture made by the user.

Step 2 As shown in Fig. 2 the input received is first morphologically filtered, wherein the image of the user's hand is eroded for better recognition of individual fingers.

Step 3 The morphologically filtered image is filtered using a rgb (Red, Blue, Green) range to detect only the elements with the specified colour (the glove) as shown in Figs. 3 and 4.

Step 4 The output processed image from Step 3 is then subjected to contour and defect detection, also the palm centre is calculated and fingers are detected individually.

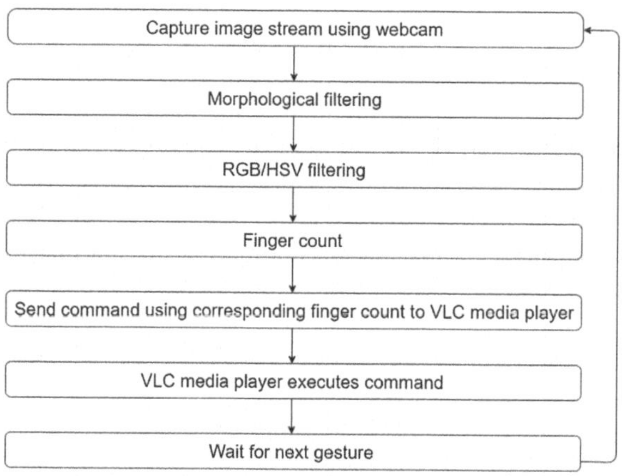

Fig. 2 Algorithm for proposed system

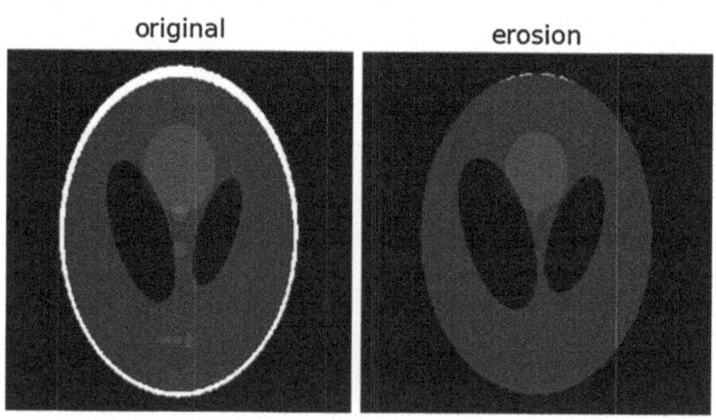

Fig. 3 Morphological filtering

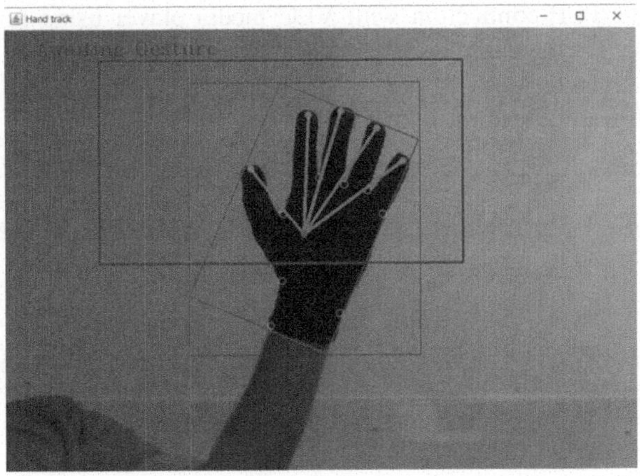

Fig. 4 Filtering of glove (RGB filtering)

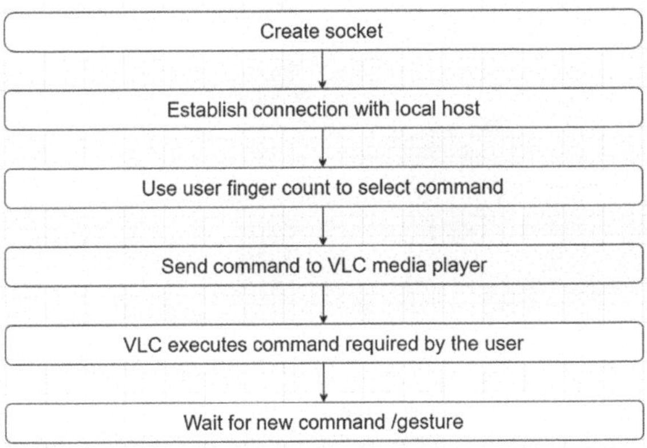

Fig. 5 Algorithm for finger count and command execution

Step 5 The system establishes a TCP Connection with VLC media player using the host (Machine on which VLC media player will run) IP and port address.
Step 6 The system waits for user to give command via gesture.

4.2 Algorithm for Finger Count and Command Execution Is Given Below

Step 1 Get finger (following above process) count (Fig. 5).
Step 2 Using above count as input to switch case goto Step 3.
Step 3 Execute predefined functionin case statement corresponding with finger count.

Step 4 Attempt TCP connection with VLC media player using host IP and port address. If connection is established, then send command.
Step 5 Close connection with VLC.
Step 6 Wait for next gesture, goto Step 1.

4.3 Following Functions Are Available with the System, Based on Number of Fingers Detected

Figure 6 shows zero (0) fingers are detected, pause command.
Figure 7 shows one finger is detected, volume up command.
Figure 8 shows two fingers are detected, volume down command.

Fig. 6 Zero (0) fingers are detected, pause command

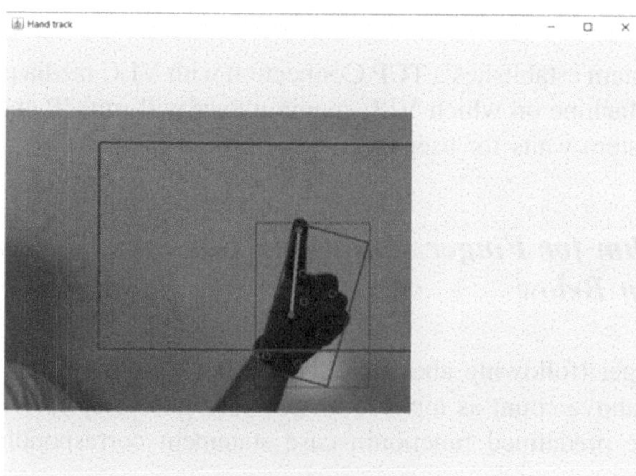

Fig. 7 One finger is detected, volume up command

Fig. 8 Two fingers are detected, volume down command

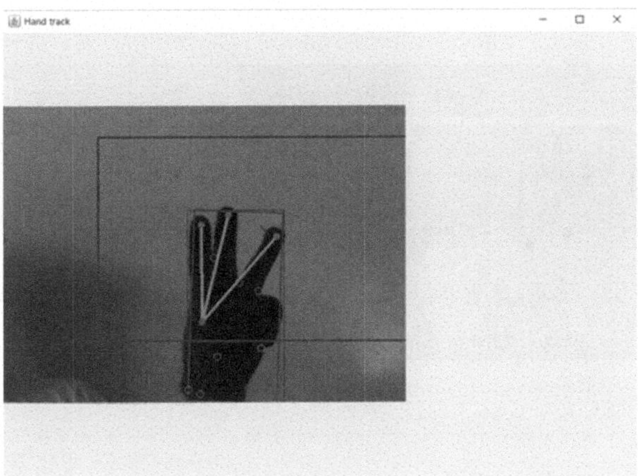

Fig. 9 Three fingers are detected, next command

Figure 9 shows three fingers are detected, next command.
Figure 10 shows four fingers are detected, previous command.
Figure 11 shows five fingers are detected, stop command.

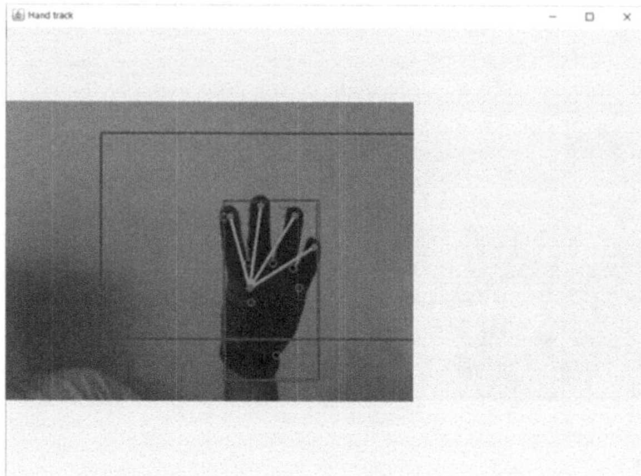

Fig. 10 Four fingers are detected, previous command

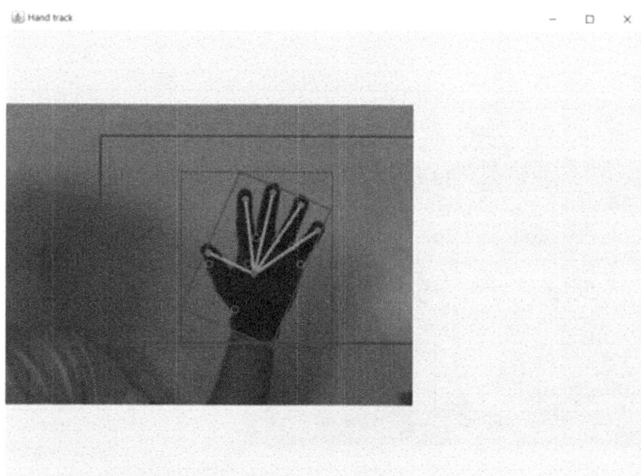

Fig. 11 Five fingers are detected, stop command

5 Performance

Performance of the system under ideal conditions is 98%. Ideal conditions are defined as Proper illumination of surroundings, no black colour background (as system is tuned to detect black gloves), Black gloves (without pattern preferred).

Processing speed is dependent on the hardware configuration. Minimum requirements are as follows Dual Core Processor (Clock speed 1.5 GHz), 1 GB RAM, On-board graphics processor and 100 MB storage memory.

Table 1 Comparison of gesture recognition system

Paper	Primary method of recognition	Number of gesture recognized	Accuracy
1	Controlling of windows media player using hand recognition system	97	–
2	Review of hand recognition techniques	–	–
3	Real time hand gesture recognition system for dynamic applications	–	–
4	MEMS accelerometer based nonspecific-user hand gesture recognition	72	91–100%
	A real-time hand gesture recognition system for daily information retrieval from internet	–	–
Proposed system	Controlling of VLC media player using hand gesture	98	98–100%

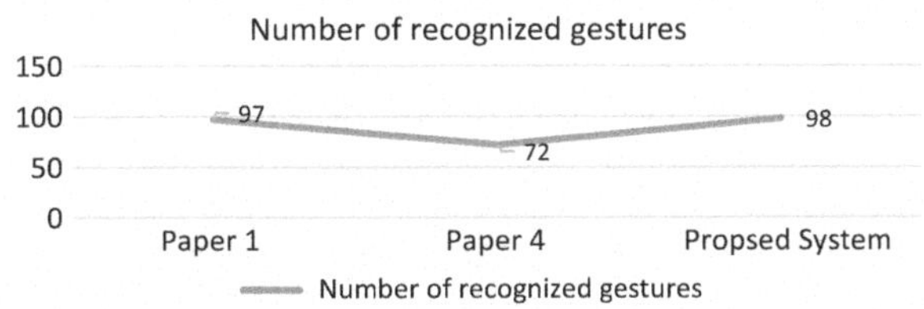

Chart 1 Number of gestures recognized

6 Efficiency

Table 1 shows comparison of various gesture recognition systems and highlights the efficiency of our system, our system gives recognition percentage of 98–100%. Efficiency is tested with 100 gestures out of which 98 were detected (Chart 1).

7 Conclusion

The implementation of this system renders easy media viewing experience for users. Differently abled individuals can use this system to operate the VLC media player from a distance, thus eliminating the need for physical operation. Easy to implement with almost zero additional cost.

8 Future Scope

The future scope of this system is to operate the mouse pointer with only gestures, thus enabling system wide access using gestures through the web-camera.

References

1. Ruize Xu, Shengli Zhou, Wen J. Li., "MEMS Accelerometer Based Nonspecific-User Hand Gesture Recognition". IEEE, 2012. Vol: 12, 1166–1173.
2. Sheng-Yu Pe'ng, Wattanachote K., Hwei-Jen Lin and Kuan-Ching Li., "A Real-Time Hand Gesture Recognition System for Daily Information Retrieval from Internet" In 4th International Conference on Ubi-Media Computing (UMedia), pages 146–151. IEEE, July 2011.
3. Ginu Thomas "Review of Various Hand Gesture Recognition Techniques" VSRD-IJEECE, Vol. 1 (7), 2011, 374–383.
4. Siddharth Swarup Rautaray and Anupam Agrawal, "A Vision based Hand Gesture Interface for Controlling VLC Media Player", International Journal of Computer Applications, 2010. Vol: 10, 0975–8887.
5. N. Krishna Chaitanya, R. Janardhan Rao, "Controlling Of Windows Media Player Using Hand Gesture Recognition System", IJES, 2014, Vol: 3, 2319–1.
6. Devikarani Patil, Varalakshmi B.D, "Hand Gesture Recognition for MP3 Player using Image Processing Technique and PIC16F8779", 2014, Vol: 4, 2231–2307.
7. Dileep Chappidi, Lavanya Thunuguntla, "MEMS Accelerometer Based Gesture Controlled Java MusicPlayer", IJERA, 2014, Vol: 2, 2248–9622.
8. Stenger B1, Thayananthan A, Torr PH, Cipolla R. "Model-based hand tracking using a hierarchical Bayesian filter", IEEE, 2006, Vol: 28, 0162–8828.
9. Xia Liu, K. Fujimura, "Hand gesture recognition using depth data", IEEE, 2004.

Part IX
Nano-Photonics, Bio-Photonics, Bio-Medical Optics

Automated Glaucoma Detection of Medical Image Using Biogeography Based Optimization

Debasree Sarkar and Soumen Das

Abstract Gradual vision loss is an important challenging issue in medical imaging system. When optic nerve is damaged, vision loss and blindness introduced as a result known as glaucoma. So, eye pressure measurement is a crucial measure for optic nerve damage which causes gradual color blindness, vision loss known as open angle glaucoma, so pressure controlling inside eye is very significant for glaucoma detection. Glaucoma detection is basically measuring by CDR value. If CDR is greater than normal value then that is glaucomatous. Correct localization of optic disc and finding cup inside disc are the Objectives of paper. Disc and cup boundaries and cup to disc ratio are shown by resulting images. Sensitivity, specificity and accuracy are the three main parameters for evaluation purpose. In our proposed method we are using RIM-ONE dataset where 290 pictures are taken for testing which gives accuracy 97.58%, sensitivity 94.44%, and specificity 99.00%.

1 Introduction

Diabetes patients have diabetic retinopathy. Gradual retinal damage is a result of this disease, which causes vision loss and also blindness [1]. Manual diagnosis is much more complex rather than an automated system for analytical measurement. Auto screening for detection of eye disease is very effective in preventing loss of sight. Detection of disease in time is important. A survey result predicted that around 80 million people will be affected by 2020. It is basically a non curable disease but it can slow down. Many people are cannot sense the until it reaches at advance level. So our aim is one to find out strategy which recognizes glaucoma

D. Sarkar (✉)
Department of CSE, UIT, Burdwan University, Burdwan, West Bengal, India
e-mail: debasree1031@gmail.com

S. Das
Department of CSE, UIT, Burdwan University, Durgapur, West Bengal, India
e-mail: getme.soumen@gmail.com

© Springer Nature Singapore Pte Ltd. 2017

I. Bhattacharya et al. (eds.), *Advances in Optical Science and Engineering*,
Springer Proceedings in Physics 194, DOI 10.1007/978-981-10-3908-9_46

before it causes blindness [2]. Optic nerve head segmentation is one of the methods to detect glaucoma. Nerve head comprise of optic disc and optic cup. Then cup to disc ratio is taken. If this ratio exceeds threshold then it is glaucoma affected otherwise healthy.

2 Methodology

See Fig. 1.

1. Input Fundus Image

To the performance measure, digital retinal images were acquired by proposed method. Retinal blood vessels, optic disc exists in the Fundus image which is actually the inner surface of the eye. In our method the Rim-one image dataset contains both normal and abnormal image.

2.1 *ROI*

In order to reduce the computational requirements the Region of Interest (ROI) localization was performed by only focusing on an appropriate region detection. In a ROI extraction initially the area of disk is located by thresholding of the selected image and image with highest brightness with largest area is then selected. Then the ROI image processed for exact disk boundary detection. The process is fast as no classification is discovered and entire region is processed together. Once if the disk area is obtained, then the cup needs to localize within this boundary.

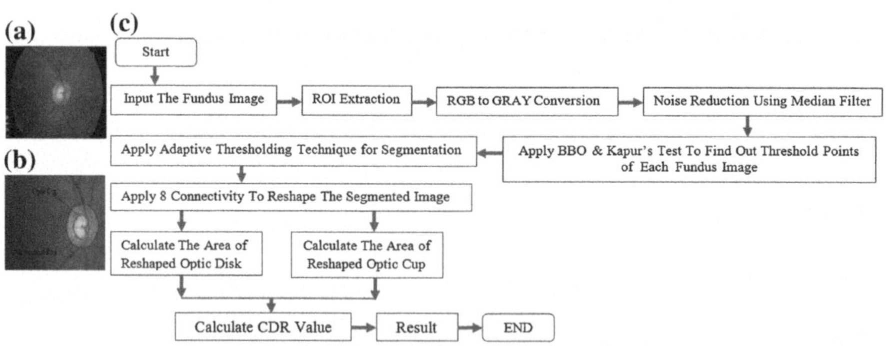

Fig. 1 **a** Structure of retinal image. **b** Optic disc and optic cup. **c** Flow chart of proposed method

2.2 *Image Intensity Conversion (RGB to GRAY)*

In image processing, images are represented as RGB color image. In the RGB images best contrast between is provided by the green channel between the blood vessels and background object while the red channel and blue channel may generate noisy result.

2.3 *Noise Reduction Using Median Filter*

Unwanted noise suppressed by Median Filtering which is added into the image. It is very much adequate to filter any outliers and is a preferable choice to select salt and pepper for removal noise.

2.4 *Threshold Points Calculation of Fundus Image Using BBO and Kapur's Test*

2.4.1 Kapur's Test

Kapur's entropy (Akay [3]; Bhandari et al. [4]) is announced as best threshold selection processes for multilevel thresholding. Multilevel thresholding of a binary image from grey-level ones finding an edge by which some pixel in the one region above the edge which is collection of one's and some are below the edge collection of zero's. An image can be described by L gray levels value. Multilevel thresholding is used by (1) in Fig. 2.

Entropy measure for the segmented histogram Kapur's Entropy [5] maximizes in such a way each separated region has more centralized distribution [6]. If an object is too much different from the background it has then the gray-level histogram is classified so that threshold value for segmentation process are chosen from the

$$\left.\begin{aligned} M_0 &= \{ g(x,y) \in I \mid 0 \le g(x,y) \le t_1 - 1 \} \\ M_1 &= \{ g(x,y) \in I \mid t1 \le g(x,y) \le t_1 - 1 \} \\ M_i &= \{ g(x,y) \in I \mid ti \le g(x,y) \le t_{i+1} - 1 \} \\ M_m &= \{ g(x,y) \in I \mid tm \le g(x,y) \le L - 1 \} \end{aligned}\right\} (1)$$

Where ti (i=1,...,m) is the ith threshold value, and the m is the number of thresholds

$$\left.\begin{aligned} H_0 &= \sum_{i=0}^{t1-1} \frac{Pi}{w0} \ln \frac{Pi}{w0}, \quad w_0 = \sum_{i=0}^{t1-1} Pi \\ H_1 &= \sum_{i=t1}^{t2-1} \frac{Pi}{w1} \ln \frac{Pi}{w1}, \quad w_1 = \sum_{i=t1}^{t2-1} Pi \\ H_j &= \sum_{i=t2}^{tj+1-1} \frac{Pi}{wj} \ln \frac{Pi}{wj}, \quad w_j = \sum_{i=tj}^{tj+1-1} Pi \\ H_m &= \sum_{i=tm}^{L-1} \frac{Pi}{wm} \ln \frac{Pi}{wm}, \quad w_m = \sum_{i=tm}^{L-1} Pi \end{aligned}\right\} (2)$$

The optimal multilevel thresholding problem is configured as an m-dimensional optimization problem

Fig. 2 Multilevel thresholding

bottom level. Later, Kapur's entropy (Akay [3]; Bhandari et al. [4]) concept reevaluate for extension as multilevel thresholding problem (Sathya and Kayalvizhi [7–9]) followed by (2) in Fig. 2 which is given above.

M-dimensional optimal thresholds are obtained by (3) which maximizes the objective function that solves the multilevel segmentation problem.

$$(t)^{*} = \arg\ \max\left(\sum_{i=0}^{m} Hi\right) \tag{3}$$

The aim of the proposed technique is to maximize the Kapur's entropy objective function using above equation.

2.4.2 Biography Based Optimization (BBO)

BBO algorithm generally is an evolutionary algorithm [10]. The emigration (μ_k) and immigration (λ_k) of each habitat are given in Fig. 3.

Where n is the habitant count, where as maximum number of habitants is defined by N. E is the maximum emigration rate where as maximum immigration rate is defined by I. The mutation is defined as follows:

$$m_n = M \times \left(1 - \frac{p_n}{p_{max}}\right) \tag{5}$$

where M is an initial value for mutation defined by user, P_n is the mutation probability of nth habitat, and $P_{max} = \arg\max\ (P_n)$. n = 1, 2, 3… (Fig. 4).

2.5 *Apply Adaptive Thresholding Technique for Segmentation*

Thresholding method for segmenting image can calculate by measuring pixel intensity value. It is a converter it converts an intensity image to a binary image.

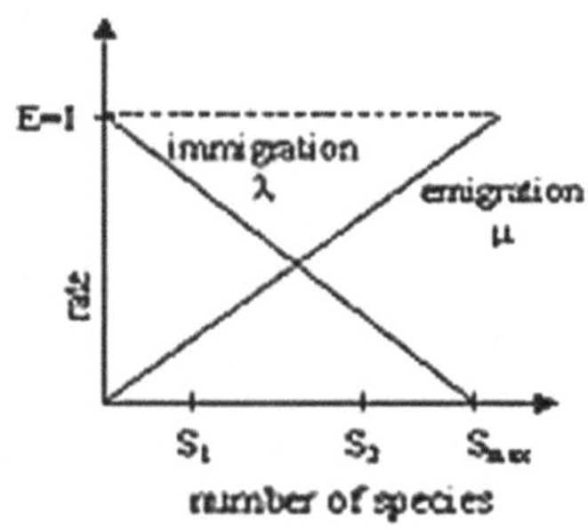

$$\mu_k = \frac{E \times n}{N} \qquad \lambda_k = I \times \frac{1-n}{N} \tag{4}$$

Fig. 3 Species model of a single habitat where S1 and S2 are good and bad solution respectively

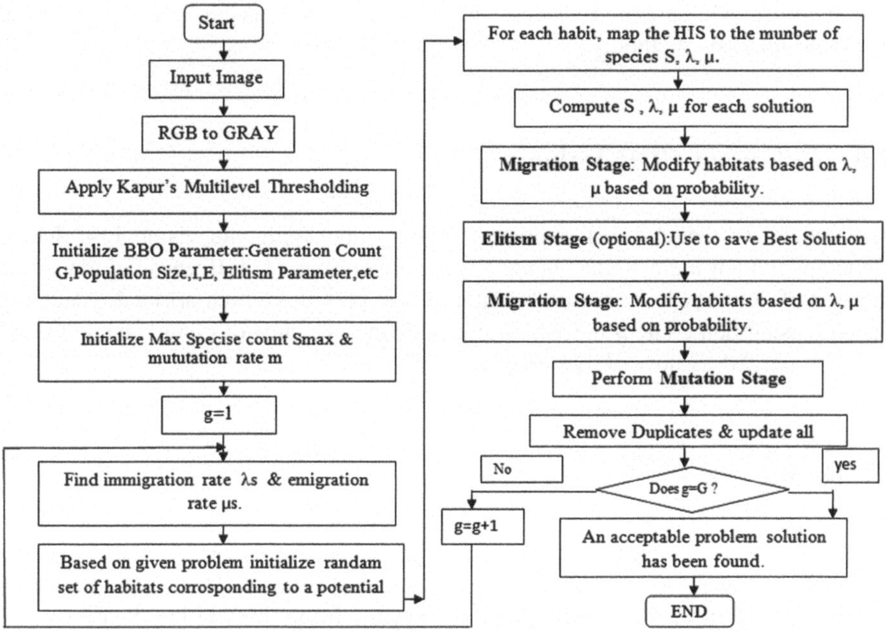

Fig. 4 Flowchart of biogeography based optimization algorithm

Here adaptive thresholding is used for histogram segmentation of an image. It finds the highest number of histogram region and tried to minimize the optimal value those are obtained from the histogram region by specific partition. Here we use Kapur's thresholding.

2.6 Apply 8 Connectivity to Reshape the Segmented Image

To get the accurate shape of cup and disk here we use 8 neighbor connectivity.

2.7 Calculate Area of Reshaped Optic Disk

The optic cup segmentation is much more difficult than optic disc. After reshaping the segmented image we can calculate the disk area near approximately.

2.8 Calculate Area of Reshaped Optic Cup

The disc and cup area could not easily be distinguishable. The component analysis method describes the location of the optic cup, although the image may exist with low contrast. This method detects the cup area more accurately than manual threshold analysis method.

2.9 *Calculate CDR Value*

The cup to disc ratio is an important feature identifying the vision loss due to glaucoma, which indicate the changes in the area of cup [11]. Because of glaucoma the area of cup slowly increased by internal eye pressure which arises dramatically vision loss. When CDR > 0.3 which indicates the presence of Glaucoma and CDR ≤ 0.3 is treated as normal eye image.

3 Result

In this paper we used MATLAB (R2014a) as an experimental tool. Here the RIMONE database is used as the source of retinal images. This method average accuracy is 97.58%. Sensitivity 94.44 and specificity 99.00% (Figs. 5 and 6).

Out of 290 images only 7 images are not classified correctly because of low quality, low contrast fundus image positive predictive value increases when Population with higher prevalence exists. Conversely, increasing prevalence results indicates decreasing negative predictive value. When we used predictive values of different tests, recognize the influence measure of the prevalence of disease. The figure below shows the relationship between disease prevalence and predictive value in a test with 99% Specificity and 94.44% sensitivity (Fig. 7).

$$\left.\begin{aligned}
\text{Sensitivity} &= \frac{TP}{TP+FN} \times 100 \\
\text{Specificity} &= \frac{TN}{TN+FP} \times 100 \\
\text{Accuracy} &= \frac{TP+TN}{TP+TN+FP+FN} \times 100 \\
\text{Precision} &= \frac{TP}{TP+FP} \times 100 \\
\text{F-score} &= 2 \times \frac{\text{Precision .Recall}}{\text{Precision + Recall}} \\
\text{Recall} &= \frac{TP}{TP+FN} \times 100 \\
\text{Positive Predictive Value} &= \frac{TP}{TP+FP} \\
\text{Negative Predictive Value} &= \frac{TN}{TN+FN} \\
\text{Prevalence of Disease} &= \frac{TP+FN}{\text{Total}}
\end{aligned}\right\} \quad (6)$$

TP= Glaucoma Effected Image / Abnormal Image
TN= Glaucoma Non-Effected Image/ Normal Image
FP= Glaucoma Non-Effected Image but Shows as Effected Image
FN= Glaucoma Effected Image but Shows as Non-Effected Image

TABLE 1

Total no of Images	Glaucoma Effected Image	Normal Image	Sensitivity	Accuracy	Specificity	Precision	Recall	Positive Predictive Value	Negative Predictive Value	Prevalence of Disease
290	85	198	94.44%	97.58%	99.00%	97.70%	94.44%	97.70%	97.53%	31.03%

Fig. 5 Result analysis

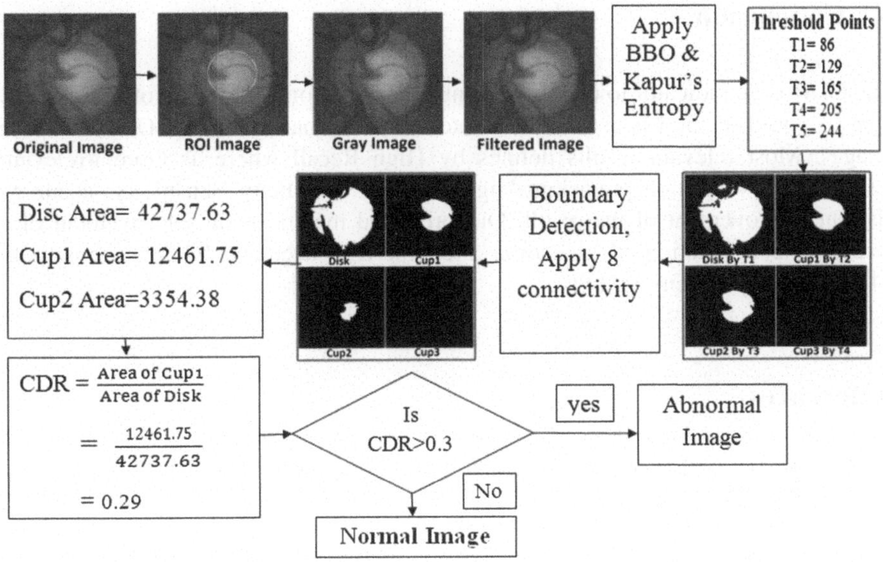

Fig. 6 Step by step evaluation process

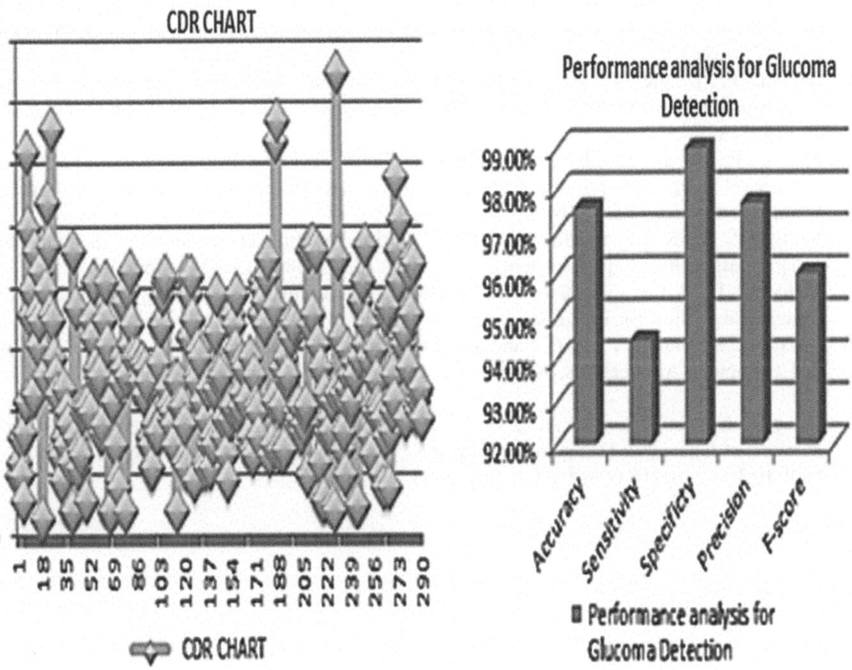

Fig. 7 CDR analysis and performance analysis

4 Conclusion

The CDR is an indicator to decide percentage of glaucoma effect ion of individuals. Our proposed method is used to calculate the performance of RIM-ONE database images. Most relevant results defines by High Recall where as more irrelevant results defines by high precision. High value of specificity, sensitivity, accuracy also an improvement of the result. Our proposed method is an improvement of a result for classification of glaucoma affection with accuracy 97.58%, sensitivity 94.44% and specificity 99.00%.

References

1. Lim, K. H., Registration of new blindness in Singapore for 1985–1995. Singap. Med. J. 39:104–106, 1998.
2. Wang, H., Hsu, W., Goh, K. G., and Lee, M. L. (2000). An effective approach to detect lesions in color retinal images. Proceedings of IEEE Conference on Computer Vision and Pattern recognition, Hilton Head Island, USA, pp. 181–186.
3. Akay, B. (2013). A study on particle swarm optimization and artificial bee colony algorithms for multilevel thresholding. Applied Soft Computing, 13(6), 3066–3091.
4. Bhandari, A. K., Singh, V. K., Kumar, A., & Singh, G. K. (2014a). multilevel thresholding using Kapur's entropy. Expert Systems with Applications, 41(7), 3538–3560.
5. Kapur, J. N., Sahoo, P. K., & Wong, A. K. C. (1985). A new method for gray-level picture thresholding using the entropy of the histogram. Computer Vision and Image Processing, 29(3), 273–285.
6. Sathya, P. D., & Kayalvizhi, R. (2010b). swarm intelligence algorithm for image segmentation using multilevel thresholding. Intelligent Learning Systems and Applications, 2, 126–138.
7. Sathya, P. D., & Kayalvizhi, R. (2010a). Tsallis entropy method based on Optimum multi-level image thresholding with bacterial for aging algorithm. International Journal of Computer Science, 7(5), 336–343.
8. Sathya, P. D., & Kayalvizhi, R. (2011a). Optimal multilevel thresholding using bacterial foraging algorithm. Expert Systems with Applications, 38(12), 15549–15564.
9. Sathya, P. D., & Kayalvizhi, R. (2011b). Based on multilevel thresholding Modified bacterial foraging algorithm for image segmentation. Applications of Expert Systems, 24, 595–615.
10. S. Gupta, K. Bhuchar, and P. Sandhu, "Color Image Segmentation Using BBO," Software and Computer Applications, Kathmandu, International Conference Nepal, pp. 79–86, July 2011.
11. Jagadish Nayak. Rajendra Acharya U. P. Subbanna Bhat. Nakkul Shetty., "Automated dtection of Glaucoma Using Fundus Images," Teik –Cheng Lim *J. Med Syst*, DOI 10.1007/s10916-008-9195-z, July 2008.

Quality Assessment of Medicinal Leaves Through Biospeckle Technique

Jitendra Dhanotia, Litesh Bopche, Vimal Bhatia and Shashi Prakash

Abstract In this paper, we report investigations undertaken towards visualization of microbiological objects in medicinal leaves using biospeckle activity approach. Both fresh and infected samples of Ficus Religiosa (peepal leaf) are illuminated by a laser source for generation of biospeckle pattern. Dynamic variation in activity images were grabbed by a Charge Couple Device (CCD) camera and quantized into 8 bit digital image using frame grabber. For image analysis revised version of conventional Fujii's method i.e. weighted parameterized Fujii method is used. It shows high contrast as compared to the other existing methods when sample is illuminated uniformly. The current study aims to distinguish between fresh and infected leaves for improvement of quality of Ayurvedic medicines.

1 Introduction

Leaves of medicinal plants have a great importance in Indian Ayurvedic pharmaceutical industry [1]. Mainly, juice and poultice of leaves are used for healing various diseases. Under certain environmental conditions, it can be affected by several microbiological organisms (fungi, bacteria, viruses, etc.) which not only reduce its medicinal attributes but also incorporate some harmful effects on human body when consumed. One of the most prevailing fungal plant syndromes is leaf spotting. Leaf spot syndromes are produced by many different types of fungi like

J. Dhanotia · V. Bhatia
Signals and Software Group, Discipline of Electrical Engineering,
Indian Institute of Technology Indore, Simrol, Madhya Pradesh 453446, India

L. Bopche · S. Prakash (✉)
Photonics Laboratory, Institute of Engineering & Technology,
Devi Ahilya University, Indore, Madhya Pradesh 452001, India
e-mail: sprakash_davv@rediffmail.com

© Springer Nature Singapore Pte Ltd. 2017

I. Bhattacharya et al. (eds.), *Advances in Optical Science and Engineering*,
Springer Proceedings in Physics 194, DOI 10.1007/978-981-10-3908-9_47

Marssonia, Septoria, etc. These comprise a large group of fungi, the affiliates of which are responsible for different diseases in human body. Powdery Mildew is another type of disease present in a leaf, for the plants raised in dry climates. General symptoms include a white or grayish powdery growth on leaves. These are also responsible for the occurrence of diseases in human body.

The biospeckle technique is an optical phenomenon based on scattering, when a biological sample is illuminated by laser light [2]. This scattered light interferes and generates a granular patterns consisting of random dark and bright areas on the observation plane. The spatial aberrance of speckle pattern is determined by using the characteristic of rough surface. When the phenomenon is used to observe biological samples, such type of pattern is called as biospeckle.

This phenomenon has been successfully used for the measurement of blood flow [3], activity and quality evaluation of fruits [4, 5], assessment of seed viability [6], drying process of non-biological sample (e.g. paint, fevicol) [7, 8], bruising in fruits [9], detection of fungi colony growth [10], etc. In the present study, leaf of Ficus Religiosa also known as 'peepal' is used as sample to demonstrate the experiment. Several leaf samples, both fresh and microbes contaminated are taken and their biospeckle activity is measured by weighted parameterized Fujii method for detection of the amount of infection in the infected leaf. Figure 1a shows original image of the peepal leaf, Fig. 1b, c shows the recorded speckle images of the fresh leaf and infected leaf, respectively.

The dynamic speckle technique is nondestructive, noncontact, and relatively new. The experimental setup used for this technique is very simple. The requirement of the component in the dynamic speckle technique is less so experimental cost of this setup is also less.

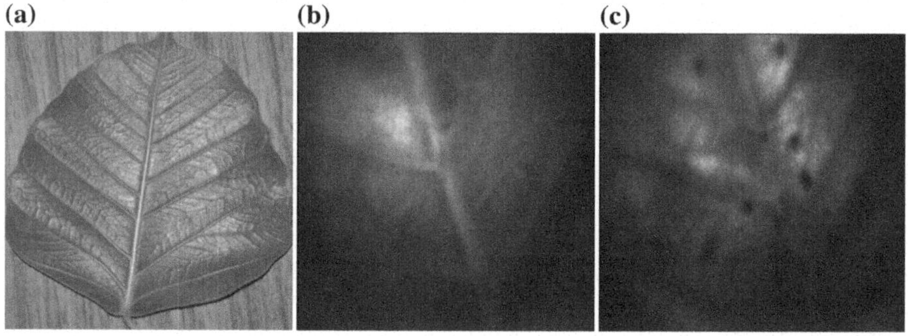

Fig. 1 **a** Original image. **b** Recorded speckle image of fresh leaf. **c** Recorded speckle image of infected leaf

2 Mathematical Equation

The dynamic speckle patterns of various kinds of specimen have been studied for evaluating their properties, quantitatively and qualitatively. First, the Fujii's method was used for the visualization purpose only. This method is based on the computations of the weighted sums of the absolute difference between consecutive image frames. It is defined by,

$$F(s) = \sum_{i=0}^{n-1} \frac{|x_i - x_{i+1}|}{(x_i + x_{i+1})} \tag{1}$$

where, F(s) is Fujii's function and n is the integer value from the image sequence s = [x0, x1 … x(n − 1)]. This method gives better results, when the light is uniformly illuminated on the sample. It shows negative effect in the results, when the light is not uniformly illuminated. For modification of results obtained by Fujii, an alternative approach known as weighted parameterized Fujii method was introduced. This method gives better results when light is non-uniformly illuminated. It is represented by,

$$F_{wp}(s, g_r) = \sum_{i=0}^{n-1} \frac{|x_i - x_{i+1}|}{(x_i + x_{i+1})}(255 - |g_r - x_i| + 255 - |g_r - x_{i+1}|) \tag{2}$$

where, $F_{wp}(s, g_r)$ is weighted parameterized function, n is integer values with image sequence s = [x0, x1 … x(n − 1)] and g_r is reference gray level. This method gives clearer image as compared to Fujii image.

3 Experimental Setup

The experimental setup used for quality assessment of medicinal leaves through biospeckle technique is shown in Fig. 2. In the experiment, we used four peepal leaves as a biological sample for biospeckle analysis. One leaf sample is fresh and other samples are infected by the microbiological organisms in different proportions, and sacrificed for laser envision immediately after its removal from the plant. Leaf surfaces were illuminated by a 15 mW He–Ne laser with an operating wavelength is 632.8 nm. Laser beam was passed through spatial filter arrangement and expanded to cover the leaf surfaces. The successive images were collected by using a high focused CCD camera with its pixel size is 4.65 μm × 4.65 μm. The

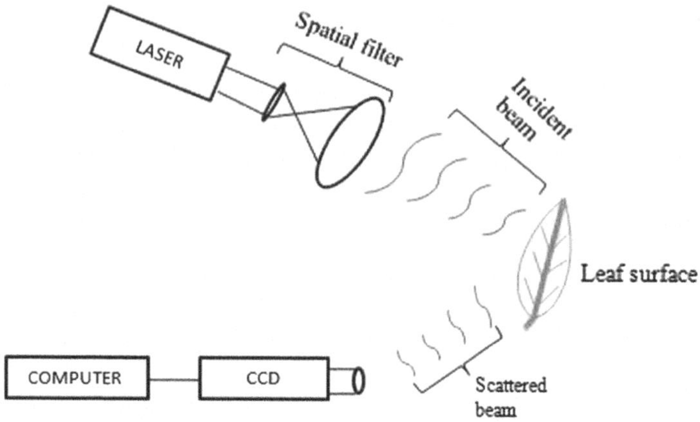

Fig. 2 Experimental setup used for quality assessment of medical leaf

angle of the laser source with the CCD camera was altered in order to retrieve high contrast images. In this technique, we analyses the recorded intensity images by various algorithm, the final result of this is also an image which shows the activity of biological sample.

4 Results and Discussion

We grabbed 50 frames of 512 × 512 pixels statistically independent laser speckle images with frame rate 20 fps. Captured images were quantized into 8 bit digital image using frame grabber card and processed by algorithm described in Sect. 2 using MATLAB image processing toolbox. In the spectral representation of biospeckle activity, red color depicts highest activity whereas blue shows lowest dynamicity. The results generated by weighted parameterized Fujii are shown in Fig. 3. The weighted sum characterized by the summation of the two consecutive images, emphasize small changes, which gives high contrast image as compared to original Fujii and alternative Fujii images. In the results, the activity areas and infected regions of the leaf are easily visible. Activity of fresh leaf is shown in Fig. 3a, which corresponds to fluid flow through inner leaf veins. Higher activity region of Fig. 3b shows presence of fungus on the sample. With the gradual passage of time, the nutrients in the leaf are consumed by the microbes. As a result, the infected region dried faster than the other regions, and the activity gradually spreads out throughout the area. This phenomenon is clearly visible in Fig. 3c, d.

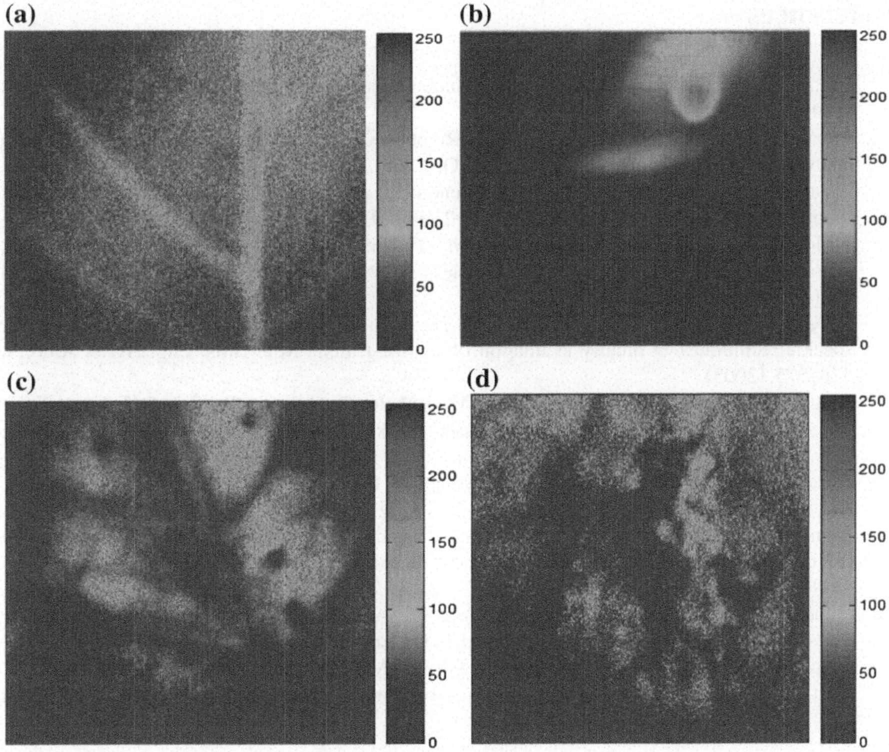

Fig. 3 Spectral activity maps of peepal leaf obtained using weighted parameterized Fujii method. **a** Activity map of fresh leaf. **b** Activity map of initially infected leaf. **c** and **d** Activity map when infected region dried gradually with respect to time

5 Conclusion

We have reported the application of intensity based technique of "weighted parameterized Fujii" for quality assessment of medicinal leaves. The technique of weighted parameterized Fujii has been found useful for qualitative measurement of speckle activities. This method has a variable contrast which can be altered with requirement by only changing reference gray level. Presence of fungus and its effect on leaves are visualized successfully using four set of samples. This makes biospeckle technique an efficient tool for sorting of quality leaves and boost up the quality of medicines manufactured by Ayurvedic pharmaceuticals.

References

1. Crellin, J. and Philpott, J., "A reference guide to medicinal plants," Duke University Press, (1990).
2. Rabal, H., Trivi, M., Arizaga, R., Cap, N., Romero, G. and Alanis, E., "Transient phenomena analysis using dynamic speckle patterns," Opt. Eng. 35(1), 57–62 (1996).
3. Aizu, Y. and Asakura, T., "Bio-speckle phenomena and their application to the evaluation of blood flow," Opt. Laser Technol. 23, 205–219 (1991).
4. Mulone, C., Budini, N., Vincitorio, F. M., Freyre, C. E., Lopez, A. J. and Ramil, A., "Biospeckle activity evolution of strawberries," SOP Transactions on Applied Physics 75–73 (2014).
5. Rabelo, G. F., Braga, R. A., Fabbro, I. M. D., Trivi, M., Rabal, H. and Arizaga, R., "Laser speckle techniques in quality evaluation of orange fruits," Rev. Bras. Eng. Agric. Amb. 9, 570–575 (2005).
6. Braga, R. A., Fabbro I. M. D., Borem, F. M., Rabelo, G., Arizaga, R., Rabal, H. J. and Trivi, M., "Assessment of seed viability by laser speckle techniques," Biosystems Eng. 86(3), 287–294 (2003).
7. Ansari, M.Z. and Nirala, A. K., "Assessment of fevicol drying process through dynamic speckle technique," Aims Bioengineering 2, 49–59 (2015).
8. Romero, G. and Rabal, H. J., "Statistics of the dynamic speckle produced by a rotating diffuser and its application to the assessment of paint drying," Opt. Eng. 39(6), 1652–1658 (2000).
9. Pajuelo, M., Baldwin, G., Rabal, H., Cap, N., Arizaga, R. and Trivi, M., "Bio-speckle assessment of bruising in fruits," Opt. and Laser Eng. 40, 13–24 (2003).
10. Vincitorio, F. M., Budini, N., Mulone, C., Spector, M., Freyre, C., Lopez, A. J. and Ramil, A., "Detection of fungi colony growth on bones by dynamic speckle," Proc. SPIE 8785 (2013).

Biospeckle Assessment of Bread Spoilage by Fungus Contamination Using Alternative Fujii Technique

Amit Chatterjee, Reena Disawal and Shashi Prakash

Abstract In present communication biospeckle phenomena has been used for assessment of spoilage of bread due to moisture level degradation and fungus colony growth. Bread sample is irradiated by a laser source and resultant boiling speckle patterns are recorded frame by frame through a charge-coupled device (CCD) camera. The three conditions, i.e. fresh bread, dried bread and its fungus contaminated variant are processed through Fujii's and alternative Fujii method. Alternative Fujii method demonstrates better contrast performance as compared to conventional Fujii's method because of modifications in its scaling factor. Obtained results conclusively establish biospeckle analysis as an efficient tool for bread spoilage detection.

1 Introduction

Bread [1], the oldest artificial food from the dawn of agriculture, is prepared from a mixture of flour, water, flavor and yeast in various proportions. This staple food has been popularized across the world due to its variety of taste and various health benefits. To satisfy consumer demand and production of bread in large scale, bakery industry is growing very rapidly. Two most common defects found in bread which causes bread spoilage are degradation of moisture level and mold. Decay in moisture level arises due to improper packaging or inappropriate storage condition and results in reduction in softness and taste. Moldiness [2], one of the most common microbiological defects found in bread industry appears due to contamination of bread by *Rhizopus* sp. and *Mucor* sp. fungi colony. Consumption of moldy bread may result in allergic reactions, respiratory problems or food poisoning. Moreover, being microscopic object, fungus could not be detected with naked eyes in early stage of development. Determination of bread spoilage period

A. Chatterjee · R. Disawal (✉) · S. Prakash
Photonics Laboratory, Department of E & I, Institute of Engineering and Technology, DAVV, Indore, Madhya Pradesh 452017, India
e-mail: reenadisawal27@rediffmail.com

© Springer Nature Singapore Pte Ltd. 2017
I. Bhattacharya et al. (eds.), *Advances in Optical Science and Engineering*, Springer Proceedings in Physics 194, DOI 10.1007/978-981-10-3908-9_48

and fungal growth can save both, the money and health for manufacturer and consumers, respectively.

Biospeckle analysis [3] is a technique based on scattering phenomena of light and used to measure activity of different samples for which illuminated pattern exhibit temporal variation. This discrepancy is particularly noticed in biological samples due to cell activity, internal fluid flow or other biological phenomena, etc. So far, freshness measurement of various fruits and vegetables (e.g. apple, pear, carrot, tomato, etc.) has been reported using biospeckle technique [4, 5]. Also it has been independently investigated by several research groups that fungi presents a dynamic speckle pattern [6] on different surfaces (e.g. bone, seed, fruits, etc.) when illuminated by a laser source. Aim of this paper is to test the veracity of biospeckle techniques for detection of spoilage due to drying and screening of fungus growth in bread. Fungi are expected to yield exacerbated activity with time which is manifested by the variation in the dynamic activity of the speckle field. Image analysis is performed by Fujii [7] and alternative Fujii method [8] using MATLAB simulation software.

2 Theory

Change in the activity dynamics of organic sample causes an alteration in intensity of the speckle pattern. For investigation of activity variation, a stack of equally time sliced images is processed using numerous image processing techniques. To calculate visual activity map Fujii's method, generalized difference (GD) algorithm and their variants are used. GD algorithm which results in summation of absolute value difference among all time sliced frames is applied mainly when object of interest is inhomogeneous in nature or the level of irradiation is non uniform. Although having a very high accuracy, this technique is not preferred due to its high computational complexity [order of $O(n^2)$]. The technique proposed by Fujii et al. [8] for processing the bank of uniformly illuminated biospeckle images is defined as a weighted summation of absolute value difference between consecutive frames and mathematically given as,

$$\mathrm{Fuji}(x, y) = \sum_{k=0}^{n-1} \frac{|I_k - I_{k+1}|}{I_k + I_{k+1}} \tag{1}$$

To improve contrast performance and efficiency under random irradiation Minz et al. introduced alternative formulation of original Fujii method, called alternative Fujii,

$$\mathrm{Fuji}_{alt}(x, y, m) = \sum_{k=0}^{n-1} \frac{|I_k - I_{k+1}|}{\sqrt[\pm m]{I_k + I_{k+1}}} \tag{2}$$

where, I is image frame, k is image sequence, n is stack size, m is integer and (x, y) represent image matrix co-ordinates. Single summation in (1) and (2) when implemented in programming environment as single loop, leads to a time complexity in the order of O(n). Consider a counterexample with high and low activity stacks are given as S1 = [7, 2, 9] and S2 = [4, 5]; then Fujii and alternative Fujii values (for m = 2) can be calculated as,

$$\mathrm{Fuji}_1 = \frac{|7-2|}{7+2} + \frac{|2-9|}{2+9} = 1.2 \quad \mathrm{Fuji}_{\mathrm{alt1}} = \frac{|7-2|}{\sqrt{7+2}} + \frac{|2-9|}{\sqrt{2+9}} = 3.78$$

$$\mathrm{Fuji}_2 = \frac{|4-4|}{4+4} + \frac{|4-5|}{4+5} = 0.11 \quad \mathrm{Fuji}_{\mathrm{alt2}} = \frac{|4-4|}{\sqrt{4+4}} + \frac{|4-5|}{\sqrt{4+5}} = 0.33$$

Notice that, $\mathrm{Fuji}_1 > \mathrm{Fuji}_2$ and $\mathrm{Fuji}_{\mathrm{alt1}} > \mathrm{Fuji}_{\mathrm{alt2}}$. This implies both the techniques can efficiently calculate biospeckle phenomena. Moreover, it is also observed that $\mathrm{Fuji}_{\mathrm{alt}} > \mathrm{Fuji}$; it models alternative Fujii as a better approach for visual inspection.

3 Experimental Arrangement

3.1 Sample Preparation

A packet of sliced bread was brought from local market. Six bread samples were selected and cut down in $6 \times 6\ \mathrm{cm}^2$ using a chopping board and knife so that it can be comparable with field of view of the optical set up. Chopped samples are then divided equally into two groups, say G1 and G2 (where G1 samples were kept in resealable plastic bags and G2 remains unpacked) and placed under room temperature. Though G1 samples were used for fungus detection, there is no need to add mold spores externally as millions of them are floating all around us which causes mold growth in bread. For detection of moisture level reduction, G2 samples were used.

3.2 Experimental Set Up

The experimental set up for dynamic laser imaging system for bread quality assessment is shown in Fig. 1. Bread surfaces were irradiated with 15 mW He–Ne laser with operating wavelength $\lambda = 632.8$ nm. To screen the sample, spatial filtering arrangement was adopted. A group of 60 time series images (having 8 bits/pixel) with a dimension of 700×700 pixels each was taken by a CCD camera. Grabbed and digitized images were processed in MATLAB simulation software using algorithms mentioned in Sect. 2.

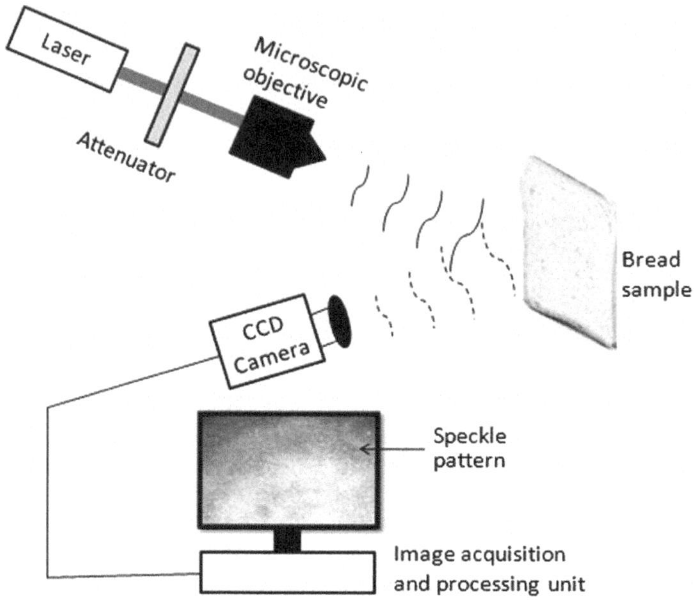

Fig. 1 Experimental schematics of biospeckle analysis for bread spoilage detection

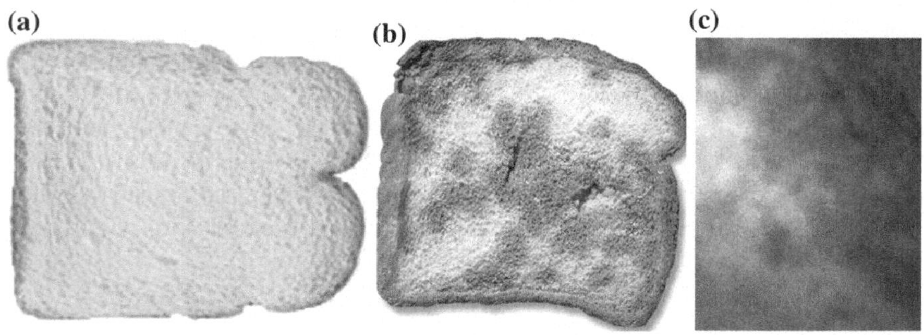

Fig. 2 Samples at different state **a** fresh, **b** fungus affected and **c** speckle pattern for bread surface

4 Results and Discussion

Fresh bread, moldy bread and its corresponding speckle pattern are shown in Fig. 2. Bread samples corresponding to different time are analyzed using visual methods i.e. Fujii and alternative Fujii analysis. Outcomes of image analysis are represented in pseudo-color with a color bar in the order: blue (lowest activity), green, yellow and red (highest activity).

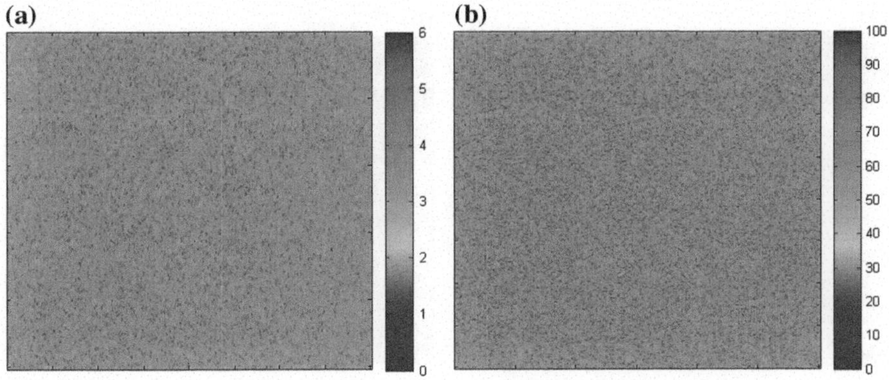

Fig. 3 Biospeckle activity of fresh bread by **a** Fujii's method, **b** alternative Fujii's method

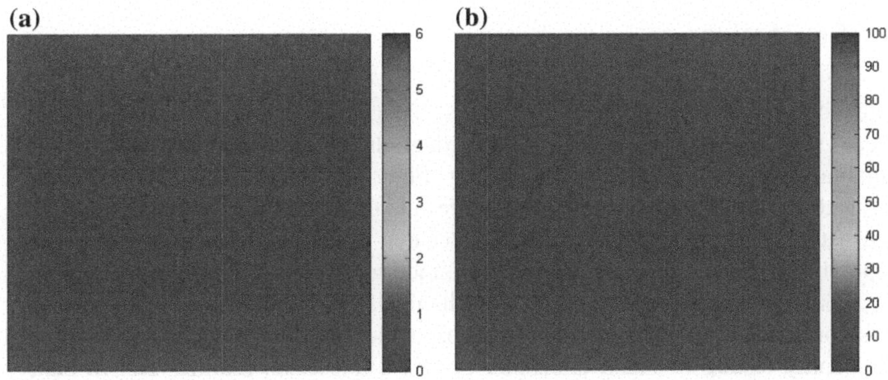

Fig. 4 Biospeckle activity of bread sample after 48 h, by **a** Fujii's method, **b** alternative Fujii's method

Figure 3 shows activity image corresponds to fresh bread sample. It is not only having high activity but also homogeneous activity. This can be good quality assessment indicator for bread industry.

G1 and G2 samples are assessed in the same time every day. It is found that after 48 h, humidity (softness) level of G2 samples reduced considerably as shown in Fig. 4.

For screening of fungus, G1 samples are analyzed further. It is observed that, after six days molds start appearing and become fully grown on tenth day. This is shown in Fig. 5. It is evident from color bar values that alternative Fujii method has noticeably higher contrast than Fujii's method.

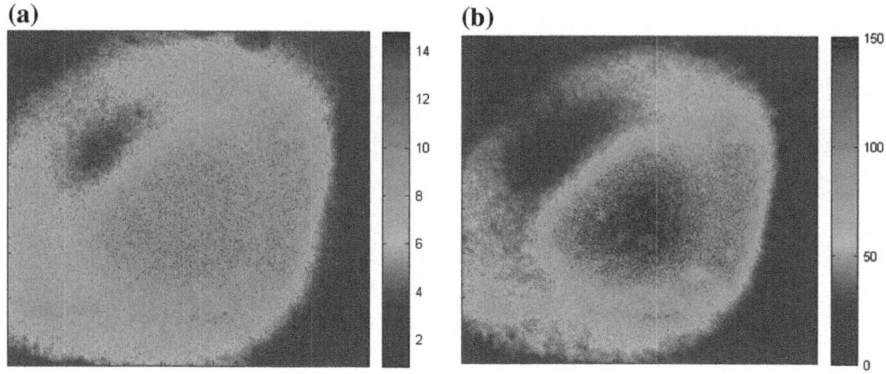

Fig. 5 Biospeckle activity of fungus affected bread sample by **a** Fujii's method, **b** alternative Fujii's method

5 Conclusion

In conclusion, we have successfully tested biospeckle technique for assessment of bread spoilage due to moisture level degradation and fungus contamination. Qualitative results are displayed using Fujii and alternative Fujii's method. Favorable correlation is achiever among results obtained by both the methods. The alternative Fujii method proves itself better due to its higher contrast and working capability when the level of illumination undergoes non uniformity.

This technique is easy, fast, non mechanical, non destructive and require very less components for implementation. Biospeckle analysis could be an useful tool for quality testing of bread and other products like cake, muffins, etc. in bakery industry.

References

1. Rubel, W., [Bread: A Global History], Reaktion Books Ltd., London (2011).
2. Pitt, J. I. and Hocking, A. D., [Fungi and Food Spoilage], Springer, US, 401–421 (1985).
3. Zdunek, A., Muravsky, L., Frankevych, L., and Konstankiewicz, K., "The biospeckle method for investigation of agricultural crops: A review," Opt. and Lasers in Engg. 52, 276–285 (2014).
4. Ansari, MDZ., Minz, P. D. and Nirala, A., "Fruit quality evaluation using biospeckle techniques," Proc. IEEE, RAIT (2012).
5. Ansari, MDZ. and Nirala, A., "Biospeckle activity measurement of Indian fruits using the methods of cross-correlation and inertia moments," Optik 124(15), 2180–2186 (2013).
6. Vincitorio, F. M., Budini, N., Mulone, C., Spector, M., Freyre, C., Díaz, A. J. L. and Ramil, A., "Detection of fungi colony growth on bones by dynamic speckle," Proc. SPIE 8785, 1–6 (2013).

7. Fujii, H., "Blood flow observed by time-varying laser speckle," Opt. Lett. 10(3), 104–106 (1985).
8. Minz, P. D. and Nirala, A., "Intensity based algorithm for biospeckle analysis," Optik 125(14), 3633–3636 (2014).

Evaluation of Aging Effect on Pea Seed Germination Using Generalized Difference Method

Amit Chatterjee, Reena Disawal and Shashi Prakash

Abstract In this work, we have applied biospeckle phenomena to study the effect of aging on pea seed germination. Dynamic speckle analysis is a non invasive technique for detection and evaluation of biological samples, when irradiated by a coherent source e.g. laser, SLD, etc. Seed samples with four different age groups were provided similar germination conditions for same time. Activity images for each group are recorded through charge-coupled device (CCD) camera and processed using generalized difference (GD) and alternative generalized difference (GD*). Obtained results conclusively establish biospeckle analysis as an efficient tool for evaluation of aging effect on seed germination.

1 Introduction

Seed aging [1, 2] referred to a process of gradual degradation in germination percentage as a result of numerous biochemical and physiological events. Maximum germination rate is achieved instantly after crop harvesting and declines with storage time. In case of seeds, aging is not only a function of time, but also moisture and temperature which reduces with uncontrolled storage environment causing considerable degradation in seed viability and vigor. Seed deterioration is a serious threat in developing and semi developing countries due to unavailability of controlled storage conditions. As quality of seed is directly related with seed aging, this effect can directly harm people associated with agricultural industry especially farmers.

Biospeckle analysis [3] is an extension of laser speckle technique that uses dynamic speckle pattern for assessment of activity in biological samples. To study the sample activity, various numerical and visual methods have been reported so

A. Chatterjee · R. Disawal (✉) · S. Prakash
Photonics Laboratory, Department of E & I, Institute of Engineering and Technology, DAVV, Indore, Madhya Pradesh 452017, India
e-mail: reenadisawal27@rediffmail.com

© Springer Nature Singapore Pte Ltd. 2017

I. Bhattacharya et al. (eds.), *Advances in Optical Science and Engineering*, Springer Proceedings in Physics 194, DOI 10.1007/978-981-10-3908-9_49

far, e.g. temporal history, co-occurrence matrix [4], absolute value of difference, Fujii's method [5], GD [6, 7], laser speckle contrast analysis [8], wavelet entropy [9] etc. For samples with different activity area (e.g. seed, animal tissue, leaf, etc.), visual methods are preferred over numerical methods due to its capability of better perception of biological activity. Aim of this paper is to show the effect of aging on its germination efficiency for *Pisum sativum* (pea seed) sample. Higher germination percentage results inflated physical and chemical movements which in turn increase its biospeckle activity. Image analysis is performed in MATLAB interface by GD and GD* method.

2 Theory

Biospeckle system produce sequence of light intensity patterns when the sample is irradiated by coherent source of light. Motion and flux of biomass generate spatiotemporal stochasticity between pixels of collected consecutive frames. This movement is able to provide useful knowledge regarding seed germination.

Visual image analysis includes evaluation of magnitude of pixel variation among dynamic image frames. Measurement techniques can be classified into two main categories: Fujii's method and GD method. Fujii's technique provides better result only when level of illumination to the sample is uniform. On the other hand, computational complexity of GD method is more, but produces highly accurate result irrespective of illumination condition. It perform absolute difference of values among all frames and mathematically given by,

$$\mathrm{GD(i,j)} = \sum_{k=1}^{N-1} \sum_{l=k+1}^{N} |I_k(x,y) - I_l(x,y)| \tag{1}$$

To improve computational efficiency of conventional GD method, concept of alternative GD (GD*) [5] was popularized and can be presented as,

$$\mathrm{GD^*(i,j)} = \sum_{k=1}^{N-1} \sum_{l=k+1}^{N} (I_k(x,y) - I_l(x,y))^2 \tag{2}$$

where, I is image, N is frame number, k and l are frame indices and (i, j) represent image matrix co-ordinates. Let high and low activity image sequences are represented with matrices $I_1 = [7, 1, 9]$ and $I_2 = [1, 1, 3]$; then we can calculate,

$$\mathrm{GD_1} = |7-1| + |1-9| + |7-9| = 16 \quad \mathrm{GD_1}* = (7-1)^2 + (1-9)^2 + (7-9)^2 = 104$$
$$\mathrm{GD_2} = |1-1| + |1-3| + |3-1| = 04 \quad \mathrm{GD_2}* = (1-1)^2 + (1-3)^2 + (3-1)^2 = 08$$

From the above calculation, $GD_1 > GD_2$ and $GD_1^* > GD_2^*$; GD* always have a higher value than GD. This implies, with GD* desired contrast level is achieved by processing lesser number of frames as compared to GD. This method has both reduced time complexity and better contrast characteristics.

3 Experimental Arrangement

3.1 Sample Preparation

Pea seeds of four different age group, (a) 2–3 months, (b) 12–13 months (c) 23–24 months and (d) 35–36 months after harvesting were collected from local seed store. All seeds were partially dipped in water and kept in room temperature for 48 h so that three essential conditions for seed germination i.e. air, water and heat can be provided to all seeds equally. One seed is randomly picked from each set and exposed under speckle analysis set up as discussed in Sect. 3.2.

3.2 Experimental Set Up

The experimental set up for boiling speckle analysis is shown in Fig. 1. A 20 mW green diode laser ($\lambda = 532$ nm), source of coherent illumination was passed through a beam expander to increase field of view. Expanded beam, without any collimating lens then exposed to the seed samples directly. A CCD camera is used to record 50 consecutive speckle images of 400 × 400 pixels each. Images are stored in host computer after digitizing to 8 bits by a frame grabber card and processed by image processing unit.

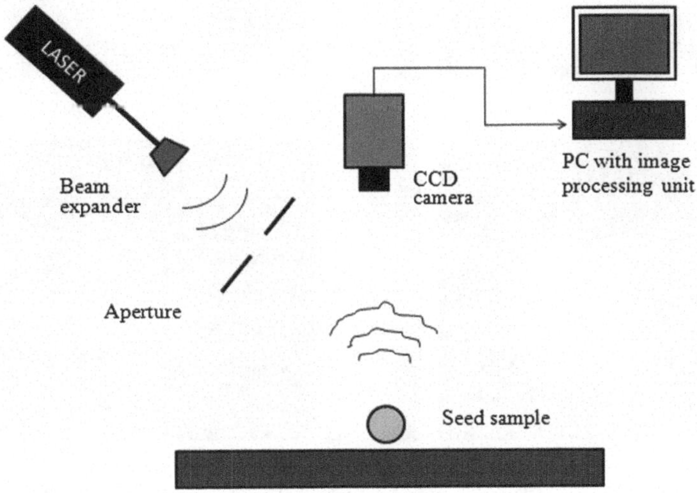

Fig. 1 Experimental schematics for biospeckle analysis

4 Results and Discussion

Figure 2 shows pea seed after 48 h germination treatment and its corresponding speckle pattern. An aperture is used between expanded laser beam and seed sample to focus on a particular portion of germinating seed.

Results obtained for seeds of four different age groups using GD and GD* method are shown in Figs. 3 and 4 respectively.

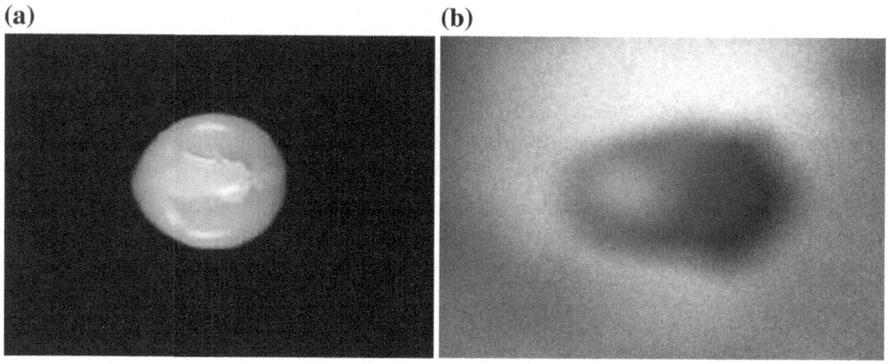

Fig. 2 **a** Arbitrary pea seed after application of germination treatment, **b** speckle pattern of (**a**)

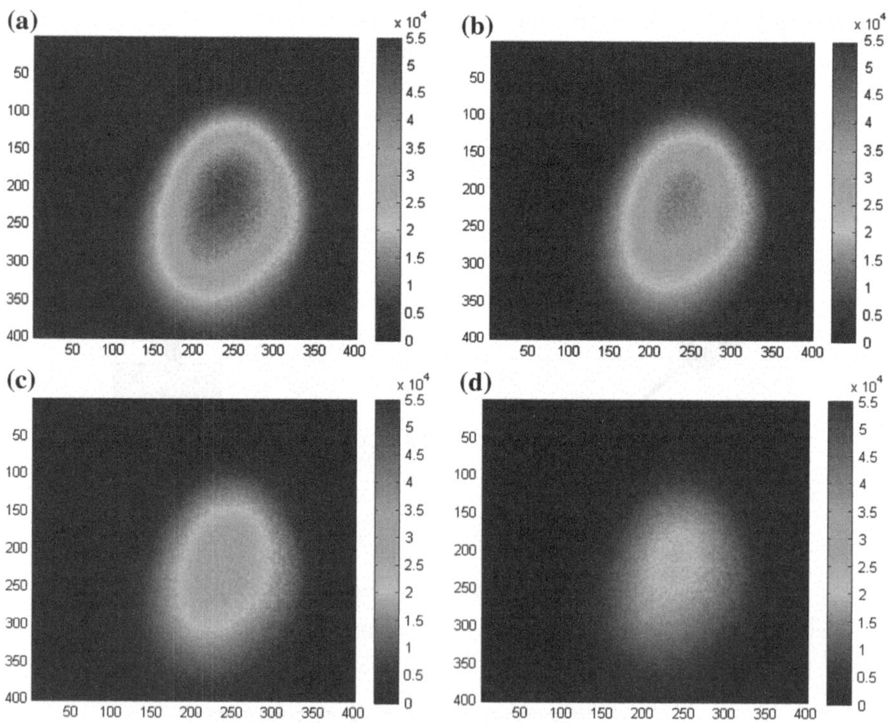

Fig. 3 Visual results using GD for **a** t = 2–3 months, **b** 12–13 months **c** 23–24 months and **d** 35–36 months

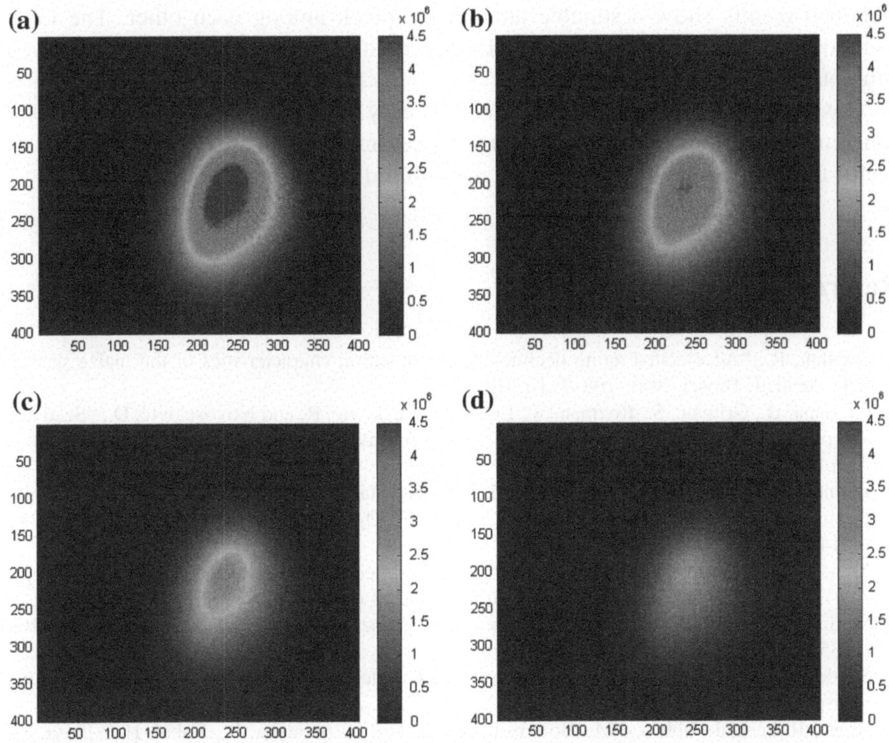

Fig. 4 Results obtained using GD* for **a** t = 2–3 months, **b** 12–13 months **c** 23–24 months and **d** 35–36 months

Outcomes are shown in false color format; intensity of sample activity is presented by red, yellow, green and blue colours respectively where red represents highest activity and blue lowest. Remarkably, it is noticed from these images, with aging cellular activity in the seed sample decays. This decay in turn leads to delayed germination and seed quality degradation with time. Furthermore, from the color bars it is clear that GD* method has remarkably higher contrast than GD method. It can be shown that, due to presence of squared term GD* can be directly correlated with variance. Although it sometime leads to temporal information loss, GD* proves itself better in two occasions: when high contrast is required and when results are to be produced with lesser number of images.

5 Conclusion

In this work, biospeckle technique was tested to study the effect of aging in germination for *pisum sativum* seeds. Qualitative assessments are displayed using generalized difference (GD) and alternative generalized difference (GD*) method.

Obtained results show a suitable amount of match among each other. The GD* method is better as compared to GD due to its higher contrast and improved computational time.

This technique is simple, non destructive, easy to implement and requires only a coherent source, a detector and an image processing unit for analysis. It could be an useful tool for evaluation of different agricultural samples and phenomena.

References

1. Kapilan, R., "Accelerated aging declines the germination characteristics of the maize seeds," Sch. Acad. J. Biosci. 3(8), 708–711 (2015).
2. Gordana, B., Grljušić, S., Rozman, V., Lukic, D., Lackovic, R. and Novoselović, D., "Seed age and ph of water solution effects on field pea (pisum sativum l.) Germination," Notul. Bot. Horti Agro. Cluj. 35(1), 20–26 (2007).
3. Zdunek, A., Muravsky, L., Frankevych, L., and Konstankiewicz, K., "The biospeckle method for investigation of agricultural crops: A review," Opt. and Lasers in Engg. 52, 276–285 (2014).
4. Men, S., Yan, L. and Qian, H.,, "Application of Bio-speckle Activity to Assess Seed Viability", Adv. Journ. of Food Sci. and Technol. 8(3), 214–218 (2015).
5. Fujii, H., "Blood flow observed by time-varying laser speckle," Opt. Lett. 10(3), 104–106 (1985).
6. Saúde, A.V., de Menezes, F. S., Freitas, P. L, Rabelo, G.F. and Braga RA Jr, "Alternative measures for biospeckle image analysis," J. Opt. Soc. Am. A 29 (8), 1648–1658 (2012).
7. Braga, R.A., Dal Fabbro, I.M., Borem, F.M., Rabelo, G., Arizaga, R., Rabal, H.J., Trivi, M. Assessment of seed viability by laser speckle techniques. Biosyst. Eng. 2003, 86(3), 287–294.
8. Briers, J.D. and Webster, S., "Laser speckle contrast analysis: a nonscanning, full-field technique for monitoring capillary blood flow," J. Biomed. Opt. 1(2), 174–179 (1996).
9. Braga Jr., R. A., Horgan, G. W., Emes, A. M., Miron, D., Rabelo, G. F. and Filho, J. B., "Biological feature isolation by wavelets in biospeckle laser images," Comp. and elect. in agriculture 58, 123–132 (2007).

Part X
Lasers, Quantum Optics and Information Technology

Electronic Band Structure of Quantum Cascade Laser

Ritabrata Chakraborty, Arpan Deyasi, Arkadeep Paul and Shrabani Nayak

Abstract Position of miniband and its separation w.r.t lowest energy band of a quantum cascade laser are analytically computed for different biasing conditions. Position-dependent effective mass following BenDaniel duke boundary conditions are considered in this simulation, which speaks in favour of realistic consideration. Findings are compared with that obtained under zero bias condition, and variation of eigenenergy of the lowest quantum state is computed as a function of electric field (applied along quantized direction). Length of the injector region is shown as a function of external bias. Result gives key information on electronic band structure of quantum cascade laser.

1 Introduction

Multiple quantum well structures are subject to theoretical and experimental research in last two decades because of the resonant tunnelling phenomenon [1], and novel photonic devices are already designed as transmitter [2] and receiver applicable in nanoelectronic domain. By virtue of quantum engineering, electronic energy states are tuned by suitable application of external excitation in this complex multilayered structure as per the requirement. Researchers have investigated the physics of resonant tunneling following the works of Esaki and Tsu [3], and series of theoretical [4] and experimental works [5] are reported thereafter. Theoretical

R. Chakraborty (✉) · A. Deyasi · A. Paul · S. Nayak
Department of Electronics and Communication Engineering,
RCC Institute of Information Technology, Kolkata, West Bengal 700015, India
e-mail: ritobroto.bms@gmil.com

A. Deyasi
e-mail: deyasi_arpan@yahoo.co.in

A. Paul
e-mail: arkadeep.paul@gmail.com

S. Nayak
e-mail: shrabani.communication@gmail.com

© Springer Nature Singapore Pte Ltd. 2017
I. Bhattacharya et al. (eds.), *Advances in Optical Science and Engineering*,
Springer Proceedings in Physics 194, DOI 10.1007/978-981-10-3908-9_50

researches are well supported by the advanced fabrication technologies [6], which allow tailoring the electronic and optoelectronic properties of quantum heterostructures. One such example is the THz laser design with low-dimensional semiconductor structures [7] with multiple layers, where miniband formation and its energy difference with lowest quantum energy state (lowermost energy band) plays crucial factor in governing the device performance [8] as optical transmitter. This is one typical semiconductor laser with room temperature operation at IR range, with good peak output power and CW mode of operation [9]. Operation of this unipolar device is based on quantum tunneling and intraband transitions; and formation of miniband critically depends on layer dimensions, their compositions and periodicity [10]. The layer thickness is essentially responsible for determining the wavelength of emitted radiation, as compared to the other semiconductor based lasers where bandgap of the material determines the wavelength. Thus accurate determination of electronic band structure of quantum cascade laser is very important for theoretical researchers from the point of view of optoelectronic application.

In the present paper, band structure of quantum cascade laser is computed in presence of moderate and high electric field, and result is compared with zero field condition. Eigenenergy is determined for a range of bias. Comparative study is carried out with the energy obtained for nonparabolic dispersion relation.

2 Mathematical Modeling

With effective mass approximation, time-independent Schrödinger equation for electron wavefunction ψ is given by

$$-\frac{\hbar^2}{2}\frac{d}{dz}\left[\frac{1}{m^*(E,z)}\frac{d}{dz}\psi(z)\right]+V(z)\psi(z)-q\xi(z)\cdot z\cdot\psi(z)=E(z)\psi(z) \tag{1}$$

where $\zeta(z)$ is the electric field applied along the direction of quantum confinement. Introducing finite difference technique, (1) can be modified as

$$\begin{aligned}-\frac{\hbar^2}{2}\left[\frac{1}{m^*(E,z)}\right]\left[\frac{\psi(z+1)+\psi(z-1)-2\psi(z)}{(\Delta z)^2}\right]\\+V(z)\psi(z)-q\xi(z)\cdot z\cdot\psi(z)=E(z)\psi(z)\end{aligned} \tag{2}$$

In order to introduce BenDaniel Duke boundary condition, we consider

$$\alpha=-\frac{\hbar^2}{(dz)^2[m^*(z)+m^*(z-dz)]} \tag{3}$$

$$\beta = -\frac{\hbar^2}{(dz)^2[m^*(z) + m^*(z+dz)]} \tag{4}$$

We introduce diagonal and off-diagonal terms of the Hamiltonian matrix as

$$t_d = -\frac{\hbar^2}{(dz)^2[m^*(z) + m^*(z-dz)]} + V(z) - q\xi(z) \tag{5}$$

$$t_{off} = -\frac{\hbar^2}{(dz)^2[m^*(z) + m^*(z+dz)]} \tag{6}$$

where effective mass is taken as position dependent. After organizing the elements of the Hamiltonian, we use the equation

$$H\psi = E\psi \tag{7}$$

Solution of this equation gives the eigenstates of the structure with appropriate boundary conditions.

3 Results and Discussions

Figure 1 depicts the electronic band structure in absence and presence of external electric field in the quantum cascade laser. It may be mentioned in this context that layer dimensions is chosen in such a way that injector and active region separations may clearly be distinguished from the band diagram. The figure in left shows for zero bias, and figure in right exhibits for very high electric field (56.8 × 10^5 V/m). From the plot, it is seen that the energy states are discrete (as expected due to

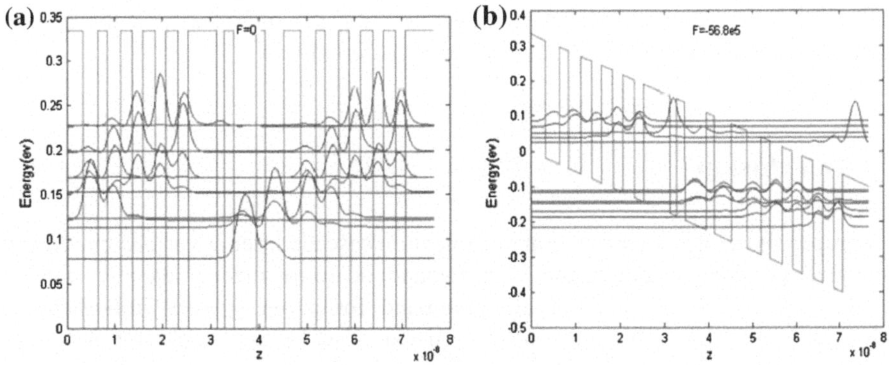

Fig. 1 **a** Band structure of QCL at unbiased condition. **b** Band structure of QCL at high bias

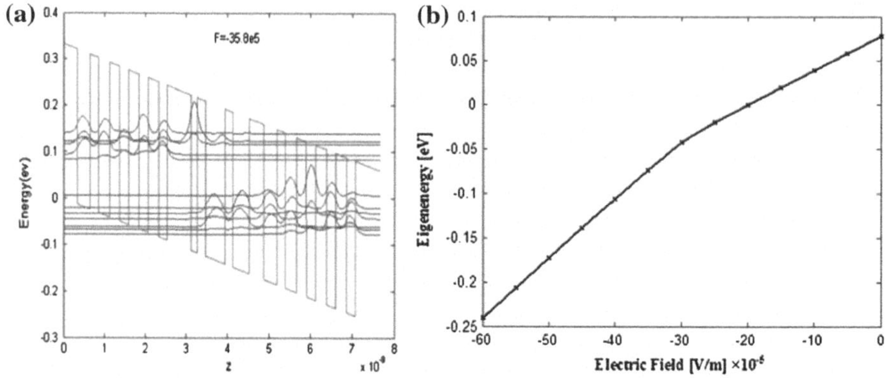

Fig. 2 **a** Band structure of QCL at moderate bias. **b** Eigenenergy variation of QCL with electric field

quantum confinement) in nature when field is not applied, which signifies the eigenstate of an otherwise multiple quantum well (MQW) structure. When strong electric field is applied, miniband is formed over the ground state energy band. It may be noted from the plot that after the injector region, the wavefunction starts to grow in the first miniband, which is the active region. Also due to the field, the structure is tilted, and thus the periodic growth of wavefuncion in the miniband appears outside the confinement region, i.e., in the quasi-continuous region. This modulation is absent if the field is moderate (35.8×10^5 V/m), which is represented in Fig. 2a (left). But moderate magnitude of electric field leads to the growth of wavefunction in the ground quantum state of the energy band. In this case, position of the miniband is found as almost inside the confinement region, i.e., miniband position is below the quasi-continuous region.

The right side of Fig. 2 shows the eigenenergy variation with the applied bias for the laser. The plot suggests that applied field is monotonically decreased with the increase of bias and the rate of increment increases when electric field is over 35×10^5 V/m. Thus stimulated emission between the miniband and the ground state energy band is effectively controlled by external bias.

4 Conclusion

The paper shows the wavefunction in a quantum cascade laser at various biased and unbiased conditions. Formation of miniband at some precise electric field is established. Result reveals that the miniband formation is very important for stimulated emission. The method showed can also be applicable for the other structures. Key factor in this calculation is that position-dependent effective mass is considered for the simulation, which is key for verification with experimental findings.

References

1. Davis, Kubis. T., Mehrotra. S. R., Klimeck. G., "Design concepts of terahertz quantum cascade lasers: Proposal for terahertz laser efficiency improvements", Applied Physics Letters, vol. 97, p. 261106 (2010).
2. Cooper. J. D., Valavanis. A., Ikonic. Z., Harrison. P., Cunningham. J. E., "Finite difference method for solving the Schrödinger equation with band nonparabolicity in mid-infrared quantum cascade lasers", Journal of Applied Physics, vol. 108, p. 113109 (2010).
3. Esaki. L, Tsu. R., "Superlattice and Negative Differential Conductivity in Semiconductors", IBM Journal of Research and Development, vol. 14(1), pp. 61–65 (1970).
4. Almansour. S. A., Hassen. D., "Theoretical Study of Electronic Transmission in Resonant Tunnelling Diodes based on GaAs/AlGaAs Double Barriers under Bias Voltage", Optics and Photonics Journal, vol. 4, pp. 39–45 (2014).
5. Toivonen. M., Jalonen. M., Pessa. M., Lefebvre. K. R., Anderson. N. G., "Experimental and theoretical studies of multi-quantum well structures for unipolar avalanche multiplication", Materials Science and Engineering: B, vol. 21(2–3), pp. 237–240 (1993).
6. Jogi. J., Verma. N., Gupta. M., Gupta. R. S., "Quantum Modeling of Electron Confinement in Double Triangular Quantum Well formed in Nanoscale Symmetric Double-Gate InAlAs/InGaAs/InP HEMT", International Semiconductor Device Research Symposium, pp. 1–2 (2011).
7. Razavipour. S. G., Dupont. E., Chan. C. W. I., Xu. C., Wasilewski. Z. R., Laframboise. S. R., Hu. Q., Ban. D., "A high carrier injection terahertz quantum cascade laser based on indirectly pumped scheme", Applied Physics Letters, vol. 104, p. 041111 (2014).
8. Hayata. H., Koshiba. M., Nakamura. K., Shimizu. A., "Eigenstate calculation of quantum well structures using finite elements", Electronics Letters, vol. 24, p. 614, 1988.
9. Bugajski. M., Kosiel. K., Szerling. A., Karbownik. P., Pierscinski. K., Pierscinska. D., Haldas. G., Kolek. A., "High performance GaAs/AlGaAs quantum cascade lasers: optimization of electrical and thermal properties", Proc. of SPIE, vol. 8432, p. 84320I (2012).

Effect of Material Parameters on the Attenuation and Amplification of an Incident Laser Beam

Rahul Basu

Abstract Laser processing is being increasingly used for treatment and processing of various materials. Etching, computer board manufacture, rapid prototyping and NC Machine tool technology have all adapted to laser use. Laser heating supplements traditional methods like nitriding and carburizing. The type of laser and its power output can vary significantly along with the depth of affected material. The moving heat source in laser melting is modeled by transformations together with a decoupling for the heat and mass transfer terms. The scale for heat diffusion is different from conduction and when small times are involved as in rapid solidification, the effect may be pronounced. Very little published work has appeared on the stability of the solid-liquid interface. A few solutions are known for certain geometries for the moving heat source. Approximate solutions incorporating the convective surface flux are obtained. It is shown that under certain conditions, a high Stefan number can attenuate an impinging laser beam and sustain thermal oscillations in the substrate. Application to thin films and amorphous material formation give criteria derived for stability in terms of surface parameters. The analysis does not include quantum effects likely in the nano region. Additionally, the surface reflectivity would influence the attenuations of the incident beam.

1 Introduction

Moving heat sources of laser melting are treated with various transformations along with a decoupling for the heat and mass transfer terms. The scale for heat diffusion is different from that of conduction and when small intervals of time are involved as in rapid solidification, the effect may be pronounced. Very little published work

R. Basu (✉)
Adarsha Institute of Technology, Off International Airport Road, Kundana, Bangalore, Karnataka 562110, India
e-mail: raulbasu@gmail.com

© Springer Nature Singapore Pte Ltd. 2017
I. Bhattacharya et al. (eds.), *Advances in Optical Science and Engineering*, Springer Proceedings in Physics 194, DOI 10.1007/978-981-10-3908-9_51

Table 1 Important thermal and non dimensional parameters for some engineering materials

Metal	Cp (J/kg K)	θ	L (kJ)	Ste	1/Ste (μ)	α
Cast iron	544	1810	272	0.33	3.03	0.134
Ni	456	1728	297	0.45	2.25	0.155
Ag	235	1235	111	0.49	1.11	1.7
Al	904	933	398	0.67	1.5	0.86
Cu	385	1358	205	0.49	2	1.14

exists on the stability of the solid-liquid interface. It is essential to understand this because instabilities in solidification processes will affect the final finish of the surface. Novel transforms specific to melting with convective surface conditions are applied to this analysis. Phase field theory notions are applied to specific cases for amorphous film transformation. The effect of material parameters to attenuate or even amplify input frequencies to nano or tetra levels is outlined.

1.1 Theoretical Considerations and Review of Past Work

Laser surface modification is through a phase change. More recently, laser processing has found application in computer controlled machine tools and Rapid prototyping, 3D printing and 3D sintering. 3D printing was described in USP 5398193 [1]. Sahoo et al. [2] describes pulsed laser treatment for TiC coating on Al. The depth of affected materials can range from nanometers to millimeters depending on types and power outputs (Table 1). Laser processing has been used to form p-n junctions and semiconductor annealing. The application of a pulsed laser gives interesting possibilities for further experimental verification due to the precision involved in focusing and controlling the intensity and time. The subsequent theoretical considerations assume that heat is not dissipated either by ablation or pore formation. Apart from the heat input, the frequency of the input laser also may affect the results. The properties of the underlying material attenuate the pulse, and it is subsequently shown that in certain cases the pulse can be sustained or even amplified.

The problem of phase change with a moving interface has been described by Carslaw and Jaeger [3], Stefan [4], Ingersoll et al. [5], as well as Crank [6], among others. A comprehensive study by Langer [7] appeared dealing with many core issues that occurred during solidification and growth of solid phases where thermal and concentration effects were combined. Many important issues are addressed with the linearization used by Mullins and Sekerka [8]. In the present work, the coupled heat diffusion and mass equations are also solved using a decoupling technique (Table 2).

Table 2 Characteristics of some laser types

Laser	Wavelength	Pulse length	$(Dt)^{0.5}$ Si	$(Dt)^{0.5}$ GaAs	Optical depth Si	Optical depth GaAs
XeCl excimer	308	30	1660	973	6.8	12.8
KrF excimer	248	30	1660	973	5.5	4.8
ArF excimer	192	30	1660	973	5.6	10.8
Nd; Y AG	1060	6	743	435	1000	N/A

Source Pulsed laser heating and melting, Sands [10]

2 Theoretical Considerations and Stability

The linearization $T = T_0 + Gz$ for the solubility phase boundary, and the similarity transformation, moving with the interface at velocity V give for a dimensionless "diffusion field" u: $\nabla^2 U + 2l\delta u/\delta z = 0$ with a thermal parameter $u = \Delta T/(L/cp)$ and the corresponding chemical potential $u = \mu/\Delta c\ (\delta u/\delta c)$, where $l = 2D/V$ (diffusive length) and L = latent heat, c_p the specific heat. The diffusion equations for both heat and mass are similar. A similar linearization was used by Davis and Schulze [9] for studying morphological instability for small Stefan numbers. Li and Beckermann [10] related growth to the critical wave number found by Mullins and Sekerka [8]. Pulsed laser heating used for surface modification was described in detail by Sands [10]. Ragas-Trigos and Calderon [11] have analysed the effects of a periodic heat source with sinusoidal modulation along with various boundary conditions by Greens function method. Kim et al. [12] have given predictions of solidification velocity in ion implanted GaAs (Gallium Arsenide). Laser spot welding with enthalpy method analysis has been described in Duggen et al. [13]. A 3-D finite element analysis for laser melting has been described by Contuzzi et al. [14]. Following Carslaw and Jaeger [3], the point source is practically modelled as a rectangular, disc shaped or strip source, for which solutions are available. All these solutions are of the form $\exp(-x^2)/(\alpha t)$, with α being diffusivity. Specific solutions for a moving source are also given in Paterson [15]. For metals, the transients are damped out very quickly, corresponding to the Fourier number $2\pi/k$. So for one wavelength, the attenuation goes as $\exp(-2\pi) = 0.0019$. Hence the solid of more than one wavelength depth may be regarded as infinite for all practical purposes. Now consider a 1-dimensional geometry for simplicity. This could also correspond to spherical symmetry after applying certain simple transformations. Applying a similarity transformation to the thermal diffusion equation with re-melting terms, results in (after suitable normalized variables are applied);

$$T'' + \eta/2T' + (\varepsilon L/\rho c_p)dc/dt = 0, \quad (1)$$

T is temperature, η similarity parameter, α diffusivity, ε porosity, c concentration, t time, ρ Density, ∇ nabla.

The transformation of the concentration derivative to non dimensional similarity parameter η, a characteristic length may be extracted from the Fourier number after knowing the characteristic time related to the frequency of the pulsed laser source. By examination of the coefficients of the equation and introducing a small parameter $\mu = g\theta c_p/(\varepsilon L) = g/(\varepsilon Ste)$, one obtains

$$\mu\theta'' + 0.5\,\mu Fo\theta' + 0.5\,Bi\theta = 0, \tag{2}$$

(where g is the thermal gradient, Ste Stefan number, Bi Biot number, and θ non dim temperature). The above equation can be solved exactly, with the inner and outer solutions easily obtained by letting $\mu \geq 0$ and $Bi \geq 0$ respectively, (assuming $\mu Fo > 0$). Accordingly, $\theta'/\theta = -(BiSte/\varepsilon)(gFo)$, $\theta''/\theta = -Fo/2$ which have immediate solutions. For small μ, the Eq. (2) can be written without the second derivative terms, thus becoming one dimensional in θ. On the other hand, if Bi is very small, the reverse holds true, and the higher terms appear. If F_0 is very small a third option is available leading to hyperbolic or sinusoidal solutions depending on the sign of Bi. The equations can still be solved exactly using a similarity transformation. The small parameter μ is calculated for some materials in Table 1. The parameter $Fo^2(gK)/(\varepsilon Ste\eta)$ can be recognized as $(Fo^2/Bi)Cp\theta$, and can be extracted from columns 1 and 2 of Table 1. (Note: this is effectively a modification of the Fourier length given by Carslaw and Jaeger [3], through the Stefan number due to the interface effect. It is seen to be a scaling through an effect on time.)

2.1 Computational Results

Two time scales which can be introduced are due to Fourier conduction and phase front motion. The scales are quite different in most cases, especially if the freezing is slow. In effect, the temperature sensed due to heat conduction of the normal boundary value problem has a superimposition of a "wave" of temperature caused by the motion of the phase front. The Dirichlet solution is modified for convective conditions by the approximations:

$$h(T_S - T_0) = kdT/dx \tag{3}$$

where

$$dT/dx = (T_S - T_m)/l \tag{4}$$

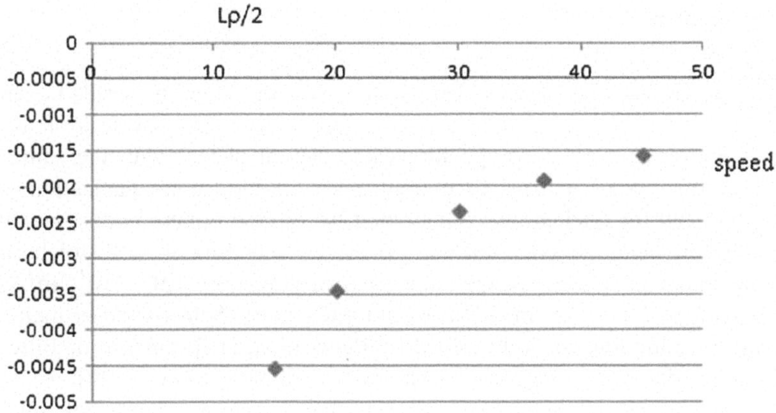

Fig. 1 Dirichlet simulation for phase change

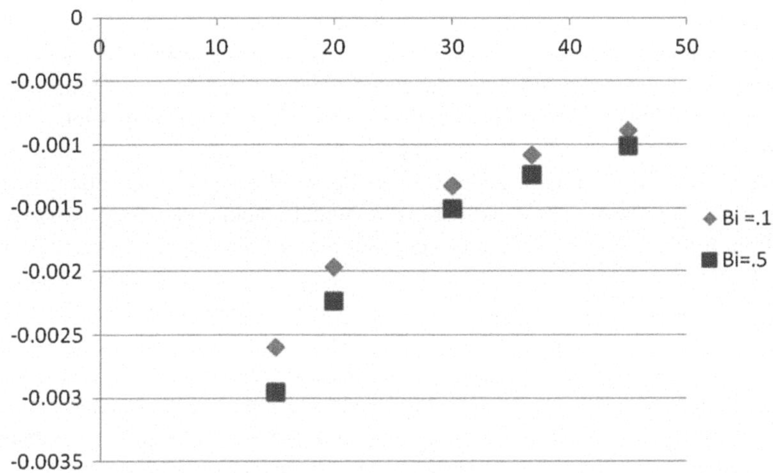

Fig. 2 Convective effects due to Bi number

Giving

$$hl/k(Ts - T_0) = (Ts - T_m) \tag{5}$$

Normalizing to Tm = 0, Ts = $BiT_0/(1 + Bi)$ where Bi = hl/k, h the convective film coefficient, l characteristic length. The Dirichlet solution is used to obtain the convective solution after replacing the boundary temperature by the modified temperature due to Bi. Simulations using WOLFRAM for the spherical case are shown in Fig. 1 along with the modification for convection in Fig. 2.

3 Discussion

According to Granasy and James [16], studies of laser treated materials began with amorphous layers in oxides, where it was found that some oxide systems with simple stoichiometric glass compositions had crystal phases with the same composition as the parent matrix. In certain cases, homogenous nucleation occurs without the need for nucleating agents aided by surface imperfections. Nucleation rates (excluding heterogeneous nucleation), were studied for several stoichiometric glasses by James [17], with steady state rates varying from 1.9×10^{12} m^3 s^{-1} for Ba_2SiO_2 glass to 1.7×10^6 m^3 s^{-1} for CaO–SiO_2 glass. Selective laser melting for processing metallic glasses is described by Pauly et al. [18]. Surface modifications of CoCe alloys have been reported by Hoekstra et al. [19], Li et al. [20], using large area electron beam irradiation of microsecond duration. Interesting results for Pt foil are reported by Xraysweb [21]. Taking typical values for Co as L = 66 cal/gm, Cp as 0.42 cal/gK, the Stefan number for a temperature differential of 400 K comes to approximately 0.4. From (1) the critical Stefan number is seen to be 0.025 g/ε. For suitable values of the Stefan number, it is possible for the laser frequency to match the resonant frequency and allow for thermal oscillations to be maintained, thus preventing homogenous nucleation. The wide range of nucleation frequencies seen for glasses indicates that other mechanisms are involved in modifying the laser action, suggested here to be an attenuation mechanism. Apart from attenuation, the high under cooling required for amorphous alloy formation may be possible by using very short laser pulses, according to Pauly et al. [18]. The effect of under cooling on amorphous alloy formation is widely acknowledged, however, the separation of this from pulse duration and initiation of homogeneous nucleation, the shape of the amorphous phase field and other factors remain an area of further work. Recently, work on high-frequency laser pulses by Mankowsky and Subedi [22], shows that due to electronic effects of THz frequency laser pulses, superconductivity and transitions from insulator to metal have been observed. The effect has been attributed to phononics and staggered dilation and contraction of Cu_2O interlayer distances in Y Ba CuO compounds. Gulian et al. [23] have also discussed evidence of superconductivity in laser processed Sr_2RuO_4 (triplet superconductor) using SQUID measurements even up to 200 K and beyond, in the top layer affected by the treatment (the "crust"). In both these examples, the presence of Oxygen and distortion of the lattice are contributing factors as are the phonon contributions. High temperature super conductors processed by a laser are described by Bauerle [24]. Laser pulses appear to shift the atoms in ceramic crystals and possible room temperature superconductivity for trillionths of a second has recently been reported.

4 Conclusion

It is found that for the usual metals, the second order terms cannot be neglected (inner solutions), and similarly for the Biot numbers encountered, the outer solution also needs to be used. The lengths involved may be seen to vary with conductivity and other parameters and thus for thin films and small dimensions the transient effects may become significant. Under certain conditions, the film can be made to sustain thermal oscillations, analogous to a mechanical spring damper system. The approximations for Fo and Bi around 0.2 and 0.1 reveal that Ste is around 0.025 g/e, and consequently the second order inertia terms varying as 1/Ste are one order of magnitude more than the damping and stiffness terms in the analogy discussed. Corrosion and oxidation due to the mass influx of foreign species from the surface layer are not considered, but would be an area of further interest in the future. Similarly the formation of an amorphous surface layer through rapid self-quenching with short heating times is an exciting possibility for development of new two phase film materials and has already been reported by several researchers. However, further work on separating the effect of the deep eutectic characteristic of amorphous glass formers from the phase field effect needs to be done. Extension of this work to the application of high temperature superconductivity seems a possibility, the data on femto-second results are however scarce and need corroboration.

References

1. deAngelis, A. "Method of three-d rapid prototyping", US Patent 5398193, 1995 Mar 14.
2. Sahoo, C.K., Sahu, J.K., Masanta, M., "Effect of Pulsed Nd:YAG Laser Parameters in Preplaced TiC Coating on Aluminium Substrate", 5th International & 26th All India Manufacturing Technology, Design, and Research Conference (AIMTDR 2014), IIT Guwahati, Assam, India, December 12th–14th, (2014).
3. Carslaw, H.S., Jaeger, J.C., [Conduction of Heat in Solids], 2nd Ed., Clarendon Press, Oxford, (1959).
4. Stefan, J., "Uber die`Theorie de Eisbildung im Polarmeer", Ann Phys Chem, NF 42, 269–86 (1891).
5. Ingersoll, L.R., Zobel, O.J., Ingersoll, A.C., [Heat Conduction], Oxford IBH, Calcutta (1960).
6. Crank, J., [Free and Moving Boundary Problems], Clarendon Press, Oxford (1984).
7. Langer, J.S., "Instabilities and Pattern Formation in Crystal Growth", Review of Modern Phys. 52, 1–28 (1980).
8. Mullins, W. W., Sekerka, R.F., "Morphological Stability of a Particle growing by Diffusion or Heat Flow", J. Appl. Phys. 34, 323–329 (1963).
9. Davis, S. H., Schulze, T.P., "Effects of Flow on Morphological Stability during Directional Solidification", Metallurgical and Mat. Trans, A, 27, 583–593 (1996).
10. Sands, D., "Pulsed laser heating and melting", Heat Transfer –Engineering Applications, V. Vikhrenko, Ed., ISBN 978-953-307-361-3, InTechopen.com (2011).
11. Rojas-Trigos, J.D., Calderon, J.A., "Heat diffusion in a homogeneous slab with an arbitrary periodical heat source: The case of heat source with square wave modulation function", Lat. Am. J. Phys. Educ. Vol.6, No. 1 March, (2011).

12. Kim, T., Pillai, M., Aziz, M.J., Scarpulla, M., "Heat flow model for pulsed laser melting and rapid solidification of ion implanted GaAs", J Appl Phys, 108, (2010) 013508-1 doi:10.1063/1.3457106.
13. Duggan, G. M., Tong, M., Browne, D.J., Proc of 3rd Intl Conf on Advances in Solidification Processes, IOP Conf Series, Mat Sci Eng 27, 012077 (2011).
14. Contuzzi, N., Campanelli S.L., Ludovico,A.D., "3D finite element analysis in the Selective Laser Melting Process", Int J Simul Model (10) 3, 113–12,1 (2011).
15. Paterson, S., " Propagation of a Boundary of Fusion", Glasgow Math Assn. Proc., 1, 42–47 (1952).
16. Granasy, L., James, P.F., "Nucleation in Oxide Glasses- Comparison of Theory and Experiment", Proc Roy Soc Lond 454, 1745–1766 (DOI:10.1098/rspa.1998.0230) (1998).
17. James, P.F., "Kinetics of crystal nucleation in lithium silicate glasses" Phys. Chem. Glasses 15, 95–105 (1974) (see ref 46).
18. Pauly, S., Lober, L., Petters,R., et al, "Processing metallic glasses by selective laser melting", Materials Today, 16,1–2, 37–41 (2013) http://dx.doi.org/10.1016/j.mattod.2013.01.018.
19. Hoekstra, J.G., Qader, S., Scully, J., Fitzgerald, J., "Laser Surface Modification of a Crystalline Al-Corrosion Co-Ce Alloy for Enhanced Resistance", Adv Eng Materials, 7, 9, 805–809 (2005).
20. Li, C., Murray, J., Volsey, K.T., McCartney, D., "Amorphous layer formation in Al86.0Co7.6Ce6.4glass-forming alloy by large-area electron beam irradiation" Appl Surface Science, 280, 431–438 (2012).
21. http://xraysweb.lbl.gov/bl1222/RESEARCH_APPLICATIONS/PtAmbLaserMelt.ppt (accessed May10 2016).
22. Mankowsky, R., Subedi, A., "Nonlinear dynamics as a basis for enhancing superconductivity in YBaCuO" Nature 516, 71–73 (4 Dec 2014) doi:10.1038/nature13875.
23. Gulian, A.M., Wood, K.S., VanVehrken, D., "Evidence for high temperature superconductivity in doped laser processed SrRuO", (2005 Sept) arxiv.org 0509313 (2005).
24. Bäuerle, D., "Laser induced formation and surface processing of high temperature superconductors", Appl. Phys. A 48: 527, doi:10.1007/BF00617854 (1989).

Distortion Analysis of 1.3 μm AlGaInAs/InP Transistor Laser

R. Ranjith, S. Piramasubramanian and M. Ganesh Madhan

Abstract We analyze the distortion characteristics of 1.3 μm AlGaInAs/InP Transistor Laser for CATV applications. The characteristics of Transistor Laser are analyzed by solving rate equations. Small signal analysis has been done for different input bias currents. Bandwidth and resonance frequency are calculated from the small signal analysis. Second harmonic distortion (2HD) is evaluated for different input bias currents. Third order intermodulation distortion (IMD3) is calculated with two input frequencies of 823.25 and 815.25 MHz. Minimum 2HD of −20.76 dBc and IMD3 of −33.61 dBc are predicted from this study.

1 Introduction

Optical communication is one of the key technologies to provide high bandwidth to users. However, optical sources such as laser diode is capable of providing only an intrinsic bandwidth up to 40 GHz [1]. So there is a need for alternate device to provide bandwidth in excess of 40 GHz. Transistor Laser (TL) operates at 980 nm was invented by Feng et al. [2, 3]. The TL is a transistor with quantum wells in its base. Quantum well in active region provides laser light output. Charge control analysis of the TL was reported by Feng et al. [4]. Electrons emitted from the emitter terminal is captured by the quantum well region because of its lower bandgap (heterostructure) with a capture time of t_{cap} and the uncaptured electrons are captured by collector terminal due to reverse bias. This reduces the carrier lifetime and increases the bandwidth when compared with conventional laser

R. Ranjith (✉) · S. Piramasubramanian · M. Ganesh Madhan
Department of Electronics Engineering, Madras Institute of Technology Campus, Anna University, Chennai, India
e-mail: ranjith4792@gmail.com

S. Piramasubramanian
e-mail: spsnanthan@gmail.com

M. Ganesh Madhan
e-mail: mganesh@annauniv.edu

© Springer Nature Singapore Pte Ltd. 2017

I. Bhattacharya et al. (eds.), *Advances in Optical Science and Engineering*, Springer Proceedings in Physics 194, DOI 10.1007/978-981-10-3908-9_52

diodes. The dynamic operation of TL has been modeled to a coupled rate equations for numerical analysis by Zhang and Leburton [5] and Faraji et al. [6–8]. The wavelengths used in the conventional optical communication system are 1.5 and 1.3 μm, due to low attenuation and dispersion in fiber respectively. An 1.3 μm AlGaInAs/InP TL was designed and demonstrated at room temperature by Shirao et al. [9]. The operation and experiments of the TL in CE and CB configuration with charge control analysis was investigated by Feng et al. [9–13]. In this work, we analyze the distortion characteristics of TL at 823.25 and 815.25 MHz. The second order harmonics and third order intermodulation products are observed and calculated for different base currents.

2 Modeling of Transistor Laser

AlGaInAs/InP TL is represented by coupled rate equations developed by Faraji et al. [6] for numerical analysis and the large signal analysis was done by Shirao et al. [9]. The charge control model is given in Fig. 1, the excess carriers flow from the emitter terminal to the collector terminal is captured by the quantum well, which is placed in the middle of the base region, with width of d (nm). The emitter and collector current in Fig. 1 flows in the direction opposite that of electron flow.

The transistor laser rate equations used for numerical analysis are given below [9].

$$\frac{dN_{vs}}{dt} = -\frac{N_{vs}}{\tau_{cap}} - \frac{N_{QW}}{\tau_{esc}} + \frac{Ivs}{dqA} \tag{1}$$

$$\frac{dN_{QW}}{dt} = \frac{N_{vs}}{\tau_{cap}} - \frac{N_{QW}}{\tau_{esc}} - \frac{g' s(N_{QW} - N_g)}{1 + \epsilon_{\text{other}} S} - \frac{N_{QW}}{\tau_s} \tag{2}$$

$$\frac{dS}{dt} = \frac{\epsilon G' S(N_{QW} - N_g)}{1 + \epsilon_{\text{other}} S} - \frac{S}{\tau_p} + C\frac{N_{QW}}{\tau_s} \tag{3}$$

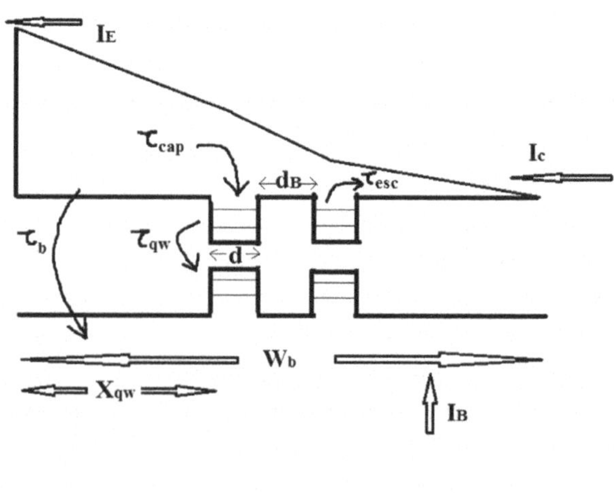

Fig. 1 Charge concentration in base region in two quantum well transistor laser [5]

Equation (1) denotes the virtual state carrier density of the region above the quantum well. I_{vs} is the current available in the higher energy state region above the QW [9]. Equation (2) represents the QW carrier density. The photon density extracted from the QW is provided in (3). The parameters used in the above coupled rate equations for 1.3 μm AlGaInAs/InP TL has been taken from Shirao et al. [9]. The above equations are numerically solved by using MATLAB®. Optical power output from the TL is formulated by the (4) [12].

$$P = \eta\, v_g(\alpha_i + \alpha_m) h\, f\, S\, (v/2\epsilon) \tag{4}$$

where v is active layer volume and v_g is the group velocity. Based on the charge continuity equation, the penetration of minority carrier in the device depends on input frequency i.e. diffusion length [9].

$$L_D = \sqrt{\frac{D_n \tau_b}{1 + j\omega\tau_b}} \quad L_D = \sqrt{D_n \tau_b} \tag{5}$$

From (5) the diffusion length is inversely proportional to the given frequency. For DC characteristics diffusion length is defined by taking $\omega = 0$.

3 Simulation Results

3.1 DC Characteristics

The DC characteristics of a transistor laser have been analyzed by giving a constant bias current as input. NPN TL is made to work in the active region by forward biasing the base emitter junction and reverse biasing the base collector junction. The coupled rate equations are solved by keeping time derivatives equal to zero. The threshold current is calculated by plotting optical power for different input DC currents. Optical power is calculated from (4) and plotted in Fig. 2. The estimated

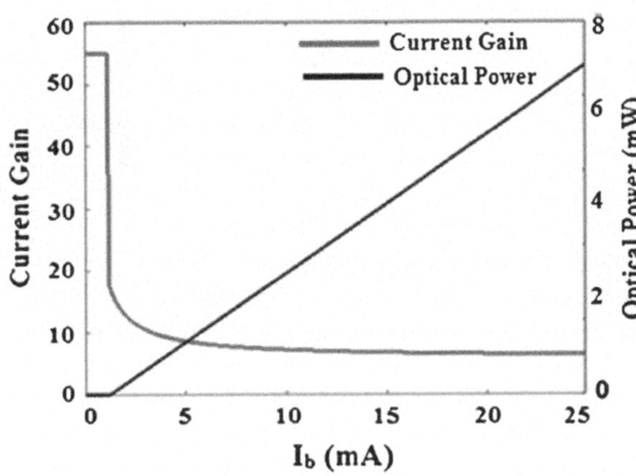

Fig. 2 Optical power and current gain variation with base current

threshold base current for two quantum well transistor laser from Fig. 2 is 1 mA. Before the threshold current, only spontaneous emission occurs. The stimulated emission is made possible by achieving population inversion by increasing the input current. The current gain β is represented in Fig. 2. It is found that the current gain decreases drastically beyond threshold current. This is due to high recombination in the base region which includes quantum well.

3.2 Small Signal Analysis

Small signal analysis is done by choosing different bias input currents after threshold and by giving input AC signal within the chosen linear region in the nonlinear transfer characteristics of TL [1].

$$I(t) = I_b + I_m \sin(\omega t) \tag{6}$$

where, I_b is the bias current and I_m is the amplitude of the time varying input signal. By varying the input current below and above the threshold, the output optical power also varies, which is called direct or intensity modulation. Increasing the frequency also the limits the performance due to the rate of depletion of stored minority carriers. Magnitude response versus frequency was determined for different input bias currents. The magnitude plot for different input bias currents are shown in Fig. 3a. From Fig. 3 it is found that the bandwidth and resonance frequency increases for increasing the input base current. From Fig. 3b the bandwidth achieved by TL in CE configuration is found as 40 GHz when the bias current is fixed as 200 mA ($200I_{th}$). Hence it is clear that direct modulation by using TL can reach up to 40 GHz bandwidth which is not the case in laser diodes. However, the bandwidth saturates at 40 GHz, even with increasing base current.

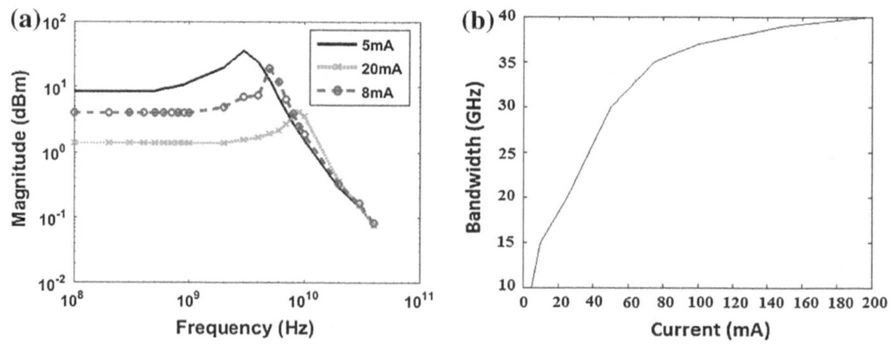

Fig. 3 **a** Magnitude response and **b** bandwidth variation with bias current

3.3 *Harmonic Analysis*

Due to non linear characteristics of the TL, the harmonics other than the fundamental component also generated. Hence an analysis of harmonics developed by the TL becomes necessary. Harmonic analysis is carried out to predict second harmonic distortion and third order intermodulation distortion. These distortion have to be reduced in analog optical link and RoF applications [13].

3.3.1 Second Order Harmonic Distortion

An 823.25 MHz signal (CH 65 of video carrier in CATV Band) is given as input to the TL. The given AC signal, intensity modulate with the optical carrier. FFT of the output optical signal is taken by 524,288 sampling points to plot the frequency spectrum of the input signal. The spectrum of input signal and corresponding optical power are shown in Fig. 4a, b respectively. The input signal spectrum is plotted in Fig. 4a with fundamental component at 823.25 MHz with magnitude of 4.22 dBm. Due to the nonlinear characteristics of the TL, the output spectrum has

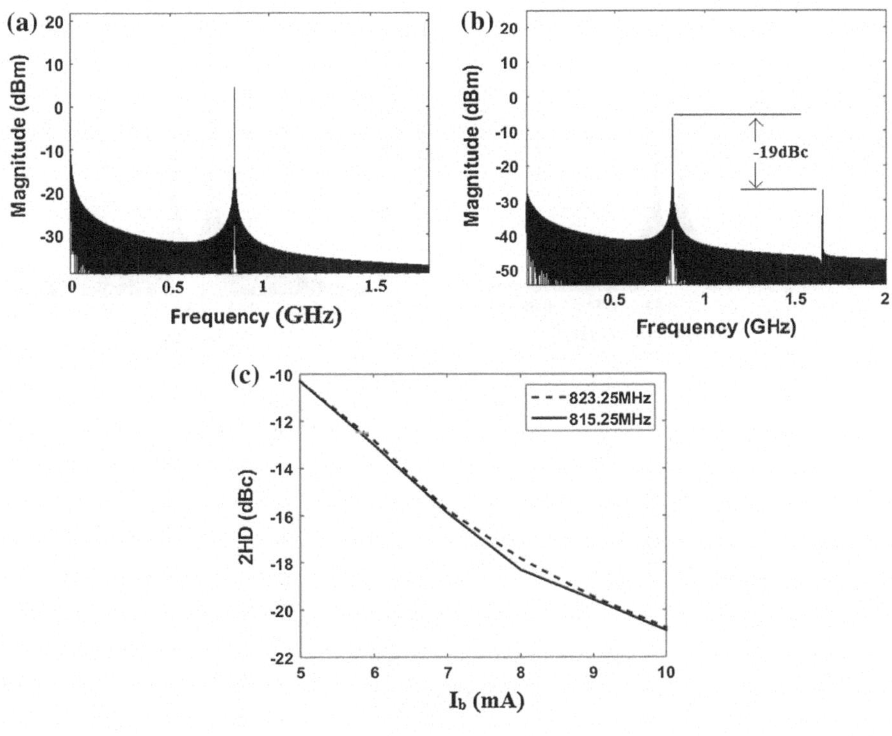

Fig. 4 **a** Spectrum of current at 823.25 MHz **b** spectrum of optical power and **c** variation of 2HD with bias current

additional harmonic components than the required fundamental one. From the small signal analysis, it has been found that the device bandwidth increases with increasing bias current. Hence the effect of bias current in second harmonic distortion (2HD) is also determined and shown in Fig. 4c. The 2HD is measured in (dBc), which is the difference between the fundamental one and the second harmonic component. As the input bias current increases, the 2HD decreases (Fig. 4c). A minimum of −21 dBc is observed for the bias current of 10 mA.

3.3.2 Third Order Intermodulation Distortion (IMD3)

In applications such as Radio over Fiber (ROF), the input signal has more than one frequency component. For simplicity we consider an input signal which contains two tones. The intermodulated harmonic components occur more closely to the fundamental frequency, than the second order harmonics [13]. For CATV applications IMD products are analyzed within the 8 MHz span of the video carrier. We have chosen two video carriers 815.25 and 823.25 MHz (CH 64 and CH 65) and the input and output spectrum are plotted in Fig. 5a, b. It has been observed that the

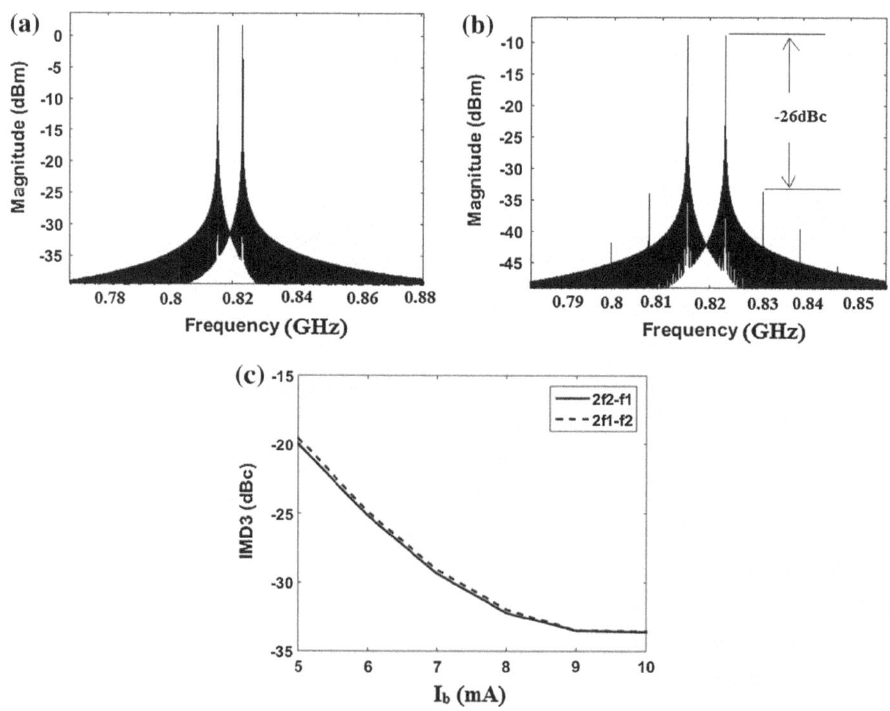

Fig. 5 **a** Spectrum of two tone input signal at f_1 = 815.25 MHz and f_2 = 823.25 MHz **b** spectrum of optical power and **c** variation of IMD3 with bias current

output has additional frequency components compared to the input. The variation of IMD3 with bias current in shown in Fig. 5c.

From this analysis, it has been found that the third order intermodulation component occurs more closely to the fundamental tone within 8 MHz span of the video carrier. The other intermodulated components closer to the fundamental tone are fifth and seventh order components. It also found that the IMD3 components have large magnitude than other IMD products. The IMD3 for different input bias currents has been calculated to study the effect of input bias current. The IMD3 products are found to decrease for increasing input bias currents and a minimum of −33.61 dBc is predicted for a bias current of 10 mA.

4 Conclusion

The DC and AC characteristics of transistor laser are analyzed by numerically solving rate equations. It has been found that the bandwidth increases for increasing the input bias current and a maximum bandwidth of 40 GHz is predicted. The second harmonic distortion and third order intermodulation distortion are analyzed for different base currents. The distortion of TL decreases with increasing the input bias currents. A minimum 2HD of −20.76 dBc and IMD3 of −33.61 dBc are predicted in our analysis for the bias current of 10 mA.

References

1. Gerd keiser, "Optical Fiber Communications" Special Indian Edition McGraw Hill (2013).
2. G. Walter, Nick Holonyak Jr., M. Feng and R. Chan, "Laser Operation of a Heterojunction Bipolar Light Emitting Transistor", Applied Physics Letters, 85, 4768 (2004).
3. M. Feng, N. Holonyak Jr., G. Walter, and R. Chan, "Room temperature continuous wave operation of a heterojunction bipolar transistor laser", Applied Physics letters 87, 131103 (2005).
4. M. Feng, N. Holonyak Jr., H. W. Then, and G. Walter, "Charge control analysis of transistor laser operation", Applied Physics letter 91, 053501 (2007).
5. Lingxiao Zhang,, and Jean-Pierre Leburton, "Modeling of the Transient Characteristics of Heterojunction Bipolar Transistor Lasers", IEEE Quantum Electronics, Vol 45, No. 4, (2009).
6. B. Faraji, D. L. Pulfrey, and L. Chrostowski, "Small-signal modeling of the transistor laser including the quantum capture and escape lifetimes", Applied Physics letters 93, 103509 (2008).
7. B. Faraji, W. Shi, D. L. Pulfrey, and L. Chrostowski. "Common-emitter and common-base small-signal operation of the transistor laser", Applied Physics letters 93, 143508 (2008).
8. Behnam Faraji, Wei Shi, David L. Pulfrey, and Lukas Chrostowski "Analytical Modeling of the Transistor Laser", IEEE Journal of selected topics in Quantum Electonics, Vol 5, No 3. (2009).
9. Mizuki Shirao, Masashi, Seunghun Lee, Nobuhiko Nishiyama, and Shigehisa Arai, "Large Signal Analysis of Transistor Laser" IEEE Quantum Electronics Vol. 47, No. 3, (2011).

10. Mizuki Shirao, Takashi Sato, Noriaki Sato, Nobuhiko Nishiyama, and Shigehisarai, "Room-temperature operation of npn-AlGaInAs/InP multiple quantum well transistor laser emitting at 1.3-μm wavelength", OSA OPTICS EXPRESS Vol 20, No 4, (2012).
11. Larry A. Coldren, "Diode Lasers and Photonic Integrated Circuits", Wiley Series in Microwave and Optical Engineering (2012).
12. S. Piramasubramanian, M. Ganesh Madhan, Jyothsna Nagella, G. Dhanapriya, 'Numerical analysis of distortion characteristics of heterojunction bipolar transistor laser', Optics Communications 357, 177–184 (2015).
13. Safwat W. Z. Mahmood, Alaa Mahmood and Mousatfa Ahmed, "Noise Performance and nonlinear distortion of semiconductor laser under two tone modulation for use in analog CATV systems", International Journal Of Numerical Modelling: Electronic Networks, Devices And Fields, Int. J. Numer. Model (2015).

Oscillator Strength of Gaussian Double Quantum Well for Intersubband Transition

Debasmita Sarkar and Arpan Deyasi

Abstract Oscillator strength and absorption cross-section of Double quantum well triple barrier structure with Gaussian geometry is analytically computed for inter-subband optical transition between ground state and first excited state. Electric field is applied along quantum confinement, and Kane-type conduction band non-parabolicity of first order is considered for near accurate computation. Result suggests that oscillator strength monotonically increases with wavelength, and is higher when nonparabolicity factor is considered. Cross-section is higher for lower well dimension. Result is also compared with parabolic overestimation. Simulated findings are important for designing optical detector.

1 Introduction

Emerging nanophotonic devices becomes one of the field of research in last decade due to various novel applications in medical [1], defense [2] or communication [3] arena. As far the application is concerned, the fundamental property of the nanostructure that needs to be evaluated is the eigenstate, which is the function of device dimension [4, 5] and material composition [6]. Solution of eigenstates for complex geometrical structures are very difficult by existing analytical method and thus different numerical methods are incorporated by theoretical researchers [7–9] for near accurate composition. In this connection, consideration of band structure of the device plays crucial role in determining energy state [10], and thus band non-parabolicity factor should be considered in mathematical analysis. Photonic properties of quantum heterostructures are heavily dependent on the eigenenergies.

D. Sarkar (✉) · A. Deyasi
Department of Electronics and Communication Engineering,
RCC Institute of Information Technology, Kolkata, West Bengal 700015, India
e-mail: sarkar.debasmita100@gmail.com

A. Deyasi
e-mail: deyasi_arpan@yahoo.co.in

© Springer Nature Singapore Pte Ltd. 2017

I. Bhattacharya et al. (eds.), *Advances in Optical Science and Engineering*,
Springer Proceedings in Physics 194, DOI 10.1007/978-981-10-3908-9_53

In low-dimensional semiconductor structure, quantization of energy states makes them characteristically different from existing bulk devices and transition in inter and intra level controls their electronic and optoelectronic properties. Oscillator strength is such a parameter which speaks the probability of transition between any two quantized energy levels, and thus its near accurate determination plays crucial role for application of the device in precise field. Absorption cross-section determines the area of the device required for photodetector application. The first one is determined form the absorption coefficient, and the later is dependent on the former. Thus computation of absorption coefficient and photoluminescence study becomes important for theoretical research in the point of view of device design. Oscillator strength for quantum wire [11–13] is already calculated for field-induced transition, and also for quantum dot [14, 15]. Absorption cross-section is also calculated for wire [13] and dot [16].

Present paper deals with the computation of oscillator strength of Gaussian double quantum well in presence of field with the consideration of Kane-type band nonparabolicity. Corresponding absorption cross-section is also calculated for different structural parameters. Results are important for the structure for optical detector application.

2 Mathematical Modeling

Absorption coefficient for a quantum well structure may be put in the following form

$$\alpha(\omega) = \frac{n_s \pi q^2 \hbar}{2\varepsilon_0 \varepsilon_r c n_r m^* L} f_{21} \delta(\Delta E - \hbar\omega) \tag{1}$$

where L is the length of quantum well, n_s is the sheet charge density, f_{21} is defined as the oscillator strength for the transition between ground state and 1st excited state. f_{21} is given by

$$f_{21} = \frac{2}{m^* \hbar\omega} \left[\frac{8\hbar}{3L}\right]^2 \tag{2}$$

For practical profile, the delta function is replaced by Lorentzian lineshape function, which modifies (1) as

$$\alpha(\omega) = \frac{n_s \pi q^2 \hbar}{2\varepsilon_0 \varepsilon_r c n_r m^* L} f_{21} \frac{\Gamma}{\pi\left[(\hbar\omega - \Delta E)^2 + \Gamma^2\right]} \tag{3}$$

where ΔE is the intersubband transition energy, Γ is the full-width at half-maximum.

Once oscillator strength is calculated, absorption cross-section may be defined as

$$\sigma_{2,1} = \frac{q^2 T}{2\varepsilon_0 m^* nc} f_{21} \left[1 + T^2 \left(\frac{E - \Delta E_{2,1}}{\hbar} \right)^2 \right]^{-1} \tag{4}$$

where T is the relaxation time.

3 Results and Discussions

Using (3) and (4), oscillator strength and absorption cross-section of double quantum well structure are computed and plotted as a function of wavelength and energy respectively. Figure 1 shows the variation of oscillator strength with wavelength for DQW structure. In Fig. 1a (in left), it is seen that oscillator strength monotonically increases with wavelength. Also with increase of well width, oscillator strength decreases. This can be directly predicted from (2). This can be explained as follows: higher well width reduces quantum confinement, which, in turn, lowers eigenenergies. This reduces the separation between energy values, so oscillator strength decreases. In Fig. 1b (in right), comparative study is made with the results obtained for parabolic overestimation, which shows that consideration of band nonparabolicity gives higher magnitude of oscillator strength for a given structural parameters with constant electric field. This speaks in favor of higher transition probability when accurate band structure is considered in modeling.

Figure 2 shows the absorption cross-section of the structure as a function of incident radiation for different well widths. It is seen form the plot that peak position of cross-section remains invariant w.r.t well dimension. But height of the profile increases as well layer width decreases. This is due to the fact that with

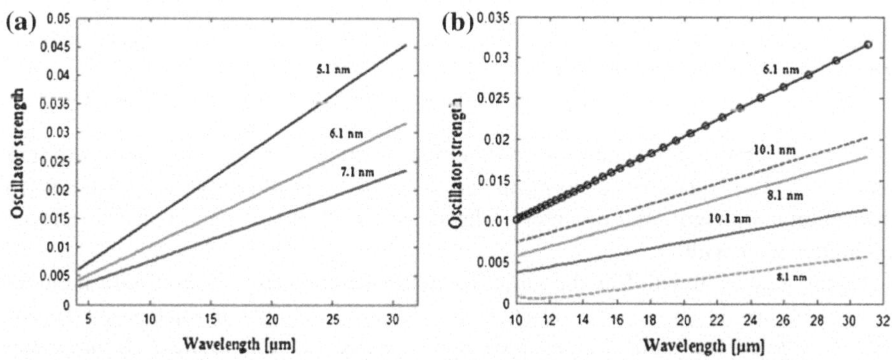

Fig. 1 **a** Oscillator strength with wavelength for different well widths for nonparabolic dispersion relation in presence of electric field; **b** Oscillator strength with wavelength for parabolic and nonparabolic dispersion relations in presence of electric field

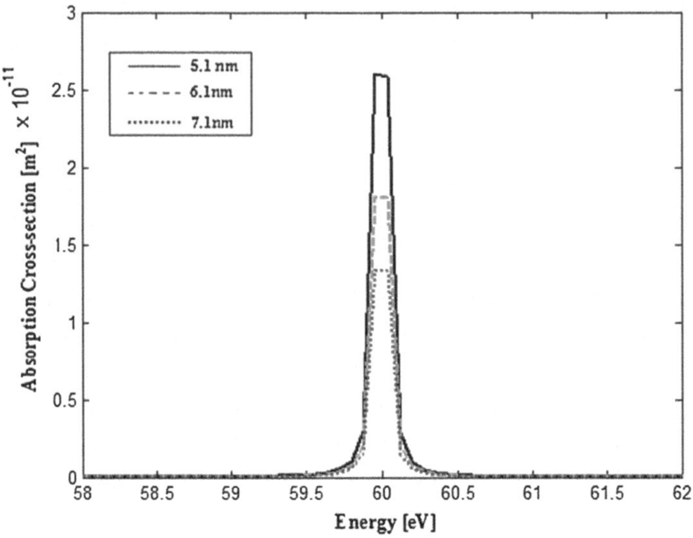

Fig. 2 Absorption cross-section with incident radiation for different well widths

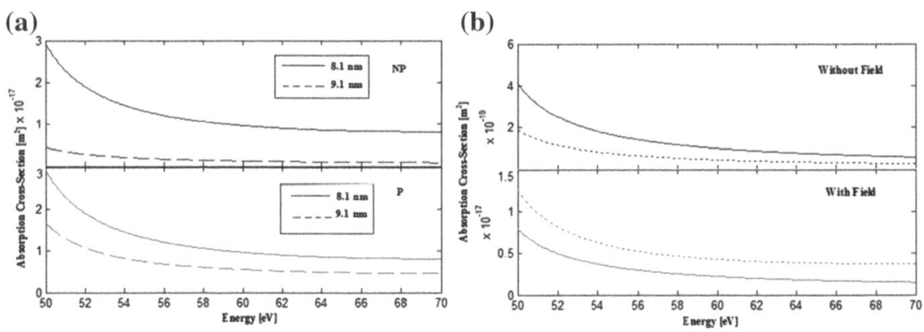

Fig. 3 Absorption cross-section with incident radiation for (a) different dispersion relations; (b) in presence and absence of electric field

increase of well width, quantum confinement decreases, hence oscillator strength decreases. The quantity absorption cross-section, is directly proportional to absorption coefficient, and inversely related with well dimension. Hence it reduces for higher well width.

Figure 3a and 3b exhibits the variation of absorption cross-section with incident energy. Figure 3a shows the variation for different dispersion relations, whereas Fig. 3b studied the same in presence and absence of electric field. It has been observed that effect of field reduces cross-section. This is due to the fact that electric field lowers the eigenstate of the device, and hence subband transition energy reduces. This reduces the absorption cross-section. Similarly, band nonparabolicity factor reduces eigenstate, more precisely it affects the higher quantum state. If Kane-type nonparabolicity is considered, then computation already revealed that eigenvalue is reduced compared to that calculated for parabolic band structure. The change is more significant for higher energy values. Thus lowermost curve of cross-section gets affected compared to the plots generated for lowermost transition energy.

4 Conclusion

Oscillator strength and absorption cross-section of a double quantum well system is analytically computed using the knowledge on eigenstates. Result suggests that cross-section decreases with increasing well width, and it is more significant when band nonparabolicity is considered for thicker well layer. Application of electric field along the quantized direction reduces the cross-section. The same nature is also exhibited for oscillator strength. Results are important for optoelectronic application of the structure, more precisely for designing of quantum detector.

References

1. Perera, A. G. U., "Quantum Structures for Multiband Photon Detection", Opto-Electronics Review, vol. 14, pp. 99–108 (2006).
2. Phillips, M. C., Taubman, M. S., Bernacki, B. E., Cannon, B. D., Schiffern, J. T., Myers, T. L., "Design and Performance of a Sensor System for Detection of Multiple Chemicals using an External Cavity Quantum Cascade Laser", Proceedings of SPIE (Quantum Sensing and Nanophotonic Devices VII), vol. 7608, p. 76080D (2010). (January 23, 2010).
3. Wei, R., Deng, N., Wang, M., Zhang, S., Chen, P., Liu, L., Zhang, J., "Study of Self-assembled Ge Quantum Dot Infrared Photodetectors", 1st IEEE International Conference on Nano/Micro Engineered and Molecular Systems, pp. 330–333 (2006).
4. Deyasi, A., Bhattacharyya, S., Das. N. R., "A Finite Difference Technique for Computation of Electron States in Core-Shell Quantum Wires of Different Configurations", Physica Scripta, vol. 89(6), p. 065804 (2014).
5. Li, Y., Voskoboynikov, O., Lee, C. P., Sze, S. M., "Electron Energy State Dependence on the Shape and Size of Semiconductor Quantum Dots", Journal of Applied Physics, vol. 90, pp. 6416–6420 (2001).
6. Deyasi, A., Bhattacharyya, S., Das, N. R., "Computation of Intersubband Transition Energy in Normal and Inverted Core-Shell Quantum Dots using Finite Difference Technique", Superlattices & Microstructures, vol. 60, pp. 414–425 (2013).
7. Yun, K., Sheng, W., Xianli, L., "Electron Energy States in a Two-Dimensional GaAs Quantum Ring with Hydrogenic Donor Impurity in the Presence of Magnetic Field", Journal of Semiconductors, vol. 36(3), p. 032003 (2015).
8. Lô, B., Gueye, S. B., "Numerical Verification of Transition's Energies of Excitons in Quantum Well of ZnO with the Finite Difference Method", Journal of Modern Physics, vol. 7, pp. 329–334 (2016).
9. Simion, C. E., Ciucu, C. I., "Triple–barrier Resonant Tunneling: A Transfer Matrix Approach", Romanian Reports in Physics, vol. 59, pp. 805–817 (2007).

10. Schaevitz, R. K., Roth, J. E., Ren, S., Fidaner, O., Miller, D. A. B., "Material Properties of Si-Ge/Ge Quantum Wells", IEEE Journal of Selected Topics in Quantum Electronics, vol. 14 (4), pp. 1082–1089 (2008).
11. Deyasi, A., Das, N. R., "Oscillator Strength and Absorption Cross-Section of Core-Shell Triangular Quantum Wire for Intersubband Transition", Springer Proceedings in Physics: Advances in Optical Science and Engineering, vol. 166, Chapter 78, pp. 629–635 (2014).
12. Tshipa, M., "Oscillator Strength for Optical Transitions in a Cylindrical Quantum Wire with an Inverse Parabolic Confining Electric Potential", Indian Journal of Physics, vol. 88(8), pp. 849–853 (2014).
13. Holovatsky, V. A., Voitsekhivska, O. M., Gutsul, V. I., "Optical Oscillator Strengths for the Electron Quantum Transitions in Elliptic Nanotubes", Romanian Journal of Physics, vol. 53 (7–8), pp. 833–840 (2008).
14. Bhattacharyya, S., Das, N. R., "Effect of Electric Field on the Oscillator Strength and Cross-Section for Intersubband Transition in a Semiconductor Quantum Ring", Physica Scripta, vol. 85(4), p. 045708 (2013).
15. Leistikow, M. D., Johansen, J., Kettelarij, A. J., Lodahl, P., Vos, W. L., "Size-dependent Oscillator Strength and Quantum Efficiency of CdSe Quantum Dots Controlled via the Local Density of States", Physical Review B, vol. 79, p. 045301 (2009).
16. Nanda, J., Ivanov, S. A., Htoon, H., Bezel, I., Piryatinski, A., Tretiak, S., Kilmov, V. I., "Absorption Cross Sections and Auger Recombination Lifetimes in Inverted Core-Shell Nanocrystals: Implications for Lasing Performance", Journal of Applied Physics, vol. 99, p. 034309 (2006).

Part XI
E. M. Radiation Theory and Antenna

Multi-slot Loaded Dual Band Compact Half-Mode SIW Triangular Antenna

Soumen Banerjee, Sombuddha Chatterjee, Sampoorna Das Mazumdar, Malay Gangopadhyaya and Biswarup Rana

Abstract An equilateral triangular antenna resembling half-mode SIW structure and operating at 5.28 GHz WLAN UNII-2 band resonating frequency is considered. The same antenna resonates at the first harmonic frequency of 9.57 GHz. The antenna is made compact and designed to support dual-band operation with insertion of inverted V-shaped slot and a horizontal rectangular slot. The new antenna exhibits dual-band operation at the frequencies of 4.3 and 9.45 GHz. A shift of 980 MHz for the fundamental frequency and 120 MHz for the first harmonic is witnessed.

1 Introduction

A very promising state-of-the-art technology in manufacturing and designing modern day sophisticated antenna is substrate integrated waveguide (SIW) technology. The choice of SIW based antenna is obvious owing to its immense advantage over conventional metallic waveguides. A typical SIW antenna bisected along the fictitious quasi-magnetic wall keeping identical field distribution results in the formation of Half-mode SIW (HMSIW) structure [1–8]. The authors have chosen a HMSIW equilateral triangular antenna operating at 5.28 GHz resonating frequency for the TE_{101} mode with its first harmonic at 9.57 GHz for the TE_{202} mode [5]. An inverted V-shaped slot and a horizontally placed rectangular slot are inserted on the antenna for

S. Banerjee (✉)
Department of ECE, University of Engineering & Management, Kolkata, West Bengal 700160, India
e-mail: prof.sbanerjee@gmail.com

S. Chatterjee · S.D. Mazumdar · M. Gangopadhyaya
Department of ECE, Institute of Engineering & Management, Salt Lake, Kolkata, West Bengal 700091, India

B. Rana
Department of ETCE, Indian Institute of Engineering Science & Technology, Shibpur, West Bengal 711103, West Bengal, India

© Springer Nature Singapore Pte Ltd. 2017

I. Bhattacharya et al. (eds.), *Advances in Optical Science and Engineering*, Springer Proceedings in Physics 194, DOI 10.1007/978-981-10-3908-9_54

miniaturization purpose. The position of the coaxial feed is optimized to achieve dual band operation. The new antenna resonates at 4.3 GHz for its fundamental frequency with its next higher mode resonating at 9.45 GHz thereby exhibiting both compactness as well as dual band operation. The antennas are designed on Arlon AD270 substrate and ANSYS make HFSS v15.0 is used to simulate various antenna parameters.

2 Half Mode SIW Triangular Antenna

2.1 The Antenna Structure

Figure 1 depicts the schematic diagram of the coaxial feed based parent HMSIW equilateral triangular antenna [5]. The overall dimension of the antenna is shown in Table 1. The mathematical expression associated with resonant frequency in connection to TE_{mnp} mode of a rectangular SIW resonator is

$$f_{mnp}^{SIW} = \frac{1}{2\pi\sqrt{\mu\varepsilon}}\sqrt{\left(\frac{m\pi}{L_{eff}^{SIW}}\right)^2 + \left(\frac{n\pi}{h}\right)^2 + \left(\frac{p\pi}{W_{eff}^{SIW}}\right)^2}$$

where, m = 1, 2, 3, …, n = 1, 2, 3, …, p = 1, 2, 3, …, substrate permeability $\mu = \mu_o\varepsilon_r$ and substrate permittivity $\varepsilon = \varepsilon_o\varepsilon_r$. The SIW resonator thickness is h while its equivalent length and width are expressed as

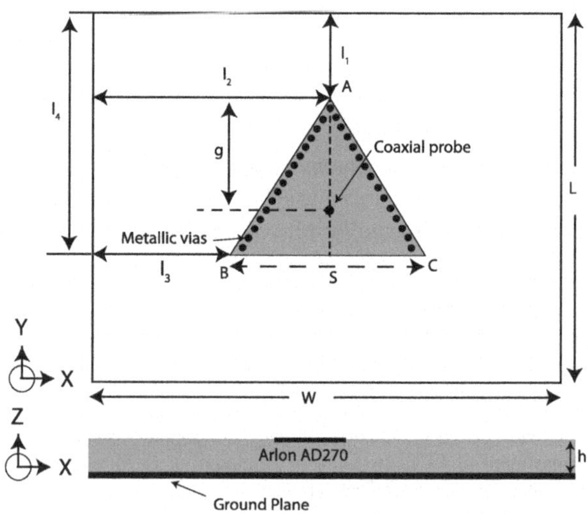

Fig. 1 Schematic diagram of SIW based equilateral triangular antenna

Table 1 Various dimensions of parent antenna (in mm)

L	W	S	h	l_1	l_2	l_3	l_4
60	60	31.18	0.79	12	30	14.41	39.01

$$L_{eff}^{SIW} = L_C - 1.08\frac{d^2}{d_1} + 0.1\frac{d^2}{L_C} \quad \text{and} \quad W_{eff}^{SIW} = W_C - 1.08\frac{d^2}{d_1} + 0.1\frac{d^2}{W_C}$$

where, the rectangular SIW resonator's length and width being L_C and W_C, the via diameter and the spacing between the vias being d and d_1 respectively. The chosen HMSIW antenna has equivalent length and width as

$$W_{eff}^{HMSIW} = \frac{W_{eff}^{SIW}}{2} + \Delta W \quad \text{and} \quad L_{eff}^{HMSIW} = \frac{L_{eff}^{SIW}}{2} + \Delta W$$

where, the additional width ΔW is estimated using the formula

$$\Delta W = h\left(0.05 + \frac{0.3}{\varepsilon_r}\right) \times \ln\left(0.79\frac{W_{eff}^{SIW}}{4h^3} + \frac{52W_{eff}^{SIW} - 261}{h^2} + \frac{38}{h} + 2.77\right)$$

The resonant frequency for the HMSIW antenna is estimated using the relation

$$f_{mnp}^{HMSIW} = \frac{1}{2\pi\sqrt{\mu\varepsilon}}\sqrt{\left(\frac{m\pi}{L_{eff}^{HMSIW}}\right)^2 + \left(\frac{p\pi}{W_{eff}^{HMSIW}}\right)^2}$$

Arlon AD270 substrate (tm) used in designing the antenna has thickness h = 0.79 mm, dielectric constant (ε_r) = 2.7 and loss tangent (tan δ) = 0.002. The side length S = 31.18 mm while the overall dimension of the antenna is 60 (L) mm × 60 (W) mm × 0.79 mm. A series of metallic vias placed along the two sides of the triangle makes the PEC wall while the surface devoid of it is PMC wall. Vias are of diameter 1 mm with inter-separation distance of 1.5 mm. The antenna is matched for exciting TE_{101} mode and the next higher TE_{202} mode. The optimized position of the coaxial feed is 22 mm. The distribution of electric field for the two different TE_{101} and TE_{202} modes are as shown in Fig. 2.

2.2 Simulated Results

At 5.28 GHz, the value of simulated S_{11} for the parent HMSIW antenna is −29 dB and the same is −17 dB at the first harmonic frequency of 9.57 GHz, as shown in Fig. 3. The simulated E plane and H plane radiation pattern of the antenna is shown in Fig. 4. From the figure it is evident that the peak values of co-pol E and H plane radiation are 5.7 and 5.1 dBi respectively, while the cross-polarization is about 30–40 dBi lower than their respective co-polarization values.

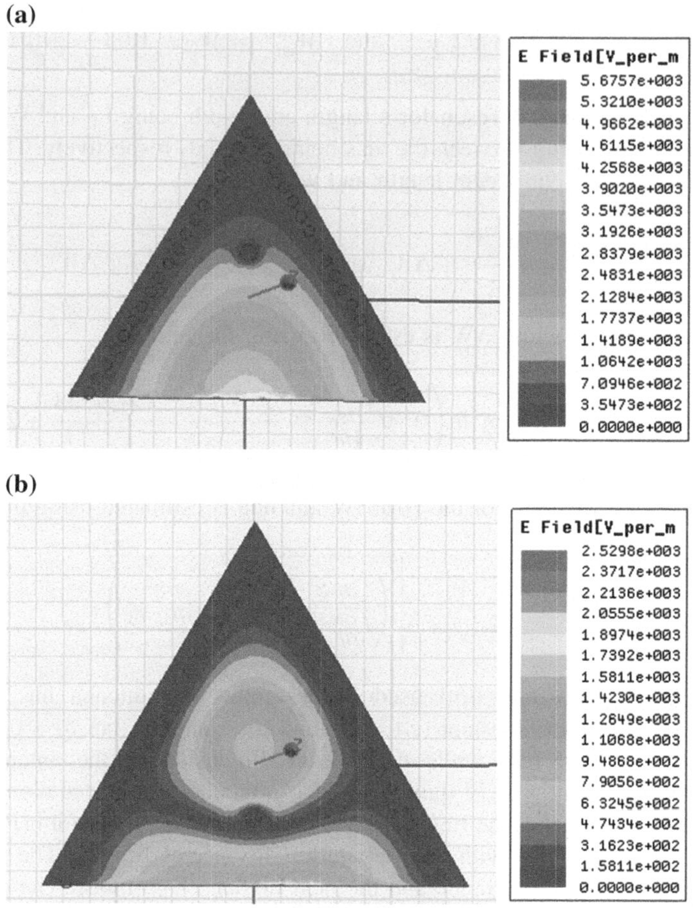

Fig. 2 Distribution of electric field for **a** TE_{101} mode. **b** TE_{202} mode for the parent antenna

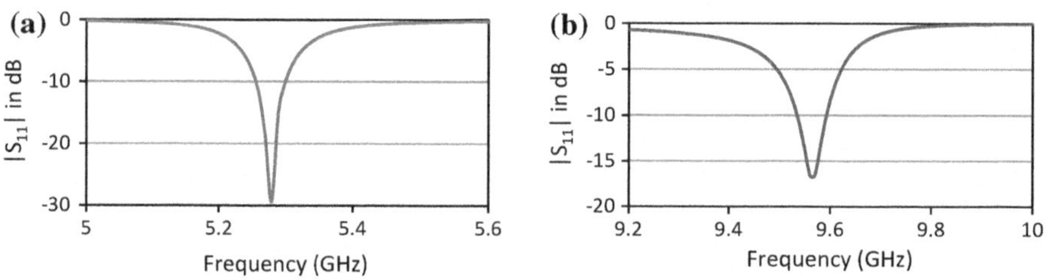

Fig. 3 Simulated return loss for **a** fundamental resonating frequency 5.28 GHz and, **b** first harmonic frequency 9.57 GHz

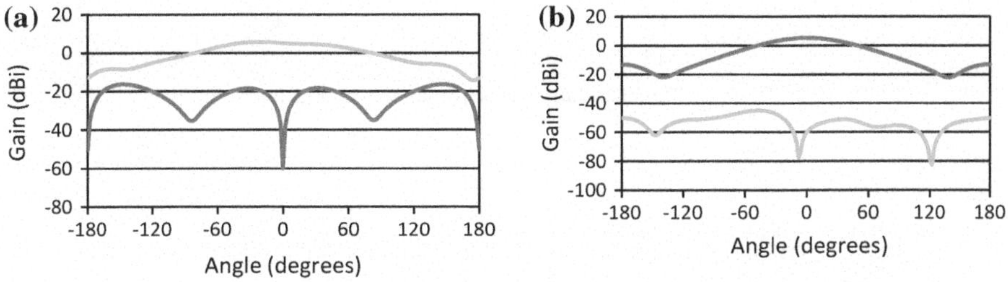

Fig. 4 Simulated **a** E plane and **b** H plane radiation patterns for frequency 5.28 GHz

Fig. 5 Schematic diagram of compact dual band antenna with multi-slots

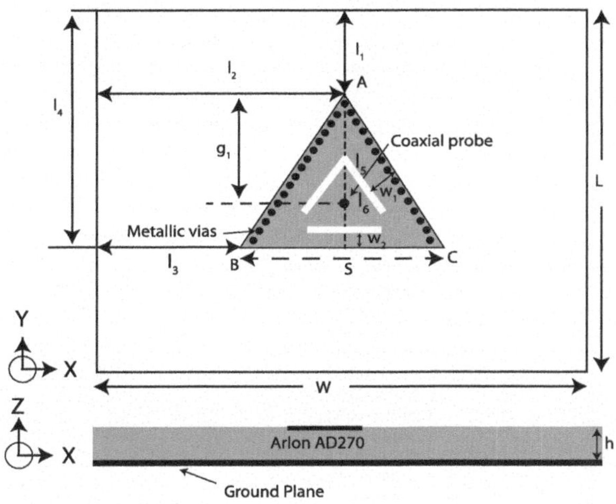

3 Dual Band Compact Antenna

3.1 The Configuration

The parent HMSIW triangular antenna is miniaturized by incorporating multi-slots for dual band operation. Figure 5 depicts the new antenna with its various dimensions shown in Table 2. The new antenna is loaded with an inverted V-shaped slot and a rectangular slot in the middle thereby causing the excited surface current to meander along its path which leads to enhancement of its equivalent capacitance and inductance, finally causing reduction in resonating frequency.

From the surface current distribution shown in Fig. 6, it is evident that the former inverted V-shaped slot lowers the higher order TE_{202} mode, while the lowering of the fundamental TE_{101} mode is caused by the presence of the rectangular slot.

Table 2 Various dimensions of compact dual band antenna (in mm)

g1	l_5	l_6	l_7	w_1	w_2
17.75	13.77	11.52	18	4.71	5

Other parameters same as the parent antenna

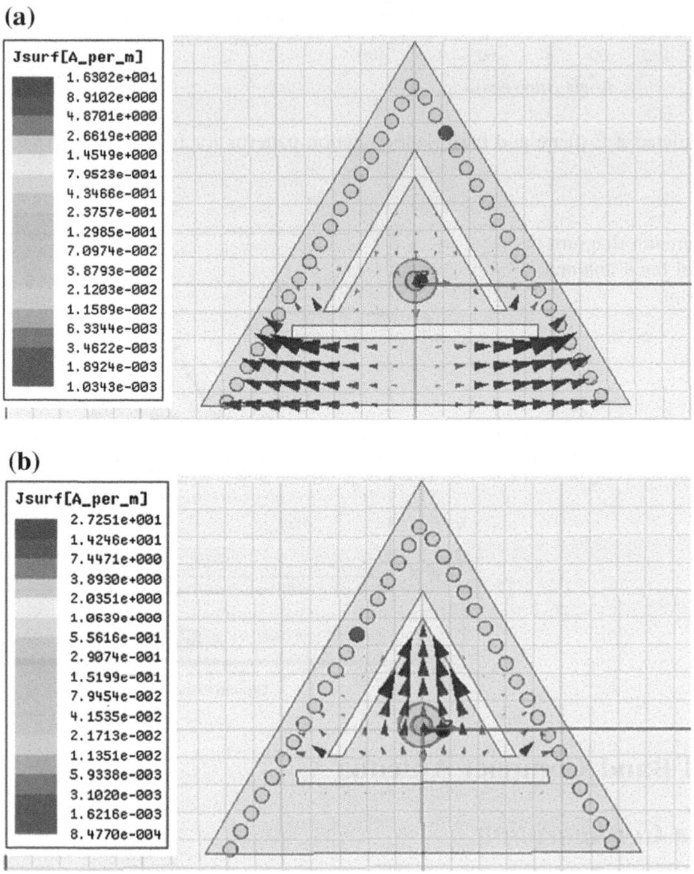

Fig. 6 Surface current distribution for **a** fundamental frequency (4.3 GHz). **b** First harmonic frequency (9.45 GHz)

Fig. 7 S_{11} for the dual band compact antenna

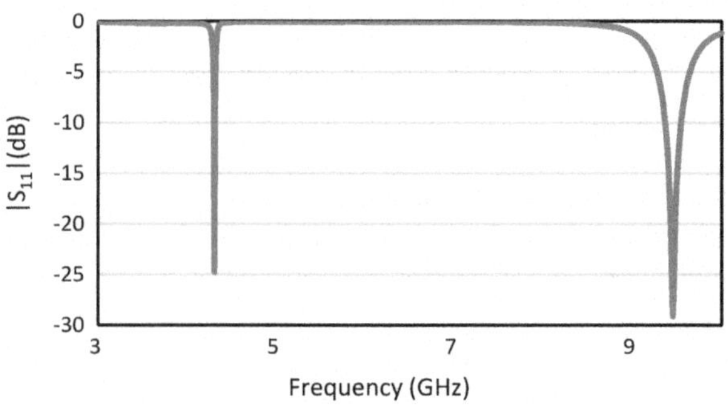

At 4.3 GHz, the simulated S_{11} for the slotted HMSIW antenna is −24.8 dB and the same is −29.14 dB at 9.45 GHz as shown in Fig. 7 and the shift in resonating frequency is shown in Fig. 8. The simulated E plane and H plane radiation pattern of the antenna is shown in Figs. 9 and 10 for frequencies 4.3 and 9.45 GHz respectively. From the figures it is evident that the peak values of co-pol E and H plane radiation are 3.53 and 3.1 dBi respectively for the fundamental frequency of 4.3 GHz and 8.21 and 6.76 dBi respectively for the first harmonic frequency 9.45 GHz, while the cross-polarization is lowered to the extent of 30 dBi as compared to the corresponding co-polarization values.

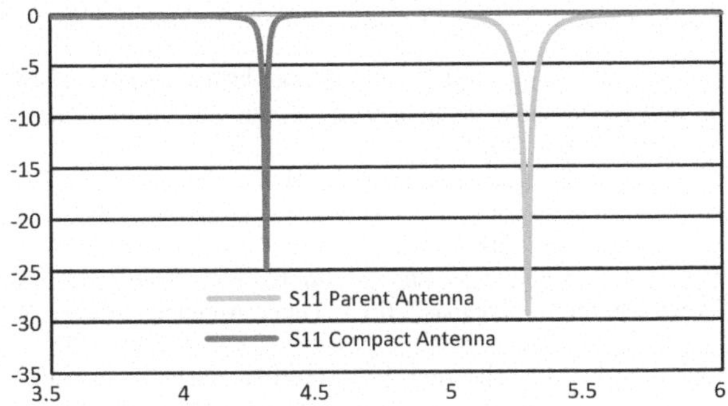

Fig. 8 Shift in fundamental resonating frequency

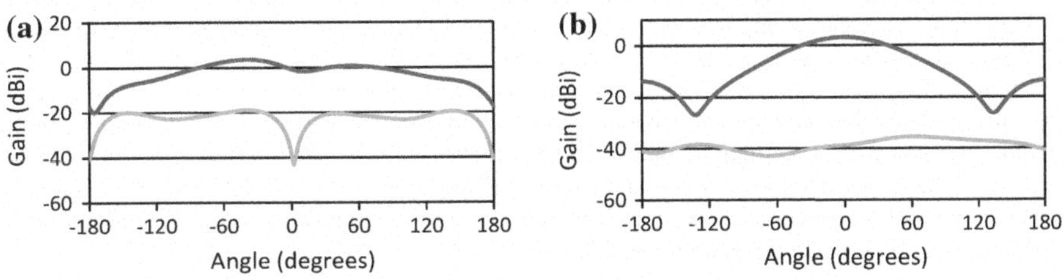

Fig. 9 Simulated **a** E plane and **b** H plane radiation patterns for fundamental frequency 4.3 GHz

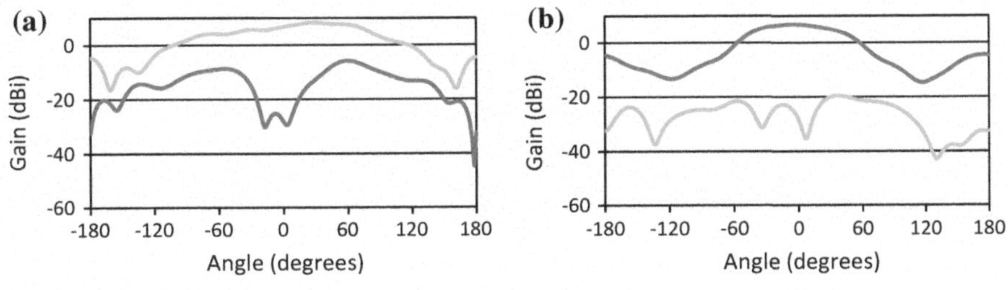

Fig. 10 Simulated **a** E plane and **b** H plane radiation patterns for first harmonic frequency 9.45 GHz

4 Conclusion

From the HMSIW triangular antenna, resonating at 5.28 GHz for TE_{101} mode and 9.57 GHz for next higher TE_{202} mode, the new compact dual-band antenna has been designed and proposed by the authors. The dual band antenna is found to exhibit a shift in resonating frequency of 980 and 120 MHz for both the modes respectively, with little degradation in its performance. Hence, arrays using such antenna may be designed to achieve higher gain, required for specific applications.

References

1. Q. Lai, C. Fumeaux, W. Hong, and R. Vahldieck, "Characterization of the propagation properties of the half-mode substrate integrated waveguide," IEEE Trans. Microw. Theory Tech. Vol. 57, pp. 1996–2004, 2009.
2. J. Xu, W. Hong, H. Tang, Z. Kuai, and K. Wu, "Half-mode substrate integrated waveguide (HMSIW) leaky-wave antenna for millimeter wave applications," IEEE Antennas Wireless Propag. Lett. Vol. 7, pp. 85–88, 2008.
3. Q.H. Lai, W. Hong, Z.Q. Kuai, Y.S. Zhang, and K. Wu, "Half-mode substrate integrated waveguide transverse slot array antennas," IEEE Trans. Antennas Propag. Vol. 57, pp. 1064–1072, 2009.
4. G.M. Zelinski, G.A. Thiele, M.L. Hastriter, M.J. Havrilla, and A.J. Terzouli, "Half width leaky wave antennas" IET Microwave AP. Vol. 1, pp. 341–348, 2007.
5. Soumen Banerjee, Biswarup Rana, and Susanta Kumar Parui, "Substrate integrated waveguide based triangular antenna arrays," Microwave and Optical Technology Letters, Vol. 58, no. 3, pp. 675–679, March 2016.
6. Soumen Banerjee, Biswarup Rana and Susanta Kumar Parui, "SIW based compact and dual-band equilateral triangular antennas," Journal of Electromagnetic Waves and Applications (Taylor & Francis), Vol. 30, no. 5, pp. 637–650, 2016.
7. Soumen Banerjee, Twinkle Mohanty, Sneha Das and Biswarup Rana, "Slot-loaded Compact HMSIW Triangular Antennas," Int. Conf. on Microelectronics, Communication and Computing, NIT Durgapur, India, January 2016.
8. Soumen Banerjee, Biswarup Rana and Susanta Kumar Parui, "A Miniaturized Semi-circular Half-mode SIW Antenna," 6th Int. Conf. on Computers and Devices for Communication (CODEC), Kolkata, India, December 2015.

A Compact Half-Mode SIW Based Semi-circular Antenna with X-Shaped Slot

Soumen Banerjee, Twinkle Mohanty, Sneha Das and Biswarup Rana

Abstract A half mode substrate integrated waveguide (HMSIW) based semi-circular antenna is chosen which operates at 5.22 GHz within the IEEE 802.11a WLAN UNII-1 band. The same antenna is miniaturized by the inclusion of x-shaped slot. The compact antenna is found to operate at the center frequency of 4.35 GHz thereby witnessing a shift of 870 MHz from 5.22 GHz.

1 Introduction

In context to development of antennas, substrate integrated waveguide (SIW) is a state-of-the-art technology providing high quality performance with low-cost fabrication. It is observed to preserve the advantages of conventional waveguides while simultaneously allowing mounting of one or more chip sets on the same substrate referred as system-on-substrate (SoS). A variation of SIW known as HMSIW has already been proposed and several such antennas have been implemented [1–11]. The authors have chosen a HMSIW semi-circular antenna operating at 5.22 GHz [12] and have miniaturized the same by introducing an x-shaped slot near the middle of the antenna. Metallic vias are placed all along the semi-circular edge making it a PEC wall. The new compact antenna is found to operate at a shifted frequency of 4.35 GHz thereby reflecting miniaturization. The antennas are designed on Arlon AD270 substrate and its various parameters are obtained through simulation using ANSYS make HFSS v15.0.

S. Banerjee (✉) · T. Mohanty · S. Das
Department of ECE, University of Engineering & Management,
Kolkata, West Bengal 700160, India
e-mail: prof.sbanerjee@gmail.com

B. Rana
Department of ETCE, Indian Institute of Engineering Science & Technology,
Shibpur, West Bengal 711103, West Bengal, India

© Springer Nature Singapore Pte Ltd. 2017

I. Bhattacharya et al. (eds.), *Advances in Optical Science and Engineering*,
Springer Proceedings in Physics 194, DOI 10.1007/978-981-10-3908-9_55

2 Methodology

The HMSIW based semi-circular shaped antenna with slot based miniaturized version is shown schematically in Fig. 1. The various dimensions of both the antennas are reflected in Tables 1 and 2 respectively. Both the antennas have radius r = 14.5 mm and they are fabricated using 0.79 mm thick Arlon AD 270 substrate with an overall size of 60 mm × 60 mm. The vias occupying the PEC wall are of 1 mm diameter with centre-to-centre distance of 1.5 mm. After optimization, the location of feed for the compact antenna is g = 5.6 mm and they excite the TM_{010} mode as shown in Fig. 2. The electric field distribution and surface current distribution of HMSIW semi-circular antenna with slot are depicted in Fig. 3.

The optimized value of simulated return loss for the semi-circular antenna and that for semi-circular antenna with slot are −25.5 and −20.2 dB respectively at the corresponding centre frequencies of 5.22 and 4.35 GHz as shown in Fig. 4. The radiation patterns of the antennas are plotted in Figs. 5 and 6. The simulated E-plane and H-plane co-polarized gain for the semi-circular antenna are 6.5 and 5.3 dBi respectively. The same for the compact antenna obtained through simulation are 3.7 and 2.3 dBi respectively. For both the antennas, the x-pol gain values are 30–40 dBi below their respective co-pol values. Table 3 depicts the antenna parameters for both the antennas.

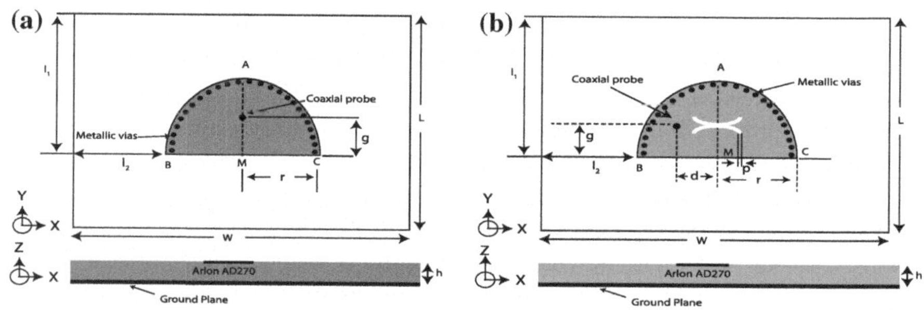

Fig. 1 Schematic diagram of **a** HMSIW semi-circular antenna and **b** HMSIW semi-circular antenna with slot

Table 1 Dimensions of HMSIW semi-circular antenna (in mm)

L	W	r	h	l_1	l_2	g
60	60	14.5	0.79	39	15.52	10.5

Table 2 Dimensions of the compact antenna (in mm)

L	W	r	h	l_1	l_2	g	p	d
60	60	14.5	0.79	39	15.52	10.5	1	9.5

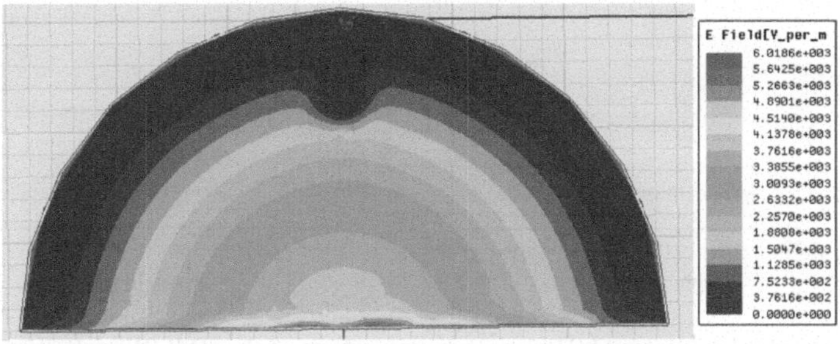

Fig. 2 Electric field distribution for TM_{010} mode at 5.22 GHz for HMSIW semi-circular antenna

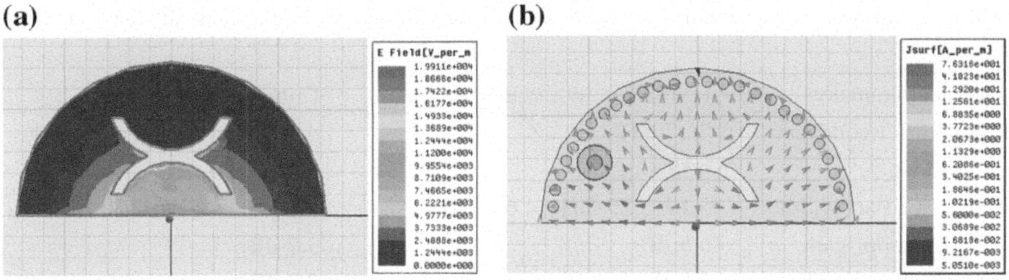

Fig. 3 **a** Electric field distribution. **b** Surface current distribution of HMSIW semi-circular antenna with slot

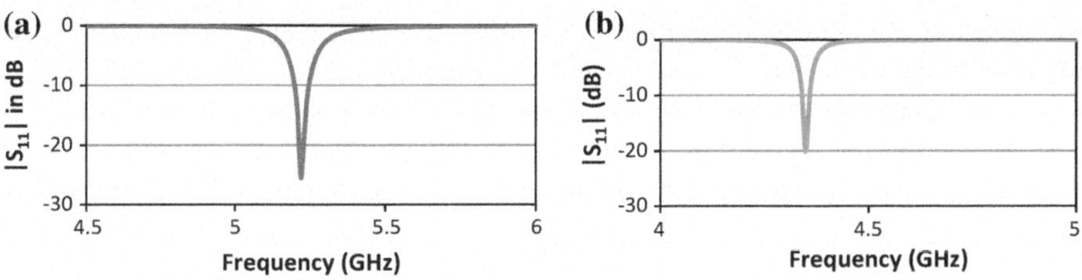

Fig. 4 Simulated S_{11} parameter of **a** HMSIW semi-circular antenna and **b** HMSIW semi-circular antenna with slot

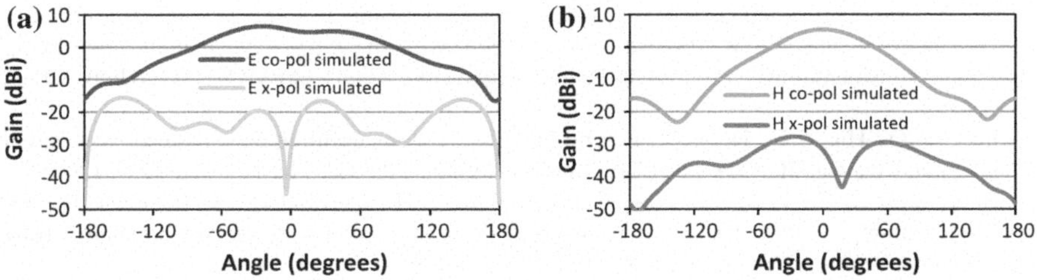

Fig. 5 Simulated **a** E-plane and **b** H-plane radiation pattern of the HMSIW semi-circular antenna at 5.22 GHz

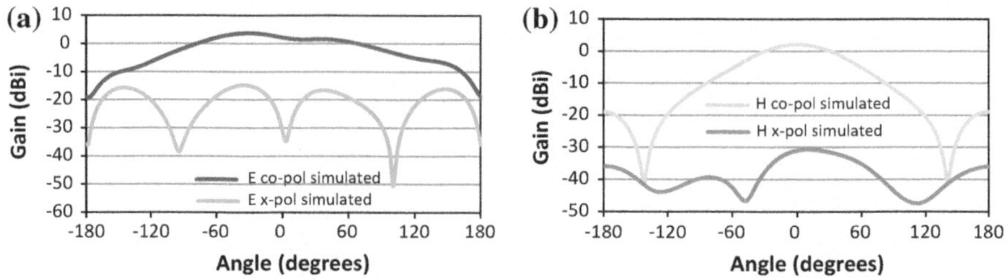

Fig. 6 Simulated **a** E-plane and **b** H-plane radiation pattern of the HMSIW semi-circular antenna with slot at 4.35 GHz

Table 3 Comparison of antenna parameters of HMSIW semi-circular antenna and compact antenna

Antenna parameters	HMSIW semi-circular antenna	Compact antenna
Resonating frequency	5.22 GHz	4.35 GHz
Return loss	−25.5 dB	−20.2 dB
Impedance bandwidth	5.19–5.25 GHz = 60 MHz	4.36–4.34 GHz = 20 MHz
E-plane co-pol gain	6.5 dBi	3.7 dBi
H-plane co-pol gain	5.3 dBi	2.3 dBi

3 Conclusion

The compact antenna exhibits high value of co-polarized gain and hence arrays using such antennas may be constructed for attaining higher gain and directivity. Moreover, such array antennas would require lesser space as they possess low value of mutual coupling. It is thus expected that the research findings would be of immense value to the researchers for further research activities in the same domain.

References

1. Yan, L., Hong, W., Hua, G., Chen, J. X., Wu, K., and Cui, T. J., "Simulation and experiment on SIW slit array antennas," IEEE Microw. Wireless Compon. Lett. 14, 446–448, (2004).
2. Deslandes, D., and Wu, K., "Substrate integrated waveguide leaky wave antenna: concept and design considerations," Proc. Asia-Pacific Microwave Conf. (APMC'05), Suzhou, China, (2005).
3. Luo, G. Q., Hu, Z. F., Dong, X. L., and Sun, L. L., "Planar slit antenna backed by substrate integrated waveguide cavity," IEEE Antennas Wireless Propag. Lett. 7, 236–239, (2008).
4. Bohorquez, J. C., Pedraza, H. A. F., Pinzon, I. C. H., Castiblanco, J. A., Pena, N., and Guarnizo, H. F., "Planar substrate integrated waveguide cavity-backed antenna," IEEE Antennas Wireless Propag. Lett. 8, 1139–1142, (2009).
5. Awida, M. H., and Fathy, A. E., "Substrate integrated waveguide Ku-band cavity-backed 2 × 2 microstrip patch array antenna," IEEE Antennas Wireless Propag. Lett. 8, 1054–1056, (2009).

6. Wang, H., Fang, D. G., Zhang, B., and Che, W. Q., "Dielectric loaded substrate integrated waveguide (SIW)—plane horn antennas," IEEE Trans. Antennas Propag. 58, 640–647, (2010).
7. Dashti, H., and Neshati, M. H., "Development of Low-profile patch and semi-circular SIW cavity Hybrid antennas," IEEE Trans. Antennas Propag. 62, 9, 4481–4488, (2014).
8. Dashti, H., and Neshati, M. H., "Comparative investigation of Half-mode SIW cavity and Microstrip Hybrid antenna using different patch shapes," IJE Transactions A: Basics, 27, 10, 1573–1580, (2014).
9. Banerjee Soumen, Rana Biswarup, and Parui, Susanta Kumar, "Substrate integrated waveguide based triangular antenna arrays," Microwave and Optical Technology Letters, 58, 3, 675–679, (March 2016).
10. Banerjee Soumen, Rana Biswarup, and Parui, Susanta Kumar, "SIW based compact and dual-band equilateral triangular antennas," Journal of Electromagnetic Waves and Applications (Taylor & Francis), 30, 5, 637–650,(February 2016).
11. Banerjee Soumen, Mohanty Twinkle, Das Sneha and Rana Biswarup, "Slot-loaded Compact HMSIW Triangular Antennas," Int. Conf. on Microelectronics, Communication and Computing, NIT Durgapur, India, (January 2016).
12. Banerjee Soumen, Rana Biswarup, and Parui, Susanta Kumar, "A Miniaturized Semi-circular Half-mode SIW Antenna," 6th Int. Conf. on Computers and Devices for Communication (CODEC), Kolkata, India, (December 2015).

Topical Survey on Daylighting System

Divya Pandey, Rajeev Ranjan, Rishabh Raj, Anukriti Tyagi and R. Navamathavan

Abstract Over ages science and technology has grown a lot. In day to day life new inventions and discoveries are made. Science and technology has reached into every sector of our life. From quite a few time there has been a need for the introduction of new technology, utilizing daylight in building and to monitor their performance for the benefit of those concerned with the building design practitioners lighting engineers, builders, product manufacturers, building owners and property managers. There are many different options and systems which serve many different purposes. Henceforth it is difficult to find the apt product. The overview and the description of the product help a lot to choose the desired daylighting system for the given conditions. In the present work theoretical design of holographic optical element has been done to filter out ultraviolet and unwanted portion of solar radiation and to allow only visible spectrum to enter inside the interior of the room. Such system provides even distribution of sunlight within the interior of the building which is hygienic and supportive to human health and activities as well as to reduce energy demands. To redirect the uniform daylight diffracted by the HOE into the remote interiors of a building, light pipes or optical fibres are quiet promising. For cost effectiveness, Plastic Optical Fibres (POFs) are used as wave guide for daylighting.

D. Pandey · R. Ranjan
Department of Physics, National Institute of Technology, Jamshedpur, India

R. Raj (✉)
SENSE, VIT University Chennai, Chennai, India
e-mail: rishabh.raj50@gmail.com

A. Tyagi
SCSE, VIT University Chennai, Chennai, India

R. Navamathavan
Division of Physics, School of Advanced Sciences, VIT University Chennai, Chennai, India

© Springer Nature Singapore Pte Ltd. 2017

I. Bhattacharya et al. (eds.), *Advances in Optical Science and Engineering*, Springer Proceedings in Physics 194, DOI 10.1007/978-981-10-3908-9_56

1 Introduction

In this ever growing technological era, people prefer easier, quicker and cheaper means to accomplish their endeavours without having concerns about the aftermath. To meet out the demands of residential as well as commercial building in urban area with growing population towering and crowded buildings are being constructed as a result sunrays do not reach every nook and corner of the building. Using artificial lights during daytime, which seems to be a trivial issue to moot, is in fact causing a major energy drain and environmental break down.

The estimation of energy consumption due to electric lighting in buildings has soared up to an appalling figure of approximately 40–50% of the total energy cost [1]. Buildings employing efficient daylighting techniques have plummeted electric lighting energy consumption by 50–80% [2]. Moreover, these light sources lack in the distribution needed for complete biological functions. The three light sources namely incandescent lamp, energy efficient fluorescent light and cool white florescent lamp is widely used in daylighting system. Presently these lamp lack blue portion of the spectrum which is very beneficial part for humans. Thus redirecting the sun rays which contains all spectrum beneficial to human health becomes necessary. Our prepared system regarding implementation of holographic optical elements to redirect sun rays into the interior of a building is capable of redirecting the entire spectrum of sun rays into interior of building [3].

In the sphere of physical sciences, daylighting points out to a system that delivers, in a given space, well administered and occupant responsive lighting with favourable environmental attribute and reasonable financial value. A given area is said to be day lit when not only it is swamped with natural light as primary source of illumination but also when it renders thermally and visually comfortable place which also manifests outdoor views. Further, scientists at the Lighting Research Centre (LRC), in Troy, N.Y. have reported the enhanced productivity and comfort levels of daylight-exposed occupants. Also, it caters to the mental and visual stimulation that tunes human circadian rhythms [4].

The obvious visual dividends of daylighting include its broad electromagnetic spectrum with excellent colour rendering property. The non-visual benefits of daylighting are too efficacious to ignore. This is particularly true in case of psychological and physiological outcomes. Daylighting significantly scales down the psychological sadness or Seasonal Affective Disorder (SAD) stimulated by the artificial light [5]. Energy savings is one of the most crucial benefactors of daylighting. The benefits run the gamut from reduction of fossil fuel consumption, reduction in greenhouse gas emission to the decreased heating and cooling load from artificial light. Above all, it avoids double conversion of energy i.e. transforming solar energy into electricity using PV cells and then into artificial light. Daylighting can be successfully employed not only in residential and commercial buildings but also in educational facilities, laboratories, workshops, healthcare facilities, photo bioreactors etc.

This work focuses on daylighting solution with the help of optical fibres and HOE's. Daylighting can be vaguely split into three categories namely, visible light, Ultra-Violet (UV) rays and Infra-Red (IR) waves. In our circumferential solar radiation, ultraviolet (UV) accounts for 7% and infrared accounts for 46% of radiation. However 46% of the solar radiation falls under visible range [6, 7]. IR radiation of the solar spectrum is responsible for producing heat inside the interior which is not suitable for hot climate. Further, prolonged exposure to solar UV radiation may result in acute and chronic health effect on skin, eyes and immune system. Therefore, visible light should reach the interior free from UV and IR to avoid spatial heating and for protection of optical fibre.

1.1 Various Daylighting Systems

While considering a daylighting system for a building three important factors should be considered are (i) solar shading, (ii) glare control and (iii) Redirecting the sunlight inside the building [8–10].

Daylighting with shading and without shading are the two major classification of daylight systems [11]. Both the categories are operational with direct and diffused sunlight. Shading strategy consists of installing Louvers and blinds, light shelves, turnable lamellas, lasercut panels are systems for direct sunlight application whereas prismatic panels, anidoliczenithal opening, mirror elements, prisms and venetian blinds for diffused skylight.

Daylighting without shading method is in cognate with the optical system modus operandi. Zenith light guiding elements with holographic optical elements, fish system, anidolic ceiling, light shelf comes under diffused light guiding system whereas prismatic panels, laser cut panels, lighting guiding glass, HOE falls under direct light guiding system. A holographic optical element is a new class of optics that operates on the principal of diffraction. Conventional optical elements use their shape to bend light but holographic recording materials changes the optical property by variation of refractive index. Mostly conventional method of daylighting techniques are bulky and require maintenance and adjustments for optimal performance unlike HOE system, which usually have no moving part and require low maintenance.

HOEs qualify to fulfill all the requirements which must be met by an ideal light delivery system. A single HOE can be used to redirect the sunlight to the interior part of a building and can also be used as solar shading and glare controlling device [12–14].

Use of as holographic windows [14] for daylighting application in buildings is quite attractive for last three decades. Holographic windows redirect sunlight from the immediate window area into the rear of the room so as to illuminate the darker regions with reduced glare. Main feature of HOEs are large optical apertures, light weight, thin film geometry, and cost effectiveness. HOEs are wavelength selective and hence they can also be made to operate over a narrow wavelength band.

Further, holographic optical elements (HOE) are used advantageously as filters [15–17] to filter out ultraviolet and unwanted portion of solar radiations while remaining portions of useful solar radiation are made to diffract into interior of buildings with reasonably good diffraction efficiency. From the architectural point of view, HOEs have certain useful properties for light transmission and radiation control.

Now in these days people are utilizing solar energy for daylighting application in various means using HOEs [13, 14], fiber optics [5, 18] and others ways [19]. Each technique has its own limitation. A comparative study has been given by Ullah and Wang [20].

2 Theoretical Background

In order to analyze the spectral characteristics of volume phase transmission holograms, we use the coupled-wave theory [21] which gives analytical equations for the diffraction efficiency assuming refractive index variation to be sinusoidal. When the wavelength of the reconstructing beam satisfies Bragg's law the diffraction efficiency is given by

$$\eta = \frac{\sin^2\left\{\left(\xi^2 + \upsilon^2\right)^{\frac{1}{2}}\right\}}{\left(1 + \frac{\xi^2}{\upsilon^2}\right)} \tag{1}$$

where, parameters ξ and υ are defined by the following relations

$$\xi = \delta \frac{2\pi n}{\lambda} d \sin\theta \quad \& \quad \upsilon = \frac{\pi n_1 d}{\lambda \cos\theta} \tag{2}$$

where n_1 is the depth of refractive index modulation, d is the film thickness, n is the average refractive index of the medium, λ is the free space wave-length of the reconstruction light beam and δ is the angular deviation in radians with respect to Bragg's angle θ.

Bragg's angle θ is related to the fringe spacing Λ recorded in the hologram through the relation given by

$$\sin\theta = \frac{\lambda}{2n\Lambda} \quad \& \quad \cos\theta = \left\{1 - \left(\frac{\lambda}{2n\Lambda}\right)^2\right\}^{\frac{1}{2}} \tag{3}$$

When the illumination is made at Bragg's angle (i.e. $\delta = 0$) we have from (1)

$$\eta = \sin^2 \nu \tag{4}$$

$$\eta = \sin^2 \left(\frac{\pi n_1 d}{\lambda \cos \theta} \right) \tag{5}$$

$$\eta = \sin^2 \left(\frac{\pi n_1 d}{\lambda \left\{ 1 - \left(\frac{\lambda}{2n\Lambda} \right)^2 \right\}^{\frac{1}{2}}} \right) \tag{6}$$

Using (6) variation in diffraction efficiency (η) for a holocon with wavelength at Bragg's angle for different values of depth of refractive index modulations of holocon recording has been plotted. While drawing the curves care has been taken to ensure that criteria for thick phase transmission holographic lens are fulfilled for which (6) holds good. A holographic lens is said to be thick if its Q parameter $\left(Q = \frac{2\pi\lambda d}{n\Lambda^2}\right)$ is greater or equal to 10 [22].

3 Experimental

3.1 Holographic Lens Recording and Reconstruction

Experimental setup for recording holographic lens is shown in Fig. 1. For present work a holographic lens has been recorded on commercially available high resolution silver halide plate PFG-01 (film thickness d = 8 μm and average refractive

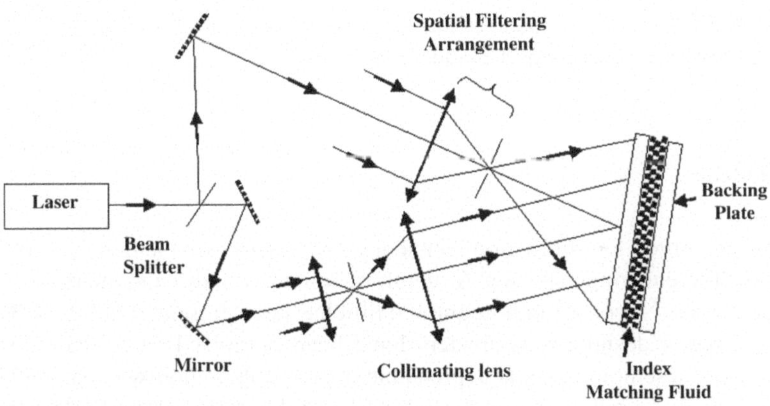

Fig. 1 Schematic of the geometry for recording holographic lens

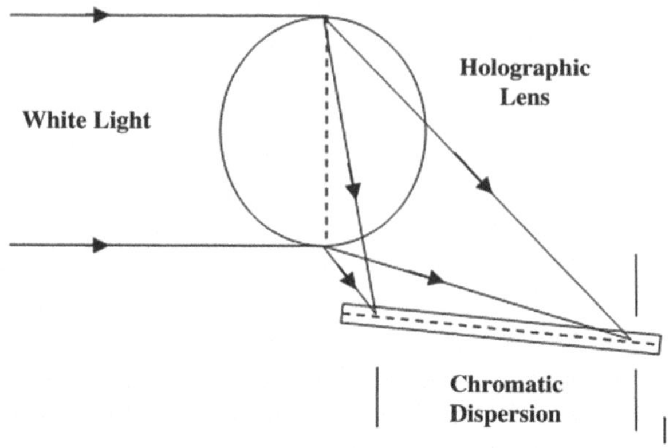

Fig. 2 Schematic of the reconstruction setup of holographic lens

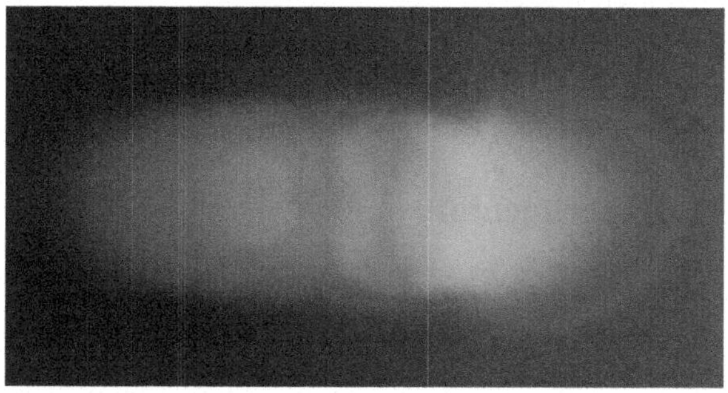

Fig. 3 Photograph of the Spectrum of white light diffracted by a typical holographic lens

index n = 1.61) using a He–Ne Laser ($\lambda = 0.6328$ μm) of power 12 mW. The recorded hologram is processed using standard procedure.

When recorded holographic lens is illuminated with white light, chromatic dispersion will occur. Schematic of chromatic dispersion through hololens is shown in Fig. 2 whereas experimental evidence is in Fig. 3.

4 Results

Theoretical curves for diffraction efficiency 'η' versus wavelength 'λ' for holograms of different fringe spacing Λ, depth of refractive index modulation 'n_1' and film thickness 'd' recorded in holographic film has been presented in Fig. 4. For the curves of Fig. 4 designing parameters have been optimized such that diffraction efficiency of these holograms is appreciable over visible region whereas diffraction efficiency of these holograms is poor for ultraviolet region. Such holograms are suitable for cold climate. Diffraction efficiency versus wavelength curves can be optimized for IR region and are suitable for hot climate condition.

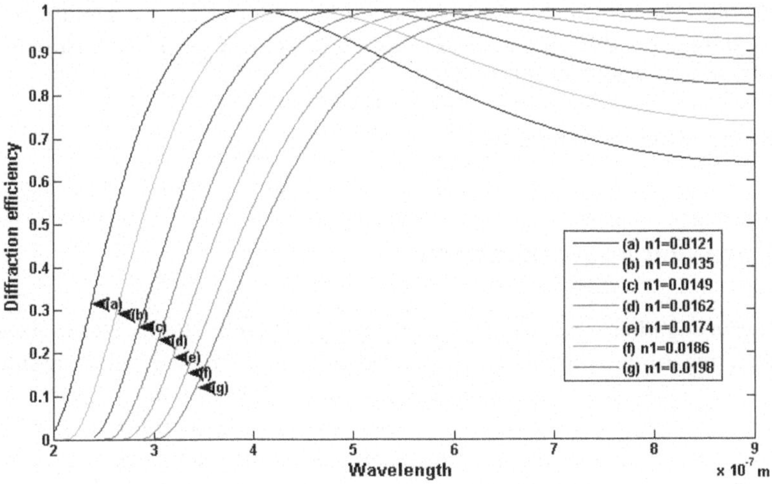

Fig. 4 Variation of diffraction efficiency with wavelength for different values of depth of refractive index modulation (n_1) at fixed value of Λ = 0.51 μm, n = 1.61 and d = 16 μm

5 Application of Holographic Filter in Daylighting Using Optical Fibre

Fibre optic cables by means of thin solid fibre transmits light with high efficiency by total internal reflection. Apart from its ease of installation, absence of mirrors at bends and uniform illumination throughout, fibre optics shows incomparable colour rending property. What makes optical fibre a strong contestant in daylighting systems is its controllability and ease of application. Apart from this, it also gains control over glare, heat gain and consistency issues that besets the conventional daylighting designs. Additionally, it steers sunlight quite deep into buildings without altering space layout or introducing complications like heat gain, glare, disruption of the intensity and variability in lighting, duration of exposure to light etc., which generally leads to fatigue, mood shifts such as Seasonal Affective Disorder.

Silica has a very good light transmission, but it is expensive, especially for bundle production and quartz fibres are fragile and rigid. Glass fibres have light attenuation higher than silica fibres, but they are considerably cheaper and more flexible.

Generally Plastic Optical Fibre are used as fibre bundles as they have low-bending radii, high levels of flexibility, low cost, higher strength and better acceptability for complex wiring in buildings [23]. A light guide material, Polymethylmethacrylate polymer, is employed in POF [24, 25]. It has good light transmission properties, especially for white lights.

An ideal daylighting system should cut down all sorts of spatial heating causes. One of them is exposure by UV and IR waves. Another reason to filter out IR waves is their prolong exposure to POFs can drastically deteriorate the performance of plastic fibre. Hence suitable filtering should be employed so that the fibre is unaffected by UV and IR radiations. In this work, we have proposed the use of HOE filter to do the required task [26, 27].

Furthermore, the light collected by HOE has to be concentrated to pass through the aperture made up of fibre ends.

6 Discussion and Conclusion

Present investigation reveals that properly designed HOEs fulfils all the necessary conditions regarding redirection of solar radiation into the interior of a building so as to ensure cost effectiveness daylighting supportive to human health. Further designing parameters of daylighting holograms can be optimized to filter out unwanted portion of solar radiation depending upon the requirement of hot or cold climate condition thereby minimizing the cost of cooling or heating the interior of a building. Such filters may advantageously be implemented to optical fibres used to redirect the filtered solar radiation to the remote interior of a building.

References

1. Green Building Rating Systems—Draft Recommendations for a U.S. Rating System; U.S. Green Building Council: Bethesda, MD, USA, (1995).
2. Ossa, A., Gottfried, D.A., Walsh, T., Simon, L.N. Sustainable Building Technical Manual: Green Building, Design, Construction and Operations; U.S. Public Technology Inc.: Los Angeles, CA, USA, (1996).
3. Franta, G., Anstead, K. "Daylighting Offers Great Opportunities" Window and Door Specifier —Design Lab, Spring; pp. 40–43 (1994).
4. Leslie, R.P., Capturing the daylight dividend in buildings: why and how? Building and Environment 38, 381–385 (2003).
5. Sapia, C., "Daylighting in buildings: Developments of sunlight addressing by optical fiber," Solar Energy 89, 113–121 (2013).
6. Benton, C.C., "Daylighting can improve the quality of life and save energy", Architectural lighting 1, 46–48 (1996).
7. Liberman, J. Light Medicine of the Future. New Mexico: Bear and company publishing. (1991).
8. KuppaswamyIyengar—"Sustainable Architectural Design; An over view",—Book, Routledge, Taylor & Francis (2015).
9. CIBSE Code for Interior lighting, Charted Institution of building Services Engineers (CIBSE); London, U.K., (1994).
10. Ruck, N., et.al, Daylight in Buildings: A Source Book on Daylighting Systems and Components, IEA SHC Task 21/ ECBCS Annex 29, Report Lawerence Berkeley National Laboratory, Berkeley, CA 94720 (2001).
11. Martin kischkoweit-lopin, an overview of daylighting system, solar energy 73(2), 77–82, (2002).
12. Ludman, J.E., "holographic solar concentrator", Appl. Opt. 21, 3057(1982).

13. Muller, H.F.O., Prof. Dr.-Ing "Application of holographic optical elements in building for various purposes like daylighting, solar shading and Photovoltaic power generation" Renewable energy, Vol. 5, part II, pp. 935–941, (1994).
14. Tholl, H.D., Stojanoff, C.G., Kubiza, R., and Willbold-Lohr, G., "Design Optimization and manufacturing of holographic windows for daylighting applications in buildings," in Optical Materials Technology for Energy Efficiency and Solar Energy Conversion XII, C.M. Lampert, ed., July 13–14, San Diego, California, Proc. SPIE 2017, 35–45 (1993).
15. Wang Bo, Chang Liang, TAO Shiquan, "Design optimization of volume holographic gratings for wavelength filters", Proceedings of SPIE Vol. 5636 (SPIE, Bellingham, WA), 267–276, (2005).
16. Shakher, C., Yadav, H.L., Dependence of diffraction efficiency of holographic concentrators on angle of illumination hologram-thickness and wavelength of illuminating light, Journal of Optics **21**(6), pp. 267–272 (1990).
17. Todorov, T., Markovski, P., Mazakova, M., Miteva, M., Razsolkov, V., Pancheva, M., Spectral Characteristics of Thick Phase Holographic Gracting, Optica Acta 379–388 (1981).
18. Ullah, I., Shin, S., "Highly concentrated optical fiber-based daylighting systems for multi-floor office buildings," Energy Build. 72, 246–61 (2014).
19. Franta, G, Anstead, K. "Daylighting Offers Great Opportunities," Window and Door Specifier- Design Lab, Spring, 40–3 (1994).
20. Irfan Ullah, Allen Jong-Woei Wang, "Development of Optical Fiber—based Daylighting System and Its Comparison", Energies 8, 7185–201 (Open Acess), ISSN 1996-1073 (2015).
21. Kogelnik, H., Coupled wave theory for thick hologram grating, Bell Syst, Tech. J., 48, 2909 (1969).
22. Gaylord, T. K., and Moharm M. C., Thin and thick gratings: Terminology Clarification, Appl. opt., **20**, 3271 (1981).
23. Deepak P. R., Madhu Sharma, Fiber Optic Daylighting System, International Journal of Engineering Research and Technology (IJERT) ISSN: Vol. 4 Issue 03, 2278–0181, (2015).
24. M. Ravi Kumar, a review of optical fibers for daylighting, International Journal of Instrumentation Electrical Electronics Engineering, Vol. 1, No. 2, (2013).
25. Ullah, I., Shin, S. Development of optical fiber-based daylighting system with uniform illumination. J. Opt. Soc. Korea 16, 247–255 (2012).
26. Optical Coating Japan, Mirrors, Optical Coatings Japan (2000) URL: http://www.techmark.
27. Kupo. (2000). URL: http://www.Dichroic.Com.tw/ (2002-04-28).

Performance Modifications of a Dielectric Ring Resonator Loaded Monopole Using a Composite Frequency Selective Surface

Rudraishwarya Banerjee, Biswarup Rana and Susanta Kumar Parui

Abstract Here, a simple frequency selective surface (FSS) is intuitively designed, and two such FSS are placed at right angle with each other, with a small predetermined air gap between them, to form a composite FSS structure, and a conical dielectric ring resonator (DRR) loaded monopole antenna is placed in between the two FSS. The DRR loaded monopole offers wide impedance bandwidth from 6 to 20 GHz, with a peak gain of 4 dBi. This composite FSS causes rejection of frequency band of nearly 2 GHz from anywhere in between 6 and 20 GHz, the wavelength of the center frequency of the rejected band being proportional to the dimension of the FSS and distance of the composite FSS from the antenna. Hence the composite FSS shows bandstop character, whereas it also causes 6–7 dB gain enhancement in its passband. The FSS structure is non complex, cost effective, and easy to fabricate.

1 Introduction

Frequency selective surfaces (FSSs) are planer periodic metallic structures etched on dielectric substrate, and it has found its various applications in the performance modification and enhancement of different antennas. FSS can be used to control the transmission and/or reflection of an incident electromagnetic (EM) wave and hence shows bandpass or bandstop characteristics [1]. Recently, FSSs with multiple independent transmission bands are required for various applications, and hence several approaches are explored by the researchers to present multi-band transmissions such as use of multiple layers [2], concentric square loops [3], anchor shaped geometries [4], lumped element resonators [5], and so on. On the other hand, dielectric resonator antenna (DRA) has attracted significant attention due to its various advantageous properties, and different shapes of dielectric ring resonator loaded monopole antenna has been already studied [6–8].

R. Banerjee · B. Rana (✉) · S.K. Parui
Indian Institute of Engineering Science and Technology, Shibpur, Howrah, India
e-mail: b.rana.in@ieee.org

© Springer Nature Singapore Pte Ltd. 2017
I. Bhattacharya et al. (eds.), *Advances in Optical Science and Engineering*,
Springer Proceedings in Physics 194, DOI 10.1007/978-981-10-3908-9_57

Here, two simple FSS are kept at right angle with each other, to form a composite FSS, and a small air gap is introduced between the two FSS. An ultra wide band (UWB) conical shaped dielectric ring resonator (DRR) loaded monopole antenna is placed at a predetermined distance from the composite FSS. It is observed that the FSS behaves like a bandstop filter, the wavelength of its stop band being directly proportional to its dimension, and distance from the antenna. On the other hand, it enhances the gain by nearly 6 dB in its passband. The bandstop characteristic of the composite FSS is verified through parametric study. The most interesting part of the work is the use of FSS along with UWB DRR loaded monopole, which is proposed here for the first time, as far our knowledge.

2 Antenna Configuration

The dimension and the top view of the small conical piece of dielectric resonator, that has been used to design the DRR is shown in Fig. 1a, b, where H = 6 mm, and rU = 2.5 mm and rL = 7.5 mm. Figure 1c shows the cross sectional view of the conical DRR loaded monopole antenna, where a = 2 mm and p = 11 mm.

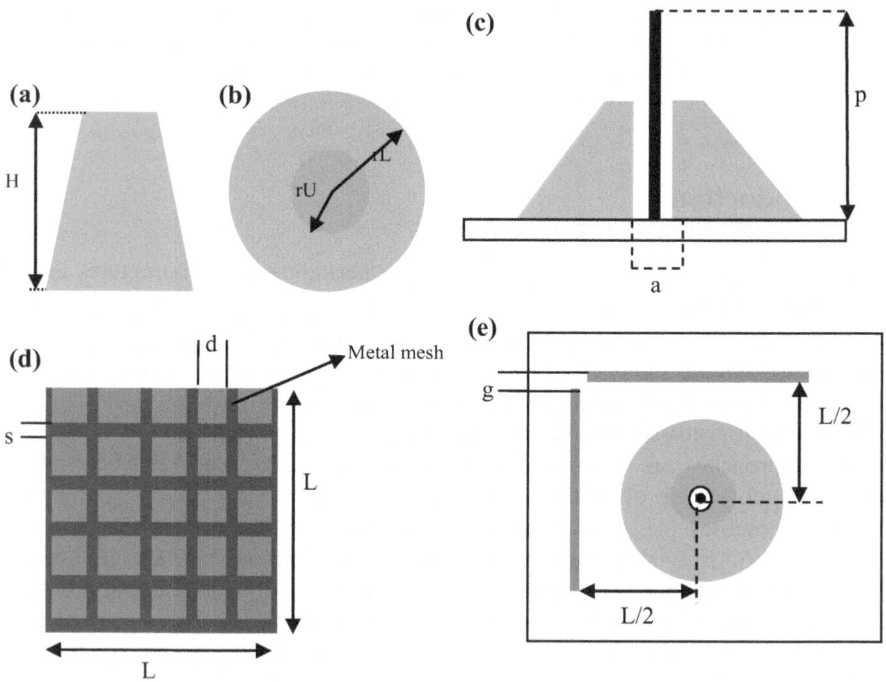

Fig. 1 **a** Cross-sectional view of the conical DRA. **b** Top view of the conical DRA. **c** Cross sectional view of the DRR loaded monopole. **d** Structure of the single FSS. **e** Top view of the proposed configuration (DRR loaded monopole with two FSS placed at right angle to each other)

The dimension of the FSS is proportional to the wavelength of the rejected frequency band. The geometry of the proposed FSS is shown in Fig. 1d, where it is observed that the FSS has square cross section with a simple metallic mesh designed on a Arlon ($\varepsilon = 2.7$) substrate. 'L' indicates the length of the FSS, and 'd' and 's', which characterize an unit cell of the FSS, are kept 5 and 1 mm respectively. Here, the FSS for a particular frequency 10 GHz, for which L = 30 mm, is shown in the figure.

The top view of the proposed configuration is given in Fig. 1e, where it is seen that two similar FSS are kept at right angle to each other, with a small gap 'g' = 1 mm between them at one corner of the ground plane. The distance between the antenna and each of the FSS is kept L/2. The rejected frequency band is a function of L. The diagram is given for L = 30 mm. Two FSS, kept at right angle to each other with a small predetermined gap of 1 mm between them, forms a composite FSS structure.

3 Simulation Results for Optimum Design Parameters

Initially in Fig. 2, the return loss characteristics of the only UWB DRR loaded monopole is shown. Then in Fig. 3 the S_{11} parameter of proposed configuration is studied for different values of L, and the observation is presented in a tabular form below in Table 1.

Hence, it is evident that the composite FSS shows bandstop behaviour, where the dimension 'L' is equal to the wavelength of the center frequency of the rejected frequency band.

Taking L = 30 mm, the radiation pattern in the X-Z plane of the proposed configuration and the standalone UWB antenna is compared at 6 and 16.3 GHz, and shown in Figs. 4 and 5. It is clearly seen that the peak gain of the UWB antenna is nearly 3.5 dBi at both 6 GHz and 16.3 GHz, whereas, after using the composite FSS, the radiation pattern become directive due to reflection from the composite

Fig. 2 Simulated S_{11} (dB) and of the DRR loaded monopole antenna (rU = 2.5 mm, rL = 7.5 mm, H = 6 mm, p = 11 mm, a = 1.5 mm)

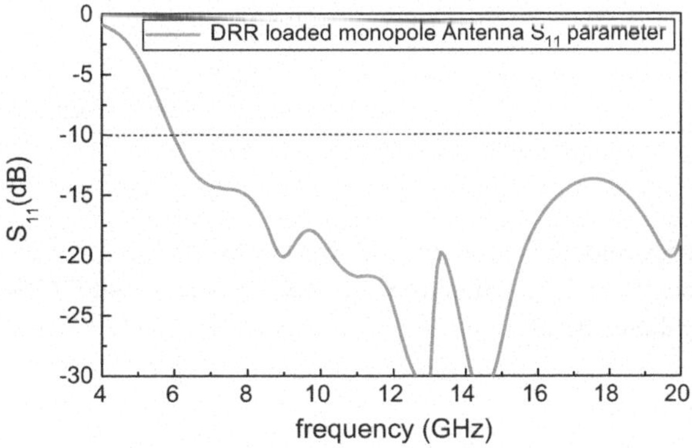

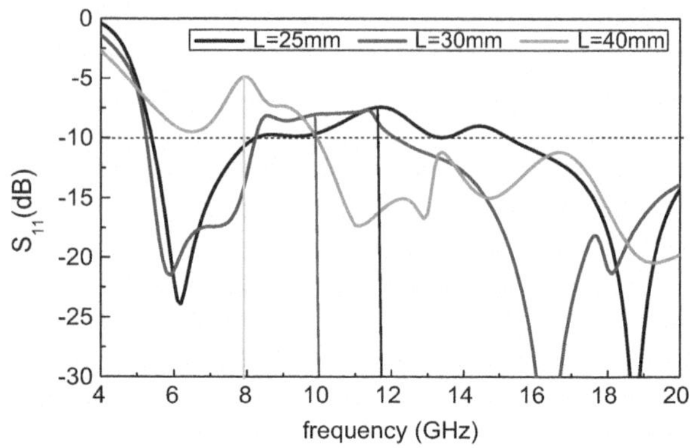

Fig. 3 Simulated S_{11} (dB) and of the proposed configuration (rU = 2.5 mm, rL = 7.5 mm, H = 6 mm, p = 11 mm, d = 5 mm, s = 1 mm, g = 1 mm)

Table 1 Stop bands for different frequencies

Calculated frequency (GHz) and wavelength	L in terms of wavelength (mm)	Stop band obtained in GHz
10 GHz, λ = 30 mm	L = λ = 30	10 GHz (8–12 GHz)
12 GHz, λ = 25 mm	L = λ = 25	11.9 GHz (10–14.5 GHz)
7.5 GHz, λ = 40 mm	L = λ = 40	8 GHz (up to 10 GHz)

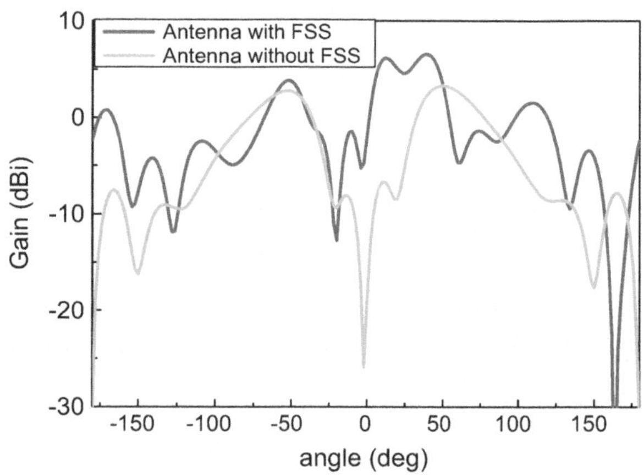

Fig. 4 Comparison of the simulated radiation pattern in X-Z plane at 6 GHz of the proposed configuration and the only UWB antenna (rU = 2.5 mm, rL = 7.5 mm, H = 6 mm, p = 11 mm, g = 1 mm, d = 5 mm, s = 1 mm, L = 30 mm)

FSS and the peak gain increases to 8 dBi at 6 GHz and 11 dBi at 16.3 GHz at nearly −55°. Thus the use of the composite FSS causes nearly 5–7 dB gain enhancement.

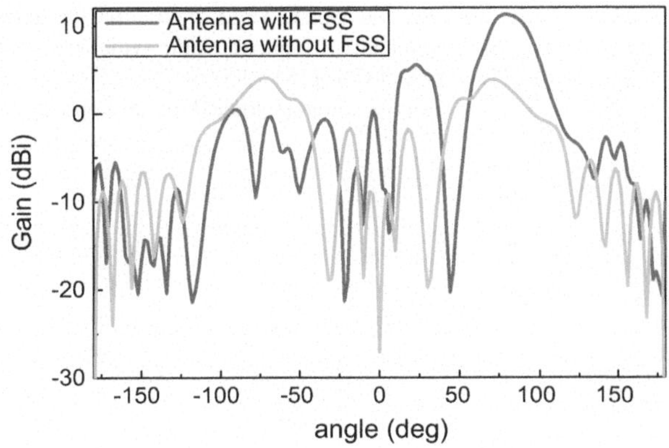

Fig. 5 Comparison of the simulated radiation pattern in X-Z plane at 16.3 GHz of the proposed configuration and the only UWB antenna (rU = 2.5 mm, rL = 7.5 mm, H = 6 mm, p = 11 mm, d = 5 mm, s = 1 mm, L = 30 mm, g = 1 mm)

4 Conclusion

It is evident that the proposed composite FSS, though quite simple in structure, but successful to eliminate any frequency band of 2 GHz from 6 to 20 GHz which is covered by this DRR loaded monopole, depending on the dimension of the FSS and distance of the composite FSS from the antenna, along with significant gain enhancement in its passband. But, there are various parameters that control the performance modification of the antenna which need detailed study and analysis. All the theoretical details will be revealed in near future and fabrication and experiment will be done to validate this concept. Moreover, the role of the gap in between the two FSS is to be investigated with greater attention.

Acknowledgements Authors like to thank IIEST, Shibpur and UGC, Govt. of India, for providing financial support.

References

1. Wu, T. K., [Frequency Selective Surface and Grid Array], Wiley, New York, USA, (1997).
2. Salehi, M., and Behdad, N., "A second-order dual X-/Ka-band frequency selective surface," IEEE Microwave and Wireless Components Letters 18, 785–788 (2008).
3. Wu, T. K., "Four-band frequency selective surface with double square loop patch elements," IEEE Transactions on Antennas and Propagation 42, 1659–1663, (1994).
4. Yan, M., Qu, S., Wang, J., Zhang, J., Zhou, H., Chen, H., and Zheng, L., "A miniaturized dual-band FSS with stable resonacne frequencies of 2.4 GHz/5 GHz for WLAN applications," IEEE Antennas and Wireless Propagation Letters 13, 895–898 (2014).
5. Xu, R. R., Zhao, H. C., Zong, Z. Y., and Wu, W., "Dual-band capacitive loaded frequency selective surfaces with close band spacing," IEEE Microwave and Wireless Components Letters 18, 782–784 (2008).
6. Guha, D., Gupta, B., and Antar, Y. M., "Hybrid monopole-DRAs using hemispherical/conical-shaped dielectric ring resonators: Improve ultra wideband designs," IEEE Transactions on Antennas and Propagation 60(1), 393–398, (2012).

7. Guha, D., Gupta, B., and Antar, Y. M., "New pawn-shaped dielectric ring resonator loaded hybrid monopole antenna for improved ultra-wide bandwidth," IEEE Antennas and Wireless Propagation Letters 8, 1178–1181 (2009).
8. Ozzaim, C., "Monopole antenna loaded by a stepped radius dielectric ring resonator for ultrawide bandwidth," IEEE Antennas and Wireless Propagation Letters 10, 843–845 (2011).

Part XII
Cryptography, Micro-Electronics and VLSI

Chaotic Elements—A Novel Physical Cryptographic Primitive for Document Authentication

Sajan Ambadiyil and V.P. Mahadevan Pillai

Abstract Today's digital revolution has also made the flexibility to produce the forged document very easily. It is possible to secure the valuable document by adding a combination of several materials based security features like substrate, design, print and ink based features. However, any manufactured material based feature is susceptible for imitation. With the advent of sophisticated technological progress, fraudsters require only the time to copy these features. Hence, to avoid all such counterfeiting threats, it is better to utilise unpredictable, disordered, latent or microscopic intrinsic unclonable elements, having confusing nature and are formed from a totally random process independent from human intervention. In the absence of a general term, this element can be termed as "CHAOTIC ELEMENTS" and this paper introduces a novel promising technology based on these chaotic elements. This shall address all the existing challenges in the conventional as well as contemporary security technologies for valuable documents without incurring extraordinary high costs. The natural disorder and entropy of the chaotic elements can be taken as cryptographic protocols and primitives for authentication whose security does not rest on the usual unproven number-theoretic assumptions

1 Introduction

Identification, authentication, and integrity checking are important tasks for ensuring the security and protection of valuable documents [1]. Earlier itself, several sophisticated security features like the watermark, intaglio printing, security fibres and window thread were introduced into the realm of document security. Until the emergence of the colour copier, most of the security features were based

S. Ambadiyil (✉)
Center for Development of Imaging Technology, Thiruvanathapuram, Kerala 695027, India
e-mail: ambadycdit@gmail.com

V.P. Mahadevan Pillai
Department of Optoelectronics, University of Kerala, Thiruvanathapuram, Kerala 695058, India

© Springer Nature Singapore Pte Ltd. 2017

I. Bhattacharya et al. (eds.), *Advances in Optical Science and Engineering*, Springer Proceedings in Physics 194, DOI 10.1007/978-981-10-3908-9_58

on Optically Invariable Devices (OIDs). As the Optically Invariable Devices are independent of the angle of illumination and observation, such devices could become easily copiable by the advanced colour copiers and other four colour reproduction systems. In order to overcome this, various optically variable devices (OVDs) such as colour changing inks, holograms and kinegrams are developed contemporaneously. In general, these devices (OVDs) are unattainable by any of the advanced colour copy or four colour reproduction systems. OVD features are reported to be the most advanced security features which are gaining more and more popularity today. This is because; a defining feature of OVDs is that their appearance in terms of colour and imagery is strongly angular dependent. Though the technology is developed day by day and the capabilities of the image processing tools and mechanised systems are enhancing, keeping valuable documents secure is a constant challenge. Nevertheless, instead of eschewing the paper/plastic document; applying effective security methodologies to the same are the feasible solutions. Combination sophisticated methods based on substrate, design, print and ink which is too expensive or not accessible for counterfeiters provides high security. However, any manufactured material based feature is susceptible for imitation due to technological progress. Fraudsters require only the time to copy these features. Also, if the material itself is available to the fraudulent individuals, such features will not be useful to protect the document. To deter all such counterfeiting threats, this investigation identified an unpredictable, disordered, latent or microscopic intrinsic unclonable elements named as CHAOTIC ELEMENTS from the base material itself and utilized the same to create a Unique ID, to authenticate the valuable document.

2 Chaotic Elements

A chaotic element on security document means the physical elements which are having completely unordered and unpredictable and confusing nature and are formed from a totally random process independent from human intervention. Generally, chaotic elements on security document have physical unclonable nature.

Physical systems with some randomness, which are produced by an uncontrolled process, are good candidates for chaotic elements on the security document. It is hard to produce an exact copy of the chaotic elements on security document because of the randomness [2]. Physical unclonability is generated from the complex unpredictable intrinsic properties and is always unique even when made in the same way. The contemporary technologies used non-reproducible unique biological traits included in fingerprints, face, hand, iris, DNA etc. for person's identification. These unique elements, which derive the randomness, can also be considered as in the category of chaotic elements in the living objects. In the non-living objects like paper and plastic, traits of paper fibres, security fibres, and surface texture can be taken as the chaotic elements which are also disordered, unique, inherent, and hard-to-forge characters in the physical objects. It is not possible to fabricate another object with the same traits with the chaotic elements, even if the exact structure of the original one is known. Macroscopic chaotic elements having physical unclonable nature can create security features with an advantage of it requires only minimal added cost to incorporate, it is recognizable and it does not require specialized tools for authentication. Chaotic elements can be used as one of the primitive as a physical source of randomness either to generate cryptographic keys or to make the element for authentication. It is better to take the chaotic elements as information to be encrypted rather generates a key for authentication of the valuable document. The technology devised from this chaotic element is named as Chaotic Element Technology (CET). This technology comprises of three functionalities such as Extraction of chaotic elements, Encryption as a digital signature and Integration to the valuable documents. In this study, the security optical fibre in the banknote is considered as a chaotic element, as these fibres are implanted into the banknote in a random manner. The spatial distribution of security fibre generating varying spacing, angle, depth and fibre colour and it forms a banknote unique having physical unclonable nature. In the encryption and decryption stage, Public key encryption [3, 4] algorithms are used to validate the owner of the document. A unique ID is extracted here from the 'fingerprint' generated by the chaotic element after secure encryption. In the Integration process, the unique ID generated is converted into QR code and printed on the valuable document as well as store in the database for verification and secure track and trace.

3 Experimental Demonstartion

3.1 Extraction of Chaotic Element

In Indian banknote, dual coloured optical fibres having fluorescent properties are used as the second level security feature which is invisible under normal light. This is visible only under UV light. Figure 1a shows a banknote under the UV light having 366 nm wavelengths. The first step in the chaotic element technology for authentication is to extract the security optical fibres pattern from the banknote. Before extracting the pattern, a Region of Interest (ROI) has to be applied and only the security optical fibre patterns inside this ROI is extracted. Here the rectangular region in white watermark portion of the banknote as shown in Fig. 1b has been selected as the ROI.

As the security optical fibres have a prominent greenish colour due to fluorescence under the UV image, colour detection algorithm using MATLAB can be used [5] to extract the security optical fibres pattern from the image. Here the colour bands are separated out from the ROI of the input UV image into 3 separate arrays, one for each colour component of the three basic colours Red, Green and Blue. Then the image histogram of the three colour bands is plotted as shown in the Fig. 1c for thresholding. Here a threshold is set such that the Red and Blue bands

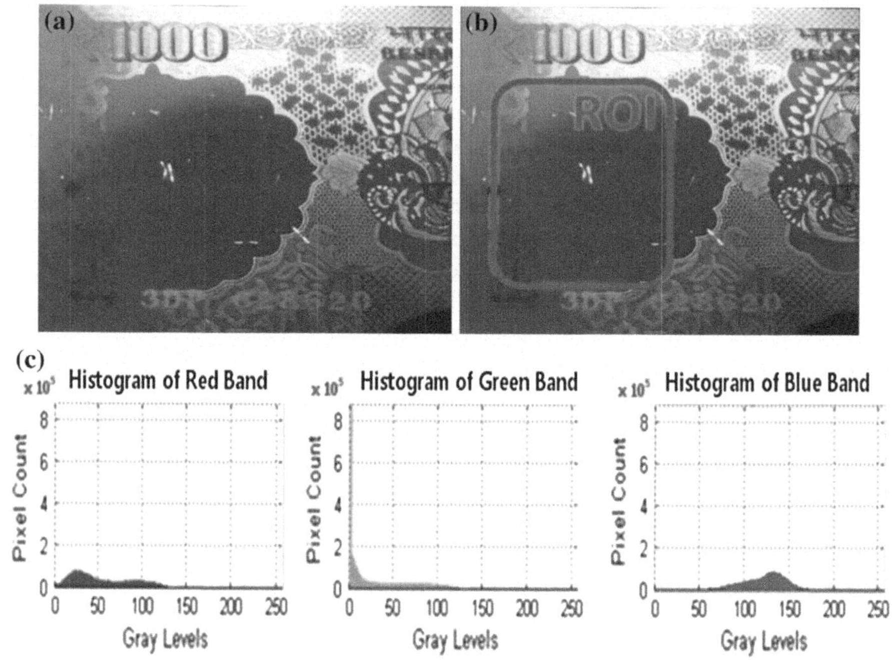

Fig. 1 **a** Banknote under UV light. **b** ROI selected in the banknote. **c** Histogram of the 3 color bands

are masked and only the green portions remain. Now the green portions of the image which are the security fibre pattern are extracted is used to generate a numerical value in the form of a corresponding matrix array [6]. Here the numerical value is taken from variable parameters such as area of each security optical fibers; X-Y coordinates of the security optical fibers, total number of security optical fibers, and shape of the security optical fibers along with value of the banknote, serial number and year of printing.

3.2 *Encryption as Unique ID*

The corresponding matrix array is then encrypted as a digital signature using an RSA cryptoservice provider. During encryption, privet key of the document issuing authority is used to generate a ciphertext whereas the public key is used to decrypt the ciphertext, which is available on the open platform. This cipher text is then encoded into QR code using QR code generation software. This QR code as a Unique ID can be easily integrated into the security documents. Flow chart for the process of UID generation is shown in Fig. 2a.

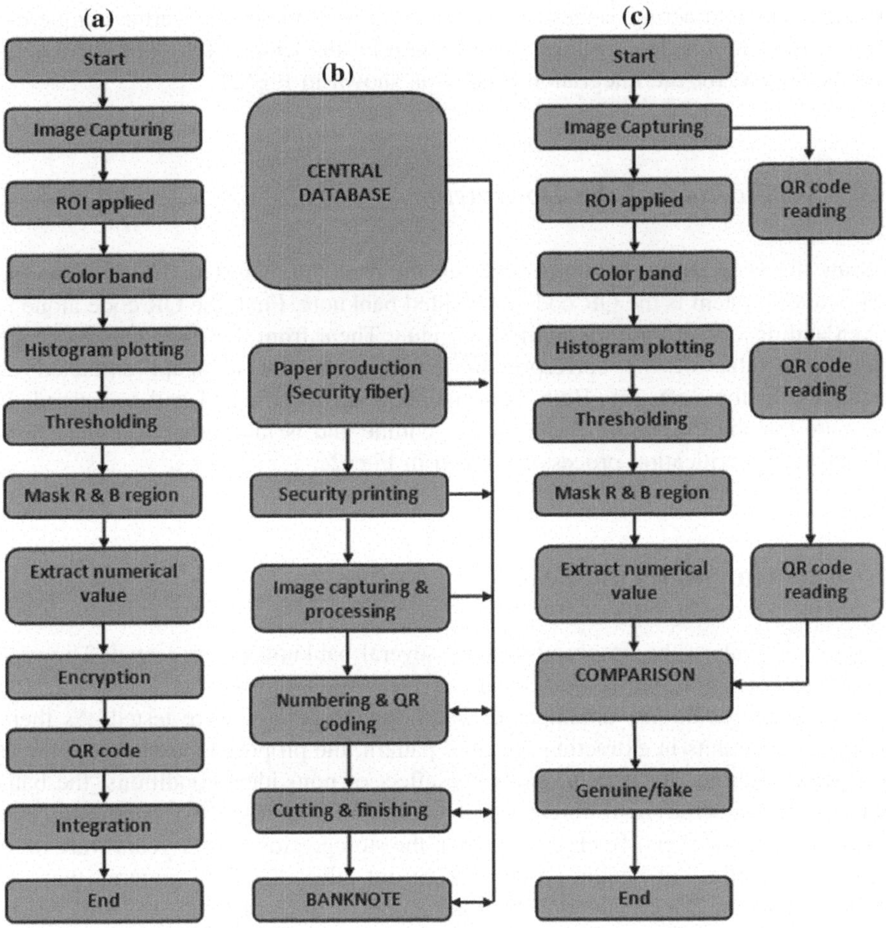

Fig. 2 Flowchart of the **a** UID generation **b** Integration and **c** verification process

3.3 *Integration of Unique ID to the Document*

In the banknote production, the major steps included are security paper production, printing and cutting and finishing the banknote [7]. During the paper preparation, the security fibres are put inside the paper not in a scalable manner; their distribution, position and number are absolutely random. Then the banknote paper is distributed several high-security printing works. After this, the numbering stage comes to incorporate the serial numbers on the banknote. Finally cutting and finishing such the final inspection and quality checking makes the banknote ready for circulation. In order to generate and embedded the Unique ID into the bank note, the image capturing and processing devices can be integrated in between the final printing and numbering stages. Then, in synchronisation with serial numbering device, the variable data printing device can print the Unique ID on the banknote. The flow chart for the integration process is shown in Fig. 2b.

3.4 Verification of the Document

During the verification, the authenticity of the banknote is tested. The input to the verification system is the QR code embedded banknote. First, the QR code alone is taken and decoded to get the numerical value. Then, from the same banknote, the numerical value of the corresponding chaotic element is again extracted as explained in the Sect. 3.1. Both information is then correlated and compared. If they are matching then the banknote is genuine and is fake otherwise. The flow chart for the verification process is shown in Fig. 2c.

4 Experimental Validation

In order to analyse the proposed system, several banknotes were tested. The proposed system was tested in both ideal and non-ideal conditions as follows. In the case of ideal conditions, untampered brand new banknotes were tested. As there were no constraints in extracting the fibre pattern, the proposed system works well in ideal conditions. In order to study the effect of non- ideal conditions, the banknote was tampered in different ways such as (a) ageing, (b) crumpling and (c) soaking in water etc. Banknotes kept in the storage closet for 4 years were used to study the effect of ageing. Ageing does not affect the fibre patterns that are embedded in the paper material. So, extraction of fibre material was still easily possible and hence the ID generated was unique even after ageing for 4 years. Also, Banknotes were crumpled thoroughly to study its effect. Crumpling only affects the surface characteristics of the paper and it will not affect the fibres embedded internally in the paper. Though the extraction of the security fibre features was possible even after crumpling of the paper, minor variation in (X, Y) coordinates has been observed. Even though, reliable authentication of the banknote is possible even after crumpling of the paper. Further, the banknote was submerged in water to study the effect of soaking with water. After drying the banknote, it was kept under UV light and the image obtained. It is observed that the fibre pattern is not affected even after soaking in water. Hence, reliable authentication of the banknote is also possible even after soaking in water.

The parameters such as (a) False Acceptance Rate (FAR), (b) True Acceptance Rate (TAR), (c) False Rejection Rate (FRR) and True Rejection Rate (TRR) [8] are also found out to appraise the performance of the proposed system.

In order to ascertain the False Acceptance Rate (FAR), ten different banknotes were taken. FAR is the ratio of the number of times the systems wrongly states a successful match between two different banknotes, to the total number of banknotes. The procedure was repeated for all the ten different banknotes. Ideally, False Acceptance Rate (FAR) of the system must be zero. False Acceptance Rate (FAR) of the current system is shown in Fig. 3a and it is observed that that the value of FAR is close to zero. In order to calculate True Acceptance Rate (TAR), ten UV images of the same banknotes were taken. True Acceptance Rate (TAR) is the ratio of the number of times the system properly states a successful match between the same banknotes, to the total number of banknotes. The procedure was repeated for all the ten banknotes to find the True Acceptance Rate (TAR). Ideally, True Acceptance Rate (TAR) must be 100%. True Acceptance Rate (TAR) of the current system is shown in Fig. 3b and it is observed that the value of True Acceptance Rate (TAR) is close to 100%. In order to ascertain False Rejection Rate (FRR), again ten different images of the same banknotes were taken. FRR is the ratio of, the number of times the systems wrongly states the difference between same banknotes, to the total number of banknotes. The above procedure was repeated for all the ten banknotes. Ideally, it should be zero. False Rejection Rate (FRR) of the current system is shown in Fig. 3c and it is observed that that the values of False Rejection Rate (FRR) is very low and close to zero.In order to obtain the True Rejection Rate (TRR), ten different banknotes were again taken. TRR is the ratio of, the number of times the system properly states the difference between two different banknotes, to the total number of banknotes. The procedure was then repeated for all ten different banknotes. Ideally True Rejection rate (TRR) must be 100%. True Rejection Rate (TRR), of the current system is shown in Fig. 3d and it is observed that that the value of True Rejection Rate (TRR) is very high and close to 100%.

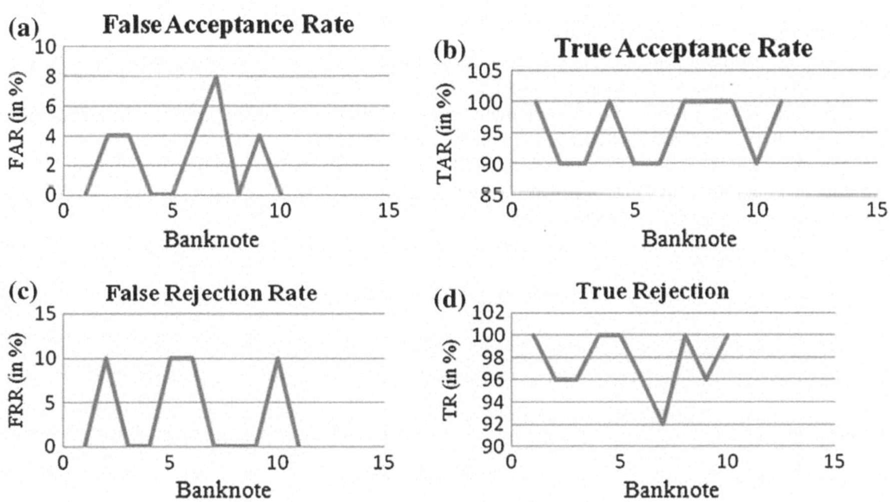

Fig. 3 Analysis of **a** FAR. **b** TAR. **c** FRR. **d** TRR

5 Conclusion

A novel cost effective chaotic element technology is introduced in this paper, which explore intrinsic characteristics of the object to prevent counterfeiting and secure track and trace of the object. The natural disorder and entropy of the chaotic elements can also be cryptographic protocols and primitives for authentication whose security does not rest on the usual unproven number-theoretic assumptions. In this investigation, it has been identified such chaotic elements for banknote, extracted the information and utilized the same to create a unique ID in the form of QR code, to authenticate the banknote. Here the randomly distributed security fibers in the banknote are taken as chaotic element for generating the unique ID. The system was analyzed in ideal and non-ideal conditions. The system performs well in both the cases. Apart from security, such unique ID can facilitate automation, secure track and trace, effective maintenance of inventory at various levels; knowledge based identification and added protection. This technology can also be integrated with the smart phone which enables the public for easy and automated verification of banknote.

References

1. Ruhrmair, U., S. Devadas, and F. Koushanfar, Security based on Physical Unclonability and Disorder, Introduction to Hardware Security and Trust, Springer, New York, 65–102 (2012).
2. P. Tuyls, B. Skoric, T. Akkermans, W. Ophey, and S. Stallinga, "Security analysis of physical uncloneable functions", Symposium On Information Theory, (2004).
3. R. L. Rivest, A. Shamir and L. Adleman, "A method for obtaining digital signatures and public-key cryptosystems", Communications of the ACM, 21(2), 120–126 (1978).
4. Jonsson, J. and B. Kaliski, "Public-Key Cryptography Standards (PKCS) #1: RSA Cryptography Specifications, Version 2.1", (2003).
5. Mathworks, Simple color detection, 20 January 2010 (updated 19 august 2010), http://www.mathworks.com/matlabcentral/fileexchange/26420-simplecolordetection.
6. Math works, Regionprops, http://www.mathworks.in/help/images/ref/regionprops.html#bqkf8ln.
7. http://www.fleur-de-coin.com/eurocoins/banknote-production.
8. Sajan Ambadiyil, K. S. Soorej and V. P. Mahadevan Pillai, "Biometric based Unique ID Generation and One to One Verification for Security documents", Elsevier Journal—Procedia Computer Science, 46, 507–516 (2015).

Analytical Modelling of Hybrid CMOS SET Rijndael Cryptography

J. Gope, S. Chowdhury, S. Chakraborty and S. Bhadra

Abstract Hybrid CMOS-SET has come up as a promising candidate for next generation ultra small, digitized, low power consuming and high speed Nano device to replace the conventional CMOS electronics. Researchers since last decade exhibitedtheir skill to develop Hybrid CMOS-SET based Nano ICs. Also contemporary research aims to incorporate the same Nano ICs in consumer electronics, health diagnostic systems as well as in cyber security. Few attempts have been made so far to fabricate nm ICs for ultra modern cyber security systems. CMOS made Rijndael cryptographic hardware is ubiquitous today, because of its excellent encryption standard. In order to model the hardware in nm region the authors here report an empirical study to implement hybrid CMOS-SET based Rijndael IC. The proposed architecture is a nm ASIC which resembles high speed during encryption beside other novelties.

1 Introduction

Post CMOS era envisaged the resurgence of numerous nano scale devices. Amid which quantum electronics plays a pivotal role in optimizing new principle of operation of molecular scale electronics. But owing to its typical material and process related limitations quantum electronics suffered serious setback. The technological shift then tends towards a new device principle where freedom of electronics was of utmost credibility. This is known as Single Electron Technology (SET) [1]. Single electronics is a fascinating technology which involves greater figure of merit deliberated from high integrating density, low power consume-ability, high processing speed, simplicity, straight forwardness, robustness and of course the potentiality to uphold one bit of information using few electrons

J. Gope (✉) · S. Chowdhury · S. Chakraborty
Department of ECE, CSET, Barasat, Kolkata, West Bengal 7000124, India
e-mail: jayanta.gope.1983@ieee.org

S. Bhadra
Department of EE, UEM, Saltlake, Kolkata, West Bengal 700156, India

© Springer Nature Singapore Pte Ltd. 2017
I. Bhattacharya et al. (eds.), *Advances in Optical Science and Engineering*,
Springer Proceedings in Physics 194, DOI 10.1007/978-981-10-3908-9_59

or a single electron. Besides, the fragility of single electronics is that the fabrication can not be obtained at room temperature operation maneuver and also it posses random background charge problem, low gain, high output impedance. This feared the Scientist that might SET alone could not replace CMOS in future VLSI/ULSI circuit design.

On the other hand CMOS is well studied since the 70s and thus a vibrant technological enhancement is omnipresent. This methodically aids a device engineer in prolific CMOS based logical designing. CMOS intrinsically foster high gain; can be operated within room temperature, does not propagate any background charge and also the speed power product lies in proximity to the Heisenberg's principle. The cosmic effects are quite relevant. Combining the goodliness of CMOS and SET, Scientist introduced a co-integrated model that diminishes the flaw of both CMOS and SET and thus Hybrid CMOS SET [2] was conceptualized. Presently Hybrid CMOS SET made novel architecture mimic the Boolean logic and categorically several Hybrid CMOS SET made logic realizations are reported [3].

On the other hand the role of internet is ubiquitous in today's era. This has augmented the electronic financial transfer system manifold. Subsequently to maintain the internet security cryptography is utmost essential. It is the technique of sending and receiving data in a concealed form so that only the specific receiver can read and process it. It facilitates inbound security to the every concerned even. It also maintains data integrity, confidentiality and authentication. In this regard all over the world, different algorithms were made available in cryptography. In 1997, the National Institute of Standards and Technology (NIST) initiated a programme to select a suitable algorithm on cryptography and in 1998, NIST announced the acceptance of 15 algorithms and after that they selected 5 as finalist. Among them Rijndael Algorithm which was designed by Joan Deemen and Vincent Rijmen, was unanimously accepted as Advanced Encryption Standard (AES) [4] in 2000. AES selects this model depending upon a number of properties such as performance, efficiency, security, flexibility and easy to implement. Additionally, this model was simulated using VHDL since its very inception. Overall this algorithm offers excellent key set up time and well key agility and also requires less memory. Owing to its unmatched advantages, authors opted to design Hybrid CMOS SET based Rijndael model.

Basically, the author here tends to incorporate Hybrid CMOS SET in a new paradigm of cyber security based cryptographic hardware realization and henceforth submit this ephemeral architecture of Hybrid CMOS SET Rijndael circuit.

2 Modelling of Rijndael Algorithm

Rijndael algorithm is unputdownable in cryptography due to some of its features like simple design technique, good speed, withstand against hacking and code compactness. For exquisite performance the algorithm uses different input block lengths like 128, 192 and 256 bits. Additionally the algorithm employs different key length and different number of rounds depending upon its block length as enunciated in the Table 1.

The entire encryption process is done subsequently in four steps [5, 6]. It initiates with 'Add Around Keys'. In this operation simple XOR function is applied between the 'State' and the 'Round Key'. The Round Key is derived from the 'Cipher Key' by Key Scheduling Technique. Besides 'State' is a simplistic but straight forward array of bytes having four rows and the number of column is equal to the block length divided by 32 intervals. The state and Round Key have the same size for this purpose. Consequently a new state is obtained following the operation

$$S(i,j) = S(i,j)\,K(i,j) \tag{1}$$

where S is the state and K is the Round Key.

This is followed by Sub Byte Transformation. Thus it is a non linear bit wise substitution of each state bit in independent arbitration. Rijndael algorithm offers only one Substitution Table defined as S-box. The designing process of S-box is such that it can prevail over attack.

Last but not least is Shift Row Transformation. Here the rows of the state are cyclically shifted by different offsets which are dependent on the block length 'Nb' and depicted in Table 2.

It is further continued using Mix Column Transformation.It is done by operating over different columns. In mix column transformation the columns of the current state are considered as polynomials of Galois Field (28). It is multiplied by modulo $x^4 + 1$ with a fixed polynomial

$$C(x) = [03]x3 + [01]x2 + [01]x + [02] \tag{2}$$

The decryption process is done by the reverse of encryption process i.e. Inverse Sub Byte, Inverse Shift Row Transformation, and Inverse Mix Column Transformation. Noticeably the Add Around Key operation is same as it perform the XOR operation, but the positions are reversed.

Table 1 Mapping of block length, key length and no. of rounds

Block length (bits)	Key length	No. of rounds
128	4	10
196	6	12
256	8	14

Table 2 Offset value of row shifting

Block length	Shift of row 1	Shift of row 2 (byte)	Shift of row 3 (byte)	Shift of row 3 (byte)
4	No change	1	2	3
6	No change	1	2	3
8	No change	1	3	4

Different types of hardware implementation has already made by different Researchers so far. CMOS based Rijndael Model [7] plays a great role in this area. But CMOSs face adverse effects from low power density, sub 10 nm physical limitations like problems. Then Jayanta and Prakash [8] attempted SET based Rijandael circuit. But it was a hypothetical approach only and also SET suffers from low current drive, background charge effect, lack of room temperature operable technology like troubles. So authors here have tried to make a Hybrid CMOS SET based Rijndael Model which compensates the drawbacks of CMOS and SET internally.

3 Hybrid CMOS Set Macromodelling of Rijndael Cryptography Technique

The authors adhered to PARTSIM simulator as the macro model of hybrid CMOS SET comprises of typical node applications. The node to node analysis is imperative for obtaining transient response and also it includes all the virtues of co tunneling phenomena within a Hybrid CMOS SET model. The versatility of PARTSIM is robustness, simplicity, easiness and above all it takes less computational time compared to other existing simulators.

Mohammad Reza Karimian et al., initiated the first macro modeling consisting of Quantizer to reinforce the SET tunneling phenomena [9]. Figure 1 depicts the proposed macro-model of the basic Hybrid CMOS SET inverter circuit using PARTSIM.

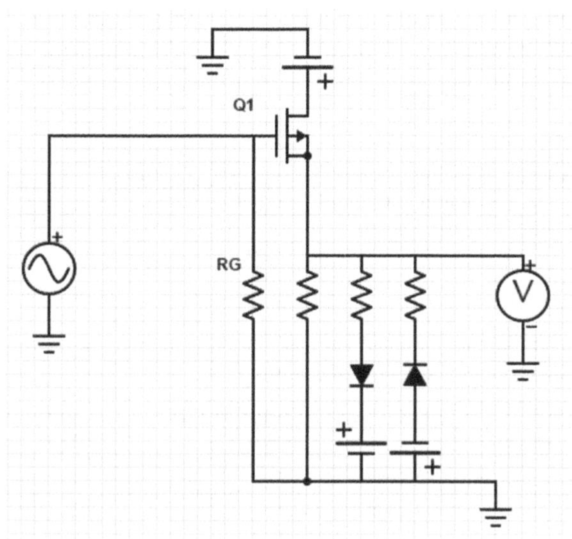

Fig. 1 Proposed macro-model of the basic Hybrid CMOS SET inverter circuit using PARTSIM

The Hybrid CMOS SET Affine Transformation module of Rijndael Cryptography Technique is demonstrated in Fig. 2, and the Inverse Affine Transformation module is followed next in Fig. 3 and subsequently both are simulated using PARTSIM.

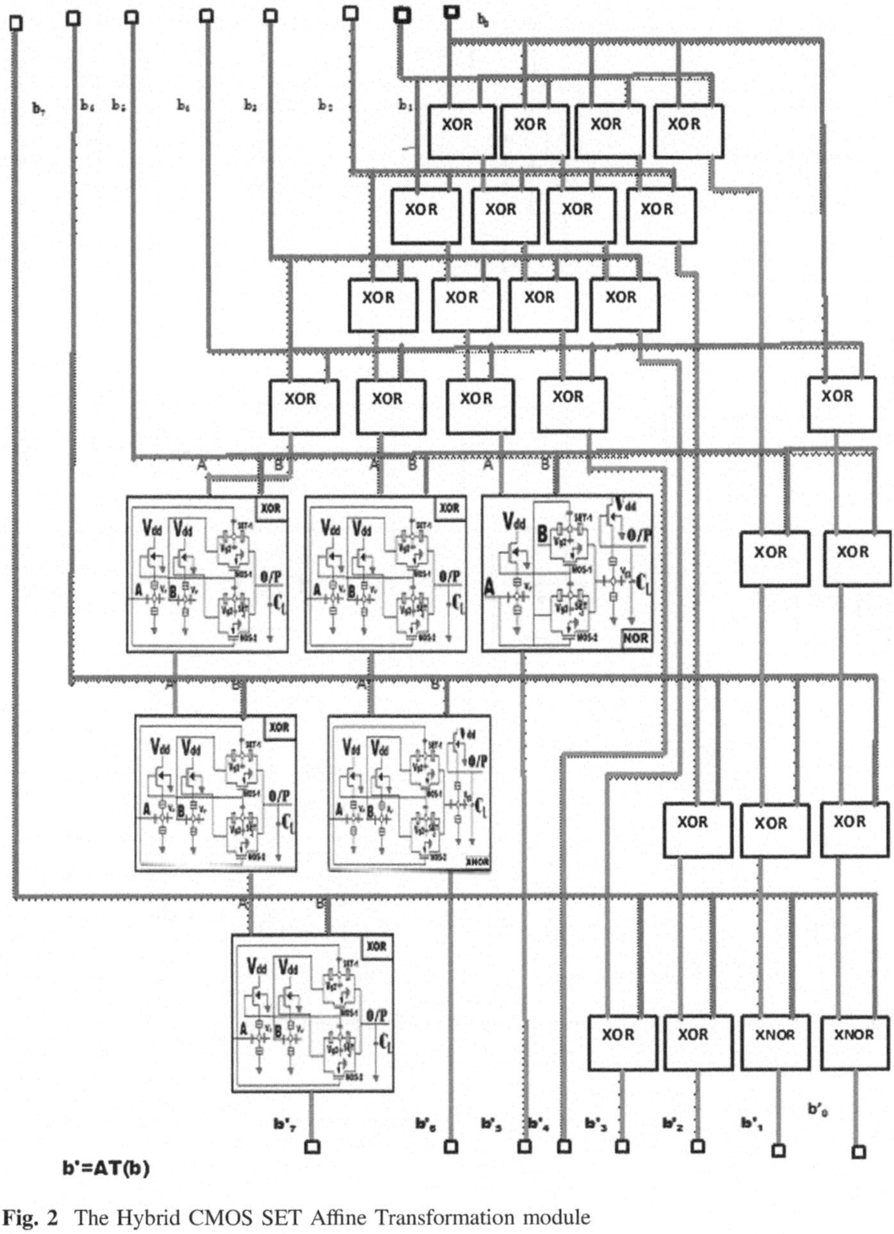

Fig. 2 The Hybrid CMOS SET Affine Transformation module

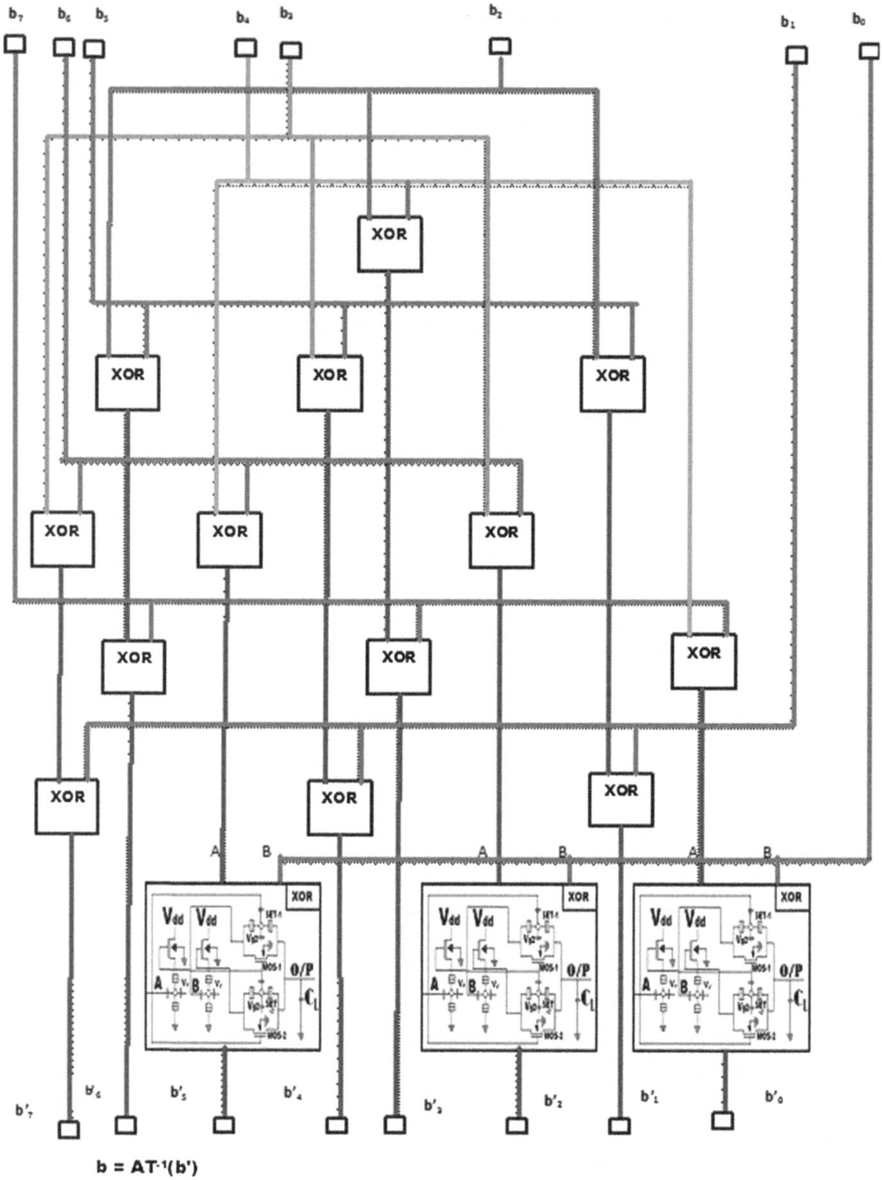

Fig. 3 The Hybrid CMOS SET Inverse Affine Transformation module

4 Conclusion

Simulation of both the structures reveal that the merits of Hybrid CMOS-SET modelling is finest and also high speed electron tunneling is omnipresent in both CMOS and SET part. It is a comprehensive ASIC designing of Rijandael Cryptography. The intrinsic Coulomb Blockade effect is controlled via exchange of regulated flow of electron supply from input V_{gs}. It is the ultimate form of device research where the authors can manipulate the flow of electrons. Empirical study aptly insights the goodliness of Hybrid CMOS SET compared to conventional CMOS. The offerings are limited to requirement of less number of gates compared to conventional CMOS Rijandael technique. As the number of gates are less so the propagation delay is also lessened which offers high speed. The integration density is also high and owing such high integrating density the circuit becomes non volatile. Furthermore, the typical aspirations of Rijandael is enunciated by true means. Thus the authors advocate for more Hybrid CMOS SET modellings in cryptography in near future.

References

1. Vinay Pratap Singh, et al., "Analytical Discussion of Single Electron Transistor (SET)," International journal of Soft Computing and Engineering (IJSCE). Papers 2(3), (2012).
2. Santanu Mahapatra, et.al., "Analytical modelling of Single Electron Transistor for Hybrid CMOS SET Analog IC Design," IEEE TRANSACTION ON ELECTRON DEVICE. Papers 51(11), (2004).
3. D. Samanta, S.K. Sarkar, "A simple SET-MOS universal hybrid circuit realization of all basic logic functions," IEEE Advances in Engineering, Science and Management (ICAESM), International Conference. Papers 336–339 (2012).
4. Dr. Reinhard Wobst, "The Advanced Enryption Standard (AES): The Successor of DES," Information Security Bulletine. Papers 31–40 (2001).
5. Prof. N. Penchalaiah et al. "Effective Comparison and Evaluation of DES and Rijndael Algorithm (AES)," International Journal on Computer Science and Engineering, Papers 02(05), 1641–1645 (2010).
6. J. Daemen and V. Rijmen, AES Proposal: Rijndael. Papers 2, (1999).
7. Joan Daemen and Vincent Rijmen, "A Specification for the AES Algorithm Rijndael," Papers 3 (7), (2003).
8. Jayanta Gope, Prakash Kumar Shah "Advanced and Secured Rijndael Hardware Realization Using Single Electron Transistor Technology," International Journal of Emerging Research in Management andTechnology, Papers 3(5), (2014).
9. Mohammad Reza Karimian et al., "A New SPICE Macro-model for the Simulation of Single Electron Circuits," Journal of the Korean Physical Society. Papers 56(4), 1202–1207 (2010).

Audio File Digitization and Encryption Using ASCII Conversion

Soumen Das and Debasree Sarkar

Abstract Audio File encryption is an advance encryption technique in real world. It is basically a technique of hybridization, combination of transformation and cryptography. To send message securely inside noisy channel through insecure internet, cryptography plays a most crucial and vital role for encryption. In this paper initially we are trying to encode and decode a .wav file, then compare both the file before encoding and after decoding that they sound technically similar or not.

1 Introduction

Encryption of .wav file is one of the challenging issue in the field of network security. Through noisy channel it is very much difficult to send such kind of file. With the help of transformation firstly we convert the file into digital format that means analog to digital conversion [1] can take place, to do so we use pulse code modulation technique. To get back the original sound in terms of .wav file apply pulse code demodulation technique. At the last we compare both the files that they sound similar or not. But there is a scope to encrypt the encoded combination of zero, one in 8 bit ASCII character then shuffling then to produce more encrypted form. To do this project we use MATLAB 2014(a).

2 Methodology

See Fig. 1.

S. Das (✉)
Department of CSE, UIT, Burdwan University, Durgapur, West Bengal, India
e-mail: getme.soumen@gmail.com

D. Sarkar
Department of CSE, UIT, Burdwan University, Burdwan, West Bengal, India
e-mail: debasree1031@gmail.com

© Springer Nature Singapore Pte Ltd. 2017
I. Bhattacharya et al. (eds.), *Advances in Optical Science and Engineering*,
Springer Proceedings in Physics 194, DOI 10.1007/978-981-10-3908-9_60

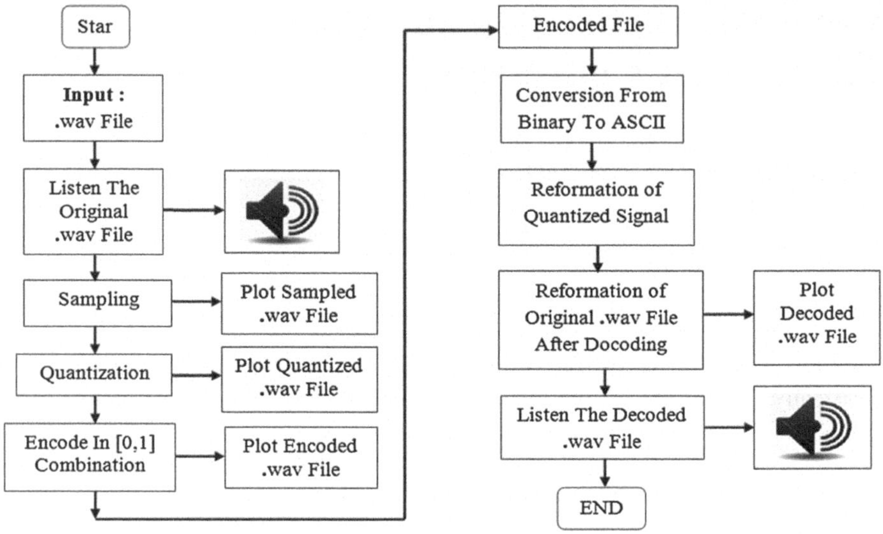

Fig. 1 Flow chart of proposed method

3 PCM Encoding and Decoding Process

Pulse code modulation (PCM) [2] is basically a 3 steps process for analog to digital conversion which includes sampling [3], quantization [3], and encoding. In this paper we use PCM modulation and PCM demodulation technique for encrypting and decrypting a .wav file in terms of combination of zeros, ones and reveres respectively.

3.1 Sampling

In case of sampling [4], the .wav file (continuous time and continuous amplitude) can be converted into discrete time continuous amplitude which is the next input to the quantization process. In practically it is impossible to digitize the .wav file which has infinite number of sample points. So we have to sample the continuous time into discrete time with time interval T.

$$F_s = \frac{1}{T} \text{ Samples per second HZ)} \quad (1)$$

where sampling interval is T and the sampling frequency is F_s.

To avoid aliasing must satisfy nyquist interval rate where $Fs \geq 2F_{max}$.

3.2 Quantization

Using quantization [5] we convert the discrete time continuous amplitude signal (output after sampling) into discrete time and discrete amplitude signal. To do so we have to perform the following

$$\Delta = \frac{\text{xmax} - \text{xmin}}{\text{L}} \tag{2}$$

$$\text{L} = 2^{\wedge}\text{m} \tag{3}$$

$$\text{i} = \text{round}\,(\frac{\text{x} \;\; - \text{xmin}}{\Delta}) \tag{4}$$

$$\text{Xq} = \text{x}_{\text{min}} + \text{i}\Delta, \quad \text{for i} = 0,\ 1\ldots\text{L} - 1 \tag{5}$$

where the maximum value is x_{max} and minimum value is x_{min} of the analog input signal x. L denotes the number of quantization levels [6]. m is the number of bits used in analog to digital conversion (ADC). Here step size the of the quantizer or the ADC resolution is delta denoted by the symbol Δ. Quantization level is denoted by X_q, and index i is corresponding to the binary code.

3.3 Encoding

After getting the discrete time and discrete amplitude values produced by quantization process converted into combination of zero and ones row vector which will produce the encoded signal.

3.4 Decoding

To decode the encoded combination of zero one we apply the reverse procedure that means initially convert the binary data to decimal data. Then we regain the quantized value by which the demodulated .wav file can re generate.

4 Encryption and Decryption [7]

For encrypting the zero one matrix convert into m × (n = 8 bit) matrix where m is the number of rows and n is the number of column [8]. So after getting the matrix with 8 bit in each row we can easily convert it in their corresponding extended 8 bit ASCII, one row at a time (Figs. 2 and 3).

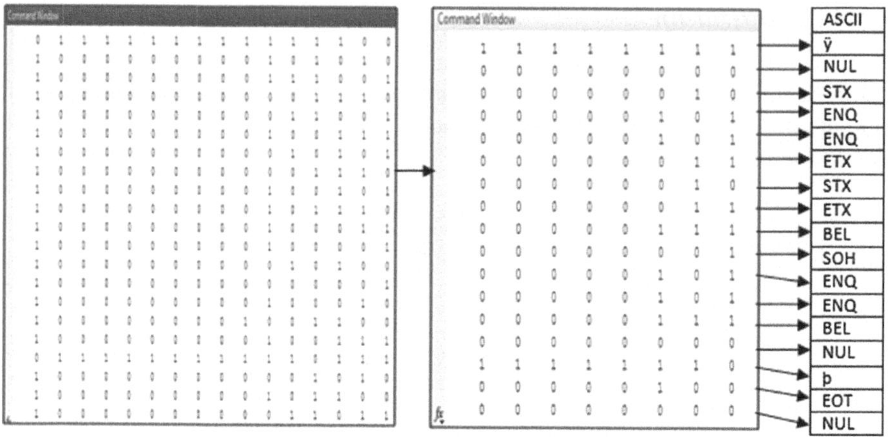

Fig. 2 16 bit to 8 bit data in each row conversation then ASCII values

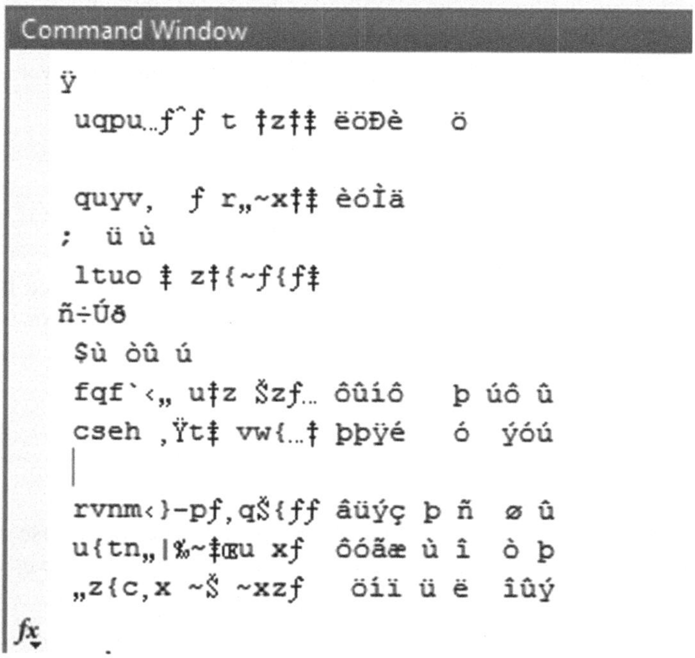

Fig. 3 Encryted form of .wav file

For decrypting the message we can apply the reverse procedure that means convert ASCII to m × (n = 8 bit) matrix, then again covert it to its original size matrix m × n. Then to regain the original .wav file we again have to transfer the matrix m × n into serial row vector (Fig. 4).

After getting the serial data given by row vector then converted into its decimal form to get back the quantized values which will reproduce the demodulated .wav signal.

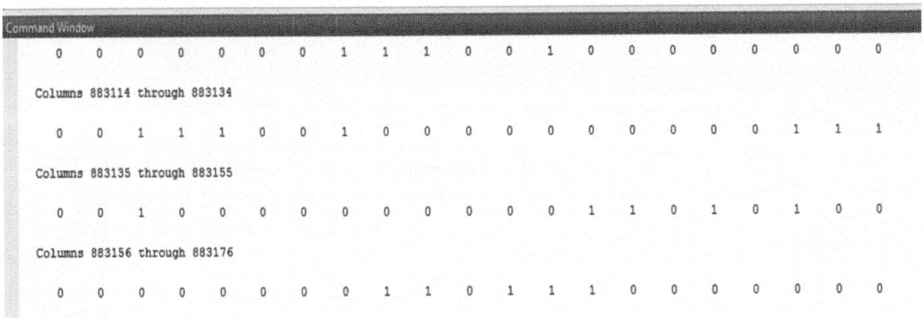

Fig. 4 Row vector

5 Result

In our proposed method initially we listen the original .wav file, then using PCM it convertes into its binary combination of zeros and ones and then encrypts. To reform the .wav file we apply the reverse procedure then again we recheck the

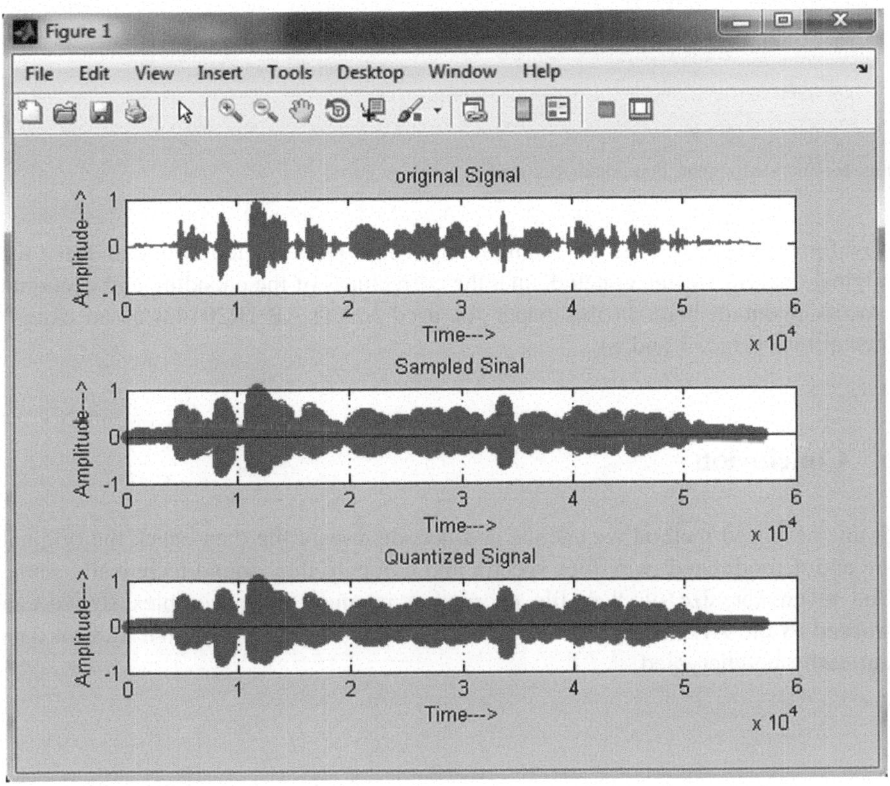

Fig. 5 Original signal, sampled signal, quantized signal

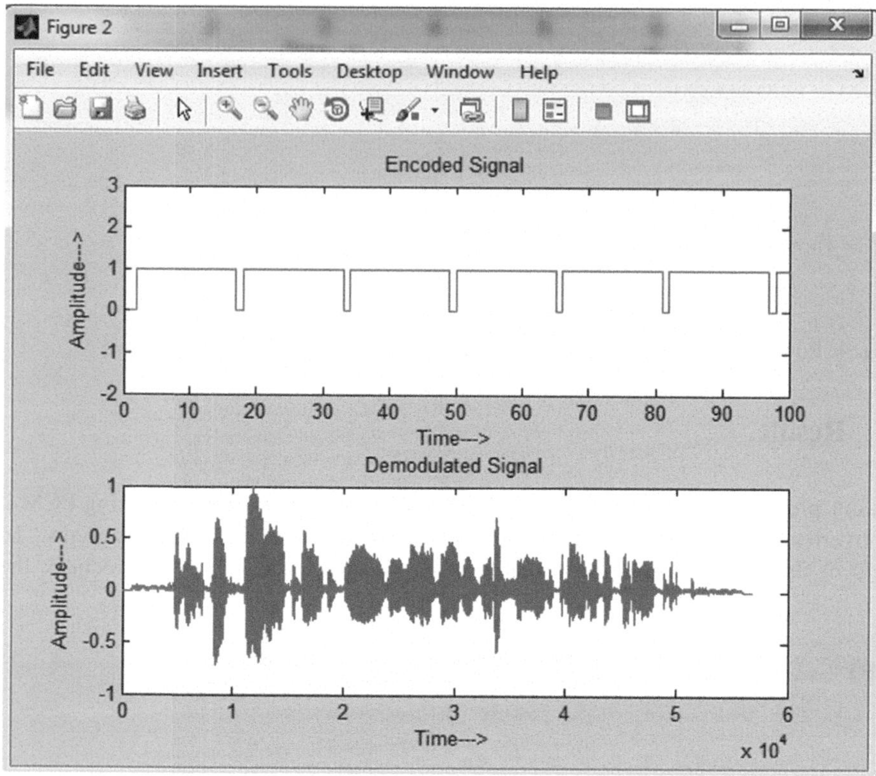

Fig. 6 Encoded signal, demodulated signal

waveform and sound it as the original one or not. Here we found it sound like the original one. So we can conclude that the correctness of the encoding and decoding process maintain here. In this paper we used MATLAB (R2014a) as an experimental tool (Figs. 5 and 6).

6 Conclusion

In this proposed method we encode and decode a .wav file then check the original file and demodulated .wav files spectra and compare they sound technically same. Due to greater size of .wav file encryption is much more complex so we can proceed to the AES encryption standard by which higher bit 16, 64, 128 bit data can easily be encrypted.

References

1. Digital signal processing by BP Lathi.
2. Digital Transmission of Analog Signals: PCM, DPCM and DM, CHAPTER 6,NPTEL Lactures, by Prof. V. Venkata Rao, Professor of Indian Institute of Technology Madras.
3. "TranterBook" — 2003/11/18 — 16:12 — page 55, Chapter 3 SAMPLING AND QUANTIZING.
4. Chapter 5 Sampling and Quantization, Lecture Notes for ELE201 Introduction to Electrical Signals and Systems, 1999-2002 by Sanjeev R. Kulkarni.
5. "Quantization and Dither: A Theoretical Survey" J.Audio Eng. Soc., Vol.40,No. 5,1992 May, Presented at the 91st Convention of the AudioEngineering 2 Society, New York, 1991 October 4-8; revised 1991 October21, by STANLEY P. LIPSHITZ, AES Fellow,ROBERT A. WANNAMAKER, AND JOHN VANDERKOOY, AES Fellow.
6. Digital signal processing fundamental and applications by LI Tan.
7. Jayaram P "Information Hiding Using Audio Steganography" The International Journal of Multimedia & Its Applications (IJMA) Vol.3, No.3, August 2011
8. Study of Encryption and Decryption of Wave File in Image Formats by Rahul R Upadhyay Department of Mechanical Engineering, BBD National Institute of Technology and Management, Lucknow.

Part XIII
Quantum and Non-Linear Optics, Opto-Electronic Devices

Effect of TiO_2 Nanoparticle Doping on the Electrical Properties of Ferroelectric Liquid Crystal

Pradeep Kumar and Aloka Sinha

Abstract In this study, the electrical properties of ferroelectric liquid crystal (W206E) doped with anatase titanium dioxide (TiO_2) nanoparticles were studied. The experimental results show that the dopant TiO_2 reduces the free ion concentration and therefore the conductivity of doped W206E samples. This reduction in the free ion concentration is due to the trapping of free ions by TiO_2 nanoparticles dispersed in W206E. The reduction in the free ion concentration is related to the doping concentration of TiO_2. The experimental results show that the higher concentration of TiO_2 is more effective in the trapping of mobile ions. This result will help in the development of fast response display devices with better contrast and low threshold voltage.

1 Introduction

Liquid crystal displays are the most widely used displays among the other existing displays in the display industry. But, nowadays the modern and sophisticated devices require displays with improvised properties such as higher contrast, low power consumption, faster response time and full color capability. The performance of the liquid crystal display devices is dependent mainly upon the purity of the used liquid crystal material. But it is a known fact that, some ionic impurities are always present in the liquid crystals that influences the various physical properties including the conductivity of liquid crystals [1]. The mobile ions in liquid crystal can originate during the synthesis of liquid crystal, due to the ionic dissociations of residual impurities present in liquid crystal, from the alignment layers and at the time of cell filling process [1]. These ionic impurities are mainly responsible for the poor performance i.e. slow response, higher operating voltage, poor contrast and

P. Kumar · A. Sinha (✉)
Department of Physics, Indian Institute of Technology, Delhi 110016, India
e-mail: aloka@physics.iitd.ac.in

© Springer Nature Singapore Pte Ltd. 2017

I. Bhattacharya et al. (eds.), *Advances in Optical Science and Engineering*,
Springer Proceedings in Physics 194, DOI 10.1007/978-981-10-3908-9_61

image sticking of display devices [2]. Therefore, to improvise the performance of liquid crystal based display devices, the free ionic impurities present in liquid crystal must be minimized before using it for any display application. Doping of different types of nanoparticles such as semiconducting nanoparticle [3, 4], metallic nanoparticle [5, 6], ferroelectric nanoparticle [7, 8] and dielectric nanoparticle [9, 10] in liquid crystal has improved the various electro-optical properties of the existing liquid crystals. The insulating nanoparticles are generally used to reduce the free ions present in liquid crystal. In this study, we have doped the FLC W206E with TiO_2 nanoparticles in two different concentrations and there after studied for different electrical properties.

2 Experimental

The phase sequence of the ferroelectric liquid crystal (FLC) W206E used in the present study is Crystal ↔ SmC*(86 °C) ↔ SmA (92 °C) ↔ N*(97.6 °C)↔ Isotropic. The TiO_2 nanoparticles (particle size 18–23 nm) procured from Sigma Aldrich USA, were used to prepare two different FLC nanocolloids, W206E + 0.5 wt% TiO_2 and W206E + 1.0 wt% TiO_2. The TiO_2 nanoparticles were taken in fixed proportion and mixed with W206E to prepare the FLC nanocolloids. The FLC nanocolloids were prepared using ultrasonication technique, where we have added chloroform to the mixture of W206E + TiO_2 nanoparticles and then sonicated till the dispersion was visibly homogeneously mixed. After that the dispersion was left at temperature $\sim$45 °C, till the chloroform evaporated completely in order to get FLC nanocolloids. The homogeneous liquid crystal cells were prepared using photolithography and conventional rubbed polyimide technique. The thickness of the cells was maintained $\sim$3.4 μm using Mylar spacer. The pure and FLC nanocolloids were introduced into the liquid crystal cells at their isotropic temperatures by means of capillary action.

These filled cells were studied for electrical properties. The conductivity measurement from 20 Hz to 25 kHz and dielectric measurements in the frequency range of 500 to 10^5 Hz, were carried out at different temperatures using Agilent E4980A LCR meter. The temperature controller INSTEC mK-1000 was used in this study.

3 Results and Discussion

3.1 Conductivity

To investigate the effect of doping of TiO_2 nanoparticles on electrical properties of pure W206E and FLC nanocolloids, W206E + 0.5 wt% TiO_2 and W206E + 1.0 wt% TiO_2, the conductivity measurements were carried out at $\sim$40 °C.

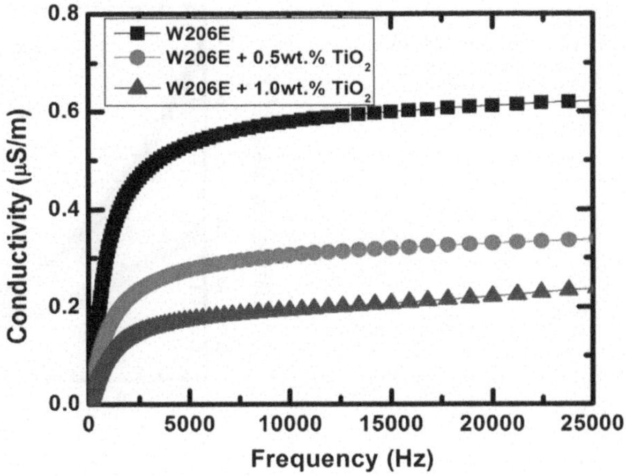

Fig. 1 Conductivity as a function of frequency

The results are shown in Fig. 1. The observed results clearly show that the conductivity of doped samples is decreased as compared to pure W206E. The decrease in the conductivity is more significant at higher doping concentration of TiO_2 nanoparticles in W206E. This can be explained on the basis of ion trapping phenomenon [11], where TiO_2 nanoparticles trapped the mobile ions present in pure W206E. Since the number of TiO_2 nanoparticles will be more in case of high doping concentration of 1 wt% as compared to the lower doping concentration of 0.5 wt%. Therefore, we can expect more trapping of ions in case of 1 wt% TiO_2 as compared to 0.5 wt% TiO_2 and hence the conductivity of W206E + 1.0 wt% TiO_2 is less than the W206E + 0.5 wt% TiO_2.

3.2 *Dielectric Measurement*

In the lower frequency region, the dielectric measurements can give valuable information about the mobile ions present in liquid crystalline material. Therefore, to further substantiate the effect of mobile ions on the observed conductivity results, we have carried out dielectric measurements in the lower frequency region in the SmC* phase of each sample from 500 to 10^5 Hz between the temperature range 30–65 °C. The observed dielectric loss (ε'') results for pure and doped samples are shown in Fig. 2. The conductivity of liquid crystal $\sigma = nq\mu$, is mainly dependent upon the concentration of ions (n), charge (q) and mobility ($\mu = qD/k_bT$, where D is diffusion coefficient, k_b is Boltzman's Constant and T is the temperature). To calculate the concentration of ions (n) and diffusion coefficient (D), we have fitted the dielectric data with Uemura formalism given in (1) and (2) [12, 13] in the SmC* phase of each sample.

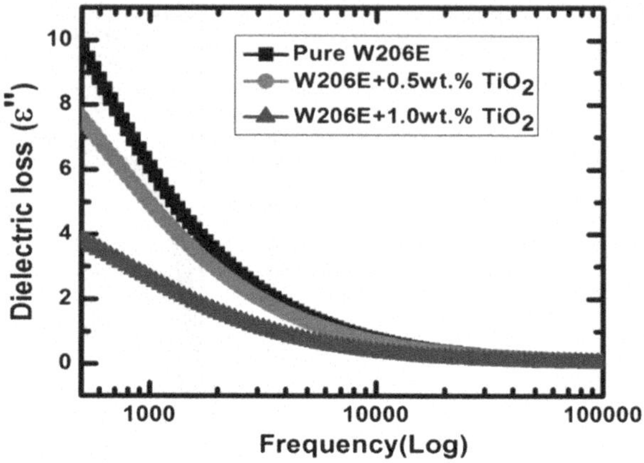

Fig. 2 Variation of dielectric loss (ε″) with frequency

$$\varepsilon' = -\left(\frac{nq^2D}{\omega\varepsilon_0 k_b T A}\right)\left[\frac{1 + 2\exp(A)\sin(A) - \exp(2A)}{1 + 2\exp(A)\cos(A) + \exp(2A)}\right] \tag{1}$$

$$\varepsilon'' = \left(\frac{nq^2D}{\omega\varepsilon_0 k_b T}\right)\left\{1 + \frac{1 - 2\exp(A)\sin(A) - \exp(2A)}{A[1 + 2\exp(A)\cos(A) + \exp(2A)]}\right\} \tag{2}$$

where, q is charge, d is the separation between electrodes and $A = d(\omega/2D)^{1/2}$. The concentration of ions and diffusion coefficient are calculated from the fitting of dielectric loss data using (2) and are shown in Fig. 3. Figure 3a, b clearly show that the concentration of ions and diffusion coefficient for W206E + 1.0 wt% TiO_2 is lower than the W206E + 0.5 wt% TiO_2 and pure W206E samples. The decrease in the concentration of ions in FLC nanocolloids is due to the trapping of mobile ions present in FLC host by TiO_2 nanoparticles. The trapping of mobile ions will be more for the higher doping concentration of 1.0 wt% TiO_2 as compared to 0.5 wt% TiO_2 in pure W206E. Further, as shown in Fig. 3b, the diffusion coefficient also decreases for the FLC nanocolloids in comparison to pure FLC. This decrease is more significant for high doping concentration of 1 wt% TiO_2 in comparison to 0.5 wt% TiO_2. The decrease in the value of diffusion coefficient may be attributed to the strong interaction between the charged particles and the increased viscosity of the FLC nanocolloids [14]. According to Einstein-Stokes theorem, the diffusion coefficient (D) is given by (3);

$$\mathrm{D} = \left(\frac{k_b T}{6\pi\eta R}\right) \tag{3}$$

where, k_b is Boltzmann's constant, T is temperature, η is viscosity of the medium and R is the radius of particle. In (3), all other quantities except η are constants. The viscosity (η) of the liquid crystal medium can be changed by the doping of nanoparticles in liquid crystal material. From (3) it is clear that, if the viscosity

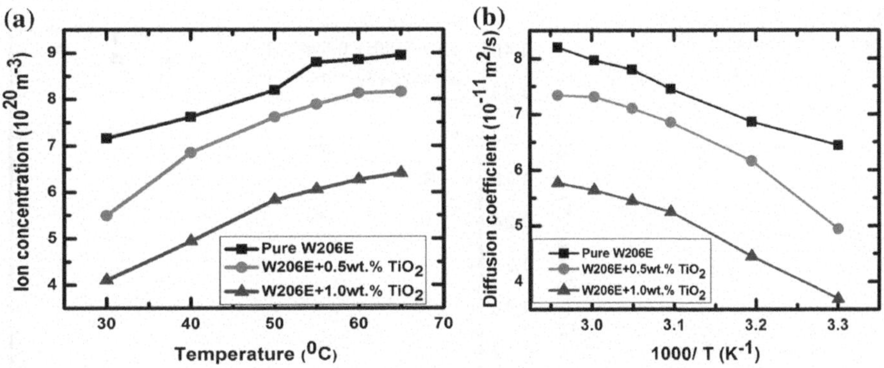

Fig. 3 **a** Concentration of ions (n) and **b** diffusion coefficient (D) as a function of temperature

Fig. 4 Variation of activation energy as a function of TiO_2 nanoparticle concentration

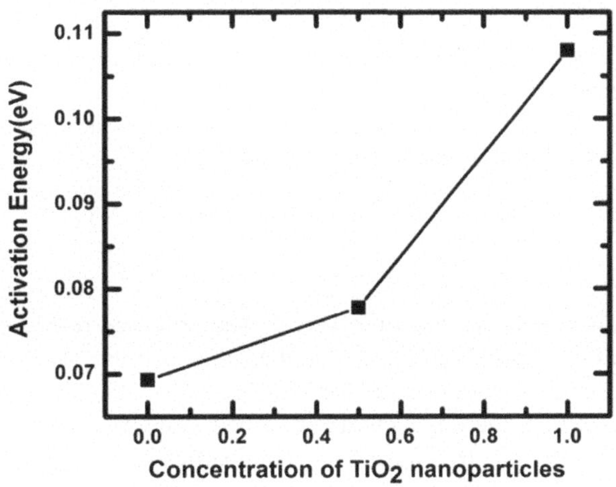

increases, then the value of diffusion coefficient will decrease and vice versa. Also, it is known that the activation energy of the system increases with increasing viscosity. To further substantiate this we have calculated the activation energy for pure W206E and FLC nanocolloids using (4).

$$\mathrm{D} = D_0 \exp\left(-\frac{E_a}{k_b T}\right) \tag{4}$$

where, D_0 is constant and E_a is activation energy. Figure 4, shows that the activation energy of nanoparticle dispersed ferroelectric liquid crystal systems is increased as compared to pure W206E and it is maximum for W206E + 1.0 wt% TiO_2. This fact suggests that the viscosity of W206E + 1.0 wt% TiO_2 is higher than W206E + 0.5 wt% TiO_2. Therefore, from these observations we believe that the strong interactions between charged particles and increased viscosity are the main

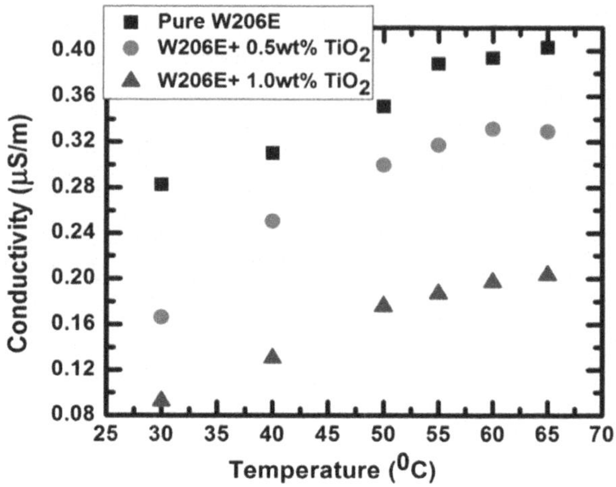

Fig. 5 Conductivity as a function of temperature

reasons for the decrement in the value of diffusion coefficient (*D*) for FLC nanocolloids.

We have also calculated the conductivity at different temperatures for pure and both doped system using (5) and

$$\sigma = \left(\frac{ne^2 D}{k_b T}\right) \tag{5}$$

the results are shown in Fig. 5. The observed results show that the conductivity of FLC nanocolloids is lower than that of pure W206E. The lowest value of conductivity was observed for W206E + 1.0 wt% TiO_2. The current study supports the fact that, the conductivity of doped systems was decreased due to the trapping of mobile ions by doped TiO_2 nanoparticles.

4 Conclusion

The pure and TiO_2 nanoparticle dispersed ferroelectric liquid crystal systems were studied for electrical properties and mobile ion concentration. The conductivity measurements show that the conductivity of the FLC nanocolloids has decreased as compared to the pure W206E. Further, with the help of dielectric measurements and Uemura formalism, the concentrations of ions and diffusion coefficient have been calculated and both of these parameters were decreased with the increase in the doping concentration of TiO_2. The decrease in ions concentration is attributed to the mobile ions trapping phenomenon while the reasons considered for the reduction in the value of diffusion coefficient are strong interactions between charged particles and increased viscosity of the FLC nanocollids. This study will help to improve the performance of liquid crystal displays by reducing the unwanted mobile ion effects.

Acknowledgements The Department of Science and Technology (EMR/2015/001897), Govt. of India is gratefully acknowledged for providing the necessary experimental facilities to conduct the experiments at IIT Delhi, India.

References

1. Basu, R. and Garve, A., "Effects of ferroelectric nanoparticles on ion transport in a liquid crystal," Appl. Phys. Lett. 105, 151905 (2014).
2. Garbovski, Y. and Glushchenko, I., "Nano-objects and ions in liquid crystals: ion trapping effect and related phenomena," Crystals, 5(4), 501–533 (2015).
3. Supreet, Kumar, S., Raina, K.K. and Pratibha, R., "Enhanced stability of the columnar matrix in a discotic liquid crystal by insertion of ZnO nanoparticles," Liquid Cryst. 40, 228–236 (2013).
4. Shukla, R.K., Galyametdinov, Y.G., Shamilov, R.R. and Haase, W., "Effect of CdSe quantum dots doping on the switching time, localised electric field and dielectric parameters of ferroelectric liquid crystal," Liquid Cryst. 41, 1889–1896 (2014).
5. Podgornov, F.V., Ryzhkova, A.V. and Haase, W., "Influence of gold nanorods size on electro-optical and dielectric properties of ferroelectric liquid crystals," Appl. Phys. Lett. 97, 212903 (2010).
6. Mandal, P.K., Lapanik, A., Wipf, R., Stuehn, B. and Haase, W., "Sub-hertz relaxation process in chiral smectic mixtures doped with silver nanoparticles," Appl. Phys. Lett. 100, 073112 (2012).
7. Kumar, P. and Sinha, A., "Effect of barium titanate nanoparticles of different particle sizes on electro-optic and dielectric properties of ferroelectric liquid crystal," Phase Trans. 88, 605–620 (2015).
8. Ouskova, E., Buchnev, O., Reshetnyak, V., Reznikov, Y. and Kresse, H., "Dielectric relaxation spectroscopy of a nematic liquid crystal doped with ferroelectric $Sn_2P_2S_6$ nanoparticles," Liq Cryst. 30, 1235–1239(2003).
9. Tang, C.Y., Huang, S.M. and Lee, W., "Electrical properties of nematic liquid crystals doped with anatase TiO_2 nanoparticles" J. Phys. D Appl. Phys. 44, 355102 (2011).
10. Kumar, P, Kishore, A. and Sinha, A., Effect of different concentrations of dopant titanium dioxide nanoparticles nn electro-optic and dielectric properties of ferroelectric liquid crystal mixture," Adv. Mater. Lett. **7**(2), 104–110 (2016).
11. Shukla, R.K., Liebig, C.M., Evans, D. R. and Haase, W., "Electro-optical behaviour and dielectric dynamics of harvested ferroelectric $LiNbO_3$ nanoparticle-doped ferroelectric liquid crystal nanocolloids," RSC. Adv. 4, 18529–18536 (2014).
12. Uemura, S., "Low-frequency dielectric behavior of poly(vinylidene fluoride)," J. Polym. Sci., Polym. Phys. Ed. 12, 1177 (1974).
13. Sawada, A., Tarumi, K., and Naemura, S., "Novel Characterization Method of Ions in Liquid Crystal Materials by Complex Dielectric Constant Measurements," Jpn. J. Appl. Phys. 38, 1423 (1999).
14. Yadav, S.P., Manohar, R. and Singh, S., "Effect of TiO_2 nanoparticles dispersion on ionic behaviour in nematic liquid crystal" Liquid Cryst. 42, 1095–1101(2015).

Improved Noise Rejection in Metamaterial based Defected Photonic Crystal Structure

Solanki Ghosh, Ruma Dutta, Varsha Shaw and Arpan Deyasi

Abstract Transmittivity of metamaterial based defected photonic crystal structure is analytically computed for optical filter design. Result suggests that if defect density lies within lower range, better noise rejection may be achieved as length of transmittance is increased compared to that obtained for ideal structure, while filter bandwidth remains unchanged. Result is computed for 1.55 μm, and is also validated for normal and oblique incidences. Structural parameters and input conditions are tuned to analyze the performance.

1 Introduction

A periodic arrangement of refractive index of two dissimilar materials (preferably both are dielectric) may be termed as photonic crystal which allows electromagnetic wave propagation in specific range [1], but other wavelengths are restricted. This property introduces a new concept termed as electromagnetic bandgap [2], whose width depends on the constituting material parameters, and layer dimensions [3]. This novel device has potential applications in antenna-embedded optical system by introducing metamaterial as a constituent. All-optical circuit design becomes the subject of research for both theoretical and experimental workers in the last few years, as it may be considered as a near perfect alternative of electronic circuits [4].

S. Ghosh (✉) · R. Dutta · V. Shaw · A. Deyasi
Department of Electronics and Communication Engineering, RCC Institute of Information Technology, Kolkata, West Bengal 700015, India
e-mail: solankighosh93@gmail.com

R. Dutta
e-mail: rumadtt5@gmail.com

V. Shaw
e-mail: varshashaw17@gmail.com

A. Deyasi
e-mail: deyasi_arpan@yahoo.co.in

© Springer Nature Singapore Pte Ltd. 2017
I. Bhattacharya et al. (eds.), *Advances in Optical Science and Engineering*,
Springer Proceedings in Physics 194, DOI 10.1007/978-981-10-3908-9_62

This structure has already proved its effectiveness in different communication systems [5, 6] with improved efficiency.

Work on 1D photonic crystal has attracted several researchers because of ease of computation and fabrication also, due to the rapid advancement of microelectronics technology. Volakis [7] showed the equivalence between in-plane anisotropy and coupled transmission lines in order to realize magnetic photonic crystal modes with degenerate band-edge and producing wideband arrays with smaller sizes. This will play important role in designing communications in RF range and also for wireless systems. Butt [8] presented modeling of silicon pillars based photonic crystals using finite element method and used in designing optical passive devices (waveguides, switches, power dividers, multiplexers). Theoretically obtained results are verified by experimental findings. Gevorgyan [9] discussed light transmission inside chiral photonic crystal structures. He showed that due to dielectric and magnetic helicities, new types of PBG are created with very large anisotropy. Behera [10] designed broadband circular polarizer inside square cell using photonic-metamaterial. Circular polarization is obtained for a wider optical range based on the design. Segal [11] experimentally showed extraordinary control over nonlinear emission generated from metamaterials by fabricating metamaterial-based photonic crystals with non-linear property, which is the pioneering work in this field. Shawon [12] proposed hollow-core fiber with multiple cladding layers using ternary one-dimensional photonic crystal. This generates zero-effective-phase bandgap for optical confinement. In the present paper, optical bandwidth of nano-fishnet (elliptic void)/air composition is computed for defected structure under TM polarization conditions, and comparative study is made with that obtained for defect-free structure.

2 Mathematical Modeling

Considering the phase factor of the field propagating through uniform medium, propagation matrix is given as the function of barrier and well widths

$$P_{1,2} = \begin{pmatrix} \exp[jk_{1,2}d_{1,2}] & 0 \\ 0 & -\exp[jk_{1,2}d_{1,2}] \end{pmatrix} \tag{1}$$

where $d_{1,2}$ is the dimension of barrier/well layer, $k_{1,2}$ is the propagation vector. Thus, transfer matrix for the elementary cell (constituting of one barrier and one well layer) is

$$M = M_1^T P_1 M_2^T P_2 \tag{2}$$

where M is the transfer matrix between the adjacent layers, given by

$$M_{1,2}^T = \frac{1}{t}\begin{pmatrix} 1 & r_{21,12} \\ r_{21,12} & 1 \end{pmatrix} \tag{3}$$

For p-polarized incident wave at angle θ_1, interface reflectivities are given by

$$r_{12} = -r_{21} = \frac{n_1 \cos(\theta_2) - n_2 \cos(\theta_1)}{n_1 \cos(\theta_2) + n_2 \cos(\theta_1)} \quad (4)$$

For s-polarized incident wave at angle θ_1, interface reflectivities are given by

$$r_{12} = -r_{21} = \frac{n_1 \cos(\theta_1) - n_2 \cos(\theta_2)}{n_1 \cos(\theta_1) + n_2 \cos(\theta_2)} \quad (5)$$

For a perfectly periodic medium composed of N such elementary cells, the total transfer matrix for such a structure is

$$M_{tot} = M_N \quad (6)$$

Transmission coefficient is given by

$$T = \frac{1}{M_{11}^2(tot)}. \quad (7)$$

3 Results and Discussions

Figure 1 exhibits transmittivity characteristics for the structure under consideration under TM mode propagation (Fig. 1a), and also for normal incidence (Fig. 1b). For both the cases, it is seen that in presence of defect, length of transmittance is increased compared to that obtained for defect-free structure. This ensures better rejection of noise as higher reflectance at either side of central wavelength

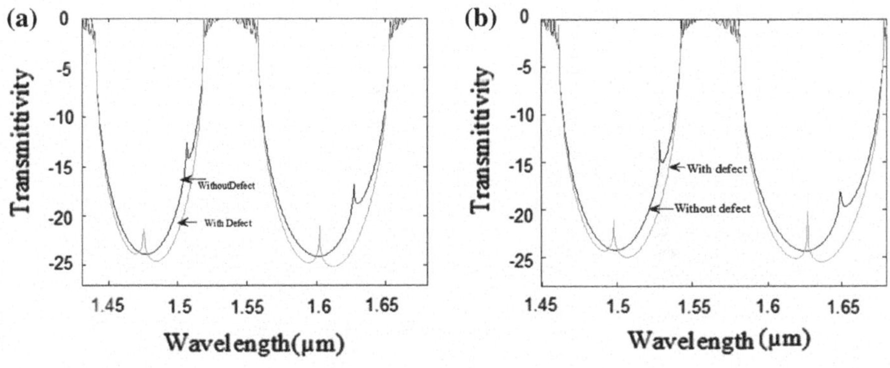

Fig. 1 Transmittivity with wavelength in presence and absence of defect for **a** TM mode propagation; **b** normal incidence

(1.55 μm) provides better filtering. This performance enhancement is made keeping optical bandwidth constant, which speaks suitability of introducing defect in otherwise perfect structure.

Variation of layer dimensions tuned the transmittance of the structure, as shown in Fig. 2. Figure 2a exhibits for different dimensions of metamaterial, where Fig. 2b shows it for different air thickness. It has been seen from Fig. 2a that increasing the thickness of elliptic void layer makes a redshift of the spectrum, but optical bandwidth of the structure remains unchanged. In contrast, Fig. 2b suggests that higher thickness of air (larger void size) increases the optical bandwidth, and redshift is also displayed. But for both the cases, noise rejection quality is detoriated with increase of dimensions, which is due to reduction of transmittance.

Incidence angle variation can tune the filter performance of the structure, as shown in Fig. 3. It is observed that higher incidence angle makes a redshift of the total spectrum. Also for higher incidence angle, noise rejection probability is

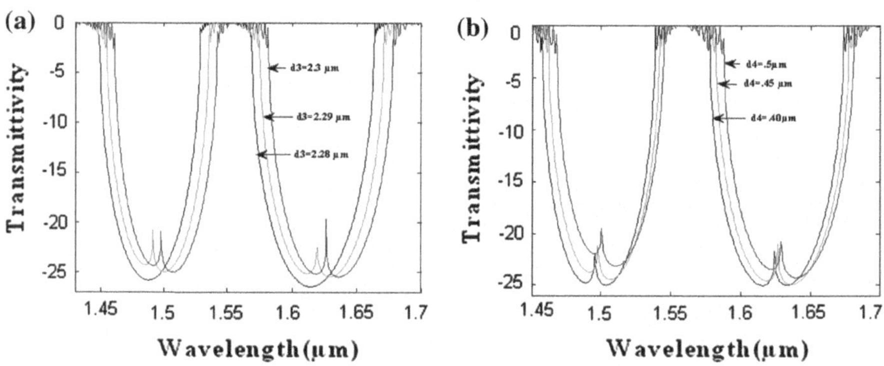

Fig. 2 Transmittivity with wavelength in presence of defect for different **a** dimensions of metamaterial; **b** different air thickness

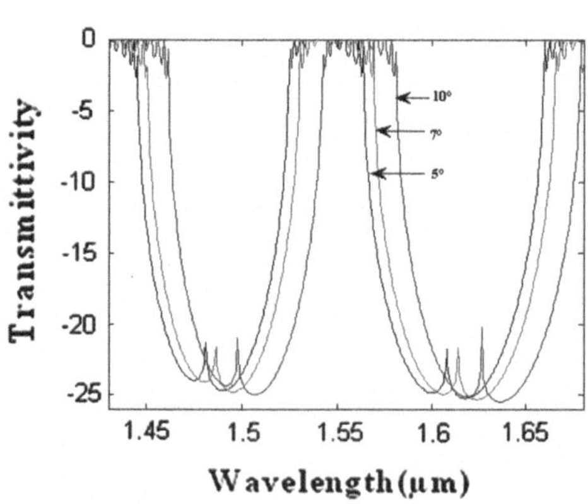

Fig. 3 Transmittivity with wavelength in presence of defect for different incidence angles

enhanced due to larger magnitude of transmittance. Thus suitable choice of external conditions prior to fabrication plays key role in design of photonic filter.

4 Conclusion

Metamaterial based defected PhC improves noise rejection keeping optical bandwidth constant than that of ideal one, and bandwidth is dependent on input conditions and structural parameters. Redshift of the bandwidth is observed by increasing either of the layer dimensions, or by increasing the incidence angle. Noise rejection quality is directly related with the length of transmittance in either side of passband, and tuning layer dimensions or incidence angle also affects noise rejection by the structure around 1550 nm. Result is important for antenna-embedded optical system.

References

1. Yablonovitch. E., Gmitter, T. J., "Photonic Band Structure: The Face-Centered-Cubic Case", Physical Review Letters, 63, 1950–1953 (1987)
2. Chen. J. C, Haus. H. A, Fan. S, Villeneuve. P. R, Joannopoulos. J. D, "Optical Filters from Photonic Band Gap Air Bridges", Journal of Lightwave Technology, 14, 2575–2580 (1996)
3. Limpert. J, Liem. A, Reich. M, Schreiber. T, Nolte. S, Zellmer. H, Tünnermann. A, Broeng. J, Petersson. A, Jakobsen. C, "Low-Nonlinearity Single-Transverse-Mode Ytterbium-Doped Photonic Crystal Fiber Amplifier", Optic Express, 12, 1313–1319 (2004)
4. Li. Y, Bhardwaj. A, Wang. R, Jin. S, Coldren. L, Bowers. J, Herczfeld. P, "All-optical ACP-OPLL Photonic Integrated Circuit", IEEE MTT-S International Microwave Symposium Digest, 1–4 (2011)
5. Srivastava. S. K, Ojha. S. P, "A Novel Design of Nano Layered Optical Filter using Photonic Band Gap Materials", 2nd National Workshop on Advanced Optoelectronic Materials and Devices, 225–230 (2008)
6. J. Limpert, T. Schreiber, S. Nolte, H. Zellmer, T. Tunnermann, R. Iliew, F. Lederer, J. Broeng, G.Vienne, A. Petersson, C. Jakobsen, "High-power air-clad large-mode-area photonic crystal fiber laser", Optic Express, vol. 11, pp. 818–823, 2003
7. Volakis. J. L, Sertel. K, "Narrowband and Wideband Metamaterial Antennas Based on Degenerate Band Edge and Magnetic Photonic Crystals", Proceedings of the IEEE, 99, 1732–1745 (2011)
8. Butt. H, Dai. Q, Wilkinson. T. D, Amaratunga. G. A. J, "Photonic Crystals & Metamaterial Filters based on 2D Arrays of Silicon Nanopillars", Progress in Electromagnetics Research, 113, 179–194 (2011)
9. Gevorgyan. A. H, Rafayelyan. M. S, "Optics of Anisotropic Metamaterial based Structurally Chiral Photonic Crystals", Journal of Optics, 15, 125103 (2013)
10. Behera. S, Joseph. J, "N-Single-Helix Photonic-Metamaterial based Broadband Optical Range Circular Polarizer by Induced Phase Lags Between Helices", Applied Optics, 54, 1212–1219 (2015)
11. Segal. N, Keren-Zur. S, Hendler. N, Ellenbogen. T, "Controlling Light with Metamaterial-based Nonlinear Photonic Crystals", Nature Photonics 9, 180–184 (2015)

12. Shawon. M. J, Mahdiraji. G. A, Hasan. M. M, Shakibaei. B. H, Gang. S. Y, Chowdhury. M. R, Adikan. F. R. M, Single Negative Metamaterial-based Hollow-Core Bandgap Fiber with Multilayer Cladding", IEEE Photonics Journal, 7, 4600812 (2015)

Computational Analysis of Carrier Mass Under Energetic Photons in Accumulation Layers of MOSFET Devices

R. Paul, S. Ghatak, S. Das, M. Mitra and T. Datta

Abstract The influence of strongly energetic photons on the carrier mass (CM) at the Fermi level in accumulation layers of MOSFET devices, has been investigated taking accumulation layers of InAs and InSb as examples. It has been observed that the CM decreases with decreasing surface electron concentration per unit area. The CM is a function of chemical potential, scattering potential and electric sub band index together with other physical variables, which is the characteristics features of such 2D systems.

The importance of CM is already well known since the inception of solid state science in general and the fundamental definition of the CM determines the Boltzmann transport equation and the different mechanisms of scattering in 3D, 2D, 1D and 0D devices. Besides the basic definition of CM offers the simplest derivation of Einstein $E = mc^2$ equation and has been extensively investigated in the literature [1–26]. In what follows we study the CM in the present case taking the materials as written in the abstract.

R. Paul
Department of CSE, University of Engineering and Management,
Kolkata, West Bengal 700156, India

S. Ghatak
Depatement of BCA and M.Sc., Institute of Engineering and Management,
Kolkata, West Bengal 700091, India

S. Das
Department of CSE, Indian Institute of Engineering Science and Technology,
Shibpur, Kolkata, West Bengal 711103, India

M. Mitra
Department of ECE, Indian Institute of Engineering Science and Technology,
Shibpur, Kolkata, West Bengal 711103, India

T. Datta (✉)
Department of Basic Science and Humanities, Institute of Engineering and Management,
Kolkata, West Bengal 700091, India
e-mail: triparna.datta@iemcal.com

© Springer Nature Singapore Pte Ltd. 2017

I. Bhattacharya et al. (eds.), *Advances in Optical Science and Engineering*,
Springer Proceedings in Physics 194, DOI 10.1007/978-981-10-3908-9_63

The 2D electron energy–electron wave vector relation for the present system can be written as

$$k_s^2 = \frac{2M_c}{\hbar^2} f_3 \tag{1}$$

where

$$f_3 = [f_1 - A_i[(2M_c)^{-1/2} \frac{|e|^2 N_{2D}}{\varepsilon_s} (f_1')]^{2/3}]$$

and the other symbols have their usual meaning.

The CM assumes the form

$$M^* = M_c \quad \text{Real Part of } f_3' \tag{2}$$

where the differentiation has to be evaluated at the Fermi level C_{pl} in which $C_{pl} = eV_g - \frac{e^2 n_s d_{ex}}{\varepsilon_{ox}} + E_{FB}$.

The E_{FB} under extremely degenerate condition, can be determined from the following equation

$$n_B = \frac{g_v}{(3\pi)^2} (\frac{2M_c}{\hbar^2})^{3/2} \quad \text{Real Part of } [f_1]^{3/2} \tag{3}$$

The CM in non parabolic band is a function of chemical potential but in this particular system the same mass in addition to chemical potential also depends on scattering potential and the sub-band index which is the characteristic feature of light aided accumulation layers. In formulating the dispersion relation we have averaged over the K.E. part and the weighting function is the impurity potential, the major features of band tails dictates the whole system in a fundamental way making the existence of CM in the band gap which is otherwise impossible. The CM increases with increasing 2D electron concentration and decreases with decreasing wavelengths with different slopes. The complex E-k relation is the consequence of the poles in the otherwise free energy-wave vector relation in bulk semiconductors in the absence of band talling effects.

Acknowledgements The authors are grateful to Prof. Dr. S. Chakrabarti, Director, Institute of Engineering and Management, Kolkata for inspiration and helpful discussion in the real sense of the term.

References

1. KP Ghatak and S Bhattacharya, "Effective Electron Mass in Low—Dimensional Semiconductors", Springer Series in Material Science 167, (Springer, Heidelberg, 2013).

2. PK Bose, N Paitya, S Bhattacharya, D De, S Saha, KM Chatterjee, S Pahari, KP Ghatak, Quantum Matter, **1**, 89 (2012).
3. N Paitya, KP Ghatak, Reviews in Theoretical Science, **1**, 165 (2013).
4. KP Ghatak, B Mitra, International Journal of Electronics, **72**, 541 (1992).
5. M Mondal, N Chattopadhyay, KP Ghatak, Journal of Low Temperature Physics, **66**, 131 (1987).
6. PK Chakraborty, GC Datta, KP Ghatak, Physica Scripta, **68**, 368 (2003).
7. B Mitra, KP Ghatak, Solid-state Electronics, **32**, 177 (1989).
8. KP Ghatak, M Mondal, Zeitschrift für Physik B Condensed Matter, **69**, 471 (1988).
9. KP Ghatak, S Bhattacharya, SK Biswas, A Dey, AK Dasgupta, Physica Scripta, **75**, 820 (2007).
10. M Mondal, KP Ghatak, Physics Letters A, **131**, 529 (1988).
11. AN Chakravarti, AK Chowdhury, KP Ghatak, S Ghosh, A Dhar, Applied physics, **25**, 105 (1981).
12. S Bhattacharya, D De, SM Adhikari, KP Ghatak, Superlattices and Microstructures, **51**, 203 (2012).
13. A Ghoshal, B Mitra, KP Ghatak, Il Nuovo Cimento D, **12**, 891 (1990).
14. KP Ghatak, B Mitra, A Ghoshal, Physica Status Solidi (b), **154**, K121 (1989).
15. AN Chakravarti, KP Ghatak, KK Ghosh, S Ghosh, A Dhar, Zeitschrift für Physik B Condensed Matter, **47**, 149 (1982).
16. M Mondal, KP Ghatak, Physica Status Solidi (b), **135**, K21 (1986).
17. KP Ghatak, A Ghoshal, B Mitra, Il Nuovo Cimento D, **13**, 867 (1991).
18. M Mondal, KP Ghatak, Physica Status Solidi (b), **129**, 745 (1985).
19. KP Ghatak, M Mondal, Zeitschrift für Naturforschung A, **41**, 881 (1986).
20. AN Chakravarti, KP Ghatak, KK Ghosh, HM Mukherjee, Physica Status Solidi (b), **116**, 17 (1983).
21. KP Ghatak, A Ghosal, Physica Status Solidi (b), **151**, K135 (1989).
22. KP Ghatak, N Chatterjee, M Mondal, Physica Status Solidi (b), **139**, K25 (1987).
23. S Debbarma, A Bhattacharjee, S Bhattacharyya, A Mondal, KP Ghatak, Journal of Advanced Physics, **1**, 84 (2012).
24. B Mitra, KP Ghatak, Solid-state Electronics, **32**, 515 (1989).
25. M Mondal, KP Ghatak, Physica Status Solidi (b), **126**, K47 (1984).
26. M Mondal, KP Ghatak, Physica Status Solidi (b), **139**, 185 (1987).

Phase Shifters in QPM Device for Domain Engineering

Nagarajan Asvin, Arvind Ganesh, Navin Infant Raj, Toijam Sunder Meetei, Shanmugam Boomadevi and Krishnamoorthy Pandiyan

Abstract We have analyzed the role of phase shifter domains in QPM devices for domain engineering. These phase shifter domains are capable of changing the phase of the second harmonics from 0 to 2π. Using this approach, we have generated multiple second harmonic peaks by introducing phase shifter domains of same width at unequal intervals in the periodic optical superlattice structure. Furthermore, we have theoretically analyzed the generation of multiple-QPM and the possibilities to engineer the intensity and bandwidth of the multiple peaks.

1 Introduction

Quasi-phase matching (QPM) technique allows efficient generation of higher harmonics by varying the polarization of a nonlinear crystal [1]. In this technique, the nonlinear optical coefficient (d_{eff}) within the nonlinear crystal is periodically reversed to 180° relative to the neighbour about the axis of propagation to form a grating of period $\Lambda = 2l_c$, where, l_c is the coherence length [2]. This kind of QPM grating structures could be easily fabricated by room temperature electric field poling method. In addition, various engineered domain structures have been proposed to achieve multiple-QPM, which includes a grating with sinusoidal modulation, phase reversal, aperiodic optical superlattice structure, continuously phase modulated superlattice structure and quasi-periodic optical superlattice structure [3]. We introduce phase shifter domains in periodic QPM device to generate multiple peaks for optical frequency conversion process [4, 5]. QPM device is useful in the

N. Asvin · A. Ganesh · N.I. Raj · T.S. Meetei · K. Pandiyan (✉)
Centre for Nonlinear Science and Engineering (CeNSE),
School of Electrical & Electronics Engineering, SASTRA University,
Thanjavur, Tamil Nadu 613401, India
e-mail: krishpandiyan@ece.sastra.edu

S. Boomadevi
Department of Physics, National Institute of Technology,
Tiruchirappalli, Tamil Nadu 620005, India

© Springer Nature Singapore Pte Ltd. 2017
I. Bhattacharya et al. (eds.), *Advances in Optical Science and Engineering*,
Springer Proceedings in Physics 194, DOI 10.1007/978-981-10-3908-9_64

current optical communication system in which the incoming optical-fiber signal is directly converted into the other channel without electrical processing in fiber based wavelength division multiplexed (WDM) systems. So, our proposed QPM design can generate desirable multiple peaks with tunable spacing, bandwidth, and intensity which could be useful to employ the same in WDM systems.

2 Theory

The coupled wave equation for basic slowly varying amplitude resulting to the growth of second harmonic generation which is obtained from Maxwell equations is given as [2],

$$\frac{dA_2}{dz} = i\kappa A_1^2 e^{-i\Delta kz} \tag{1}$$

where, $A_{1(2)}$, is the electric field amplitude of the second-harmonic wave, κ is proportional to the effective nonlinear optical coefficient and the wave-vector mismatch. The wave vector-mismatch for the QPM interaction Δk is given as,

$$\Delta k = k_2 - 2k_1 \pm G = 0 \tag{2}$$

where, G is the grating vector factor. With the assumption of negligible depletion of the fundamental wave, the amplitude of the second harmonic wave can be calculated by integrating the (1) to obtain a *sinc* form of the spectrum, κ is constant in the medium. But, κ is not constant. It is a function of z, i.e., $\kappa(z)$. Let $\Delta\kappa = 2\pi q$, then the second harmonic amplitude can be expressed as,

$$A_2(q) = iA_1^2 \int \kappa(z) e^{-i2\pi qz} dz = iA_1^2 \Im[\kappa(z)] \tag{3}$$

3 Results and Discussion

We have analyzed the second harmonic spectrum of periodic QPM device as depicted in Fig. 1a. Moreover, we introduce phase shifter domains in the periodic structure to generate multiple peaks. This phase shifter domains can vary the phase of second harmonics from 0 to 2π.

In this approach, we have distributed three phase-shifter domains of same width at unequal intervals in the periodic structure. This concept has generated four equally spaced second harmonic peaks of different intensities as shown in Fig. 1b. Similarly, we have distributed more numbers of phase shifter domains such as n = (five, seven, nine etc.,) of same width at unequal intervals in a periodic QPM

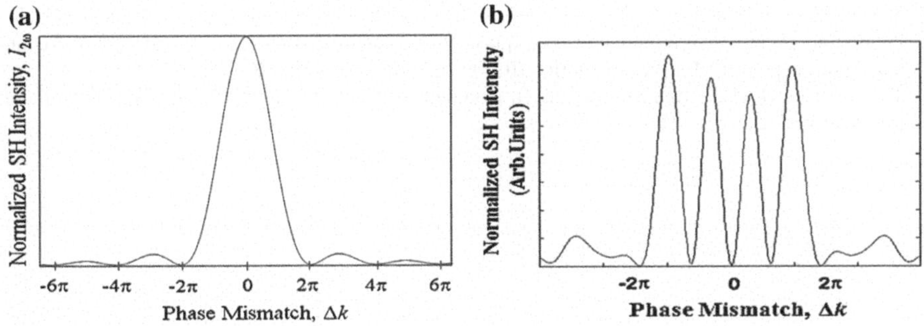

Fig. 1 **a** Normalized SH intensity of periodic QPM device without phase shifters. **b** Four SH peaks generated by introducing three phase shifter domains in the periodic structure

device. We could observed $n + 1$ numbers of second harmonic peaks which is a function of phase-shifter domains. Moreover, from our theoretical analysis we observed that the peak intensity is also a function of width and location of the phase shifter domains in the QPM device. Further works are going on to engineer the number of peaks, spacing and intensity of the second harmonic peaks to utilize in WDM systems.

4 Conclusion

Multiple SH peaks have been generated by introducing phase-shifter domains of same width at unequal position in the periodic QPM device. The proposed QPM device with phase shifter domains of same widths is capable of tailoring the intensity, spacing and number of SH peaks.

Acknowledgements The authors wish to express their sincere gratitude to the Department of Science & Technology (DST), New Delhi, India for their financial support (Ref. No.: SB/FTP/PS-089/2013).

References

1. D. S. Hum, and M. M. Fejer, "Quasi-phasematching," C. R. Physique 2, 180–198 (2007).
2. M. M. Fejer, G. A. Magel, D. H. Jundt, and R. L. Byer, "Quasi-Phase-Matched Second Harmonic Generation: Tuning and Tolerances," IEEE J. Quantum Electron. 28 (11), 2631–2654 (1992).
3. M. Asobe, O. Tadanaga, H. Miyazawa, Y. Nishida and H. Suzuki, "Multiple Quasi-Phase-Matched Device using Continuous Phase Modulation of $\chi(2)$ Grating and Its Application to Variable Wavelength Conversion," IEEE J. Quantum Electron 41 (12), 1540–1547 (2005).

4. S. Hari Hara Subramani, K. Karthikeyan, A. Mirunalini, R.K. Prasath, S. Boomadevi, and K. Pandiyan, “Analysis of a phase reversal quasi-phase matching device for the dual peak second harmonic response,” Journal of Optics (IOP) 15(5), 055205–055210 (2013).
5. T. S. Meetei, S. Hari Hara Subramani, S. Boomadevi, K. Pandiyan, “Advances in Optical and Engineering,” Springer Proceeding in Physics, 166, 461–463 (2015).

Simulated Annealing: An Approach for Multiple QPM

Siva Chellappa, Shiva Prabhakar, Narayanan Balaji, Toijam Sunder Meetei and Krishnamoorthy Pandiyan

Abstract All optical wavelength conversion is made possible with the advent of engineered Quasi-phase matching (QPM) devices. Simulated annealing is an algorithm can be used in QPM technology for designing second harmonic wave of any chosen amplitude and phase profile. In this article, we have used simulated annealing routine for achieving multiple QPM second harmonic generation with equal intensity and channel spacing. Further, the influence of the algorithm temperature and number of crystal domains on designing the multiple channel QPM response is discussed in detail.

1 Introduction

The techniques for achieving wavelength conversion by Quasi-Phase-Matching (QPM) nonlinear optics have been developing quickly in the past few decades. QPM method can be used to compensate the wave vector mismatch caused by the different phase velocities of the interactive waves involved in a nonlinear process [1]. The wavelength conversion needs efficient nonlinear optical interaction which depends on the phase vector mismatch compensation. The periodically poled crystals (e.g. Lithium Niobate) are made by changing the sign of the nonlinear co-efficient effectively at every π radian using the room temperature electric field poling process [2, 3]. In case of periodically poled QPM crystals, they can provide a series of reciprocal vectors, each of which is an integral multiple of the fundamental vector. These reciprocal vectors provide phase mismatch compensation for single interaction. While giving the complete solution to the single frequency interaction as shown in Fig. 1, periodic super-lattices failed to give solution for multiple

S. Chellappa · S. Prabhakar · N. Balaji · T.S. Meetei · K. Pandiyan (✉)
Centre for Nonlinear Science and Engineering (CeNSE),
School of Electrical & Electronics Engineering, SASTRA University,
Thanjavur, Tamil Nadu 613401, India
e-mail: krishpandiyan@ece.sastra.edu

© Springer Nature Singapore Pte Ltd. 2017
I. Bhattacharya et al. (eds.), *Advances in Optical Science and Engineering*,
Springer Proceedings in Physics 194, DOI 10.1007/978-981-10-3908-9_65

arbitrary interactions. Multiple interactions are used in many nonlinear devices such as optical routing, wavelength division multiplexing and so on.

2 Theory

The second harmonic field is governed by slowly varying amplitude equation given by,

$$\frac{dE_{3i}}{dz} = \pm ikd_{ijk}\exp(i\Delta kz) \tag{1}$$

where, E_{3i} and n_3 are the field amplitude and index of the second harmonic (SH) wave respectively at frequency ω_3 and polarized in the z-direction, d_{ijk} is the absolute value of the nonlinear co-efficient, z is the propagation distance in the crystal, and Δk is the wave vector mismatch between the SH and fundamental waves. E_{1j} and E_{2k} are the constant input fields at the fundamental wavelength with frequencies $\omega_1 = \omega_2$ and polarization directions j and k. As stated in the introduction, for a design of device that can vary all optical parameters it is important to control the intensity and phase of the SH wave. In order to design the SH wave the proposed simulated annealing routine is used. Simulated annealing algorithm used for grating design uses the summation value of E_{3i} for each domain as provided [4, 5].

$$E_{3i} = -\frac{kd_{ijk}}{\Delta k}\left[1 - (-1)^n\exp(i\Delta kQ_n) + \sum_{m=1}^{n} 2(-1)^m\exp(i\Delta kQ_m)\right] \tag{2}$$

where, $Q_n = [q_1, q_1 + q_2, q_1 + q_2 + q_3, \ldots, q_1 + \cdots + q_n]$, Qn is a vector that contains the tails of the domains. Now, the domain lengths are perturbed for a small variation in length and the corresponding energy E_{3i} is calculated. Then, the difference is determined between obtained E_{3i} and absolute E_{3i} (equal field intensity) in order to

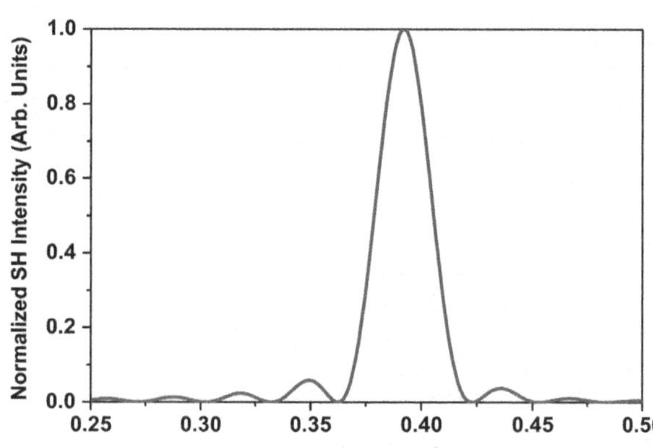

Fig. 1 Shows typical SH spectral response of grating structures designed by simulated annealing algorithm

calculate error. The perturbations are performed till we minimize the error to negligible.

3 Results and Discussion

The most important thing to understand in this algorithm is it gives an optimal solution which can be considered as one of the best solutions. The algorithm does this by running iterative loops for several hundred times to yield the best result. This algorithm does not get stuck with the local minima and maxima peaks where the randomness of the algorithm plays a vital role in choosing the domain of the grating. The convergence speed of the algorithm also depends on the type of the grating and the initial domain chosen. The state with the highest maxima function is chosen when two states are compared by using an acceptance probability term ***exp*** (E/T). The algorithm temperature which we defined as a function of temperature is generally multiplied by a decaying factor α. The algorithm temperature also determines the convergence speed of the algorithm. As a result of multiplying the temperature with a decaying factor the algorithm is most likely to accept even the worst case results and then refining its solution again and again as the loop iteration occurs. The acceptance probability dictates the algorithm how to switch to the better solution based on the following conditions.

exp (E/T) = 1; switch over to the new state
exp (E/T) = 0; do not switch over to the new state
exp (E/T) = 0–1; compare this with a random number between 0 and 1 and switch over if it is greater.

Here, the number of domains and algorithm temperature are the key factors in designing desired spectral response. In order to design a QPM crystal with any peak intensity profile, the desired profile from now on is denoted as D (Δk). Initially periodically poled domain is chosen and we have shown above the profile of a single peak QPM of grating size 8 μm with side lobes suppression using simulated annealing. Although it may seem easy for the design of single peak profiles, there are various factors like fixing the error threshold and the initial temperature which decide the convergence time of the algorithm to be considered while designing the grating of multiple peak profiles [6, 4]. And these domains are perturbed randomly up to 1%. The perturbation is chosen very small so that multiple changes can be done in a series of iterative loops preventing a very large change in domain size in a single iteration. The implementation of simulated annealing algorithm is shown in Fig. 2.

Fixing the error threshold determines how well the obtained peak response is similar to the desired peak response. Generally for a dual peak, triple peak profiles the intensities are half, one third of the maximum value of intensity in the single peak profile. In this way we have fixed the threshold values for different kinds of profiles and the resulting SH spectrum is shown in Fig. 3.

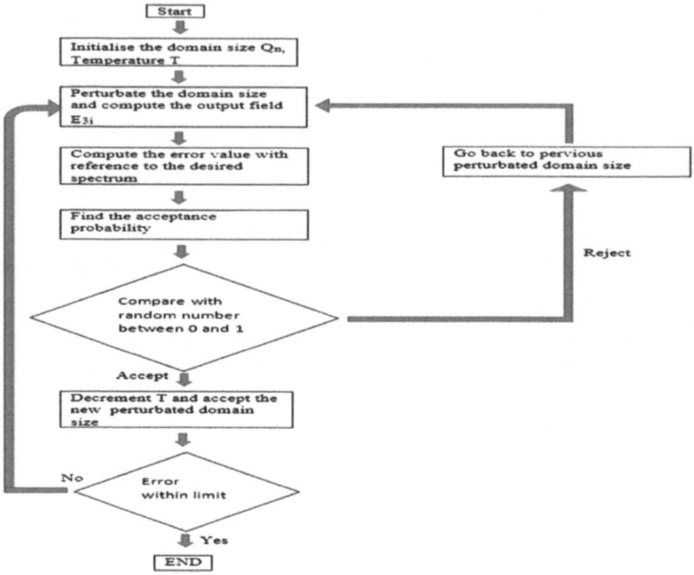

Fig. 2 Flow chart of simulated annealing algorithm used for generating SH spectral profiles containing maxima at more than one wavelength with equal intensity

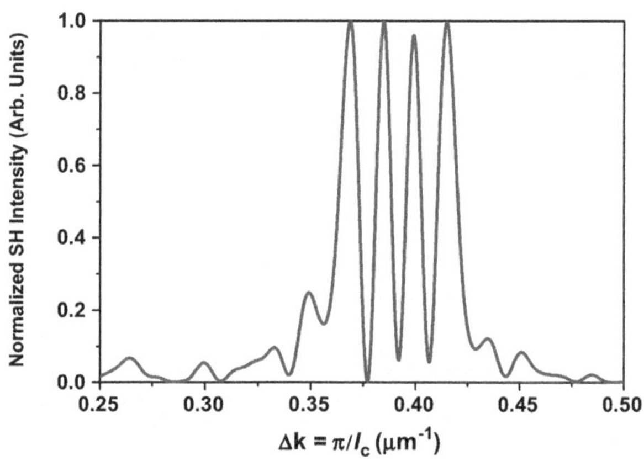

Fig. 3 SH spectral profiles containing maxima at more than one wavelength with equal intensity generated by simulated annealing algorithm

4 Conclusion

We have utilized the simulated annealing algorithm to design desired second harmonic spectral response from the QPM device. With this algorithm it is possible to have a control on intensity and phase profile in the SH wave. The algorithm temperature and number of crystal domains play a crucial role in designing desired spectral response. Using this algorithm we have demonstrated multiple channel QPM devices.

Acknowledgements The authors wish to express their sincere gratitude to the Department of Science & Technology (DST), New Delhi, India for their financial support (Ref. No.: SB/FTP/PS-089/2013).

References

1. M. M. Fejer, G. A. Magel, D. H. Jundt, and R. L. Byer, "Quasi-Phase-Matched Second Harmonic Generation: Tuning and Tolerances," IEEE J. Quantum Electron. 28 (11), 2631–2654 (1992).
2. M. Ahlawat, A. Tehranchi, K. Pandiyan, M. Cha, and R. Kashyap, "Tunable all-optical wavelength broadcasting in a PPLN with multiple QPM peaks" Optics Express 20 (24), 27425–27433, (2012).
3. K. Pandiyan, Y. S. Kang, H. H. Lim, B. J. Kim, and M. Cha, "Nondestructive quality evaluation of periodically poled lithium niobate crystals by diffraction" Optics Express. 17(20), 17862–17867 (2009).
4. J. Wu, T. Kondo, and R. Ito, "Optimal design for broadband quasi-phase-matched second-harmonic generation using simulated annealing" Journal of Light wave Technology. 13(3), 456–460 (1995).
5. D. T. Reid, "Engineered Quasi-Phase-Matching for Second-Harmonic Generation," Journal of Optics A: Pure and Applied Optics 5 (4), S97 (2003).
6. S. Hari Hara Subramani, K. Karthikeyan, A. Mirunalini, R.K. Prasath, S. Boomadevi, and K. Pandiyan, "Analysis of a phase reversal quasi-phase matching device for the dual peak second harmonic response," Journal of Optics (IOP) 15(5), 055205–055210 (2013).

Part XIV
Non-Linear Waveguides, Optical Fibre Devices, Photonic Crystal

Spectral Bandwidth Tuning at Telecommunication Wavelength by Specific Intensity Modulation in Nonlinear Plasmonic Cavity Coupled Waveguide

Sushmita Paul and Mina Ray

Abstract A single mode plasmonic waveguide coupled to a cavity, filled with third order nonlinear Kerr medium is designed to provide bandwidth tuning at 1.55 μm. This is achieved by controlling the intensity of a pump beam which modulates the resonance response of the output beam. Usually, the intensity variation causes a shift in the resonance wavelength. However, for certain intensity values of the pump excitation, the system is found to resonate at the same wavelength with varying spectral bandwidths. Quality factor tuning resulting from spectral width variation has been demonstrated quantitatively.

1 Introduction

A subwavelength plasmonic waveguide enables photon confinement at nanoscale due to strong localization of surface plasmon polaritons (SPP) at metal-dielectric interface [1, 2]. The enhanced intensity of these localized surface plasmons facilitates nonlinear optical phenomenon in nanoplasmonic structures [3]. Plasmonic resonances are highly sensitive to the permittivity of the medium surrounding the metal. This sensitivity is explored in nonlinear plasmonic systems by using a control light whose intensity perturbs the permittivity of the nonlinear medium finally modifying the resonance response. Such phenomenon enables nonlinear switching and bistability, which have been widely demonstrated in several plasmonic waveguide coupled cavity based devices [4–7].

S. Paul (✉) · M. Ray
Department of Applied Optics and Photonics, University of Calcutta,
Technology Campus, JD-2, Sector-3, Salt Lake, Kolkata, West Bengal 700106, India
e-mail: paul86.sushmita@gmail.com

M. Ray
e-mail: mraphy@caluniv.ac.in

© Springer Nature Singapore Pte Ltd. 2017
I. Bhattacharya et al. (eds.), *Advances in Optical Science and Engineering*,
Springer Proceedings in Physics 194, DOI 10.1007/978-981-10-3908-9_66

In the present study, we demonstrate bandwidth tuning in a nonlinear cavity coupled subwavelength plasmonic waveguide at a fixed wavelength (1.55 μm). This is achieved by launching the control beam at specific intensities. Resonant dips are obtained at a fixed chosen wavelength but overall resonance response shows variable spectral width for a fixed set of geometrical parameters and material properties. The cause of spectral width tuning with intensity modulation is explained with graphical analysis. Quality factor of these resonances is also evaluated quantitatively.

2 Device Description and Numerical Modelling

The structure under consideration is depicted in Fig. 1. It comprises of a metal-dielectric-metal waveguide (plasmonic waveguide) coupled to a cavity. The width of the waveguide and the cavity are of subwavelength dimension (i.e. $w_w = w_c \ll \lambda$). The subwavelength plasmonic waveguide supports only a single antisymmetric mode. The waveguide and the cavity is embedded in a metallic medium, whose permittivity is calculated using Drude model, given by,

$$\varepsilon_m = 1 - \frac{\lambda^2 \lambda_c}{\lambda_p^2(\lambda_c + i\lambda)} \tag{1}$$

where λ_p and λ_c denotes the plasma wavelength and collision wavelength respectively. In our work, the metal is considered to be silver (Ag) for its low absorption property. The effective refractive index of this single mode plasmonic waveguide is given by [8]:

$$n_{eff} = \sqrt{\varepsilon_d}\left(1 + \frac{\lambda}{\pi w_w \sqrt{-\varepsilon_m}}\sqrt{1 + \frac{\varepsilon_d}{-\varepsilon_m}}\right)^{1/2} \tag{2}$$

where ε_d is the permittivity of the dielectric medium within the waveguide. A third order Kerr medium is filled within the cavity to achieve the nonlinear operation. For

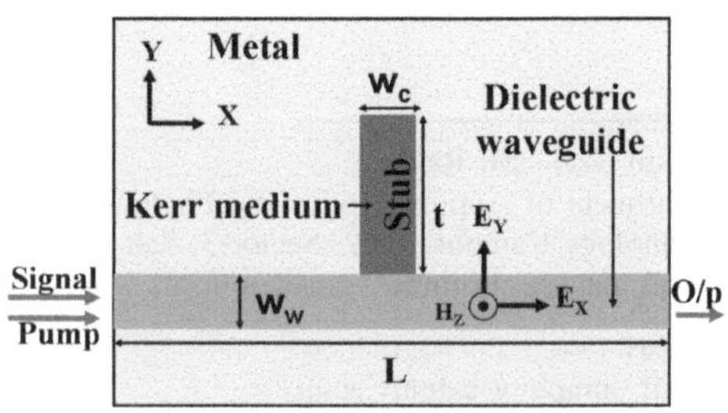

Fig. 1 Schematic diagram of the plasmonic cavity coupled structure

calculating the effective refractive index of the nonlinear cavity, ε_d in (2) is replaced by ε_{nl} given by:

$$\varepsilon_{nl} = \varepsilon_0 + \varepsilon_2 \quad (3)$$

where ε_0 refers to the linear permittivity and $\varepsilon_2 = \chi^{(3)}\overline{E}^2$ represents the intensity dependent term. The pump excitation (control beam) modulates this term resulting in change of the transmission response of the device. Here $\chi^{(3)}$ is the third order nonlinear susceptibility and $\overline{E}$ is the total average electric field within the stub.

The transmission response of the plasmonic structure can be evaluated from (4) which is derived [9, 10] using transmission line theory and scattered field technique and is expressed as:

$$T = \left|1 + \frac{Zc}{2Z_{wg}}\right|^{-2} \exp\left(-\frac{L}{L_{spp}}\right) \quad (4)$$

where Z_c denotes the effective impedance of the cavity and Z_{wg} represents the waveguide characteristic impedance. Propagation length of the SPPs in the dielectric waveguide is denoted by $L_{spp.}$ The length L of the device or the waveguide weakly affects the transmission response. Change of L affects only the magnitude of transmitted amplitude (or intensity) without altering the resonating wavelength. The resonance wavelength depends on the impedances Z_c and Z_{wg} which strongly depends on the value of stub height.

3 Results

Modulation of control light in a nonlinear system usually causes the device to operate as an on-off switch, where on-off states generally correspond to the presence or absence of the control (pump) pulse. While the 'on' state is characterized by signal with higher transmission value, the 'off' state is characterized by a dip in the transmission spectrum depicting lower transmission value. The transmission dip at a certain frequency (wavelength) is caused when a light pulse of the same frequency is launched which superimposes with the continuous wave signal used for plasmonic excitation of the system. This superposition causes all the propagating surface plasmons to be trapped within the cavity, resonating inside it, and prohibiting any signal at the output of the waveguide at that frequency. This is manifested as a resonant dip in the transmission spectrum of the system. Trapping of the electromagnetic field within the cavity at the resonating wavelength is shown in Fig. 2a. By modulating the intensity of the control beam, the refractive index of the Kerr nonlinear medium can be changed, which in turn causes a shift in the resonance dip of the transmission spectrum. Shift of the transmission dip by varying the pump

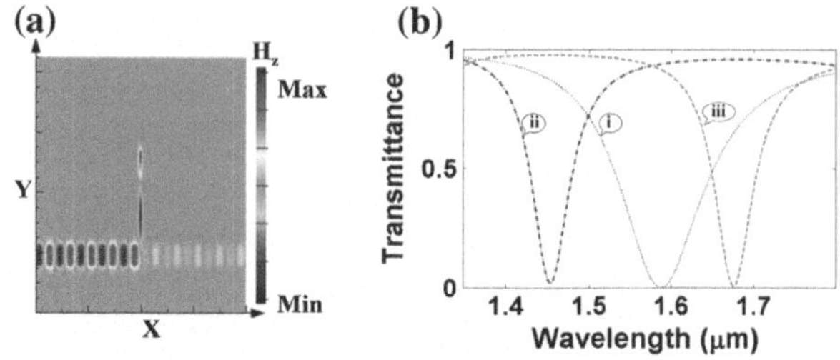

Fig. 2 **a** Magnetic field distribution in the plasmonic structure showing trapping of surface plasmons within the cavity at 1.55 μm. The field profile is simulated in Lumerical FDTD software. **b** Transmission spectra for varying intensity values: (*i*) 2.25×10^3 V^2/μm^2 (*ii*) 4.41×10^3 V^2/μm^2 and (*iii*) 6.25×10^3 V^2/μm^2

intensity is depicted in the plot of Fig. 2b. The Kerr material chosen for the simulation has a susceptibility of 1.95×10^{-3} μm^2/V^2 or 1.4×10^{-7} esu [11] and the refractive index of the waveguide is 1.52 which is same as the linear refractive index of the Kerr medium filing the cavity. The geometrical parameters considered are: t = 630 nm, $w_w = w_c$ = 100 nm and L = 400 nm. The pump intensities for which the plot is obtained are mentioned in the figure caption. However, unlike that of the switching phenomenon, it is observed from the simulated result of Fig. 3a that, at certain specific intensities, for certain fixed geometrical parameters of the structure, the system provides transmission dip at a single wavelength and the spectral bandwidth is different for each intensity value. In our case (Fig. 3a), the system is designed to produce transmission dips at 1.55 μm (telecom wavelength) for three selected specific intensities of the pump excitation. Spectrum is seen to be narrower for the highest pump intensity. The cause for this bandwidth variation can be attributed to the fact that one of the multiple dips of the transmitted spectrum for each intensity value merges at the operating wavelength (1.55 μm). This is depicted in Fig. 3b. It clearly depicts, that one transmission dip for each intensity value coincides at 1.55 μm and the spectral width varies for each of them. The spectrum width widens as the intensity value reduces. Thus, it confirms that spectral width tuning can be achieved at a particular wavelength with properly chosen optical pump intensity in a nonlinear plasmonic switch. Bistability or multistability is also one of the causes for bandwidth modulation which leads to fixed number for intensity values for which spectral width tuning can be achieved at the operating wavelength.

Quality factor (Q-factor) provides a Figure of Merit for quantitative analysis of resonator or cavity based devices. Q-factor is usually a measure of energy trapped or stored within a cavity during resonance. Q-factor can be mathematically defined as:

$$\text{Q-factor} = \frac{\lambda_O}{\text{FWHM}} \tag{5}$$

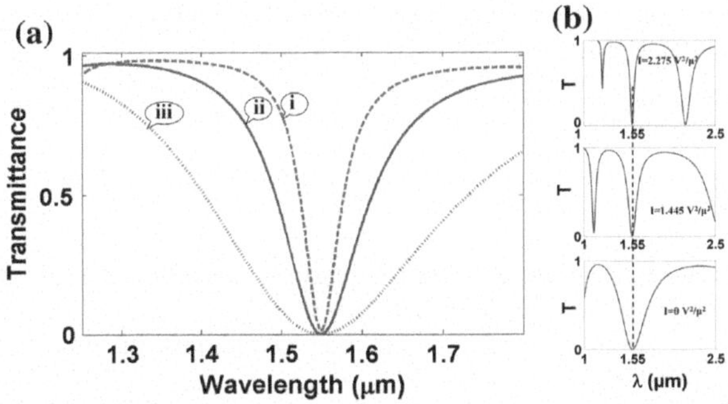

Fig. 3 **a** Transmission spectra for specific intensity values: (*i*) 5.18 $V^2/\mu m^2$ (*ii*) 2.088 × 10^3 $V^2/\mu m^2$ and (*iii*) 0 $V^2/\mu m^2$. **b** Transmission spectra for same intensity values shown separately depicting occurrence of multiple dips and merging of dips at 1.55 μm. Geometrical and material parameters are similar to that considered for Fig. 2

Table 1 Quality factor calculation for plasmonic excitations at 1.55 μm for different specific intensities

Specific intensity values ($V^2/\mu m^2$)	FWHM (nm)	Q-factor
5.18 × 10^3	54	28.70
2.088 × 10^3	120	12.92
0 × 10^3	328	4.72

Where λ_O is the operating or resonant wavelength and FWHM denotes the full width half maxima of the resonance profile. The FWHM and the corresponding Q-factor of transmission resonances are calculated from Fig. 3a for each specified intensity and are tabulated in Table 1. Quality factor is seen to be highest corresponding to the highest intensity confirming higher energy confinement for narrower spectrum.

4 Conclusion

We have designed a nonlinear plasmonic cavity coupled waveguide device in which bandwidth tuning at a particular wavelength can be achieved by application of an external control beam. The spectral width at the operating frequency is tuned only at certain specific intensities of the pump excitation. As the FWHM of the spectrum can be modulated keeping the center frequency fixed, the device can serve as tunable band-stop filter. This is advantageous over similar plasmonic waveguide coupled devices where band-pass or band-stop filtering is obtained by structural alteration, unlike in our case where an external control beam is sufficient for spectral width tuning [12, 13]. Moreover, this filter application is more appropriate where only bandwidth tuning is required without altering the operating frequency.

Acknowledgements The author S. Paul acknowledges Department of Science and Technology, Government of India, for providing INSPIRE Fellowship.

References

1. Zhang, J., Zhang, L. and Xu.W., "Surface plasmon polaritons: physics and applications," J. Phys. D: Appl. Phys. 45(11), 1–19 (2012).
2. Stockman, M. I., "Nanoplasmonics: past, present, and glimpse into future," Opt. Express 17 (22), 22029–22106 (2011).
3. Kauranen, M. and Zayats, A.V., "Nonlinear Plasmonics," Nat. Photonics 6, 737–748 (2012).
4. Lu, H., Liu, X., Wang, L., Gong, Y. and Mao, D., "Ultrafast all-optical switching in nanoplasmonic waveguide with Kerr nonlinear resonator," Opt. Express 19(4) 2910–2915 (2011).
5. Lin, X., Yan, J., Zheng, Y., Wu, L. and Lan, S., "Bistable switching in the lossy side-coupled plasmonic waveguide-cavity structures," Opt. Express 19(10), 9594–9599 (2011).
6. Pannipitiya, A., Rukhlenko, I. D. and Premaratne, M., "Analytical theory of optical bistability in plasmonic nanoresonators," J. Opt. Soc. Am. B. 28(11), 2820–2826 (2011).
7. Paul, S. and Ray, M., "Analysis of plasmonic subwavelength waveguide-coupled nanostub and its application in optical switching", Appl. Phys. A 122(22), 1–9 (2016).
8. Collin S., Pardo F. and Pelouard J. L., "Waveguiding in nanoscale metallic apertures," Opt. Express 15(7), 4310–4320 (2007).
9. Pannipitiya, A., Rukhlenko, I. D., Premaratne, M., Hattori, H. T. and Agarwal G. P., "Improved transmission model for metal-dielectric-metal plasmonic waveguides with stub structure", Opt. Express 18(6), 6191–6204 (2010).
10. Pannipitiya, A., Rukhlenko, I. D. and Premaratne, M., "Analytical modelling of resonant cavities for plasmonic-slot-waveguide junctions," IEEE Photonics journal, 3(2), 220–233 (2011).
11. Wang, X., Jiang, H., Chen, J., Wang, P., Lu, Y. and Ming, H, "Optical bistability effect in plasmonic racetrack resonatorn with high extinction ratio," Opt. Express 19(20), 19415–19421.
12. Lu, H., Liu, X., Wang G. and Mao, D., "Tunable high-channel-count bandpass plasmonic filters based on an analogue of electromagnetically induced transparency," Nanotechnology 23, 444003 (2012).
13. Xu, Q., Sandhu, S., Povinelli, M . L., Shakya, J., Fan, S. and Lipson, M., "Experimental realization of an on-chip all optical analogue to electromagnetically induced transparency," Phy. Rev. Lett 96, 123901 (2006).

Investigation on Fiber Loop Ring Resonator for Chemical Sensor Application

Arun V.S. Nair, Jijo V. Ittiarah and Tarun Kumar Gangopadhyay

Abstract An investigation on fiber loop ring resonator has been accomplished for analysing the performance of the chemical sensor application. A comparative study of direct coupled and cross coupled design of fiber loop has been performed. The phase fluctuation on circulating intensity component is critically analysed for distinctive configuration. It is predicted that a resonance peak obtained for a fiber loop length of 13.85 m will shift 10.85 mm/RIU. The fiber loop ring resonator system is practically implemented and resonance is obtained.

1 Introduction

In the past few years, the advancement in the field of fiber optic sensors, interferometers and filters [1–4] have increased the attention towards the fiber loop ring resonator due to its practical significance. Interference of light within the fiber loop is the underlying phenomena trailing the fiber loop resonance. The phase variation between different optical intensity components can produce the interference effect which relies upon the fiber loop length, the input laser wavelength and refractive index of the analyte. The interaction between the chemical and the evanescent field of the light can produce a fluctuation in the output intensity pattern. This shift in resonance is proportional to the refractive index of the chemical in the sensing region, which can be used for chemical and biochemical sensing. The resonance peak of fiber loop ring resonator is theoretically and practically obtained with a theoretical sensitivity of 10.85 mm loop length shift per refractive index unit.

A.V.S. Nair · J.V. Ittiarah · T.K. Gangopadhyay (✉)
Fiber Optics & Photonics Division, CSIR-CGCRI, Kolkata, India
e-mail: tkg@cgcri.res.in
URL: http://www.cgcri.res.in

© Springer Nature Singapore Pte Ltd. 2017

I. Bhattacharya et al. (eds.), *Advances in Optical Science and Engineering*,
Springer Proceedings in Physics 194, DOI 10.1007/978-981-10-3908-9_67

2 Mathematical Modeling of Fiber Loop Resonator

The simple structure of a fiber loop ring resonator consists of a directional coupler and a fiber loop which is shown in Fig. 1a. Based on the arrangement of the directional coupler, there are two kinds of configurations are possible with in the fiber loop ring resonator.

2.1 Direct Coupled Configuration

In direct coupled configuration, the coupled port is connected in the fiber loop and output is taken from direct port. The intensity at the output port can be calculated as [5],

$$\left|\frac{E_o}{E_i}\right|^2 = (1-\gamma_0)\left[1 - \frac{k(1-k-A)/(1-k)}{1+A-2A^{1/2}\cos\beta L}\right] \tag{1}$$

where E_O is the electric field associated with the output intensity component. E_i is the electric field associated with the input intensity component, γ_O is the coupler insertion loss, k is the coupling coefficient, β is the propagation constant, L is the fiber loop length and A is a parameter defined as [5],

$$A = (1-k)(1-a)(1-\gamma_0)e^{-2\alpha L} \tag{2}$$

where a is the splice loss in the fiber loop and α is the attenuation coefficient.

At resonance the output intensity must be minimum. Then the condition for the resonance can be derived as,

$$\beta L = 2\pi p \tag{3}$$

where p is an integer.

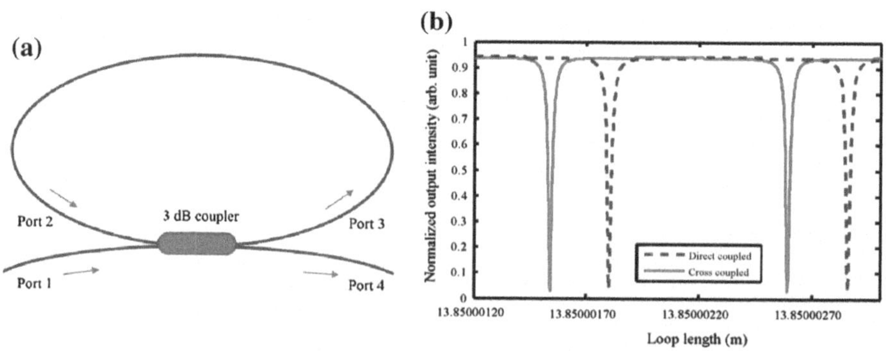

Fig. 1 **a** Schematic diagram for fiber loop ring resonator, **b** resonating pattern for direct coupled and cross coupled configuration

2.2 Cross Coupled Configuration

In cross coupled configuration, the coupled port is considered as an output and the direct port is connected to the fiber loop. The intensity at the output port can be calculated as,

$$I_o = (1-\gamma_0)\left(1 - \frac{(1-k)(1-\alpha(1-\gamma_0))}{1+\alpha k(1-\gamma_0)+2\sqrt{\alpha k(1-\gamma_0)}\sin\beta L}\right) \tag{4}$$

Similarly the condition for the resonance can also be written as

$$\beta L = (2\pi - \pi/2)p \tag{5}$$

From (1) and (4), it is clear that the output intensity of the system depends on the loss factors with in the proposed setup. The variations in the wavelength of the laser beam, fiber loop length and effective refractive index are influences only on the shift in resonating patterns. Equations (3) and (5) shows that direct coupled and cross coupled configuration have only a phase shift of 90° in resonance pattern where all other parameters remain same. This result is shown in Fig. 1b.

The loss inside the fiber loop resonator will influence the intensity level at output port and the performance. The coupler insertion loss, splice loss and attenuation in the fiber loop are the main loss in the ring resonator. The variation in the resonance curve for different coupler insertion loss (C.I.) in direct coupled and cross coupled fiber loop ring resonators are shown in Fig. 2a, b respectively.

Both direct coupled and cross coupled configuration shows that the coupler insertion loss should be minimum for good performance. Then only the fiber loop returns a sharp peak. Among the three conditions of coupler insertion loss 0.3, 3 and 5%, slope of the curve at the resonating condition is much larger for 0.03% insertion loss. So the lower insertion loss will produce a better intensity transition, when the resonating condition is reached.

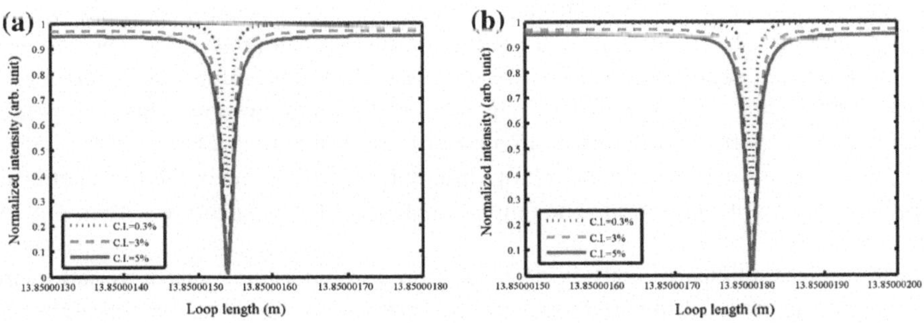

Fig. 2 Influence of coupler insertion loss on **a** direct coupled configuration and **b** cross coupled configuration

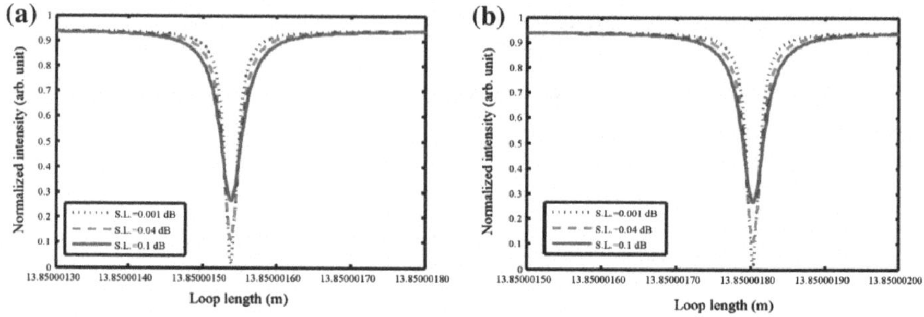

Fig. 3 Influence of splice loss on **a** direct coupled configuration and **b** cross coupled configuration

The variation in the output intensity for different splice loss (S.L.) for cross coupled and direct coupled fiber loop ring resonators are shown in Fig. 3a, b respectively.

Here, both configurations shows that the splice loss in the fiber loop ring resonator should be minimum to get a high Finesse resonator. From the curves, it can be clear that the better performance of the resonator is obtained at the splice loss in the fiber is at 0.001 dB. The splice loss in the fiber determines the output intensity level at the resonance. If the splice loss is reduced to its minimum value (~ 0 dB) then the output intensity at resonance becomes zero and the resonating curve becomes ideal.

The influence of the absorption coefficient of the fiber also affect the performance of ring resonator. From the analysis of output intensity pattern for different absorption coefficient, it is obtained that the fiber loop resonance is sharp when the absorption coefficient of the fiber is minimum.

Figure 4a, b shows the comparative study of the phase change in port 2, 3 and 4 for the direct coupled and cross coupled fiber loop ring resonator respectively. The curve shows that a constant rate of change in fiber loop length will provides a uniform change in the phase value for the confined field components at different port. The change in the phase value is in periodic nature. One period is considered as the interval between two adjacent resonance minima. During one period the circulated intensity component will produces a phase change of $\sim 2\pi$. The coupler in the proposed setup can provides a 90° phase shift. The influence of this additional phase shift at each port will depends on the configuration. Then the phase value at same port for different configurations are different.

The simulation has been done for different refractive index of the chemical applied on the sensing region. The output intensity of the resonator cavity is shown in Fig. 5.

The output intensity analysis of fiber loop resonator indicates a variation in output intensity for different loop lengths. Intensity is found to be minimum at certain loop length of the cavity and it is considered as the resonance loop length. A shift in resonance loop length has been observed for different external refractive

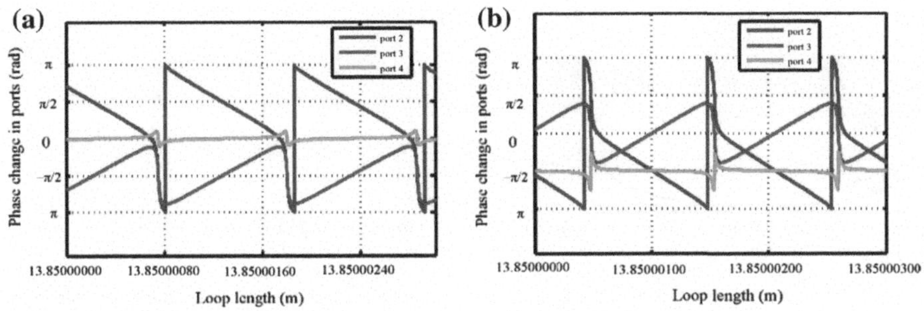

Fig. 4 The variation in the phase of the intensity component for **a** direct coupled fiber loop ring resonator and **b** cross coupled fiber loop ring resonator at port 2, 3 and 4

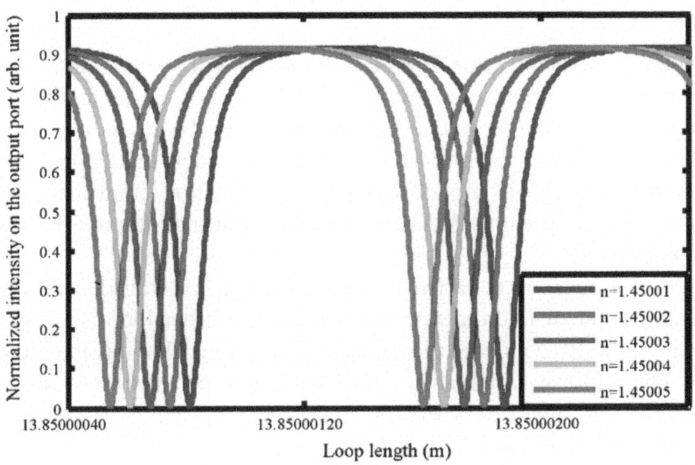

Fig. 5 The intensity at the output port of fiber loop as a function of the loop length for different chemicals

index of the analyte. For a change in refractive index of 0.00004, a shift in resonating loop length of 4.34×10^{-7} m is obtained. So the sensitivity is calculated as 10.85 mm/RIU.

3 Experimenal Setup of Ring Resonator

Fiber loop ring resonator with a loop length of 13.85 m has been demonstrated at CGCRI. The schematic diagram for the experimental setup is shown in Fig. 6a. The resonating pattern obtained from the fiber loop resonator is shown in Fig. 6b.

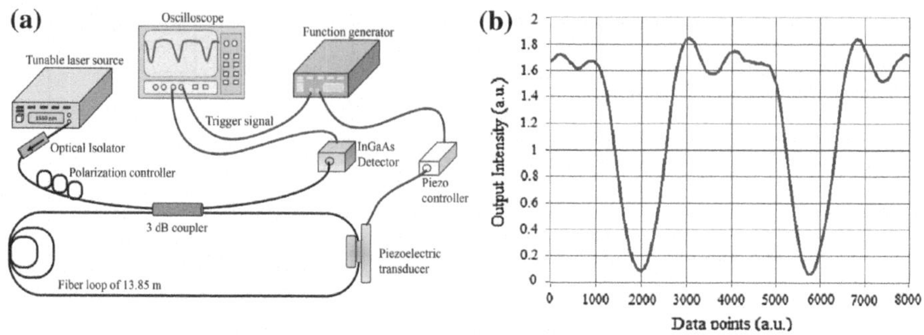

Fig. 6 **a** The experimental setup for fiber loop ring resonator having a fiber loop length of 13.85 m with a 3 dB coupler and a piezo electric optical fiber stretcher, **b** resonance obtained at the output port of the fiber loop ring resonator

The system consist of a laser source, optical isolator, polarization controller, a coupler, piezoelectric transducer, piezo driver, function generator, detector and oscilloscope. The system is excited using the laser beam at a wavelength of 1550 nm. Isolator is used to protect the laser from back reflection. The state of polarization of the beam is rotated by using polarization controller and laser is coupled to the fiber loop with a 3 dB coupler. The splitting ratio of 50:50 is selected for initial experiments.

The length of fiber loop is varied using a piezo electric transducer by stretching the fiber in the order of micrometer. The variation of fiber loop is controlled by a function generator. It produces a triangular wave signal having the frequency of 38.77 kHz. The uniform variation in fiber loop length produces a resonance on the output intensity. The intensity variation is captured by using an InGaAs detector and it is visualized using an oscilloscope.

The characteristic nature of the resonating pattern is expressed using Free Spectral Range, Full Width Half Maximum and Finesse. In the demonstrated experimental setup having a Finesse 3.46. The presence of loss in the fiber loop influences the performance of resonator. The loss will definitely reflects on the resonating curve and can easily be identifies from curve. This property of resonator is utilized for the chemical sensing applications.

4 Results and Conclusion

The loop length at which the minimum output intensity is known as the resonating length. There is a shift in the resonating loop length depending up on the refractive index at the sensing region.

The result shows a linear relation between the resonating loop length and the refractive index of the analyte. A resonating loop length shift is observed with a theoretical sensitivity of 10.85 mm loop length shift per refractive index unit and an additional 90° phase shift is observed in the cross coupled as compared to the directly coupled configuration. The phase change in each intensity component with respect to the input component also varies with the loop length. The resonance is obtained with experimental setup which can be further extended for high sensitive chemical sensing application.

Acknowledgements The research work supported by the sponsored project (No. GAP0141) from DST, Govt. of India. The author would like to acknowledge the support of the Director, CGCRI, Dr. Mukul Paul, Dr. M. Pal and Dr. Shyamal Das of FOPD, CGCRI, Kolkata for their help.

References

1. Stokes, Loren F., Marvin Chodorow, and Herbert J. Shaw. "All-single-mode fiber resonator." Optics Letters, 7.6, 288–290 (1982).
2. Gangopadhyay, Tarun Kumar, et al. "Detection of chemicals using a novel fiber-optic sensor element built in fiber loop ring-resonators." Sensors and Actuators B: Chemical, 206, 327–335 (2015).
3. Harun, S. W., Lim, K. S., Damanhuri, S. S., & Ahmad, H. "Microfiber loop resonator based temperature sensor". Journal of the European Optical Society: Rapid Publications, 6, 11026 (2011).
4. Vollmer, F., & Fischer, P. "Frequency-domain displacement sensing with a fiber ring-resonator containing a variable gap". Sensors and Actuators A: Physical, 134(2), 410–413 (2007).
5. Zhang, Feng, and John WY Lit. "Direct-coupling single-mode fiber ring resonator." JOSA A, 5.8, 1347–1355 (1988).

Spectrally Efficient SSB Optical OFDM Signal Using Balanced Detection

Gokul Boro and Subhash C. Arya

Abstract Orthogonal frequency division multiplexing (OFDM) has already gained importance in the field of optical communications with the increasing popularity of Radio-over-Fiber (RoF) technique. In this paper, generation of optical OFDM (OOFDM) signal using both direct and external intensity modulation that can be detected using direct-detection receivers is presented. Direct intensity modulation is performed using Vertical Cavity Surface Emitting Laser (VCSEL) whereas external modulation is done using Mach-Zehnder Modulator (MZM). These modulation schemes are tested using vector modulation format, i.e., quadrature amplitude modulation (QAM) that are used to modulate the orthogonal subcarriers, thus generating an OFDM signal. The performance of the system with different orders of QAM is compared based on the error vector magnitude (EVM) of the received constellation. The change in EVM is studied for the increase in the transmission distance of the signal. OOFDM systems using MZM showed better results compared to the system using VCSEL.

1 Introduction

Orthogonal frequency division multiplexing (OFDM) is a special form of multi-carrier modulation (MCM), where a high-rate data stream is split into a number of lower-rate data streams that are transmitted simultaneously over a number of subcarriers [1, 2]. One of the main reasons to use OFDM is to increase the robustness against frequency selective fading and narrow band interference.

G. Boro · S. C. Arya (✉)
Lightwave Communication System Laboratory,
Department of Electronics & Communication Engineering,
North-Eastern Hill University, Shillong, Meghalaya 793022, India
e-mail: aryasubh@yahoo.co.in

© Springer Nature Singapore Pte Ltd. 2017
I. Bhattacharya et al. (eds.), *Advances in Optical Science and Engineering*,
Springer Proceedings in Physics 194, DOI 10.1007/978-981-10-3908-9_68

On the basis of the detection, OOFDM can be classified as coherent-detection and direct-detection. Coherent-detection based OOFDM shows better performance in terms of receiver sensitivity, spectral efficiency and robustness against polarization mode dispersion (PMD). However, the receiver for CO-OFDM is complex since another local oscillator laser is required in addition to the digital signal processing (DSP) for dealing with its sensitivity to frequency offset, phase noise and I/Q imbalance [2]. Moreover, lasers at both transmitter and receiver fronts need to have a much narrow bandwidth [3]. To overcome all those complexities in CO-OOFDM, another alternative detection scheme is direct-detection OOFDM (DD-OOFDM). This receiver is simpler compared to CO-OOFDM in terms of design and cost required for its implementation. DD-OOFDM requires a simpler receiver structure because the electrical OFDM signal is recovered by detecting the optical carrier and signal beat products via a square-law photodiode [3–6]. This makes DD-OOFDM more suitable for metro or access networks where cost is the primary concern. In conventional intensity modulation, the optical carrier is modulated to generate an optical field with the carrier and double sidebands (DSB) [5, 7, 8]. When the signal is sent over fiber, chromatic dispersion causes each spectral component to experience different phase shifts depending on the fiber link distance, modulation frequency, and the fiber dispersion parameter. If the phase difference between the two optical sidebands at the photodetector is π, destructive mixing will cancel all power at the RF frequency [2]. As the RF frequency increases, the effect is even more pronounced and the fiber link distance becomes severely limited [9]. If the DSB signal is detected by the photodiode, the presence of two sidebands in DSB signal may sometimes null the power of the subcarrier, if there are phase shifts in the upper or lower sidebands thereby resulting in symbols allocated in the real axis of the constellation/signal-space diagram [2]. If there is only one sideband, there will be no null effect and hence the constellation point will appear at the point where it is destined to be, provided other distortion effects are not there. In intensity modulation schemes, dispersion effects can be reduced by the elimination of one sideband to produce an optical single-sideband (SSB) [5, 8].

In [10], a DD-OOFDM system employing VCSEL as direct-intensity modulation device was proposed at a bit rate of 3 Gbps and its performance was compared with DD-OOFDM transmission using MZM as an external modulator. In our work, DD-OOFDM system with a data rate of 10 Gbps is simulated using both external modulation and direct modulation and their performance analysis are compared based on the percentage of EVM. This paper is arranged as follows. In Sect. 2, the general process of generation of OOFDM and its direct-detection is discussed. Sections 2.1 and 2.2 discuss the results of the simulation employing both VCSEL and MZM as modulators. In Sect 2.3 the change in EVM (%) with the change in transmission distance of is discussed. In Sect. 3, the conclusions of this work are drawn.

2 Optical OFDM Generation and Direct Detection

The process of generation of Optical OFDM signal can be divided into three steps. Firstly, the OFDM signal is generated in electrical domain and mixed with RF frequency. Secondly, the RF-OFDM signal is converted to an optical signal using electro-optic devices, transmitted through the optical fiber and then re-converted to an electric signal using photodetectors. The third step is demodulating the OFDM signal, which is done coherently.

The simulation was done using OptiSystem 13.0.1. The Fig. 1 shows the general simulation set up. The PRBS generator in the figure generates pseudo-random bit sequence (PRBS) of the length specified by the user. In this work, the sequence length was taken as 16384, i.e. 2^{14}. This PRBS sequence is then fed to a QAM sequence generator. With the QAM sequence generator, the bit sequence is split into two parallel sub-sequences; each can be transmitted in two quadrature carriers when building a QAM modulator. This is achieved by using a serial-to-parallel converter. Then the next is the OFDM block which generates two outputs, in phase and quadrature phase. This OFDM spectrum is very broad which is incapable of being processed further due to the limited bandwidth processing capability of the components. Hence, this signal is reshaped using two separate pulse shaping filters for the in-phase and quadrature outputs of the OFDM modulator. In this work, a low-pass cosine roll-off filter is used for both. The cut-off frequency is set equal to 0.62 times the symbol rate of the system. The filtered signal is applied to the quadrature modulator for generating a guard band. The guard band is the band between the optical carrier and the optical OFDM signal. This guard band plays a crucial role in the direct-detection systems using square law devices and its width must be equal to the bandwidth of the OFDM signal so as to avoid the OFDM signal to be contaminated by signal-signal mixing interference (SSMI) products after photodetection [5, 11]. The RF-OFDM signal is then applied to the optical modulator to modulate the intensity of light generated by a laser. The laser is operated in continuous wave (CW) mode and its center frequency is set to be at 193.1 THz (approximately 1550 nm) because this frequency corresponds to the lowest power attenuation figure when light propagates throughout the fiber link [9].

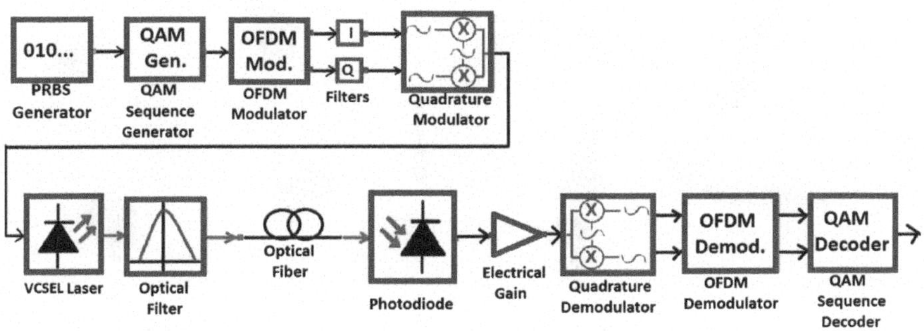

Fig. 1 Simulation setup for DD-OOFDM transmitter and receiver [10]

The optical modulator generates a double-sideband OOFDM signal [12]. A filter is used to suppress the lower sideband, thereby generating a single-sideband (SSB) OOFDM signal. The optical filter used in all the simulations performed in this work is a second order optical filter with a Gaussian frequency transfer function. This signal is then transmitted through an optical fiber channel. The optical fiber has an attenuation of 0.2 dB/km when operated at 1550 nm (approximately 193.1 THz) window [2, 9]. The signal from the fiber is directly applied to a square law photodiode, which performs the optical-to-electrical conversion. This scheme, since a single photodiode is used to detect the received signal without using any laser as a local oscillator, is called direct detection [4]. The signal is detected using a photodiode having a responsivity of 1 A/W. The result of photodetection is the RF-OFDM spectrum in the electrical domain. The OFDM demodulation is then done to recover the information transmitted.

2.1 *Results of OOFDM Signal Generation Using VCSEL*

In this simulation, VCSEL is used as an external modulator to generate OOFDM signal. The RF signal is applied directly to the VCSEL to modulate its optical intensity that is centered at a frequency of 193.1 THz. The parameters of the optical filter following the VCSEL are adjusted in order to implement a pass band compatible with the optical carrier and the upper sideband. The single sideband signal is transmitted through an optical fiber comprising of 2 loops of 50 km single-mode optical fiber with an optical amplifier at the end of each loop. The signal is then processed in the reverse order as that in the transmitter. In this simulation, the number of IFFT points taken is 1024, out of which 512 subcarriers are used to carry data and the rest are zero padded to achieve 2 × oversampling, thereby making it feasible for analog filtering. The number of cyclic prefix points taken is 100. Figure 2a shows the constellation diagram at the output of the simulated design

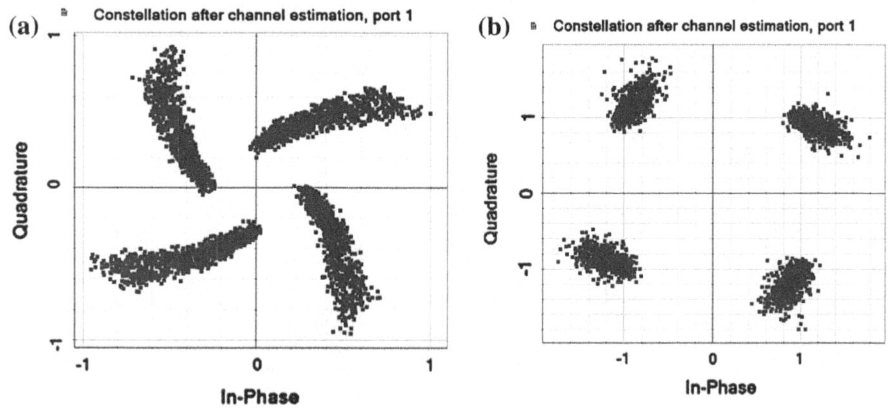

Fig. 2 **a**, **b**: Constellation of 4-QAM DD-OOFDM using direct modulation

before channel estimation. It can be seen that the constellations are moving towards the center although they are not completely concentrated at the center. This can be corrected by channel estimation [4, 5]. Channel estimation can be done by using training sequences. Training symbols are a set of symbols that a system uses to construct a transfer function to compensate for chromatic dispersion, fiber non-linearity, polarization mode dispersion and polarization dependent loss by comparing the received training symbols (at the training symbols locations) to the original training symbols [2]. Figure 2b shows the constellation diagram that is obtained after channel estimation. A decision block calculates the percentage error vector magnitude (EVM). The EVM (%) in this design was found to be 20.81% for back-to-back connection. However, the EVM (%) was found to be higher, i.e. 22.34% for a propagation distance of 100 km. Hence, this direct modulation scheme using VCSEL is not suitable for transmitting OFDM signals over longer distance fiber. It is also to be noted that there is also a limitation in the data rate because of the band limitation of the VCSEL. If the bandwidth of the OFDM signal exceeds the bandwidth of the VCSEL the performance of the system gets degraded even more.

2.2 *Results of OOFDM Signal Generation Using MZM*

In this simulation, all the components are same as that were used in the simulation of direct modulation discussed in Sect. 2.1, except the VCSEL which is replaced by a Lithium Niobate Mach-Zehnder Modulator (MZM). The RF-OFDM signal is applied to the MZM to modulate the laser output having a center frequency of 193.1 THz. The MZM is operated in the push-pull pattern. The MZM generates a double-sideband OOFDM signal [12]. A filter is used to suppress the lower sideband, thereby generating a single-sideband (SSB) OOFDM signal. This signal is then transmitted through an optical fiber channel that consists of a span of 50 km optical fiber in 6 loops with the signal amplified after the end of each span. The signal from the fiber is directly applied to a square law photodiode, which performs the optical-to-electrical conversion. The signal is then demodulated coherently in electrical domain to recover the transmitted data. The system parameters used for the simulation setup of 4-QAM OOFDM signal generation is similar to that of Sect. 2.1 except for the data rate which is increased to 40 Gbps in this case. The received constellations are shown in Fig. 3a, b. As seen in Fig. 3a, the constellations seem to be rotating. This is due to chromatic dispersion of the optical fiber channel. When transmitted over a shorter distance, chromatic dispersion does not affect the signal. But for longer distances, this dispersion cause shift in the position of the constellation points [13]. This shift can be compensated by using training symbols. Here, two training symbols were used for the purpose of channel estimation. It can be seen from Fig. 3b that, the constellation points are clear after

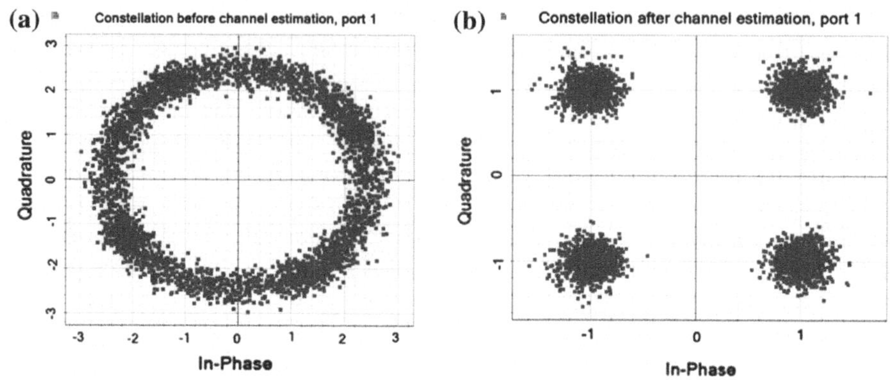

Fig. 3 **a**, **b**: Constellation of 4-QAM DD-OOFDM using external modulation

channel estimation. The percentage of error vector magnitude was 11.643%. The same system was simulated for a lower data rate, that is, 10 Gbps. The bandwidth of the RF-OFDM signal decreases with the decrease in the data rate, if all other parameters of OFDM block are unchanged. Hence, for 10 Gbps data transmission, the OFDM signal has a bandwidth of 2.5 Gbps with 4-QAM symbol mapping. The guard band required for the optical OFDM signal is equal to the bandwidth of the baseband OFDM signal. So, in this case, the RF frequency required for the RF up-conversion is 3.75 GHz. The RF spectrum at the transmitter and the receiver of this simulation looks similar to that of the spectrum of the 40 Gbps data transmission. But they do not lie at the same center frequency. The center frequency of RF OFDM signal for the 10 Gbps data transmission is at 3.75 GHz due to up-conversion by a frequency of 3.75 GHz. Since the RF OFDM spectrum is centered at 3.75 GHz, the sidebands after optical modulation will appear at (193.1 $\pm$ 0.00375) THz. The parameters of the optical filter are adjusted to suppress one of the two sidebands, the lower sideband in this case. The EVM for this case was found to be 9.372% which is less compared to 40 Gbps data transmission.

2.3 EVMVS Transmission Distance of OOFDM

The effect of the increase in fiber length on the performance of the system is studied. Figure 4 shows the plot of the change in the EVM (%) with the change in the length of optical fiber. It can be seen from the figure that the EVM (%) rises with the increase in the length of the fiber. This plot was based on the parameter set for 4-QAM and 16-QAM with 128 subcarriers. Note that the EVM obtained is without finding any optimum bias point for the MZM.

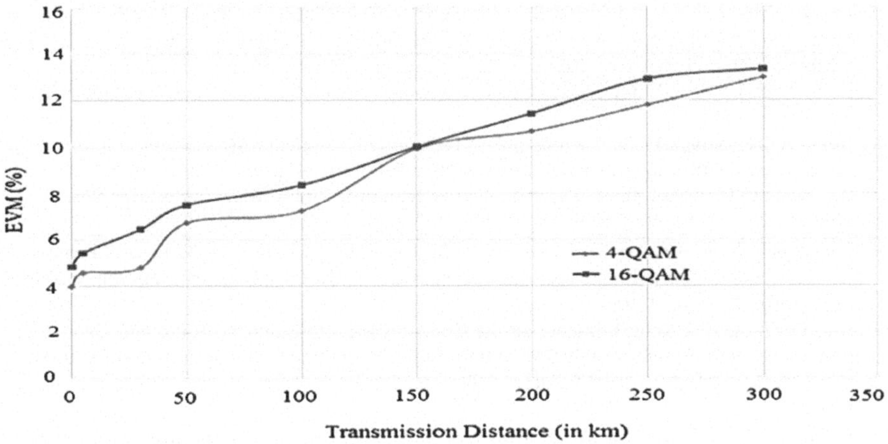

Fig. 4 Variation in EVM (%) with change in transmission distance

3 Conclusions

After simulating, both the direct and external modulation, it was found that the percentage error vector magnitude (EVM) differs for both the cases. The EVM (%) for direct modulation of 4-QAM DD-OFDM was found to be around 22.34% when transmitted through 100 km optical fiber, while the system using external modulation showed the better result with an EVM (%) of 4.038%. From this result, it can be concluded that the system using external modulator performs better.

Acknowledgements Authors are thankful to Prof. Rajan and Dr. Pankaj Sarkar for their valuable suggestions and having discussions to improve the manuscript.

References

1. Bingham, J. A. C., "Multicarrier Modulation for Data Transmission: An Idea Whose Time Has come", IEEE Communications Magazine, 2(1), 5–14 (1990).
2. Shieh, W., and Djordjevic, I, [Orthogonal frequency division multiplexing for optical communications], Mass Academic Press, Burlington, (2010).
3. N. Cvijetic, OFDM for Next-Generation Optical Access Networks, Journal of Lightwave Technology, vol. 30, no. 4, pp. 384–398, 2012.
4. Schmidt, B., Lowery, A., and Armstrong, J., "Experimental Demonstrations of Electronic Dispersion Compensation for Long-Haul Transmission Using Direct-Detection Optical OFDM", Journal of Lightwave Technology 26(1), 196–203 (2008).
5. Nezamalhosseini, S., Chen, L., Zhuge Q., Malekiha, M., Marvasti, F., and Plant, D., "Theoretical and Experimental Investigation of Direct Detection Optical OFDM Transmission using Beat Interference Cancellation Receiver", Optics Express 21(13), 15237–15246 (2013).

6. Ma, J., "Simple Signal-to-Signal Beat Interference Cancellation Receiver Based on Balanced Detection for a Single-Sideband Optical OFDM Signal with a Reduced Guard Band", Optics Letters 38(21), 4335–4338 (2013).
7. Armstrong, J., "OFDM for Optical Communications", Journal of Lightwave Technology 27(3), 189–204 (2009).
8. Smith, G.H., Novak D., and Ahmed, Z., "Technique for optical SSB generation to overcome dispersion penalties in fiber-radio systems", IEEE Photonics Journal 33(1), 74–75 (1997).
9. Agarwal, G.P. [Fiber-Optic Communication Systems], John Wiley & Sons, New York, (2002).
10. Khwandah, S. A., Cosmas, J. P., Glover, I. A., Lazaridis, P. I., Prasad, N. R., and Zaharis, Z. D., "Direct and External Intensity Modulation in OFDM RoF Links", IEEE Photonics Journal 7(4), 7902710, 1–11 (2015).
11. Ma, J., Zhou, W., "Joint Influence of the Optical Carrier-to-Sideband Ratio and Guard Band on Direct- Detection SSB-OOFDM System", IEEE Photonics Journal 7(5), 1–13 (2015).
12. Seimetz, M., [High-Order Modulation for Optical Fiber Transmission], Springer, Berlin, (2009).
13. Kikuchi, K., "Fundamentals of Coherent Optical Fiber Communications", Journal of Lightwave Technology 34(1), 157–179 (2016).

Design of 1 × 4 All Optical Splitter Based on 2D Photonic Crystal

Bani Gandhi, Anil Kumar Shukla and G.N. Pandey

Abstract In this study 1 × 4 optical splitter has been designed for TE-polarized light. The design comprises of 2D Hexagonal lattice where elliptical Si-rods are arranged in the air. Power splitting device with four outputs (1 × 4) has been designed and simulated with Y-Junction. Finite Difference Time Domain (FDTD) method was used for carrying out the simulations and testing of splitting properties of the beam splitter was done. The band gap of the structure was realized through the Plane Wave Expansion (PWE) band solver simulator. The overall size of the 2D Y-junction splitter device designed is 21 μm * 15 μm. The objective was to attain equal power at each output channel with the least amount of loss. The losses were measured at each port and were calculated to be −6.02 db. Three band gaps were measured and their gaps were calculated as 0.20382, 302364; 0.397845, 0.513458; 0.540259, 0.548335 with the tolerance as 0.001.

1 Introduction

Regardless of strained market circumstances, optical communication systems still persists to advance at an expeditious momentum. The prominence of industry research has shifted from long haul to metro to fiber to home networks in the recent times. These different networks require very different solutions. Varieties of new

B. Gandhi (✉) · A.K. Shukla
Amity Institute of Telecom Engineering and Management, Amity University, Noida, Uttar Pradesh, India
e-mail: bani.gandhi@gmail.com

A.K. Shukla
e-mail: akshukla2@amity.edu

G.N. Pandey
Department of Physics, Amity Institute of Applied Science, Amity University, Noida, Uttar Pradesh, India
e-mail: gnpandey2009@gmail.com

© Springer Nature Singapore Pte Ltd. 2017
I. Bhattacharya et al. (eds.), *Advances in Optical Science and Engineering*, Springer Proceedings in Physics 194, DOI 10.1007/978-981-10-3908-9_69

optical components have been evolved and persist to be developed to address these requirements [1].

One of the important and necessary optical device is an optical splitter, it is a device which is used to split the power or the fiber optic light into several parts by a certain ratio for example in a 1 × 4 power divider or splitter the light will be divided into 4 optic-light by a ratio of ¼ or 25% of the original source in the ideal case. The splitters are passive devices i.e. they do not require any external source of energy to perform any activity and are often used in radio technology. The various applications of a power splitter comprises of providing a signal model for feedback monitoring or measurement, coupling feeds to and from antenna, also antenna beam forming provide taps for systems that are for cable distributed like cable TV, and also used to split the transmitted and received signals on telephone lines (Fig. 1).

Since the Photonic Crystals have been in discussion for a long time now due to their potential to provide with the ultra compact structure that will then lead to miniaturization of optical circuits [3, 4]. A Photonic crystal is sporadic dielectric crystal which has a periodic division of refractive index where the span of period is of the range of wavelength of light. The most simple form of a photonic crystal that exists has 1D periodical configuration that is because of the periodic structure and band gap wave nature of light. Due to diffraction, in an optical band gap a specific segment of the spectrum is not able to proliferate, this means the wave which has the wavelength in band gap is not able to localize light in crystals. Photonic crystals occur in nature and control how photons are absorbed or reflected off them. These are made of dielectric nanostructures that permit specific wavelengths to bypass whereas hinder the rest of the wavelengths [5] (Fig. 2).

Fig. 1 Symbol for a 1 × N power splitter [2]

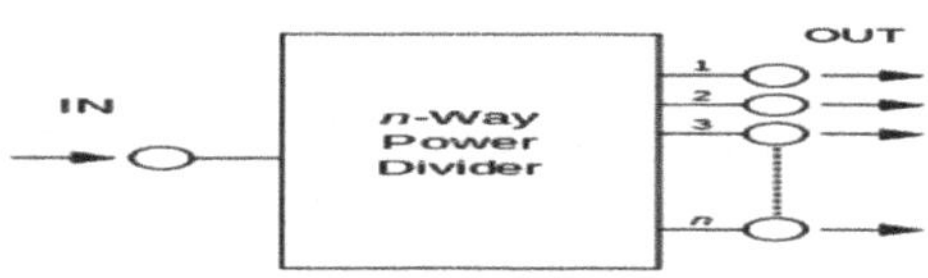

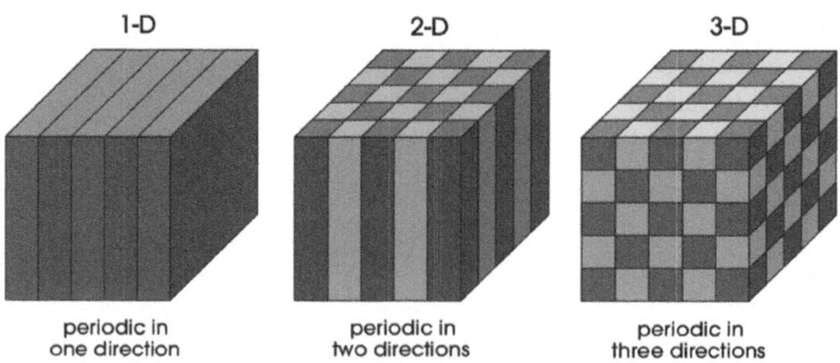

Fig. 2 1D, 2D, 3D photonic crystals [4]

2 Layout Design and Simulation

The wafer size of SiO_2 is chosen to be 21 μm * 15 μm in length and width respectively. Air with refractive index 1 was decided as the dielectric material. The input signal was chosen to be Gaussian modulated continuous wave with the wavelength of 1.55 μm. The planar PC structure is simulated on a silicon-on-insulator substrate. A 2D Hexagonal lattice is selected for the design. The elliptical waveguide is selected for the same and the atom radius is 0.3 μm (Fig. 3).

The simulation of the light was studied by 2D Finite-difference time-domain (FDTD) method that is a standard simulator. FDTD is a time-domain numerical method which is used to model the propagation of EM waves in an optical based medium, which uses Maxwell's equations for the calculations. The time domain method is found to be very proper and exact for the calculations and analysis of the photonic crystal [6].

The simulation was performed for the TE mode, definite Mesh used is −0.041 μm (delta X) × 0.041 μm (delta Z) and the number of mesh cells were 400 (X) and 500 (Z). For result finalization the simulation has been done for 4000 time steps. The boundary condition used is Anisotropic Perfectly Matched Layer (APML). The valid reasons for using Photonic Crystal waveguides instead of conventional dielectric waveguides for optical interconnects includes better routing capabilities with lower

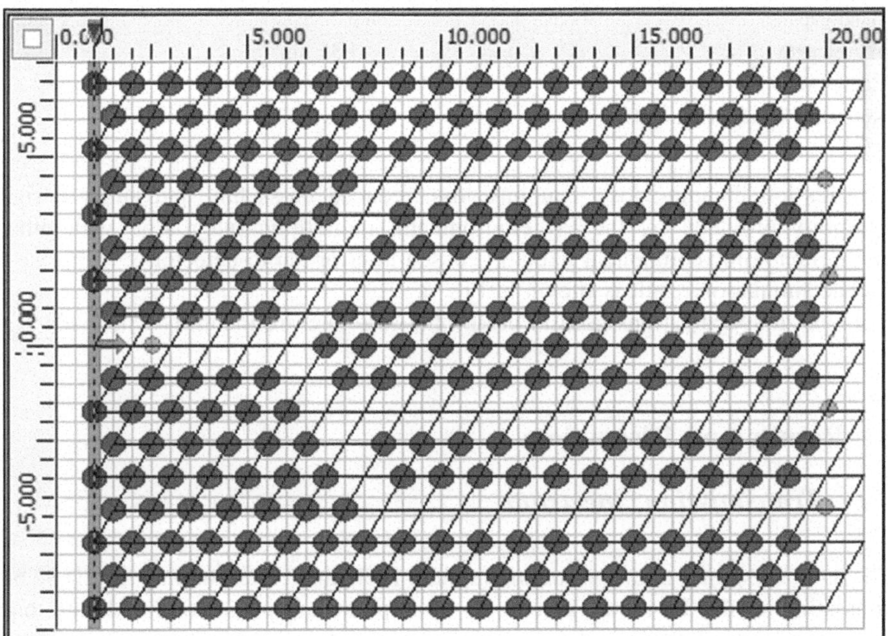

Fig. 3 1 × 4 Y-junction splitter

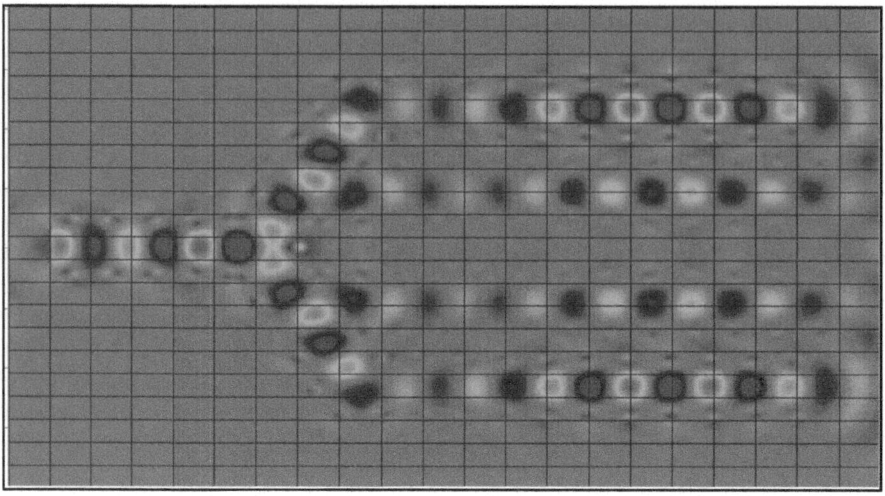

Fig. 4 Simulation result of 1 × 4 Y-junction splitter

Table 1 Physical specifications of the splitter

Parameters	Y-junction splitter
Configuration	Holes with hexagonal lattice
Device size (Length (μm) X * Width (μm))	21 * 15
Band gaps	3
Wavelength	1.55 μm
Radius of rods	0.3 μm
Refractive index of the rods	3.4

transmission, bending loss, coupling loss, reduced cross talk and larger design flexibility provided by the PC environment, monolithic integration with other on-chip photonic component like a PC based laser or photo-detector.

The simulation result is shown in Fig. 4.

Physical specifications used to design the structure are mentioned in Table 1.

3 Results and Discussions

3.1 Transmission Spectrum

The output response of the pre-defined splitter is obtained after running the 2D 32 bit simulation for the TE mode. Five observation points are placed on the design, one for the input and the other four to calculate the response at the output ports.

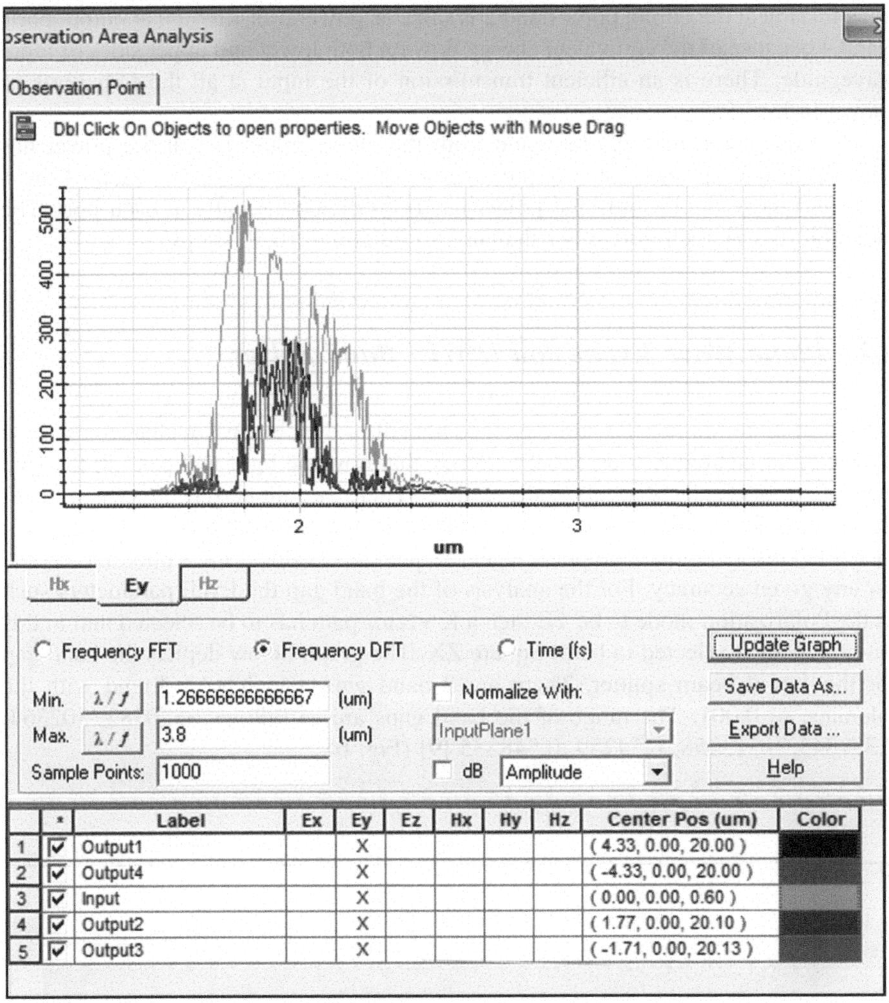

Fig. 5 Power distribution or transmission spectrum of all the outputs in 1 × 4 Y-junction splitter

Discrete Fourier Transform (DFT) calculations are used by the FDTD method to obtain the transmission spectrum of this 1 × 4 beam splitter. Post simulation process the FDTD analyzer window opens up to view the output of this 1 × 4 power divider. The amplitude or the power can be computed by setting the observation points at the input and the output and can be analyzed by the graph which is obtained at the analyzer [7] (Fig. 5).

As it can be observed from Fig. 5 that the transmitted power is uniformly spread into the four outputs which is due to the proportional structure. The power transmission at the output ports 1 and 2 is same as power transmission at output ports 3 and 4 because of the equivalent energy flow on both lower and upper sides of input waveguide. There is an efficient transmission of the input at all the four ports as shown in Fig. 5.

The conclusion that can be made from the above graph is that the power has been divided in a ratio that is approximately 25% of the original at each of the 4 ports and most significantly the power has been divided equally at each port. The radiation loss at each port was calculated to be about −6.02 db [8].

3.2 *Plane Wave Expansion (PWE) Band Solver*

Design of photonic crystals and then take benefit of their exclusive characteristics, a computation technique is essential to verify that how the light will pass all the way through a particular crystal configuration. There are numerous proficient techniques, but Plane Wave Expansion method is the most studied and consistent method. This technique computes eigen frequencies (values) for a photonic crystal for any given accuracy. For the analysis of the band gap the PWE parameters such as the Polarization mode to be TE then a K-vector path has to be selected and in this case the path is selected to be as Square ZX. The graph below depicts the band gap for the 1 × 4 beam splitter. There are 3 band gaps which were found with the tolerance of 0.001. The range of the band gaps are as follows: 0.20382, 302364; 0.397845, 0.513458; 0.54259, 0.548335 [9] (Fig. 6).

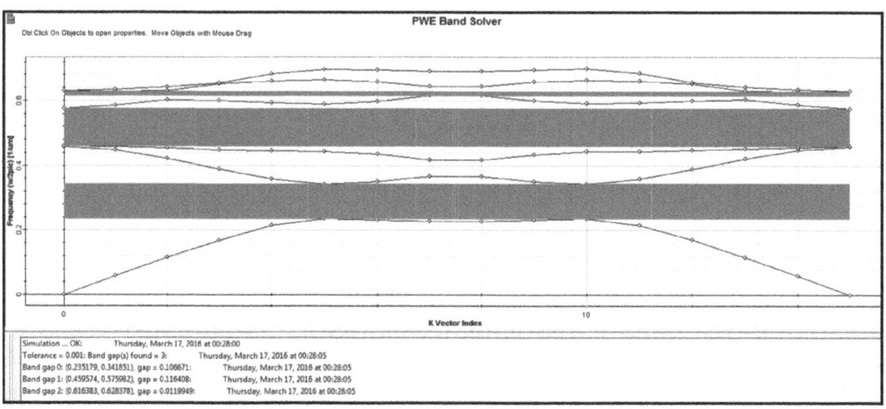

Fig. 6 Band gap diagram for 1 × 4 beam splitter

4 Conclusion

In this study, assessment and investigation of photonic crystal beam splitter has been carried out. The transmission spectrum or power distribution among the four output ports of this 1 × 4 beam splitter were analyzed by using FDTD method. The Photonic Band Gap (PBG) was calculated by PWE technique. It has been observed that there is almost an equivalent division of power at each output port of this 1 × 4 beam splitter. The power transmission at the output ports 1 and 2 is same as power transmission at output ports 3 and 4 because of the equivalent energy flow on both lower and upper sides of input waveguide. This configuration helps in providing high transmission efficiency. Added improved performance may be obtained by optimizing the range of design parameters conditions. The device may be helpful in optical communication applications and photonic integrated circuits.

Acknowledgements The authors gratefully acknowledge the unconditional support provided by the family, friends and mentors. We would also thank the reviewers for their insightful comments which have helped improve this paper.

References

1. Juhi Sharma, Ritu Sharma and Lalit Kumar Dusad, "Review and Analysis of Photonic Crystal Beam Splitters for Optical Communication Applications", GECA, India (2015).
2. http://www.wow.com/wiki/Combiner
3. Rajat Dey, "Optical Power Splitting Techniques Using Photonic Crystal Line Defect Waveguides," Ph.D Thesis, The University of Western Ontario, London, Canada, 2011
4. J. D. Joannopoulos, R. D. Meade and J. N. Winn, "Photonic Crystals", Princeton University Press, Princeton (1995).
5. Myhrvold, N., "Confessions of a cybershaman," Slate, 12 June 1997, http://www.slate.com/CriticalMass/97-06-12/CriticalMass.asp (19 October 1997). http://www.optics4yurresearch.com/7752.html
6. Jones, C.J., Director, Miscellaneous Optics Corporation, interview, Sept. 23 2011
7. H. Hojo and N. Uchida, "Control of Electromagnetic waves by 2-D plasma Photonic Crystals", plasma and industrial applications, IEEE, (2005)..
8. K M. Leung and Y.F. Liu "Photon band structures: The plane-wave method", Phys. Rev. B, vol. 41, pp. 10188–10190, 1990.
9. Peiyuan Zhang, Hao Guo, Hongjun Chen, Cancan Wang, Xiong Zhang, "Novel composite beam splitter with directional coupler and Y-junction using photonic crystal," Optik 124, pp. 3384–3386, 2013

Analysis of a Novel Grating-Assisted Directional Coupler by Using the Normal Mode and Coupled Mode Analyses

Parvinder Kaur and M.R. Shenoy

Abstract We present a novel structure of the directional coupler which consists of two non-identical waveguides, with a periodic refractive index perturbation along the direction of propagation. We have compared the coupling characteristics obtained by using the approximate coupled mode analysis and the exact normal mode analysis for the grating-assisted directional coupler. We find that, for the proposed structure, using the modes of the isolated waveguides in the coupled mode theory does not give accurate results, and one should apply the normal mode analysis to obtain the coupling characteristics.

1 Introduction

There are several different designs of the grating-assisted directional couplers (GADC) [1–4]. The coupled mode theory (CMT) has been widely used to determine the power transfer characteristics in GADCs [5–8], because an exact analytical treatment is difficult. The GADCs considered earlier have the periodic refractive index variation only at the boundary of any one waveguide. We have proposed a new GADC in which periodic perturbation of the refractive index is across the complete width of the directional coupler. In order to study the characteristics of the presented directional coupler, we have used exact normal mode theory (NMT) and approximate CMT. We have carried out a comparative study of the results obtained from the NMT (using the normal modes of the structure) and CMT. We show that the power transfer characteristics between the two waveguides, calculated using NMT and CMT differ significantly.

P. Kaur (✉) · M.R. Shenoy
Department of Physics, Indian Institute of Technology Delhi, Hauz Khas, New Delhi 110016, India
e-mail: parvinder.iitd@gmail.com

© Springer Nature Singapore Pte Ltd. 2017

I. Bhattacharya et al. (eds.), *Advances in Optical Science and Engineering*, Springer Proceedings in Physics 194, DOI 10.1007/978-981-10-3908-9_70

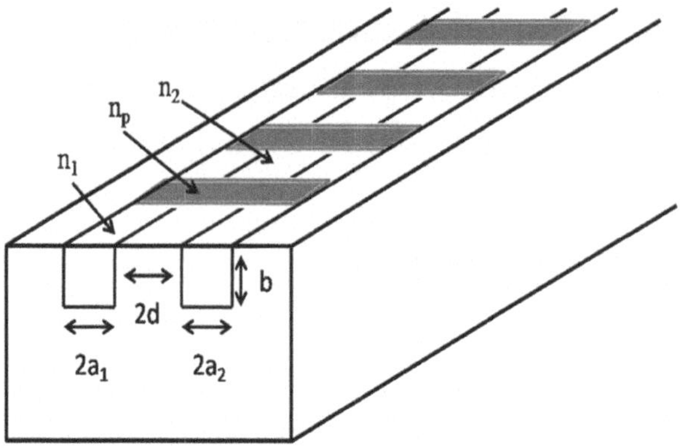

Fig. 1 Schematic of the GADC. $2a_1$ and $2a_2$ are the widths of the core of waveguide 1 and 2, respectively; $2d$ represents the spacing between the two waveguides. n_1 and n_2 are the refractive index of the core and the cladding of each waveguide; n_p indicates the periodic refractive index perturbation along the direction of propagation

2 Configuration of Directional Coupler

The basic configuration of the directional coupler consists of two non-identical waveguides placed adjacently; a periodic grating structure is embedded all along the length of the directional coupler, and with the high index segments across the whole width (see Fig. 1). The grating period is chosen such that the spatial harmonic of the periodic grating provides phase matching between the two non-synchronous waveguides.

3 Theoretical Analysis

To investigate the power transfer characteristics of the directional coupler, we have used coupled wave equations.

$$\frac{\partial A}{\partial z} = -i\kappa_{ii}A + \kappa_{ij}Be^{i(\beta_i-\beta_j-K)z} \quad (1)$$

$$\frac{\partial B}{\partial z} = -i\kappa_{jj}B - \kappa_{ji}Ae^{-i(\beta_i-\beta_j-K)z} \quad (2)$$

$$\Delta\beta = \beta_i - \beta_j; \quad K = \frac{2\pi}{\Lambda} \quad (3)$$

$$\kappa_{ij} = \frac{k_0^2}{2\beta_i}\left(\frac{\int_0^b \int_{-\frac{d}{2}-2a_1}^{2a_2+\frac{d}{2}} \psi_i(x,y)(n_p^2 - n^2(x,y))\psi_j^*(x,y)dxdy}{\int_{-\infty}^{\infty}\int_{-\infty}^{\infty} \psi_i^*(x,y)\psi_i(x,y)dxdy}\right) \quad (4)$$

with $i = 1, j = 2$

The coupling length is given by

$$L = \frac{\pi}{2\sqrt{\kappa_{ij}\kappa_{ji}}} \quad (5)$$

In the above equations ψ_i and ψ_j are electric fields of the two interacting modes, $k_0 = 2\pi/\lambda$, where λ is the wavelength; β_i and β_j are the propagation constants of the two modes; K represents the spatial frequency; Λ is the period of the grating; A and B are normalized amplitudes of the two mode fields; $n_1 = 1.668$ and $n_2 = 1.662$ are the refractive indies of core and cladding, respectively; $n_p = 1.6686$ is the refractive index of the high-index region of the periodic grating; $2a_1 = 4$ μm and $2a_2 = 2$ μm are widths of the core of the two non-identical waveguides, and $2d$ is the spacing between the two cores of the directional coupler. Equations (1)–(5) provide the mathematical tools to describe the GADC. We have computed numerical values of the coupling coefficients and studied its characteristics. In an identical waveguide directional coupler, the two normal modes of the directional coupler resemble with the individual modes of the two waveguides of the directional coupler. In CMA to simplify the calculation the two normal modes replace with more easily computed individual modes of the two waveguides. For identical waveguide, directional coupler this approximation gives the correct order of magnitude of the coupling coefficients but, it is not adequate for the precise study of the behavior of the directional coupler. For comparison of the exact NMA and the approximate CMA, we have computed coupling length of the directional coupler as a function of the spacing 2d and compare the coupling characteristics of the directional coupler using both theories. To calculate the exact coupling coefficients, we have used normal modes ψ_1 and ψ_2 of the non-identical directional coupler in (4). Figure 2a shows the symmetric and antisymmetric modes of the non-identical waveguide directional coupler, which is used in NMA. On the other hand, to calculate approximate coupling coefficients, we have used the modes of the isolated waveguides of the directional coupler in (4). Figure 2b shows the isolated fields of the two non-identical waveguides of the directional coupler used in CMA. Here, we can see the difference in mode fields of the directional coupler in both cases. The approximation of the CMA will not make much difference, if the two waveguides are slightly different. Because in that case the two isolated modes of the waveguides will be approximately similar to the normal modes of the directional coupler. Hence, the CMA theory and the NMA, estimate the same order of coupling coefficients. But in our structure the two waveguides are strongly different; due to this the power transfer characteristics from the NMA and the CMA theory have large differences.

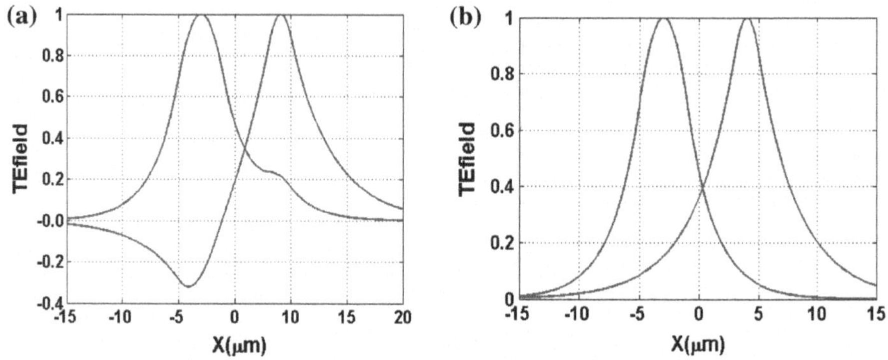

Fig. 2 **a** Field profiles of the symmetric and antisymmetric mode of the non-identical waveguide directional coupler. **b** Is isolated mode fields profiles of the two nonidentical waveguides of the directional coupler

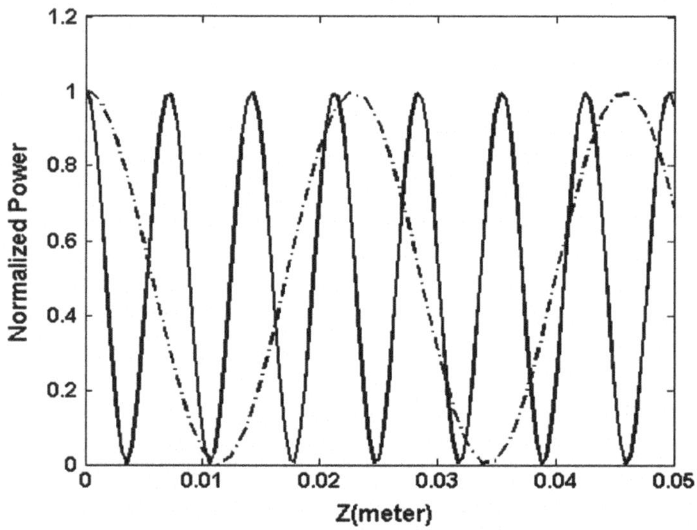

Fig. 3 Transmitted normalized power in waveguide 1, along the direction of propagation for a GADC. *Dotted line* represents exact NMA and *solid line* corresponds to CMA. The spacing between the two waveguides of the directional coupler is $2d = 4.5$ μm

Power transfer characteristics of the GADC are shown in Fig. 3 for $2d = 4.5$ μm. the power transfer characteristics are significantly different for proposed structure from both theories. However, difference in coupling length estimated from power characteristics graph is not exactly equal to the difference in the coupling length calculated from coupling coefficients shown in Fig. 4. This is due to including the self-coupling coefficients in the two coupled wave equations of the directional coupler. Figure 4 shows the variation of coupling length with the

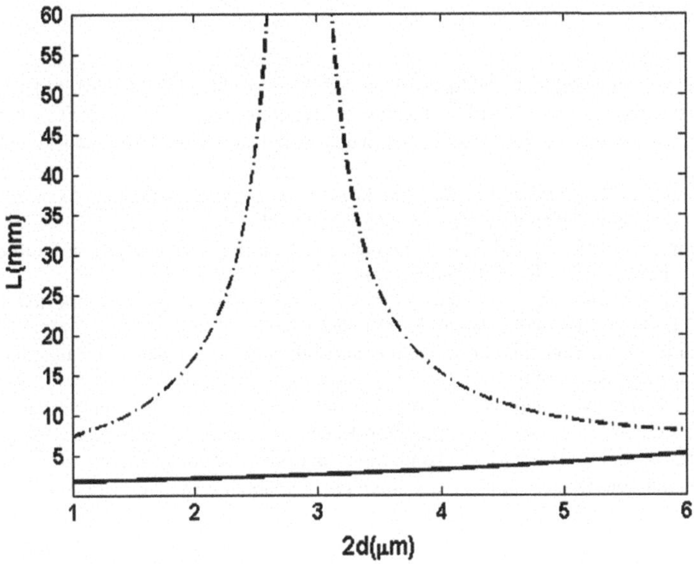

Fig. 4 Variation of coupling length 'L' as a function of the spacing (2*d*) between the two waveguides of the directional coupler. *Dotted line* represents results from exact NMA and *solid line* corresponds to approximate CMA

spacing between the two waveguides. Since the normal modes have a zero of the field at one transverse point, at that point the overlap integral used to calculate the coupling coefficients vanishes, and hence the coupling length becomes very large. When that happens, the NMA and CMA have large discrepancies [1].

4 Conclusion

We have presented a study of novel non-identical waveguide directional coupler with a periodic grating in the direction of propagation. Coupling characteristics of GADC have been studied. The exact analysis of a GADC requires the use of the normal modes of the complete structure. Since the analytical derivation of this problem is not simple, the widely used approximate analytical expressions for the coupling coefficients are derived, by approximating the normal modes by the modes of the individual, isolated waveguides. Here, we have shown a comparison of the exact and approximate results which are not in agreement. Hence, for description of the proposed structure, one should use the exact NMT of the structure instead of approximate CMT.

References

1. D. Marcuse, “Directional coupler made of non-identical asymmetric slabs. Part II: Grating-Assisted couplers,” IEEE J. Lightwave Technol., Papers 5, 268–273 (1987).
2. Y. Chen and A.W. Snyder, “Grating-Assisted couplers,” Opt. Lett., Papers 16, 217–219 (1991).
3. W. P. Huang, B. E. Little and S. K. Chauduri, “A new approach Grating-Assisted couplers,” IEEE J. Lightwave Technol., Papers 9, 721–727 (1991).
4. B. E. Little, W. P. Huang and S. K. Chauduri, “A multiple-scale analysis of Grating-Assisted couplers,” IEEE J. Lightwave Technol., Papers 9, 1254–1263 (1991).
5. W. P. Huang and H. A. Haus, “Power exchange in grating assisted couplers,” IEEE J. Lightwave Technol., Papers 7, 920–924 (1989).
6. D. Marcuse, “Radiation loss of grating-assisted directional coupler,” J. Quantum Electron, Papers 26, 675–684 (1990).
7. G. Griffel, M. Itzkovich and A. A. Hardy, “Coupled mode formulation for directional couplers with longitudinal perturbation,” J. Quantum Electron., Papers 27, 985–994 (1991).
8. W. P. Huang and J. W. Y. Lit, “Non orthogonal coupled-mode theory of grating-assisted co-directional couplers,” J. Light wave Technol., Papers 9, 845–852 (1991).

Part XV
Optical and Digital Image Processing Devices

1-D Photonic Crystal Based Dynamic Encoder/Decoder for 2D W-T OCDMA System

M. Bala Subramanian, K.S. Resmi and Prita Nair

Abstract We propose a novel low power and simple incoherent 2D optical code division multiple access (OCDMA) system comprising of reconfigurable encoder based on voltage dependent wavelength selection in the electro optic (EO) polymer infiltrated silicon photonic crystal and a tunable decoder based on silicon 1-D photonic crystal structures. Using simulation, the power efficient and error free system functioning of the encoding/decoding devices is verified. Dynamic provisioning of code patterns are possible using a single physical implementation of optical encoder/decoder and it can support flexible data rates of about 200 Gbps by using only 4 pair of wavelengths.

1 Introduction

The internet traffic rate is growing in the order of exabytes per month [1] due to combined requirements of cloud computing and storage, distributed computing, escalating mobile backhaul and emerging IoT applications. A consequence of this is the requirement of flexible dynamic data-rate provisioning even down in the access and last-mile networks where the high density of components imposes a more stringent requirement of ensuring low operational and capital expenses per device while giving performance compatible to the costlier high-end more flexible optical systems of core networks. One of the schemes for providing flexible low cost easily upgradable multiple access compatible to even legacy systems and those using conventional 10 Gbps systems is optical code division multiple access (OCDMA).

M. Bala Subramanian (✉)
Department of Electronics and Communication Engineering,
SSN College of Engineering, Rajiv Gandhi Salai, Kalavakkam,
Chennai, Tamil Nadu 603110, India
e-mail: balasubramanianm@ssn.edu.in

K.S. Resmi · P. Nair
Department of Physics, SSN College of Engineering, Rajiv Gandhi Salai,
Kalavakkam, Chennai, Tamil Nadu 603110, India

© Springer Nature Singapore Pte Ltd. 2017
I. Bhattacharya et al. (eds.), *Advances in Optical Science and Engineering*,
Springer Proceedings in Physics 194, DOI 10.1007/978-981-10-3908-9_71

Incoherent schemes are more suitable for access systems as they avoid the non-linearities, dispersion effects and high cost complex architecture associated with coherent encoding/decoding systems [2]. The key requirements for incoherent OCDMA access system are a dynamically reconfigurable, low cost high cardinality, fast encoding/decoding device coupled with a simple decoding scheme. Existing optical encoder/decoder [3–5] implementations have unique physical implementations per code with limited re-configurability. This paper introduces and analyses a dynamically reconfigurable 2D OCDMA system architecture based on single common physical encoder/decoder that can be programmed to generate all the different codes and a new scheme for simple decoding. The device implementation of both encoder and decoder part are possible by silicon fabrication technology on SOI wafer. The W/T (wavelength-time) encoder is a silicon-electro-optic (EO) organic hybrid 1D photonic crystal waveguide with a tunable defect mode. Silicon platform leverages all the advantages of CMOS processing, with $\chi(2)$ nonlinearity of linear electro-optic organic polymers ensuring fast low voltage low power tuning. The simple decoder employs a two step process of wavelength decoding using tunable filters based on photonic band gap structure followed by time chip identification.

Illustration of the design, capabilities and limitations of full architecture of this new 2D OCDMA system is presented using 10 Gbps data stream with 100 ps time period chipped in to four 25 ps time chips and a maximum of 2 wavelengths/code.

2 The Optical Encoder/Decoder System

The optical encoder is an EO polymer–Si 1D photonic crystal with central silicon fabry-perot cavity embedded in a silicon waveguide as illustrated in the block diagram shown in Fig. 1. The silicon ridge waveguide is 6 μm × 240 nm × 250 nm in dimension with embedded 3 layers of EO polymer infiltrated holes with 400 nm periodicity (doped polymer simulated with EO coefficient 80 pm/volt), on either side of central Si cavity. The voltage dependent wavelength selection through refractive index tuning of EO polymer is made possible by a pair of electrodes placed parallel to the length of the waveguide optimally positioned at a separation of 325 nm ensuring minimal deleterious impact on wave guiding effect at the same time ensuring sufficient field strength required for nonlinear EO tuning of polymer refractive index. The device is thus designed to work as a fast electro-optically tunable filter capable of transmitting a single tunable wavelength in the C band based on applied voltage. The tuning curve of the encoder (depicted in Fig. 1), obtained through its FEM simulation incorporating the dispersion of both the EO polymer and silicon [6], and voltage dependent change in refractive index of EO polymer [7], shows that, under zero volt drive voltage, there will be no transmission provided the optical source bandwidth is limited to the C-band. This ensures that '0' bit in the coded data generates no output at the end of the encoder for the duration of the '0' bit. For voltages up to 5 V, the transmitted wavelength can be tuned within the C band. The applied voltage thus decides the wavelength selected for

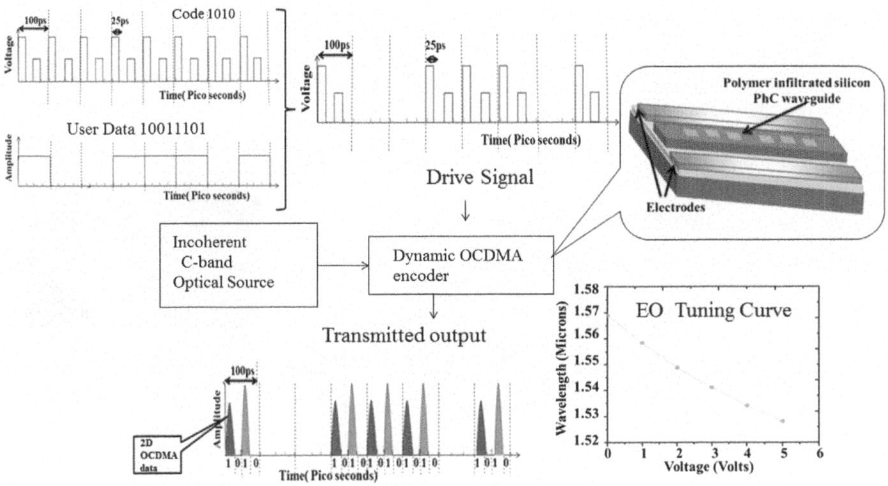

Fig. 1 Block diagram of the encoder with schematic of the device

transmission from the input bandwidth of the broadband source. Tuning sensitivity of 7.5 nm/V has been demonstrated. The novelty of this encoder is that when driven by a time chipped, pulse amplitude modulated data (as shown in Fig. 1), the encoder combines wavelength selectivity based on amplitude of pulse, time gating of selected wavelength based on duration for which this amplitude is held (duration of the chip) and time delay by ensuring nil transmission when voltage is zero. This single device is thus capable of implementing any code sequence adaptively and can hence be used for dynamic OCDMA encoding.

The 2D OCDMA decoder decodes the incoming code in two consecutive stages. In the first stage it performs wavelength decoding and the output of the wavelength decoder is used to do time decoding. This separation of the 2D decoding ensures flexible low cost simple decoding. Figure 2 shows a block diagram of the 2D decoder. Four tunable PBG (photonic band gap) filters function as the wavelength decoder. The use of 4 filters ensures fast parallel decoding and also flexibility to program the 2D wavelength pair independently. The design of each filter consists of 3 silicon layers 800 nm wide separated by 1 μm air gap on either sides of a defect layer of variable thickness. The electrostatically actuated comb structure is connected to the set of Si-Air bilayers on either sides of the defect layer. The voltage dependent change in defect layer thickness gives rise to different transmitted wavelengths and electrostatic actuation ensures low power consumption. The dimensions of the device are so chosen that at zero bias the transmitted wavelength falls outside the C-band. At higher voltages, the device transmits modes that fall within the C-band and hence choosing the wavelengths that are expected during the high bits of the code sequence. Tuning curve simulated using FEM software shows that a decrease in defect length of 0.2 μm tunes the wavelength over the entire C band. To realize this decrease each set of bi layers need move only by 0.1 μm each.

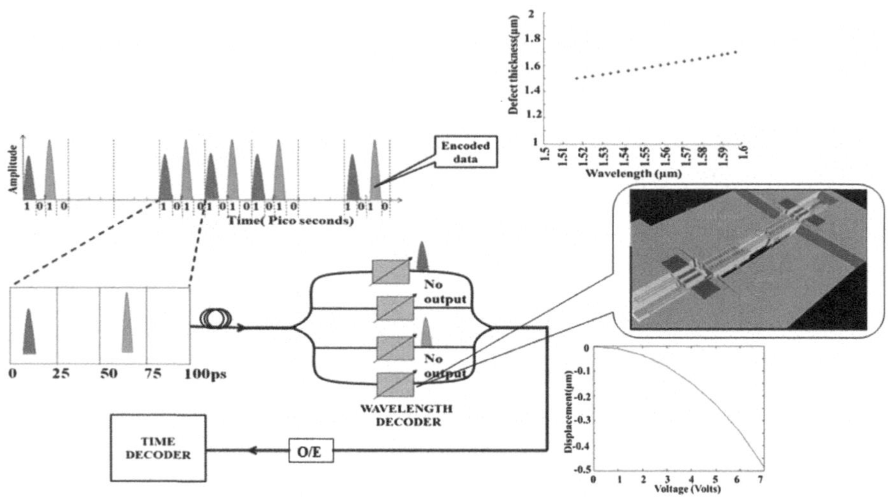

Fig. 2 Block diagram of the wavelength decoder with Intellisuite image of the device

Thus the output of the four bits will appear only for the matched pair of wavelengths for each code. The time decoding section performs O to E conversion of the combined output of the wavelength decoder and matches the TS coding based on identification of the time gaps between the arrival of the wavelength bits at the output of the wavelength decoder and the duration for which they appear. If the time gaps and durations match that of the expected code, then the output is accepted and the data is reconstructed. This 2 step decoding process ensures cross-correlation free decoding for non-cyclic codes.

3 MATLAB Modeling of the 2D OCDMA System

The system level modeling in MATLAB, starts with specifying inputs in the form of time chip coding sequence (1010 in this demonstration) and the wavelengths to be used (1550 and 1535 nm) for the coding. To simulate this 2D OCDMA system with 4 time chips and 2 wavelengths, two 25 ps 10 Gbps pulse streams are generated with amplitudes mapped to required carrier frequencies using a polynomial fit to the voltage tuning curve of the encoder and are interleaved as per code sequence as shown in Fig. 3. This interleaved time encoded signal is then combined (by AND operation) with pseudo random NRZ data to generate the drive signal. This drive signal can be seen to transmit the code when user data is high. An FFT of the drive signal is also shown in Fig. 3 illustrating that it has only the 2 wavelengths specified by the user, 1535 and 1550 nm. The modulated wavelength-time signal is such that, the code sequence gets transmitted when 100 ps data bit of the user data is high.

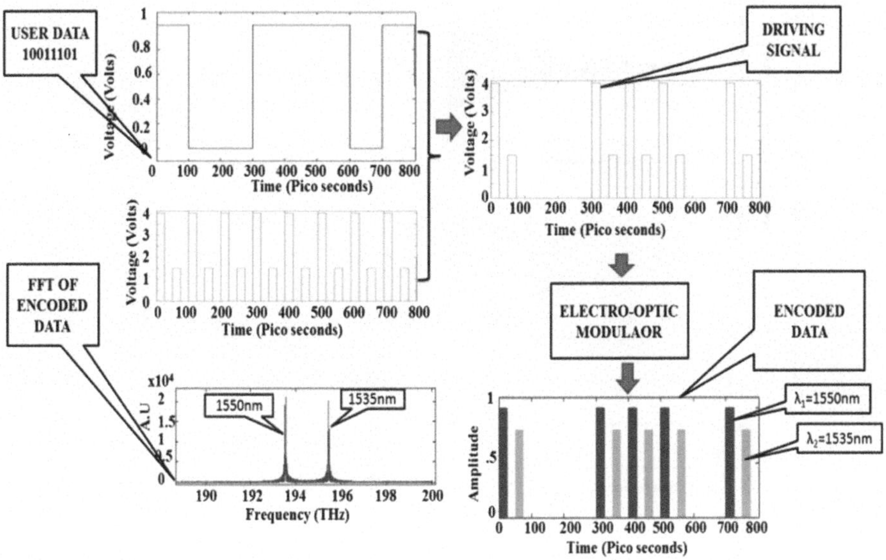

Fig. 3 MATLAB results of wavelength-time encoding of user data

At the decoder end, (Fig. 4) the wavelength filtering of the 4 bit encoded data signal is simulated in MATLAB by multiplying the FFT of the incoming signal by transfer function of each filter at appropriate drive voltage setting. The wavelengths identified FFT of the filter are shown in Fig. 4a, b illustrating that filter 3 and filter 1 transmits only the 1535 and 1550 nm wavelengths respectively and their FFTs are shown in the right. After the O/E conversion the obtained signal has 2 pulses, the initial one generated by the 1535 nm and the second pulse by 1550 nm. These electrical signals are differentiated and a timing signal is switched ON on reception of the first positive going pulse and switched OFF at the consecutive positive pulse. If the time period of this signal is 50 ps, only then the code is accepted as the correct one. If the timing signal resets before 50 ps or does exceeds 50 ps then data is dropped or else a 100 ps long pulse is generated after a wait period of an additional 25 indicating reception of a valid high bit user data.

The code identification at the decoder is shown in Fig. 4. The time circuitry identifies the time of arrival of the pulse as in Fig. 4c. This confirms that the assigned frequency pair is 195.4 and 193.5 THz detected with a delay of 50 ps. Therefore the decoded code sequence is identified as 1010 and its time signal for 100 ps is shown in Fig. 4d.

In this decoding scheme, for a particular code word to be decoded successfully, it should match both the wavelength decoding conditions and time gap circuitry condition. When any one of the condition is not satisfied the code word is not accepted as the correct code. For instance, in the example the illustrated code word is B0R0 (Two wavelengths blue and red). Suppose Y0G0 (Two wavelengths yellow and green) is the code word that arrives at the decoder, the code word gets

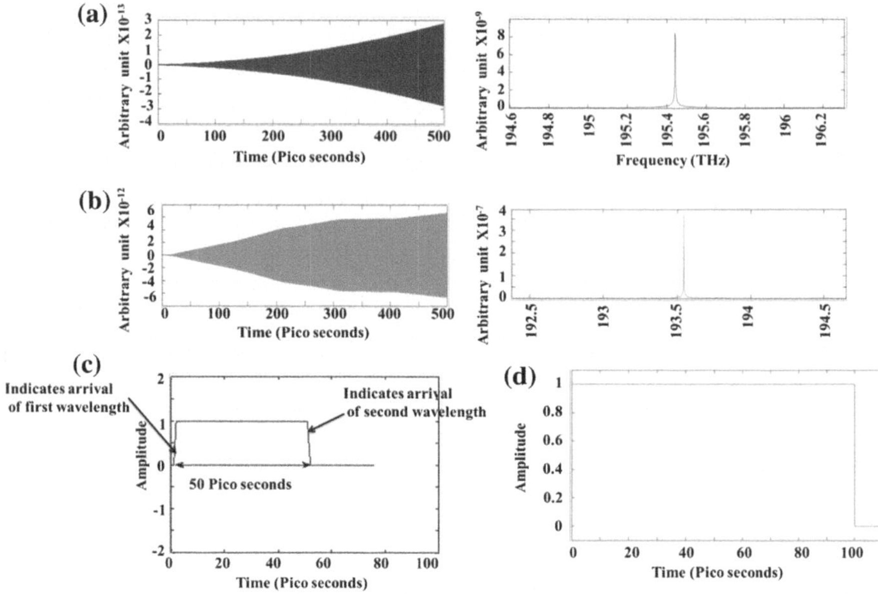

Fig. 4 MATLAB simulation results of the decoder. Decoded time domain signal and its corresponding FFT **a** at output of filter set to detect λ_1 **b** at output of filter set to detect λ_2 and **c** time delay detection between λ_1 and λ_2 **d** decoded user data for first 100 ps

blocked at the wavelength decoder itself. Or suppose BR00 is the code word at the decoder, even if the code word satisfies the wavelength selection criteria, the code gets discarded at the time circuitry. Hence a simple error free decoding system is ensured. The only problem occurs when a cyclic code of B0R0 (say R0B0) arrives at the decoder. The code word gets accepted by both the wavelength condition and time gap conditions. Hence to ensure error free detection and to make the decoding simpler cyclic codes are not used. If the O/E conversion is done individually for each filter, the filter the wavelength that pings first can be identified and hence for a pair of wavelengths, we have six available non-cyclic codes that can be used. So, the existing 10 Gbps system can be effectively increased upto 60 Gbps. If four pair of wavelengths in the C band are used then, the capacity of the system can be increased four times i.e. more than 200 Gbps.

4 Conclusion

We demonstrate through simulation, a novel low voltage (<5 V) completely reconfigurable optical 2D CDMA system based on Si/EO hybrid PhC devices. The chip rate and power budget is dependent on the efficiency of EO polymer and can hence be improved further. The decoding scheme is simple as it does not need any time synchronization or external local oscillator for its functioning. This system can be used for secure flexible line-rate provisioning in access networks. MEMS decoder filters along the lines have already been demonstrated by our group [8] and realization of the encoder can lead to actual experimental demonstration of this concept for providing flexible bandwidth provisioning.

References

1. Korotky, S. K., "Semi-empirical description and projections of Internet traffic trends using a hyperbolic compound annual growth rate," Bell Labs Tech. J. 18, (2013).
2. Yin, H. and Richardson, D. J., [Optical Code Division Multiple Access Communication Networks]. Springer-Verlag Berlin Heidelberg, Tsinghua University Press and Springer (2007).
3. Zhang, Y., Chen, H., Si, Z., Ji, H. and Xie, S., "Design of FBG Encoder/Decoders in Coherent 2D Time Wavelength OCDMA systems," IEEE Photonics Technology Letters 20(11), 891–893 (2008).
4. Kitayadama, K., Wang, X. and Wada, N., "OCDMA Over WDM PON—Solution Path to Gigabit-Symmetric FTTH," J. Lightwave Technol. 24, 1654–1662 (2006).
5. Yengnanarayanan, S., Bhushan, A. S. and Jalali, B., "Fast Wavelength-Hopping Time-spreading Encoding/Decoding for Optical CDMA," IEEE Photonics Technology Letters 12(5), 573–575 (2000).
6. Lindroos, V., Tilli, M., Lehto, A. and Motooka, T., [Handbook of Si Based MEMS materials and technologies], UK: ElsevierInc (2010).
7. Roussey, M., Baida, F. I. and Bernal, M. P., "Experimental and theoretical observations of the slow light effect on a tunable photonic crystal," J. Opt. Soc. Am. B 24, 1416–1422 (2007).
8. Renilkumar M, Prita Nair, "Low Voltage Widely Tunable Photonic Crystal Channel Drop Filter in SOI Wafer" IEEE Journal of Micromechanical Systems, Vol. 21, No. 1, pp. 190–197, (2012).

Full-Duplex 20 Gbit/s Fiber-Optic Link Design Based on Dual-Polarization Differential Quaternary Phase-Shift Keying

Dhiman Kakati and Subhash C. Arya

Abstract Low loss and ultra-high bandwidth available in optical domain has led to the shift from electrical signal processing such as modulation, bandwidth allotment etc. toward all optical processing. Implementing millimeter-wave (mm-wave) signals along with advanced modulation techniques to modulate the light has led to the development of technologies such as Radio over Fiber (RoF), Passive Optical Network (PON) and finds application in present and future broadband communication. In this paper a full-duplex 20 Gbit/s RoF system is presented for a transmission distance up to 100 km and overall system performance is studied in terms of Q-factor, Bit Error Rate (BER) and Error Vector Magnitude based on simulation platform. It is observed that the system maintains good performance in terms of coherency (i.e. single laser source) and spectral efficiency (bits/s/Hz) after transmission over Standard Single Mode Fiber (SSMF). Findings are useful to design high speed, lesser bandwidth, and cost-efficient optical communication system. The same can also be implemented in WDM system to get a much higher data rate of 320 Gbps for system using 16 channels.

1 Introduction

Recently, the information processing and communication society highly rely on broadband communication solutions such as interactive multimedia services, network games, High Definition Television (HDTV) distribution which require very high bandwidth and with the increasing number of users the network traffic is

D. Kakati (✉) · S.C. Arya
Lightwave Communication Systems Laboratory,
Department of Electronics and Communication Engineering,
North-Eastern Hill University, Shillong, Meghalaya 793022, India
e-mail: dhiman.kakati@gmail.com

© Springer Nature Singapore Pte Ltd. 2017
I. Bhattacharya et al. (eds.), *Advances in Optical Science and Engineering*,
Springer Proceedings in Physics 194, DOI 10.1007/978-981-10-3908-9_72

growing exponentially [1]. So for comparing the technologies available at present, one way is to consider the system parameters such as bit rate, regeneration-free transmission distance and cost of the implementation. One way of implementing cost effective transmission is to make the remote base stations as simple as possible and network functions such as signal processing, frequency allocation is carried out at the center station. The other way is to implement advanced modulation schemes owing to their high data rate and tolerance to fiber impairments such as attenuation and dispersion. In this paper, we have implemented Differential Quaternary Phase Shift Keying (DQPSK) so that with increased data rate, direct detection is also possible. There are two advantages, (1) multiple data is transmitted per sample and (2) in terms of coherency because the uplink carrier signal is preserved in the downlink signal and single laser source. Furthermore, differential phase modulation technology is found to have excellent transmission characteristics and 3 dB enhanced reception accuracy [2]. The mm-wave along with higher data rate capability is used for avoiding the spectral congestion in the lower frequency bands. The RoF technology allows a dynamic radio resource configuration and capacity allocation. It also eliminates the need for local oscillator at the radio access unit. In the later part, the same system may be incorporated to design a Wavelength Division Multiplexing (WDM) system and have a higher bit rate and more number of carriers transmitted through a SSMF. The polarization multiplexing technology increases the capacity of the single carrier to two times. The demodulation procedure includes Mach-Zehnder Interferometer (MZI) demodulator, followed by balanced Photo Detector (PD) and decision logic [3]. To eliminate the issue of generating electrical mm-wave of required range 40–60 GHz as this exceeds the limit of electrical devices, we preferred to generate it in optical domain by means of optical heterodyne beating [4].

The paper is organized as follows. Section 2 describes the working principle of the designed full-duplex wired communication system theoretically. Section 3 presents the simulation of the proposed system along with the results.

2 Principle of Operation

Figure 1 shows the schematic of the proposed full-duplex link. The lightwave emitted from the CW Laser has a central frequency of $f_0 = \omega_0/2\pi$ and the emitted lightwave can be expressed as $E_0(t) = E_0 e^{j\omega_0 t}$. Then it is provided to both the DQPSK modulator where the data rate is half of the input data rate. The Serial to Parallel (S/P) Converter converts the input bit stream into two output sequence of half the input bit rate. The drive waveform of the $LiNbO_3$ Mach-Zehnder Modulator (MZM) is in Non-Return to Zero (NRZ) format pulses $p(t) = \sin(\pi t/T_S)$, where T_S is the symbol period [5].

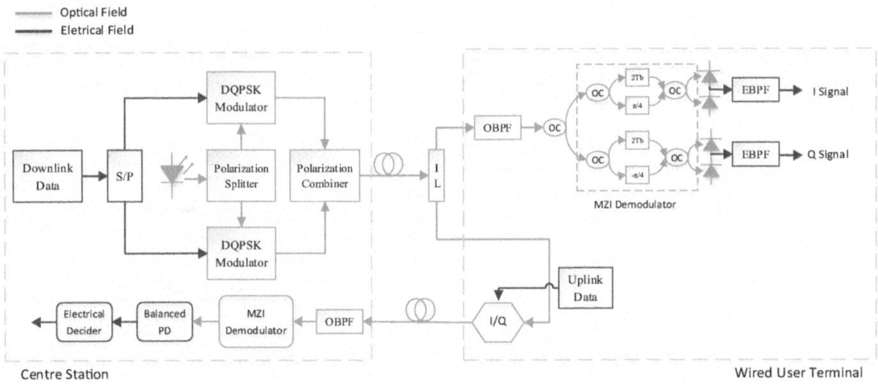

Fig. 1 Design of the proposed full-duplex optical fiber link [5], where, serial to parallel converter (S/P), delay interleaver (IL), in-phase and quadrature modulator (I/Q), electrical band-pass filter (EBPF), optical band-pass filter (OBPF), Mach-Zehnder interferometer (MZI), balanced photo-detector (PD)

At the transmitter output the downlink data can be expressed in terms of In-phase and Quadrature components as below [3]:

$$\begin{bmatrix} I(t) \\ Q(t) \end{bmatrix} = \begin{bmatrix} \sum_n a_{I,n} p(t - kT_s) \\ \sum_n a_{Q,n} p(t - kT_s) \end{bmatrix} \tag{1}$$

Here, $a_{I,n}$ and $a_{Q,n}$ are the data symbol transmitted in-phase and quadrature components at time kT_s. The in-phase and quadrature components are multiplied with mm-wave sinusoidal carrier of $V_m(t) = V_m \cos \omega_m t$ and $V_m(t) = -V_m \sin \omega_m t$ represented as [3]:

$$\begin{aligned} E_{DQPSK}(t) &= V_m[I(t) \cos \omega_m t - Q(t) \sin \omega_m t] \\ &= V_m \sqrt{I(t)^2 + Q(t)^2} \cos[\omega_m t + \varphi_i(t)] \\ &= \sqrt{2} V_m \cos[\omega_m t + \varphi_i(t)] \end{aligned} \tag{2}$$

where, $\varphi_i(t) = \arctan\{Q(t)/I(t)\} = \pi/4, 3\pi/4, 5\pi/4$ and $7\pi/4$ for i = 1, 2, 3 and 4 respectively. If we represent fiber length by z, the transmission function can be represented as:

$$H(\omega) = e^{-[\alpha + j\beta(\omega)]z} \tag{3}$$

where, α is the attenuation coefficient and $\beta(\omega)$ is propagation constant at angular frequency of ω. The DQPSK mm-wave can be represented in optical domain as [3]:

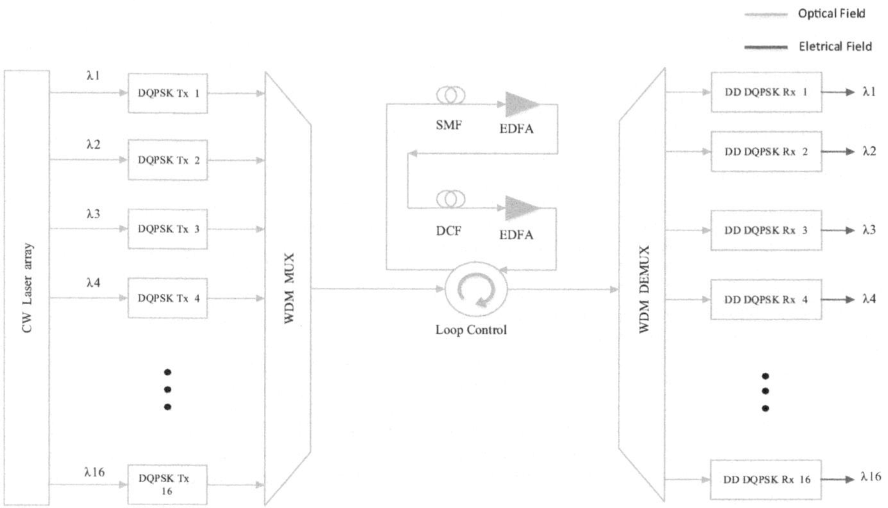

Fig. 2 Wavelength division multiplexing implementation of the system

$$\begin{aligned} E_{DQPSK}(z,t) &= F^{-1}\{H(\omega)F[E_{DQPSK}(0,t)]\} \\ &\approx E_C e^{-\alpha z} e^{j[\omega_0 t-\beta(\omega_0)z]} + E_S e^{-\alpha z} e^{j\{(\omega_0+\omega_m)t-\beta(\omega_0+\omega_m)z-\varphi[t-(\omega_0+\omega_m)^{-1}\beta(\omega_0+\omega_m)z]\}} \\ &= E_C(z,t)+E_S(z,t) \end{aligned} \tag{4}$$

Here, $\omega_0 + \omega_m$ represents 1st order sideband, $F\{\}$ is forward Fourier transform and $F^{-1}\{\}$ is inverse Fourier transform. E_C and E_S correspond to carrier and baseband signals.

The same system can be implemented in WDM System for increasing the data rate. Moreover, it is useful in the context of frequency reuse environments where multiple frequency generation and simple base station is a major factor. The function of the WDM multiplexer (MUX) is to combine all the channel wavelengths and transmit over the fiber. Figure 2 shows a simplex WDM implementation of the proposed system.

3 Simulation Setup and Results

To check the system performance, the design is implemented on OptiSystem V13 simulation platform. The full-duplex link uses two wavelengths that makes the system to communicate in both directions simultaneously. The laser has a central wavelength of 193.1 THz and linewidth of 0.1 MHz which is injected to the $LiNbO_3$ MZM that has a half wave voltage of 4 V. The oscillator produces a sine

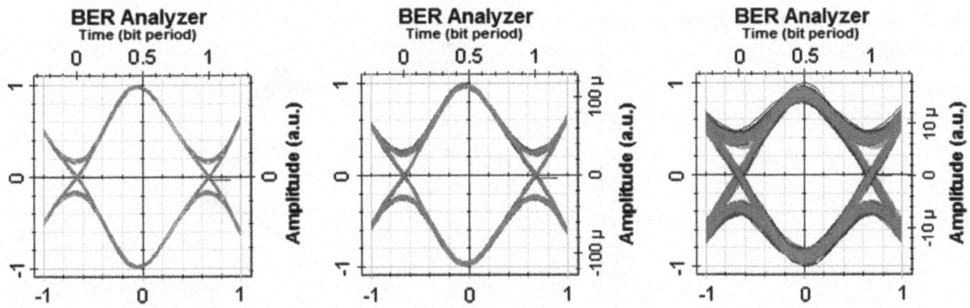

(a) 20 km (b) 60 km (c) 100 km

Fig. 3 The eye diagram of the 20 Gbps transmitted signal for different fiber lengths

Table 1 Q-factor, bit error rate, eye height with respect to fiber lengths

Distance (km)	Q-factor	Bit error rate	Eye height (μV)
0	59.687	0	0.00209
40	38.019	1.893×10^{-51}	0.0003
80	28.692	1.467×10^{-27}	4.22×10^{-5}
100	18.641	1.893×10^{-12}	1.057×10^{-5}

wave of frequency of one fourth of bit rate (i.e. bit rate/4) and it modulates the optical carrier to produce the required mm-wave. The resulting signal is applied to the two parallel Dual Drive $LiNbO_3$ MZM where the electrical converted data stream is imprinted to the optical carrier. Then modulated data is transmitted over the SSMF having attenuation of 0.2 dB/km and a dispersion of 16.75 ps/nm-km. Figure 3 shows the eye diagram for various fiber lengths of 20, 60, 80 and 100 km respectively. It is clearly visible that the system maintains a good performance which have been shown through Q-factor and BER.

Table 1 shows various parameters that represents transmission quality of the system in terms of Eye opening (Height peak to peak) which measure the additive noise in the signal, Quality factor (vertical eye opening in comparison to the noise present), and BER.

The Error Vector Magnitude (EVM) is basically a standardized error magnitude between the experimentally observed constellation and with difference to ideal constellation and can be calculated from optimized constellation diagram. The magnitude error, phase error and EVM can be expressed as [6] (Fig. 4):

$$\text{Magnitude Error} = \sqrt{I^2 + Q^2} - \sqrt{I_{ideal}^2 + Q_{ideal}^2} \tag{5}$$

$$\text{Phase Error} = \tan^{-1}\left(\frac{Q}{I}\right) - \tan^{-1}\left(\frac{Q_{ideal}}{I_{ideal}}\right) \tag{6}$$

$$\text{EVM} = \sqrt{(I - I_{ideal})^2 + (Q - Q_{ideal})^2} \tag{7}$$

Fig. 4 Basic concept of error vector magnitude (EVM)

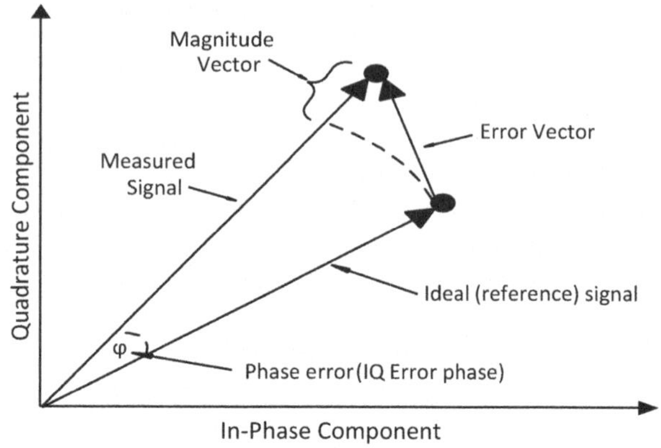

Fig. 5 EVM (%) versus received optical power

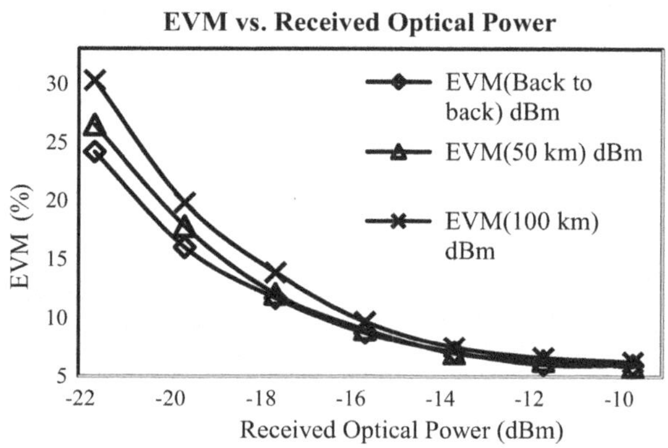

A graph is plotted for various received power versus EVM (%) for various fiber lengths as in Fig. 5. A maximum distance of 100 km is achieved and the system performance is tested for 0, 50 and 100 km. It is seen that when the received power becomes less the system is more prone to error. The receiver sensitivity is found to be −20 dBm for achieving a BER of 10^{-6} i.e. an EVM of 20%. Below this power level the BER of the system changes abruptly. Then a simple implementation for WDM simplex system is done and its performance is evaluated for wavelengths λ_1 = 1552.4 nm to λ_{16} = 1564.67 nm in the C-band. The frequency separation for different wavelength is set at a separation of 0.8 nm (100 GHz). The dispersion effects have been compensated with the implementation of Dispersion Compensating Fiber (DCF) which has a negative dispersion co-efficient of −85 ps/nm/km and the fiber attenuation is combated with the help of Erbium Doped Fiber Amplifier (EDFA) with a gain of 4 dB at regular intervals. With attenuation and dispersion compensation the transmission distance can be extended to 140 km over SSMF.

4 Conclusion

This paper implements a full-duplex RoF network architecture design with 20 Gbps DP-DQPSK down link and 5 Gbps up-link data rates. The laser free user terminal with direct detection makes it useful in lesser complexity and low cost base stations, reduced phase noise and also good coherence due to same optical source. From the performance analysis based on simulation platform the designed system can be realized for bidirectional transmission with 100 km fiber transmission and for WDM system with attenuation and dispersion compensation a maximum data rate of 320 Gbps for 140 km is achieved.

References

1. Winzer, P. J. and Essiambre, R. J., "Advanced Optical Modulation Formats," proc. of the IEEE, Vol. 94, No. 5 (2006).
2. Li, L., Jin-ling, C. and Ji-jun, Z, "Research of 100 Gbit/s DP-QPSK Based on DSP in WDM-PON System," International Journal of Signal Processing, Vol. 8, No. 3, 121–130 (2015).
3. Zhang, R., Ma, J. and Xin, X., "Full-Duplex Fiber-Wireless link for alternative wired and 40 GHz wireless access based on Differential Quaternary Phase Shift Optical Single sideband millimeter wave signal", Optical Engineering vol. 54, No. 2, 026101–10 (2015).
4. Ma, J., Zhan, Y., Zhou, M., Liang, H., Shao, Y. and Yu, C., "Full-Duplex Radio Over Fiber with a Centralized Optical Source for a 60 GHz Millimeter-Wave System with a 10 Gbps 16-QAM Downstream Signal Based on Frequency Quadrupling," J. Opt. Commun. Netw., Vol. 4, No. 7, 557–564 (2012).
5. Chen, A. and Murphy, E. J., [Broadband Optical Modulators], CRC press, Taylor and Francis Group, New York (35–54).
6. Chang, P. H. and Cho, G. R., "Enhanced feedforward control of non-minimum phase systems for tracking predefined trajectory," International Journal of Control, Vol. 83, No. 12, 2440–2452 (2010).

Development of MATLAB Based Image Stitching Tool for Detection of Hidden Objects at 89 GHz

Triveni Keskar, Vijay R. Dahake, Kshitij Mittholiya, Archana Hegde, A.M. Basil and Anuj Bhatnagar

Abstract Our system uses a 108 mW IMPATT diode source operating at 89 GHz, a 4.5 cm × 4.5 cm detector camera from TeraSense and a motorized XY stage covering 34 cm × 22 cm area. Hidden objects having large dimensions can be imaged using small area detector using this technique in 10–11 min. Image stitching and enhancement is done using MATLAB and imageJ software programming.

1 Introduction

Closely spaced 89 GHz band emulates characteristics of sub-terahertz Extremely High Frequency. Non-ionizing radiation and transparency to non-metallic-non-polar dielectrics highlights the applicability of this frequency band. It is highly reflective when obstructed by metals and has absorption signatures whilst passing through most liquids and explosives. Consequently, it is applicable for security scanning, product composition and quality checking [1, 2]. The system is capable of giving a resolution up to 3.37 mm $\approx$ 3 mm which adds to the precision of imaging achieved.

T. Keskar (✉) · V.R. Dahake
Department of Electronics and Telecommunication Engineering, Ramrao Adik Institute of Technology, Mumbai, India
e-mail: trivenikeskar@gmail.com

V.R. Dahake
e-mail: vijayr_dahake@rediffmail.com

K. Mittholiya · A. Hegde · A.M. Basil · A. Bhatnagar
Department of Photonics, Society of Applied Microwave Electronics Engineering and Research (SAMEER), Mumbai, India
e-mail: anuj@sameer.gov.in

© Springer Nature Singapore Pte Ltd. 2017

I. Bhattacharya et al. (eds.), *Advances in Optical Science and Engineering*, Springer Proceedings in Physics 194, DOI 10.1007/978-981-10-3908-9_73

2 Procedure for Object Detection

2.1 Video Scanning by TeraSense Camera and Movable X-Y Stage Motion Using LabVIEW

- A stepper motor based movable XY stage controlled by Holmarc MicroMotion software is coded on LabVIEW platform which enables it to travel from initial Home position at (0, 0) to the Y direction full sweep in 4 loops. At the end of each Y-directional full sweep iteration, we increment the X direction loop spanning a virtual S shape during wait time. The Stage dimensions are 34 cm × 22 cm.
- To achieve the whole scan in minimum time, we set the motors at maximum speed = 8000 μm/s. With these parameters, we are able to record a full stage scan in 10 min. The arrangement of developed equipment system and source detector orientation can be observed in schematic sketch Fig. 1.
- The camera captured image shows low intensity dark region where the radiation gets obstructed by a metallic surface (highly reflective), pharmaceutical or narcotic drugs (highly absorptive). The TeraSense camera captures each 4.5 cm × 4.5 cm frame as a 128 × 128 pixel video onto system. The parameters such as Exposure, Accumulation, Working area and Rec. Normalized are used to enhance clarity in the normalized intensity image of the object (Figs. 2 and 3).

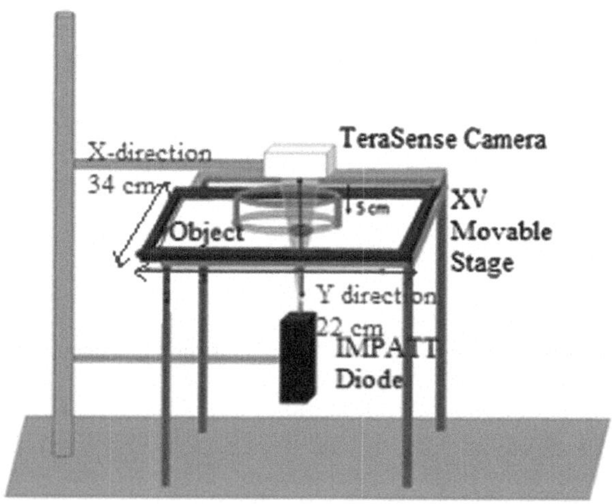

Fig. 1 Schematic of imaging system

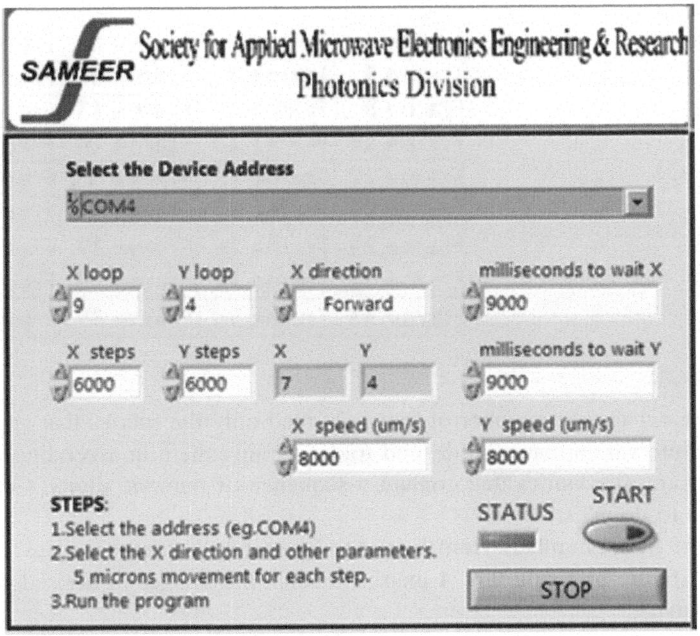

Fig. 2 LabVIEW UI to control stage positioned

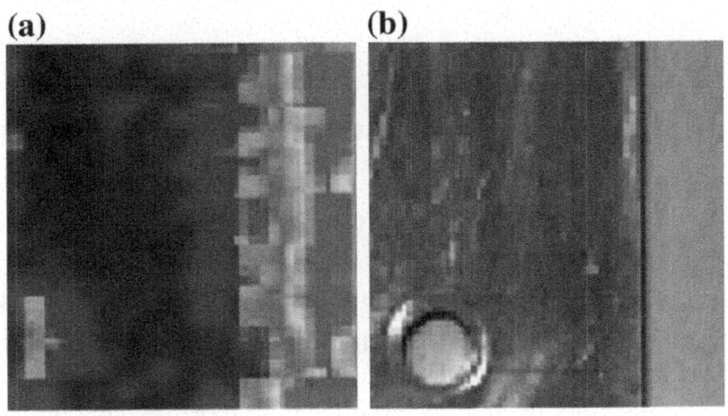

Fig. 3 **a** Intensity image of **b** part of the object

2.2 *Image Construction Using MATLAB*

Equal interval frames are chosen out of the captured video and stitched in S-scan direction using MATLAB program. We have developed MATLAB code for image stitching, enhancement and edge detection according to the following algorithm:

(1) Read the video file from 'save video' location on the system.
(2) Count the total number of frames (b) and choose an interval 'm' to pick 36 equally spaced images: m = b/36.

Fig. 4 Desired frame arrangement

Frame 1	Frame 2	Frame 3	Frame 4
Frame 8	Frame 7	Frame 6	Frame 5
Frame 9	Frame 10	Frame 11	Frame 12
Frame 16	Frame 15	Frame 14	Frame 13
Frame 17	Frame 18	Frame 19	Frame 20
Frame 24	Frame 23	Frame 22	Frame 21
Frame 25	Frame 26	Frame 27	Frame 28
Frame 32	Frame 31	Frame 30	Frame 29
Frame 33	Frame 34	Frame 35	Frame 36

(3) From t = 1 to total number of frames b, read only the frames that appear after given interval and store in desired folder naming them in ascending order.
(4) Match any file names that contain a sequence of numeric digits. Convert the strings to doubles.
(5) Sort the frame numbers from lowest to highest and arrange indices.
(6) Create figure by using 9×4 matrix to stitch the images in desired order.
(7) Save image.

Using our algorithm, we obtain the desired frame arrangement for montage function which stitches equally spaced video snapshots to get the final image of entire stage (Fig. 4).

3 Image Enhancement and Edge Detection

As the resolution obtained with 89 GHz is 3.37 mm i.e., $\approx$3 mm. The rays may get diffracted from the thinner edges or corners of objects, still allowing some rays to pass shearing the reflective surface. Consequently, the images that we obtain after stitching are usually distorted. Practically in our stitched images, the edges of thin objects are prone to show a relatively higher intensity. Thus further improvement on contrasting, averaging of the edge surface intensities and filtering of the image is necessary.

I. Our MATLAB program first converts the RGB intensity image to grayscale and then applies median filtering technique. We can also enhance the contrast of grayscale image using deviation based thresholding of the image.
II. Open source imageJ tools are used for filters such as 'Butterworth, Floyd Steinberg Dithering, Enhance contrast or Dilation' to enhance properties of the RGB image for a better edge detection.
III. 'Canny' edge detection algorithm [3] is used to detect the edges of the object in a simpler format. A channel separation with edge detection in imageJ gives better quality.

4 Results

4.1 Stitching and Edge Detection Outputs

In our experimentation, stitched images of a metallic plate and a hidden plier with image processing are shown. A metallic plier hidden inside a 2 mm thick optically opaque cardboard box has been detected. The same object's intensity image obtained using stitching can be improved in quality for edge detection using inbuilt tools of open source application imageJ. We have used the math class functions for multiplying image intensities, background subtraction, color channel separation, conversion to binary and dilation of pixels thereafter to withdraw the noise from the picture (Figs. 5, 6 and 7).

The timing profile of MATLAB stitching and detection program is ≤ 4 s as shown in the Table 1.

4.2 Image Quality Analysis

The resultant scan images show high level of closeness to the original object's optical images. Using intensity color-map images for comparison, we have quantified the parameters such as Peak Signal to noise ratio (PSNR), Signal to noise ratio (SNR) and structural similarity matrix (SSIM) with the help of built in psnr, snr and ssim functions of MATLAB. The PSNR for plate shown in our results is 14.2881,

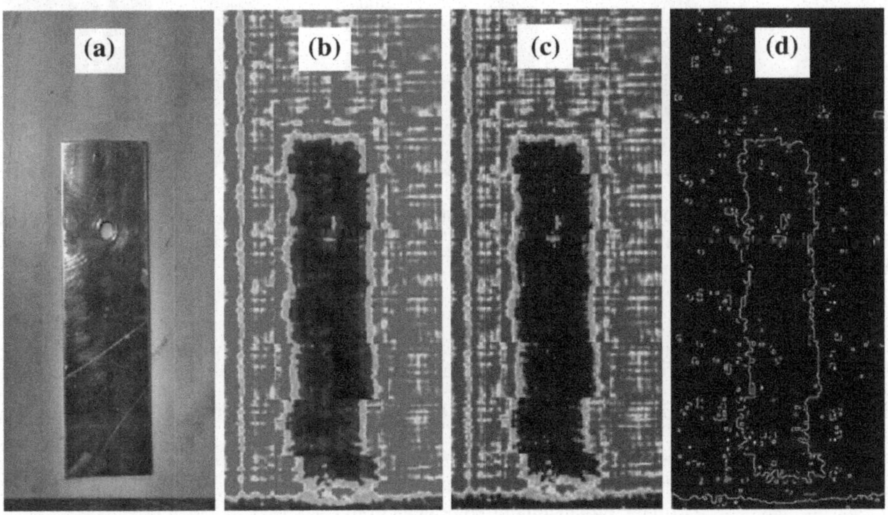

Fig. 5 **a** A steel plate object (optical image), **b** its scanned intensity image, **c** contrast enhanced grayscale image of (**b**) and **d** edge detected when scanned through our system

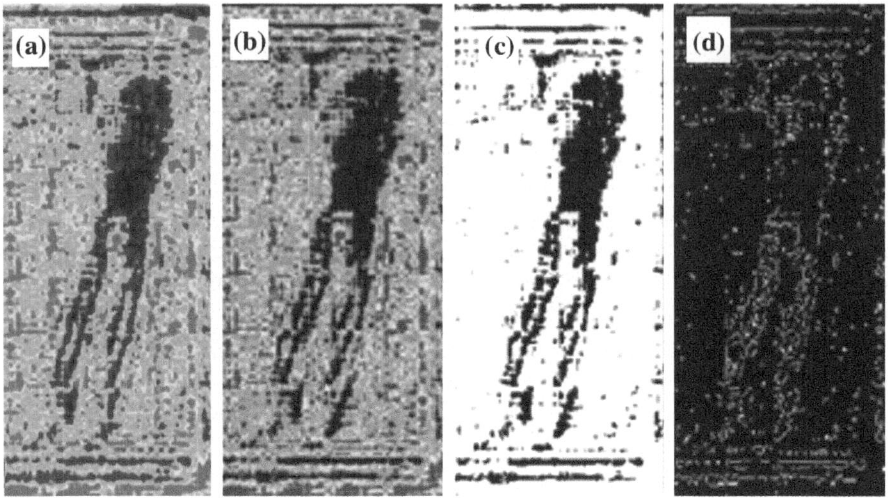

Fig. 6 **a** Stitched intensity image of a hidden plier recorded by the system, **b** contrast enhanced grayscale image, **c** thresholding enhancement image and **d** edge detected image

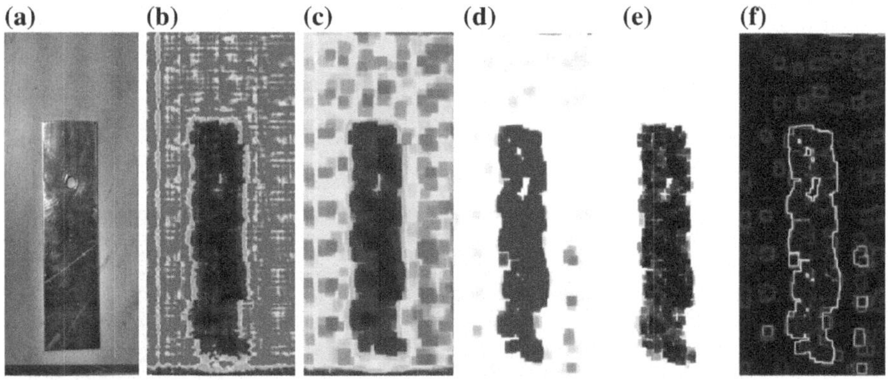

Fig. 7 **a** A steel plate object (optical image), **b** its intensity image, **c** 25% background subtracted output of (**b**), **d** triple multiplied intensity image, **e** red channel separated image of (**d**) and **f** edge detection

Table 1 MATLAB profiler timing stitching image and edge detection program

Profile summary	Generated on 19-Feb-2016 14:24:04 using CPU time		
Sr.	Function names	Calls	Total timing (s)
1	Image stitching	1	2.2997
2	Image enhancement edge detection	1	1.775

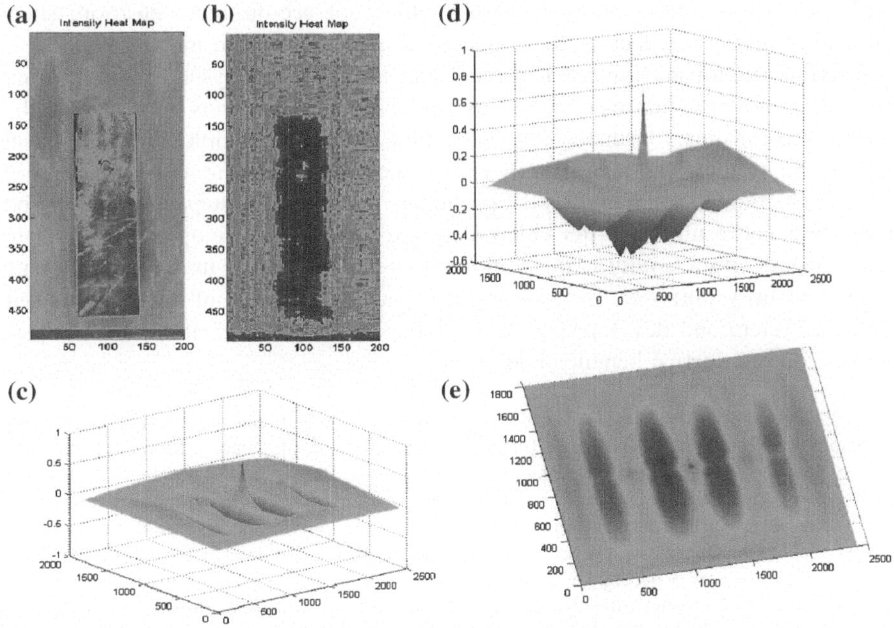

Fig. 8 Intensity heat map of **a** Optical image, **b** terahertz scan image, correlation matrix 3D plots **c** upper side view, **d** lower side view **e** top view for scanning result of metallic plate

with SNR value of 13.3679 and similarity matrix score of 0.6559. Ideally, PSNR and SNR values higher than 1 are considered as indicators of low noise, whereas ssim values range from −1 to 1, closeness to 1 signifying a good structural matching. The normalized correlation matching calculated for plate between the scanned and optical images is shown in (Fig. 8).

5 Concluding Discussion

Development of a programming user interface and analytical tool is our proposed way to utilize low cost resources to build a system for large hidden object scanning to detect materials opaque or absorptive to 89 GHz such as metals, liquids, drugs and explosives. The low intensity portion obtained from camera is essential to detect such objects which can be hidden to the naked eye due to covering of a non-dielectric substance [4]. The design of XY scanner gives a wide area to use in sub-terahertz applications. Advantages of this tool include a higher safety level for prolonged human interaction as compared to conventional scanning methods, adherence to the safety norm standards, ability of see-through detection, short computing timing and lack of dependence on human error and intervention.

There is significant limitation of recording timing in our system which we are trying to better as a future scope to this application. As test runs have shown, the mechanical systems consume 98% of the time. Therefore, employing high speed mechanical systems will significantly reduce the amount of time required to record scan. Reduction in wait time and optimizing the speed of stepper motor based translation can result in a faster video recording. As far as developmental tools are concerned, the responsivity of this tool is sufficiently high to be used for wide range of vital security purposes. Considering the trade-off between all the aspects, our imaging system and developed tools for intensity image acquisition and analysis present a cost effective beneficial solution.

Acknowledgements The authors would like to express their gratitude towards two immensely cooperative institutions 'Ramrao Adik Institute of Technology', affiliated to Mumbai University and 'Society for Applied Microwave Electronics Engineering Research', IIT Bombay (Department of Electronics Information Technology Ministry of Communications Information Technology, Govt. of India) for providing with an inclusive, encouraging environment. Special thanks to our colleagues Saurabh Bharadwaj and Suyog Chowdhari for their active timely contribution in project and peer reviews were a pacemaker to the reporting aspects of this paper. On behalf of all the authors, I would like to acknowledge our friends and families who have directly and indirectly joined hands to make this application development successful.

References

1. "Camera looks through clothing", BBC News 24, 10 March, 2008, retrieved 10 March, 2008.
2. Basil A.M, Triveni Keskar, Kshitij Mittholiya, Archana Hegde, A. Bhatnagar, "Terahertz Imaging System for Scanning Concealed Objects", Journal of Instrumentation Society of India, Vol.46, No.1, pp. 57–58, March 2016.
3. J. Canny, "A Computational Approach to Edge Detection", in IEEE Transactions on Pattern Analysis and Machine Intelligence, vol. PAMI-8, no. 6, pp. 679–698, Nov. 1986. Doi:10.1109/TPAMI.1986.4767851.
4. Basil A. M., Kshitij Mittholiya, Archana Hegde, Dr. Anuj Bhatnagar, "Development of terahertz imaging system", Journal of Instrumentation Society of India, Vol. 45 No. 2, 30 June 2015.

Modelling for Spectral Domain Optical Coherence Tomography (SD-OCT) System

Suyog Choudhari, Mukesh Patil and Roshan Makkar

Abstract Optical Coherence Tomography (OCT), a revolutionary technology was initially developed for imaging retinas to investigate various eye diseases like glaucoma, diabetic retinopathy etc. It later found acceptance for its non-invasive nature in other bio medical applications like cardiology, dermatology etc. It is very similar to ultrasound in nature; the only difference is usage of light waves in place of sound waves and it offers higher axial resolution in comparison with most of other existing technologies. The technology behind SD-OCT is Michelson interferometry, which offers the usage of Super Luminescent Diode (SLD) source as low temporal coherent source with broad bandwidth to measure its absorption/scattering through the sample specimen. In the following paper, we depict the model of SD-OCT system. The spectral response of the source and the output generated waveforms show the response of the model in respect to the actual hardware. LabVIEW software could generate the graphical user interface. The samples were used on the basis of their refractive indices. The maximum depth obtained depends on the number of pixels. Depth of up to 0.5 mm can be obtained with 127 pixels and 1.7 mm with 400 pixels with axial resolution as 7.75 μm for 100 nm bandwidth 1310 nm SLD source. Simulation results at 632.8, 840 and 1310 nm are compared and discussed.

S. Choudhari (✉) · M. Patil
Department of Electronics and Telecommunication Engineering,
Ramrao Adik Institute of Technology, Mumbai, India
e-mail: suyogc7@gmail.com

M. Patil
e-mail: mukesh.rait@gmail.com

R. Makkar
Photonics Division, Society of Applied Microwave Electronics Engineering
and Research (SAMEER), Mumbai, India
e-mail: roshan@sameer.gov.in

© Springer Nature Singapore Pte Ltd. 2017

I. Bhattacharya et al. (eds.), *Advances in Optical Science and Engineering*,
Springer Proceedings in Physics 194, DOI 10.1007/978-981-10-3908-9_74

1 Introduction

In order to enhance the diagnostics and clinical administration of diseases, new therapeutic innovations are coming up. The current advancements as the Ultrasound, X-Ray, Computed Tomography (CT filter), Magnetic Resonance Imaging (MRI), Radioisotope Imaging have changed symptomatic medication amid the most recent decades. Computed tomography can accomplish high resolutions yet utilizes ionizing radiation and bears an inherent danger. Magnetic Resonance Imaging does not include such a danger, but rather in return it can't resolve objects smaller than 0.3 mm. Ultrasound has a depth resolution of around a couple of centimeters however poor image resolution, therefore even high-frequency ultrasound of 50 MHz is constrained to 30 μm resolution. Besides, it needs a decent transport medium, for example, gel since sound waves are highly weakened in air [1]. OCT performs cross-sectional imaging by measuring the magnitude and echo time delay of back-scattered light. Measuring the echo time delay of several reflections generates the axial scans or the A-scans. This will produce the two-dimensional data set, which represents the back scattering in a cross-sectional plane. Images or B-scans can be shown by the false shading or dark scale to imagine tissue pathology.

Three-dimensional OCT information will contain the volumetric basic data, which can be compared with the CT scan or MRI pictures. OCT is a non-invasive optical imaging methodology that procures depth resolved information of organic specimens in both two and three dimensions [1]. OCT was initially presented by Dr. James Fujimoto [2]. It relies on low-coherence interferometry and makes use of low coherence light sources or ultra-short pulsed lasers [3]. OCT was first performed in time domain which involved manual movement of the reference mirror. This type of OCT is called Time Domain OCT (TDOCT).

Section 2 highlights the basic principle of OCT in detail. Mathematical model for OCT is described in Sect. 3 i.e. signal processing in SDOCT. The simulation flow and the simulation results in LabVIEW (Laboratory Virtual Instrument Engineering Workbench) are discussed in Sect. 4. Section 5 gives the conclusion derived from this paper followed by acknowledgement.

2 Principle of Operation

A.A. Michelson imagined the optical interferometer usually called as the Michelson interferometer [4]. The mechanical assembly of the most well-known design comprises of the beam splitter and a light source as the major components. The light from the source is part into two arms; sample and reflected, utilizing a beam splitter. Each of the arms reflects back the light towards the beam splitter which then adds their intensities interferometrically. The subsequent interference pattern is directed towards the photoelectric detector or CCD (Charge Coupled Device) camera.

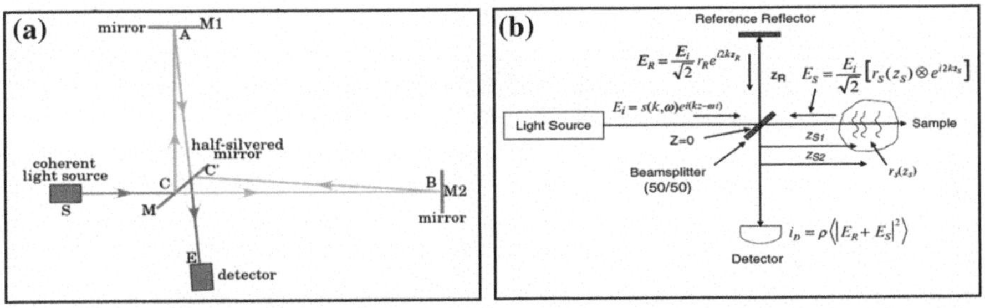

Fig. 1 **a** Michelson interferometer [4] and **b** schematic representation of Michelson interferometer [6]

Shown in Fig. 1 is Michelson interferometer which consists of a half silvered mirror acting as a beam splitter. Mirrors M1 and M2 are placed equidistant to follow the principle of path difference as required for interference. The mathematical representation of Michelson interferometer is shown in Fig. 1b.

3 Signal Processing in SDOCT

The source spectrum in Spectral Domain Optical Coherence Tomography (SD OCT) is Gaussian in nature [5]. Hence, for analysis and simulation we assume a bell shaped Gaussian spectrum. OCT can be mathematically illustrated as [6];

$$\begin{aligned} I_D(k) &= \frac{\rho}{4}\{S(k)[R_R + R_{S1} + R_{S2} + \cdots]\} \\ &+ \frac{\rho}{2}\{S(k)\sum_{n=1}^{N}\sqrt{R_R R_{Sn}}\cos[2k(Z_R - Z_{Sn})]\} \\ &+ \frac{\rho}{4}\{S(k)\sum_{n\neq m=1}^{N}\sqrt{R_{Sm} R_{Sn}}\cos[2k(Z_{Sn} - Z_{Sm})]\} \end{aligned} \tag{1}$$

Here, S(k) represents the Gaussian spectrum, R_R and Z_R are the reflectivity and physical distance of the reference mirror respectively, R_{S1}; R_{S2}; … ; Z_{S1}; Z_{S2};… are the reflectivities and physical distances of each sample respectively. ρ is the responsivity of the detector. The first term in (1) is the 'Constant' or the 'Direct Current (DC) Component' which is the largest component of the detector current. It is dependent on the reference reflectivity, sample reflectivity and responsivity of the detector. The second term in (1) is the 'Cross Correlation' term. It is dependent upon the light source spectrum and the path length difference between the sample arm and the reference arm. The third term of (1) is referred to as the 'Auto Correlation' term. This term represents the interference between different sample reflectors. Axial resolution is an essential parameter in OCT frameworks [7].

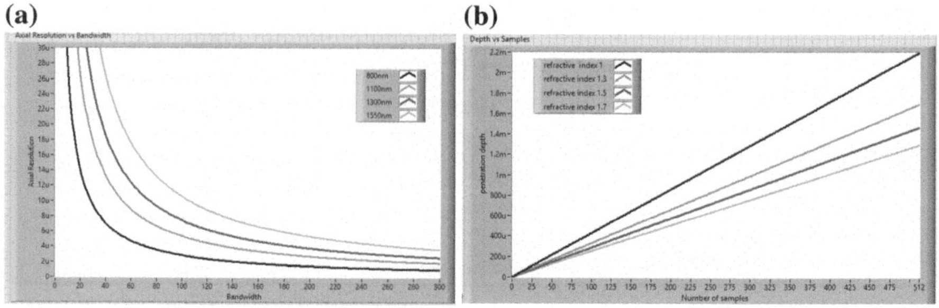

Fig. 2 **a** Relation between axial resolution and bandwidth for different wavelengths **b** relation between penetration depth and pixel number for different refractive indices

Axial resolution is half of the coherence length. The coherence length is given by Tomlins and Wang [8] as shown in (2):

$$lc = \frac{4 \ln 2}{\pi} * \frac{\lambda^2}{\Delta\lambda} \tag{2}$$

The length of coherence is the function of λ (central wavelength) and Δλ (bandwidth of the source).

The relation between axial resolution and bandwidth is depicted in Fig. 2a whereas Fig. 2b depicts the relation between penetration depth and pixel number. The above relation was obtained in LabVIEW.

The maximum depth Zmax is given by Tomlins and Wang [8] as shown in (3):

$$Zmax = \frac{\lambda^2}{4 * \Delta\lambda * RI} * N \tag{3}$$

where RI is the refractive index of the sample and N is the pixel number.

4 Results

In the first step we enter the login credentials. A person with a valid username and password only can access this simulation. On entering the valid credentials, the user enters the parameters. The parameters are converted to appropriate units as required in (1). Calculations of coherence length, linear interpolation are the steps which follow. After interpolation, the data is converted to k space, the matrix is then flipped. Inverse Fast Fourier Transform (IFFT) is applied on the flipped matrix. Plane swapping is performed to get the double sided spectrum.

Single sided spectrum obtained after taking the absolute values and dc subtraction, the spectrum thus obtained is mapped to depth. Finally, false coloring gives the final image. The simulation interferogram and depth image for 632.8 nm

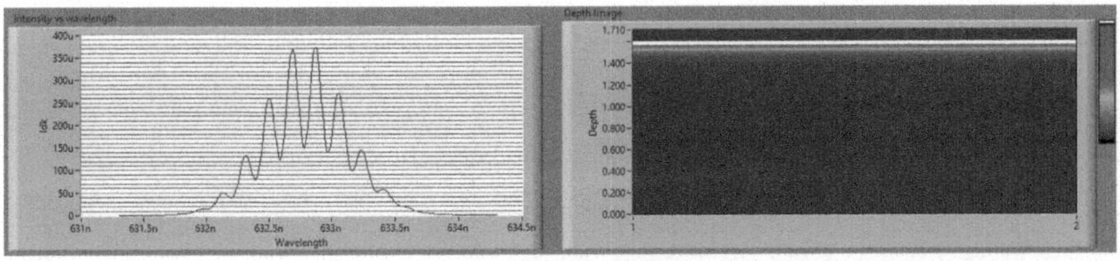

Fig. 3 Simulation of interferogram and depth results at 632.8 nm

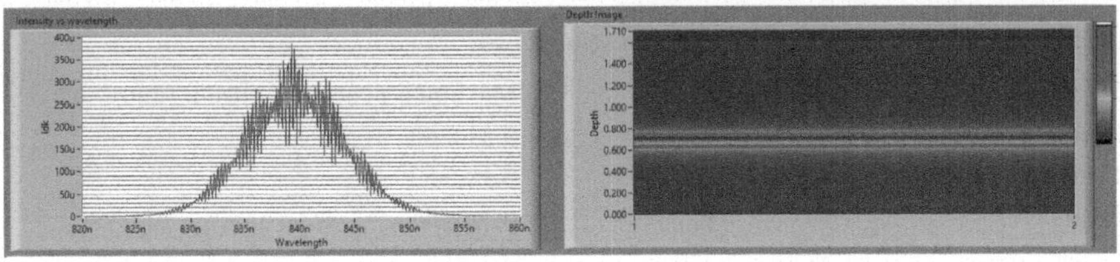

Fig. 4 Simulation of interferogram and depth results at 840 nm

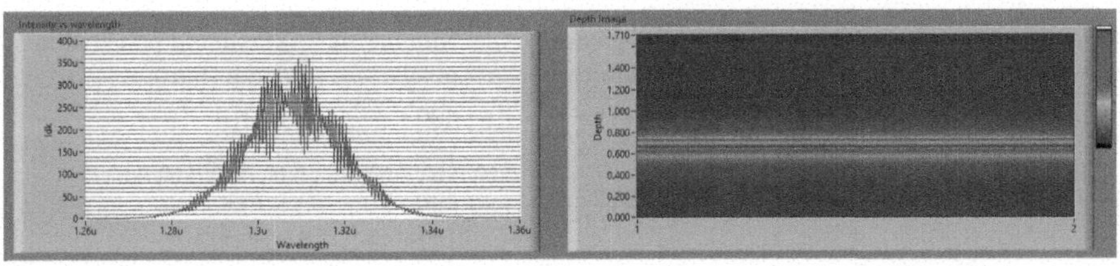

Fig. 5 Simulation of interferogram for two reflectors at 1310 nm central wavelength

source is shown in Fig. 3. The depth results were acquired for two reflectors. But since the coherence length for 632.8 nm laser source is 58.902 μm, the two reflectors merge into one. Hence we are not able to section the sample appropriately. In the next analysis, we acquire simulations for 840 nm SLD and 40 nm bandwidth. For 840 nm SLD source, the maximum depth acquired is 1.764 mm as shown in Fig. 4. Here, two reflecting surfaces can be seen clearly as the axial resolution is 7.784 μm. The spectrum is evenly spaced around 840 nm.

Figure 5 depicts the simulation screen for 1310 nm. The Inverse Fast Fourier Transform (IFFT) graph is a representation of number of samples and intensity. The cross correlation term contain the layers detected from the interferogram. The final depth image is mapped to scale by the Zmax formula.

Figure 6a, b depicts the depth analysis for three and four reflectors respectively. The results shown are simulated for 1310 nm central wavelength and 100 nm bandwidth. Here, the sample reflectors are spaced 0.1 mm apart and are resolved effectively as the axial resolution is 7.57265 μm. Table 1 shows the simulation

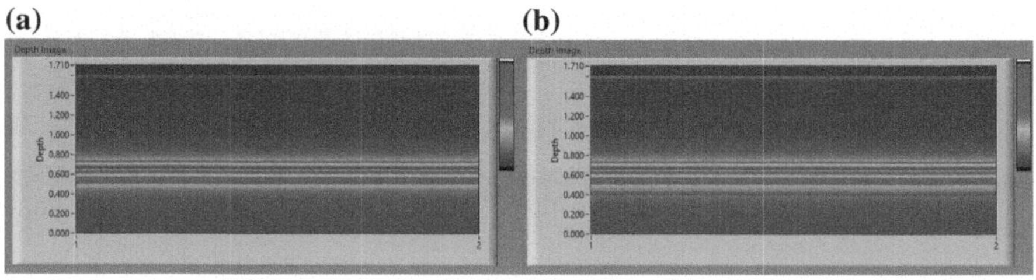

Fig. 6 **a** Depth analysis at 1310 nm central wavelength with 3 reflectors and **b** 4 reflectors

Table 1 Simulation results for variation in sample used

Sample	Refractive index	Maximum penetration depth (mm)
Mirror	1	0.544862
Leaf	1.425	0.382352
Glass	1.5	0.363241
Cover slip	1.52	0.358462

Table 2 Simulation results for variation in central wavelength, half bandwidth and pixel number

Central wavelength (nm)	Half bandwidth (nm)	Coherence length (μm)	Number of pixels	Maximum penetration depth (mm)
1310	50	15.1453	400	1.7161
			127	0.544862
840	20	15.568	400	1.764
			127	0.56007
632.8	1.5	117.801	400	13.3479
			127	4.23795

results for different input parameters. In this table, 1310 nm source and 100 nm bandwidth is kept constant along with 127 pixels.

In Table 2, the refractive index is constant at 1, the responsivity of the detector is 0.001 A/W. The results for variation in number of pixels, central wavelength and bandwidth are obtained. The coherence length is a function of the SLD used and hence becomes a crucial parameter in SDOCT imaging system.

5 Conclusion

Here, we have simulated a model for SDOCT with central wavelength 1310, 840 and 632.8 nm. These wavelengths are selected for simulation as they represent the vast difference in results obtained on increasing the central wavelength and bandwidths. Different samples are simulated by varying their refractive indices as shown in Table 2. This model aims to provide an insight into the practical model of SDOCT. The signal processing steps of SDOCT are explained with the help of a simulation model.

Acknowledgements The authors would like to express their gratitude towards the affiliated institutions 'Society for Applied Microwave Electronics Engineering and Research' (Department of Electronics Information Technology Ministry of Communications Information Technology, Govt. of India) and 'Ramrao Adik Institute of Technology', affiliated to Mumbai University for providing an inclusive and encouraging environment. The authors are thankful to Dr. Ramesh Vasappanavara, Principal, Ramrao Adik Institute of Technology for financial assistance to publish this work. The authors would also like to thank Ms. Prachi Bhatia for her immense contribution to make this work a success. Also the authors are highly grateful to Director, SAMEER and Photonics Division, SAMEER, Mumbai for providing LabVIEW software and other equipments. On behalf of all the authors, I would like to acknowledge our friends and families who have directly or indirectly joined hands to make this application development successful.

References

1. James G. Fujimoto, Costas Pitris, "OCT-An Emerging Technology for Biomedical and Biopsy", Neoplasia Vol. 2, Nos. 1–2, January-April 2000.
2. W. Drexler, J. G. Fujimoto, C. Pitris, "In-vivo Ultrahigh Resolution Functional OCT", CLEO 2000/Thursday, MIT Cambridge, Massachusetts.
3. Wolfgang Drexler, "Ultrahigh-Resolution OCT", Journal of Biomedical Optics 9(1), 4774 (Jan/Feb 2004).
4. http://physics.gmu.edu/~ellswort/p263/michelson.pdf.
5. Murtaza Ali, Renuka Parlapalli, "Signal Processing Overview of Optical Coherence Systems for Medical Imaging", Texas Instruments, June 2010.
6. Joseph Izatt, M. A. Choma, "Introduction to Optical Coherence Tomography", Springer, Heidelberg, 03/2008.
7. Peng Xi, Kai Mei, Tobias Bruler, Chuanqing Zhou and Qiushi Ren, "Evaluation of spectrometric parameters in Spectral-Domain Optical Coherence Tomography", Appl. Opt. 50, 366–372 (2011).
8. P. H. Tomlins and R. K. Wang, "Theory, Developments and Applications of Optical Coherence Tomography", Journal of Applied Physics, 38 (2005) 25192535.

Characterization of Memory Effect of Polarization Speckles from a Birefringent Scatterer

Abhijit Roy, Rakesh K. Singh and Maruthi M. Brundavanam

Abstract Propagation of a coherent light through a random birefringent scatterer generates a speckle pattern with spatially varying polarization called polarization speckle. The spatial polarization distribution of the polarization speckle is random in nature making zero net polarization. In this paper, we experimentally characterize the memory effect of the polarization speckle from a birefringent scatterer for different orientations of an analyzing polarizer using far-field intensity correlation measurements.

1 Introduction

The spatial coherence of a coherent light is completely lost due to modification of the wavefront after passing through a random scattering medium. The unaffected temporal coherence allows the scattered light to interfere to produce randomly fluctuating grainy intensity pattern known as speckle. Although the speckle is treated as noise in coherent imaging system, it has been found that the object information is not completely lost in the speckle pattern rather scrambled into it and is retrievable [1]. Subsequently different non-invasive imaging techniques for object information retrieval have been proposed and successfully demonstrated [2].

It has been observed that although the speckle pattern is visually random in nature, any kind of modification in the input beam affects the generated speckle pattern accordingly which is treated as the memory effect of the random field [3]. The study of the memory effect in terms of intensity correlation at the far-field provides better understanding of the effect. The change in the angle of incidence of

A. Roy · M.M. Brundavanam (✉)
Department of Physics, Indian Institute of Technology
Kharagpur, Kharagpur, West Bengal 721302, India
e-mail: bmmanoj@phy.iitkgp.ernet.in

R.K. Singh
Department of Physics, Indian Institute of Space Science and Technology,
Thiruvananthapuram, Kerala 695547, India

© Springer Nature Singapore Pte Ltd. 2017
I. Bhattacharya et al. (eds.), *Advances in Optical Science and Engineering*,
Springer Proceedings in Physics 194, DOI 10.1007/978-981-10-3908-9_75

the incident beam on scattering medium displaces the generated speckle pattern by the same angle [4]. The displaced speckle pattern remains correlated with the original pattern up-to certain angle of the tilt and starts de-correlating. The far-field angular correlation function of the light scattered from a one-dimensional rough conducting [5] and dielectric surface [6] for different input polarizations have been studied experimentally. The existence of two peaks in the angular correlation function has been observed: one due to the auto-correlation function and another because of the cross-correlation function. It has also been found that the angular correlation function contains the contribution of short and long range correlation function [7]. In another separate work, the existence of the memory effect for double passage of light through a one-dimensional random phase screen has been established both theoretically and experimentally [8, 9]. No significant changes in the rotational memory effect have been observed by rotating the incident beam at different angles from 0° to 360° [10]. Recently, the theoretical and experimental validation of the existence of translational memory effect has been reported [11]. All these reported studies on the memory effect are confined to non-birefringent scatterer.

The speckle from a non-birefringent scatterer has uniform spatial polarization distribution and is same as the incident beam. In contrast, the speckle generated from a birefringent scatterer has random spatial polarization distribution that makes the net polarization zero called, polarization speckle. Recently it has been reported that imaging using polarization speckle has potential applications in skin cancer detection [12]. Although there has been some effort to characterize the polarization speckle [13, 14], no attention has been paid to study the memory effect to the best of our knowledge. In the present paper, we study the memory effect of the polarization speckle by modifying the spatial polarization of the generated speckle using a polarizer. The memory effect of polarization speckle for different orientations of the analyzing polarizer for example two mutually orthogonal orientations of the polarizer is presented.

2 Mathematical Modeling

Speckle pattern is characterized using intensity correlation at the far-field under the assumption of spatial ergodicity and Gaussian statistics. The degree of coherence (DoC), γ of the speckle field is related to the two point fourth order field or second order intensity correlation at the far-field following the relation

$$\gamma^2(\mathbf{r}_1, \mathbf{r}_2) = \frac{\langle \Delta \mathrm{I}(\mathbf{r}_1)\Delta \mathrm{I}(\mathbf{r}_2)\rangle}{\langle \mathrm{I}(\mathbf{r}_1)\rangle\langle \mathrm{I}(\mathbf{r}_2)\rangle} \tag{1}$$

where $\mathbf{r}$ is the spatial position in the transverse plane, $\Delta \mathrm{I} = \mathrm{I} - \langle \mathrm{I}\rangle$ is the fluctuation of the intensity from its mean value and '$\langle\ \rangle$' is the ensemble average of the variable. The degree of polarization (DoP), P that provides information about the spatial polarization distribution of the speckle can be calculated using the following relation

$$P^2(0) = 2\gamma^2(0) - 1 \quad (2)$$

The characterization of the speckle pattern is done following the (1) and (2). The memory effect is studied by calculating the correlation between two different speckle patterns using (1) where $\mathbf{r}_1$ and $\mathbf{r}_2$ are the two spatial positions in two different patterns.

3 Experiment and Result

The schematic diagram of the experimental setup used for the present study is shown in Fig. 1. A horizontally polarized He–Ne laser source of wavelength 632.8 nm is made diagonally polarized using a half-wave plate, HWP. The diagonally polarized beam of size of 2 mm, adjusted with the help of an aperture is passed through a birefringent ground glass, BGG (composed of OHP sheet and ground glass) placed at the front focal plane of a Fourier transforming biconvex lens (L) of focal length of 100 mm. An analyzer is placed just after the BGG to make the spatial polarization distribution of the speckle pattern uniform. The uniformly polarized far-field speckle pattern is recorded by a CCD camera placed at the back focal plane of the lens. The CCD camera is from Thorlabs having pixel dimension (1024 × 1280). The analyzer is oriented at different orientations from −90° to +180° in separation of 20° and data is recorded. The unperturbed speckle pattern i.e. without the analyzer is also recorded to characterize the speckle pattern generated from the BGG.

The two point second order intensity correlation technique is applied following (1) to characterize the speckle pattern recorded without the analyzer and for analyzer orientation of $\theta = 0°$ and $90°$. The result is shown in Fig. 2. It can be observed that without the analyzer, the value of $\gamma^2(0)$ is very close to 0.5. The calculated value of DoP, P(0) using (2) comes out to be zero that indicates the generated speckle is polarization speckle and has random spatial polarization distribution.

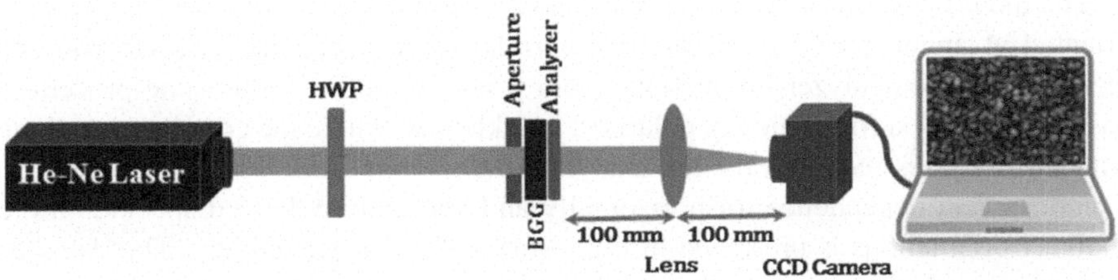

Fig. 1 The Schematic diagram of the experimental set up is presented. *HWP* is the half-wave plate, *BGG* is the birefringent ground glass, and *Lens* is the Bi-convex lens of 100 mm focal length

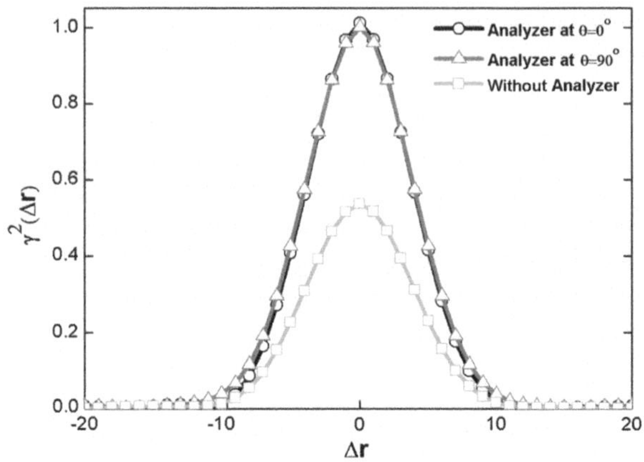

Fig. 2 The plot of correlation function for three different cases: without analyzer, analyzer oriented at 0° and 90°. The maximum values of the correlation functions are very close to 0.5 and 1.0 respectively

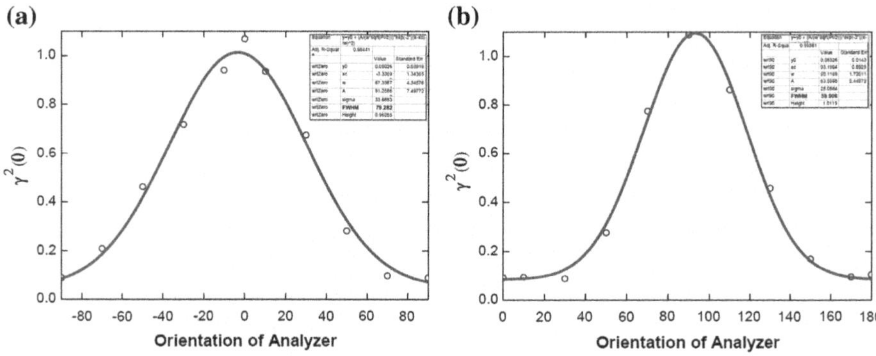

Fig. 3 The dependence of maximum value of cross-correlation function on the analyzer orientation for horizontally and vertically polarized component is presented in **a** and **b** respectively. The dependence is fitted with the Gaussian function. The *dots* are the experimental data and the *solid line* is the fitted curve

The value of $\gamma^2(0)$ is found to be very near to 1.0 for analyzer orientation of $\theta = 0°$ and 90° and calculated P(0) is also near to 1.0 which indicates that the speckle has uniform spatial polarization distribution.

The memory effect of the horizontally polarised speckle i.e. when the analyzer is oriented at angle $\theta = 0°$ is studied by correlating it with other speckle patterns recorded for the analyzer orientations from $\theta = -90°$ to 90°. The same procedure has been followed for vertically polarised speckle also where the correlation is done with the speckles recorded for the analyzer orientations from $\theta = 0°$ to 180°. The memory effect dependence for horizontally and vertically polarized speckle on the analyzer orientations is presented in Fig. 3 (a) and 3 (b) respectively. The data has been fitted with the Gaussian distribution.

It is observed from the Fig. 3 that the information hidden in horizontally or vertically polarized components is stored in other polarization components also and is retrievable significantly up to certain orientation of the analyzer, taken as the FWHM of the Gaussian fitted curve. The FWHM of the fitted curve for horizontally and vertically polarized speckle is 79.28 and 59.00 respectively. So, it can be concluded that the information stored in the horizontally polarized speckle, is retrievable significantly for more orientation of the analyzer compared to the vertically polarized one.

In conclusion, we have studied the polarization based memory effect of the polarization speckle for different orientations of the analyzer. We have shown that the information stored in the two polarized components is retrievable significantly up to certain orientations of the analyzer and this is valid for any polarized component. It has also been found that the memory effect exists more for horizontally polarized speckle than vertically polarized one. This work will surely provide a deep insight in the polarized memory effect and imaging through polarization speckle.

References

1. Leith, E.N. and Upatnieks, J., "Holographic imagery through diffusing media," J. Opt. Soc. Am. **56**, 523 (1966).
2. Katz, O., Heidmann, P., Fink, M. and Gigan, S., "Non-invasive single-shot imaging through scattering layers and around corners via speckle correlations," Nat. Photonics **8**, 784 (2014).
3. Feng, S., Kane, C., Lee, P.A., and Stone, A.D., "Correlations and fluctuations of coherent wave transmission through disordered media," Phys. Rev. Lett. **61**, 834 (1988).
4. Freund, I., Rosenbluh, M. and Feng, S., "Memory effects in propagation of optical waves through disordered media," Phys. Rev. Lett. **61**, 2328 (1988).
5. Knotts, M.E., Michel, T.R. and O'Donnell, K.A., "Angular correlation functions of polarized intensities scattered from a one-dimensionally rough surface," J. Opt. Soc. Am. A **9**, 1822 (1992).
6. Lu, J.Q. and Gu, Z.H., "Angular correlation function of speckle patterns scattered from a one-dimensional rough dielectric film on a glass substrate," Appl. Opt. **36**, 4562 (1997).
7. Malyshkin, V., McGurn, A.R., Leskova, T.A., Maradudin, A.A., Nieto-Vesperinas, M., "Speckle correlations in the light scattered from a weakly rough one-dimensional random metal surface," Opt. Lett. **22**, 946 (1997).
8. Escamilla, H.M., Mndez, E.R. and Hotz, D.F., "Angular intensity correlations in the double passage of waves through a random phase screen," Appl. Opt. **32**, 2734 (1993).
9. Lin, Z.Q. and Gu, Z.H., "Memory effect in the double passage of waves through a one-dimensional random phase screen," Proc. SPIE **3141**, 232(1997).
10. Amitonova, L.V., Mosk, A.P. and Pinkse, P.W.H., "Rotational memory effect of a multimode fiber," Opt. Express **23**, 20569 (2015).
11. Judkewitz, B., Horstmeyer, R., Vellekoop, I.M., Papadopoulos, I.N. and Yang, C., "Translation correlations in anisotropically scattering media," Nat. Phys. **11**, 684 (2015).
12. Tchvialeva, L., Dhadwal, G., Lui, H., Kalia, S., Zeng, H., McLean, D.I. and Lee, T.K., "Polarization speckle imaging as a potential technique for in vivo skin cancer detection," J. Biomed. Opt. **18**, 061211 (2013).

13. Singh, R.K., Naik, D.N., Itou, H., Miyamoto, Y. and Takeda, M., "Characterization of spatial polarization fluctuations in scattered field," J. Opt. **16**, 105010 (2014).
14. Vinu, R.V. and Singh, R.K., "Experimental determination of generalized Stokes parameters," Opt. Lett. **40**, 1227 (2015).

Part XVI
Micro-Electronics and VLSI

Analytical Study of High Speed Low Power Consuming Reversible Nano Device CPLD

Jayanta Gope, Sanjay Bhadra and Shantanu Bhadra

Abstract In modern context Reversible Logic evolved as one of the promising technologies owing to its wide spread application in quantum computing and low dimensional system architecture. Largely, the Reversible Logic circuit configurations are CMOS made. However, scaling down of CMOS technology is facing tremendous challenges owing to its inherent physical limitations. In such situations Scientists have come up with the idea of new Nano scale device technologies. The authors here render Application Specific design of Reversible Complex Programmable Logic Devices (CPLD) using CMOS and Single Electronics Device combination which is unique of its kind. In precise a Hybrid CMOS-SET made Reversible PLA (HRPLA) modelled here. The designed IC is analytically studied using TANNER environment tool. The results obtained shows greater trade off amid conventional and commercially available CPLDs and the newly proposed one.

1 Introduction

Downscaling of device sizes is the driving force of device research. The key element in the field of device research is the downscaling of device sizes. But the limiting factors in downsizing are: (i) what would be the minimum gate oxide thickness? (ii) what would be the minimum gate length? (iii) what would be the doping placement effect? Thus to venture out the future attributes of VLSI technology in post CMOS era technological shift was much awaited. In persuasion to this up

J. Gope (✉)
Department of ECE, CSET, MAKAUT, Kolkata, West Bengal 7000124, India
e-mail: jayanta.gope.1983@ieee.org

S. Bhadra
Department of EE, University of Engineering and Management Kolkata, Kolkata, West Bengal 700160, India

S. Bhadra
Department of Computer Application, RAMT, MAKAUT, Kolkata, West Bengal 700052, India

© Springer Nature Singapore Pte Ltd. 2017

I. Bhattacharya et al. (eds.), *Advances in Optical Science and Engineering*, Springer Proceedings in Physics 194, DOI 10.1007/978-981-10-3908-9_76

gradation several novel technologies including Quantum Electronics and Single Electronics evolved and subsequently ushered new horizon in the device research.

Low power consuming attributes and its maximized integration made Single Electronics an indispensable element in post CMOS era. This prioritized the transformation of FETs to Single Electron Transistors (SETs) thereby coining a newly modeled nano scaled switching device. It retains the scalability even at atomic state as well as it controls the motion of a single electron or few a number of electrons. Evidently, SET's have matured enough to be considered as a prolific composite of modern ULSI circuits [1]. However, SETs lack in few device criterions such as low gain, background charge problem and extremely low room temperature operation. Thus complete replacement of CMOS using SET is probably not feasible with the available lithography technique. To the contrary, CMOS possesses high operating speed and low static power supply; efficient uses of energy and high degree of noise immunity. Besides, CMOS has vibrant Look Up Table (LUT) to be incorporated for IC designing. Co-integrating both the advantages of CMOS and SET—Researchers in Delft University in Netherlands conceptualized amalgamation of these two technologies only to substitute the predecessor CMOS or the novel SET. Concurrently Prof. S. Mahapatra and his team of co Researchers proposed Mahapatra-Ionescu-Banerjee (MIB) [2] Model by hybridizing CMOS and SET. Later on Sarkar et al., facilitated excellent SET-MOS bidirectional switching device [3].

Reversible logic gates transport information in lossless mode and thus information loss is eventually made to zero in any stage of computation. Computing systems during information processing dissipate heat when voltage discharges from positive to negative. But in reversible engineering this loss is minimized to obtain absolute identical input output correlation.

Undoubtedly the Hybrid CMOS SET is a fascinating but challenging field of research awaiting for some elegant solutions. The authors here modelled Hybrid CMOS SET made complex programmable logic device. The same device is first realized using Reversible Logic [4–7] orientation and of late the same is compared with its conventional counterpart. The proposed reversible logic is optimized in terms of reliability and robustness.

2 Modus Operandi of HRPLA Modeling

To facilitate the randomness, quantum arithmetic is pivotal and should be designed using Reversible Logic which ensures cosmic relevance. The HRPLA consists of reversible AND array designed from reversible gates. This is to done to generate the compliment of input and avoid the problem of fan out. HRPLA is a specific reversible programmable logic array designed in a way to implement complex combinational logic circuits. The proposed model is designed for multi output minimization with few unique min term combinations where many min terms are shared among the output functions. It also supports multilevel implementation. The

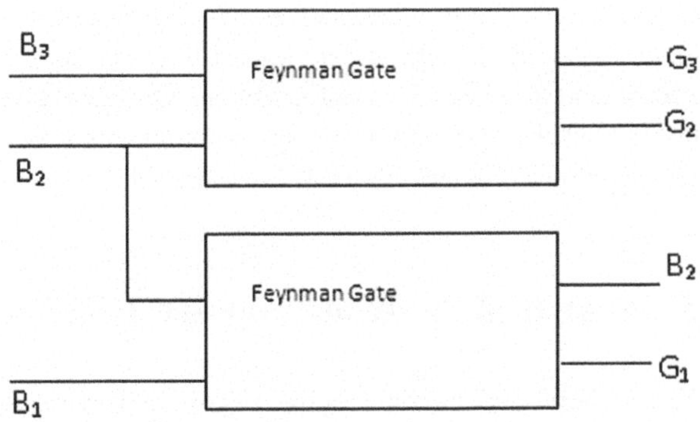

Fig. 1 Block diagram representation of RCPLD using reversible logic

most remarkable feature of the model is the presence of on chip nonvolatile memory which is commonly used for boot loader function before handing to other devices that does not have its own permanent program storage. The block diagram realization is shown in Fig. 1 and the proposed model is depicted in Fig. 2.

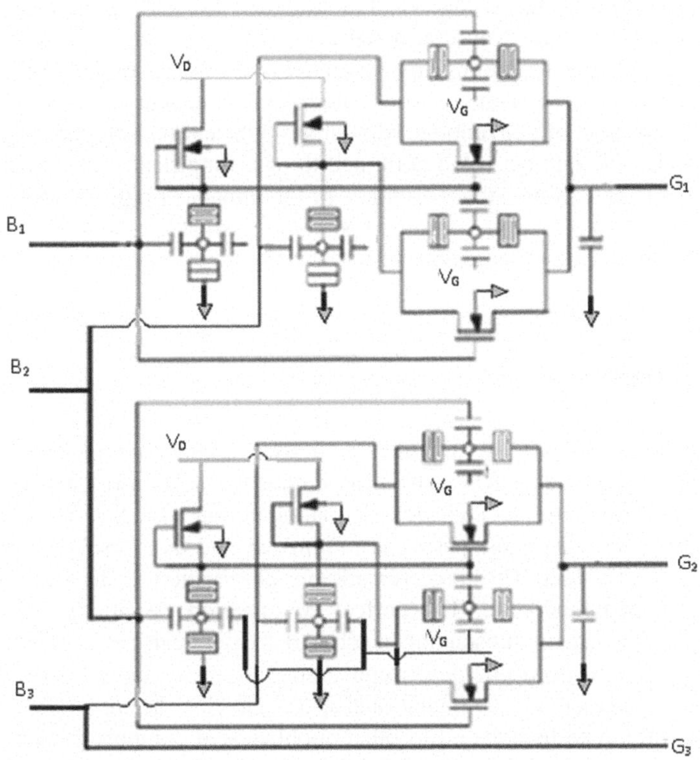

Fig. 2 Proposed model reversible nano device CPLD

The product term generated in each AND gate is listed along the output of the gate. The product term is determined from the inputs whose cross points are connected and marked. The output of an OR gate gives the logic sum of selected product terms. The output may be complemented off late in its true form depending on the connection for one of the XOR gate inputs.

3 Analytical Study of Proposed HRPLA

The authors' insight on Double Feynman Reversible Logic Gate which ensures three inputs and three outputs. The input as shown in Fig. 1 is B_1, B_2 and B_3 and the system output is G_1, G_2 and G_3. The outputs in ephemeral are obtained as $G_1 = B_1 \oplus B_2$; $G_2 = B_2 \oplus B_3$; $G_3 = B_3$. Notably the Quantum cost of this gate lies in proximity to 2.

The proposed HRPLA is straight forward but simple circuit that comprises of four MOS, eight SET and sixteen capacitors. The three inputs B_1, B_2 and B_3 form the left side of circuit where the input voltages are applied in the range of $\approx$0.02 mV. The circuit counts on the given Gate Voltage $\approx$ 0.06 mV and the output is obtained from G_1, G_2 and G_3. For simplicity the garbage value is excluded in Fig. 2. The circuit has been optimized using TSPICE and BSIM 4.0. The proposed model ensures minimum propagation delay using less number of gates with increased efficiency. The elegant part of the research endeavor is that array need not be defined as in case of conventional digital systems and thereby interconnects are diminished and the garbage value is limited to one. Empirical study reveals that as HRPLA is a simply Reversible Complex Programmable Logic Device (RCPLD) and nearly less than 5% of the consumed power is dissipated as leakage power. This intrinsically minimizes propagation delay.

4 Conclusion

The authors here successfully implemented a novel architecture of next generation advanced Nano modeled HRPLA/RCPLD circuit using Hybrid CMOS-SET. The speed is increased to a considerable amount due to the co integration. Consequently, the power dissipation is minimized and as a result the heat dissipation is also very less. The most remarkable observation is the fact that it can operate in room temperature. The perfect combination essentially delivers high speed output. Two main substantial benchmarks have been achieved here in this endeavor—(1) high speed low power consuming arithmetic circuit is designed and (2) it possess extremely fast computational speed. Hereby, the authors conclude that such composition can be adhered to other sophisticated complex circuits.

References

1. K.K. Likharev, "Single-electron devices and their applications" – IEEE Proc. (87), 606–632 (1999).
2. Santanu Mahapatra and Adrian MihaiIonescu "Realization of multiple valued logic and memory by hybrid SETMOS architecture" IEEE transactions on Nano-technology, 4(6) (2005).
3. A. Jana, N. Basanta Singh, J.K. Singh and Subir Kumar Sarkar, "Design and simulation of hybrid CMOS-SET circuits,". Papers 53(4), 592–599 (2013).
4. Feynman, "Quantum mechanical computers," Optics News. Papers 11–20 (1985).
5. E. fredkin, T. Toffoli, "Conservative Logic", International Journal of Theory of Physics. Papers (21), 219–253 (1982).
6. Bhagyalakshmi, H.R, Venkatesha, M.K, "An improved design of a multiplier using reversible logic gates," International Journal of Engineering Science and Technology. Papers (2), 3838–3845 (2010).
7. Tanay Chattopadhyay, "All-optical modified Fredkin gate" Selected Topics in Quantum electronics, IEEE Journal. Papers (99), 1–8 (2012).

Reversible Logic Gates Based on Single Spin Logic

J. Gope, S. Mondal, M. Kundu, S. Chowdhury and S. Bhadra

Abstract Reversible Logic Gates catered substantial anticipations because of its low power consumption and dissipation and furthermore because of its non-volatility in nature. Several attempts have been reported so far to earnestly fabricate reversible logic based synthesis. Meanwhile, spintronics unlike other charge couple devices yield revolutionary features from its very inception such as non-volatility, high speed information processing, low power consuming, highest integration density; they are not easy vulnerable and the power dissipation is merely $1/10^6$ when compared to conventional CMOS topology. This technical documentation in ephemeral enunciates the prospect ability of improvising spin based logical interventions for reversible circuitry.

1 Introduction

Epitaxial beam lithography has augmented Small Scale Integration of CMOS devices in 21st century. The much desired cramping of electrons transistor size has crossed the 60′ nm land mark. The research trend categorically emphasized in shrinking of device sizes, but apathetically device miniaturization is not a 'never ending process'. Apparently, the search for new principle of operation for small scale devices is long pressed. Amid such nanometric devices like Carbon Nano Tubes (CNT), Resonant Tunneling Devices (RTD), Rapid Single Flux Quantum (RSFQ) and Quantum Dots (QD) are the few alternatives that evolved as a pioneering substitute of CMOS [1].

J. Gope (✉) · S. Chowdhury
Department of ECE, CSET, Barasat, Kolkata, West Bengal 7000124, India
e-mail: jayanta.gope.1983@ieee.org

S. Mondal · M. Kundu
Department of EEE, CSET, Barasat, Kolkata, West Bengal 7000124, India

S. Bhadra
Department of EE, UEM, Saltlake, Kolkata, West Bengal 700156, India

© Springer Nature Singapore Pte Ltd. 2017
I. Bhattacharya et al. (eds.), *Advances in Optical Science and Engineering*,
Springer Proceedings in Physics 194, DOI 10.1007/978-981-10-3908-9_77

Until the late 90s, device research chronologically centered upon charge based quantum computing. But the fragility is that such charge based devices catastrophically fall apart owing to its inherent physical limitations. The leakage power factor is omnipresent in every charge coupled devices. It intrinsically fatigue the idea of complete replacement of CMOS using SET and/Hybrid CMOS-SET and so on.

Prof. Supriyo Bandyopadhyay of Virginia Commonwealth University in late 90s conceptualized the spin orientation property of electron and subsequently advocated for spintronics [2]. Soon after Researchers typically adhered in mobilizing the spin phenomena into quantum dot thereby processing information using the spin degree of freedom. Concurrently, Sarkar et al. [3] enunciated the spin properties in logic synthesizing, there by coining the term 'Single Spin Logic (SSL)' that mimics logic synthesis of conventional CMOS topology.

By the time being efforts were on to inhibit the heat dissipation through dropping the fan out. Several new logic appeared alike one reversible logic [4, 5]. The impetus of reversible logic lies in the fact that aptly zero information is lost during processing. Moreover, the number of input and output remains same. Such physiognomies of reversible logic augments the concurrent optical computing, DNA computing, computer graphics designing and secures the data path for DSP research. This technology is more likely to headway nanotechnology as it is indispensable for designing arithmetic circuits. Indeed such aspirations inspired the authors to account on reversible logic based analytical study.

On the other hand the unputdownable journey of SSL has rolled enough mileages such as NAND gates (H. Agarwal et al.), Full Adder (Soumitra Shukla et al.), Multiplexer (T.K. Bhattacharyya) and other few reported significant contributions [6]. Authors here have motivated themselves to incorporate SSL in realizing Reversible Logic Gates only to extract the maximum outcomes of quantum cost. The structures are compared to its conventional counterpart and they lie in proximity to each other and off course they are in good agreement.

2 SSL Orientation of Reversible Logic Gates

The reversible gates currently have several offspring and the authors here intend to realize few of them using SSL. Additionally and for simplicity we have obtained $\overline{A}$ identical to green and likewise $\overline{B}$, $\overline{C}$, $\overline{D}$ is pinpointed using sky blue, yellow and purple respectively.

2.1 Feynman Gate

Figure 1a, b represent quantum diagram and spin realization of a 2 * 2 Feynman gate [7] which is further referred as controlled NOT (C-NOT) or Quantum XOR. It

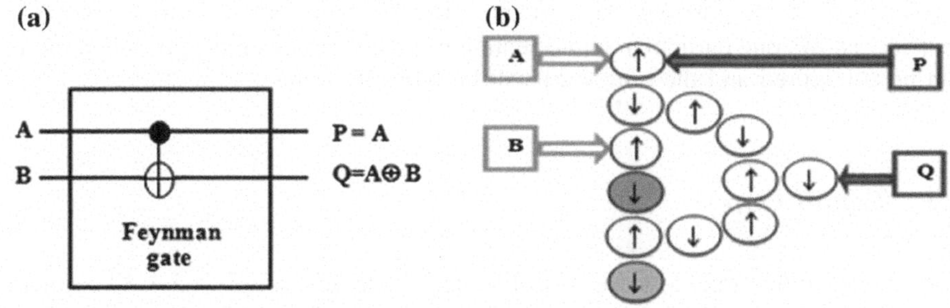

Fig. 1 **a** Block diagram of Feynman Gate. **b** SSL realization of Feynman Gate

is mainly used as a fan-out (Coping Gate). The inputs in green are A, B and the defined outputs are P = A, Q = A ⊕ B. Its Quantum Cost is 1. Here a notable feature is that realizing Feynman with spin attributes zero garbage value.

2.2 *Double Feynman Gate*

Advancing a bit further the Fig. 2a, b depicts the quantum structure and SSL realization of a 3 * 3 Double Feynman Gate [8]. The three inputs are A, B and C respectively and the corresponding corollary outputs are P, Q and R. The outputs P = A, Q = A ⊕ B, R = A ⊕ C are obtained from the spin attribution of Double Feynman Gate or Double C-NOT gate. While spinning this gate emphasis were

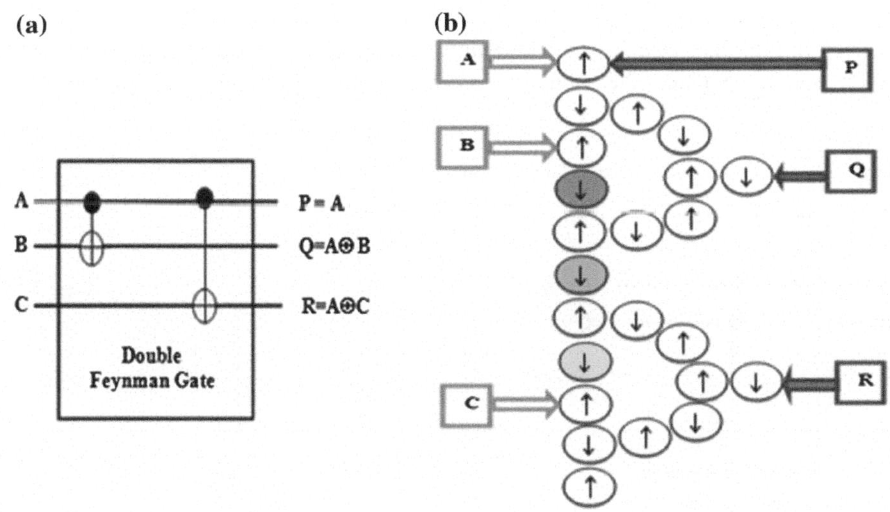

Fig. 2 **a** Block diagram of Double Feynman Gate. **b** SSL realization of Double Feynman Gate

given to reduce the garbage value to zero. The Up Spin Down Spin combination basically shows one particular spin orientation. Likewise any other spin orientations can be configured and the outcomes will be typically same.

2.3 Toffoli Gate

The authors further tend towards Toffoli Gate [9] which is simply a 3 * 3 Feynman Gate. The inputs A, B, C categorically construct the output P = A, Q = B, R = AB ⊕ C. Empirical study reveals that its quantum cost is 5. Figure 3a enunciates the quantum structure of Toffoli Gate and its SSL representation is provided in proximity in Fig. 3b. The SSL orientation reveals zero garbage value for Toffoli Gate compared to conventional Toffoli Gate.

2.4 Fredkin Gates

Diverse to other typical Feynman gates the Fredkin [10] gate although has three inputs i.e., A, B and C but the outputs are typically derived as P = A, $Q = \overline{A}B + AC$, $R = \overline{A}C + AB$. The quantum structure and its SSL realization of a 3 * 3 Fredkin Gate is shown in Fig. 4a, b respectively. Its Quantum Cost is 5.

2.5 Peres Gate

Beneath in Fig. 5a is one advanced quantum structure of Peres Reversible Gate [11] which is a 3 * 3 gate. The inputs are A, B and C respectively and the obtained

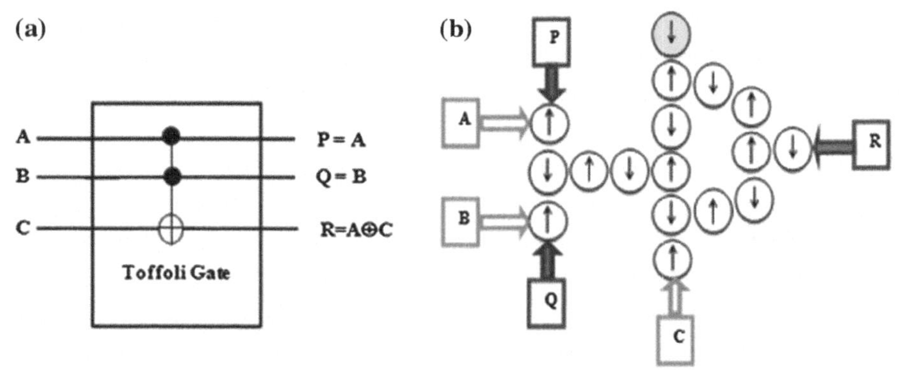

Fig. 3 **a** Block diagram of Toffoli Gate. **b** SSL representation of Toffoli Gate

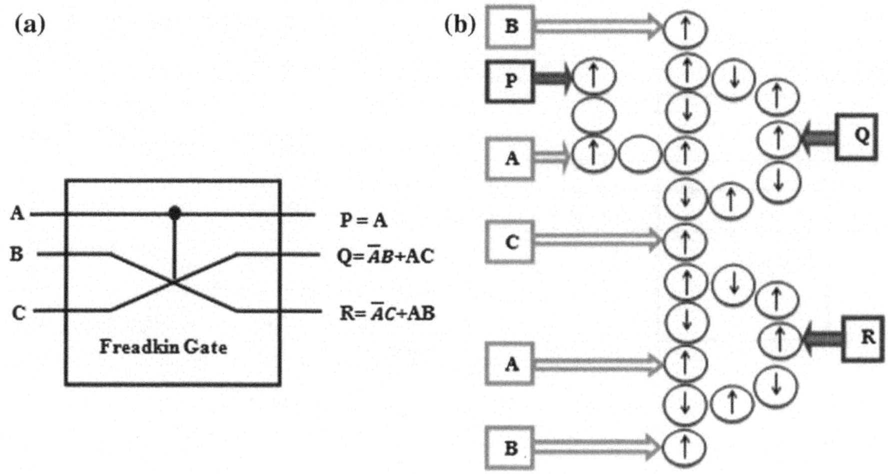

Fig. 4 **a** Block diagram of Fredkin Gate. **b** SSL realization of Fredkin Gate

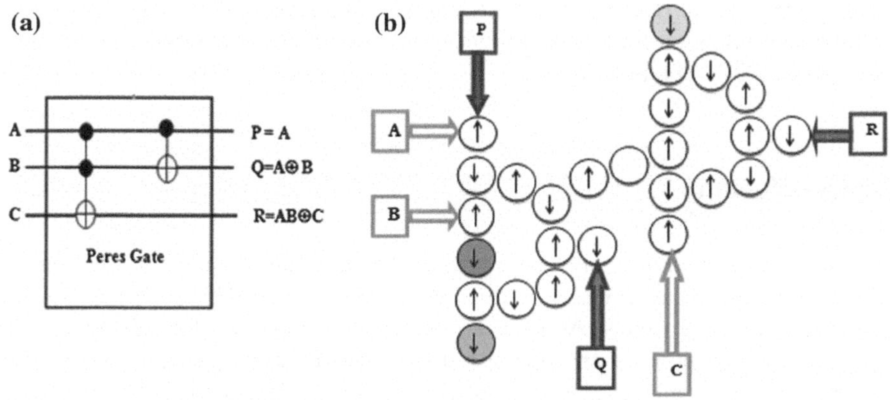

Fig. 5 **a** Block diagram of Peres Gate. **b** SSL realization of Peres Gate

outputs are P = A, Q = A ⊕ B, R = AB ⊕ C. Its Quantum Cost is 4. The SSL realization of Peres Gate is depicted in Fig. 5b. The garbage value when realized using SSL is zero for Peres Gate.

2.6 *BVF Gate*

A 4 * 4 BVF Gate [12] with four inputs A, B, C and D is revealed in Fig. 6a with its quantum structure and the outputs are P = A, Q = A ⊕ B, R = C, S = C ⊕ D

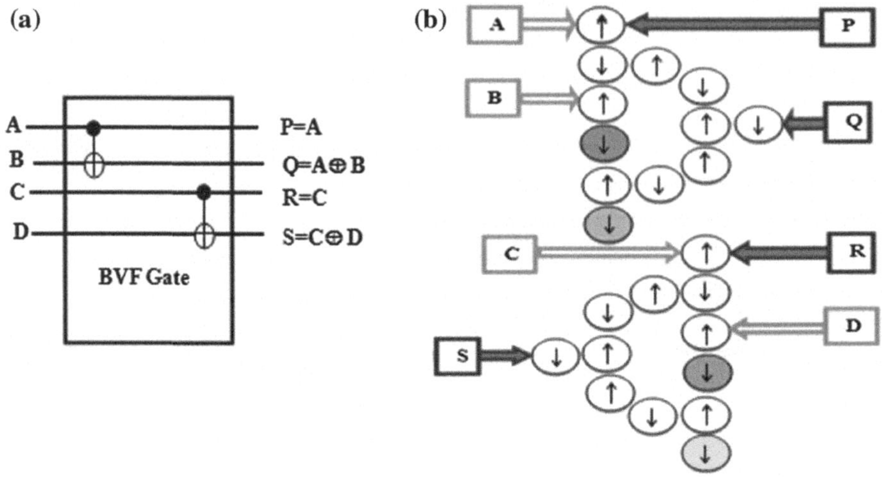

Fig. 6 **a** Block diagram of BVF Gate. **b** SSL realization of BVF Gate

as shown in Fig. 6b. Basically this particular gate is a combination of two Feynman Gates and can be used for fan-out purpose. Its Quantum Cost is limited to 2 only if SSL realization is obtained.

3 Conclusion

The authors profoundly emphasized upon lossless information processing to introduce a new paradigm in communication engineering. This is benevolent in next generation electronics. As heat dissipation increases with fan out, reversible logic can be an optimum solution for advanced encryption of information. But conventional reversible circuits reasonably surmounts the garbage value. This is why SSL realization is adhered in this category of research. Incorporation of SSL in reversible circuitry pragmatically lessened the garbage value. An ephemeral note at the end would be to improvise SSL technique for reversible circuitry as it has no limits of potential entities when compared to its conventional counterpart. The obtained qualities are unique of its kind. Thus the authors truly embrace the framework of SSL based reversible circuit designing.

References

1. Dr. Jayanta Gope et.al., "Hybrid CMOS-SET Decision Making Nano IC: A Case Study, "International Journal of Science, Engineering and Technology Research (IJSETR)". Papers 4 (6), (2015).

2. Marc Cahay, Supriyo Bandyopadhyay, "An electron's spin—Part I", Potentials, IEEE, (2009).
3. Subir Kumar Sarkar et.al., "SPINTRONICS DEVICE BASED POWER EFFICIENT VLSI CHIP DESIGN FOR UNIVERSAL CODE CONVERTER," Canadian Journal of pure & applied science, SENRA Academic Publishers, Burnaby, British Columbia. Papers 2(3), 595–600 (2008).
4. R. Bhagyalakshmi, MK Venkatesha, "An improved design of a Multiplier using Reversible logic gates," International Journal of Engineering Science and Technology. Papers 2(8), 3838–3845 (2010).
5. Prashant. R. Yelekar et al., " Introduction to Reversible Logic Gates & its Application," 2nd National Conference on Information and Communication Technology (NCICT) Proceedings published in International Journal of Computer Applications® (IJCA). Proc. (2011).
6. Dr. Jayanta Gope et.al, "Exploration of Single Spin Logic (SSL) Based Expandable Hardware Design," IJSER. Papers 7(4), 146–148 (2013).
7. H. R. Bhagyalakshmi, MK Venkatesha, "An improved design of a Multiplier using Reversible logic gates," International Journal of Engineering Science and Technology. Papers 2(8), 3838–384 (2010).
8. Md. Selim Al Mamun et al., "Quantum Cost Optimization for Reversible Carry Skip BCD Adder," International Journal of Science and Technology. Papers 1(10), (2012).
9. T. Toffoli, "Reversible Computing||Tech memo," MIT/LCS/TM-151, MIT Lab. (1980).
10. E. Fredkin, T Toffoli, "Conservative Logic," International Journal of Theoretical Physics. Papers 2, 219–253(1982).
11. Nagamani A N et al., "Novel Low Power Comparator Design using Reversible Logic Gates," Indian Journal of Computer Science and Engineering (IJCSE). Papers 2(4), 566–574(2011).
12. Jayashree H V et al., "Modified TOFFOLI GATE and its Applications in Designing Components of Reversible Arithmetic and Logic Unit," International Journal of Advanced Research in Computer Science and Software Engineering. Papers 2(7), 207–210 (2012).

Hypothetical Modeling of Single Spin Logic Based Booth's Multiplier IC

Jayanta Gope, Snigdha Chowdhury (Kolay), Sanjay Bhadra and Shantanu Bhadra

Abstract The longstanding CMOS device exploration is pinpointed on the hypothesis of the flow of electrons. The same ideology is copycatted in post CMOS devices similarly. Contrariwise, Spintronics marked the device academics and research as a typically distinguished occurrence since it solely relies on the spinning of an electron instead on the charge of an electron. This innate virtues of electron was long unheeded. Even in its very sprouting phase Spintronics manifested the notion of small scale device integration. Intrinsically the spin lies in a quantum dot and it is typically informative owing to its spin up and spin down attributes. Unlike any charge coupled devices Spintronics are non-volatile; neither has it necessitated any interconnections nor do any refreshing; thus the power efficiency of spin based devices increases vividly. Chronologically since the last decade several research attempts have been published to integrate the spin properties in Boolean logic. This manuscript precisely explore one such Single Spin Logic (SSL) phenomenon in Booth's Multiplier (BM) circuitry in order to blueprint a high speed, low power consuming, non-volatile exquisite SSL-BM ICs of merely nm dimension.

1 Introduction

Researchers nearsighted the end of CMOS era in late 9 s owed to physical limitations in CMOS. They being pragmatic, dreaded the potential growth of VLSI/ULSI circuits in post CMOS era. The pursuit for next generation technologies ushered

---

J. Gope (✉) · S.C. (Kolay)
Department of ECE, CSET, MAKAUT, Kolkata, West Bengal 7000124, India
e-mail: jayanta.gope.1983@ieee.org

S. Bhadra
Department of EE, University of Engineering and Management,
Kolkata, West Bengal 700160, India

S. Bhadra
Department of Computer Application, RAMT,
MAKAUT, Kolkata, West Bengal 700052, India

© Springer Nature Singapore Pte Ltd. 2017
I. Bhattacharya et al. (eds.), *Advances in Optical Science and Engineering*,
Springer Proceedings in Physics 194, DOI 10.1007/978-981-10-3908-9_78

charge based Quantum Electronics, Single Electronics like promising post CMOS device aspirants. ITRS 2003 [1] also divulged the path of several new topologies afar CMOS to pave the new horizon of technological shift. This technological shift is the propelling potency of today's Nanoelectronics. But the ardent fact is that all charge coupled devices encounters leakage power loss like deep-down limitations.

Further, Prof. S. Bandyopadhyay projected 'Spintronics' that is governed by the 'spin degree of freedom of electrons'. This Spin degree of freedom of electronics has led to the resurgence of new technology called Spintronics. It rendered Researchers to explore SSL circuits in order to design and supersede various CMOS built Boolean circuits. Petty few circuits such as Arithmetic circuits which includes Half Adder [2], Full-Adder [3], Half Subtractor [4], Full Subtractor [5]; Combinational circuits which includes Multiplexer [6], Demultiplexer [7], Decoders [8], Encoders [9], Converters [10] etc. and Sequential circuits such as Flip-Flops [11], Registers [12], Counters [13] are realized by SSL.

Booth multiplier circuit has reserved it's space low power consuming and low dimensional CPLD application. Accordingly, Booth radix sequential multiplier has designed few years back. The circuit reduces the number of partial product to minimize the complexity of the CPLD thereby it has become a common features in modern FPGAs. The circuit follows simple booth multiplication algorithm and takes into account several shift, register, counter and multiplexer.

This manuscript drafts the designing of a complex nm Booth's Multiplier IC using SSL. It is to be considered as an endeavor to enterprise computational nm tools for future logic realizations. This article also elucidate the various circuit components of Booth's Multiplier which are realized by SSL. Last but not least, the article exhibit the comparative study between conventional Booth's Multiplier and SSL oriented Booth's Multiplier with respect to processing speed, device scaling, device integrity, power consumption, power dissipation.

2 Layout of Proposed Hands on Training Tool

Initially each part of the circuit is realized using spin logic and extreme care is taken to articulate anti-ferromagnetic behavior of the spin in the circuit. The proposed circuit follows the minimum energy dissipation from the laws of thermodynamics derived by Landauer-Shannon limit. For simplicity a linear array of three single electron units containing quantum dots are considered. Here the following figure are enlisted as below, Fig. 1. NOT gate; Fig. 2. AND gate; Fig. 3. OR gate; Fig. 4. XOR gate, Fig. 5. Full Adder; Fig. 6. ADD/SUB Unit (4-Bit); Fig. 7. 4-Bit Register; Fig. 8. 2-Bit Counter; Fig. 9. Multiplier, Fig. 10. SSL BM IC. For justification of circuit following multiplier and multiplicand is used.

Fig. 1 NOT gate

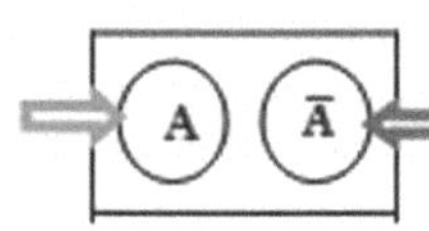

Fig. 2 AND gate

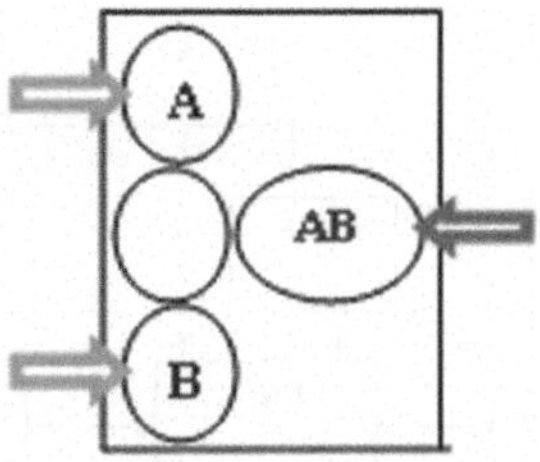

Fig. 3 OR gate

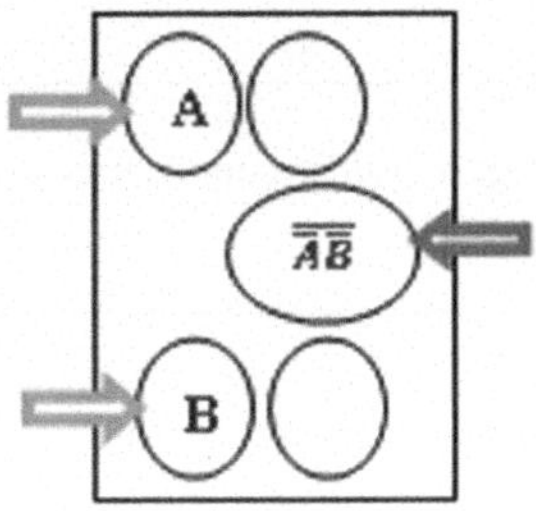

Fig. 4 XOR gate

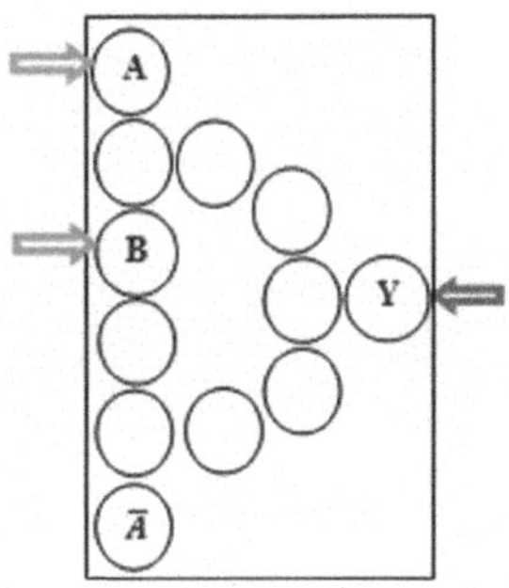

Fig. 5 Full adder

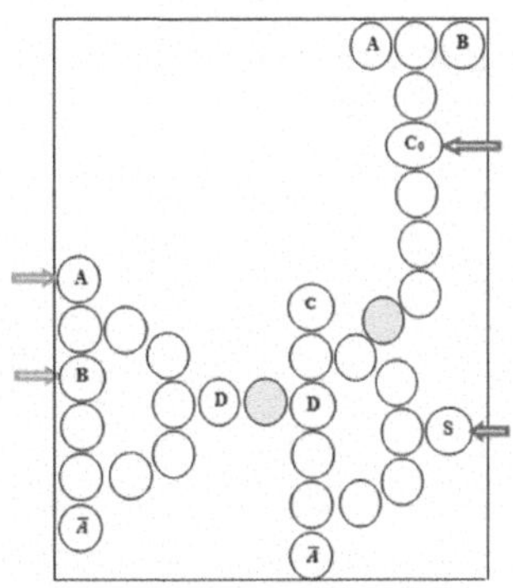

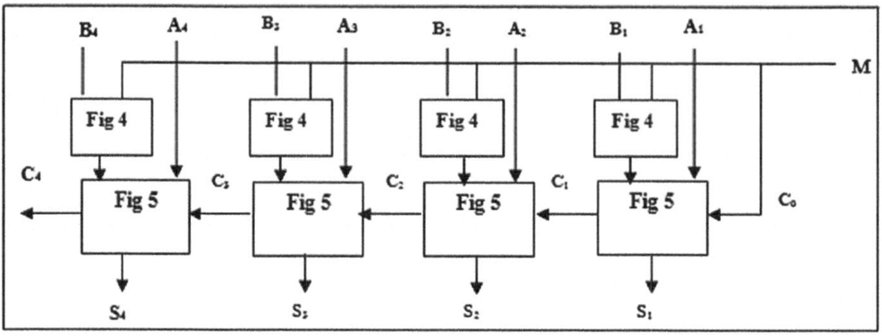

Fig. 6 Adder cum subtractor

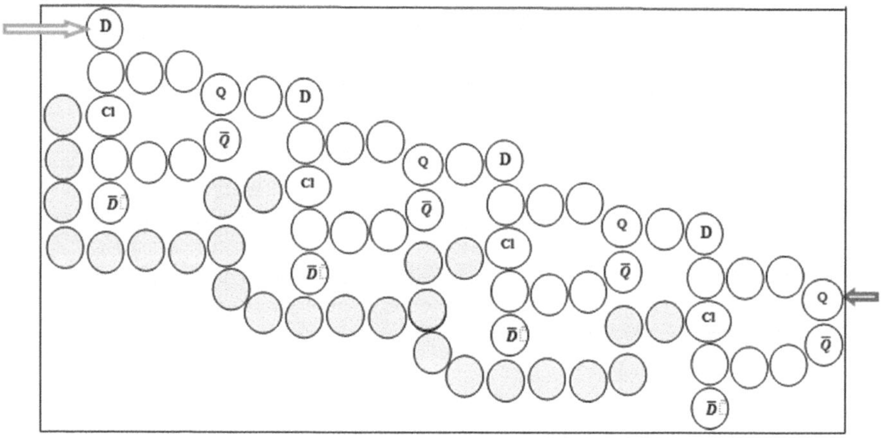

Fig. 7 4 bit register

Fig. 8 2 bit counter

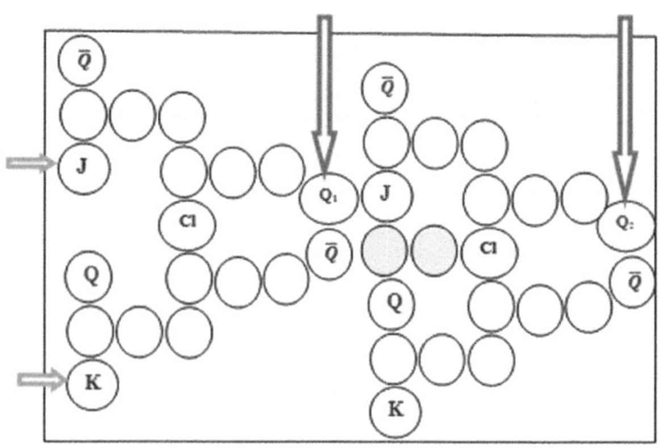

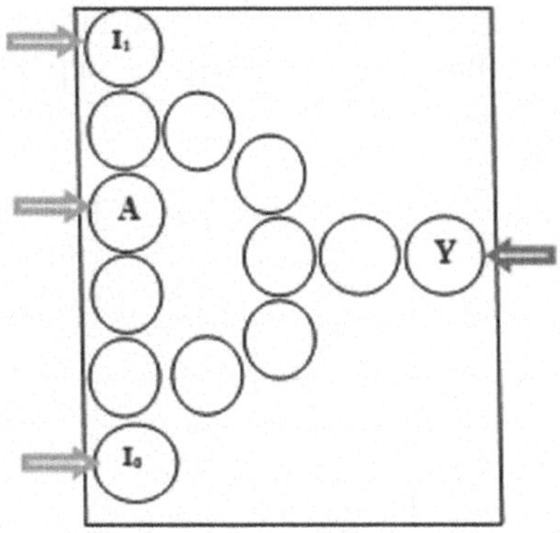

Fig. 9 Multiplexor

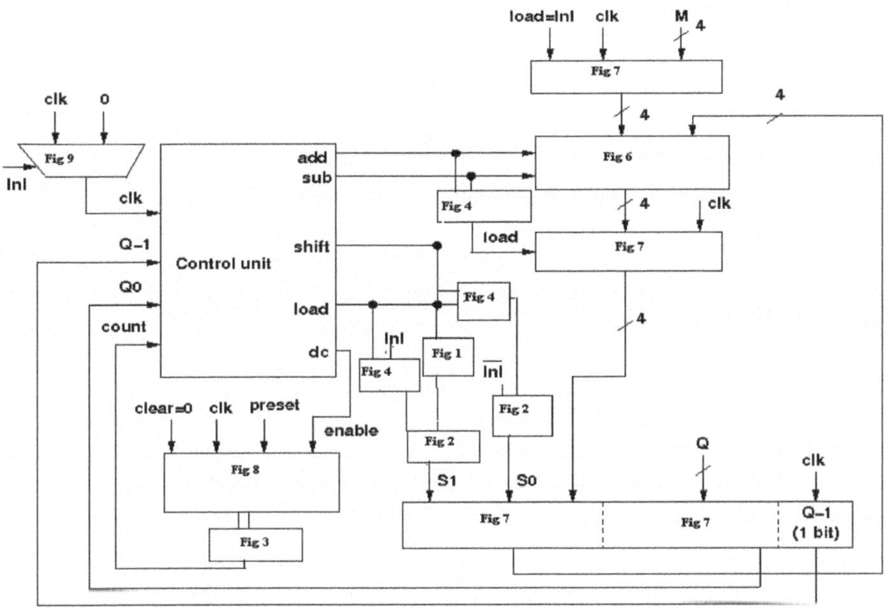

Fig. 10 SSL based Booth's multiplier IC

3 Application of Booth's Multiplier

To illustrate the method of Booth's Multiplier we consider, the multiplier $Q = 12 = (1100)_2$ and the multiplicand $M = 5 = (0101)_2$ F = Flip Flop, A = Accumulator. In Step 1 If Q[0] = 0 then bits will be right shifted and the size of multiplier will be decremented by 1. The Step 2 follows the same sequence as Step 1. In Step3 Q[0] = 1 the content of the accumulator is added with the multiplicand and the right shift is materialized and the size of multiplier is decremented by 1. The Step 4 follows the same step as Step 3.

Initial Configuration	M	F	A	Q	Size
	0101	0	0000	1100	4
Step1:					
As Q[0]=0	0101	0	0000	1100	—
RS (FAQ)	0101	0	0000	0101	
Size--					3
Step2:					
As Q[0]=0	0101	0	0000	0110	—
RS (FAQ)	0101	0	0000	0011	
Size--					2
Step3:					
As Q[0]=1	0101	0	0000	0011	—
A=A+M	0101	0	0101	0011	—
RS(FAQ)	0101	0	0010	1001	
Size--					1
Step4:					
As Q[0]=1	0101	0	0010	1001	1
A=A+M	0101	0	0111	1001	1
RS(FAQ)	0101	0	0011	1100	
Size--					0

Mathematical representation of Booth's Multiplier

4 Conclusion

In this presentation the authors proposition ensures light weight, low cost, high integration density, low power dissipation, non volatile long duration battery SSL based Booth's Multiplier circuit which is compatible for future generation Nano device. All such potentials are to be integrated in future as SSL is anticipated as a promising technology for future embedded system.

References

1. S Bandyopadhyay, B Das, and A E Miller, "Supercomputing with Spin polarized single electrons in a quantum coupled architecture" Nanotechnology, 113–133,5(1994).
2. T.K. Bhattacharya et.al., "Single-Spin implementation of a multiplexor", Elsevier Journal, Physica E 41(2009) 1184–1186.
3. Soumitra Shukla et al, "1-Bit Full Adder implementation using single spin logic paradigm", World Scientific Journal, Spin, 2(2), 10.1142/S2010324712500129.
4. T. Basu, S.K. Sarkar,S. Bandyopadhyay, "Arithmetic logic unit of a computer based on the spin-polarised single electrons", 1(3), 194–199, (2007).

5. Subir Kumar Sarkar et.al., "Spintronics device based power efficient vlsi chip design for universal code converter", Canadian Journal of pure & applied science, SENRA Academic Publishers, Burnaby, British Columbia, 2(3),595– 600,(2008)
6. S. K. Sarkar, T. Basu and S. Bandyopadhyay, "Single Spin Logic Circuits," Phys. of low dim. Struct., (2005).
7. H. Agarwal et al., "Single spin universal Boolean logic gates", IOP Publishing and Deutsche Physikalische Gesellschaft, New Journal of Physics, (10),(2008).
8. Dr. Jayanta Gope et al., "Hybrid CMOS-SET Decision Making Nano IC: A Case Study", International Journal of Science, Engineering and Technology Research (IJSETR), 4(6), 1768–1772, (2015).

Designing Comprehensive Tool for Analytical Modeling of Single Spin Logic

J. Gope, Shantanu Bhadra, Sanjay Bhadra and Koustuv Sarkar

Abstract Exploration of device Research manifested typically in two distinctive arenas such as device fabrication and development of device logic. The latter one counts largely on soft computational analysis for which analytical tools are inevitable. Spintronics, a post CMOS device initiative largely attracted the Researchers owing to its significant attributes. Alike its predecessors, Spintronic fabrication augmented in Research labs keeping Single Spin Logic (SSL) orientation far lagging owing to unavailability of robust soft computational tools for SSL. This has motivated the authors to implement a soft computational tool for SSL in situ with all the spin properties of electron in a quantum dot. This attempt is unique of its kind and continuous relentless effort is made to integrate the spin properties.

1 Introduction

The propagation of 'spin based electronics' has conquered the blockage of charge-coupled logical circuits. From its very inception single spin logic evolved as a driving force in post CMOS era device research. The composite materials of SSL encapsulate quantum dot and several ferromagnetic semiconductor and insulator layer. Paradigm of SSL is logically realized by up and down spin that are encoded in orthogonal spin polarization confined in a quantum dot. SSL although in its embryonic stage renders numerous figure of merits like low power consumption, non-volatility, high integration density, non interconnected (i.e., has no wire) Robustness, low power dissipation, high reliability, etc. [1]. This has motivated the

J. Gope (✉)
Department of ECE, CSET, Barasat, Kolkata, West Bengal 700124, India
e-mail: jayanta.gope.1983@ieee.org

S. Bhadra
Department of Computer Application, RAMT, Kolkata, West Bengal 700052, India

S. Bhadra · K. Sarkar
Department of EE, University of Engineering and Management,
New Town, Kolkata, West Bengal 700160, India

© Springer Nature Singapore Pte Ltd. 2017
I. Bhattacharya et al. (eds.), *Advances in Optical Science and Engineering*,
Springer Proceedings in Physics 194, DOI 10.1007/978-981-10-3908-9_79

Researchers to explore SSL in fundamental digital circuits like NAND, MUX, AND, ALU [2] etc. In order to realize next generation spin-based VLSI/ULSI circuits typical analytical tool for SSL is indispensable. Apathetically, such tools are not omnipresent. Thus there is an ample dearth in this type of soft computation analytical study. In order to substantiate this computational fragility the authors here render a heuristic approach to develop a soft computing tool for SSL. The computational tool is initially subdivided into several segments based on its cosmic relevance. Few of such are reported here mostly through flow diagram, design and interfacing.

2 Modus Operandi of Tools

The authors initialize the simple logic realization using SSL depicted in Fig. 1. Here Up spin = logic 1 = ↑ (Up arrow), Down spin = logic 0 = ↓(Down arrow).

For operational maneuver:

1. Read Spin A, Y where A is the input and Y is the output.
2. If input A has ↑ and if condition is YES then output Y has ↓. On other hand if the condition is NO i.e., A has ↓ then output Y has ↑.
3. The required output Y is subsequently ousted and then after
4. Stop.

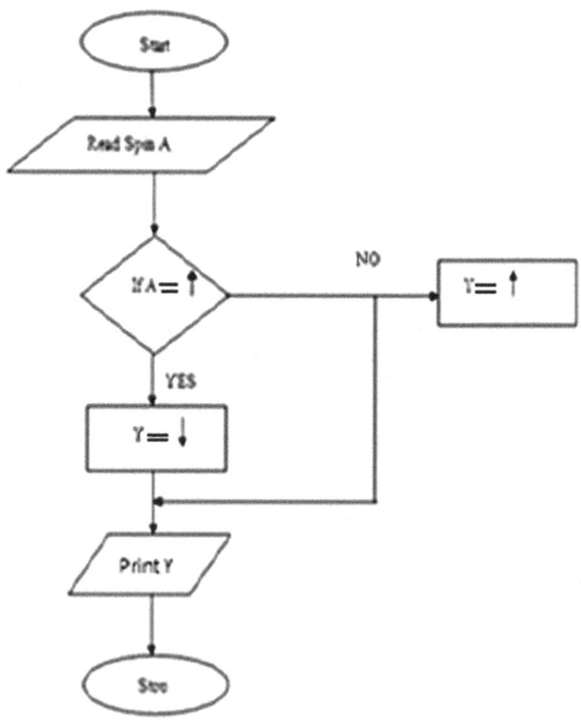

Fig. 1 Flow diagram of NOT gate

3 Operational Maneuver of SSL Simulator

The operational aspect of the simulator is based on a simplified theory of spin quantization. It is a unilaterally consider that each dot comprised of a single size quantized level and there remains on average one electron per quantum dot. The interaction are quite anti-ferromagnetic in nature amide the electrons in adjacent quantum dots. For strong intra dot the tunnel coupling serves as quantum wires and there by the simulator operates without any wiring or in other words no inter-connection required by the simulator. The authors cognitively adhere to intrinsic properties of spin and thus for intra dot Coulomb reaction Hubbard model [3], Heisenberg model [4] and Hamilton model [5] are considered. The Hamiltonian model is based on the Pauli's matrices [6] and is derived as,

$$\begin{aligned}\widehat{H} &= J\sum\nolimits_{(ij)} \hat{\sigma}_{zi}\hat{\sigma}_{zj} + J\sum\nolimits_{(ij)} (\hat{\sigma}_{xi}\hat{\sigma}_{xj} + \hat{\sigma}_{yi}\hat{\sigma}_{yj}) \\ &= J\sum\nolimits_{(ij)} \hat{\sigma}_{zi}\hat{\sigma}_{zj} + 2J\sum\nolimits_{(ij)} (\hat{\sigma}_{+i}\hat{\sigma}_{-j} + \hat{\sigma}_{-i}\hat{\sigma}_{+j})\end{aligned} \tag{1}$$

For covalent systemization, the entire simulator further follows the Hamiltonian interaction given by

$$\widehat{H}_{int} = \sum\nolimits_{(input)} \hat{\sigma}_{zi} h_{zi}^{input} \tag{2}$$

Each and every part of the simulator follows the same hypothesis and typically the basic logic gates as well as complex combined logic gates are realized. Precisely the Eigen values are the controlling factor of the Hamiltonian interaction and eventually this forms the chassis of the simulator (Fig. 2).

The Cognitive Modeling requires extensive adaptation of virtual tools. The tool comprises of window pane and typically offers several Single Spin Logic (SSL) orientation of gates, up spin downspin buttons, clock pulses, buffers, and few simplified models. For revelation of the working of tools Figs. 3, 4, 5, 6 and 7 are included in the documentation here. Few noteworthy physiognomies of the tools are the user can learn the metaphor of the spin logic gates, the waveform of the simulated models, the orientation of up spin downspin and its exploration during modeling, application of buffer for perfect logicalizing and also offers a segment to connect to the developers through web.

4 Pedagogy of Spintronic Simulator

From the developers perspective the authors intends to provide unbounded opportunity to the user through open simulation platform. In this regard orientation assistant (OA) [7] which happens to be a pedagogical web agent is tied up from the very inception of the formation of this said simulator. The problem solving/create

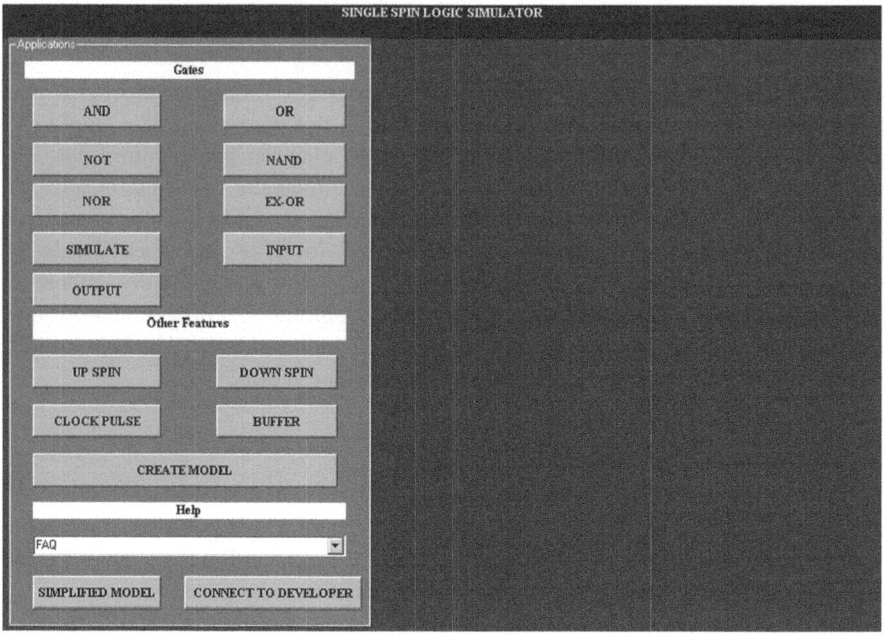

Fig. 2 Designing view of single spin logic simulator

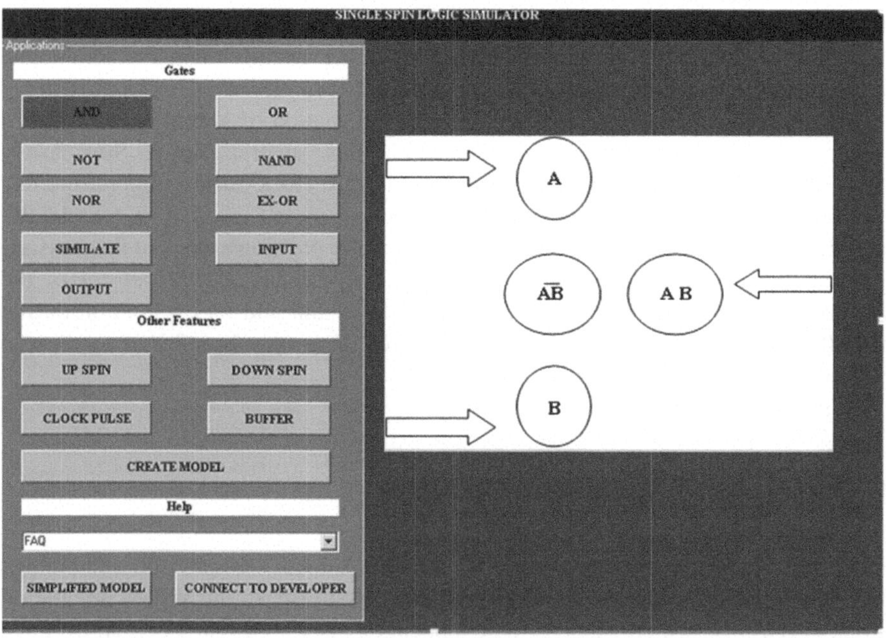

Fig. 3 Circuit diagram of AND gate

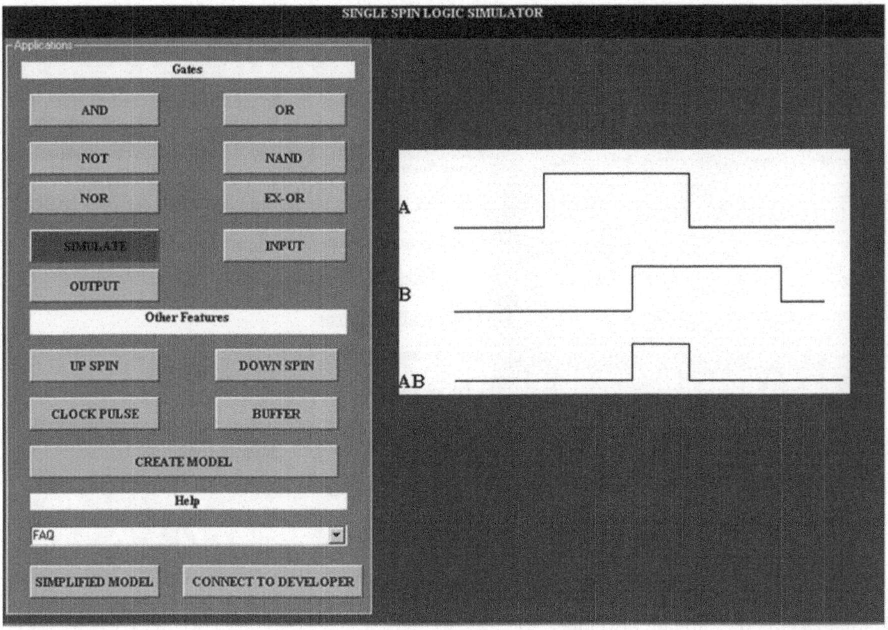

Fig. 4 Simulation of AND gate

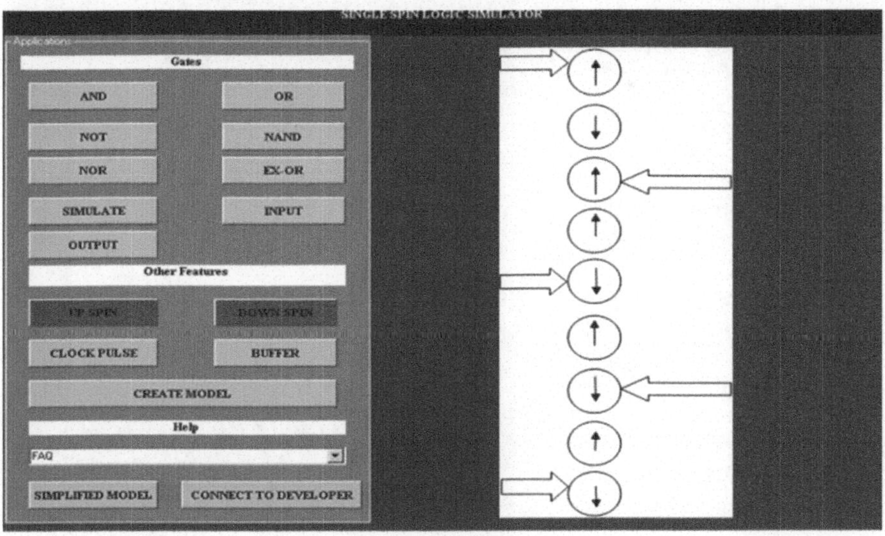

Fig. 5 Designing encoder using up spin and down spin buttons

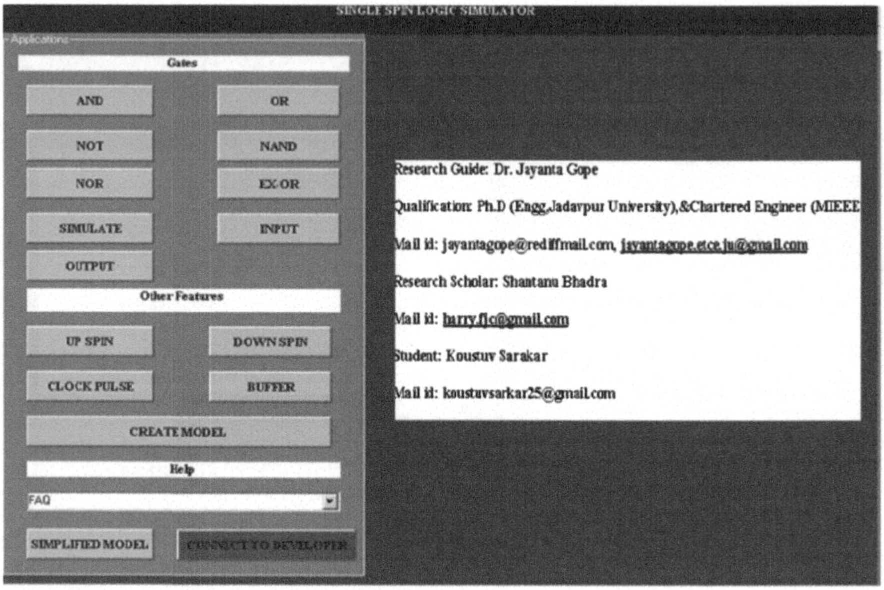

Fig. 6 Details of software developer

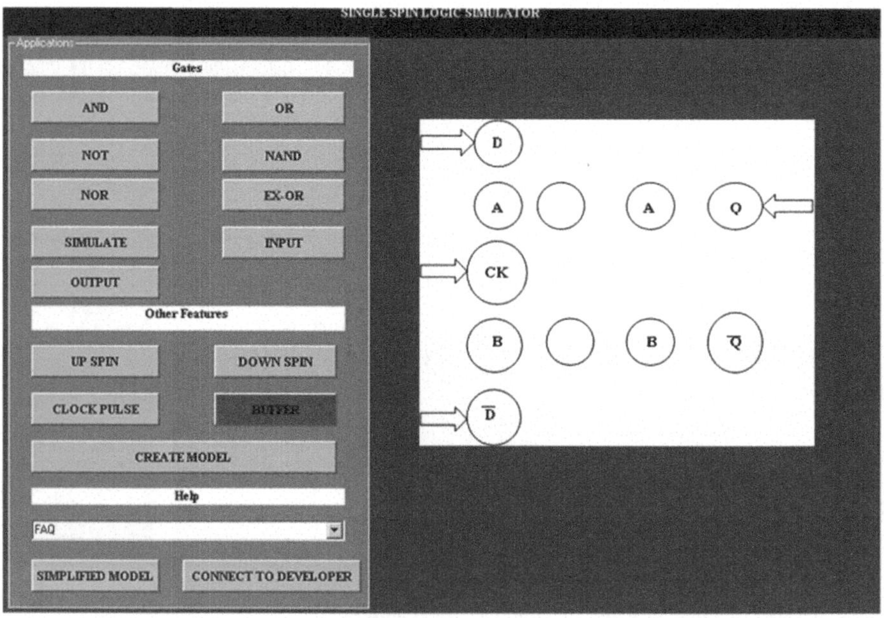

Fig. 7 Designing D flip-flop using buffer buttons

new models is a approximated understanding of the knowledge base of the user and it simultaneously aids in creating future spin based Nano IC's and ASICS. The interactive navigation area is 'mouse click' operated and it follows the Hamilton equation smaller popup window appear on the screen after simulation is executed which is provides possible output consequences in a wave form like structure. Eventually this is a trial and error process which reduces computational timing and diminishes failure rate of the user. Of late the pen paper method is driven into more sophisticated spin logic oriented less time consuming efficient simulation technique.

5 Conclusion

Nano device simulation research adheres to metacognitive understanding of the nano devices. In this regard Spintronics logicalizations doesn't fall apart. The authors here intend to art craft the Hamilton process for spinning the simple logics and there by proposed Single Spin Simulator is unique of its kind. The authors are keen to develop the nano ICs with spin and consequently they put forward this Spintronics simulator. The intrinsic merits of such simulator is that logic realization will be much easier and less time consuming. Furthermore, the simulator is simplistic and user friendly. Thus the authors advocates for the incorporation of the proposed Single Spin Logic simulator in very recent years.

References

1. S Bandyopadhyay, B Das, and A E Miller, "Supercomputing with Spin polarized single electrons in a quantum coupled architecture" Nanotechnology, 5, 113–133 (1994).
2. Dr. J. Gope (MIEEE, CE), S. Mondal, M. Kundu, S. Chowdhury (Kolay). "Spin Realization of Reversible Logic Gate," IJSAR, 7(4), 143–145 (2016).
3. S. Bandyopadhyay, B. Das, and A. E. Miller, "Nanotechnology" 5, 113 (1994).
4. S. Bandyopadhyay, V. P. Roychowdhury, and X. Wang, "Phys. Low-Dim Struct". 28,819, (1995).
5. S. N. Molotkov, S. S. Nazin, "Single-electron computing: Quantum dot logic gates," JETP 83 (4), 794–802 (1996).
6. S. Bandyopadhyay and V. P. Roychowdhury, "Computational Paradigm in Natwelectronics: Single-Electron Logic and Neuromorphic Networks", invited paper presented at SSDM'95, Osaka, Japan, (1995).
7. Lawrence E. Carlson and Jacquelyn F. Sullivan, "Hands-on Engineering: Learning by Doing in the Integrated Teaching and Learning Program", Int. J. Engng Ed. 15(1), 20–31 (1999).

Part XVII
Interdisciplinary: Lasers, Interferometry and Devices

Talbot Interferometry for Focal Length Measurement Using Linear and Circular Gratings

Rahul K. Choudhary, Sunit M. Hazarika and Rajpal S. Sirohi

Abstract Several methods for measurement of focal length of a positive lens involve the measurement of distances of the object and image planes from principal planes, which are mathematical planes; the positions of which are difficult to figure out accurately. The Talbot method for focal length measurement is a method, which does not require the distances from the principal planes. Further Talbot interferometry can be used to measure focal length variation over the surface of a progressive power lens. A technique for the measurement of focal length using Talbot interferometry and moiré effect has been described. The Talbot image of a Ronchi grating placed just after a test lens is superimposed on the second grating to give us the moiré fringes. The angle of tilt of the moiré fringes can be used to find out the focal length. The effect of unequal pitches of the gratings on moiré formation is also investigated. The investigation is also carried out using binary gratings having equidistant circular rings (circular gratings). The theory and experimental results obtained with both types of gratings have been discussed.

1 Introduction

A number of methods have been proposed and used for the measurement of focal length [1]. Combination of Talbot [2] and moiré phenomena [3] provides a method that is suited for the measurement of long focal length of a lens and large radius of curvature of a mirror. There has been large number of publications on this subject [4–12]. Use of both linear and circular gratings is made for the measurement of focal length.

R.K. Choudhary (✉) · S.M. Hazarika · R.S. Sirohi
Department of Physics, Tezpur University, Tezpur, Assam, India
e-mail: rkc.girija@gmail.com

© Springer Nature Singapore Pte Ltd. 2017
I. Bhattacharya et al. (eds.), *Advances in Optical Science and Engineering*,
Springer Proceedings in Physics 194, DOI 10.1007/978-981-10-3908-9_80

2 Theory

When a linear grating is placed in a collimated beam of monochromatic light of wavelength λ, it images itself, due to diffraction, at equidistant planes from the grating: the separation between two consecutive self-image planes is the Talbot distance given by d^2/λ, where d is the pitch of the grating. When the illumination of the grating is by a spherical wave, the Talbot planes are not equidistant. The position of the nth Talbot plane is given by Sirohi [1]

$$z_n^s = \frac{n}{\frac{\lambda}{d^2} \pm \frac{n}{R}} \tag{1}$$

where R is the radius of curvature of the wave at a plane where the linear grating is placed: here + sign stands for convergent spherical wave and − sign stands for divergent spherical wave. For the convergent wave illumination, the separation between consecutive Talbot planes keeps on decreasing while for divergent illumination, it keeps on increasing. Further the pitch at the Talbot plane is such as if the grating is projected, implying that the pitch decrease in convergent illumination and increases in divergent illumination.

Figure 1 shows an experimental set-up for measuring focal length of a lens. There are two situations: in one the grating is kept in contact with the lens, while in the second it is kept at some distance away from the lens.

2.1 *Grating in Contact with Lens*

It is assumed that the principal plane of the lens coincides with the grating. In other words the radius of curvature of the spherical wave at the grating plane is f, where f is the focal length of the lens. The locations of the Talbot planes is given by

$$z_n^s = \frac{n}{\frac{\lambda}{d^2} + \frac{n}{f}} \tag{2}$$

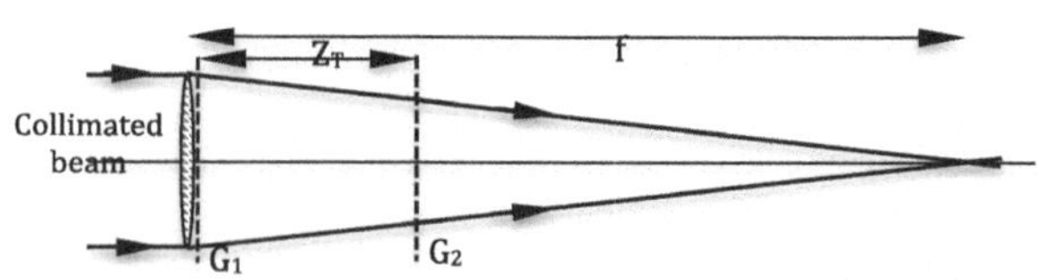

Fig. 1 Experimental arrangement

Further the pitch d′ of the grating at its nth image position, using projection geometry, is given by,

$$d' = d\frac{f - z_n^s}{f} \rightarrow \frac{d'}{d} = 1 - \frac{z_n^s}{f} = \frac{\lambda f}{\lambda f + nd^2} \tag{3}$$

2.2 Grating Is Located a Distance a_0 from the Lens

Defining the radius of curvature of the converging wave at the grating plane as f′ = f − a_0, we can write for the location of Talbot planes and the grating pitch as

$$z_n^s = \frac{n}{\frac{\lambda}{d^2} + \frac{n}{f'}} = \frac{n}{\frac{\lambda}{d^2} + \frac{n}{f - a_0}} \tag{4a}$$

$$d' = d\frac{f' - z_n^s}{f'} = d\frac{f - a_0 - z_n^s}{f - a_0} \tag{4b}$$

2.3 Moiré Fringe Formation

Let the two gratings G_1 and G_2 be defined as

$$y = \cot\theta\, x - \frac{n\, d_1}{\sin\theta} \tag{5a}$$

$$y = -\cot\theta\, x + \frac{m\, d_2}{\sin\theta} \tag{5b}$$

The lines in these gratings make angles of ±θ with the y-axis: alternately the grating vectors make angles of ±θ with the x-axis. Further, d_1 and d_2 are the pitches of the gratings G_1 and G_2 respectively. The indicial equation that leads to the moiré fringe equation is given by

$$n - m = p \tag{6}$$

Eliminating n and m from the grating equations, we obtain

$$y = \frac{d_2 - d_1}{d_2 + d_1}\cot\theta\, x - \frac{d_2 d_1}{d_2 + d_1}\frac{p}{\sin\theta} = \tan\phi\, x - \frac{p\, d_0}{\cos\phi} \tag{7}$$

where ϕ is the angle that a moiré fringe makes with the x-axis and d_0 is the period (pitch) of the moiré fringes. When the pitches of the gratings G_1 and G_2 are equal $(d_2 = d_1 = d)$, the moiré fringe equation is given by

$$y = -\frac{dp}{2\sin\theta} \tag{8}$$

The fringes run parallel to the x-axis and the fringe spacing $d_0 = d/2 \sin \theta$. For the case of unequal pitches, the angle ϕ with the x-axis is given by

$$\tan\phi = \frac{d_2 - d_1}{d_2 + d_1}\cot\theta \tag{9}$$

From this

$$\sin\phi = \frac{(d_2 - d_1)\cos\theta}{\sqrt{(d_2^2 + d_1^2 - 2d_2 d_1 \cos 2\theta)}} \tag{10a}$$

$$\cos\phi = \frac{(d_2 + d_1)\sin\theta}{\sqrt{(d_2^2 + d_1^2 - 2d_2 d_1 \cos 2\theta)}} \tag{10b}$$

The moiré fringe pitch is obtained by equating

$$\frac{d_2 d_1}{d_2 + d_1}\frac{1}{\sin\theta} = \frac{d_0}{\cos\phi} \rightarrow d_0 = \frac{d_2 d_1}{d_2 + d_1}\frac{\cos\phi}{\sin\theta} = \frac{d_2 d_1}{\sqrt{(d_2^2 + d_1^2 - 2d_2 d_1 \cos 2\theta)}} \tag{11}$$

2.4 Talbot Imaging and Moiré Phenomenon

The gratings G_1 and G_2 are identical i.e. having the same pitch d but are oriented at angle of $\pm\theta$ with the vertical. The grating G_1 is illuminated with a collimated beam of wavelength λ and the grating G_2 is placed at the first Talbot plane of grating G_1 resulting in the formation of moiré pattern with fringes running horizontally. Next the test lens is placed in contact with grating G_1 as shown in Fig. 1, and the moiré fringes between the Talbot image of G_1 and the grating G_2 is observed at the Talbot plane. Since the grating G_1 is in the convergent beam, the Talbot plane shifts towards the lens and its pitch changes. This result in the formation of moiré pattern with fringes oriented at an angle ϕ with the horizontal where

$$\tan\phi = \frac{d - d\frac{f - z_n^s}{f}}{d + d\frac{f - z_n^s}{f}}\cot\theta = \frac{z_n^s}{2f - z_n^s}\cot\theta \rightarrow \frac{\cot\theta}{\tan\phi} = \frac{2f}{z_n^s} - 1 \tag{12}$$

Therefore

$$\frac{\cot\theta}{\tan\phi}+1=\frac{2f}{z_n^s}=\frac{2f(\lambda f+nd^2)}{n\,d^2}=\frac{2\lambda f}{n\,d^2}+2\rightarrow f=\left(\frac{\cot\theta}{\tan\phi}-1\right)\frac{n\,d^2}{2\,\lambda} \tag{13}$$

Here it may be noted that the moiré fringes lie in first and third quadrant when G_1 makes an angle θ while G_2 makes an angle of −θ with the vertical. If G_1 makes an angle −θ while G_2 makes an angle of +θ with the vertical, the moiré fringes will lie in second and fourth quadrant.

The expression for focal length in (13) is different than the reported expression [4], which is obtained when one of the gratings G_1 is vertical and the other grating G_2 is inclined at an angle θ with the vertical. The moiré fringes in the collimated beam illumination are not horizontal but are inclined at an angle θ/2 with the horizontal. The moiré pattern makes an angle of ϕ with the horizontal when a lens of focal length f is placed in contact with the grating G_1, where

$$\tan\phi=\frac{1+\frac{nd^2}{\lambda f}-\cos\theta}{\sin\theta} \tag{14}$$

The focal length f is obtained as

$$f=\frac{1}{\cos\theta+\ \tan\phi\ \sin\theta-1}\frac{n\,a^2}{\lambda} \tag{15}$$

If the grating G_1 is vertical and the grating G_2 is inclined at angle θ with the vertical, the moiré fringes lie in first and third quadrant. If grating G_2 is vertical and grating G_1 makes an angle of −θ with the vertical, the moiré fringes lie in second and fourth quadrant.

3 Experimental

3.1 Moiré Between Two Linear Gratings

The gratings G_1 and G_2 have a pitch of 0.356 mm each. These are placed in the collimated beam of wavelength 633 nm and inclined at an angle ±θ with the vertical. The Talbot distance is 20.02 cm. Figure 2a is the photograph of the moiré pattern obtained in collimated illumination. The angle between the gratings is calculated to be 2.125°. We use the method to measure the focal length of two ordinary lenses with nominal focal lengths of 2 and 4 m respectively. The lens is placed in contact with grating G_1 and the grating G_2 is moved to the first Talbot plane and then the second Talbot plane. Figure 2b, c shows the moiré pattern recorded at these planes with a lens of nominal focal length of 2 m. The angles of the moiré fringes at first Talbot and second Talbot planes are measured to be 52° and 68° respectively.

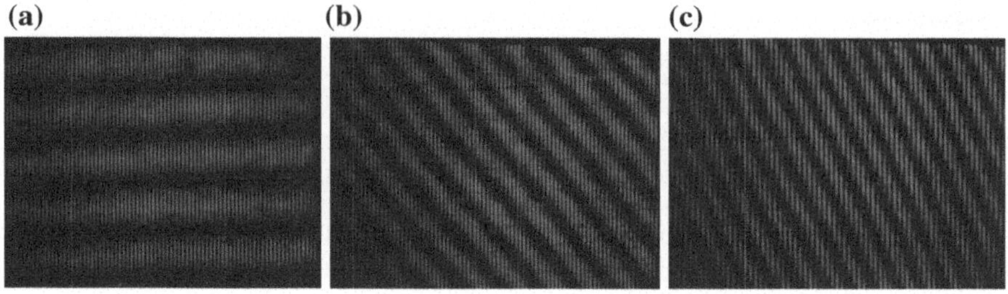

Fig. 2 **a** Moiré fringes at 1st Talbot plane in collimated illumination, Moiré fringes with lens (f ~ 2m) in contact with G_1 **b** at 1st Talbot Plane and **c** at 2nd Talbot Plane

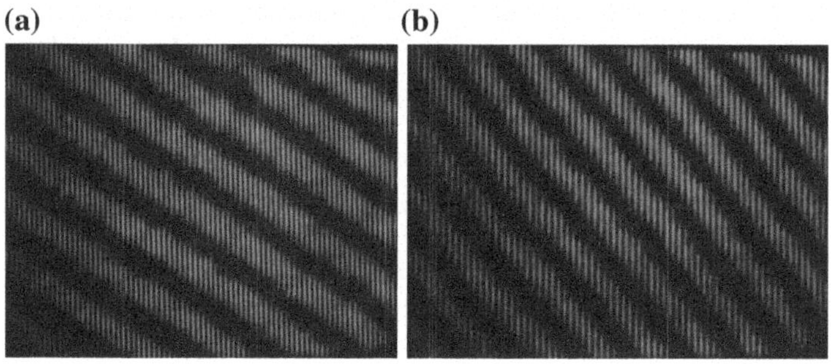

Fig. 3 Moiré fringes with lens (f ~ 4m) in contact with G_1 **a** at 1st Talbot Plane and **b** at 2nd Talbot Plane

The accuracy of the angle measurement is $\pm 1°$. The angle can, however, be measured with much better accuracy approaching $\pm 0.1°$ using the curser. The calculated values of the focal length at 1st and 2nd Talbot planes are focal length values 2.008 and 1.98 m respectively. The uncertainty in focal length measurement depends on the absolute values of the angles and their measurement uncertainties. Assuming that the uncertainty in the calculated value (calculated from fringe width) of θ is $\pm 0.01°$ and that of ϕ is $\pm 1°$, the calculated uncertainties in focal length values could be a few percentages.

Figure 3 shows the moiré fringes at first Talbot and second Talbot planes when a lens of nominal focal length of 4 m is placed in contact with grating G_1. The measured angles of the moiré fringes at these planes are 34° and 52° respectively. The calculated values of the focal lengths are 3.9999 and 4.016 m respectively.

3.2 *Measurement of Variable Focal Length*

Talbot interferometry along with moiré phenomenon provides a possibility to map the power distribution on an optical element. Straight lime moiré fringes are

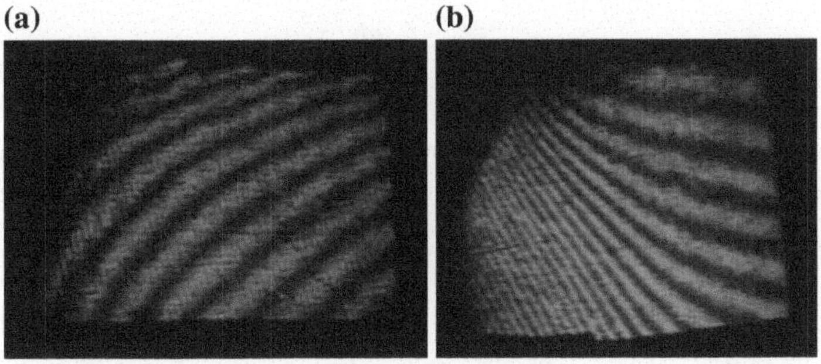

Fig. 4 **a** Moiré pattern of a progressive spectacle lens and **b** Moiré pattern of the same spectacle lens in a different orientation

obtained only when the power is constant over the whole surface under investigation. As is seen from the expression (13) focal length and the angle ϕ are related in a complex way. In order to prove this point, progressive power spectacle lens was placed in contact with grating G_1 and the moiré pattern is observed on a plane that may still be called a Talbot plane. This is due to the fact that the Talbot plane is not defined due to the variable focal length. Nevertheless a moiré pattern can be observed. Figure 4 shows two moiré patterns corresponding to two orientations of the spectacle lens. Notice the curved nature of fringes, signifying the variable power of the lens.

3.3 Moiré Between Two Circular Gratings

The gratings, again represented as G_1 and G_2 are described by

$$x^2 + y^2 = b^2 n^2$$

$$x^2 + y^2 = a^2 m^2$$

where *an* and *bm* are the radii of the circular rings in the two gratings respectively. The indicial equation is again written as n − m = p.

Eliminating n and m from the equations we have

$$\frac{\sqrt{x^2 + y^2}}{b} - \frac{\sqrt{x^2 + y^2}}{a} = p$$

The equation of moiré fringes can be written as

$$x^2+y^2=\left(\frac{ab}{a-b}\right)^2 p^2$$

The radii of moiré fringes are given by $\frac{ab}{a-b}p$. Here as well the gratings G_1 and G_2 are identical, i.e. a = b. When used for measuring focal length, the radii *nb* scale as

$$\frac{nb}{na}=\frac{f-z_s}{f}=1-\frac{z_s}{f}\rightarrow 1-\frac{b}{a}=\frac{z_s}{f}$$

The radii of the pth moiré fringe is given by

$$r^p=\frac{ab}{a-b}p=a\frac{\frac{b}{a}}{1-\frac{b}{a}}p=a\frac{1-\frac{z_s}{f}}{\frac{z_s}{f}}p=a\frac{\frac{\lambda f}{\lambda f+na^2}}{\frac{na^2}{\lambda f+na^2}}p=\frac{\lambda f}{na}p \tag{16}$$

Similarly

$$\mathrm{r}^{\mathrm{p}-1}=\frac{\lambda f}{na}(p-1) \tag{17}$$

Therefore the focal length is obtained by measuring the radii of the circular moiré fringes on subtracting (16) and (17)

$$f=\frac{na\Delta r^p}{\lambda p}$$

It may be noted that $\Delta \mathrm{r}^{\mathrm{p}}$ is the pitch of the moiré fringes. Strangely it is linear relationship. The accuracy therefore depends on how accurately one could determine the radius of the various moiré fringes. Further, it is advised to ignore first few moiré fringes for finding the average pitch.

4 Conclusion

Talbot interferometry in conjunction with moiré phenomenon has been used for the measurement of focal length of a lens and radius of curvature of spherical surfaces. We have worked out a formula for the focal length assuming that the gratings are symmetrically oriented with respect to the vertical. This formula is simple and convenient. We have then used this technique to measure focal length of lenses. It is also shown that circular gratings could also be used for the measurement of focal length albeit poor accuracy. Moreover, a possibility to map the power distribution on an optical element with variable focal length has been shown using a spectacle lens.

References

1. Sirohi, Rajpal S., *Introduction to Optical Metrology*, CRC Press (2015)
2. Talbot, H. F., "Facts relating to optical science", Philos. Mag., No. IV, 9 (1836)
3. Bryngdahl, Olof, "Moiré: Formation and Interpretation, JOSA, 64(10), 1287–1294 (1974)
4. Nakano, Y. and Murata, K., "Talbot interferometry for measuring the focal length of a lens," Appl. Opt. 24, 3162–3166 (1985)
5. Chang, C. W. and Su, D. C., "An improved technique of measuring the focal length of a lens," Opt. Commun. 73, 257–262 (1989)
6. Su, D. C. and Chang, C. W., "A new technique for measuring the effective focal length of a thick or a compound lens," Opt. Commun. 78, 118–122 (1990)
7. Glatt, I. and Kafri, 0., "Determination of the focal length of nonparaxial lenses by moiré deflectometry," Appl. Opt. 26, 2507–2508 (1987)
8. Bernardo, L. M. and Soares, O. D. D., "Evaluation of the focal distance of a lens by Talbot interferometry," Appl. Opt. 27, 296–301 (1988)
9. Sriram, K. V., Kothiyal, M. P. and Sirohi, R. S. "Talbot interferometry in non-collimated illumination for curvature and focal length measurements," Appl. Opt.31, 75–79 (1992)
10. Sriram, K. V., Kothiyal, M. P. and Sirohi, R. S., Use of a non-collimated beam for determining the focal length of a lens by Talbot interferometry, J. Opt. (India), *22*, 61–66 (1993)
11. Hao Chen, Yong He,* Jianxin Li, and Heng Lu, Measurement of long focal lengths with a double-grating interferometer, Appl. Opt. 52(27), 6696–6702 (2013)
12. Jia Luo, Jian Bai,* Jinchun Zhang, Changlun Hou, Kaiwei Wang, and Xiyun Hou, Long focal-length measurement using divergent beam and two gratings of different periods, Optics Express, 22(23), 27921–27931 (2014)

Simulations Studies for Femtosecond Laser Inscribed Bragg Grating Structures on Polymer

Sanyogita, U. Das and P.K. Panigrahi

Abstract The present study reports the simulation results on the reflectance of the uniform polymer Bragg gratings in the visible spectral range for fabrication by direct writing using femtosecond laser. The simulations are based on the coupled-mode theory. Simulations studies have been carried out for a Bragg grating with a grating period of 190 nm and a reflection up to 99.9% with varying bandwidth and grating length. The reflectivities increase rapidly with increasing grating length or/and refractive index modulation in the optical waveguide. However, this also leads to an increase in the side lobes strength and the reflectance bandwidth. For the BG of grating length 1 mm used for tuning purposes along with other gain medium, a reflectance of ~46% for a bandwidth of 1.1809×10^{-10} m is found to be suitable for integrated photonics applications. Potential applications of this study are in field of optofluidic lasers, sensor technology and bio photonics.

1 Introduction

Bragg gratings (BG) in the visible spectral range find widespread use in bio photonics [1], spectroscopy [1], laser applications [1, 2] and sensors. Bragg grating fabrication in the visible range is more challenging as it required grating period in the range of few 100 nm. Femtosecond laser micromachining emerged as new technique to fabricate nano/micro structures on polymer material which gives the ability to fabricate strong and stable Bragg gratings on transparent material including glass or polymer materials with control on grating period over a few 100 nm [3]. The density change brought about by nonlinear multiphoton ionization is a possible explanation for the mechanisms behind induction of refractive index change. Femtosecond laser created phase gratings with refractive index changes of $\Delta n = 1 \times 10^{-4}$ to 5×10^{-4} on polymer material [3]. Focus spot of the laser beam plays an important role during nano fabrication, Specific objective lenses can be

Sanyogita (✉) · U. Das · P.K. Panigrahi
Centre for Lasers and Photonics (CELP), IIT Kanpur, Kanpur, Uttar Pradesh 208016, India
e-mail: sanyogi@iitk.ac.in

© Springer Nature Singapore Pte Ltd. 2017

I. Bhattacharya et al. (eds.), *Advances in Optical Science and Engineering*,
Springer Proceedings in Physics 194, DOI 10.1007/978-981-10-3908-9_81

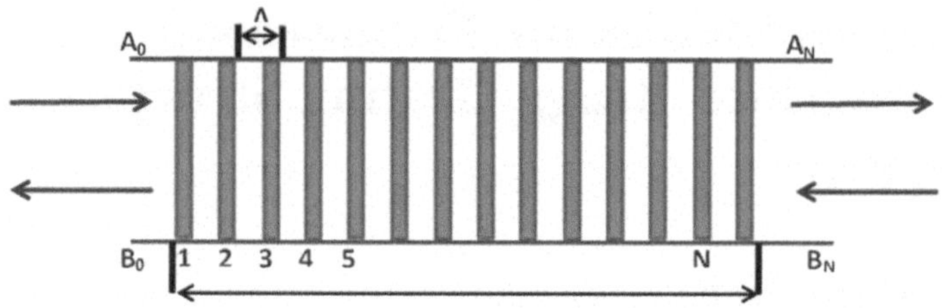

Fig. 1 Uniform Bragg grating

used to reduce diffraction limited nanometer spot size [4]. The feature sizes after the fabrication is mainly depend upon the incident laser power and scan speed of laser focus. At certain laser powers, with specified scan speed, fabrication in the nanometer regime is possible [4]. In this article, modulation in refractive index of polymer material (PMMA) for grating calculations has been considered by femtosecond laser direct writing process [3]. In a simple Bragg grating, the refractive index of the substrate varied periodically along the length of light propagation, the schematic of an uniform Bragg Grating and light propagating is shown in Fig. 1. Here Uniform indicates constant grating period and change in the refractive index for total grating length. Here, the grating period is Λ. The equation relating the grating spatial periodicity and the Bragg resonance wavelength λ_B is given as:

$$\lambda_B = 2n_{eff}\Lambda \quad (1)$$

where n_{eff} is the refractive index of the waveguide in which the grating is formed.

In this paper we discuss simulation results for the reflectance of the uniform polymer Bragg grating in the visible range to obtain grating parameters for its fabrication using femtosecond laser micromachining. This paper is divided into following sections. Section 1 introduces direct writing on polymer (femtosecond fabrication) in the visible spectral range. Section 2 covers the coupled mode theory in order to describe working principle of Bragg gratings. Section 3 deals with the results and discussion about the model and simulation for typical specifications, using MATLAB. Lastly, Sect. 4 described the conclusion of the work done.

2 Couple Mode Theory for Uniform Bragg Grating

Femtosecond laser based micromachining fabrication technique is planned to get a periodic modification of the refractive index along the definite length of optical waveguide in order to fabricate the Bragg grating structure. When light passes though modified region of optical waveguide, it is reflected by coherent scattering due to periodic variation of refractive index. The periodic variation in the refractive index is given as [5–8]:

$$\delta n_{eff}(z) = \overline{\delta n_{eff}}(z)\left\{1 + v\cos\left(\frac{2\pi}{\Lambda}z + \varnothing(z)\right)\right\} \quad (2)$$

where, z is the coordinate of light propagation along the length of BG. $\delta n_{eff}(z)$ is the spatially dc index change over a grating period, v is the fringe visibility of the index change and $\varnothing(z)$ is the grating chirp. Input and output characteristics of Bragg grating can be determined by using couple mode theory and Transfer Matrix Method (TMM) [5, 9]. For a grating with uniform index modulation and period the reflectivity is given by Erdogan [5]

$$R(L,\lambda) = \frac{k^2\sinh^2(\gamma L)}{\Delta\beta^2\sinh^2(\gamma L) + k^2\cosh^2(\gamma L)} \quad (3)$$

R: Grating reflectivity as a function of both grating length and wavelength, L: total length of grating.

Here, the coupling coefficient k(z) is defined by the equation

$$k(z) = \frac{\pi}{\lambda}\cdot\Delta n\cdot g(z)\cdot v \quad (4)$$

In case of uniform Bragg grating Δ_{neff} is constant, g(z) = 1, v being the fringe visibility usually 1.

$\Delta\beta$: wave vector tuning, given by $\Delta\beta = \beta - \pi/\Lambda$.

Waveguide propagation constant β is given by, $\beta = (2\,\pi n_0)/\lambda$, $\gamma = \sqrt{k^2 - \Delta\beta^2}$.

3 Simulation Results and Discussion

We have been done simulation for a first-order grating over the visible wavelength range of 565–567 nm suitable for femtosecond laser fabrication on PMMA. The parameters used for simulation are refractive index of PMMA substrate of 1.49 along with a change in the refractive index (Δn_{eff}) of 1.5×10^{-4}, 2×10^{-4}, 2.5×10^{-4} and 3×10^{-4} for grating periods (Λ) 190 nm. The grating length has been varied from 0.5 to 5 mm. For different values of grating lengths (as shown in Table 1), the reflection spectra has been obtained. From the obtained spectra it is seen as expected that the spectral properties of the uniform grating comes out to be similar to that of a Sinc function. The reflection spectra for different grating lengths 0.5, 1, 1.5 and 2 mm shown in Figs. 2, 3, 4 and 5.

Table 1 Variation in reflectivity for grating length

Grating length (mm)	Reflectivity (%)
0.5	15.49
1	46.47
1.5	71.88
2	86.63
2.5	93.95
3	97.28
3.5	98.82
4	99.49
4.5	99.78
5	99.9

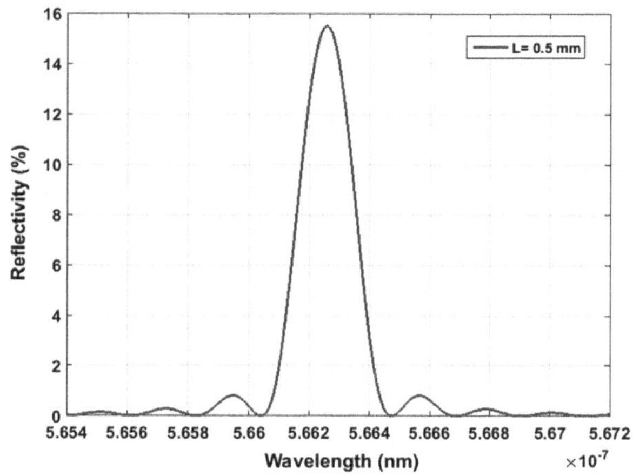

Fig. 2 Reflection spectrum at L = 0.5 mm

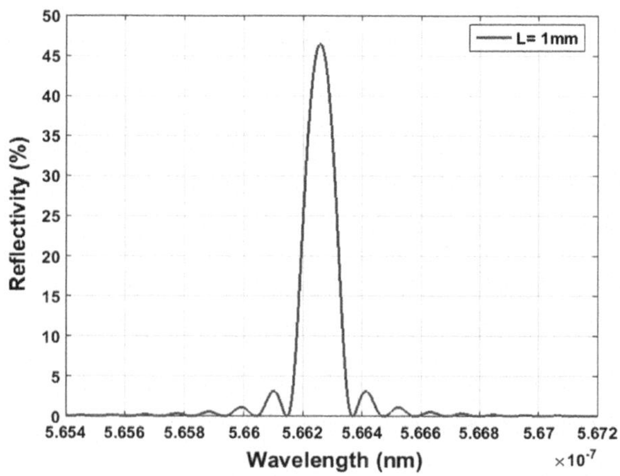

Fig. 3 Reflection spectrum at L = 1 mm

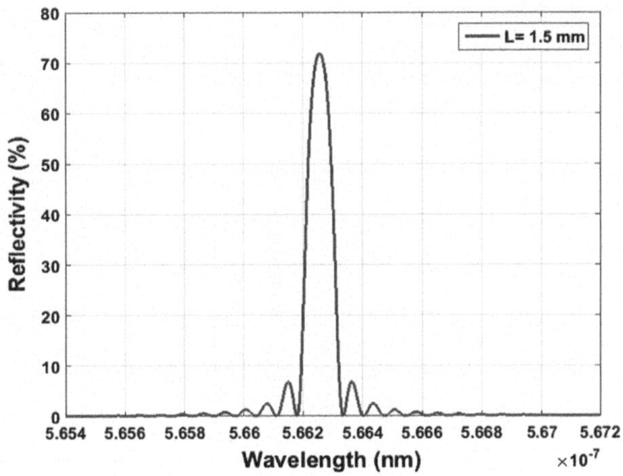

Fig. 4 Reflection spectrum at L = 1.5 mm

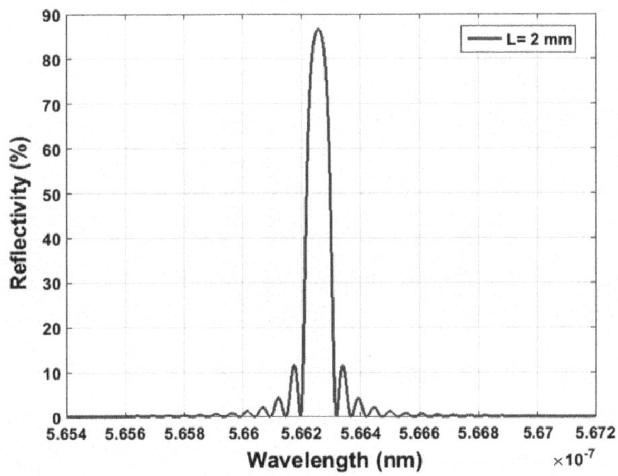

Fig. 5 Reflection spectrum at L = 2 mm

3.1 Dependence of Reflectivity of the BG on Grating Length

Reflectivity as function of grating length and refractive index of the waveguide is discussed for Bragg gratings on PMMA as shown in Fig. 6. From the above figs. it is clear that when the length of grating is increased, bandwidth of grating is decreased, when the refractive index modulation uniform over the grating length. At L = 1, 1.5, 2, and 2.5 mm respectively the maximum reflectivity is 46.47, 71.88, 86.63 and 93.95%. At L = 5 mm, the reflectivity reaches 99.98% but increased in the sides lobes reflectivity.

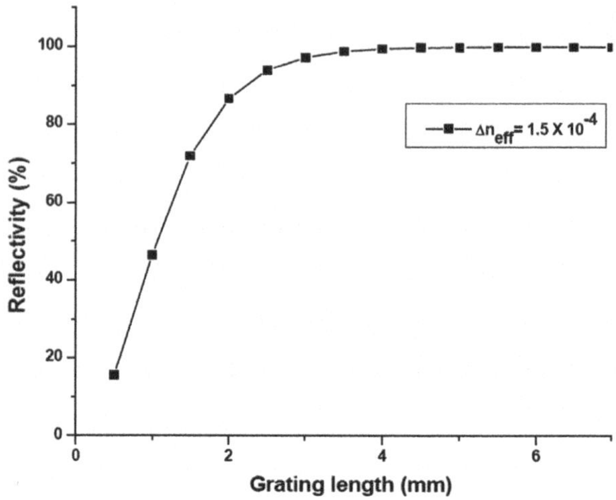

Fig. 6 Reflectivity as a function of grating length

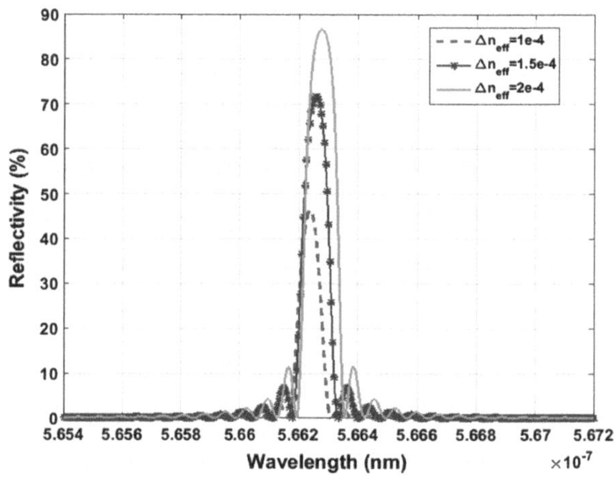

Fig. 7 Dependence of reflectivity and spectral bandwidth refractive index modulation at L = 1.5 mm

3.2 Dependence of Reflectivity and Spectral Bandwidth on Refractive Index Modulation

After that, if the length is further increased for a lossless waveguide, it is observed that the asymptotically reached the maximum reflectivity. Figures 7 and 8 show spectral reflectivity of the grating as a function of the refractive index modulation, the change in the refractive index Δn_{eff} of the polymer material is causes change in the amplitude of the reflectivity of the spectrum. The wavelength of maximum reflectivity is found to be not equal to λ_B. It moves to longer wavelengths with an increase of the refractive index modulation as shown in Figs. 7 and 8. When the length of the grating changes, the reflectivity is found to be increases for higher changes in refractive index modulation as demonstrated in Fig. 9 and Table 2.

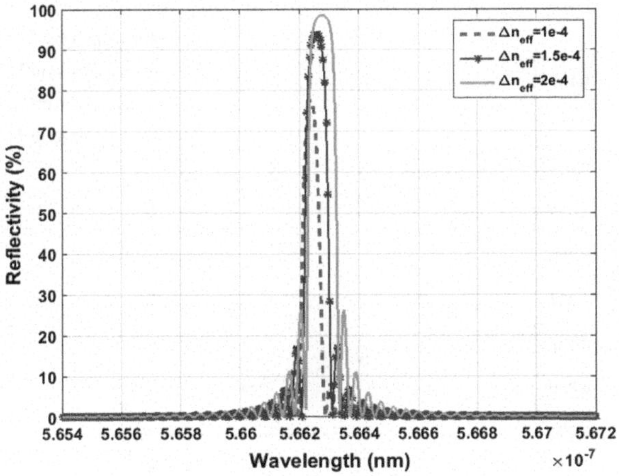

Fig. 8 Dependence of reflectivity and spectral bandwidth on refractive index modulation at L = 2 mm

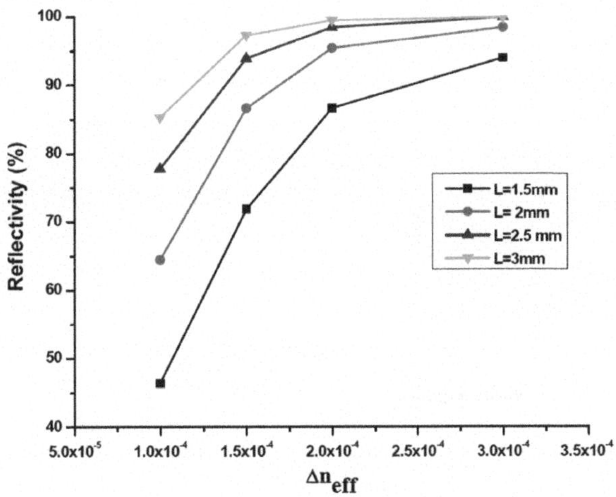

Fig. 9 Relationship between reflectivity versus refractive index modulation

Table 2 Reflectivity of Bragg grating for different length and refractive index modulation

Δn_{eff}	Reflectivity (%)			
	L = 1.5 mm	L = 2 mm	L = 2.5 mm	L = 3 mm
1×10^{-4}	46.41	64.21	77.77	85.29
1.5×10^{-4}	71.83	86.6	93.88	97.28
2×10^{-4}	86.63	95.38	98.45	99.49
2.5×10^{-4}	93.95	98.42	99.9	99.99

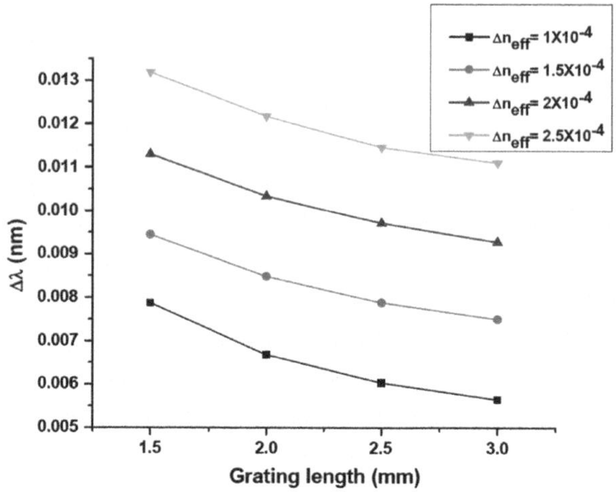

Fig. 10 Effect of grating length on bandwidth

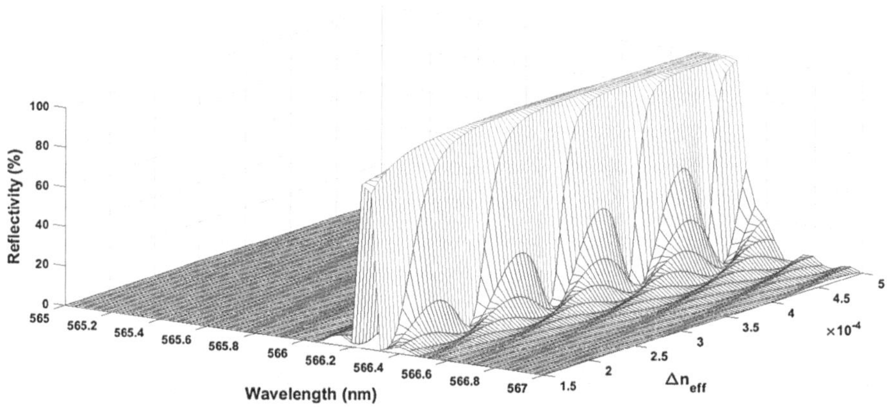

Fig. 11 Calculated reflectivity spectrum over range of refractive index modulation for polymer Bragg grating

When change in the refractive index decreases to a certain point, as shown in Fig. 10, the bandwidth appears to have reached a minimum value and remains constant for further reduction in the refractive index modulation.

The effect of grating length and refractive index modulation on bandwidth is shown in Fig. 10. It shows that the increase of the grating length causes the bandwidth to decrease and an increase of the reflectance at the Bragg wavelength. When the refractive index modulation increases, the reflectance and the bandwidth increase too. Numerical simulation results obtained from Figs. 9, 11 and 12 shows that polymer grating reflectivity and bandwidth are affected by all of grating length, period and refractive index modulation. Polymer grating fabrication by

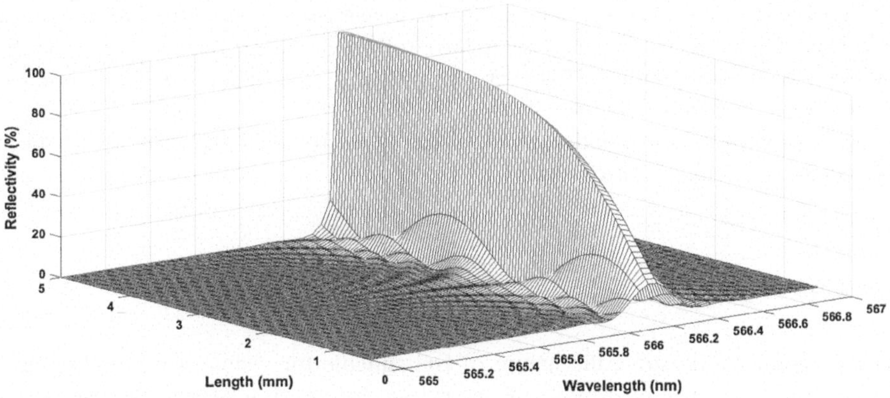

Fig. 12 Calculated reflectivity spectrum over range of grating length for polymer Bragg grating

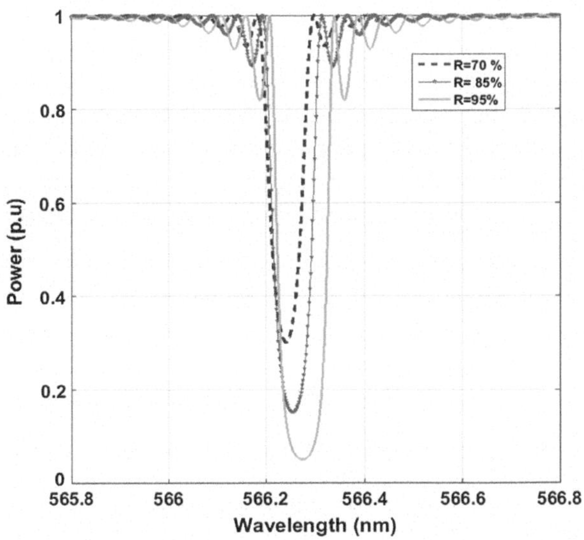

Fig. 13 Simulated transmission spectrum of Bragg grating on PMMA substrate

femtosecond laser radiation is known to increase the PMMA refractive index during interaction of femtosecond light radiation with polymer material [3].

However larger reflectivity results in larger bandwidth and this is a constraint where the BG is used for tuning purposes.

It is seen from this work that even for low bandwidth a reflectivity approximately 46% would be sufficient along with other gain materials for applications in integrated photonics These analyses give better understanding of the relationships of BG characterization and suggest possible ways of designing high performance visible BG for bio photonics applications. Figure 13 shows that any changes in the

BG reflectivity will lead to a change in the transmission spectrum. Theoretically, to prepare a good BG, it is necessary to have reflectivity larger than 50%, in order to produce a strong grating. However, in practical, the reflectivity is not easy to control, due to other factors, such as the laser system, the femto pulse energy, femto laser exposure time and scanning speed of micromachining, which will affect the quality of grating fabrication.

4 Conclusion

In this work, we illustrate the nature of BG suitable for visible wavelengths on a PMMA substrate. Coupled mode theory is used for the simulations. We organized the connotative analyses on the BG reflectivity with changes in grating length and the refractive index modulation (Δn_{eff}). It achieves 99.99% reflection at a grating length of 5 mm at a fixed refractive index modulation of $\Delta n_{eff} = 1.5 \times 10^{-4}$ and assumed to be lossless. In conclusion it is found that for a BG of 190 nm period, because of fs laser/polymer interaction, the reflectivities increase rapidly with increasing grating length or/and refractive index modulation in the optical waveguide. However, this also leads to an increase in the side lobes strength and the reflectance bandwidth. For the BG of grating length 1 mm used for tuning purposes along with other gain medium, a reflectance of $\sim$46% for a bandwidth of 1.1809×10^{-10} m is found to be suitable for integrated photonics applications.

References

1. Gary E. Carver, "High speed multispectral confocal biomedical imaging", J. Biomedical Optics, vol. 19(3), 2014.
2. G. B. Hocker, "Fiber-optic sensing of pressure and temperature," Applied Optics, vol. 18, pp. 1445–1448, 1979.
3. A Baum, "Mechanisms of femtosecond laser induced refractive index modification of poly (methyl methacrylate)", J. Opt. Soc. Am. B, Vol. 27, No. 1, January 2010.
4. D. Tan, "Reduction in feature size of two-photon polymerization using SCR500", Appl. Phys. Lett.90, 071106 (2007).
5. T. Erdogan, "Fiber grating spectra," Journal of Lightwave Technology, vol. 15, pp. 1277–1294, 1997.
6. K. Okamoto, Fundamentals of optical waveguides. Academic press, 2006.
7. Agarwal G.P, "Non Linear Fiber Optics", Academic, New York, 2nd edition, 1995.
8. Ghatak Ajoy, Thyagarajan.K, "Introduction to Fibre Optics", Cambridge University Press, 1999.
9. Ugale Sunita et al. "Fiber Bragg Grating Modeling characterization and optimization with different index profiles", International Journal of Engineering Science and Technology vol. 2(9), 4463–4468, 2010.

Focal Length Measurement Using Modified Bessel's Method

Sunit M. Hazarika and Rajpal S. Sirohi

Abstract The focal length of a lens can be measured by a number of methods. Some methods involve the measurement of object and images distances from the respective principal planes of the lens. The principal planes are mathematical planes and may lie within or outside the lens. It is thus not possible to get the values of these distances accurately. There are, however other methods, which do not require distance measurement from the principal planes. The Bessel method is one such method. We have made a modification to the conventional Bessel method. This paper discusses the focal length measurements using the modified form of the Bessel's method. This new expression is derived from the paraxial equation and involves only the displacement and difference in the magnifications.

1 Introduction

Measurement of focal length or effective focal length is an important parameter of an imaging optical system. A number of methods have been proposed and used since long times. These include (i) auto-collimation method [1], (ii) nodal slide method [2], (iii) magnification method [2], (iv) Talbot interferometry [3, 4], (v) moiré interferometry [4, 5], (vi) reflective Fresnel zone plate [6], (vii) grating diffraction [7], (viii) grating shearing interferometry [8], (ix) multiple beam-shearing interferometry [9], (x) Fourier transform method [10] and (xi) imaging conjugates [11].

Several methods require the distance measurements from the respective principal planes. There are, however, some methods, which do not require distance measurement from the respective principal planes. One such method is the Bessel's method [2, 12]. In this method, a lens is moved along the optical axis between a fixed object and a fixed image planes. For the method to work, the distance L between the object and image planes must be greater than four times the focal

S.M. Hazarika (✉) · R.S. Sirohi
Department of Physics, Tezpur University, Tezpur, Assam, India
e-mail: sunit@gonitsora.com

© Springer Nature Singapore Pte Ltd. 2017
I. Bhattacharya et al. (eds.), *Advances in Optical Science and Engineering*, Springer Proceedings in Physics 194, DOI 10.1007/978-981-10-3908-9_82

length f of the lens being measured. Two positions of the lens are found for which an image is formed on the screen, magnified in one case and reduced in the other. This follows from the possible exchange of image and object positions. If the two lens positions differ by distance d, the focal length f of the lens is given by

$$f = \frac{L^2 - d^2}{4L} \tag{1}$$

Though the distance, d, is obtained as a difference between the two positions, the distance L is not so accurately known. In the modified method the distances of image from the focal point is observed for multiple positions of equal spacing. This can be measured accurately. The successive distances between the object and image and their corresponding magnification is used to find out the focal length. In another modification of the method, the distance between two lens positions and difference between the corresponding magnifications are used to obtain the value of the focal length. Experiments are conducted to validate the theory. The paper discusses both the theory and experimental results.

2 Theory

2.1 *Image Plane Stationary, Object Displaced by Equal Amount, Lens Displaced Accordingly to Get Image*

The focal length of a lens is obtained by measuring object distance p and image distance q using the formula

$$\frac{1}{p} + \frac{1}{q} = \frac{1}{f} \tag{2}$$

The distances p and q are measured from their respective principal planes. In practice, the principal planes may not be accessible. If the distance between the object plane and the image plane is L, (2) can be recast as

$$\frac{1}{p} + \frac{1}{L-p} = \frac{1}{f} \Rightarrow Lf = pL - p^2 \tag{3}$$

Therefore the distance p is obtained as

$$p = \frac{L \pm \sqrt{L^2 - 4Lf}}{2} \tag{4}$$

For the distance p to have real values, the distance L must be equal or greater than four times the focal length of the lens $(L \geq 4f)$. This gives two values of p say

$p_1 = \left(L + \sqrt{L^2 - 4Lf}\right)/2$ and $p_2 = \left(L - \sqrt{L^2 - 4Lf}\right)/2$. The corresponding image distances are $q_1 = \left(L - \sqrt{L^2 - 4Lf}\right)/2$ and $q_2 = \left(L + \sqrt{L^2 - 4Lf}\right)/2$.

The essence of the Bessel method is to measure the magnifications m_1 and m_2 corresponding to two object distances p_1 and p_2 respectively. The magnification m_1 and m_2 are given by

$$m_1 = \frac{L + \sqrt{L^2 - 4Lf}}{L - \sqrt{L^2 - 4Lf}} \tag{5a}$$

and

$$m_2 = \frac{L - \sqrt{L^2 - 4Lf}}{L + \sqrt{L^2 - 4Lf}} \tag{5b}$$

Further it can be shown that the focal length is obtained from the measured magnifications as

$$f = \frac{m_1}{(1 + m_1)^2} L = \frac{m_2}{(1 + m_2)^2} L \tag{6}$$

Here $\frac{m_1}{(1+m_1)^2}$ or $\frac{m_2}{(1+m_2)^2}$ is a scale factor when multiplied by the distance between object and image planes give the focal length of the lens. For the focal length f to be measured accurately, the distance L must be known accurately. However, if we could modify the procedure so that it involves measurement of distance as the difference between the positions, the focal length can be measured more accurately. The distance, which is the difference between the two positions, can be measured with an accuracy determined by the scale on the optical bench. Let us then consider that the separation between the object and image plane is L. If the magnifications measured are m_1 and m_2, then

$$f = \frac{m_1}{(1 + m_1)^2} L = \frac{m_2}{(1 + m_2)^2} L \tag{6}$$

In order to use this equation, we cast it as

$$L = \frac{(1 + m_1)^2}{m_1} f \tag{7}$$

We will now vary the distance L by equal distances say ΔL_n and measure the corresponding magnifications. Thus we have

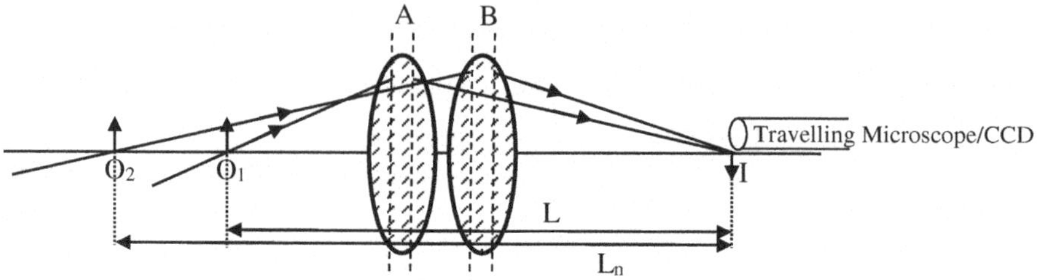

Fig. 1 Image formation by a positive lens

$$L + \Delta L_n = \frac{(1 + m_{1n})^2}{m_{1n}} f \tag{8}$$

Here L is the initial distance, which is also taken greater than or equal to 4f. This distance need not be known but should be greater than or equal to 4f. The magnification m_{1n} is measured for each ΔL_n and a graph between ΔL_n and $(1 + m_{1n})^2/m_{1n}$ is drawn. From the slope of this plot we can obtain the focal length of the lens. It may be noted that ΔL_n is known with the precision of the scale on the optical bench.

Figure 1 shows the experimental method. An object in the form of a mesh either created on a glass slide or a wire mesh is illuminated and its image formed by the test lens is examined. Let the object be at the plane that intersects the optical axis at O_1 and its image is formed at the plane that intersects the optical axis at I. The travelling microscope is focused to receive this image and the magnification m_1 is obtained. The lens is at position A and the distance between the object and the image plane is L. The distance between the object plane and the image plane is adjusted to be equal to or greater than 4 times the focal length. Now the object plane is shifted to position O_2 by ΔL_{1n} distance and the lens is adjusted to position B, such that the object is imaged on the same plane, I, and the magnification m_{1n} is obtained. The object plane is again shifted by ΔL_n, i.e. by the same amount and the lens is adjusted so that its image is formed at the plane I, and the magnification is obtained. This procedure is carried out for several object positions. Finally a graph between ΔL_{1n} and $(1 + m_{1n})^2/m_{1n}$ is drawn and its slope is determined. The focal length is obtained from the slope.

2.2 Modified Method

If we examine (6), it is to be noted that the focal length is given in terms of the measured magnifications and the distance between the object and image planes L. Since the length L is not accurately known, the measured focal length will have error. However, if the focal length is expressed in terms of the distance d, which is the difference between two lens positions and hence is measured to an accuracy given by the scale on the optical bench, the measured value of the focal length will be more accurate.

Using the imaging condition, we can obtain the following equations

$$1 + m_1 = \frac{L + d}{2f} \tag{9a}$$

$$1 + m_2 = \frac{L - d}{2f} \tag{9b}$$

where d is the distance between the two lens positions which give the magnifications m_1 and m_2 respectively. Subtracting (9b) from (9a), we obtain

$$m_1 - m_2 = \frac{d}{f} \tag{10}$$

Alternately the focal length f is obtained as

$$f = \frac{d}{m_1 - m_2} \tag{11}$$

This is a very simple formula. Since the difference in magnifications appears in the denominator, magnifications are to be measured accurately in order to get focal length accurately.

The experimental setup is organized on an optical bench such that the distance between the object plane and the image plane is more than four times the nominal focal length of the lens under test. The magnification of the object, a Ronchi grating or a mesh of known pitch or spacing is measured with lens near the object. This magnification is m_1. The lens is moved towards the image plane to obtain the second imaging position and the magnification m_2 is obtained. The distance between the two lens positions is d. Therefore all the three measured values, needed to calculate the focal length, are available.

2.3 *Special Case*

It is to be noted that magnified image of the object is formed when the lens is closer to the object and diminished image is formed when the lens is closer to the image plane. Under the restricted situation, when the distance between the object plane and the image plane is equal to four times the focal length (L = 4f), only one imaging condition is possible and the corresponding magnifications are equal ($m_1 = m_2 = 1$). When the distance between the object plane and the image plane is greater than four times the focal length, the sum of the magnifications will be greater than two ($m_1 + m_2 > 2$). For some separations, the sum of the magnifications will be an integer number. Let us take this number as N. We can now find an interesting relationship between the focal length, f, and the distance, L. Again assuming the distance, L, being greater than 4f, the magnifications m_1 and m_2 are expressed as

$$m_1 = \frac{L+d}{L-d} \tag{12}$$

$$m_2 = \frac{L-d}{L+d} \tag{13}$$

Since we would like the sum of magnifications as an integer, i.e. ($m_1 + m_2 = N$), we have

$$m_1 + m_2 = N \Rightarrow L^2(N-2) = d^2(N+2) \tag{14}$$

This gives

$$\frac{d}{L} = \sqrt{\frac{N-2}{N+2}} \tag{15}$$

Obviously for d to be positive, $N \geq 2$. At $N = 2$, we obtain only one imaging condition as the distance L is exactly 4f. Substituting for d^2 from (1) into (14), we obtain

$$\frac{f}{L} = \frac{1}{N+2} \tag{16}$$

This is quite an interesting result. This means that the distance between the object plane and the image plane is exactly four times if the sum of the magnifications is two and it is five times if the sum of magnifications is three. Obviously this method requires continuous iteration of distance L to be meaningful experimentally. Nevertheless the relationship is quite interesting.

2.4 *Magnification Method*

This method also is based on the measurement of difference between two conjugate positions and measurement of corresponding magnifications. Figure 2 shows the schematic of the imaging and also the methodology for making measurements.

An object in the form of a fine mesh or a line grating is first placed at position O_1 and its image formed at location I_1 is measured with a travelling microscope (in this case) or a CCD sensor. The ratio of the length of rectangle in the image mesh to that in the object mesh gives the magnification M_1. The object is now moved to the new

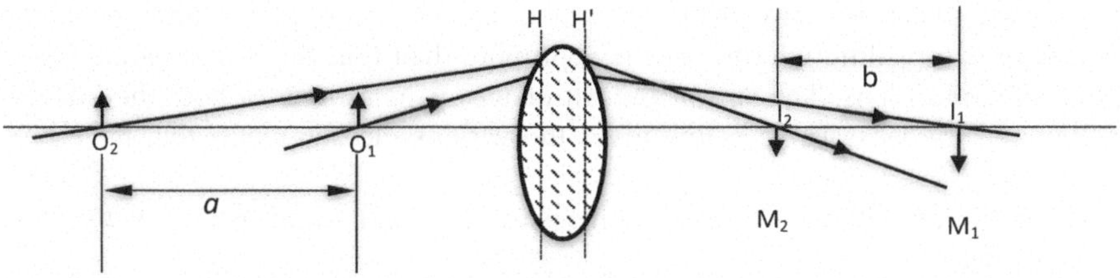

Fig. 2 Focal length measurement by magnification method: lens is stationary

position O_2 and its image is obtained by sliding the screen by b to its new position: the lens remains stationary. Essentially the travelling microscope/CCD sensor is moved by b when it receives the image of object at O_2. The length of the mesh rectangle in this image is measured. The ratio of this measurement to that of the object earlier measured gives the magnification M_2. The focal length is obtained from the measured M_1, M_2 and b values. This is shown using the imaging conditions

$$1 + M_1 = \frac{q}{f'} \tag{17}$$

and

$$1 + M_2 = \frac{q - b}{f'} \tag{18}$$

From the equations we obtain

$$f' = \frac{b}{M_1 - M_2} \tag{19}$$

A set of data of b's versus corresponding differences in magnifications can be generated and plotted. The slope gives the value of the focal length.

This requires three independent measurements. Further, the magnifications appear in difference in the expression for focal length, these needs to be measured very accurately.

3 Experimental

An achromate of nominal focal length of 10 cm is used for the measurement. Slit of width 0.102 mm is an object mounted on the rider. A traveling microscope is used to measure the width of the slit. A test lens is mounted on a rider. Object and lens can be moved on the optical bench, which has a scale graduated in millimeters.

Initially the distance between the slit plane and its image plane (also the object plane for the travelling microscope) is kept more than four times the nominal focal length of the test lens. The size of the slit images is measured for both the imaging positions of the lens. Table below gives the various measurements that have been made.

The data thus obtained is used to measure the focal length using all the three methods, namely

(i) $f = \frac{L^2 - d^2}{4L}$: the conventional Bessel method
(ii) $f = \frac{m_1}{(1+m_1)^2} L = \frac{m_2}{(1+m_2)^2} L$: the magnification method
(iii) $f = \frac{d}{(m_1 - m_2)}$: modified method

The focal length is obtained as a slope from L versus $\frac{(1+m_1)^2}{m_1}$ plot in method (ii) and d versus $(m_1 - m_2)$ plot in method (iii). In the table, m_1 is the magnification when the lens is closer to the object.

Table for all three methods

Object-slit: 0.102 mm width							
L (cm)	d (cm)	m_1	m_2	$f = \frac{L^2-d^2}{4L}$	$f = \frac{m_1}{(1+m_1)^2} L$	$f = \frac{m_2}{(1+m_2)^2} L$	$f = \frac{d}{(m_1-m_2)}$
42.5	10.5	1.454	0.690	9.98	10.26	10.27	13.73
43.5	12.6	1.709	0.582	9.96	10.13	10.11	11.18
44.5	14.2	1.873	0.542	9.99	10.10	10.15	10.68
45.5	16.0	1.997	0.507	9.97	10.12	10.15	10.74
				f_{av} = 9.97 cm	f_{av} = 10.15 cm	f_{av} = 10.17 cm	
Slope (cm)					= 10.16	= 10.16	= 10.52

Figure 3a, b are the plots of L versus $\frac{(1+m_1)^2}{m_1}$, and d versus $(m_1 - m_2)$. The straight line fit is forced to pass through the origin. It is seen that the conventional Bessel method gives consistent values of the focal length. It may also be noted that the focal length values remain consistent even though the length L may differ by a

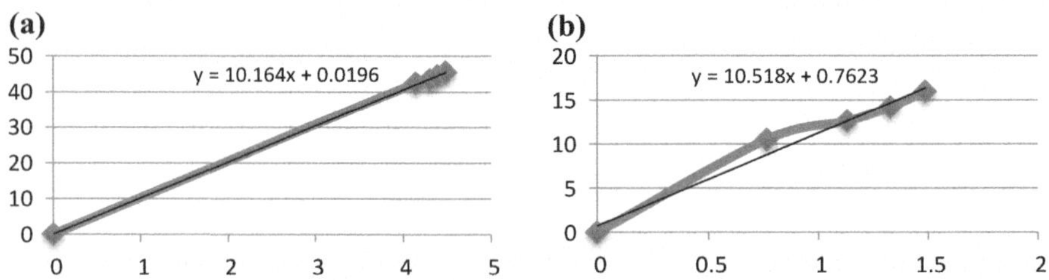

Fig. 3 **a** L versus $\frac{(1+m_1)^2}{m_1}$ plot giving a slope of 10.16 cm, **b** d versus $(m_1 - m_2)$ plot giving a slope of 10.52 cm

centimeter from the correct L value. Further the focal length obtained from this method is a paraxial focal length that assumes the lens to be thin. The magnification methods, however, give the focal length as measured from the principal plane. The average value f_{av} obtained with the magnification method is close to the nominal value. However the modified method, due to the difference in magnification appearing in the expression does not give consistent values. When the value of d is small, the two magnifications are nearly equal and hence any small error in the measurement gets amplified in the focal length value. If the magnification could be obtained with higher accuracy, the focal length measured by this method will be closer to the true value of the focal length. However, the slope obtained by a linear fit gives a value closer to the nominal value.

The main conclusion of this research is that the conventional method is quite good. However, if the magnification values are accurately obtained, then the simple formula $f = \frac{d}{(m_1 - m_2)}$ should give good results.

References

1. Ilev, I. K., "Simple fiber-optic auto-collimation method for determining the focal length of optical elements," Optics Letters, 20(6), 527–529 (1995).
2. Meyer-Arendt, J. R., [Introduction to Classical and Modern Optics] (2nd Edition), New Jersey, Prentice Hall, Inc., Englewood Cliffs, (1984).
3. Sriram, K. V., Kothiyal, M. P. and Sirohi, R. S., "Talbot interferometry in non-collimated illumination for curvature and focal length measurements," Applied Optics, 31(1), 75–79 (1992).
4. Nakano, Y. and Murata, K., "Measurements of phase objects using the Talbot effect and Moiré techniques," Applied Optics, 23(14), 2296–2299 (1984).
5. Keren, E., Kreske, K. and Kafri, O., "Universal method for determining the focal length of optical systems by moiré deflectometry," Applied Optics, 27(8), 1383–1385 (1988).
6. DeBoo, B. and Sasian, J., "Precise focal-length measurement technique with a reflective Fresnel-zone hologram," Applied Optics, 42(19), 3903–3909 (2003).
7. Sirohi, R. S., Kumar, H. and Jain, N. K., "Focal length measurement using diffraction at a grating," Proc. SPIE1332, 50–55 (1990).
8. Lei, F. and Dang, L. K., "Measuring the focal length of optical systems by grating shearing interferometry," Applied Optics, 33(28), 6603–6608 (1994).
9. Matsuda, K., Barnes, T. H., Oreb, B. F. and Sheppard, C. J. R., "Focal-length measurement by multiple-beam shearing interferometry," Applied Optics, 38(16), 3542–3546 (1999).
10. de Angelis, M., "A new approach to high accuracy measurement of the focal lengths of lenses using a digital Fourier transform," Opt. Commun., 136(5–6), 370–374 (1997).
11. Liao, Lin-Yao, Bráulio de Albuquerque, F. C., Parks, R. E. and Sasian, José M., "Precision focal-length measurement using imaging conjugates," Opt. Eng., 51(11), 113604 (2012).
12. LD Physics leaflets, "Optics: Geometrical Optics *Laws of Imaging*", P5.1.2.3.

Automatic Vehicle Detection and Motion Path Tracking Based on Gaussian Mixture Model

Kyamelia Roy, Soham Saha, Tanmoy Mondal
and Sheli Sinha Choudhury

Abstract Automated traffic surveillance plays a vital role to build up a smart transport and communication system of any modern city. This paper deals with three primary aspects of any electronic traffic control system viz., identifying moving vehicle intruding a specified traffic zone, its count and tracking of the motion path. Gaussian mixture model (GMM) is used to detect the foreground object followed by blob analysis which in turn gives the moving vehicle identification. Finally a new concept of weighted mean is deployed for motion path tracking.

1 Introduction

The lifeline of any city is its road network and so monitoring this vital infrastructural resource is very important. Traffic jams and congestion are regular problems faced by most cities of the world. For supervision of traffic, both real-time and historical data about the traffic conditions on the road should be maintained. Hence, in such a scenario, a system with monitoring and reporting ability regarding the condition of roads and estimation of traffic on different road segments is very useful. There are certain operations, whose implementation paves the path for a well maintained and robust traffic network. The operation management includes tasking like vehicle counting, average speed detection, parking violations at 'No Parking' zone and distinguishing vehicles on distinct road sections (e.g. bus lane). The quest

---

K. Roy (✉)
Department of ECE, University of Engineering & Management,
Kolkata, West Bengal 700160, India
e-mail: kyamelia_rain@yahoo.co.in

S. Saha · T. Mondal
Department of CSE, Kalyani Government Engineering College,
Kalyani, West Bengal 741235, India

S.S. Choudhury
Department of ETCE, Jadavpur University, Kolkata, West Bengal 700032, India

© Springer Nature Singapore Pte Ltd. 2017
I. Bhattacharya et al. (eds.), *Advances in Optical Science and Engineering*,
Springer Proceedings in Physics 194, DOI 10.1007/978-981-10-3908-9_83

for more traffic information has led to the increase in the dependence on traffic surveillance.

Coifman et al. [1] proposed a feature based tracking system for detecting vehicles when there is occlusion under various lighting conditions. The system supports the detection of vehicles by tracking its features, instead of tracking the complete vehicle. However, the drawback with feature based detection is the accurate and reliable location of features which is an important task to be carried out. The feature matching stage has well known correspondence problem of ambiguous potential matches occurring. Betkeet et al. [2] proposed a technique in which recognition of the cars is done by matching the templates, cropped from the input data online. Detection of cars is also performed by temporal differencing and tracking motion parameters of cars. During night hours on city expressways when lights are present in the background, distinguishing vehicles on the road from obstacles is a tricky task in this method. The system could find only the cars that are present directly in front of the car assisted with the camera. In [3] the authors proposed a method where the contribution for detection and tracking of vehicles is deployed by Wald Boost (WB) detector [4] and the TLD tracker [5]. The method poses a challenge with high percentage of trucks, occasional density of traffic and high speed of some vehicles. Sochor [6] presented a traffic surveillance system which is able to detect and simultaneously count the number of cars. The idea is enriched with the motion detection method and Kalman filter, used for tracking the same. The detection is basically based on the background subtraction method. The speed in detecting moving objects makes the user attracted to opt for this process. Unfortunately the global thresholding for inter frame differencing disclose itself as being percepture to the phenomenon of the basic hypothesis of background subtraction.

The novel work presented here counts the number of vehicles that passes through the surveillance area and marks the track of their motion. As the first step, foreground detection is done using Gaussian mixture model by pointing the change of state of the corresponding pixels from one frame to another. Foreground detection is followed by foreground filtering with morphological structural element. Finally detection of the moving vehicles is focused with the tracing of their motion paths, estimated for the first time by the authors using weighted mean of the moving vectors inside the blobs.

2 Methodology

Detection of foreground object by using Gaussian Mixture Model (GMM) [7, 8] is done, which performs background subtraction. The foreground detection, as the first step, is applied on the frames of an input video. In the next step, a structural morphological object is declared and is applied on the video frames. Following the preceding step, blob analysis is performed and the track of the number of moving vehicles in the current frame is kept and marking is done with a rectangle in green

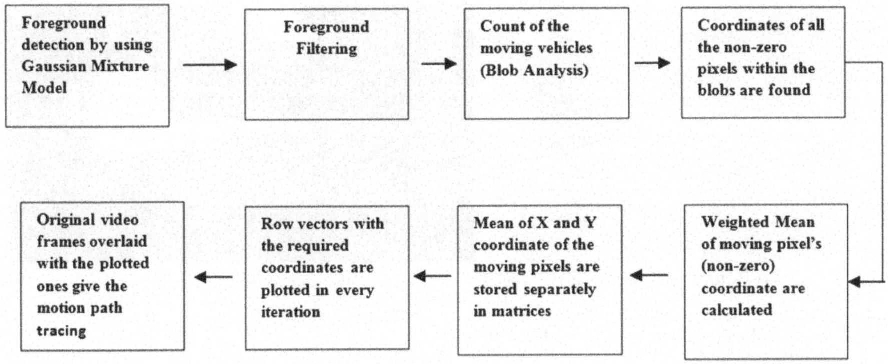

Fig. 1 Block diagram of the proposed work

color. The coordinates of the pixels falling inside the green coloured rectangle are considered and the weighted mean of these pixel's coordinates is calculated. The mean of the coordinate of the moving pixels for each foreground moving object is thus obtained. Since blob analysis gives us the foreground object so, the object is more likely to move with stationary background in consideration. Thus with the moving object the weighted mean also changes in each frame otherwise the mean remains constant thereby indicating the foreground object as stationary. Each of the mean coordinates is stored separately in different matrices and is augmented at every iteration. It is noteworthy that each row corresponds to a moving object individually. The resulting plots help to track the motion of the entire moving foreground object as is shown, when the plot has been superimposed on the original video. Figure 1 depicts the typical block diagram of the methodology followed for the proposed work.

2.1 Foreground Detection Using Gaussian Mixture Model

Segmentation of the foreground object (moving object) from the background frame image is very essential in the image series. The foreground detection in this case is done by using the foreground detector system object, which compares a gray scale or color video frame with a background model. The individual pixels are then determined, whether they are part of the background or the foreground and effectively a foreground mask is computed. The GMM is a mixture of K Gaussian distributions which points to the change of state of corresponding pixels from one frame to another. Each pixel in the GMM is characterized by its intensity in the RGB color space and is formulated as:

(a) Before applying GMM

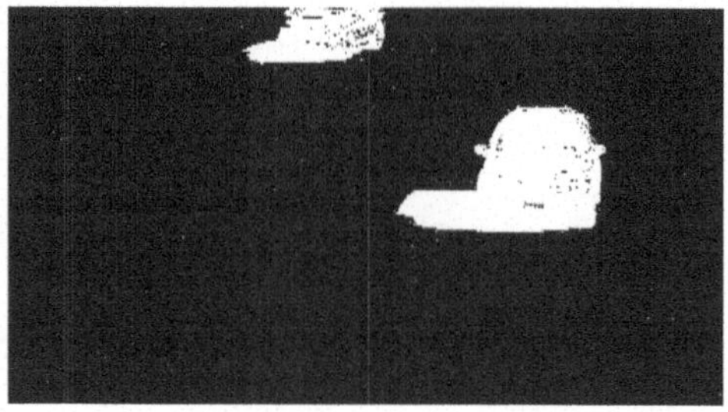

(b) After applying GMM

Fig. 2 Application of GMM on a video frame

$$P(X_t) = \sum_{i=1}^{K} \omega_{i,t}\eta(X_t; \mu_{i,t}, \sum_{i,t}) \quad (1)$$

where, K is the number of distribution; $\omega_{i,t}$ is the weight associated with the i-th Gaussian at time t with mean $\mu_{i,t}$ and standard deviation is $\Sigma_{i,t}$; η is a Gaussian probability density function.

The application of GMM changes the colored image to a binary image. The pixels in the binary image that do not undergo any state of change, is attributed as the value 0 (black), correspondingly the pixels which undergoes change is attributed as 1 (white). Thus, from a video frame the moving objects, which are labeled as foreground object, can be segmented out and exemplified as white. In Fig. 2, the result of the GMM on a video frame is shown distinctively. The pixels which are stationary have turned black (0) and the pixels which have undergone change have attribute 1 (white), clearly stating the foreground object. In the present framework cars have been detected as foreground object.

2.2 *Foreground Filtering Using Structural Morphological Element*

Image processing is contributed by a wide range of operators from mathematical morphology, all based around a few simple mathematical concepts from set theory. The operators are particularly useful for the analysis of binary images including edge detection, noise removal, image enhancement and image segmentation. A morphological filter may be used to selectively remove image structures or objects while preserving the other ones. In the present frame as shown in Fig. 4 foreground is filtered using masking. The masking is accomplished with the structural elements. The structural elements are specified by a shape and in this case, a square structuring element is used as shown in Fig. 3. As is seen from the figure that after the Gaussian Mixture Model is applied the image needs to be filtered i.e. the foreground needs to

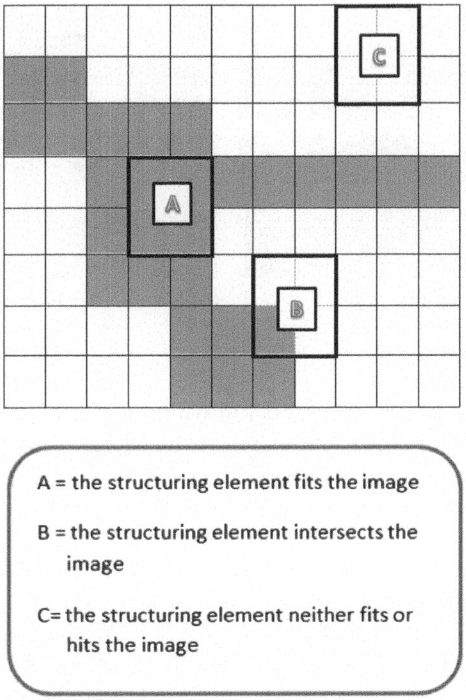

Fig. 3 Square structuring element

(a) Before using structural morphological element

(b) Foreground filtered image after using structural morphological element

Fig. 4 Foreground filtering using masking

be filtered. So the square structural morphological element is applied to filter the image which gives a more smooth result (Fig. 4b).

2.3 Moving Object Detection Using Blob Analysis

Blob is defined as a region of connected pixels and its analysis discerns pixels by their values and places them in either of the two categories i.e. foreground

(a) Original frame (b) Frame after blob analysis applied

Fig. 5 Moving vehicle detection

(usually the non-zero pixels represented in white) and background (the zero pixels in black). In the present paper, blob analysis block has been used to calculate the statistics for labeled regions in a binary image. For each frame, the number of foreground objects on the screen is generated. If the number of such objects is 'n' (say) then an $n \times m$ matrix is declared where m is dynamic in nature and is incremented in every step. For each of the 'n' foreground objects the weighted mean of the coordinates of that object is generated and stored in the matrix. An effort is being made to establish each row of the matrix into series of x coordinates through which the object is moving in the x direction. A similar approach for storing the movement in the y coordinates completes the total movement information for each frame. The block returns quantities such as the centroid, bounding box, label matrix and blob count. The Blob count in turn gives the number of vehicles count as shown in Fig. 5.

2.4 Plotting of the Vectors

Once it has been established that, storing is indeed movement information, an inner loop has been run for each frame whose limit is the number of foreground objects in the frame. In every loop iteration for each blob, the x and y coordinate information is taken from the matrices which have been already declared. For x and y, with each update the number of points increases and hence, each row vector X and Y are plotted individually at every single iteration and on the completion of the track path of the vehicle the final motion path is plotted as shown in Fig. 6c. The plotted frame can be stored to be shown later together as a movie.

Let X[] and Y[] be the two dynamic row vectors written as:

$$X[x_1, x_2, x_3, \ldots x_n, x_{n+1}] \quad \text{and} \quad Y[y_1, y_2, y_3, \ldots y_n, y_{n+1}]$$

The row vectors are initialized to '0' and the size increases with every iteration for each frame. Say, at time instant 't_0' for 0th iteration the row vectors are X[] and Y[].

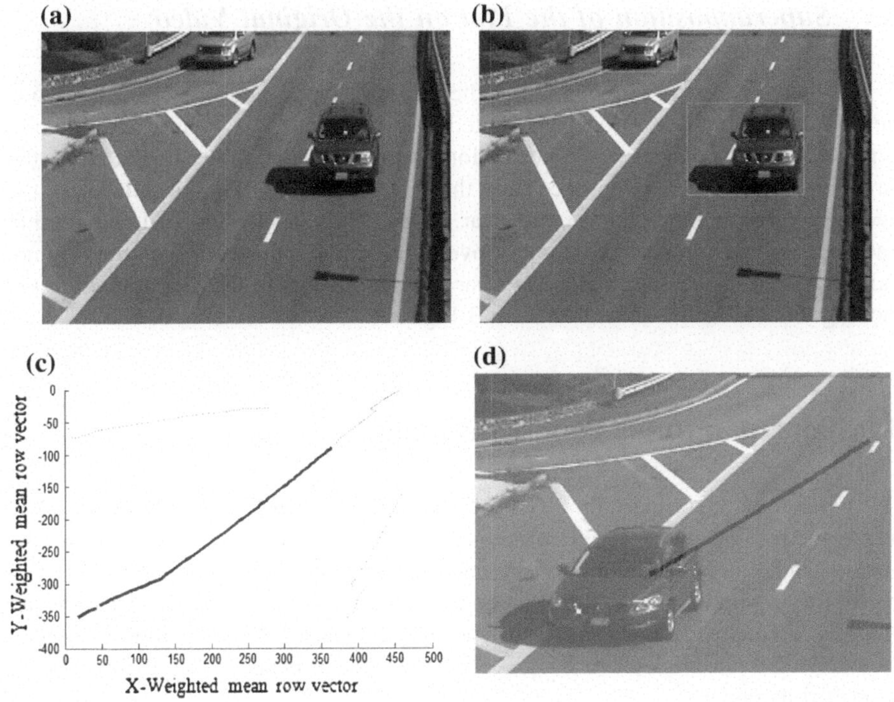

Fig. 6 **a** Original video frame, **b** moving vehicle detection, **c** plotting of the row vectors, **d** tracking of the motion path with overlaying of fig (**c**) on fig (**a**)

At time instant 't_1' at iteration 1, the row vectors are:

$$X[x_1] \quad \text{and} \quad Y[y_1]$$

At time instant 't_n' and at 't_nth' iteration, the row vectors are:

$$X[x_1, x_2, x_3, \ldots x_n] \quad \text{and} \quad X[y_1, y_2, y_3, \ldots y_n]$$

Similarly, for 't_{n+1}' we get,

$$X[x_1, x_2, x_3, \ldots x_n, x_{n+1}] \quad \text{and} \quad X[y_1, y_2, y_3, \ldots y_n, y_{n+1}]$$

The above written vectors can be generalized as:

$$X = [x_1, x_2, \ldots x_i] \quad \text{and} \quad Y = [y_1, y_2, \ldots y_i]$$

where 'i' denotes the number of iteration for the total number of frames for the considered blobs.

2.5 Superimposition of the Plot on the Original Video

A plot has already been traced along the moving foreground object in the above step and backed up by storing each plotted frame separately. If the movie of the plot is played then the trajectory of the motion path shall be smoothly traced along the direction of movement. But since only the plotting is able to be visualized so, the tracing of the motion path is not clear. Thus, to meet the need of the motion tracking, a simple process of fusion or overlaying the plotted frames on the original video is done. The result obtained is clear with the tracing of the motion path of the moving vehicle and is shown in Sect. 3 Fig. 6d.

3 Simulation Results and Discussion

The performed experiment on a typical video is shown in Fig. 6. Stepwise in the figure, the results are displayed as moving vehicle identification, plotting of the row vectors and finally the motion path tracking is marked. In plotting of the row vectors the weighted mean of the coordinates of the pixels inside the blob are considered as already mentioned. Instead of taking the commonly followed mean the weighted mean is taken to give a better performance. The justification is described as under.

Let us consider a sample 0-1 matrix where 1 represents movement of the pixel of the object at that coordinate and 0 represents that no movement of that coordinate has taken place. The matrix obtained is [A].

Taking Simple Mean

Taking the mean simply we get

Mean of Row = (1 + 1 + 1 + 1 + 1 + 1 + 1 + 1)/Number of non-zero elements in the rows of the matrix = 8/8 = 1
Mean of Column = (1 + 1 + 1 + 1 + 1 + 1 + 1 + 1)/Number of non-zero elements in the columns of the matrix = 8/8 = 1

Therefore, Mean of row and Column elements = (1, 1) which represents the first element of the matrix, i.e. making the (1, 1) coordinate 1 and the rest 0, which signifies the corner pixel but as mean value the middle pixel is appropriate.

Taking Weighted Mean
Weighted Mean of all Row elements = $(1 \times 1 + 1 \times 1 + 1 \times 1 + 2 \times 1 + 2 \times 1 + 3 \times 1 + 3 \times 1 + 3 \times 1)$/Number of non zero row coordinates = 16/8 = 2
Weighted Mean of all Column elements = $(1 \times 1 + 1 \times 1 + 1 \times 1 + 2 \times 1 + 2 \times 1 + 3 \times 1 + 3 \times 1 + 3 \times 1)$/Number of non zero column coordinates = 16/8 = 2

Weighed Mean of Coordinates of Row and Column = (2, 2). Therefore, the movement shall be represented as shown in the matrix, i.e. making the (2, 2) coordinates as 1 and the rest 0, signifying the middle pixel as the mean. Thus, it has been explored that weighted mean finds efficient result over simple mean calculation.

$$\begin{matrix} & 1\;2\;3 \\ \begin{matrix}1\\2\\3\end{matrix} & \begin{bmatrix}1&1&1\\1&0&1\\1&1&1\end{bmatrix} \\ & \text{Matrix [A]}\end{matrix} \qquad \begin{matrix} & 1\;2\;3 \\ \begin{matrix}1\\2\\3\end{matrix} & \begin{bmatrix}1&0&0\\0&0&0\\0&0&0\end{bmatrix} \\ & \text{Simple mean}\end{matrix} \qquad \begin{matrix} & 1\;2\;3 \\ \begin{matrix}1\\2\\3\end{matrix} & \begin{bmatrix}0&0&0\\0&1&0\\0&0&0\end{bmatrix} \\ & \text{Weighted mean}\end{matrix}$$

The typical video chosen is of 28.521 s with a total number of 531 frames whose individual processing time is 0.132 s. The playing time taken for the output video with motion path tracking takes 0.0715 s/frame. The same experiment is performed with another video clip and the results are shown in Fig. 7.

In Fig. 8a a typical video frame is chosen in which some vehicles are stationary while that others are missing. In Fig. 8b as shown, it can be noticed that when vehicle detection is performed then only one vehicle is detected, the vehicles present at the left side of the frame image are not detected. This gives the authentication of the moving vehicle detection. As those two vehicles are actually immobile in the video clip, so they are not detected. The second video clip is of 11 s. The total number of frames present is 290. So, the processing time for each frame is 0.0984 s and the playing time is 0.0489 s.

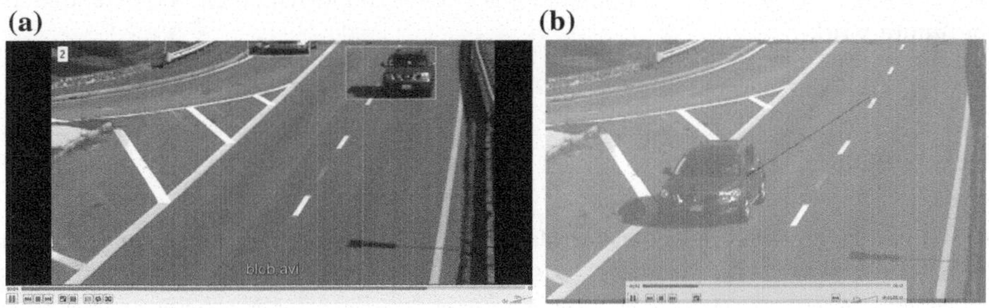

Fig. 7 **a** Screen shot of moving vehicle identification and the count of vehicles in a video clip **b** screen shot of motion tracking in a video clip

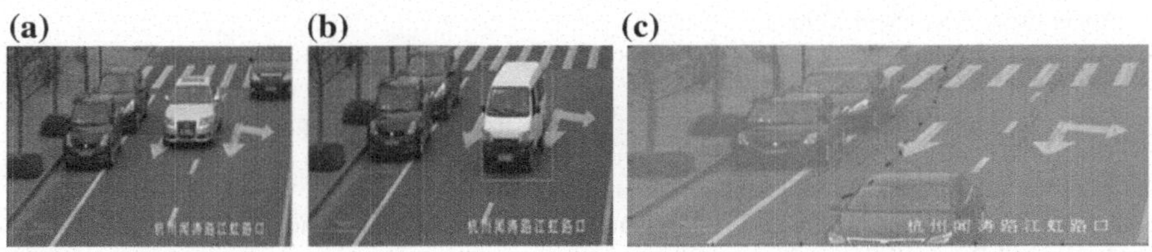

Fig. 8 **a** Original video frame **b** moving vehicle detection **c** plotting of the motion path and tracing it on the video

The proposed method was performed on a personal computer which is equipped with Intel Core i5-2410 M CPU2.30 GHz, 4 GB RAM and windows 7, 64-bit operating system.

4 Conclusion

Real time vehicle detection focuses on intelligent transportation. In this paper, a system is defined for fully automated traffic monitoring and control. The system can be used for online traffic control with a camera positioned in a specified traffic zone. The camera is an immobile one and hence the video that is captured, endures stationary background.

In the present paper, the task of moving vehicle detection, the count of moving vehicles and the motion path tracking is done successfully. The method of extraction of moving vehicle trajectories fulfills the need of detailed traffic monitoring, particularly in case of multiple lane roadways.

The tracking of the motion path and tracing it is based on the weighted mean of the pixel's coordinates inside the blob. The process is continued for a number of frames for which the particular vehicle exists. The idea of the work presented here, is well-matched for permanent as well as temporary traffic studies, such as investigating vehicle movements in the interlaced sections. Unlike the paper where the experimental procedure is applied on a video with stationary background, similar work can also be done with a non-stationary background which can be extended as the future work.

References

1. B. Coifman, B., Beymerb, D., McLauchlanb, P., and Malikb, J., "A real-timecomputer vision system for vehicle tracking and traffic surveillance," Transportation Research Part C, 6, 271–288, (1998).
2. Betkel, M., Haritaoglu, E., and Davis, L. S., "Real-time multiple vehicle detection and tracking from a moving vehicle," Machine Vision and Applications, 12, 69–83, (2000).
3. Caraffi, C., Vojır, T., Trefny, J., Sochman, J., and Matas, J., "A System for Real-time Detection and Tracking of Vehicles from a Single Car-mounted Camera," Proc. of 15th International IEEE Conference on Intelligent Transportation Systems (ITSC), 975–982, Anchorage, AK, (Sept. 2001).
4. Sochman, J., and Matas, J., "WaldBoost - Learning for Time Constrained Sequential Detection," CVPR, 2, 150–157, (2005).
5. Kalal, Z., Matas, J., and Mikolajczyk, K., "P-N Learning: Bootstrapping Binary Classifiers by Structural Constraints," Proc.IEEE Conference on Computer Vision and Pattern Recognition (CVPR), 49–56, San Francisco, CA, (June 2010).
6. Sochor, J., "Fully Automated Real-Time Vehicles Detection and Tracking with Lanes Analysis," Proc. CESCG, 59–66, (2014).

7. Fradi, H., and Dugelay, J., "Robust Foreground Segmentation Using Improved Gaussian Mixture Model and Optical Flow," Proc. IEEE ICIEV, 248–253, (May 2012).
8. Kumar Chauhan, A., and Krishan, P., "Moving Object Tracking using Gaussian Mixture Model and Optic Flow," International Journal of Advanced Research in Computer Science and Software Engineering, 3, 243–246, (April 2013).

Stability Improvement of Captive Generator Sets Utilizing FACTS Device

Utpal Goswami, Madhuria Chanda, Arijit Ganguly, P.K. Sadhu and Suprava Chakraborty

Abstract Power demand in last two decades has increased substantially increasing the complexity in power system network. Therefore to maintain the transient stability of captive power plant (CPP) during any grid fault or disturbance presents a new challenge to researchers. In case of grid fault, captive power plant immediately isolates its captive generator sets from grid to ensure safety. In this paper, an attempt has been made to improve transient stability of captive generator sets during grid fault using FACTS (Flexible AC Transmission System) controller. The performance of FACTS controllers likes Static Var Compensator (SVC) is observed during grid fault condition. The simulation analysis was performed using MATLAB/Simulink software.

1 Introduction

The modern power system network consists of several complex electrical components such as generator, automatic voltage regulator (AVR), exciter, power transformer, distribution transformer, transmission line, electrical motors and combination of variety of linear and nonlinear loads etc. Combination of these components makes the power system network more complex [1]. De-regularization in power industry emphasizes the liberal provision with respect to build up of captive power plant with a view of not only providing dependable power but also to provide high quality and economic power as well. Current industry practice suggests that after any grid fault captive power plant needs to be disconnected from the

U. Goswami (✉) · M. Chanda · A. Ganguly
Department of Electrical Engineering, University of Engineering & Management, Kolkata, West Bengal 700160, India
e-mail: utpal.goswami@uem.edu.in

P.K. Sadhu
Indian School of Mines, Dhanbad, India

S. Chakraborty
National Institute of Solar Energy, Gurugram, India

© Springer Nature Singapore Pte Ltd. 2017
I. Bhattacharya et al. (eds.), *Advances in Optical Science and Engineering*, Springer Proceedings in Physics 194, DOI 10.1007/978-981-10-3908-9_84

grid within 200–300 ms [2], if the system has not regained its stability. Stabilization of CPP, under grid disturbance condition, can be done implementing load shedding and anti-islanding scheme [3]. Active and passive filter array system is one of the useful system to manage CPP [4, 5].

Recent power electronics research suggests that FACTS device is a useful tool to enhance the transient performance index during any major grid disturbance [6]. First operational feature of FACTS device is to help the network to regain its transient stability. Static Var Controller (SVC) is one of the reliable FACTS controller which control the voltage of power network using reactive power compensation technique. SVC also acts as a shunt compensator to control the transient voltage and damp out oscillation during any grid fault [10, 11]. FACTS controller is considered to be the most versatile one and used for improving the dynamic stability of power system network. FACTS devices consist of Static Synchronous Compensator (STATCOM), Series Synchronous Compensator (SSSC), Unified Power Flow Controller (UPFC) and SVC to control many electrical parameters simultaneously. This paper presents a novel approach to improve transient stability by connecting SVC inside the process plant network system [7–11]. Result shows that the transient stability index of CPP generator improve significantly in terms of time duration to reduce generator output voltage oscillation as well as rotor angle variation of CPP.

2 Methodology

To study the transient performance of proposed network model a detailed network diagram is prepared using MATLAB/Simulink software. The network block diagram is depicted in Fig. 1.

In the network shown below, the alternator rated 30 MVA, 11 kV is connected to the grid. A part of the generated power is used to supply the plant itself to sustain the process of generation. After supplying the internal plant load of 5 MW, the

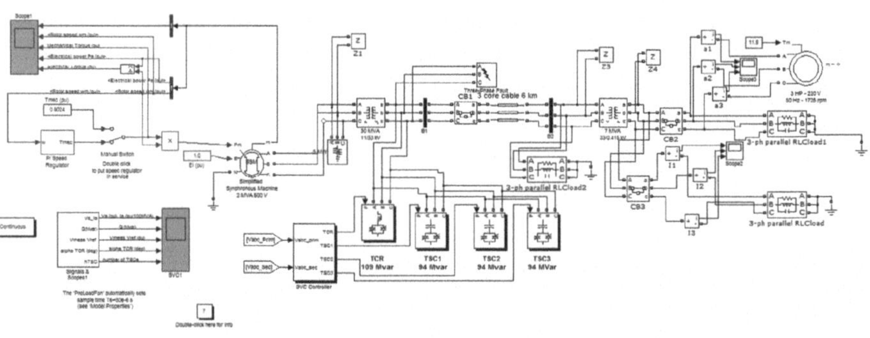

Fig. 1 Network diagram consider for analysis with SVC

generated voltage is stepped up to 33 kV. The rating of the transformer considered as 30 MVA and the primary side voltage of the transformer is 11 kV and secondary side voltage is considered as 33 kV. The power is transmitted through a line of considerable length to supply the grid. The power transmitted at this voltage is ultimately stepped down to supply the load at the distribution end.

Now the interesting area is to deal with a sudden three phase fault occurs in 33 kV bus as shown in the Fig. 1. The voltage profile of the grid shows abnormality. To maintain the safety rule as mention in IEC standard the generator immediately needs to be isolated from the grid within critical fault clearing time. The system behavior is analyzed in two domains one incorporates FACTS devices in the network and other without it. SVC and UPFC is considered as FACTS device to improve stability.

Basic points that are considered for the study:

(a) Loads as shown in various buses are motor load; static load, etc.
(b) Short circuit fault level of existing 11 kV system considered as 30 kA.
(c) Tolerance of impedances not considered except generator transformer.
(d) Tolerance of impedances for 30 MVA GT is considered as per IEC standards.
(e) Overhead 33 kV line fault level consider as 50 kA.

3 Results

The Transient stability study of the proposed network is performed in the following conditions.

(a) For a fault in 33 kV grid side without connection of SVC at generator bus.
(b) For a fault in 33 kV grid side and SVC is connected at generator bus.

Event of operations performed for the transient stability study is indicated in Table 1.

The simulation results of Transient stability study for case-1, case-2

Table 1 Event of operation performed for transient stability study

Event	Time (s)	Action	Status of SVC
Case-1	0.5	3Ph fault on grid bus	Without SVC
Case-2	0.5	3 Ph fault on grid bus	With SVC

4 Discussion

From the simulation result Figs. 2, 3, 4, 5 and 6, the summary output is depicted in Table 2.

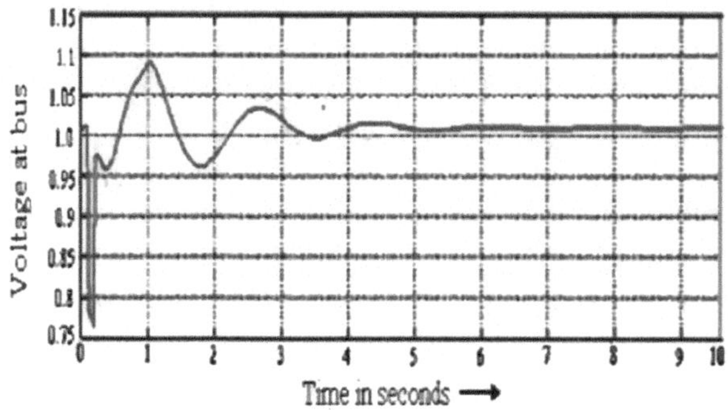

Fig. 2 Voltage profile at generator bus without SVC

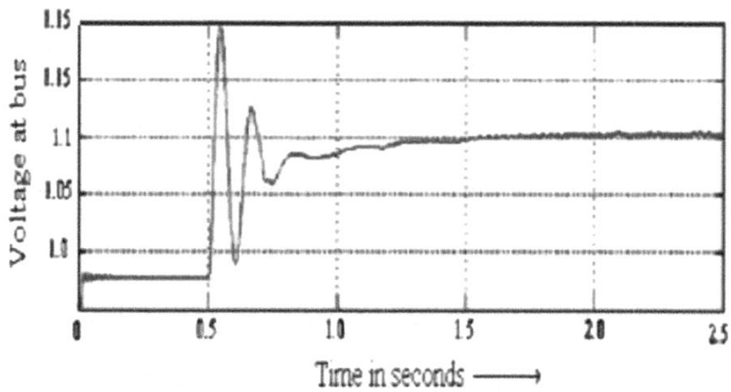

Fig. 3 Voltage profile at generator bus with SVC

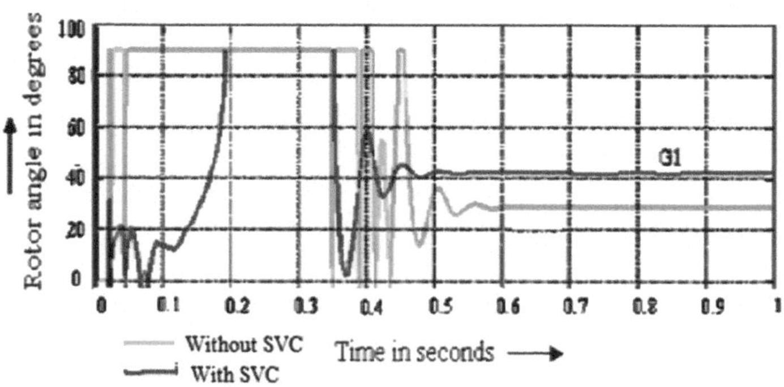

Fig. 4 Variation of rotor angle during fault with and without SVC

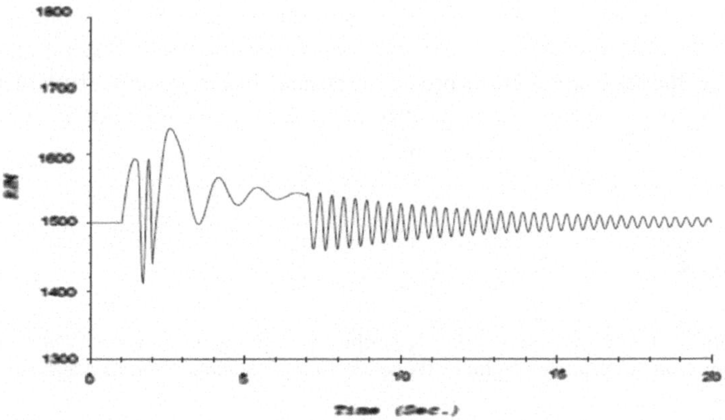

Fig. 5 Speed variation of captive generator during grid fault without SVC

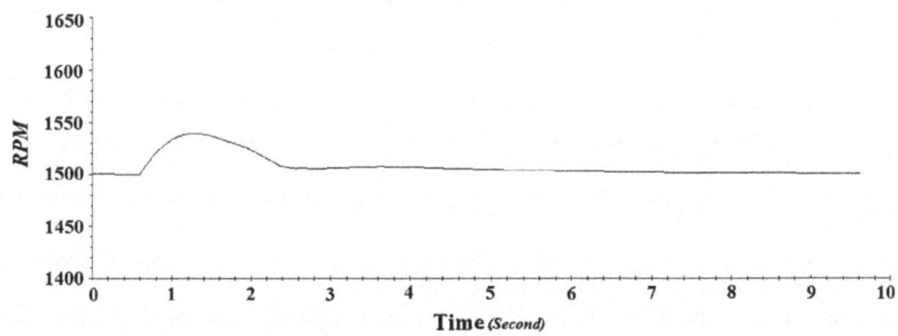

Fig. 6 Speed variation of captive generator during grid fault with SVC

Table 2 Summary of event of operation performed for transient stability study

S. No.	Status of SVC	Rotor angel stabilization time (s)	Final speed attended (rpm)	Time for stabilization (s)
1.	Without SVC	0.6	Oscillates 1550–1520	5
2.	With SVC	0.5	1500	2

5 Conclusion

As per the analysis it is observed that, during grid disturbance and after removal of fault, first acting turbine governor system along with exciter and SVC plays an important role to maintain the transient stability of the system. From the transient stability study it is very clear that turbine governor model with SVC performs better in terms of improving transient stability index i.e. voltage variation, damping oscillation and generator speed variation during a grid fault condition. With the use of SVC, time duration of voltage oscillation reduced from 5–2 s during 33 kV grid bus fault. With the use of SVC, reduction in generator speed oscillation results in reduction of thermal stress in turbine system and improvement of plant reliability significantly. Improvement of reliability also helps plant operator to perform safe islanding operation of captive generator.

References

1. Mukhopadhyay. S and Singh. B, "Distributed Generation - Basic Policy, Perspective Planning, and Achievement", India Power & Energy Society General Meeting, 2009. PES '09. IEEE.
2. Utpal Goswami, Tapas Kumar Sengupta, Arabinda Das, "Improvement of Transient Stability Performance of Captive Power Plant during Islanding Condition", TELKOMNIKA Indonesian Journal of Electrical Engineering, Vol. 12, No. 12, 2014, pp. 8001–8007. DOI: 10.11591/telkomnika.v12i12.6730.
3. Rajamani. K, Hambarde. U.K, "Islanding and Loadshedding Scheme for Captive Power Plant Power Delivery", IEEE Transactions, Vol. 14, Jul 1999.
4. NEMA Standards Publication ICs 1-1988, General Standards for Industrial Control and Systems; ICs 2-1988, Industrial Control Devices, Controllers and Assemblies; ICs 3-1988.
5. Bindon. R.E, "Emergency operation of large steam turbine generator", presented at South Eastern Electric Exchange, Atlanta, Georgia, October 13–14, 1966.
6. IEEE Std 141-1986, IEEE Recommended Practice for Electric Power Distribution for Industrial Plants (ANSI).
7. IEEE Std 242-1986, IEEE Recommended Practice for Protection and Coordination of Industrial and Commercial Power Systems (ANSI). 19458.
8. Sengupta T.K., "Studies on Assessment of Power Frequency in Interconnected Grid – Its Computer based Control and Protection", Research Paper, The Faculty of Engineering and Technology, Department of Electrical Engineering, Jadavpur University, 2008.
9. Krishnamurti. P, "Captive Power Plant Quality Journal" of TCE Limited, Vol.4-I-pp. 10, April 2006.
10. Fishow, A.G, "Transient ans Steady State Stability margin to state a power system stability standardization", published in UPS of Russia.
11. Canizarea. C. A, Bhattacharrya. K, Haghighat. H, Pan. J, Tang. C and Samahy. E. I, "Reactive Power Despatch Problem in the context of Captive Electricity Market Generation, Transmission and Distribution", Vol. 4, IET issue, February 2010.

Author Index

© Springer Nature Singapore Pte Ltd. 2017

I. Bhattacharya et al. (eds.), *Advances in Optical Science and Engineering*,
Springer Proceedings in Physics 194, DOI 10.1007/978-981-10-3908-9

The manufacturer's authorised representative in the EU is Springer Nature Customer Service Centre GmbH, Europaplatz 3, 69115 Heidelberg, Germany. If you have any concerns regarding our products, please contact ProductSafety@springernature.com

Printed and bound by CPI Group (UK) Ltd, Croydon, CR0 4YY

25/09/2026

02222881-0014